PHOTOREACTIVE ORGANIC THIN FILMS

PHOTOREACTIVE ORGANIC THIN FILMS

Edited by

Zouheir Sekkat
Osaka University, Japan

Wolfgang Knoll
Max-Planck Institut für Polymerforschung,
Germany

ACADEMIC PRESS
An imprint of Elsevier Science

Amsterdam   Boston   London   New York   Oxford   Paris   San Diego
San Francisco   Singapore   Sydney   Tokyo

Academic Press
An imprint of Elsevier Science
525 B Street, Suite 1900, San Diego, California 92101-4495, USA
http://www.academicpress.com

Academic Press
An imprint of Elsevier Science
84 Theobald's Road, London WC1X 8RR, UK
http://www.academicpress.com

Library of Congress Catalog Card Number: 2002104260

International Standard Book Number: 0-12-635490-1

PRINTED IN THE UNITED STATES OF AMERICA
02 03 04 05 06 07 MB 9 8 7 6 5 4 3 2 1

CONTENTS

2 Ultrafast Dyamics in the Excited States of Azo Compounds

TAKAYOSHI KOBAYASHI AND TAKASHI SAITO

3 Photo-Orientation by Photoisomerization

ZOUHEIR SEKKAT

II   PHOTOISOMERIZATION IN ORGANIC THIN FILMS

6 Photoisomerization in Langmuir-Blodgett-Kuhn Structures

HENNING MENZEL

7 Electronic and Optical Transduction of Photoisomerization Processes at Molecular- and Bimolecular-Functionalized Surfaces

EUGENII KATZ, ANDREW N. SHIPWAY, AND ITAMAR WILLNER

III PHOTOCHEMISTRY AND ORGANIC NONLINEAR OPTICS

8 Photoisomerization Effects in Organic Nonlinear Optics: Photo-Assisted Poling and Depoling and Polarizability Switching

ZOUHEIR SEKKAT

9 Photoisomerization in Polymer Films in the Presence of Electrostatic and Optical Fields

ANDRÉ KNOESEN

12 Photoinduced Third-Order Nonlinear Optical Phenomena in Azo-Dye Polymers

VICTOR M. CHURIKOV AND CHIA-CHEN HSU

IV OPTICAL MANIPULATION AND MEMORY

13 Photoinduced Motions in Azobenzene-Based Polymers

ALMERIA NATANSOHN AND PAUL ROCHON

14 Surface-Relief Gratings on Azobenzene-Containing Films

OSVALDO N. OLIVEIRA, JR., LIAN LI, JAYANT KUMAR, AND SUKANT K. TRIPATHY

15 Dynamic Photocontrols of Molecular Organization and Motion of Materials by Two-Dimensionally Arranged Azobenzene Assemblies

TAKAHIRO SEKI AND KUNIHIRO ICHIMURA

16 3D Data Storage and Near-Field Recording

YOSHIMASA KAWATA AND SATOSHI KAWATA

17 Synthesis and Applications of Amorphous Diarylethenes

TSUYOSHI KAWAI AND MASAHIRO IRIE

■ CONTRIBUTORS

Numbers in parentheses indicate the pages on which the authors' contributions begin.

Aleksandra Apostoluk (331) Université d'Angers, Laboratoire POMA, UMR-CNRS 6136, ERT Cellules Solaires PhotoVolataïques Plastiques, 2, Boulevard Lavoisier 49045 Angers Cedex 01 France

Lev M. Blinov (145) Institute of Crystallography, Russian Academy of Sciences 117333, Moscow, Russia

Victor M. Churikov (365) Department of Physics National Chung Cheng University, Ming Hsiung, Chia Yi 621, Taiwan, ROC

Jacques A. Delaire (305) PPSM, CNRS UMR 8531, Ecole Normale Superieure de Cachan, 61 , avenue du Président Wilson, 94235 Cachan Cedex, France

Céline Fiorini-Debuisschert (331) CEA Saclay, DRT-LIST-DECS-SE2M, Laboratoire Composants Organiques, 91191 Gif sur Yvette Cedex, France

Wolfgang Haase (145) Institute of Physical Chemistry, Darmstadt University of Technology, D-64287, Darmstadt, Germany

Chia-Chen Hsu (365) Department of Physics National Chung Cheng University, Ming Hsiung, Chia Yi 621, Taiwan, ROC

Kunihiro Ichimura (487) Research Institute for Science and Technology, Science University of Tokyo, 2641 Yamazaki, Noda, Chiba 278-8510, Japan

Masahiro Irie (541) CREST, Japan Science and Technology Corporation, JST, Hakozaki 6-10-1, Higashi-ku, Fukuoka 812-8581, Japan

Elena Ishow (305) PPSM, CNRS UMR 8531, Ecole Normale Superieure de Cachan, 61 , avenue du Président Wilson, 94235 Cachan Cedex, France

Eugenii Katz (219) Institute of Chemistry and the Farkas Center for Light-Induced Processes, Hebrew University of Jerusalem, Jerusalem 91904, Israel

Yoshimasa Kawata (513) Shizuoka University, Department of Mechanical Engineering, Johoku, Hamamatsu 432-8561, Japan

Satoshi Kawata (513) Department of Applied Physics, Osaka University, Suita, Osaka 565-0871, Japan

Tsuyoshi Kawai (541) Department of Applied Chemistry; Graduate School of Engineering, Kyushu University, Hakozaki 6-10-1, Higashi-ku, Fukuoka 812-8581, Japan

André Knoesen (289) Department of Electrical and Computer Engineering, University of California Davis, Davis, CA 95616-5294, USA

Wolfgang Knoll (107) Max-Planck Institut für Polymerforschung, Ackermannweg 10, 55128 Mainz, Germany

Takayoshi Kobayashi (49) Department of Physics, University of Tokyo, Hongo 7-3-1, Bunkyo-ku, Tokyo 113-0033, Japan

Mikhail V. Kozlovsky (145) Institute of Physical Chemistry, Darmstadt University of Technology, D-64287, Darmstadt, Germany

Jayant Kumar (429) Center for Advanced Materials, University of Massachusetts Lowell, 1 University Avenue, Lowell, MA 01854, USA

Lian Li (429) Center for Advanced Materials, University of Massachusetts Lowell, 1 University Avenue, Lowell, MA 01854, USA

Henning Menzel (179) Institut für Technische Chemie, Technische Universität Braunschweig, D-38106 Braunschweig, Germany

Keitaro Nakatani (305) PPSM, CNRS UMR 8531, Ecole Normale Superieure de Cachan, 61 , avenue du Président Wilson, 94235 Cachan Cedex, France

Almeria Natansohn (399) Department of Chemistry, Queen's University, Kingston, ON K7L 3N6, Canada

Jean-Michel Nunzi (331) Université d'Angers, Laboratoire POMA, UMR-CNRS 6136, ERT Cellules Solaires PhotoVolataï ques Plastiques, 2, Boulevard Lavoisier 49045 Angers Cedex 01, France

Osvaldo N. Oliveira, Jr. (429) Instituto de Física de São Carlos, USP, CP 369, 13560-970 São Carlos, SP, Brazil

Hermann Rau (3) Institut für Chemie, Universität Hohenheim, D-70593 Stuttgart, Germany

Paul Rochon (399) Department of Physics, Royal Military College, Kingston, ON K7K 5L0, Canada

Takashi Saito (49) Department of Physics, University of Tokyo, Hongo 7-3-1, Bunkyo-ku Tokyo 113-0033, Japan

Takahiro Seki (487) Chemical Resources Laboratory, Tokyo Institute of Technology, 4259 Nagatsuta, Midori-ku, Yokohama 226-8503, Japan

Zouheir Sekkat (63, 107, 271) Department of Applied Physics, Osaka University, Suita, Osaka 565-0871, Japan

Andrew N. Shipway (219) Institute of Chemistry and the Farkas Center for Light-Induced Processes, Hebrew University of Jerusalem, Jerusalem 91904, Israel

Sukant K. Tripathy (429) Center for Advanced Materials, University of Massachusetts Lowell, 1 University Avenue, Lowell, MA 01854, USA

Itamar Willner (219) Institute of Chemistry and the Farkas Center for Light-Induced Processes, Hebrew University of Jerusalem, Jerusalem 91904, Israel

◼ PREFACE

Research on optically sensitive chromophores containing polymeric films (photoreactive organic thin films, POTF) has exploded during the past decade, and it is continuing to be a very active area of science. Several important phenomena have been discovered, especially at the contact of photochemistry (photoisomerization) and nonlinear optics by the photoisomerization of nonlinear optical molecules in polymers. The material published in the book covers photoreactivity in organic films from the most elementary to the most sophisticated backgrounds, and discusses possible technological uses in views of applications for fast electro-optic modulation, optical data storage, and holography. This book is a comprehensive volume that encompasses the known knowledge on POTF research and applications.

POTF have been investigated intensively in the past few years due to requirements in the areas of optoelectronics and photonics; and linear and nonlinear optical (NLO) effects induced by photo-orientation of photoisomerizable NLO chromophores in polymers have attracted much attention. Light can manipulate the chromophores' orientation by photoisomerization via polarized transitions such that centrosymmetry and isotropy are alleviated, and anisotropy and quadratic and cubic optical nonlinearities are induced. Optical poling techniques, e.g. induced molecular polar orientation, which result in second order NLO effects have been reported, and the coupling between photochemistry and organic nonlinear optics has emerged as a new discipline in the past decade. Surface relief gratings created by photo-induced mass movement of polymers have also been reported. Inasmuch as optical ordering of photoisomerizable molecules is being intensively studied, its theoretical quantification helps bridge independent studies in the areas of nonlinear optics and photochemistry.

Photoisomerization was studied from a purely photochemical point of view in which photo-orientation effects can be disregarded. While this feature can be true in low viscosity solutions where photo-induced molecular orientation can be overcome by molecular rotational diffusion, in polymeric environments, especially in thin solid film configurations, spontaneous molecular mobility can be strongly hindered and photo-orientation effects are appreciable. The theory that coupled photoisomerization and photo-orientation processes was also recently developed, based on the formalism of Legendre Polynomials, and more recent further theoretical developments have helped quantify coupled photoisomerization and photo-orientation processes in films of polymer.

A number of polymers containing photoisomerizable chromophores have been reported, and several authors reported studies in Langmuir-Blodgett-Kuhn azo-polymers as multilayer structures and alignment layers for liquid crystal molecules, self-assembled monolayers, amorphous and liquid crystalline polymers and so on. In recent years, studies of the role of inter-chromophore interactions and molecular addressing have been reported, and questions have begun to arise concerning the relationship of optical ordering processes in amorphous polymers to the Tg and polymer structure, including the main chain rigidity, the free volume, and the nature of the connection of the chromophore to a rigid or a flexible main-chain. These studies correlated the optical ordering (nonpolar and polar) to the polymer structure in a series of very high Tg (up to 350 °C) rigid or semirigid NLO polymers, and demonstrated a new way of probing sub-Tg polymer dynamics in photoreactive NLO polymers.

These intensive studies on POTF have been performed by a large number of scientists from a variety of communities, including polymer scientists, photochemists and photophysicists, chemists and chemical engineers, optical physicists and optical engineers, and researchers in the field of organic nonlinear optics. Each of these scientists approached POTF research from the point of view of his or her own field. This book unifies the various sub-themes of photoisomerization research and bridges different disciplines. Leading experts have encapsulated their work on POTF research in comprehensive and self-consistent chapters. Both fundamental and application issues are discussed; and the readers, including non-specialists, can not only appreciate that this book represents the largest collection of information on this topic published in a single book, but also see and eventually acknowledge the interdisciplinary nature of POTF research and applications. We expect this book to be the essential reference on POTF science for both students and for researchers in adjacent fields.

Zouheir Sekkat
Wolfgang Knoll

I

PHOTOISOMERIZATION AND PHOTO-ORIENTATION OF AZOBENZENES

1
PHOTOISOMERIZATION OF AZOBENZENES

HERMANN RAU
Institut für Chemie, Universität Hohenheim, D-70593 Stuttgart, Germany

1.1 INTRODUCTION

Most of the phenomena described in this monograph on photoreactive organic thin films are based on the isomerization of units deliberately built into molecules, molecular assemblies, or polymers. Most especially, the spectroscopic and isomerization behavior of these units determines the switching and triggering properties of the photoreactive systems and devices. Information storage and nonlinear optical properties, as well as photo-control of equilibria and of polymer, membrane, and other properties are exploited in applications.

The majority of the systems outlined in this monograph contain the azobenzene moiety as the photoswitchable unit. Therefore, the first chapter of this monograph deals with the properties of this simple molecule and its simple derivatives. Applications for various purposes are covered in some of the following chapters.

Azobenzene **1** (Figure 1.1A) is particularly useful for these applications for the following reasons:

1. The azobenzene unit is chemically stable at moderate temperatures and against UV/VIS radiation;
2. On E-Z (trans-cis) conversion,[1] it changes its absorption spectrum considerably (Figure 1.1B);
3. On E-Z conversion, it changes its molecular shape, reducing the distance between the p-positions from 1.0 to 0.59 nm (Figure 1.1A) and increasing the dipole moment from 0 to ca. 3 Debye;[2]
4. Donor/acceptor substituted azobenzenes, which have large second- and third-order nonlinear optical properties, show a fast thermal Z-E (cis-trans) conversion.

Items 2 to 4 warrant this chapter on photo-induced and thermal E-Z isomerization of the parent molecules, which are incorporated in numerous macromolecular and supramolecular assemblies as the photo-active element.

Azo compounds are systematically addressed as "diazenes." This has to be borne in mind when one conducts a literature research. The azo (diazene) group is isosteric with the ethene group; stilbene and azobenzene have many related features, but they also possess relevant different properties that make azobenzenes superior for use as photo-switches.

To understand the photoresponsive properties of azobenzene and its molecular family, it is necessary to discuss their spectroscopy and the mechanistic options of isomerization. A review of the spectroscopic properties of azo compounds appeared in 1973.[3] The isomerization properties of azobenzene were reviewed for several periods. Wyman[4] covered the literature up to 1954, Ross and Blanc[5] up to 1969, and Rau[6,7] up to 1988. The present stand-alone review is restricted to the spectroscopy and isomerization of simple aromatic azo compounds. It is meant to serve as a basis for the detailed treatments of complex photoresponsive systems in the following chapters of this monograph.

The spectroscopic and photochemical, especially isomerization, features of the azoaromatics warrant separation into three classes according to the relative energy of the lowest lying $^1(n,\pi^*)$ and $^1(\pi,\pi^*)$ states: the azobenzene type, the aminoazobenzene type, and the pseudo-stilbene type.[6] This determines the structure of this chapter. Section 1.2 provides basic information on

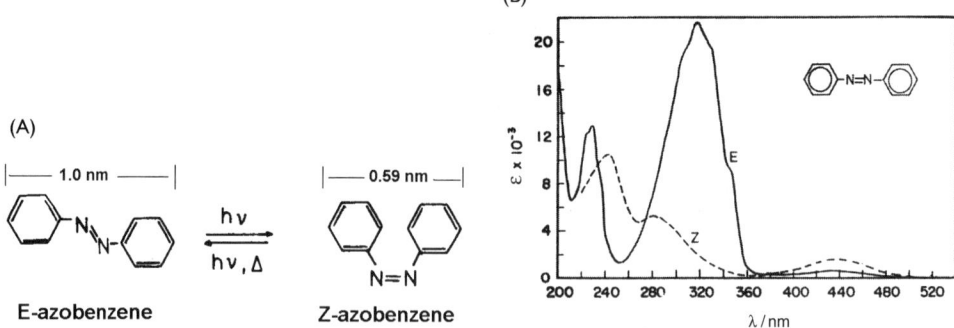

FIGURE 1.1 (A) The E/Z isomerization system of azobenzene. (B) Absorption spectra of E- and Z-azobenzene in EtOH solution.

the azo group, the kinetics of the E-Z isomerization, and the use of kinetic evaluation methods. Thereafter, Section 1.3 covers the spectroscopic and isomerization data of the azobenzene type azo compounds; Section 1.4 includes those of the aminoazobenzene type, and Section 1.5 presents those of the pseudostilbene type azo compounds. In Section 1.6, I discuss the mechanistic aspects of the E-Z isomerization, and I make concluding remarks in Section 1.7.

Few publications on the spectroscopic and isomerization properties of simple azo compounds have appeared in the last 15 years, as compared to the decades before. There is, however, one exception: Ultrashort time-resolved spectroscopy of azobenzene and its relatives has opened new access to the dynamics following pico- and femtosecond excitation. The results are most relevant for the mechanisms of the photophysical and photochemical processes, which in azoaromatic compounds primarily are isomerizations. There is, however, a host of newer investigations into the isomerization of azobenzene and its family that are directed to applications in photoswitchable systems and devices. Some of them are relevant for the understanding of the parent molecules and therefore are included in this chapter.

1.2 THE AZO GROUP

1.2.1 Spectroscopic Properties

The azo group is planar and observed in the E- and Z-configurations (Fig. 1.1A). It is characterized by the ethenic π-electron system which has antisymmetric wave functions relative to the molecular plane and by the unique n-electron system. The n-orbitals centered at the two adjacent nitrogen atoms are symmetric in relation to the molecular plane. At a distance of 123 pm,[8] they interact strongly and split into an n_+ and n_- molecular orbital with a large energy separation (photoelectron spectra give 3.3 eV $\approx$ 25000 cm^{-1} in azomethane 2[9]). The n- and π-systems are orthogonal for symmetry reasons.

The states built of these orbitals determine the spectroscopic and photochemical behavior. The features of the azo group are best represented in the spectra of the aliphatic azo compounds diazene H-N=N-H, azomethane CH_3-N=N-CH_3[10] or 2,3-diazabicyclo[2.2.1]hept-2-ene (DBH, **3**).[11] In azomethane, a floppy E-azo compound, the forbidden n → π* band in the region of 350 to 400 nm is very weak ($\varepsilon \approx 10$ l mol^{-1} cm^{-1}) and continuous. In DBH and the homologous DBO (diazanorbornene, **4**),[11,12] rigid Z-azo compounds, the n → π* band is weakly allowed ($\varepsilon \approx 300$ l mol^{-1} cm^{-1}), structured sharply in the gas phase and weakly in solution. A large energy gap of ca. 20,000 cm^{-1} separates the weak n → π* bands from the intense allowed π → π* bands around 215 nm in E-azomethane as well as in Z-DBO (here also sharply structured in the gas phase).[12]

2 **3** **4**

Although there are some "reluctant"[13] aliphatic azo compounds, in general these molecules are photochemically not very stable.[14] Thus, they are not used in the systems covered in this book and will not be reviewed in this contribution.

In aromatic azo compounds, the π system is extended. X-ray data[2,15] for E-azobenzene give the N=N distance as 124.7 pm, not much different from that in azomethane, and the C-N distance as 142.8 pm. The NNC angle is 114.1^0, somewhat off the sp^2 hybridization angle, and the molecule is planar (>CNNC = 180^0). The corresponding values for Z-azobenzene are: N=N 125.3 pm, C-N 144.9 pm, >NNC 121.9^0, >CNNC 172^0, and the twist angle of the phenyl rings is 53.3^0. This is in agreement with earlier work.[16] Electron diffraction data[17] for E-azobenzene do not differ more than 2 pm from the X-ray results, but they indicate a small twist angle >C-N of 30^0.

The extension of the conjugation system increases the photostability of the molecules and lowers the excitation energies compared to those of the aliphatic compounds: The n → π* absorption of aromatic azo compounds occurs in the visible region; they are colored (Figure 1.1B). The (n,π*) energy is moderately lowered by 5000 to 7000 cm^{-1}, and the (π,π*) state energy strongly by about 20,000 cm^{-1}; thus the band gap is reduced to about 10,000 cm^{-1} in azobenzene. This energy gap is very sensitive to substitution, which influences the spectroscopic and photochemical features of different azoaromatics. Therefore, we group the azoaromatics into three classes according to the relative energy of the lowest-lying $^1(n,\pi^*)$ and $^1(\pi,\pi^*)$ states (Figure $1.2^{18})^7$: an azobenzene type with a low-lying $^1(n,\pi^*)$ state, an aminoazoben-

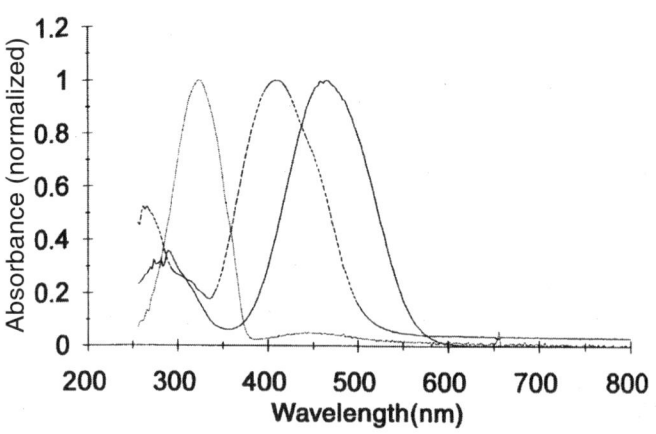

FIGURE 1.2 Absorption spectra of azobenzene ·······, 4-aminoazobenzene - - - -, and pseudostilbene type (4-nitro-4'-aminoazobenzene) ——— molecules in a polar environment. (Adapted from reference 18, by permission.)

zene type where $^1(n,\pi^*)$ and $^1(\pi,\pi^*)$ are at comparable energies (see also Figure 1.11), and a pseudo-stilbene type with a $^1(\pi,\pi^*)$ state as the lowest-excited state (see also Figure 1.13). The assignment of an azo molecule or an azobenzene-containing macromolecule or system to one of these classes can be made by a simple inspection of the absorption spectrum. The spectroscopic properties of these types of azo compounds will be covered in Sections 1.3 to 1.5 of this chapter.

1.2.2 Isomerization

A major feature of the azo group is its capability to isomerize; this is the property used widely in photoresponsive organic thin films. The E-Z photoisomerization of azobenzene was detected by Hartley,[1] who created the Z-form by irradiation of E-azobenzene. Generally, the E-forms of azo compounds are more stable than the Z-forms. The parent E-azobenzene is by ca. 50 kJ mol^{-1} the more thermodynamically stable isomer.[19,20] Isomerization is the main photoreaction of most aromatic azo compounds. Other thermal- and photoreactions lose the competition with isomerization.

If the processes occurring during thermal- and photoisomerization are to be analyzed and understood a kinetic analysis is very useful. This can help give answers to the following questions: How fast and how effective is the isomerization reaction? What is the molecular mechanism? and Do all units isomerize according to the same mechanism and kinetics? Some tools for this analysis are presented here.

1.2.2.1 Concentration/Time Relations: How Fast?

The general scheme of geometric photoisomerization is given by three elementary reactions of Scheme I (see also Figure 1.1)

$$E \underset{h\nu,\Delta}{\overset{h\nu}{\rightleftarrows}} Z$$

(Scheme I)

Isomerization can be induced by light in both directions or by heat in the Z → E direction. The reverse thermal reaction is not observed at normal temperatures. Any one of the elementary reactions can be missing. Z-azobenzene in solution [1,21] has a thermal Z → E activation enthalpy $\Delta H^{\#} \approx 96$ kJ mol^{-1} and a half life time of 2 to 3 days at room temperature. Thus, the thermal reaction is irrelevant for the photoisomerization at usual irradiation intensities (for comparison: Z-stilbene[22] has $E_a \approx 180$ kJ mol^{-1}, is liquid, and is kinetically stable). On the other hand, one of the photoreactions may not be active (e.g., when an irradiation wavelength is selected where one form does not absorb or when the quantum yield is too small). Inspection of Figure 1.1B shows that E- and Z-azobenzene have virtually no spectral region without overlapping absorption.

Photoisomerization can be induced by direct irradiation and by triplet sensitization. Monitoring by UV/VIS spectroscopy is a convenient means of following the kinetics. According to Scheme I, a photostationary state of different E/Z compositions is reached in which the composition is determined

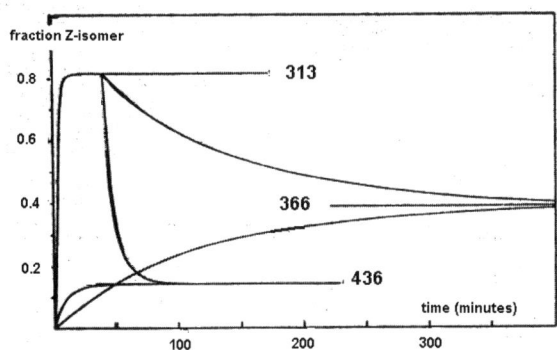

FIGURE 1.3 Time development of the Z-isomer of azobenzene on 313, 366, and 436 nm irradiation. The respective photostationary states are reached regardless of the starting point, pure E-isomer, or the 313 nm pss. (Adapted from reference 32, by permission.)

by the magnitude of the absorption coefficients of E and Z at the irradiation wavelength and of the quantum yields of the two isomerizations reactions and also by the rate constant of the thermal reaction (Figure 1.3). The photostationary state is independent of irradiation intensity if there is no thermal Z → E isomerization.

For photoresponsive systems, the kinetics of the change of a property induced by isomerization or photoisomerization is an important characteristic. Therefore, I will work this point out in some detail.

The kinetic evaluation of an isomerization according to Scheme I is subject to the condition $c_0 = c_E + c_Z$. Then the rate equation in homogeneous solution for c_Z is

$$\frac{dc_Z}{dt} = + 1000\, I_0^\lambda \frac{1 - 10^{-A^\lambda(t)}}{A^\lambda(t)}\, \varepsilon_E^\lambda \phi_E^\lambda c_E - 1000\, I_0^\lambda \frac{1 - 10^{-A^\lambda(t)}}{A^\lambda(t)}\, \varepsilon_Z^\lambda \phi_Z^\lambda c_Z - k^{therm} c_Z$$

$$= 1000\, I_0^\lambda \frac{1 - 10^{-A^\lambda(t)}}{A^\lambda(t)} (\varepsilon_E^\lambda \phi_E^\lambda c_0 - (\varepsilon_E^\lambda \phi_E^\lambda + \varepsilon_Z^\lambda \phi_Z^\lambda) c_Z) - k^{therm} c_Z \qquad (1.1)$$

where λ is the irradiation wavelength, I_0^λ is the irradiation intensity at this wavelength, ϕ_E^λ is the quantum yield of photoisomerization of E → Z, and ϕ_Z^λ is the yield of Z → E, and ε_E^λ is the absorption coefficient of E, and ε_Z^λ that of Z at the irradiation wavelength. (Note that ϕ_E^λ and ϕ_Z^λ are wavelength dependent for azobenzene (see Figure 1.9). $A^\lambda(t)$ is the time-dependent absorption of the solution due to all molecules absorbing at the irradiation wavelength. The fraction $(1 - 10^{-A^\lambda}(t))/A^\lambda(t)$ is called the *photokinetic factor*, which takes into account that the absorbance changes during the photoreaction and may be rationalized by the concept that it transforms the irradiation time axis to an axis of "photons absorbed." If A^λ is >0.01, which may well be the case in thin films, then $(1 - 10^{-A^\lambda})/A^\lambda$ is a constant of 2.3 within 1%.

For many azobenzene derivatives, the thermal back reaction can be neglected at ambient temperatures; in that case, the last term in Equation 1.1 can be omitted. If this is not the case, the photostationary state is dependent on the irradiation intensity. At very high irradiation intensities, the thermal back reaction becomes relatively unimportant. Thus, the thermal back reaction can be eliminated by choosing different irradiation intensities and extrapolating to $I_0^\lambda \to \infty$.[23]

For the development of the following procedures, the thermal reaction is considered slow compared to the photoreaction. From fundamental kinetics it is known that opposing reactions end in an equilibrium, opposing photo-reactions in a photostationary state (pss). The rate equations of neither c_E nor c_Z in Equation 1.1 are of first order. However, that equation can be trans-formed into an equivalent one describing the rate as a function of $(c_{Z\infty} - c_Z)$, i.e., the approach to the pss, which, indeed, is of first order, not in the irradia-tion time axis but in the axis "photons absorbed":

$$\ln \frac{c_{Z\infty} - c_Z}{c_{z\infty} - c_{Z0}} = -1000 \, I_0^\lambda \, (\varepsilon_E^\lambda \phi_E^\lambda + \varepsilon_Z^\lambda \phi_Z^\lambda) \int \frac{1 - 10^{-A^\lambda(t)}}{A^\lambda(t)} \, dt \qquad (1.2)$$

Remember, for $A < 0.01$ the integral is 2.303 t. Another useful equation derived from Equation 1.1 is for the pss with $k^{therm} = 0$:

$$\frac{c_{Z\infty}}{c_{E\infty}} = \frac{\varepsilon_E^\lambda \phi_E^\lambda}{\varepsilon_Z^\lambda \phi_Z^\lambda} \qquad (1.3)$$

From Equation 1.3 it is evident that the pss for different irradiation wave-lengths has varying compositions. Figure 1.3 shows E-azobenzene that was irradiated with 313 nm leading to a pss with 80% Z-isomer. Then the irradi-ation wavelength was changed to 366 nm, which creates a new pss with 48% Z-form. The same pss is reached when starting from pure E-form. 436 nm gives ca. 15 % of Z-form in the pss. The concentration/time relation in Figure 1.3 is not purely exponential, because the irradiation time is the variable.

For the evaluation of Equation 1.2, the irradiation intensity I_0^λ is needed. This is determined by actinometers,[24,25] which may be either physical in nature or chemical reaction systems with known quantum yields. Because the values of $(\varepsilon_E^\lambda \phi_E^\lambda + \varepsilon_Z^\lambda \phi_Z^\lambda)$ for azobenzene are well documented,[26] azobenzene itself is used as a convenient actinometer.[27]

Any physical property proportional to concentrations may be used for the analysis according to Equation 1.2. But the concentrations, especially in thin films, are usually determined by means of absorbance measurements at a wavelength ξ where only the Z- and/or E-forms of the azobenzene derivative change their absorbances. Then the irradiation (λ) and analyzing (ξ) wave-lengths must be held apart:

$$\ln \frac{A_\infty^\xi - A^\xi}{A_\infty^\xi - A_0^\xi} = -1000 \, I_0^\lambda \, (\varepsilon_E^\lambda \phi_E^\lambda + \varepsilon_Z^\lambda \phi_Z^\lambda) \int \frac{1 - 10^{-A^\lambda(t)}}{A^\lambda(t)} \, dt \qquad (1.4)$$

and

$$\frac{A_\infty^\xi - \varepsilon_E^\xi c_0}{A_\infty^\xi - \varepsilon_Z^\xi c_0} = \frac{\varepsilon_E^\lambda \phi_E^\lambda}{\varepsilon_Z^\lambda \phi_Z^\lambda} \qquad (1.5)$$

Equation 1.4 can determine whether the observed photoreaction is of first order or not, i.e., whether Scheme I is valid. Cases where Scheme I may not be sufficient for a description of the photoisomerization include systems with differing sites of isomerizing probe molecules with different quantum yields. An example is azobenzene in monomeric methylmethacrylate and poly-methylmethacrylate with a linear plot according to Equation 1.2 in the monomer, but nonlinear plots in polymers.[28,29,30] In this case, Scheme I, which for photoreactions requires equal values for ε and ϕ for all probe molecules is not sufficient. Different sites may be present (see Section 1.2.2.3).

1.2.2.2 The Quantum Yield: How Effective?

One of the most important features of a photoreaction is the value of the quantum yield ϕ_i^λ of compound i,[31] which is the quantifying answer to the question: "How effective?" In principle, the quantum yield is the ratio of the number of reacting molecules to the number of quanta absorbed. *In praxi*, there are several definitions of the quantum yield: true (only light absorbed by the reactant is considered) and apparent (there are other absorbers present), differential (at the moment) and integral (mean). In the previous rate equation, ϕ_E and ϕ_Z are the true differential yields. The monoexponential kinetics of Equation, 1.2 or 1.4 allow one to determine the yields in systems where the starting solution is already a mixture of E- and Z-forms (which can happen easily if the E-form is not prepared under strict exclusion of light). It turns out, however, that the values of the Z $\rightarrow$ E quantum yield are especially sensitive to small errors in the ε values.

For the determination of quantum yields in photoisomerizations without thermal back reaction, several situations regarding the availability of data of E- and Z-isomers are considered.

- The spectra and absorption coefficients of both E- and Z-forms are available. This is the case when both isomers can be isolated in sufficient quantities to allow the determination of their absorption coefficients. In this situation, sufficient information for the determination of ϕ_E^λ and ϕ_Z^λ is available via Equations 1.4 and 1.5.[32]
- The spectrum and absorption coefficients of the E-form are known, but for the Z-form only a spectrum is available. This is the case when the amount of Z-isomer is sufficient to isolate it (e.g., by TLC) and to take a spectrum but not enough material is available to weigh it. In this situation, there is one piece of information lacking for the evaluation according to Equations 1.4 and 1.5. This can be obtained by using two analysis wavelengths.[33,34]
- The spectrum and absorption coefficients of the E-form are available, but there is no spectral information for Z. This is the case where the isolation of Z is impossible—a case often encountered with thin films. In this situation, two pieces of information are lacking: the value and the spectral dependence of the absorption coefficient ε_Z. Fischer[35] presented a solution to this problem (1) by doing two isomerization experiments employing two different irradiation wavelengths: λ_1 and λ_2, and (2) by assuming that $\phi_Z^{\lambda_1}/\phi_E^{\lambda_1} = \phi_Z^{\lambda_2}/\phi_E^{\lambda_2}$, i.e., the ratio of the E $\rightarrow$ Z and Z $\rightarrow$ E quantum yields at the two irradiation wavelengths should be equal, which is only partly true for azobenzene (see Section 1.3.2.1). Fisher's article is not easy to read; one must be careful to discriminate strictly between irradiation and analysis wavelengths.

A review of these methods, with the mathematics of the kinetics, is given in a technical report.[36]

1.2.2.3 Solid Matrices: One or More Independent Photoisomerizations?

The kinetic analysis presented in Section 1.2.2.1 should result in a first order of the reaction when all photoresponsive units are equivalent (uniform reaction). This is generally the case in homogeneous solution, but it may be different in solid matrices, especially if these consist of different molecular components, and even more so if there is some kind of complexing of the photoresponsive and a matrix molecule. Then there may be nonequivalent sites that influence the absorption coefficient and/or the quantum yield of the photoresponsive unit. This leads to deviations from the first order behavior.

Mauser developed a method for the mathematical analysis of the kinetics of complex reaction systems,[37,38,39,40] which can apply here because the reactions at different sites are kinetically independent. As simple graphic instruments, the Mauser diagrams are useful for determining how many kinetically independent reactions (*s*) occur in the reaction system (the three elementary reactions in Scheme I are not independent).

- *s* = 1: In a series of spectra taken during the reaction, two arbitrary wavelengths are selected, and their absorbance changes (or the absorption readings themselves) at different times are plotted versus one another. If these AD-diagrams (<u>a</u>bsorbance <u>d</u>ifference) are linear, then exactly one independent reaction is present. If the plots are bent, more than one reaction occurs (Figure 1.4, parts A and B).
- *s* = 2: If the AD-diagrams are nonlinear, absorbance changes at three arbitrary wavelengths are selected. The changes at two wavelengths are divided by the absorbance change at the third and these <u>a</u>bsorbance <u>d</u>ifference <u>q</u>uotients are plotted in an ADQ2-diagram. If this diagram is linear, there

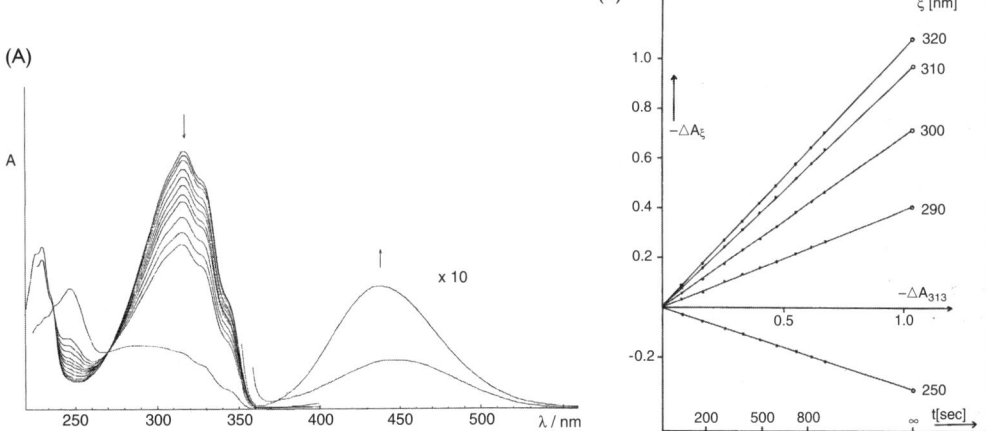

FIGURE 1.4 (A) Reaction spectra of E-azobenzene in hexane under 313 nm irradiation of $I_0 \approx 1.5 \cdot 10^{-9}$ Einstein cm^{-2} s^{-1}: two intervals of 5 s, five of 10 s, three of 20 s and a period of 1 h 25 min for reaching the pss. (B) Mauser AD-plots ΔA_ξ vs ΔA_{313} values for different wavelengths) for 313 nm photoisomerization of azobenzene.

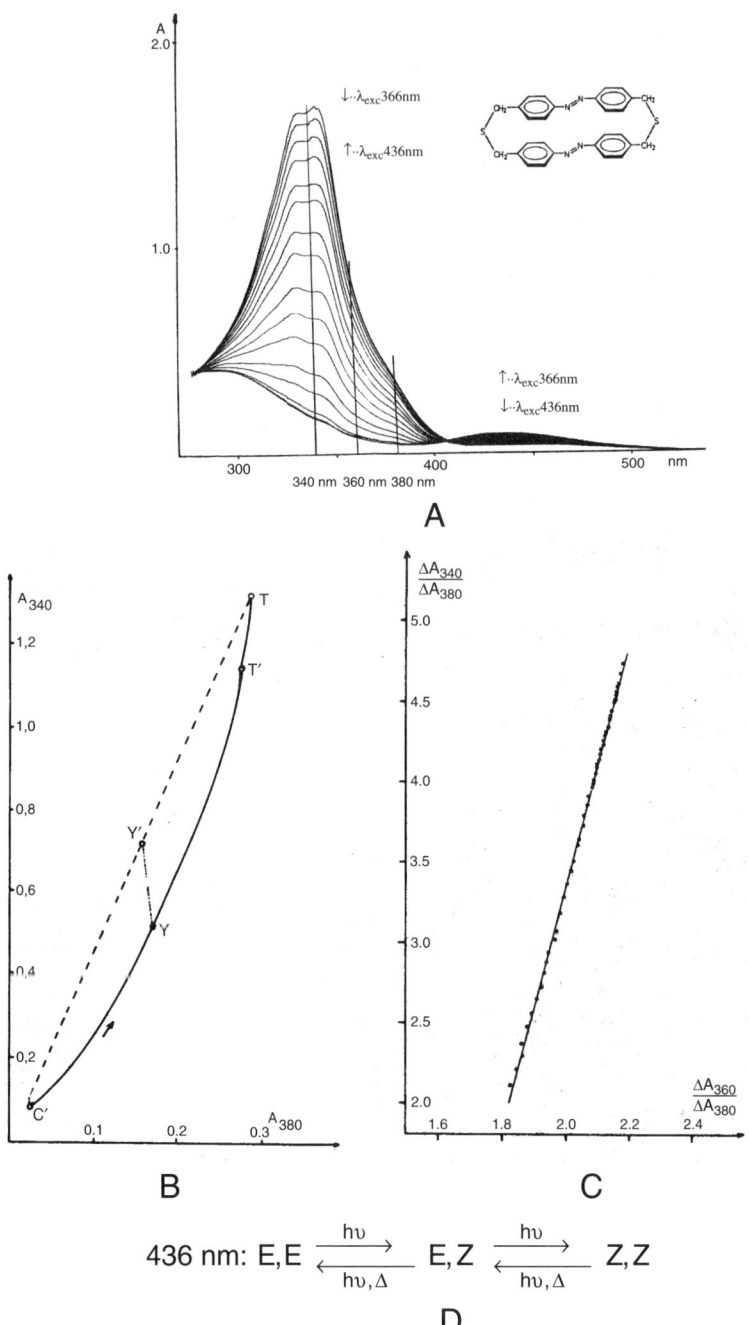

$$436 \text{ nm}: \quad E,E \; \underset{h\upsilon,\Delta}{\overset{h\upsilon}{\rightleftarrows}} \; E,Z \; \underset{h\upsilon,\Delta}{\overset{h\upsilon}{\rightleftarrows}} \; Z,Z$$

D

FIGURE 1.5 (A) Reaction spectra for 436 nm photoisomerization of an azobenzenophane, pre-irradiated by 313 nm to the photostationary state. (B) Nonlinear Mauser A plot of this photoisomerization. T: pure E,E-form; T′: pss at 436 nm irradiation; C′: pss at 313 nm irradiation, nearly pure Z,Z form. —— 436 nm photoreaction: C′ → T′ in ca 2 h. - - - - slow thermal reaction: C′ → T in ca. 10 d. ⋯⋯ Thermal reaction: Y → Y′ in 10 min. After interrupting irradiation at Y. The slow thermal reaction brings the system from Y′ to T within days. (C) Linear Mauser ADQ2 plot and (D) the reaction scheme (Adapted from reference 40, by permission.)

are exactly two independent reactions in the system; if not, there are more than two (Figure 5.1, parts A through D).

- s > 2: By choosing four or more wavelengths, higher-order ADQs-diagrams can be constructed with similar diagnostic value.[41] However, the scatter resulting from the propagation of experimental errors in dividing jeopardizes the usefulness of this procedure in most cases.

Mauser diagrams are very versatile. Every wavelength-dependent property can be used for their construction, e.g., the refractive index or, for optically active photoresponsive molecules, the ellipticity, a property quite sensitive to asymmetric perturbations of the environment. An example is given in Figure 1.6,[42] which shows that on fixation of the azo steroid in PMMA, an optically inactive matrix, the sites are not equivalent.

Mauser diagrams are useful for all sorts of reactions. For instance, for acid dissociation, the pH value takes the place of reaction time, and the pK values of different carboxy groups in a molecule can be determined.[41] Another example is complex formation. From the concentration dependence of the spectra, the degree of aggregation can be determined (dimers or higher aggregates?). At constant concentration, the aggregation as a function of temperature can be studied. The diagrams are one order higher, however, because solvent contraction mimics an independent reaction.

1.3 AZOAROMATICS OF THE AZOBENZENE TYPE

Spectroscopic and isomerization properties are relevant for the isomerization mechanism, a major discussion point on azo compounds. It is not easy, there-

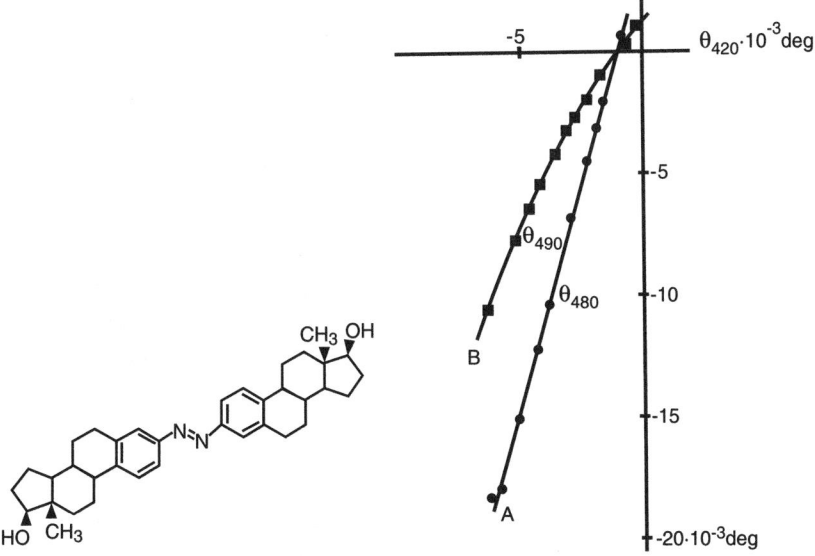

FIGURE 1.6 A Mauser θ- diagram of the photoreaction of an optically active azobenzene. A in diethyleneglycol, B in PMMA matrix. (Adapted from reference 42.)

fore to separate the data and mechanism sections, and as a result some redundancy will be encountered.

1.3.1 Spectroscopic Properties

1.3.1.1 UV/VIS Spectroscopy

The absorption spectra of E- and Z-azobenzene in EtOH solution were presented in Figure 1.1B. In these spectra, a weak, low-energy band that is well separated from the intense, higher-energy bands is identified. This constitutes the azobenzene type. The lowest-lying excited state is of the $^1(n,\pi^*)$ type; there is a large energy gap between this lowest (n,π^*) state and the next higher state that is of the (π,π^*) type.

1.3.1.1.1 E-Azobenzene

In the spectrum of E-azobenzene, the unstructured low intensity and low energy $n \to \pi^*$ band is identified in the 400 to 500 nm region (n-hexane: λ_{max} = 449 nm, ε_{max} = 405 l mol^{-1} cm^{-1}).[43] This band corresponds to an $n_- \to \pi^*$ excitation and is forbidden under the symmetry C_{2h} of E-azobenzene. Compared to $n \to \pi^*$ transitions in other molecules that of azobenzene is very intense. This may be due to nonplanar distortion and vibrational coupling.[44] The $n \to \pi^*$ intensity is stolen from the relatively far-off first $\pi \to \pi^*$ band, as shown by their common direction of polarization.[45,46] ε of the (n,π^*) state decreases by about 20% when the solution is frozen at 77K[3], which may be interpreted as better planarity at low temperature. The $n \to \pi^*$ band of the E-compound is continuous; a case with vibrational features has never been found. This band extends to 620 nm, where ε becomes smaller than $5 \cdot 10^{-3}$ l mol^{-1} cm^{-1}, which gives a state energy of ca. 17 500 cm^{-1} (2.1 eV, 205 kJ mol^{-1}).[47]

The energy gap between the low-lying (n,π^*) and the next (π,π^*) state is about 9000 cm^{-1} if the band origin of the continuous $n \to \pi^*$ band is taken, as usual, at 10% intensity—perhaps even greater. The intense $\pi \to \pi^*$ band is at λ_{max} = 316 nm (ε_{max} = 23,000 l $mol^{-1}cm^{-1}$), and a second $\pi \to \pi^*$ band appears at 229 nm (n-hexane: ε_{max} = 14,400 l $mol^{-1}cm^{-1}$)[43]. In the short wavelength region, the spectrum of E-azobenzene is very similar to that of stilbene, but red shifted by some 2000 cm^{-1}. The $\pi \to \pi^*$ bands show weakly expressed vibrational structures at room temperature, much less strongly expressed than in stilbene. But when solutions are cooled to 77 K in a rigid solvent, a well-defined structure with vibrational spacings of 1200 to 1400 cm^{-1} appears on both bands.[3,46] Less expressed vibrational structure can be induced by very viscous environments.[3]

1.3.1.1.2 Z-Azobenzene

In the spectrum of Z-azobenzene, the $n \to \pi^*$ transition is allowed under the symmetry of C_{2v}. The $n \to \pi^*$ band maximum is 440 nm with ε = 1250 l mol^{-1} cm^{-1}, much higher than that of E-azobenzene.[43] This is relevant to the E-Z isomerization reaction. This band is also continuous, as in all nonrigid Z-azo compounds.

The $\pi \rightarrow \pi^*$ band in Z-azobenzene has an odd appearance. It does not have an expressed maximum but seems to be situated at 274 nm with $\varepsilon = 5000$ l mol^{-1} cm^{-1}.[43] This indicates nonplanarity, which is also found in the X-ray crystal analysis.[16] When the structure is rigidified, such as by fixing two ortho-positions by a -CH$_2$- bridge (o,o'-azo-diphenylmethane), the band shifts to near 340 nm (indicating also more planarity) and gains intensity and some structure.[3] This indicates that on $\pi \rightarrow \pi^*$ excitation of Z-azobenzene, a Franck-Condon state that is on a slope of the potential energy surface of the $^1(\pi,\pi^*)$ state is reached.

1.3.1.1.3 Influence of Substitution

Many substituted azobenzenes belong to the azobenzene type, as well. Substitution may shift the n $\rightarrow \pi^*$ band from 440 nm (azobenzene), to 465 nm (hexamethylazobenzene), to 480 nm (2,2'-dimethyl-4,6,4',6'-tetra-tert.butylazobenzene), and even to 520 nm (hexaphenylazobenzene). Hydrocarbon, halogen, nitro, carboxy, acetyl, hydroxy, m-amino[48] and even 2,2'4,4',6,6' hexaphenyl substitution[49] influences the (π,π^*)-(n,π^*) state energy gap, but not so much as to shift the molecule into the aminoazobenzene group. In the context of this book, it is important that the long-chain and polymeric molecules containing azobenzene units coupled by means of hydroxy and carboxy substituents are of the azobenzene type.

Distortion of the azobenzene unit from planarity influences the spectral properties. In E,E-[2.2](4,4')-azobenzenophane 7, the N=N distance is the same as it is for azobenzene itself, but the tilt angle (dihedral angle C-N=N-C) of the benzene units is 169°, which is unusually far from the usual 180°. This causes an increase in the wavelength of the absorption maximum (462 nm) and the value of the absorption coefficient (1200 l mol^{-1} cm^{-1}) of the n $\rightarrow \pi^*$ band, and a decrease of the intensity of the $\pi \rightarrow \pi^*$ band ($\varepsilon = 20700$ l mol^{-1} cm^{-1}) without a band shift.[50]

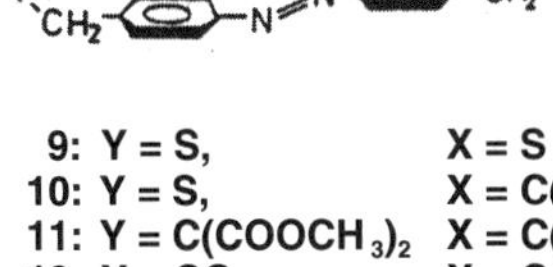

7

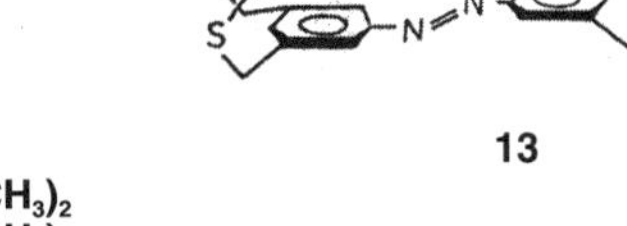

8

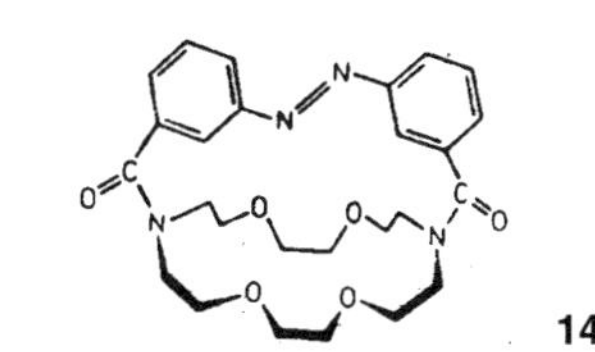

9: Y = S, X = S
10: Y = S, X = C(COOCH$_3$)$_2$
11: Y = C(COOCH$_3$)$_2$ X = C(COOCH$_3$)$_2$
12: Y = CO X = CO

13

14

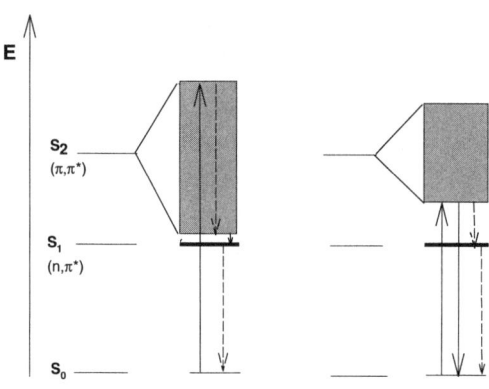

FIGURE 1.7   Energy state scheme with exciton splitting (two states for dimers, bands for polymers).

1.3.1.1.4 Influence of Solvent

The influence of solvent on the spectra is not very expressed. The $n \rightarrow \pi^*$ band is shifted approximately 5 to 10 nm. The position of the $\pi \rightarrow \pi^*$ band is also not very solvent dependent.

1.3.1.1.5 Aggregation

In the context of this book, aggregates of azobenzene are of interest. Azobenzenophanes[50,51,52] are the prototypes of dimers in fixed geometry. Figure 1.5 showed the absorption spectra of azobenzenophane **9** with two -CH$_2$-S-CH$_2$- bridges in the para-positions. One can see that the $n \rightarrow \pi^*$ band of the azobenzenophane is not much influenced but that there is a weak "phane-band" at the long wavelength tail of the $\pi \rightarrow \pi^*$ band around 380 nm. (This band was used for the construction of the AD-diagram in Figure 1.5B.) This feature also appears in the comparable spectrum of an analog fourfold bridged stilbenophane **13**.[53] The arrangement of the transition moments μ of the units in these dimers is parallel and like a card pack, and the spectral features suggest an exciton interaction,[54] which should amount to a state splitting of $\Delta E = 2\ \mu^2/R^3$ (Figure 1.7). Exciton splitting should be visible if the transition moment is large, which is the case only for the $\pi \rightarrow \pi^*$ band, and the dimer separation is small, which is true for the phanes with contact of the units (d = 308 pm). In this arrangement, the transition to the low energy state is forbidden, which explains the weakness of the "phane band."

$$CH_3\text{-}(CH_2)_{n-1}\text{-}O\text{---}\langle\bigcirc\rangle\text{---}N{=}N\text{---}\langle\bigcirc\rangle\text{---}O\text{-}(CH_2)_m\text{---}\overset{\overset{\displaystyle CH_3}{|}}{\underset{\underset{\displaystyle CH_3}{|}}{N}}\text{-}CH_2\text{-}CH_2OH \quad Br^-$$

15

Higher aggregation modifies the state diagram in the direction of a band structure. Such higher aggregates occur in many supramolecular systems, such as ordered structures created in Langmuir-Blodgett films or in vesicles by the intermolecular forces of large molecules. Kunitake and coworkers[55,56]

lead a very detailed investigation on bilayer assemblies of different amphiphiles of the type of **15**. They[55] have used the effect of temperature on aggregation and bring their spectral results, which are obtained for different
azobenzene amphiphiles of the p,p'-dihydroxy series, in line. The maximum
of the monomeric unit is around 355 nm; card-packed aggregates absorb at
lower wavelengths, down to 300 nm; head-to-tail arrangements, as verified in
the tilted bilayers, lead to high-intensity bands at longer wavelengths up to
400 nm. The influence of packing also becomes evident in the structure that
appears on the exciton spectra of the head-to-tail aggregates.[56]

1.3.1.1.6 Emission

Azobenzene-type azoaromatics generally do not emit with reasonable
quantum yields after excitation either to the $^1(n,\pi^*)$ or to the $^1(\pi,\pi^*)$ states.
There are only very few reports, and in most of them powerful lasers that
allow detection of quantum yields down to $5 \cdot 10^{-7}$ are used. Struve reported a
short-lived fluorescence of azobenzene in cyclohexane from the (n,π^*) state
$(\tau = 25 \text{ ps})$[57] and even from the second excited (π,π^*) state $(\tau = 5 \text{ ps})$,[58] a finding that violates Kasha's rule, but gains support from the fact of the large
$^1(n,\pi^*) - ^1(\pi,\pi^*)$ state separation. In a conference contribution, Hamai and
Hirayama[59] have presented confirmative results and given a $(\pi,\pi^*) \rightarrow S_0$
quantum yield of $1.7 \cdot 10^{-5}$ and a calculated lifetime of 0.06 ps. Azuma et al.[60]
report that the liquid crystalline *trans*-4-butyl-4'-methoxyazobenzene diluted
in hexane, which according to its absorption spectrum clearly is of the
azobenzene type, emits from the $^1(\pi,\pi^*)$ state (S_2) at 400 nm with about
0.25 ps. Although the residuals are not too good, the result is supported by
femtosecond absorption measurements which show that the $^1(n,\pi^*)$ state (S_1)
is populated with the same time constant. This state lives about 2.3 ps,
according to transient absorption data. In the femtosecond studies of
Fujino et al.,[61] the excitation intensity is obviously so high that a spectrum of
$S_2 \rightarrow S_0$ and $S_1 \rightarrow S_0$ fluorescence can be taken (Figure 1.8), with quantum
yields of $2.52 \cdot 10^{-5}$ and $7.54 \cdot 10^{-7}$.

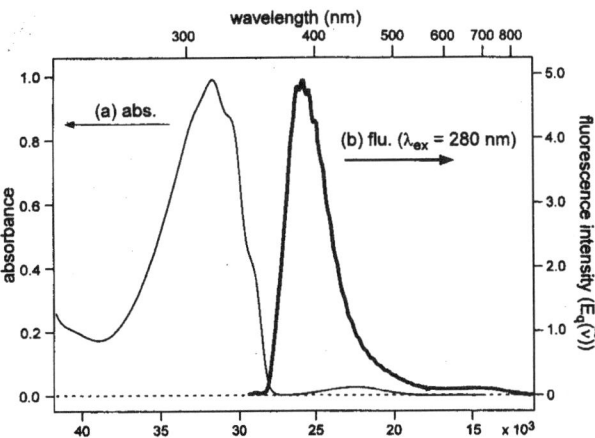

FIGURE 1.8 Absorption and emission of azobenzene under high intensity excitation. (Adapted from reference 61, by permission.)

Some azobenzenes that are locked against rotation by bulky substituents in all four ortho positions may show fluorescence when frozen rigidly at 77 K: 2,2',4,4',6,6'-hexaisopropyl; 2,2'-dimethyl-4,4'6,6'-tetra-tert-butyl azobenzene belong to this series.[62] Azobenzenophanes **7** to **13** do not emit, even at 77 K; this is the expectation for card-packed dimers.

Some reports on fluorescence occurring in, for instance, porous materials such as Nafion[63] or aluminophosphates,[64] do not refer to azobenzene but to protonated azobenzene, which is classified as a pseudostilbene (see Section 1.5). Emission from nonprotonated, isolated azobenzene-type molecules is still very rare. Aggregated systems, however, seem more prone to show fluorescence emission. Shinomura and Kunitake[56] have detected fluorescence bands with a maximum of near 600 nm in bilayer systems built from the monomers of **15**. They have shown that the ability to emit is tied to the type of aggregation: Head-to-tail aggregates emit relatively strongly, with quantum yields of up to $\phi = 10^{-3}$ and lifetimes below 2 ns. Their prototype of card-packed dimers does not emit at all. This is expected because of the low transition probability at the lower band edge, which favors radiationless deactivation, probably via the S_1 state (see Figure 1.7).

Tsuda et al.[65] found the same fluorescence at $\lambda_{max} = 600$ nm in a giant vesicle in a card-packed azobenzene arrangement. The noisy appearance of their fluorescence trace (if not due to the technique of confocal laser microscopy), however, suggests a very low emission intensity. Both Shinomura and Kunitake[56] and Tsuda et al.[65] report time-dependent orientation phenomena on Z-E isomerization in the supramolecular arrangement, which is reflected in the fluorescence intensity. So the former general statement that azobenzene-type azo molecules do not emit[3] needs to be modified.

On the other hand, azobenzene can quench the fluorescence of other molecules. This has been investigated in molecules containing both a fluorescing and an azobenzene unit.[66] It was found that the E-form is about 3 times,[67] or even up to 13 times,[68] as effective a quencher as the Z-form. The influence of the environment on such bichromophoric molecules was studied by Eisenbach et al.[69]

o-hydroxyazobenzene and other o- or p-hydroxy substituted azo compounds show emission at low temperature. Although this seems to be an unexpected n ← π* fluorescence, in reality it is not the fluorescence of an azo compound but that of the tautomeric hydrazone form.[70]

1.3.1.1.7 The Triplet State

Triplet state data for azobenzene-type azo compounds are very limited. Direct absorption of a 0.51 mol l^{-1} solution in $C_7H_{15}J$ in 5 cm cells has not been detectable.[47] Neither has phosphorescence been detected. The energy of triplet states has been located only by "chemical spectroscopy," i.e., the quenching of other molecules' triplet states by azobenzene. Ronayette et al.[71,72] found two relevant triplet states at about 196 and 180 kJ mol^{-1} (E-azobenzene) and about 192 and 142 kJ mol^{-1} (Z-azobenzene). Monti et al.[73] located triplet states at 146 (E-azobenzene) and 121 kJ mol^{-1} (Z-azobenzene). From their kinetic results, they inferred that the azobenzene acceptor should be twisted (phantom triplet) when accepting the energy and calculated the distortion energy to be 56 kJ mol^{-1} for the E-form and 77 kJ mol^{-1} for the Z-form. Pragst[74] used triplet sensitizers created in the recombination reaction of electrochemically produced radicals and reported 170 ± 10 kJ mol^{-1}.

1.3.1.1.8 Electronic State Calculations

Electronic state calculations for azobenzene in early papers suffered from the inability of older methods to take into account the mixing of (n,π^*) and (π,π^*) states. New calculations using ab-initio methods are successful, even in mastering donor/acceptor substituted azobenzenes.[75] A survey of calculations in connection with the isomerization mechanism will be given in Section 1.6.

1.3.1.1.9 Conclusion

The spectroscopy of azobenzene-type azo compounds is characterized by the large energy gap between the low-lying $^1(n,\pi^*)$ state and the next higher $^1(\pi,\pi^*)$ state. A certain floppiness of the molecular shape leads to state interactions, which, together with the ability of the molecules to isomerize, leads to fast, radiationless deactivation of azobenzene and its derivatives. This photostability is an asset for practical use. Aggregation rigidifies the molecular structure, as clearly demonstrated by the increase of vibrational structure of the bands. It also lowers the (π,π^*) energy states with longer (head-to-tail) or shorter (card-packed arrangements) lifetimes. An investigation of the photostability of aggregated molecules compared to nonaggregated or dispersed molecules is needed to assess possible disadvantages of the aggregates.

1.3.1.2 IR and Raman Spectroscopy

The relevant vibrations for this review are the N=N and C-N (Ph-N) stretching vibrations and, perhaps, torsional vibrations around the C-N bond. The E-azobenzene molecule has a center of inversion, and therefore the N=N vibration is infrared-inactive, but Raman-active, and has been found to be at 1442 cm^{-1}.[76] By IR spectroscopy, Kübler et al.[77] located the symmetric C-N stretching vibration at 1223 cm^{-1} in E- and at 866 cm^{-1} in Z-azobenzene. The N=N vibration in Z-azobenzene is at 1511 cm^{-1} (in KBr pellets). These numbers are confirmed by newer work; Biswas and Umapathy report 1439 and 1142 cm^{-1} for the N=N and C-N vibrations (in CCl$_4$),[78] and Fujino and Tahara[79] found nearly identical results (1440 cm^{-1} and 1142 cm^{-1}). A thorough vibrational analysis of the E-isomer is given by Amstrong et al.[80] The vibrations in the $^1(n,\pi^*)$ excited state are very similar: 1428 cm^{-1} and 1130 cm^{-1}.

1.3.1.3 Picosecond and Femtosecond Spectroscopy

The short pulse duration combined with the high photon density of ps- and fs-lasers have provided the means to study the properties of the excited states by emission and transient absorption measurements. Fluorescence of the lowest and higher excited states of azobenzene can be detected, but most work is being directed toward the dynamics of isomerization. Because questions about the isomerization mechanism are prominent in this field, this work will be discussed in Section 1.6: The Isomerization Mechanism.

1.3.2 Isomerization of Molecules of the Azobenzene Type

1.3.2.1 Thermal Isomerization

The E-form of azobenzenes is by 50 to 55 kJ mol^{-1} thermodynamically more stable than the Z-form.[19,20,81] For azobenzene, the heats of formation are[19] $\Delta H^0_{298}(E) = 311$ kJmol^{-1} and $\Delta H^0_{298}(Z) = 367$ kJmol^{-1}. The Z-form can be isolated[1] and crystallized, it is kinetically rather stable and is protected by an activation barrier that depends on the environment. In homogeneous[1] and liquid crystalline solution,[82] $E_a \approx 95$ kJ mol^{-1} is reported, a bit more in the melt[81] ($E_a = 105$ kJ mol^{-1}) and a value more than doubly as high in the crystal[81] ($E_a = 233$ kJ mol^{-1}). Andersson et al.[21] have determined the activation parameters in all three phases and find a linearity in the $\Delta H^{\neq}$ vs. $\Delta S^{\neq}$ plot, which leads them to conclude that the mechanism is the same in all three phases (inversion; see Section VI.) The thermal Z $\rightarrow$ E isomerization can be accelerated by iodine[83,84] and electron donating and accepting catalysts.[85,86] For a series of substituted azobenzene-type azobenzenes in solution, Talaty and Fargo[87] give activation energies between 85 and 100 kJ mol^{-1} and pre-exponential factors of 10^{12} to $3 \cdot 10^{13}$ s^{-1}. The dependence of the thermal isomerization rate on pressure is small. It increases less than 20% at 2100 bar and is not very dependent on the solvent ($\Delta V^{\neq} = -3.3$ ml mol^{-1} in hexane, -0.7 ml mol^{-1} in methanol).[88]

The thermal Z $\rightarrow$ E isomerization of azobenzene has been widely used to determine free volume in polymers at room[89] and temperatures as low as 4 K.[90,91] The thermal reaction is also important in the context of photo-response, as an information written or a signal or state produced by switching E to Z is slowly fading. However, the Z-lifetime is strongly modified by strain in the molecule: Z-azobenzene in solution at room temperature has a half life of about 2 days; the Z,E $\rightarrow$ E,E isomerization in the [3.3](4,4')azo-benzenophane 9 has a half life of ca. 4 min.;[52] the [2.2](4,4')azobenzenophane 7 has a half life of ca. 15 seconds;[50] and in dibenzo[2.2][4.4')-azobenzeno-phane 8 the life of the E,Z isomer drops to 1 s.[92] On the other hand, the Z,Z $\rightarrow$ Z,E isomerization in these phanes is slowed down enormously: Z,Z-7 lives 2.5 days, Z,Z-9 about 5 days, and Z,Z-10 about 1 year[93] at room temperature. Activation energies are available in the publications. The Z,E $\rightarrow$ E,E isomerization in most azobenzenophanes is very fast. However, in 2,19-Dioxo[3.3](3,3')azobenzolophane 12, the Z,E-form is relatively stable.[94] The remarkable differences in these and other structures[92,95] are not due to different activation enthalpies but to different activation entropies.

Of course, the surrounding matrix influences the lifetimes. Double and multiple decays of the Z-form concentration are found in polymer matrices[29] and monolayers.[96]

Electrochemical reduction of azobenzene leads to very fast isomerization of the resulting radical anion. Thus, there is no difference in the polarographic behavior of the two isomers.[97]

1.3.2.2 Photoisomerization

According to the absorption coefficients, irradiation in the UV drives the E-Z ratio toward Z and irradiation in the visible toward the E-form (Figure

1.4). At irradiation wavelengths where both forms absorb—and there is no wavelength region where only one absorbs—photostationary states are reached at a sufficiently long irradiation period that should not depend on the irradiation intensity when the thermal reaction is slow. (Figure 1.3 and Equation 1.3). Isomerization can be effected by powerful laser irradiation at wavelengths as long as 633 nm,[98,99] where $\varepsilon > 10^{-2}$ l mol^{-1} cm^{-1}. It is not clear whether the $^1(n,\pi^*)$ or the $^3(n,\pi^*)$ state is populated under these conditions.

The photoinduced and thermal isomerization reactions are nearly perfectly reversible, and side reactions are virtually absent. In de-aerated hydrocarbon solution, azobenzene can be irradiated for days with near UV or visible radiation without any change of absorbance after the photostationary state is established. Under air, the only side reaction is a very slow oxidation to azoxybenzene.[100] This can be checked without much effort by Mauser diagnostics (Section 1.2.2.3). For most azobenzenes, the application of absorbance diagrams gives perfectly straight lines, indicating that the isomerization is the only reaction (Figure 1.4). This fact warrants the use of azobenzene as a convenient actinometer.[24,25,101,102]

1.3.2.2.1 Quantum Yields

Not every photon absorbed by azo compounds induces isomerization. Table 1.1 shows a series of quantum yields of azobenzene collected from different authors. The spread of the values reflects the experimental problems of a seemingly simple system.

The quantum yields for azobenzene in solution are not concentration dependent.[103]

One feature of the data in Table 1.1 is that $\phi_{E \to Z}$ and $\phi_{Z \to E}$ do not add up to unity. This is an indication that more than one potential surface is involved in the isomerization mechanism (see Section 1.6). The most unusual fact disclosed in Table 1.1, however, is that the quantum yields are wavelength dependent. $\phi_{E \to Z}$ is about twice as large for low-energy excitation of the $^1(n,\pi^*)$ state as it is for high-energy excitation of the $^1(\pi,\pi^*)$ state. Obviously, for azobenzenes Kasha's rule—that fluorescence and photochemistry take

TABLE 1.1   Isomerization Quantum Yields of Azobenzene in Hydrocarbon Solution

$\pi \to \pi^*$		$n \to \pi^*$		Reference
$\phi_{E \to Z}$	$\phi_{Z \to E}$	$\phi_{E \to Z}$	$\phi_{Z \to E}$	
0.11	0.42	0.27	0.75	(32)
0.09	0.4	0.25	0.4	(113)
0.10	0.42	0.28	0.55	(71)
0.10	0.44	0.20	0.68	(103)
	0.40		0.55	(104)
0.11	0.27	0.25	0.56	(112)
0.13	0.27	0.31	0.52	(105)
0.12		0.24		(106)

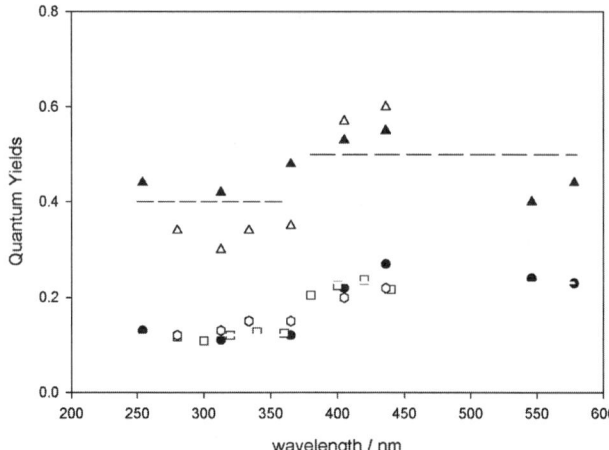

FIGURE 1.9 Wavelength dependence of the photoisomerization quantum yields. ● , o and □ are $\phi_{E \to Z}$ values from references 32, 106 and 105, ▲ and △ are $\phi_{Z \to E}$ values from references 32 and 105.

their start from the lowest excited state—is violated. The quantum yields are constant at excitation within one of the absorption bands (Figure 1.9)[32,106]: $\phi_{E \to Z}(\pi,\pi^*) = 0.12$ and $\phi_{E \to Z}(n,\pi^*) = 0.23$. A similar, but less expressed, difference is observed in the quantum yields of $Z \to E$ photoisomerization: $\phi_{Z \to E}(\pi,\pi^*) = 0.40$ and $\phi_{Z \to E}(n,\pi^*) = 0.55$.

There are, however, azobenzenes that have wavelength-independent isomerization quantum yields and thus obey Kasha's rule. The structure of these molecules inhibits rotation. Rau and Lüddecke[52] investigated azobenzenophane **9** and Rau[34] the azobenzene capped crown ether **14**, and these researchers found identical E,E → E,Z, and E → Z quantum yields respectively, regardless of which state was populated. The photoisomerization of azobenzenophanes **7**[50] and **13**[52] could not be evaluated in the same way because the photoisomerization is intensity-dependent. A series of azobenzenes substituted in all ortho positions to the azo group has equal quantum yields for n → π* and π → π* excitation if the substituents are ethyl, isopropyl, tert.butyl, or phenyl.[49] This provides clues for the elucidation of the isomerization mechanism (Section 1.6).

Surprisingly, 4,4'-dimethoxyazobenzene and 2,2'-dimethoxyazobenzene,[107] which show well-developed n → π* bands, also have equal E → Z quantum yields for 365 nm ($\phi_{E \to Z} = 0.35$ and 0.46, $\phi_{Z \to E} = 0.40$ and 0.52) and 436 nm irradiation (4,4'-DMOAB:$\phi_{E \to Z} = 0.36$, $\phi_{Z \to E} = 0.52$). Polyoxiethylene macrocycles containing the 4,4'-DMOAB unit are quite similar.

If more than one azobenzene unit is present in larger molecules or supramolecular systems, stepwise isomerization is observed. No cooperative effect in isomerization is noted.[50,52,108,109]

1.3.2.2.2 Influence of Temperature

E → Z quantum yields decrease slightly with decreasing temperature ($E_a \approx 3$ kJ mol-1).[103] Malkin and Fischer[110] confirmed that the difference in $\phi_{E \to Z}$ on n → π* and π → π* excitation exists down to 115 K. However, the

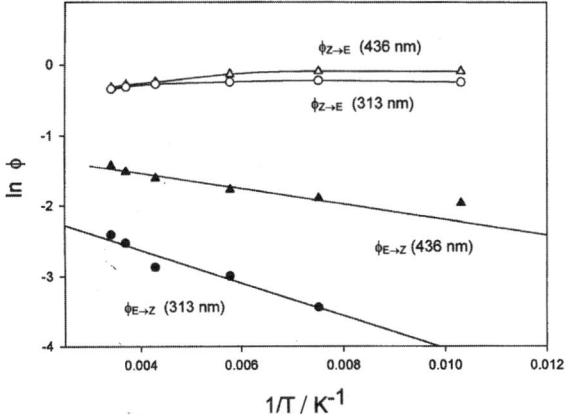

FIGURE 1.10 Temperature dependence of the quantum yields of azobenzene. (Data from reference 110.)

$Z \rightarrow E$ isomerization yield is nearly temperature-independent (Figure 1.10) or increases at low temperature, with only a small difference for excitation to the two lowest-excited states. So obviously, the $E \rightarrow Z$ photoisomerization—after irradiation to the (n,π^*) state as well as the $Z \rightarrow E$ isomerization—proceeds even at low temperature and in frozen solvents. In solid matrices, fast and slowly isomerizing molecules are observed on $\pi \rightarrow \pi^*$ excitation.[28,90] The fast process has a quantum yield of $\phi \approx 0.14$[90] that is temperature independent down to 4 K. With strong lasers, photoisomerization in the $E \rightarrow Z$ direction have been exploited, even at 4 K in hole burning experiments.[111] Thus, azobenzene photoisomerization cannot be frozen out.

1.3.2.2.3 Influence of Environment

Bortolus and Monti have determined the quantum yields of azobenzene isomerization in different solvents.[112] They found an increase of $E \rightarrow Z$ isomerization but a decrease of $Z \rightarrow E$ isomerization when solvents with high dielectric constant are used. This phenomenon is independent of the irradiation wavelength. Table 1.2 shows the special feature of wavelength-dependent quantum yields of azobenzene.

TABLE 1.2 Solvent Dependence of Isomerization Quantum Yields of Azobenzene[112]

Solvent	$\phi_{E \rightarrow Z}$		$\phi_{Z \rightarrow E}$	
	$n \rightarrow \pi^*$	$\pi \rightarrow \pi^*$	$n \rightarrow \pi$	$\pi \rightarrow \pi^*$
n-Hexane	0.25	0.11	0.56	0.27
Ethyl Bromide	0.26	0.11	0.58	0.25
Ethanol	0.28	0.25	0.51	0.24
Acetonitrile	0.31	0.15	0.46	0.21
Water/EtOH (80/20% v:v)	0.35	0.32	0.41	0.15

Gegiou et al.[113] found "only a very slight viscosity effect, both in the n-π* and in the π-π* absorption bands" on the isomerization quantum yield. They used glycerol as a viscous solvent, but the result may also be transferred to polymer matrices. In solid matrices, several photoisomerization modes are observed (see the preceding section on the influence of temperature). A comparison between azobenzene isomerization in liquid methylmethacrylate and the slow mode in poly(methylmethacrylate) showed that the difference in quantum yields on S_1 (0.17) and S_2 excitation (0.03) is retained in the solid matrix.[28] The fast process is not observed in n → π* excitation.[28] These data are important in relation to the use of the azobenzene isomerization method for the determination of the free volume in a polymer.

Confinement of azobenzene in defined structures changes the quantum yields. In the cavity of ß-cyclodextrine, the $\phi_{E \to Z}$ are reduced and become nearly wavelength-independent, whereas the $\phi_{Z \to E}$ are practically unaffected by the inclusion.[114] Photoisomerization is found for azobenzene-type molecules in ZSM-5 and sicalite-1 zeolites.[115] E-stilbene, in comparison, does not isomerize on direct excitation in the channels of narrow pore zeolites (the fluorescence yields increase by a factor of 10),[116,117] and the triplet-sensitized isomerization of Z-stilbene in zeolites is inhibited.[118] These findings are relevant to the discussion of the isomerization mechanism in Section 1.6.

In monolayers of azobenzene containing amphiphiles, the characteristics of photoisomerization are dependent on the chain length.[119]

There is one report of a concentration dependence of the 313 nm photostationary state of azobenzene and 4-methoxyazobenzene in cyclohexane—not in benzene or CCl_2F-$CClF_2$—in the literature.[120] A bimolecular excimer intermediate was postulated. Further work is needed to elucidate whether the absorption coefficients or the quantum yields are concentration-dependent, for instance by ground or excited-state association (cf. Equation 1.3).

1.3.2.2.4 Triplet-State Isomerization

Isomerization of azobenzenes may also be sensitized by triplet sensitizers. Jones and Hammond[121] and Fischer[122] came up with different results: 2% of Z-form versus 25% in the photostationary state. A thorough study by Lemaire and coworkers[71,72] showed that two triplet states at 195 and 180 kJ mol^{-1} in the E- and at 190 and ca. 140 kJ mol^{-1} in the Z-isomer are involved in the reaction. According to Bortolus and Monti,[112] sensitizers with high (>190 kJ mol^{-1}) triplet energy transfer their energy efficiently (diffusion controlled) to both E- and Z-azobenzene. Still, the isomerization yield is small— $\phi_{E \to Z}$ = 0.015—in agreement with Jones and Hammond.[121] On the other hand, the sensitized Z → E isomerization has $\phi_{Z \to E}$ = 1.0. Azobenzenophane **9** also undergoes triplet-sensitized isomerization.[52]

1.3.2.2.5 The Azobenzene Radical Anion

One-electron reduction of azobenzene yields the azobenzene radical anion. Its Z-form thermally isomerizes fast. This explains the cleavage/recombination mechanism reported for azosulfides[123] and the reduction/oxidation mechanism of azobenzene derivatives in Langmuir-Blodgett monlayers

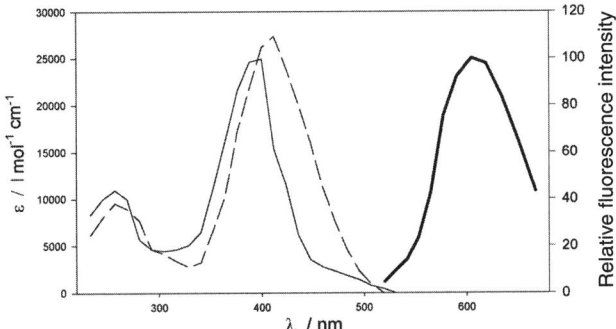

FIGURE 1.11   Absorption spectra of 4-dimethylamino azobenzene in ——— hydrocarbon and - - - - - in ethanol solution. Emission spectrum in hydrocarbon solution at 77 K ———. (Adapted from reference 7, by permission.)

deposited on an electrode.[124] Liu et al. use the combination of reduction, isomerization, and reoxidation as an analytical instrument in thin films.[125]

1.4 AZOAROMATICS OF THE AMINOAZOBENZENE TYPE

1.4.1 Spectroscopic Properties

Substitution of azobenzene by good electron-donating groups, (such as o- or p-amino groups) shifts the $^1(\pi,\pi^*)$ state far to the red, whereas the $^1(n,\pi^*)$ is nearly constant in energy. Thus, azo compounds of the aminoazobenzene type are characterized by a close energetic proximity of the $^1(\pi,\pi^*)$ and $^1(n,\pi^*)$ states, which may interchange their relative energetic position on outside influences (Figure 1.11). Thus, they may appear to be of the azobenzene type in hydrocarbons but of the pseudo-stilbene type in hydrogen-bonding or acidic solutions.

The absorption bands of aminoazobenzene-type molecules have some charge transfer characteristics. So the vibrational structure is weakly expressed and the spectra are sensitive to the polarity of the solvent[48] (Figure 1.11). Note that the intensity of the n → π^* band is increased relative to the π → π^* compared to the azobenzene type molecules. This may indicate increased state mixing.

Thus, fluorescence may occur in aminoazobenzene-type compounds, but it is not prominent. Low-temperature, glassy solvents[48] or other methods of external rigidification, such as adsorption to a surface[126] at liquid nitrogen temperature, are necessary. For 4-dimethylaminoazobenzene, fluorescence in hydrocarbon solution at 77 K is on the red side of the n → π^* band and assigned to n←π^* emission. With the high-power lasers available today, one would anticipate emission of many molecules of this class.

Phosphorescence has not been observed with these compounds. However, using energy transfer experiments, Monti et al.[127] have determined that the energy of the triplet state is near 140 kJ mol^{-1} for 4-diethylamino-, 4-diethylamino-4'-methoxy- and 4-nitroazobenzene. They assign the triplet state to the (n,π^*) type in contrast to what one would expect; normally the singlet-

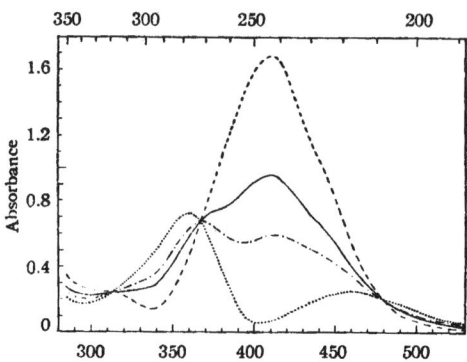

FIGURE 1.12 Photoisomerization of 4-dimethylamino azobenzene. The dotted line is the extrapolated Z-spectrum. (From reference 128, by permission.)

triplet split of (π,π^*) states is greater than that of (n,π^*) states. Moreover, a twist which lowers the energy and may be the reason that no triplet-triplet absorption is observed is inferred in the triplet state.

The lifetime of the Z-isomers of aminoazobenzene-type molecules is generally short. Approximate spectra are gained by extrapolation of a series of photoisomerization reaction spectra[128] (Figure 1.12).

The azo group, whose n-electrons are orthogonal to the π-system, is more basic than the amino group, whose n-electrons are part of the π-system. It is protonated before the amino group[129] in acid/ethanol mixtures of increasing acid concentration.

1.4.2 Isomerization of Molecules of the Aminoazobenzene Type

1.4.2.1 Thermal Isomerization

Although the activation energies of aminoazobenzene-type compounds (E_a between 75 and 88 kJ mol^{-1}) are not very different from those of azobenzene-type molecules, thermal Z → E isomerization of aminoazobenzene-type molecules is in general much faster than that of the azobenzene-type compounds. Conventional flash experiments are necessary to monitor the changes.[130] The half-life of the Z-form of dimethyl-aminoazobenzene in toluene at 298 K is 220 s.[131] A Linear Free Energy Relationship and Hammett relation is established, which includes azobenzene- and aminoazobenzene type compounds.[130] A linear ln k_{isom} vs. π^*, the Taft parameter of solvent polarity, is also observed.[132] The dependence of the isomerization rate on pressure is weak: In most solvents, it increases less than 35% at 2100 bar, $\Delta V^{\#} \approx -1.65$ ml mol^{-1}. Methanol is exceptional, with $\Delta V^{\#} = -17$ ml mol^{-1}.[88,133]

1.4.2.2 Photoisomerization

The photoisomerization of aminoazobenzene-type compounds is complicated by the presence of the rapid thermal Z → E isomerization. Pure Z-isomers generally cannot be isolated, and their spectra thus cannot be measured directly. Extrapolation is possible, however[128] (Figure 1.12). So for 4-dimethylamino-azobenzene flash excitation,[130] a rotating-shutter technique[128] or low-temperature techniques[131] have been applied. Another technique is based on the intensity-dependent photostationary state of the reaction; this technique extrapolates to infinite light intensity, where the thermal reaction becomes unimportant.[23]

1.4.2.2.1 Quantum Yields

The method of variation of the irradiation intensity gives $\phi_{E \rightarrow Z}$ as well as $\phi_{Z \rightarrow E}$. For 4-dimethyl-aminoazobenzene in n-hexane irradiation at 366 nm or 405 nm, both of which are wavelengths of the $\pi \rightarrow \pi^*$ band, the results are $\phi_{E \rightarrow Z} = 0.17$ and $\phi_{Z \rightarrow E} = 0.25$.[23] Albini et al.,[134] who calculated the extent of Z $\rightarrow$ E isomerization by an extrapolation of this conversion back to zero time of irradiation, found somewhat higher $\phi_{E \rightarrow Z}$ values for 4-diethylamino-azobenzene in de-aerated cyclohexane solution (Table 1.3).

The quantum yield at 436 nm is significantly higher than it is at shorter wavelength irradiation (Table 1.3) which is like in azobenzene-type compounds. Indeed, in hydrocarbon solution, the absorption spectrum shows the azobenzene-type feature of a long wavelength n $\rightarrow \pi^*$ band (Figure 1.11).

The photostability of aminoazobenzene-type azo compounds in hydrocarbon solution is high, at least for irradiation in the visible and near UV ($\lambda > 313$ nm) and is comparable to the azobenzene-type compounds. For short wavelength (254 nm) irradiation, photoreduction is detected, but with quantum yields of less than 1%.[134]

1.5 AZOAROMATICS OF THE PSEUDO-STILBENE TYPE

1.5.1 Spectroscopic Properties

Azoaromatics of the pseudo-stilbene type are characterized by the feature that is of the (π,π^*) the lowest-excited singlet state type (Figure 1.13). Compared to the azobenzene type, the $^1(n,\pi^*)$ and $^1(\pi,\pi^*)$ states are rearranged. This reordering can happen by lowering the energy of the (π,π^*) state (e.g., by donor/acceptor substituents) and/or by increasing the energy of the (n,π^*) state (e.g., by asymmetric[135,136] protonation or by complexing the n-electrons of the azo group).

TABLE 1.3 Quantum Yields for E $\rightarrow$ Z Isomerization of 4-Diethylaminoazobenzene (4-DEAMAB) and 4-Diethylamino-4'-methoxyazobenzene (4-DEAM-4'-MOAB) in De-arerated Cyclohexane (Adapted from reference 134.)

Irradiation wavelength	254 nm	313 nm	366 nm	436 nm
4-DEAMAB	0.23	0.25	0.21	0.72
4-DEAM-4'-MOAB	0.34	0.31	0.27	0.84

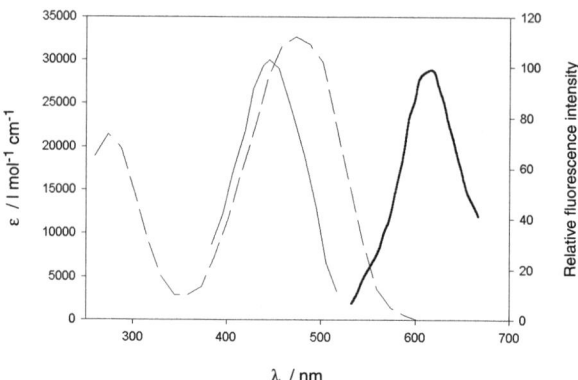

FIGURE 1.13 Absorption spectra of 4-dimethylamino-4'nitro azobenzene in ——— hydrocarbon and - - - - - in ethanol solution. Emission spectrum in hydrocarbon solution at 77 K ———. (Adapted from reference 7, by permission.)

1.5.1.1 Absorption

All azoaromatics, be they of the azobenzene, aminoazobenzene, or donor/acceptor pseudo-stilbene type, experience considerable spectral changes on protonation[137,138] or complexation.[107,139,140] Ortho metallation has the same effect.[141,142] Usually the $\pi \rightarrow \pi^*$ band is red-shifted, possibly due to the localized charge at the N-atom. By the same token, the $^1(n,\pi^*)$ state is shifted to higher energies. Minor band shifts and intensity changes indicate double protonation[143] of azobenzene.

The prototype molecule for donor/acceptor substitution is 4-dimethyl-amino-4'-nitroazobenzene. Here, the $\pi \rightarrow \pi^*$ band is shifted far to the red due to the charge transfer character of the transition. The band has few vibrational features, and its energy is influenced by the polarity of the solvent. The weak $n \rightarrow \pi^*$ band cannot be seen under the intense $\pi \rightarrow \pi^*$ band (Figure 1.13). Most commercial azo dyes are pseudo-stilbenes rather than azobenzene-type molecules.

1.5.1.2 Emission

Pseudo-stilbenes may emit fluorescence that is, contrary to true stilbenes, generally weak at room temperature and often weak even at low temperatures. Protonated azobenzene-type molecules and many protonated azo dye molecules emit strong fluorescence in sulfuric acid at 77 K[44,144] with quantum yields of about 0.1. Inclusion of azobenzene in the channels of $AlPO_4$-5 crystals provides complexation of the n-electrons and space confinement. This leads to emission by protonated azobenzene at room temperature.[64] For their cyclopalladated azobenzenes, Ghedini et al.[141] report quantum yields of ca. $1 \cdot 10^{-4}$ and lifetimes of ca. 1 ns. In contrast, donor/acceptor pseudo-stilbenes, if emitting at low temperatures[48] or when adsorbed to surfaces,[126] are weak emitters. In textile chemistry, it has long been known that azo dyes adsorbed to fibers may show fluorescence.[145]

1.5.1.3 The Tripet State

The energy of the triplet states of pseudo-stilbenes cannot be detected by phosphorescence. In some cases, transient absorption or energy transfer gives some information. Görner et al.[146] investigated the temperature-dependent kinetics of population and decay of the triplet state of 4-dimethylamino-4'-nitroazo-benzene by laser flash experiments and found that its energy is higher than 165 kJ mole^{-1}. Monti et al.[130] used the energy transfer method and found a $^3(\pi,\pi^*)$ state at 140 kJ mole^{-1}; they inferred nonplanarity by twisting of the azo bond ("phantom triplet state"). Higher triplet states may play a role in biphoton holographic storage;[147] this proposition, however, has been challenged.[148]

Because the thermal Z → E isomerization is very fast, the Z-isomers cannot be isolated. The spectra are available only from an extrapolation of different E/Z mixtures produced by conventional flash irradiation[149] and have to be considered qualitative.

1.5.2 Isomerization of Molecules of the Pseudo-Stilbene Type

1.5.2.1 Thermal Isomerization

There is a great difference between the two types of pseudo-stilbenes—donor/acceptor-substituted and protonated azo compounds—with regard to the stability of the Z-forms. The former isomerize quickly, whereas the latter may be relatively stable.

1.5.2.1.1 Protonated Azo Compounds

The rate of Z-azobenzene-H$^+$ isomerization is dependent on acid strength:[138,140] It is low below 70% of H_2SO_4 (by weight), the half lives at room temperature at 83 wt% H_2SO_4 are 20 min, at 90% 7 min.[143] This is rationalized by the formation of an intermediary ABH_2^{2+} species in concentrated acids, which indeed has been detected in $SbF_5/FSO_3/SO_2$ ("magic acid") at $-80°C$.[150] Surprising are the high negative entropies of activation of the reaction. Unfortunately, there are no studies of the temperature dependence of the thermal isomerization of complexed azobenzenes. This type of compound may be of higher importance in complex matrices than hitherto suspected.

1.5.2.1.2 Donor/Acceptor-Substituted Azo Compounds

Contrary to the Z-forms of protonated azobenzenes, the Z-forms of donor/acceptor di-substituted azobenzenes isomerize very quickly at room temperature. To investigate this reaction, the Z-form is produced by flash excitation. The isomerization of these pseudo-stilbenes is strongly dependent on the polarity of the solvent.[130,133,149,151] For instance, 4-diethylamino-4'-nitroazobenzene isomerizes with $k_{Z\to E} = 0.007$ s^{-1} in hexane, but with $k_{Z\to E} = 600$ s^{-1} in N-methylformamide.[149,152] 4-Anilino-4'-nitroazobenzene in cyclohexane obeys the relation ln $k_{Z\to E} = 22(\pm3) - 72(\pm9)\cdot10^3/(8.314\cdot T)$.[153] Schanze et al.[154,155] have established two linear relationships between the free-activation enthalpy of isomerization and the Kosower Z values of aprotic and protic solvents. Sanchez and de Rossi[156,157] report a strong pH sensitivity of

Z → E isomerization of 4-dimethylamino-azobenzene and its o'- and p'-carboxylated derivatives (o- and p-methyl red) in the alkaline realm, with a warning about "the interpretation of results when the rate of isomerization is measured in organized media if the pH is not appropriately controlled."

1.5.2.1.3 Influence of Pressure

Pressure dependence was thoroughly investigated by Asano and his group.[88,132,133] It turns out that the partial volumes of the Z-forms of 4-dimethylamino-4'nitroazobenzene and related molecules are ca. 250 cm^3 mol^{-1} in all solvents. Those of the E-forms are smaller and solvent-dependent.[132] Thermal isomerization rates are weakly dependent on pressure in nonpolar solvents, but contrary to azobenzene- and aminoazobenzene-type compounds, they are strongly dependent in polar solvents:[88] in hexane 10%, in acetone 475% for 2100 bar ($\Delta V^{\#}$ = -0.7 and -25.3 cm^3 mol^{-1}, respectively). This has implications for the discussion of the mechanism of isomerization (Section 1.6).

One concludes from these facts that pseudo-stilbenes are not suitable for persistent switching of the molecular form. Any information based on E-Z isomerization is quickly lost. If, however, fast interconversion of E- and Z-forms is the aim, as it is in the alignment of the higher-order polarizability tensor of donor/acceptor azobenzenes, then thermal isomerization supports the photoisomerization process.

1.5.2.2 Photoisomerization

1.5.2.2.1 Protonated Azo Compounds

In a reaction analogous to that of stilbene, protonated azobenzene isomerizes at room temperature under irradiation with $\phi_{E \rightarrow Z}$ = 0.27 and $\phi_{Z \rightarrow E}$ = 0.25.[158] Subsequently, the dehydro-photocyclization starts from the photostationary E/Z mixture to give protonated benzo[c]cinnoline[137,138] with a yield of ϕ = 0.02.[158] This is parallel to the stilbene-to-phenanthrene reaction, and it proceeds not in concentrated sulfuric acid where the thermal Z → E reisomerization is fast, but best in ca. 66% acid. Protonated benzo[c]cinnoline and hydrazobenzene are formed, and the latter undergoes the benzidine rearrangement.[137,138,158] The complexation with metal ions in azobenzene crown ethers under participation of the n-electrons of the azo group leads to an increase of the E → Z quantum yield from 0.25 of the metal free compound to 0.4 to 0.6 in the Ba^{2+} and Ca^{2+} complexes. The Z → E yield decreases from 0.18 to 0.13 and 0.05. If the crown is larger, the values increase to $\phi_{E \rightarrow Z}$ ≈ 0.35 and $\phi_{Z \rightarrow E}$ ≈ 0.45.[107]

1.5.2.2.2 Donor/Acceptor-Substituted Azo Compounds

At room temperature, the isomerization of e.g. 4-dimethylamino-4'-nitroazobenzene cannot be observed reasonably, because the thermal back isomerization of the Z-form is too fast.[159] To determine the spectrum of the Z-form, Gabor and Fischer[160] applied low-temperature and extrapolation techniques. The photoisomerization quantum yields are $\phi_{E \rightarrow Z}$ ≈ 0.20 and $\phi_{Z \rightarrow E}$ ≈ 0.75 at 163 K. They decrease to $\phi_{E \rightarrow Z}$ ≈ 0.06 to 0.09 at 143 K, whereas $\phi_{Z \rightarrow E}$ does not change much. Newer data regarding quantum yields are not available.

1.5.2.2.3 Influence of Solvent

The dependence of the photoisomerization process on the polarity of the environment varies greatly for differently substituted compounds. King et al.[161] found that p-nitro- or cyanosubstituted azobenzenes had photostationary states with similar E/Z ratios, but that an additional p'-amino function stopped photoisomerization in acetonitrile. This was not true, however, in methylcyclohexane, where lifetimes of Z-isomers were determined to be in the order of seconds at room temperature. These findings are retained at -35⁰C, which is taken as proof that it is not the fast thermal Z → E isomerization that fakes the lack of photoisomerization. This conclusion may be questionable, however, considering the weak temperature dependence for azobenzene-type molecules (Figure 1.10).

Pseudo-stilbene type azobenzenes are not as stable under UV irradiation as the azobenzenes.[159,162,163,164] Irick and Pacifici found that in de-aererated alcoholic solvents, photoreduction gives the hydrazo compound with a yield of ca. 10^{-4} at 254 nm irradiation.[159] Albini et al[165] found yields of one order higher, but at 313 nm and longer wavelength irradiation, the decomposition is virtually absent. Some caveat is necessary when triplet-sensitizing additives are present, because irradiation under air does not lead to photoreduction.[165] Benzophenone-sensitized reduction in benzene proceeds with a yield of 0.17.[164]

1.6 THE ISOMERIZATION MECHANISM

1.6.1 Azobenzene-Type Molecules

The close similarity between isosteric azobenzene and stilbene suggested that their isomerization mechanisms might be the same and that this would be by rotation around the central double bond. As early as 1941, however, Magee et al.[166] speculated about a different mechanism for azobenzene. They proposed a planar transition state with a single bond between the N-atoms of the azo group. The "lateral shift mechanism," today called "inversion," was proposed by Curtin et al.[167] a rehybridization of the n- and σ-electrons of the azo group should create a planar transition state that would only weakly influence the π-system. Such a mechanism is established in the ground state isomerization of imines.[168,169] The two alternative paths from Z- to E-azobenzene are visualized in Figure 1.14. For the two routes, the simple coordinates: twist angle or N-N-C-bond angle, should be good approximations of the true reaction coordinates. They have different potential energy profiles.

Photoreactions proceed on the potential energy surfaces of both the excited and the ground state.[170] Interesting features of the potential energy diagrams are the maxima and minima as well as the loci where the system changes from one potential energy surface to another, usually from an excited-state curve to the ground-state curve. Note that the gradient of the potential energy curves gives the force exerted on the molecular system along the respective coordinate.

Even after the advent of femtosecond spectroscopy, a potential energy

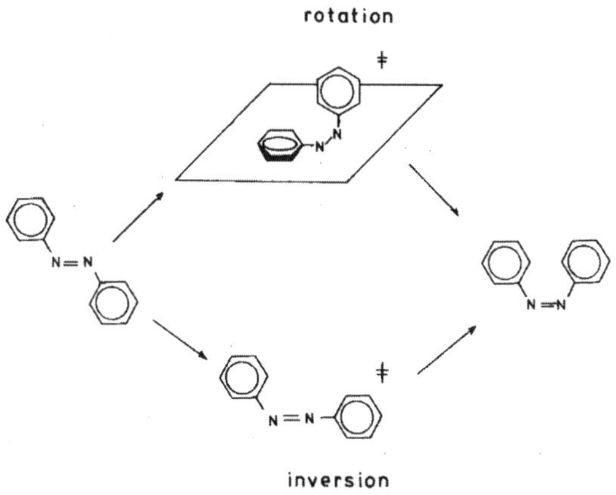

FIGURE 1.14 The rotation and inversion mechanisms of isomerization of azobenzene. (Adapted from reference 7, by permission.)

surface cannot be mapped out fully by experimental probing. We a re still restricted to the two possibilities for creating a potential energy diagram. One is by experimentally guided intuition, which constructs the unobservable parts of the curves by exploiting the experimental data, such as steady-state and transient absorption and emission spectra and their activation energies, and principles like the correlation of reactant and product states, symmetry considerations, the principle of avoided crossings of energy curves, and so on. The other possibility is by calculation, which scans the surfaces point by point for different geometrical configurations. With the development of computing capacity and improved software, the latter method has become increasingly reliable. One may address this as "experimental" theoretical chemistry.

1.6.1.1 The Thermal Isomerization Mechanism

For the thermal ground-state $Z \rightarrow E$ isomerization of azobenzene, the inversion mechanism was readily accepted, mostly on the grounds of the much lower activation energy as compared to stilbene (96 versus 180 kJ mol^{-1}) and the parallel reaction in the imines.[168] A direct proof of inversion has been given by Rau and Lüddecke[52] and by Tamaoki et al.,[50] who found that azobenzenophanes that cannot rotate still isomerize. Chemical intuition would indicate that the inversion starts with a symmetrical molecular vibration leading finally to a linear transition state. Calculations have shown convincingly, however, that the inversion takes place at only one N-atom in a semilinear transition state (Figure 1.14);[171,172,173] a linear transition state would be energetically too unfavorable. This also holds for an inversion mechanism in the excited state.

For pseudo-stilbene–type molecules, the question of the mechanism of thermal isomerization was taken up again in the early eighties by Whitten et al.[149] and later by Kobayashi et al.,[174] who, on the basis of their isomerization experiments with donor/acceptor-substituted azobenzenes in polar solvents, postulate rotation. Asano and coworkers[88,133,151,175] infer from the isomerization activation volumes in their experiments under pressure, a dependence of the mechanism on solvent and even parallel reaction paths along these mechanistic coordinates. A close proximity of the calculated energies of the rotation and inversion transition states was pointed out by Cimiraglia and Hofmann.[176] They draw attention to the strong lowering of the energy barriers of rotation on substitution.

Isomerization needs some extra sweep volume. The volumes for the two mechanisms of azobenzene should be quite different—ca. $0.25 \, nm^3$ for rotation and ca. $0.12 \, nm^3$ for inversion.[177,178] This bears out in restricted spaces. In some zeolites azobenzene can isomerize whereas stilbene does not. Kuriyama and Oishi found that there are two separate $\Delta H^{\#}$ versus $\Delta S^{\#}$ lines for azobenzenes isomerizing by inversion (azobenzene type) and rotation (pseudo-stilbene type).[179]

1.6.1.2 The Photoisomerization Mechanism

In their 1971 review, Ross and Blanc[5] expressed doubts as to the operation of the inversion mechanism in the *excited* states. This opened another round of heated discussion. The rotation/inversion controversy invoked much theoretical and experimental work.

The use of Walsh diagrams,[180] based on one-electron molecular orbitals, shows that on $n \rightarrow \pi^*$ excitation the azobenzene molecule is stretched, which is the beginning of inversion. All calculations and suggestions for an inversion mechanism agree that the potential energy curve for inversion has a relatively steep slope at the E- and the Z- geometries. This is corroborated by the experimental evidence of a continuous $n \rightarrow \pi^*$ absorption band in both isomers. In fact, a structured $n \rightarrow \pi^*$ band in an azo compound that can isomerize has never been observed.

The crucial finding was that in the case of azobenzene, the isomerization quantum yields for excitation of the higher $^1(\pi,\pi^*)$ state are about one-half those on excitation to the lower $^1(n,\pi^*)$ state (Table 1.1). Similar facts hold for aminoazobenzene 1.1×2 type molecules with low-lying $^1(n,\pi^*)$ state (Table 1.3). On the other hand, for azobenzene-type molecules whose structure inhibits rotation, the quantum yields become equal. To rationalize their results, Rau and Lüddecke[52] suggested that two different isomerization pathways in the excited states in azobenzene-type molecules should be active: rotation in the high energy $^1(\pi,\pi^*)$ state and inversion in the low energy $^1(n,\pi^*)$ state. This concept was extended to a potential energy diagram[34] (Figure 1.15B). At the geometry of E-azobenzene the $^1(n,\pi^*)$ state has a relatively steep slope in the direction of the inversion coordinate leading to a minimum from which the deactivation to the ground state occurs in the vicinity of a ground state maximum providing for 25% of isomerization. In contrast, in the $^1(\pi,\pi^*)$ state, the slope of inversion is uphill, rotation is favored and leads to a "bottleneck" state near the geometry of the maximum of the twisted ground state, and the partition of the population toward the Z-isomer is different from that of $^1(n,\pi^*)$ inversion. The crucial point in this model is the vanishing internal conversion $^1(\pi,\pi^*) \rightarrow ^1(n,\pi^*)$ near the E-geometry. This channel, however, seems to be disfavored by the large energy gap if the rotation channel is open.

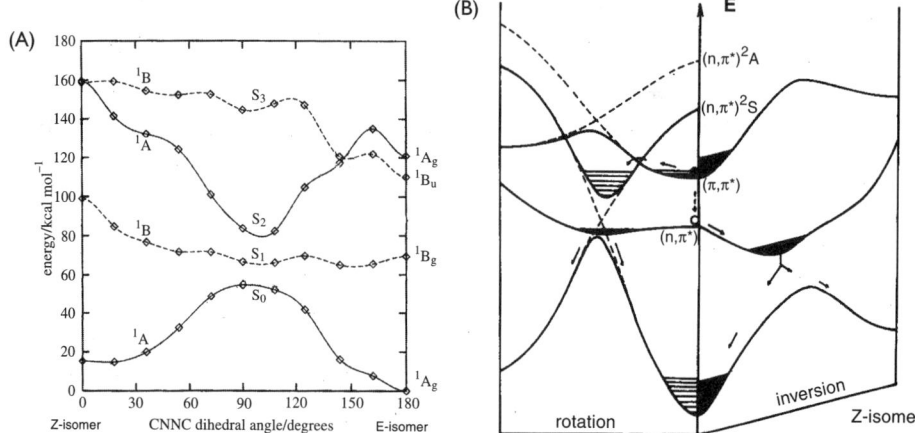

FIGURE 1.15 (A) The calculated singlet states relevant for isomerization by rotation according to Cattaneo and Persico. (Adapted from reference 184, by permission.) (B) The potential energy curve system for rotation and inversion according to Rau (From reference 34, by permission.)

Besides solving the quantum yield enigma, this concept also rationalizes some other results. If rotation is inhibited by, say, structural design as, for instance, in azobenzenophanes or constraint from outside as, for instance, in restricted spaces as in ß-cyclodextrin[114] or zeolites[116,117] or in solid matrices or low temperature down to 4 K,[111] then the internal conversion from the $^1(\pi,\pi^*)$ to the $^1(n,\pi^*)$ state provides a virtually barrierless path of isomerization. The fact that the stilbenophane analogue of Tamaoki's azobenzenophane shows isomerization[181,182] does not invalidate this reasoning—the azobenzenes choose the easiest isomerization path.

This suggestion, based on intuitive penetration into then nonvisible geometries of azobenzene, has met critical theoretical examination by calculations of the potential energy surfaces and critical experimental examination that comes from ultra-short spectroscopy.

1.6.1.3 Calculations of Potential Energy Curves

Some older calculations suggest rotation in the $^1(n,\pi^*)$ state[171] or the activity of both mechanisms in the S_1 and S_2 state.[172,173] They were of the highest quality in their time. In a newer paper, Monti, Orlandi, and Palmieri[183] reported an ab-initio calculation of the state energies at four special geometries for azobenzene: E and Z, planar semilinear, and 90° twisted azobenzene configurations. These authors found that the twisted $^1(n,\pi^*)$ is higher in energy than the twisted $^1(\pi\pi^*)$ state and even the $^1(\pi\pi^*)$ state of the E-isomer, whereas the semilinear $^1(n,\pi^*)$ state should be at lower energy than the two corresponding $^1(n,\pi^*)$ states. With these cornerstones, the authors conjectured a system of potential energies for rotation and inversion (Figure 1.16), which differs from that of Figure 1.15 for the rotation pathway. According to the diagrams of Monti, Orlandi, and Palmieri, on

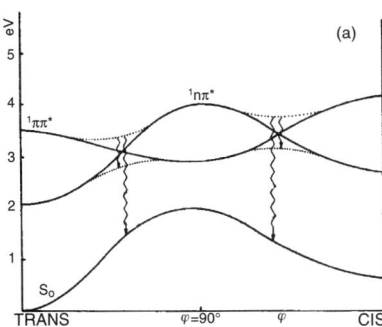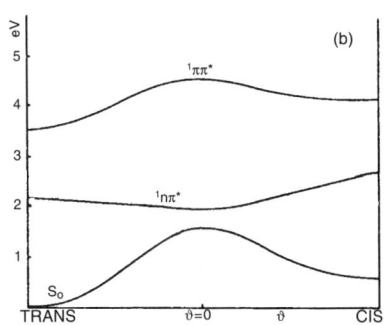

FIGURE 1.16 The singlet potential energy curve system for rotation (a) and inversion (b) according to Monti, Orlandi, and Palmieri. (Adapted from reference 183, by permission.)

excitation of E-azobenzene to the $^1(\pi,\pi^*)$ state, isomerization by rotation is not expected because of the energy barrier toward the Z-side of the diagram. But after some rotational motion, internal conversion occurs and a portion of the molecules finally isomerize via inversion on the $^1(n,\pi^*)$ surface. Figure 1.16 was misinterpreted by many later authors as indicating rotation in the $^1(\pi,\pi^*)$ state. Cattaneo and Persico[184] state: "Although these results give only a qualitative picture of the potential energy surfaces (PES) of azobenzene, they constitute for more than 15 years the leading theoretical reference for the interpretation of azobenzene photochemistry." In a recent ab-initio study, Cattaneo and Persico extended the calculations to 12 geometrical configurations along the rotation (Figure 1.15A) and 7 geometries along the inversion coordinates. Their potential energy curves are very similar to those in Rau's scheme (Figure 1.15B), and they find that the S_3/S_2 transition near the E-geometry on the rotational coordinate is very fast, because in C_2 geometry, the crossing is a conical intersection. Thus, isomerization in the $^1(\pi,\pi^*)$ state occurs by rotation. For $n \rightarrow \pi^*$ excitation, these authors predict an inversion mechanism but cannot clearly exclude rotation.

On the other hand, very new calculations by Ishikawa et al.[185] reject the possibility of inversion. They predict fast isomerization on $n \rightarrow \pi^*$ excitation via a conical intersection with the ground state near 90° on the *rotational* coordinate. These authors agree with Cattaneo and Persico: On $\pi \rightarrow \pi^*$ excitation, isomerization should occur by rotation via a doubly excited state, which at the 90° geometry interacts with the (n,π^*) and ground state. But there is no conclusive discussion about the difference in $n \rightarrow \pi^*$ and $\pi \rightarrow \pi^*$ quantum yields.

Thus, the newest theoretical results challenge the picture of the double inversion and rotation mechanism that was accepted for two decades.

1.6.1.4 Ultrashort Transient Spectroscopy

In the experimental realm, picosecond and femtosecond experiments contribute to the understanding of isomerization. A review of the ultrafast dynamics of photochromic systems appeared in 2000.[186] A study of the photochemistry of E-azobenzene on excitation of the S_2 state [$^1(\pi,\pi^*)$] was conducted by Lednev et al.[187] They find a transient absorption in the region red to the $\pi \to \pi^*$ absorption band with two lifetimes: $\tau_1 \approx 1$ ps and $\tau_2 = 11–15$ ps. The ground state recovers (in part) with a single time constant of $\tau_3 = 13$ ps. This indicates that the S_2 state does not deactivate directly to the ground state, but rather that a "bottleneck" state $S^\dagger$ with 13 ps lifetime is passed. The authors find their results to be compatible with rotation in the S_2 state. An extended study by Lednev et al.[188] also included $n \to \pi^*$ excitation. This creates transient absorptions near 400 nm (strong) and 550 nm (weak). Both bands decay with a single time constant of $\tau = 2.5$ ps when the excitation is near the origin of the absorption band. On $\pi \to \pi^*$ excitation, the improved time resolution of the equipment allowed a modification of the earlier results: An immediate transient at 475 nm decays very fast ($\tau > 0.2$ ps) and creates the band at 400 nm with the same time constant. This 400 nm band in turn decays biexponentially with the times of ca. 0.9 and ca. 15 ps. In a third paper, Lednev et al.[189] investigate the molecule 14 where rotation is inhibited. They find, regardless of excitation wavelength, a transient absorption with a decay time of ca. 2.6 ps and the same time constant of the (partial) recovery of the ground state. From the persistent ground-state absorption change, they estimate $\phi_{E \to Z} = 0.26$ for S_2 excitation, in agreement with the steady-state result.[34] This is in accord with Figure 1.15. Nägele et al.[190] give an idea about the relative slopes of the $^1(n,\pi^*)$ inversion curve at the Z- and E-geometries by the time constants of the disappearance of the first transient absorption in the 450 to 550 nm region: 170 fs for the Z- and 320 fs for the E-isomer. After the intramolecular processes, the vibrational cooling of the ground state by energy transfer to the solvent occurs in about 20 ps,[191] in agreement with the results of Terazima et al.[192]

Picosecond time-resolved Raman studies give information about the early time dynamics of the isomerization process. For $n \to \pi^*$ excitation, the N=N bond of E-azobenzene shrinks by 3 pm, and the C-N bond is elongated by 1 pm within 5 to 30 fs,[78,193] which is in accord with an inversion mechanism. Fujino and Tahara[79] see the properties of the (n,π^*) and ground state after $\pi \to \pi^*$ excitation. In the picosecond realm, the N=N vibrations in the ground (1440 cm^{-1}) and the (n,π^*) states (1428 cm^{-1}) are nearly the same, indicating the retention of the N=N double bond, which suggests the retention of a planar geometry. The authors infer internal $S_2 – S_1$ conversion from their findings.

In a very new report, Fujino et al.[61] challenge the two-isomerization-mechanism concept on the basis of their time-resolved and time-integrated femtosecond fluorescence measurements of E-azobenzene following excitation of the $^1(\pi,\pi^*)$ state. They use the extremely weak fluorescence (cf. Figure 1.8) as an indicator for the population of the emitting state. From the ratios of their measured fluorescence lifetimes (S_2: 0.11 ps; S_1: 0.5 ps) and the radiative lifetimes deduced from the (absorption-spectra-based) oscillator strengths, they determine the fluorescence quantum yields: $2.53 \cdot 10^{-5}$ for the $S_2 \to S_0$ emission and $7.54 \cdot 10^{-7}$ for the $S_1 \to S_0$ emission. By comparison with the integrated $S_2 \to S_0$ and $S_1 \to S_0$ fluorescence intensities, they derive an efficiency of unity ($\phi = 1.07 \pm 0.15$) for the $S_2 (\pi,\pi^*) \to S_1 (n,\pi^*)$ internal conversion. On this basis, they exclude any isomerization in the (π,π^*) state.

The given error margin would exclude compatibility with the undoubtedly wavelength-dependent isomerization yields found in steady-state experiments (Figure 1.9). The authors seek a not very convincing explanation in the internal conversion of vibrationally excited states of S_1 directly to hot vibrational states of the ground state, whose cooling time has been determined to be ca. 16 ps. To comply with the results for rotationally blocked molecules, these vibrations should experience rotational features that do not lead to isomerization. Vibrationally excited states may be important, because the lifetime of the $^1(n,\pi^*)$ state is much smaller than the cooling time in a solvent.[191]

With Fujino and Tahara's proposition,[61] based on experimental results, that no isomerization should be possible in the high energy $^1(\pi,\pi^*)$ state and that exclusively inversion in the $^1(n,\pi^*)$ state should lead to isomerization, and with Ishikawa's proposition,[185] based on computational results, that all isomerization should occur on a rotational pathway even in the (n,π^*) state, another round of discussion of the isomerization mechanism in azobenzene seems to have been opened. Further work will be necessary, but both single-route mechanisms will need to rationalize the wavelength-dependent quantum yields.

1.6.1.5 The Triplet State

Isomerization in the triplet manifold seems to be totally separate from the singlet route. Here also, however, isomerization quantum yields are reported that are dependent on the sensitizer's triplet energy.[71] No implications for the isomerization mechanism have been discussed.

1.6.2 Pseudo-Stilbene–Type Molecules

1.6.2.1 The Thermal Isomerization Mechanism

Based on their isomerization experiments in polar solvents, Whitten et al.[149] stated that thermal $Z \rightarrow E$ isomerization of donor/acceptor-substituted azobenzenes proceeds by rotation. Later, Shin and Whitten[194] modified this point of view and saw a dual mechanism active dependent on solvent and donor strength. Asano and coworkers,[88,133,151,175] from their experiments under pressure, inferred a dependence of the mechanism on solvent and even parallel reaction paths along these mechanistic coordinates.

1.6.2.2 The Photoisomerization Mechanism

The photoisomerization mechanism in aminoazobenzene- and pseudo-stilbene–type compounds has attracted far less attention than the mechanism for azobenzenes. In pseudo-stilbenes, the $^1(n,\pi^*)$ state is buried under the intense $\pi \rightarrow \pi^*$ band and cannot be populated selectively. No state-specific quantum yields are available because the yields are independent of the exciting wavelength.[160] There is only a very narrow experimental basis for a discussion of these two mechanisms. However, this may change when pseudo-stilbenes are subjected to ultrashort-time experiments.

Monoprotonation of azobenzenes is asymmetric.[135,136] a $^1(n,\pi^*)$ state still exists, but it is expected to shift to high energy, and the molecules are very stilbene-like. Rotation would be the isomerization mode. This concept is supported by the strong temperature dependence of fluorescence: Azobenzene-

H^+, for instance, does not emit at room temperature, but does emit strongly at 77 K in rigid H_2SO_4.[126]

In the donor/acceptor-substituted azobenzenes, the π conjugation system is predominantly affected. A decrease in energy of the (π,π^*) states and a weakening of the central double bond is previewed.[176] The close energetic proximity of the (π,π^*) and (n,π^*) states, combined with easy distortions of the molecule from planarity, suggest vibrational coupling of the two states. Then one can envisage fast transitions between them, and it may be the special conditions of substitution, environment, temperature, pressure, etc. that determine the route of isomerization. Rotation is predicted as the preferred path, but inversion may also be possible, calculations predict energetic proximity of the transition states of the two isomerizations paths.[195]

1.6.2.3 The Triplet State

As in the azobenzene type systems, the triplet pathway seems to be decoupled from the singlet route in pseudo-stilbenes. Little is known about the mechanism in the triplet state. The only information comes from calculations, and these show that the triplet surfaces are frequently similar in shape to the singlet surfaces. Thus, both mechanisms may be operative in the triplet state, too.

1.7 CONCLUDING REMARKS

The photoisomerization of all types of azobenzenes is a very fast reaction on either the singlet or triplet excited-state surfaces according to the preparation of the excited state, with nearly no intersystem crossing. "Bottleneck states" have lifetimes on the order of 10 ps. The molecules either isomerize or return to their respective ground states with high efficiency. So photoisomerization is the predominant reactive channel, and the azobenzenes are photochemically stable. Only aminoazobenzene-type molecules and pseudo-stilbenes have small quantum yields of photodegradation.

Thus, azobenzene-type compounds especially have the potential to survive many isomerization cycles. Therefore, they are preferred in photoresponsive devices where bistability is a goal. If substitued azobenzenes have to be considered for such a device—usually for synthetic reasons—care should be taken to find a compound with the characteristic $n \rightarrow \pi^*$ band in the absorption spectrum. Substituents meeting this demand are, *cum grano salis*: halogenalkyl, and hydroxy or carboxy groups.[48,87] There are many examples of such devices in the following chapters of this monograph.

Pseudo-stilbenes are better suited for the creation of devices with a high anisotropic molecular order and large macroscopic higher-order polarizabilities. To reach this, a fast thermal $Z \rightarrow E$ isomerization is required. For this purpose, molecules with a long wavelength intense $\pi \rightarrow \pi^*$ band should be selected. The selection of p-alkyamino and p'-nitro or cyano groups gives the best donor/acceptor combinations. There are also many examples for such devices in the following chapters.

From the knowledge of the isomerization data of Sections 1.3 through

1.5. of this chapter and the considerations on the isomerization mechanism in Section 1.6 guidelines can be established for the construction of new photo-responsive systems.

REFERENCES

1. Hartley, G. S. (1938). The cis-form of azobenzene and the velocity of the thermal cis → trans conversion of azobenzene and some derivatives. *J. Chem. Soc.*, 633–643.
2. De Lange, J. J., Robertson, J. M., and Woodward, I. (1939). X-ray crystal analysis of azobenzene. *Proc. Roy. Soc. London, Sect. A* **171**, 398–410.
3. Rau, H. (1973). Spektroskopische Eigenschaften organischer Azoverbindungen. *Angew. Chem.* **85**, 248–258; *Angew. Chem. Intl. Ed. Engl.* **12**, 224–235.
4. Wyman, G. (1955). The cis-trans isomerization of conjugated compounds. *Chem. Rev.* **55**, 625–667.
5. Ross, D. J., and Blanc, J. (1971). Photochromism by cis-trans isomerization. In: *Photochromism* (G.H. Brown, Ed.) Wiley-Interscience, New York.
6. Rau, H. (1990). Azo compounds. In *Photochromism. Molecules and Systems* (H. Dürr and H. Bouas-Laurent, Eds.) Chapter 4, pp. 163–192, Elsevier, Amsterdam.
7. Rau, H. (1990). Photoisomerization of azobenzenes. In *Photochemistry and Photophysics* (J. F. Rabek, Ed.) Vol. II, pp. 119–141, CRC Press, Boca Raton.
8. Landolt-Börnstein. *Zahlenwerte und Funktionen aus Naturwissenschaften und Technik*, Bd.1, 4. Teil, p. 104, Springer Verlag, Berlin, 1955.
9. Houk, K. N., Chang, Y.-M., and Engel, P. S. (1975). Photoelectron spectra of azo compounds. *J. Am. Chem. Soc.* **97**, 1824–1832.
10. Robin, M., Hart, R., and Kuebler, N. (1967). Electronic states of the azoalkanes. *J. Am. Chem. Soc.* **89**, 1564–1572.
11. Solomon, B. S., Thomas, T. F., and Steel, C. (1968). Primary processes in the photochemistry of bicyclic azo compounds. *J. Am. Chem. Soc.* **90**, 2249–2258.
12. Wall, J. (1997). *Spektroskopische Untersuchungen am 2,3-Diazabicyclo[2.2.2]oct-2-en (DBO)*. Wissenschaftliche Arbeit, Hohenheim.
13. Engel, P., Horsey, D., Keys, D., Nalepa, C., and Soltero, L. (1983). Photolysis of reluctant azoalkanes. Effect of structure on photochemical loss of nitrogen from 2,3-diazabicyclo[2.2.2]oct-2-ene derivatives. *J. Am. Chem. Soc.* **105**, 7108–7114.
14. Turro, N. J. (1978). *Modern Molecular Photochemistry* Chapter 13.4, pp. 544–550, Benjamin Cummings, Menlo Park.
15. Bouwstra, J. A., Schouten, A., and Kroon, J. (1983). Structural Studies of the System trans-Azobenzene/trans-Stilbene. I. A Reinvestigation of the Disorder in the Crystal Structure of trans-Azobenzene $C_{12}H_{10}N_2$. *Acta Crystallogr. Sect. C.* **39**, 1121–1123.
16. Hampson, C. G., and Robertson, J. M. (1941). Bond lengths and resonance in the cis-azobenzene molecule. *J. Chem. Soc.* 409–413.
17. Traetteberg, M., Hilmo, I., and Hagen, K. (1977). A gas electron diffraction study of the molecular structure of trans-azobenzene. *J. Mol. Struct.* **39**, 231–239.
18. Ho, M.-S., Natansohn, A., Barrett, C., and Rochon, R. (1995). Azo polymers for reversible optical storage 8. The effect of polarity of the azobenzene groups. *Can. J. Chem.* **73**, 1773–1778.
19. Schulze, F.-W., Petrik, H.-J., Camenga K. H., and Klinge, H. (1977). Thermodynamic properties of the structural analogues benzo[c]cinnoline, Trans-azobenzene, and Cis-azobenzene. *Z. Phys. Chem. (Wiesbaden) NF*, **107**, 1–19.
20. Adamson, A. W., Vogler, A., Kunkely, H., and Wachter, R. (1978). Photocalorimetry. Enthalpies of photolysis of trans-azobenzene, ferrioxalate and cobaltioxalate ions, chromium hexacarbonyl, and dirhenium decacarbonyl. *J. Am. Chem. Soc.* **100**, 1298–1300.
21. Andersson, J. A., Petterson, R., and Tegnèr, L (1982). Flash photolysis experiments in the vapour phase at elevated temperatures. I: Spectra of azobenzene and the kinetics of its thermal cis-trans isomerization. *J. Photochem.* **20**, 17–32.
22. Kistiakowski, G. B., and Smith, W. R. (1935). Kinetics of thermal cis-trans isomerization. III. *J. Am. Chem. Soc.* **56**, 638–642.

23. Rau, H., Greiner, G., Gauglitz, G., and Meier, H. (1990). Photochemical quantum yields in the A (+hv) ⇔ B (+hv, Δ) system when only the spectrum of A is known. *J. Phys. Chem.* **94**, 6523–6524.

24. Kuhn, H., Braslavsky, S. E., and Schmidt, R. (1989). Chemical Actinometry. *Pure Appl. Chem.* **61**, 187–210.

25. Gauglitz, G. (1990). Actinometry. In *Photochromism. Molecules and Systems* (H. Dürr and H. Bouas-Laurent, Eds.), Chapter 25, pp. 883–902. Elsevier, Amsterdam.

26. Gauglitz, G., and Hubig, S. (1985). Chemical actinometry in the UV by azobenzene in concentrated solution: a convenient method. *J. Photochem.* **30**, 121–125.

27. Gauglitz, G., and Hubig, S. (1981). Azobenzene as a convenient actinometer: evaluation values for UV mercury lines and for the N_2 laser line. *J. Photochem.* **15**, 255–257.

28. Shen, Y.-Q., and Rau, H. (1991). The environmentally controlled photoisomerization of probe molecules containing azobenzene moieties in solid poly(methylmethacrylate). *Makromol. Chem.* **192**, 945–957.

29. Lamarre, L., and Sung, C. S. P. (1983). Studies of physical aging and molecular motion by azochromophoric labels attached to the main chains of amorphous polymers. *Macromolecules* **16**, 1729–1736.

30. See Chapter 2 of this monograph.

31. (1983) Glossary of Terms Used in Physical Organic Chemistry. *Pure Appl. Chem.*, **55**, 1281–1371.

32. Zimmerman, G., Chow, L.-Y., and Paik, U.-J. (1958). The photochemical isomerization of azobenzene. *J. Am. Chem. Soc.* **80**, 3528–3531.

33. Rau, H. (1984). Some useful formulae for the evaluation of the A ⇄ B photoreaction system. *EPA Newsletter*, **21**, 31–32.

34. Rau, H. (1984). Further evidence for rotation in the π,π*- and inversion in the n,π*-photoisomerization of azobenzenes. *J. Photochem.* **26**, 221–225.

35. Fischer, E. (1967). The calculation of photostationary states in systems A ⇄ B when only A is known. *J. Phys. Chem.* **71**, 3704–3706.

36. Rau, H., and Greiner, G. (1991). Determination of Quantum Yields of the X/Y Isomerization System from Spectroscopic Data. *EPA Newsletter* **41**, 40–55.

37. Mauser, H. (1974). *Formale Kinetik*. Bertelsmann Universitätsverlag, Düsseldorf.

38. Mauser, H. (1968). Zur Theorie der isosbestischen Punkte. *Z. Naturforsch.* **23b**, 1021–1025.

39. Mauser, H. (1968). Zur spektroskopischen Untersuchung der Kinetik chemischer Reaktionen. II. Extinktionsdifferenzen-Diagramme. *Z. Naturforsch.* **23b**, 1025–1030.

40. Ritter, G., Häfelinger, G., Lüddecke, E., and Rau, H. (1989). Tetrazetidine: Ab-initio calculations and experimental approach. *J. Am. Chem. Soc.* **111**, 4627–4635. Appendix.

41. Blume, R., and Polster, J. (1975). New methods for spectrophotometric determination of pK values. III. Titration of trivalent protolytes. *Z. Naturforsch.* **30b**, 358–372.

42. Kölle, U., Schätzle, H., and Rau, H. (1980). Molecular probes: Photoisomerization of optically active azo compounds. *Photochem. Photobiol.* **32**, 305–311.

43. *DMS UV Atlas of Organic Compounds*. Verlag Chemie, Weinheim. 1966–1971. Spectrum C7/5. Solvent is hexane.

44. Rau, H. (1968). Über die strahlungslose Desaktivierung von Azoverbindungen und die Lichtechtheit von Azofarbstoffen. *Ber. Bunsenges. Physik. Chem.* **72**, 408–414.

45. Hochstrasser, R. M., and Lower, S. K. (1962). Polarization of the spectra of crystalline azobenzene and mixed crystals of azobenzene in stilbene at 77° and 4.2° K in the region of the lowest n → π* transition. *J. Chem. Phys.* **36**, 3505–3506.

46. Rau, H. (1961). Über die Bandenstruktur der Spektren von Azoverbindungen. *Diplomarbeit, Tübingen.*

47. Rau, H. (1970). Über einige spektroskopische Eigenschaften der Azogruppe. *Habilitationsschrift Tübingen.*

48. Bisle, H., Römer, M., and Rau, H. (1976). Der Einfluß der Kopplung von ^{1}(n,π*)- und ^{1}(π,π*) Zuständen auf die Fluoreszenzfähigkeit von Azobenzolen. *Ber. Bunsenges. Physik. Chem.* **80**, 301–305.

49. Rau, H., and Shen, Y. Q. (1988). Photoisomerization of sterically hindered azobenzenes. *J. Photochem. Photobiol. A: Chemistry* **42**, 321–327.

50. Tamaoki, N., Ogata, K., Koseki, K., and Yamaoka, T. (1990). [2.2](4,4')Azobenzeno-

phane. Synthesis, structure, and cis-trans isomerization. *Tetrahedron* **46**, 5931–5942.

51. Gräf, G., Nitsch, H., Ufermann, D., Sawitzki, G., Patzelt H., and Rau, H. (1982). Phane des Azobenzols. *Angew.Chem.* **94**, 385–386. *Angew. Chem. Intl. Ed. Engl.* **24**, 313–314.

52. Rau, H., and Lüddecke, E. (1982). On the rotation-inversion controversy on photoisomerization of azobenzenes. Experimental proof of inversion. *J. Am. Chem. Soc.* **104**, 1616–1620.

53. Rau, H., and Waldner, I. (2002). A non-rotatory isomerization path in ethene derivatives? Investigation of a stilbenophane and protonated azobenzenophanes ("pseudo-stilbenes"). *Phys. Chem. Chem. Phys.* in press.

54. Davidov, A. S. (1962) *Theory of molecular excitons.* Transl. by M. Kasha and M. Oppenheimer Jr. McGraw-Hill, New York.

55. Shinomura, M., Ando, R., and Kunitake, T. (1983) Orientation and spectral characteristics of the azobenzene chromophore in the ammonium bilayer assembly. *Ber. Bunsenges. Physik. Chem.* **87**, 1134–1143 .

56. Shinomura, M., and Kunitake, T. (1987). Fluorescence and photoisomerization of azobenzene-containing bilayer membranes. *J. Am. Chem. Soc.* **109**, 5175–5183.

57. Struve, W. S. (1977). Emission from the $^1(n,\pi^*)$ state of azobenzene: Spectrum and ultrashort decay time. *Chem. Phys. Lett.* **46**, 15–19.

58. Morgante, C. G., and Struve, W. S. (1979). $S_2 \to S_0$ fluorescence in trans-azobenzene. *Chem. Phys. Lett.* **68**, 267–271.

59. Hamai, S., and Hirayama, F. (1984). $S_2 \to S_0$ fluorescence of azobenzene. *Book of Abstracts, Annual Symposium on Photochemistry (Japan)*, 315–316.

60. Azuma, J., Tamai, N., Shishido, A. and Ikeda, T. (1998). Femtosecond dynamics and stimulated emission form the S_2 state of a liquid crystalline trans-azobenzene. *Chem. Phys. Lett.* **288**, 77–82.

61. Fujino, T., Arzhantsev, S. Y., and Tahara, T. (2001). Femtosecond time-resolved fluorescence study of photoisomerization of trans-azobenzene. *J. Chem. Phys. A,* **105**, 8123–8129.

62. Bisle, H., and Rau, H. (1975). Fluorescence of noncyclic azo compounds with a low-lying $^1(n,p^*)$ state. *Chem. Phys. Lett* **31**, 264–266.

63. Tung, C.-H., and Guan, J.-Q. (1996). Modification of photochemical reactivity by Nafion. Photocyclization and photochemical cis-trans isomerization of azobenzene. *J. Org. Chem.* **61**, 9417–9421.

64. Lei, Z., Vaidyalingam, A., and Dutta, P. K. (1988). Photochemistry of azobenzene in microporous aluminophosphate $AlPO_4$-5. *J. Phys. Chem. B* **102**, 8557–8562.

65. Tsuda, K., Dol, G. C., Gensch, T., Hofkens, J., Latterini, L., Weener, J. W., Meijer, E. W., and DeSchryver, F. C. (2000). Fluorescence from azobenzene functionalized poly(propylene imine) dendrimers in self-assembled supramolecular structures. *J. Am. Chem. Soc.* **122**, 3445–3452.

66. Autret, M., Le Plouzennec, M., Molnet, C., and Sirnonneaux, G. (1994). Intramolecular Fluorescence Quenching in Azobenzene-substituted Porphyrins. *J. Chem. Soc., Chem. Commun.* 1169–1170.

67. Asakawa, M., Ashton, P. R., Balzani, V., Brown, C. L., Credi, A., Matthews, O. A., Newton, S. P., Raymo, F. M., Shipway, A. N., Spencer, N., Quick, A., Stoddart, J. F., White, A. J. P., and Williams, D. J. (1999). Photoactive azobenzene-containing supramolecular complexes and related interlocked molecular compounds. *Chem. Eur. J.* **5**, 860–875.

68. Zacharias, P. S., Ameerushina, S., and Korupoju, S. R. (1998). Photoinduced fluorescence changes on E-Z isomerization in azobenzene derivatives. *J. Chem. Soc. Perkin. Trans. II*, 2055–2059.

69. Eisenbach C. D., Dimberger, K., and Ficht, K. (1998). Matrix relaxation and local free volume effects in mono- and bichromophoric azobenzene polymers. *Polym. Prepr.* **39**, 279–280.

70. Rau, H. (1968). Über die Fluoreszenz der Hydroxyazoverbindungen. *Ber. Bunsenges. Physik. Chem.* **72**, 637–643.

71. Ronayette, J., Arnaud, R., Lebourgeois, P., and Lemaire, J. (1974). Isomérisation photochimique de l'azobenzène en solution. I. *Can. J. Chem.* **52**, 1848–1857.

72. Ronayette, J., Arnaud, R., and Lemaire, J. (1974). Isomériation photosensibilisée par des colorants et photoréduction de l'azobenzène en solution. II. *Can. J. Chem.* **52**, 1858–1867.

73. Monti, S., Gardini, E., Bortolus, P., and Amouyal, E. (1981). The triplet state of azobenzene. *Chem. Phys,. Lett.* 77, 115–119.

74. Pragst, F. (1980). Über die Chemilumineszenz beim Elektronentransfer zwischen den Radikalionen aromatischer Azoverbindungen und polycyclischer aromatischer Kohlenwasserstoffe [1] Korrektur zur Triplettenergieabschätzung von Azobenzol. *Z. Phys. Chem.(Leipzig)* 261, 791–792.

75. Astrand, P.-O., Ramanujam, P. S., Hvilsted, S., Bak, K. L., and Sauer, S. P. A. (2000). Ab-initio calculation of the electronic spectrum of azobenzene dyes and its impact on the design of optical data storage materials. *J. Am. Chem. Soc.* 122, 3482–3487.

76. Stammreich, H. (1950). Raman spectrum of azobenzene. *Experientia* 6, 225.

77. Kübler, R., Lüttke, W., and Weckherlin, S. (1960). Infrarotspektroskopische Untersuchungen an isotopen Stickstoff verbindungen. 1. Mitteilung: Die Lokalisierung der Valenzfrequenz der N=N-Doppelbindung. *Z. Elektrochem.* 64, 650–658.

78. Biswas, N., and Umapathy, S. (1997). Early time dynamics of trans-azobenzene isomerization in solution from resonance Raman intensity analysis. *J. Chem. Phys.* 107, 7849–7858.

79. Fujino, T., and Tahara, T. (2000). Picosecond time-resolved Raman study of trans-azobenzene. *J. Phys. Chem. A,* 104, 4203–4210.

80. Armstrong, D. R., Clarkson, J., and Smith, W. E. (1995). Vibrational analysis of trans-azobenzene. *J. Phys. Chem.* 99, 17825–17831.

81. Wolf, E., and Camenga, H. K. (1977). Thermodynamic and kinetic investigation of the thermal isomerization of cis-azobenzene. *Z. Phys. Chem. (Wiesbaden) NF,* 107, 21–28.

82. Ortruba III, J. P., and Weiss, R. G. (1983). Liquid crystalline solvents as mechanistic probes. 11. The syn → anti thermal isomerization mechanism of some low-"bipolarity" azobenzenes. *J. Org. Chem.* 48, 3448–3453.

83. Yamashita, S. (1961). Iodine-catalyzed cis-trans isomerization of azobenzene. *Bull. Chem. Soc. Japan* 24, 842–845.

84. Arnaud, R., and Lemaire, J. (1974). Isomérisation cis-trans de l'azobenzène par l'iode. III. *Can. J. Chem.* 52, 1869–1871.

85. Schulte-Frohlinde, D. (1958). Über den Mechanismus der katalytischen cis → trans-Umlagerung von Azobenzol. *Liebigs Ann. Chem.* 612, 131–138.

86. Hall, C. D., and Beer, P. (1991). Trico-ordinate phosphorus compounds as catalysts for the isomerization of (Z)- to (E)-azobenzene. *J. Chem. Soc. Perkin Trans. II,* 1947–1950.

87. Talaty. E. R, and Fargo, J. C. (1967). Thermal cis-trans-isomerization of substituted azobenzenes: a correction of the literature. *J. Chem. Soc., Chem. Commun.* 65–66.

88. Asano, T., Yano, T., and Okada, T. (1982). Mechanistic study of thermal Z-E isomerization of azobenzenes by high-pressure kinetics. *J. Am. Chem. Soc.* 104, 4900–4904.

89. Eisenbach, C. D. (1980). Relation between photochromism of chromophores and free volume theory in bulk polymers. *Ber. Bunsenges. Phys. Chem.* 84, 680–690.

90. Yoshii, K., Yamashita, T., Machida, S., Horie, K., Itoh, M., Nishida, F., and Morino, S. (1999). Photo-probe study of siloxane polymers. 1. Local free volume of an MQ-type silicone resin containing crosslinked nanoparticles probed by photoisomerization of azobenzene. *J. Noncryst. Sol.* 246, 90–103.

91. Kondo, T., Yoshii, K., and Horie, K. (2000). Photoprobe study of siloxane polymers. 3. Local free volume of polymethylsilsesquioxane probed by photoisomerization of azobenzene. *Macromolecules* 33, 3650–3658.

92. Tamaoki, N., and Yamaoka, T. (1991). Light-intensity dependence in the photochromism of dibenzo[2.2][4.4']- azobenzenophane. *J. Chem. Soc. Perkin Trans. II,* 873–878.

93. Rau, H., and Röttger, D. (1994). Photochromic azobenzenes which are stable in the trans and cis forms. *Mol. Cryst. Liq. Cryst.* 246, 143–146.

94. Schmiegel, J., and Grützmacher, H.-F. (1990). A macrocyclic 2,19-dioxo[3.3](3,3')azobenzolophane by transition metal carbonyl complex-mediated CO insertion and cyclization. *Chem Ber.* 123, 1749–1752.

95. Moss. R. A., and Jiang, W. (1997). Modulation of photoisomerization in double-azobenzene-chain liposomes. *Langmuir* 13, 4498–4501.

96. Zhao, W., Wu, C.-X., and Iwamoto, M. (1999). Rate-equation theory of azobenzene monolayer isomerization. *Chem. Phys. Lett.* 312, 572–577.

97. Klopman, G., and Boddapaneni, N. (1974). Electrochemical behavior of cis and trans-azobenzene. *J. Phys. Chem.* 78, 1825–1828.

98. Sekkat, Z., Morichere, D., Dumont, M., Loucif-Saibi, R., and Del, J. A. (1992). Photoisomerization of azobenzene derivatives in polymeric thin films. *J. Appl. Phys.* **71**, 1543–1548.

99. Ramanujam, P. S., Hvilsted, S., Zebger, I., and Siesler, H. W. (1995). On the explanation of the biphotonic processes in polyesters containing azobenzene moieties in the side chain. *Macromol. Rapid Commun.* **16**, 455–461.

100. Mauser, H. (1969). personal communication.

101. Gauglitz, G., and Hubig, S. (1984). Photokinetische Grundlagen moderner chemischer Actinometer. *Z. Phys. Chem. N. F.* **139**, 237–246.

102. Gauglitz, G. (1975). Azobenzene as a Convenient Actinometer for the Determination of Quantum Yields of Photoreactions *J. Photochem.* **5**, 41–46.

103. Yamashita, S., Ono, H., and Toyma, O. (1962). The cis-trans isomerization of azobenzene. *Bull. Chem. Soc. Japan* **35**, 1849–1853.

104. Siampiringue, N., Guyot, C., Monti, S., and Bortolus, P. (1987). The cis → trans photoisomerization of azobenzene: An experimental reexamination. *J. Photochem.* **37**, 185–188.

105. Hubig, S. (1984). *Chemische Aktinometrie. Photokinetische Grundlagen, Entwicklung und Kalibrierung neuer chemischer Aktinometer.* Ph. D. Thesis, Tübingen.

106. Bihler, O. (1976). Die Wellenlängenabhängigkeit der Quantenausbeuten der photochemischen Cis-transisomerisierung bei Azobenzol. *Wiss. Arbeit für das Höhere Lehramt, Hohenheim.*

107. Tahara, R., Morozumi, T., Nakamura, H., and Shinomura, M. (1997). Photoisomerization of azobenzenecrown ethers. Effect of alkaline earth metal ions. *J. Chem. Phys. B* **101**, 7736–7743.

108. Junge, D. M., and McGrath, D. V. (1999). Photoresponsive azobenzene-containing dendrimers with multiple discrete states. *J. Am. Chem. Soc.* **121**, 4912–4913.

109. Tauer, H., Grellmann, K.-H., and Heinrich, A. (1991). Photochemistry of tetrabenzo[c,e,i,k]tetraazacyclodecine in 2-propanol. *Chem. Ber.* **124**, 2053–2055.

110. Malkin, S., and Fischer, E. (1962). Temperature dependence of photoisomerization quantum yields of cis ⇄ trans isomerizations in azo-compounds. *J. Phys. Chem.* **66**, 2482–2486.

111. Yoshii, K., Machida, S., and Horie, K. (2000). Local free volume and structural relaxation studied with photoisomerization of azobenzene and Persistent Spectral Hole Burning in poly(alkylmethacrylate)s at low temperatures. *J. Polym. Sci. B: Polymer Physics* **38**, 3098–3105.

112. Bortolus, P., and Monti, S. (1979). Cis-trans photoisomerisation of azobenzene. Solvent and triplet donor effects. *J. Phys. Chem.* **83**, 648–652.

113. Gegiou, D., Muszkat, K. A. and Fischer, E. (1968). Temperature dependence of photoisomerization VI. The viscotiy effect. *J. Am. Chem. Soc.* **90**, 12–18.

114. Bortolus, P., and Monti, S. (1987). cis ⇄ trans Photoisomerization of azobenzene-cyclodextrin inclusion complexes. *J. Phys. Chem.* **91**, 5046–5050.

115. Hoffman, K., Marlow, F., and Caro, J. (1997). Photoinduced switching in nanocomposites of azobenzene and molecular sieves. *Adv. Mat.* **9**, 567–570.

116. Ramamurthy, V. (1992). Photochemical and photophysical studies within zeolites. *Chimia* **46**, 359–376.

117. Ramamurthy, V., Caspar, J. V., Corbin, D. R., Eaton, D. F., Kauffman, J. S., and Dybowksi, C. (1990). Modification of photochemical reactivity by zeolites: arrested molecular rotation of polyenes by inclusion in zeolites. *J Photochem. Photobiol. A:Chemistry* **51**, 259–263.

118. Gessner, F., Olea, A., Lobaugh, J. H., Johnston, L. J., and Sciano, J. C. (1989). Intrazeolite photochemistry 5. Use of zeolites in control of Photostationary ratios in sensitized cis-trans isomerizations. *J. Org. Chem.* **54**, 259–261.

119. Hong, J.-D., Park, E.-S., and Jung, B.-D. (1999). Studies on photochemical isomerization of azobenzene in self-assembled boloamphiphilic monolayers. *Mol. Cryst. Liq. Cryst., Sci. Technol. Sect. A*, **327**, 119.

120. Kojima, M., Tagaki, T., and Karatsu, T. (2000). Concentration dependent photodimerizsation of azobenzenes in solution. *Chem. Lett.* 686–687.

121. Jones, L. B., and Hammond, G. S. (1965). Mechanisms of photochemical reactions in solution. XXX. Photosensitized isomerization of azobenzene. *J. Am. Chem. Soc.* **87**, 4219–4219.

122. Fischer, E. (1968). Photosensitized isomerization of azobenzene. *J. Am. Chem. Soc.* **90**, 796.

123. Giurec, P., Hapiot, P., Moiroux, J., Neudeck, A., Pinson, J., and Tavani, C. (1999). Isomerization of azo compounds. Cleavage recombination mechanism of azosulfides. *J. Phys. Chem. A*, **103**, 5490–5500.

124. Enomoto, T., Hagiwara, H., Tryk, D. A., Liu, Z.-F., Hashimoto, K., and Fujishima, A. (1997). Electrostatically induced isomerization of azobenzene derivatives in Langmuir-Blodgett Films. *J. Phys. Chem. B*, **101**, 7422–7427.

125. Liu, Z.-F., Hashimoto, K., and Fujishima, A. (1991). Thermal cis-trans isomerization kinetics of azo compounds in the assembled mono-layer film: an electrochemical approach. *Chem. Phys. Lett.* **185**, 501–504.

126. Rau, H. (1971). Über die Fluoreszenz p-substituierter Azoverbindungen. *Ber. Bunsenges. Physik. Chem.* **75**, 1343–1347.

127. Monti, S., Dellonte, S., and Bortolus, P. (1983). The lowest triplet state of substituted azobenzenes: An energy transfer investigation. *J. Photochem.*, **23**, 249–256.

128. Brode, W. R., Gould, J. H., and Wyman, G. (1952). The relation between the absorption spectra and the chemical constitution of dyes. XXI. Phototropism and cis-trans isomerism in aromatic azo compounds. *J. Am. Chem. Soc.* **74**, 4641–4646.

129. Gerson, F., and Heilbronner, E. (1962). Physikalisch-chemische Eigenschaften und Elektronenstruktur der Azo-Verbindungen. Teil XI: Bemerkungen zur Struktur des Azonium-Kations des p,p'-Bis-dimethylamino-azobenzols. *Helv. Chim. Acta*, **45**, 51–59.

130. Nishimura, N., Sueyoshi, T., Yamanaka, H., Imai, E., Yamamoto, S., and Hasegawa, S. (1976). Thermal cis-to-trans isomerization of substituted azobenzenes. II. Substituent and solvent effects. *Bull. Chem. Soc. Japan* **49**, 1381–1387.

131. Fischer, E., and Frei, Y. (1957). Photoisomerization equilibira in azodyes. *J. Chem. Phys.* **27**, 328–330.

132. Nishimura, N., Tanaka, T., Asano, M., and Sueishi, Y. (1986). A volumetric study on the thermal cis-to-trans isomerization of 4-(dimethylamino)-4'-nitroazobenzene and 4,4'-bis(dialkylamino)azobenzenes: Evidence of an inversion mechanism. *J. Chem. Soc., Perkin Trans. II* , 1839–1845.

133. Asano,T., and Okada, T. (1984). Thermal Z-E isomerization of azobenzene. The pressure, solvent, and substitution effects. *J. Org. Chem.* **49**, 4387–4391.

134. Albini, A., Fasani, E., and Pietra, S. (1983). The photochemistry of azo dyes. Photo-isomerisation versus photoreduction from 4-diethylaminoazobenzene and 4-diethylamino-4'-methoxyazobenzene. *J. Chem. Soc. Perkin Trans. II*, 1021–1024.

135. Haselbach, E. (1970). Physikalisch-chemische Eigenschaften und Elektronenstruktur der Azo Verbindungen. Teil XV. Über die Struktur der protonierten Azobrücke in Azobenzolderivaten. *Helv. Chim. Acta,* **53**, 1526–1543.

136. Haselbach, E., and Heilbronner, E. (1967). The structure of the protonated azo-link in 2,2'-azo-isobutane. *Tetrahedron Lett.* 4531–4535.

137. Badger, G. B.,. Drewer, R. J., and Lewis, G. E. (1963). Photochemical reactions of azo compounds. II. Photochemical cyclodehydrogenation of methyl- and dimethylazobenzenens. *Aust. J. Chem.* **16**, 1042–1050.

138. Lewis, G. E. (1960). Photochemical reactions of azo compounds. I. Spectroscopic studies of the conjugate acids of *cis-* and *trans-*azobenzene. *J. Org. Chem.* **25**, 2193–2195.

139. Gutmann, V., and Steininger, A. (1965). Komplexbildung von *trans-*Azobenzol und seinen Derivaten mit Akzeptorhalogeniden in Acetonitril. *Monatsh. Chem.* **96**, 1173–1182.

140. Lewis, G. E., and Mayfield, R. J. (1966). Photoreactions of azo compounds. VII. Studies of the photochemical reactions of azobenzene under various acidic conditions. *Aust. J. Chem.* **19**, 1445–1454.

141. Ghedini, M., Pucci, D., Calogero, G., and Barigelletti, F. (1997). Luminescence of azobenzene derivatives induced by cylcopalladation. *Chem. Phys. Lett.* **267**, 341–344.

142. Barigelletti, F., Ghedini, M., Pucci, D., and LaDeda, M. (1999). A mercurated azobenzene complex for photoswitching between trans- and cis forms. *Chem. Lett.* 297–298.

143. Rau, H., Crosby, A. D., Schäufler, A., and Frank, R. (1981). Triplett-sensibilisierte Photoreaktionen von Azobenzol in schwefelsaurer Lösung. *Z. Naturforsch.* **36a**, 1180–1186.

144. Rau, H. (1967). Über die Fluoreszenz von Azoverbindungen. *Ber. Bunsenges. Physik. Chem.*, **71**, 48–53.

145. Pringsheim, P. (1949). "Fluorescence and Phosphorescence," Interscience, New York, p. 422.

146. Görner, H., Gruen, H., and Schulte-Frohlinde, D. (1980). Laser flash photolysis study of substituted azobenzenes. Evidence for a triplet state in viscous media. *J. Phys. Chem.*, 84, 3031–3039.

147. Fei, H., Wei, Z., Wu, P., Han, L., Zhao, Y., and Che, Y. (1994). Biphoton holographic storage in methyl orange and ethyl orange dyes. *Opt. Lett.* 19, 411–418.

148. Bach, H., Anderle, K., Fuhrmann, T., and Wendorff, J. H. (1996). Biphoton-induced refractive index change in 4-amino-4'-nitroazobenzene/polycarbonate. *J. Phys. Chem.* 100, 4135–4140.

149. Wildes, P. D., Pacifici, J. G., Irick Jr., G., and Whitten, D. G. (1971). Solvent and substituent effects on the thermal isomerization of substituted azobenzenes. A flash spectroscopic study. *J. Am. Chem. Soc.* 93, 2004–2008.

150. Olah, G. A., Dunne, K., Kelly, D. P., and Mo, Y. K. (1972). Stable Carbocations. CXXIX. Mechanism of the benzidine and Wallach rearrangements based on direct observation of a dicationic reaction intermediate and related model compounds. *J. Am. Chem. Soc.* 94, 7438–7447.

151. Asano, T., Okada, T., Shinkai, S., Shigematsu, K., Kosano, Y., and Manabe, O. (1981). Temperature and pressure dependence of thermal cis-to-trans isomerization of azobenzenes which evidence an inversion mechanism *J. Am. Chem. Soc.* 103, 5161–5165.

152. Gille, K., Knoll, H., and Quitsch, K. (1999). Rate constants of the thermal cis-trans isomerization of azobenzene dyes in solvents, acetone/water mixtures, and in microheterogeneous surfactant solutions. *Int. J. Chem. Kinet.* 31, 337–350.

153. Hair, S. R., Taylor, G. A., and Schultz, L. W. (1990). An easily implemented flash photolysis experiment for the physical chemistry laboratory. The isomerization of 4-anilino-4'-nitroazobenzene. *J. Chem. Educ.* 67, 709–712.

154. Schanze, K. S., Mattox, T. F., and Whitten, D. G. (1982). Correlation of the rate of thermal cis-trans isomerization of p-nitro-p'-dialkylaminoazobenzenes with solvent Z value applied to study polarity in aqueous surfactant solutions. *J. Am. Chem. Soc.* 104, 1733–1735.

155. Schanze, K. S., Mattox, T. F., and Whitten, D. G. (1983). Solvent effects on the thermal cis-trans isomerization and charge transfer absorption of 4-diethylamino-4'-nitroazobenzene *J. Org. Chem.* 48, 2808–2813.

156. Sanchez, A., and de Rossi, R. H. (1993). Strong inhibition of cis-trans isomerization of azo compounds by hydroxide ion. *J. Org. Chem.* 58, 2094–2096.

157. Sanchez, A., and de Rossi, R. H. (1995). Effect of hydroxide ion on the cis-trans thermal isomerization of azobenzene derivatives. *J. Org. Chem.* 60, 2974–2976.

158. Mauser, H., Francis, D., Niemann, H. (1972). Zur kinetischen Analyse von Photoreaktionen. III: Zur Kinetik der photochemischen Cyclodehydrierung von Azobenzol in Schwefelsäure. *Z. Phys. Chem. NF* 82, 318–333.

159. Irick Jr., G., and Pacifici, J. G. (1969). Photochemistry of azo compounds. I. Photoreduction of 4-diethylamino-4'-nitroazobenzene. *Tetrahedron Lett.* 1303–1306.

160. Gabor, G., and Fischer, E. (1971). Spectra and cis-trans isomerism in highly bipolar derivatives of azobenzene. *J. Phys. Chem.* 75, 581–583.

161. King, N. R., Whale, E. A., Davis, F. J., Gilbert, A., and Mitchell, G. R. (1997). Effect of media polarity on the photoisomerization of substituted stilbene, azobenzene and imine chromophors. *J. Mater. Chem.* 7, 625–630.

162. Rehak, V., Novák F., and Cepciansky, L. (1973). Photochemie von Farbstoffen und Zwischenprodukten. I. Photoreduktion monosubstituierter Azobenzolderivate in Alkoholen. *Coll. Czechoslov. Chem. Commun.* 39, 697–705 .

163. Blaisdell, B. E. (1949). The photochemistry of aromatic azo compounds in organic solvents. *J. Soc. Dyers Colourists*, 618–628.

164. Irick Jr., G., and Pacifici, J. G. (1969). Photochemistry of azo compounds. II. Effect of ketonic sensitizers on photoreduction of 4-dimethylamino-4'-nitroazobenzene. *Tetrahedron Lett.* 2207–2209.

165. Albini, A., Fasani E., and Pietra, S. (1982). The photochemistry of azo-dyes. The wavelength-dependent photoreduction of 4-diethylamino-4'-nitroazobenzene. *J. Chem. Soc. Perkin Trans. II* , 1393–1395.

166. Magee, J. L., Shand Jr., W., and Eyring, H. (1941). Non-adiabatic reactions. Rotation about the double bond. *J. Am. Chem. Soc.* 63, 677–688.

167. Curtin, D. Y., Grubbs, E. J., and McCarthy, C. G. (1966). Uncatalyzed *syn-anti* isomerization of imines, oxime ethers and haloimines. *J. Am. Chem. Soc.* **88**, 2775–2776.

168. Kessler, H. and Leibfritz, D. (1970). NMR spectroscopic detection of intramolecular mobility. XIX. Magnetic nonequivalence as proof of nitrogen inversion in imines. *Tetrahedron Lett.* 1423–1426 .

169. Knorr, R., Ruhdorfer, J., Mehlstäubl, J., Böhrer, P., and Stevenson, D. S. (1993). Demonstration of the nitrogen inversion mechanism of imines in a Schiff base model. *Chem. Ber.* **126**, 747–754.

170. Turro, N. J. (1978). *Modern, Molecular Photochemistry.* Chapter 4, Benjamin/ Cummings, Menlo Park.

171. Baird, N. C., and Swanson, J. R. (1973). Quantum organic photochemistry. IV. The photo-isomerization of diimide and azoalkanes. *Can. J. Chem.* **51**, 3097–3101.

172. Camp, R. N., Epstein, I. R., and Steel, C. (1977). Theoretical Studies of the Photochemistry of Acyclic Azoalkanes *J. Am. Chem. Soc.* **99**, 2453–2459.

173. Olbrich, G. (1978). INDO-SCF and CI calculations on the trans-cis isomerization of azomethane in the ground and in excited states. *Chem. Phys.* **27**, 117–125.

174. Kobayashi, S., Yokoyama, H., and Kamei, H. (1987). Substituent and solvent effects on electronic absorption spectra and thermal isomerization of pull-push-substituted cis-azobenzenes. *Chem. Phys. Lett.* **138**, 333–338.

175. Asano, T. (1980). Pressure effects on the thermal cis → trans isomerization of 4-dimethylamino-4'-nitroazobenzene. Evidence for a change of mechanism with solvent. *J. Am. Chem. Soc.* **102**, 1205–1206.

176. Cimiraglia, R., and Hofmann, H.-J. (1994). Rotation and inversion states in thermal E/Z isomerization of aromatic azo compounds. *Chem. Phys. Lett.* **217**, 430–435.

177. Victor, J. G., and Torkelson, J. M. (1987). On measuring the distribution of local free volume in glassy polymers by photochromic and fluorescence techniques. *Macromolecules* **20**, 2241–2250.

178. Naito, T., Horie, K., and Mita, I. (1991). Photochemistry in polymer solids. 11. Effects of the size of reaction groups and the mode of photoisomerization on photochromic reactions in polycarbonate film. *Macromolecules* **24**, 2907–2911.

179. Kuriyama, Y., and Oishi, S. (1999). Mechanism of thermal isomerization of azobenzene in zeolite cavities. *Chem. Lett.* 1045–1046.

180. Walsh, A. D. (1953). The electronic orbitals, shapes, and spectra of polyatomic molecules. Part III. HAB and HAAH molecules. *J. Chem. Soc.*, 2288–2296.

181. Thulin, B., and Wennerström, O. (1976). Synthesis of [2.2](3,6)phenanthrenophanediene. *Acta Chem. Scand. B*, **30**, 369–371.

182. Tanner, D., and Wennerström, O. (1981). [2.2](4,4')-*trans*-stilbene. *Tetrahedron Lett.* **22**, 2313–2316.

183. Monti, S., Orlandi, G. and Palmieri, P. (1982). Features of the photochemically active state surfaces of azobenzene. *Chem. Phys.* **71**, 87–99.

184. Cattaneo, P., and Persico, M. (1999). An *ab-initio* study of the photochemistry of azobenzene. *Phys. Chem. Chem. Phys.* **1**, 4739–4743.

185. Ishikawa, T., Noro, T., and Shoda, T. (2001). Theoretical study on the photoisomerization of azobenzene. *J. Chem. Phys.* **115**, 7503–7512.

186. Tamai, N., and Miyasaka, H. (2000). Ultrafast dynamics of photochromic systems. *Chem. Rev.* **100**, 1875–1890.

187. Lednev, I. K., Ye, T.-Q., Hester, R., and Moore, J. N. (1996). Femtosecond time-resolved UV-visible absorption spectroscopy of trans-azobenzene in solution. *J. Phys. Chem.* **100**, 13338–13341.

188. Lednev, I. K, Ye, T.-Q., Matousek, P., Towrie, M., Foggi, P., Neuwahl, F. V. R., Umapathy, S., Hester, R. E., and Moore, J. N. (1998). Femtosecond time-resolved UV-visible absorption spectroscopy of trans-azobenzene: dependence on excitation wavelength. *Chem. Phys. Lett.* **290**, 68–74.

189. Lednev, I. K., Ye, T.-Q., Abbott, L. C., Hester, R. E., and Moore, J. N. (1998). Photoisomerization of a capped azobenzene in solution probed by ultrafast time-resolved electronic absorption spectroscopy. *J. Phys. Chem. A* **102**, 9161–9166.

190. Nägele, T., Hoche, R., Zinth, W., and Wachtveitl, J. (1997). Femtosecond photoisomerization of cis-azobenzene. *Chem. Phys. Lett.* **272**, 489–495.

191. Hamm, P., Ohline, S. M., and Zinth, W. (1997). Vibrational cooling after ultrafast photo-isomerization of azobenzene measured by femtosecond infrared spectroscopy. *J. Chem. Phys.* **106**, 519–529.

192. Terazima, N., Takezaki, M., Yamaguchi, S., and Hirota, N. (1998). Thermalization after photoexcitation to the S_2 state of trans-azobenzene in solution. *J. Chem. Phys.* **109**, 603–609.

193. Biswas, N., and Umapathy, S. (1995). Wavepacket dynamical studies on trans-azobenzene: absorption spectrum and resonance Raman excitation profiles in the n-π* transition. *Chem. Phys. Lett.* **236**, 24–29.

194. Shin, D.-M., and Whitten, D. G. (1988). Solvent induced mechanism change in charge-transfer molecules. Inversion versus rotation paths for the $Z \rightarrow E$ isomerization of donor-acceptor substituted azobenzenes. *J. Am. Chem. Soc.* **100**, 5206–5208.

195. Kikuchi, O., Azuki, M., Inadomi, Y. and Morihashi, K. (1999). Ab-initio GB study of solvent effect on the cis-trans isomerization of 4-dimethylamino-4'-nitroazobenzene. *J. Mol. Struct. (Theochem)* **468**, 95–104.

2 ULTRAFAST DYNAMICS IN THE EXCITED STATES OF AZO COMPOUNDS

TAKAYOSHI KOBAYASHI
TAKASHI SAITO
Department of Physics, University of Tokyo, Hongo 7-3-1, Bunkyo-ku, Tokyo 113-0033, Japan

ABSTRACT

The photoisomerization dynamics of two azo compounds, one with and the other without intramolecular hydrogen bonding, were studied using picosecond and femtosecond spectroscopy apparatus. Isomerization time constants in 1-phenylazo-2-hydroxynaphthalene (1PA2N, Sudan) with hydrogen bonding at room temperature were determined to be 14±3, 28±5, and 110±30 ps in solvents with viscosities of 0.729, 2.06, and 9.63 cP, respectively, by using 6-ps pulsed laser. The results were discussed in terms of the relation between the viscosity and the association of isomerization and tautomerization in the relaxation mechanism. The real-time dynamics of molecular vibration in the excited state of *trans*-4-(dimethylamino)azobenzene (DMAAB, methyl yellow) in solution was also studied by sub-5fs pump-probe spectroscopy, using visible pulsed laser developed by our group.[1,2] It was found for the first time that the vibration frequencies of N=N and C–N stretching modes are modulated quasi-periodically. This is explained in terms of coupling between the two modes via an ~80cm^{-1} torsion mode. More detailed analysis indicated that there is a small frequency difference between the two modulations. From the results, we can conclude that the reaction does not proceed via either the pure *rotation* or pure *inversion* mechanism.

2.1 INTRODUCTION

Trans-cis photoisomerization is one of the most important and widely investigated unimolecular photochemical reactions. Molecular systems showing the isomerization contain double bonds such as C=C, N=N, and C=N bonds. Most representative examples of the molecules are stilbene, azobenzene, and anile. There are many other molecular systems, including biological chromophores. Photoisomerization triggers conformational changes in retinoid proteins such as rhodopsin and bacteriorhodopsin, relevant to vision[3] and ATP synthesis,[4] respectively. Although the spectra of both isomers in many molecular systems are well known, the dynamics of these processes have not been sufficiently studied because the rate of conversion from one isomer to the other is quite often very fast and at room temperature, and in nonviscous solvents, it frequently occurs in the picosecond range. Azobenzene and its derivatives are known to form a group of molecules showing the *trans-cis* photoisomerization. They are studied to gain basic understanding of the fundamental chemistry. These are also widely investigated for various photonics applications, such as optical switches and high-density optical memory storage devices.[5,6] To determine the rate of ultrafast isomerization, it is important to search for azo compounds that are best-suited to such applications.

In the first part of this article, we present a kinetic study on the isomerization of 1-phenylazo-2-hydrozynaphthalene (1PA2N) measured by picosecond spectroscopy. The selection of this compound was based on its known photochemical properties of *cis-trans* isomerization and tautomerization by proton transfer. The phototautomerization of 1PA2N is evidenced by the fact that in solution this compound is a mixture of the azo tautomer A and the hydrazo tautomer H,[7–10] as shown in Figure 2.1. It is thought that the internal hydrogen bond between the oxygen and a nitrogen atom is a strong factor in the tautomerization process. In fact, all o-hydroxyazo compounds exhibit an internal hydrogen bond between the oxygen and the nitrogen, and the presence of tautomerization enforces the thought that this internal bond facilitates tautomerization. In this article, we shall present the results of the time constant measurement of the isomerization of 1-phenylazo-2-naphthol and correlate this rate with the viscosity of the solvent. We shall also discuss the relaxation channel in the system in comparison with other hydrogen-bonded systems of indigo and tioindigo.

There is another aspect of the dynamics of isomerization in azo compounds, namely the rotation versus inversion mechanism. Since azobenzene has lone-pair electrons on the nitrogen atoms, the n-π^* electronic transition is observed in addition to the π-π^* transition, which may result in a different isomerization mechanism from that of stilbene. It has been proposed that the photoisomerization mechanism of *trans*-azobenzene depends on the excitation wavelength.[11,12] Figure 2.2A shows the schematics of a generally accepted model. According to this model, isomerization proceeds with a rotation of the phenyl ring(Φ) around the N=N double bond after π-π^*(S_2) excitation in the *rotation* mechanism. On the other hand, n-π^*(S_1) excitation induces the in-plane bending of the phenyl ring through a configuration with a straight $-N^+=N^-=\Phi$ group (like an allene structure) in the *inversion* mecha-

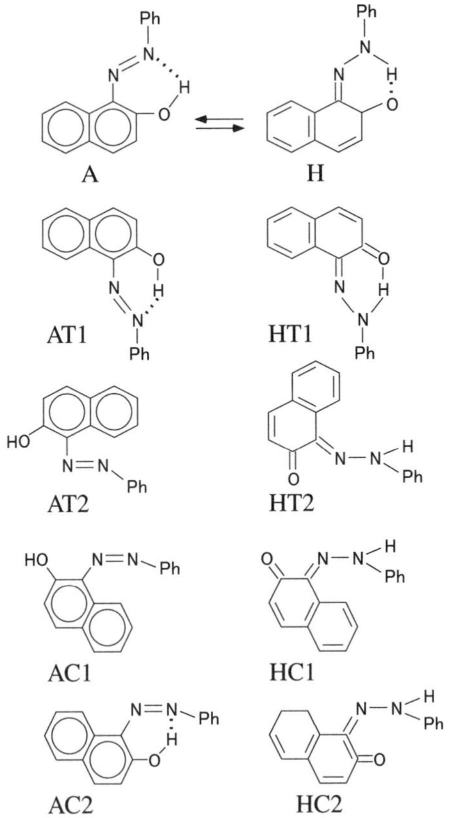

FIG. 2.1 Azo tautomers (A) and hydrazo tautomers (H) of 1PA2N.

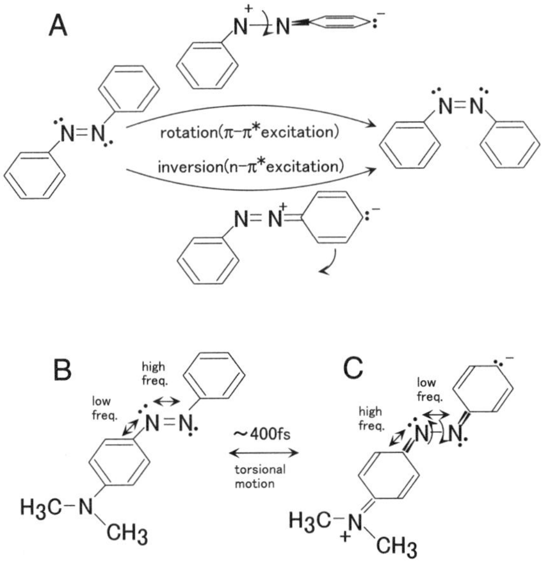

FIG. 2.2 Rotation and inversion pathways of isomerization (A). Dynamic mode-coupling model of DMAAB in the excited state (B and C).

nism. There is still an argument about the mechanism.[13] In the second part of this paper, the reaction mechanism in relation to the real-time dynamics of molecular vibration in the excited state of *trans*-4-(dimethylamino)azobenzene (DMAAB, methyl yellow) in solution is studied by pump-probe spectroscopy, using a sub-5fs visible pulsed laser recently developed by our group.[1,2]

2.2 EXPERIMENTAL SECTION

1PA2N was purchased from Aldrich and used without further purification. Experiments were performed with 1PA2N in methylcyclohexane, glycerin, or mixed solvents of methylcyclohexane and cyclohexanol. Output pulses from an Nd glass laser with 6-ps width and 40-mJ energy were used as a basic pulsed light source. The 532-nm with 6-ps width was used for both the pump and the generation of a picosecond broadband continuum for the interrogation of the transient spectral change. All data of picosecond spectroscopy of 1PA2N were obtained at room temperature.

DMAAB (guaranteed reagent grade, Tokyo Kasei) and dimethyl sulfoxide (DMSO; special reagent grade of Japan Industry Standard, Kanto Chemicals) used as solute and solvent, respectively, were utilized without further purification. Output pulses from the pulse-front-matched noncollinear optical parametric amplifier (NOPA)[1,2] with 4.7-fs width and 5-μJ energy at 1-kHz repetition rate were used as both pump and probe pulses. The spectrum of the pulses covers from 520 to 730nm with a nearly constant phase. Laser pulse energies of the pump and probe pulses are about 20nJ and 2nJ, respectively. The pump beam was mechanically chopped at 500Hz in synchronization with the laser pulse at 1kHz. The probe pulse intensity was measured with a Si photodiode attached to a monochromater with 4-nm resolution. Transmitted probe intensity change induced by the pump was detected with a lock-in amplifier. All measurements of sub-5-fs spectroscopy of DMAAB were performed at room temperature.

2.3 RESULTS AND DISCUSSION

2.3.1 1PA2N

2.3.1.1 Absorption Spectrum

The absorption spectrum of 1-phenylazo-2-naphthol in various solvents and the spectra of the photoinduced colored species at low temperature generated by the exposure of this compound to light at various wavelengths was studied in detail.[14] For any o-hydroxyazo compound, e.g., 1-phenylazo-2-naphthol, the combined results of tautomerization and rotation around single and double bonds lead to the eight configurations AT1, AT2, AC1, AC2, HT1, HT2, HC1, and HC2,[10,15] where H and A represent the hydrazone and azo forms, respectively, and T and C denote the trans and cis configurations, respectively.

The study by Gabor et al.[16] showed that the absorption spectrum of 1PA2N is a superposition of A, absorbing in the region 350–440 mm, and of

H, having its strong absorption bands of π-π* transition in the region 450–550 mm. This strong π-π* absorption band overlaps the weak n-π* band of A. In the present experiments, the excitation wavelength is 530 nm. It is safely assumed that only the H form is selectively excited. From the absorption spectrum of 1PA2N in methylcyclohexane and mixed solutions of methyl-cyclohexane-cyclohexanol, it can be clearly seen that when changing from nonpolar methylcyclohexane solution to more polar mixed solvents, the absorbance in the region 480–600 nm increases relative to that in the region of 350–460 nm. This indicates that the concentration ratio of H to A increases with the polarity of the solvent. The samples were placed in a 2-mm thick cell, and the absorbance was adjusted and maintained between 0.25 and 0.30. Conditions that were other than these are specifically pointed out.

2.3.1.2 Absorbance Changes and Decay Kinetics

The absorbance changes, ΔA, in the spectrum of 1PA2N dissolved in neat methylcyclohexane induced by 6-ps pulses, indicate that the kinetics of the recovery of the bleached band do not follow a simple monotonic rate but rather are composed of two time constants.

The effect of pulse excitation energy was studied on the 480-nm bleaching kinetics of 1PA2N dissolved in pure methylcyclohexane and a 5:2 mixture of methylcyclohexane-cyclohexanol. For the methylcyclohexane solution, we observed that, when the excitation energy was increased by a factor of 2 from 70 to 140 mJ/cm^2, the decay time constant of the short component was found to decrease from 14±3 to 9±1 ps. This decrease is caused by a self-damping process similarly observed for DODCI.[17] The short time constant was also observed to decrease from 28±5 to 22±8 ps for the 5:2 mixed solvent when the excitation energy was increased from 70 to 140 mJ/cm^2. In addition, we found that when the concentration of the sample was increased by a factor of 1.7, i.e., absorbance at 530 nm is increased from 0.3 to 0.5, the decay time constant of the short component is again decreased from 28±5 to 22±8 ps. To avoid nonlinear effects such as self-damping and saturation, ΔA vs. concentration and vs. energy were plotted for 1PA2N in various solutions. It was found that slight saturation occurs at higher excitation density (~30 mJ/cm^2). Hence, only the data obtained in the linear region are presented and discussed in the following sections.

2.3.1.3 Effect of Solvent Viscosity on the Decay Time Constant

The time dependence of ΔA after 6-ps pulse excitation of 1PA2N in methylcyclohexane was monitored at 480 nm, which is near the maximum of the visible absorption band. A negative absorbance change, i.e., bleaching, was observed at 480 nm. This observation is in agreement with the work of Gabor et al.,[14] showed that the absorbance of the H tautomer at 480 nm is much higher than that of the A tautomer. The bleaching we observed was indicated by a decrease of the H form of 1PA2N. The recovery kinetics induced upon the A form by a 10-mJ/cm^2 pulse were found to be governed by two distinct rates, one with a lifetime of more than 1 ns and the other a fast component with a time constant of 14±3 ps. The short lifetime component was found to decrease to 9±1 ps by increasing the excitation pulse density from 10 to 20 mJ/cm^2. Further experiments were performed on the same molecules dissolved in a mixture of solvents with the purpose of examining the effect of viscosity on the rate of isomerization of 1PA2N as revealed by the kinetics of the recovery of the 480-nm band after excitation with a non-saturating 5-ps, 530-nm pulse.

We present here only the kinetics of the 480-nm band, which is representative of the transient kinetics. We also make the obvious and trivial assumption that a fraction of the excited singlet state population remains there during the relaxation process. The factor of 10 changes in the viscosity of the solvent were achieved by using methylcyclohexane (η=0.729 cP) and methylcyclohexane-cyclohexanol 5:2 and 5:9 mixtures (with respective viscosities of 2.09 and 9.63 cP).

For every solution, we found that the recovery of the bleaching 480-nm band follows biphasic kinetics composed of two exponential functions. The time constants correspond to the 5:9 methylcyclohexane-cyclohexanol (η=9.63 cP) solution, excited with a 10-mJ/cm^2 pulse. The long-lifetime components have more than 1 ns time constant, while the short component exhibits a time constant, of 107±24 ps. We believe that the variation in the 480-nm recovery time constant observed with the two solutions, 14±4 ps vs. 107±20 ps, which were excited and monitored with identical pulses, is caused by the difference in viscosity.

This proposal regarding the dependence of the short decay time component on viscosity is supported by the fact that the 530-nm pulse excites preferentially the H, hydrazone, form of the 1PA2N molecule and specifically the trans isomer of the H form (HT). In addition, the cis form is known to be unstable, quite possibly because of the steric hindrance between the naphthalene and benzene components of the 1PA2N, and therefore of the very low ground state population.

If we assume that the HT form is excited mainly with the 530-nm pulse, there are three possible candidates for the mechanism that governs the decay rate of the fast component: (1) intersystem crossing from the excited singlet state of HT to HT triplet; (2) *trans-cis* isomerization; and (3) internal conversion to the ground state. The first possibility is excluded, based on the absence of a reasonable yield of phosphorescence, even at low temperatures.[14] Possibility 3 is not very plausible because of the existence of the additional long component and the strong dependence on the viscosity of the solvent. *Trans-cis* isomerization is the most reasonable possibility.

2.3.1.4 *Trans-cis* Isomerization

The assignment of the observed fast kinetic rate to *trans-cis* isomerization is strongly supported by our experimental data, which show that the time constant is related to the viscosity by approximately $\eta^{2/3}$. The Forster and Hoffman model[18] was developed originally to explain the Q= $\eta^{2/3}$ relationship where Q is the fluorescence quantum yield and η is the viscosity of the medium for triphenylmethane dyes. In addition, it was predicted that the fluorescence lifetime, τ, should follow a similar relationship: τ = c $\eta^{2/3}$. According to this model, absorption of light produces a vertically excited Franck-Condon state with the phenyl rings still at a ground state equilibrium configuration, forming an angle θ_0. The rings subsequently rotate to a new equilibrium angle θ, and nonradiative deactivation of the excited state now depends upon $(\theta-\theta_0)^2$, following the generalized Hook's model.

It was suggested by Gegiou et al.[19] that the photoisomerization of azobenzene probably proceeds via a mechanism different than the *cis-trans* isomerization of stilbene. Such an isomerization process in azobenzene could involve a pyramidal inversion of a nitrogen atom, in contrast to stilbene and its derivatives, where rotation about the central double bond is required. Based on the data, the AT→AC isomerization mechanism may be similar to that of the triphenylmethane dyes, whose ground state structure is known to resemble a three-dimensional, propeller-shaped D_3 structure with the phenyl rings rotated 32 degrees from the central plane.

The viscosity dependence of the internal-conversion time constant has been measured previously for two kinds of triphenylmnethane dyes: crystal violet[20] and malachite green.[21] The recovery time constant of the ground-state depletion of crystal violet and the $1/e$ point for complete recovery were reported[8] to obey an $\eta^{1/3}$ relationship rather than the $\eta^{2/3}$ relationship that we observed in 1PA2N. The reason for the $\eta^{1/3}$ dependence is not apparent to us. We have also found transient absorption in the 545-nm region for 1PA2N dissolved in each of the three solvents discussed. Because of the low absorption cross-section of the intermediate at this wavelength, it was very difficult to investigate the decay constant dependence on concentration and excitation energy. The results indicate, however, that the same relaxation processes are present in the 545-nm intermediate as were observed at 480 nm at the same concentration. On the basis of this data, we can assign the cross section to the following order: $\sigma(HT, S_0, 480 \text{ nm}) > \sigma(HT, S_1, 480 \text{ nm}) > \sigma(I, 480 \text{ nm})$, where I is the intermediate and S_0 and S_1 are the ground and lowest-excited singlet states, respectively.

2.3.1.5 Relaxation Mechanism

It has been stated[22] that o-hydroxy derivatives are the only aromatic azo compounds exhibiting luminescence. The reason for this phenomenon may be the tautomeric phenylhydrazones, which exist in at least these few compounds. Gabor et al.[14] showed that among o- and p-hydroxyphenylazobenzenes and o-hydroxyazobenzene, only 1-phenylazo-2-hydrozy-naphthalene, 2-phenylzao-o-hydrozynaphthanele, and diphenylhydrazone o-naphthoquinone emit any appreciable luminescence—i.e., above an estimated quantum yield of 0.001. The fluorescence spectra are practically mirror images of the absorption spectra in their respective hydrazone forms; they are also both found independent of the wavelength of excitation from 360 to 546 nm. However, the approximate quantum yields were ~0.05 and decreased sharply at higher temperatures. These experimental results (Gabor et al.[14] indicate that the A form does not emit fluorescence. Yet after excitation with 360–480-nm light, this molecule relaxes to the hydrazone form via a proton transfer process, and this hydrazone form emits at temperatures higher than −180°C, most probably because of the more efficient *trans-cis* isomerization.

Experimental data obtained by Gabor et al.[14] have provided evidence that the lowest excited state of the H form is lower than that of the A form.

Therefore, the 530-nm excited HT cannot relax via the AT form, but it must necessarily remain within the HC manifold. As in the case of stilbene, we believe that for 1PA2N there is an intermediate I that might have a pyramidal form[19] different from the perpendicular form of the stilbene intermediate. Relaxation from the HT form takes place via the I intermediate, with the rate being dependent upon the viscosity of the solvent. In methylcyclohexane, the time constant is found to be about 20 ps, and therefore fast enough to quench the fluorescence. In more viscous solvents, such as mixed methyl-cyclohexane-cyclohexanol, the decay process becomes slower and the probability of fluorescence increases.

In the case of stilbene two mechanisms of cis-trans isomerization have been proposed. One suggests that the isomerization proceeds via the singlet, and the other invokes a triplet-state intermediate. There are cases, however, where only one mechanism is operative, such as with azonaphthol, where *cis-trans* photoisomerization is thought to take place via the singlet mechanism only.[23] Our present study on 1PA2N is consistent with the singlet mechanisms in contrast to indigo and 6,6'-dimethoxyindigo, which do not show photo-chromism. Their singlet lifetimes are ~50 ps for 6,6'-dimethoxyindigo and 150 ps for indigo. This highly efficient internal conversion to the ground state[24] provides a strong reason for the photostability of indigo and 6,6'-dimethoxyindigo. Thus, it can be argued that the excited singlet states of these indigos are quenched by fast internal conversion enhanced by hydrogen bonding, and that therefore they do not exhibit luminescence or isomeriza-tion.[24] On the other hand, 1-phenylazo-2-naphthol is hydrogen-bonded, but it shows fluorescence at low temperature. Therefore, we propose that the very fast decay from the singlet state of 1PA2N hydrazone proceeds via a *trans-cis* isomerization path.

2.3.2 DMAAB

2.3.2.1 Absorption Spectrum

DMAAB sample of DMSO solution was prepared in a 0.1 mm–thick handmade cell with a microscope cover-glass plate and a slide-glass plate as front and back windows, respectively, to prevent the reduction of the time resolution due to pulse broadening by the group-velocity dispersion. Figure 2.3 shows the absorption spectrum of the DMAAB sample solution (0.01 and 0.2 mol dm^{-3}). A band with a peak around 420 nm in the spectrum is assigned to a strongly allowed π-π^* transition. The substitution of azobenzene by *p*-amino groups is known to shift the π-π^* band to longer wavelengths[25,26] because of the electron-donative nature of the group. The tail extending to 650 nm is mainly due to the weak n-π^* absorption, the wavelength of which is less dependent on substituents than that of the π-π^* transition.[27] This absorption band is not clearly seen because of the intense π-π^* band existing close to the n-π^* transition.

The absorption spectral shape does not show any concentration depend-ence between 0.01 and 0.2 mol dm^{-3}, indicating that there is no aggregate formation. The sample of 0.2 mol dm^{-3} was used mainly for the femtosecond

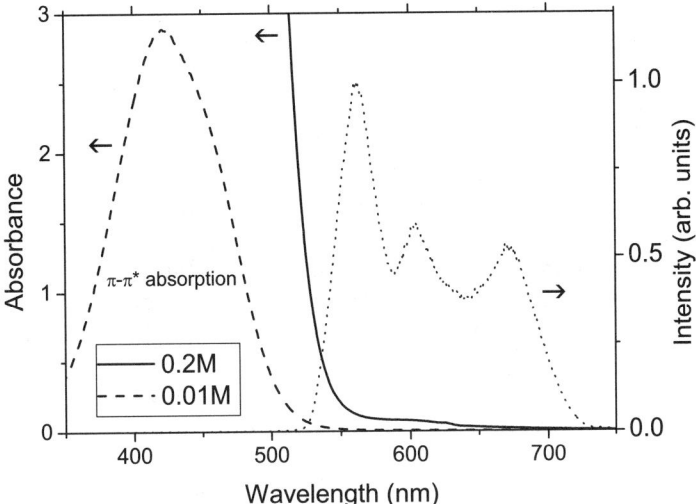

FIG. 2.3 Absorption spectra of DMAAB/DMSO at 0.2 (solid curve) and 0.01 (dashed curve) mol dm^{-3}. The spectrum of NOPA (noncollinear optical parametric amplifier) output (dotted curve) was used for both pump and probe pulses.

pump-probe experiments. The pulse spectrum has a small but finite overlap with the sample's π-π* absorption band tail. The absorption coefficients at 420nm(π-π*) and 600nm(n-π*) are 30,000 and 280mol^{-1} dm^2, respectively. By separating the absorption spectrum into π-π* and n-π*, the fraction of the former excitation is estimated as 70%.

2.3.2.1 Wave Packet in the Ground and Excited States

It is well known that a coherent wave packet can be generated both in the ground and excited states and that the ratio between the amounts of the wave packets excited on the two potential curves is sensitive to the chirp direction.[28–30] Positively chirped (PC) and Fourier-transform limited (TL) pulses excite a wave packet in the excited state much more efficiently than in the ground state, whereas negatively (NC) chirped pulses with an appropriate rate induces a wave packet more efficiently in the ground state. This is because the damping of the wave packet photogene rated by the leading edge of the pulse takes place by the trailing edge of the NC pulse in cases where the chirp rate is close to the lowering rate of the molecular system's potential energy in the excited state. We prepared NC, PC, and TL pulses for sample excitation by changing the insertion amount of a prism pair used as a pulse compressor in NOPA. The second-harmonic generation frequency-resolved optical gating (SHG-FROG) measurement[31,32] was used for the pulse characterization. The quadratic phase terms $\Phi''(\nu)$ of the NC and PC pulses were determined from the FROG traces to be −930 and 990 fs^2, respectively. The TL pulse is slightly down-chirped, but the absolute value of the residual chirp is smaller than 330 fs^2 in the main part of the pulse. The full width at half maximum of the NC, PC, and TL pulses are 9.6, 8.4, and 7.4fs, respectively.

These are slightly broader than 4.7fs, the FWHM of the pulse just at the output of the NOPA. This is due mainly to the smearing effect induced by the finite angle of pump and probe pulses (2.5 degrees) and the higher-order chirp caused by cell-wall glasses or the thin beam splitter that separates pump and probe beams.

The transient transmittance changes ($\Delta T/T$ (t)) of DMAAB/DMSO at 570 nm (π-π^* breaching is dominant) induced by the pump pulse were measured up to 1.8ps at the three previously mentioned chirp rates. They show very complicated but clear oscillating features in all cases. The positive transmittance change, averaged over the several oscillation periods due to the ground-state bleaching, does not decay in the measured delay-time range because of the decay of the excited state followed by the ground-state thermalization, which takes place for about 50ps.[33] The average transmittance change over the 50–1800 fs decay-time range is nearly proportional to the pump-pulse energy, guaranteeing negligible contribution of nonlinear photoexcitation, such as two-photon absorption.

Figure 2.4 shows the Fourier transform of oscillating components of the real-time spectra of $\Delta T/T$ (t) (the time dependence of the normalized transmittance change). In both PC- and NC-pulse excitations, two strong vibrating components are clearly observed in the $\Delta T/T$ (t) trace. These peaks, at 1140 and 1410cm^{-1}, are attributed to the C–N and N=N stretching modes, respectively, in DMAAB molecules from the Raman data in literature.[13,34–36]

2.3.2.2 Spectrogram Analysis

To investigate the frequency modulation of these modes, we performed time-frequency analysis by calculating the spectrograms of these two oscillating components corresponding to the C–N and N=N stretching. First, these two components were separated by frequency filter to avoid interference with each other. Then, time-frequency distributions (TFD) were obtained from the signals containing the single-frequency components. TFD indicate the energy

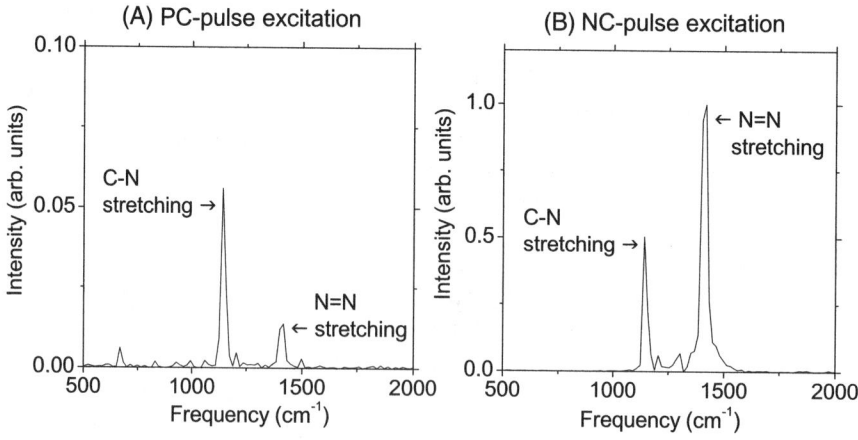

FIG. 2.4 Fourier transform of the oscillating components of the $\Delta T/T(t)$ traces using (A) positively chirped(PC) and (B) negatively chirped (NC) pulses.

content of a signal as a function of both time and frequency. There are many methods of constituting TFD.

In this paper, we utilized the method of short-time Fourier transform (spectrogram). The time-frequency $(t\text{-}\omega)$ distributions $P(t, \omega)$ of the time-dependent signal $s(t)$ is given as follows:

$$P(t,\omega) = |\textstyle\int e^{-i\omega\tau} s(\tau)h(\tau-t)\,d\tau|^2 \tag{2.1}$$

where $h(t)$ is the time window function given by

$$h(t) = \frac{1}{2} - \frac{1}{2}\cos(2\pi t/T) \tag{2.2}$$

Figure 2.5 shows the instantaneous frequencies $\omega(\tau)$ at delay time τ for C–N and N=N stretching modes, calculated by integrating the TFD along the lines at τ parallel to frequency axis. In the case of PC-pulse excitation (Figure 2.5A), the frequencies of both molecular vibration modes are modulated at the period of about 400fs. Furthermore, the phase difference between the two modulations in Figure 2.5A is about π, in contrast to Figure 2.5B, which shows less correlation between two frequency modulations. The phase difference between the two modulations can be even more clearly seen in

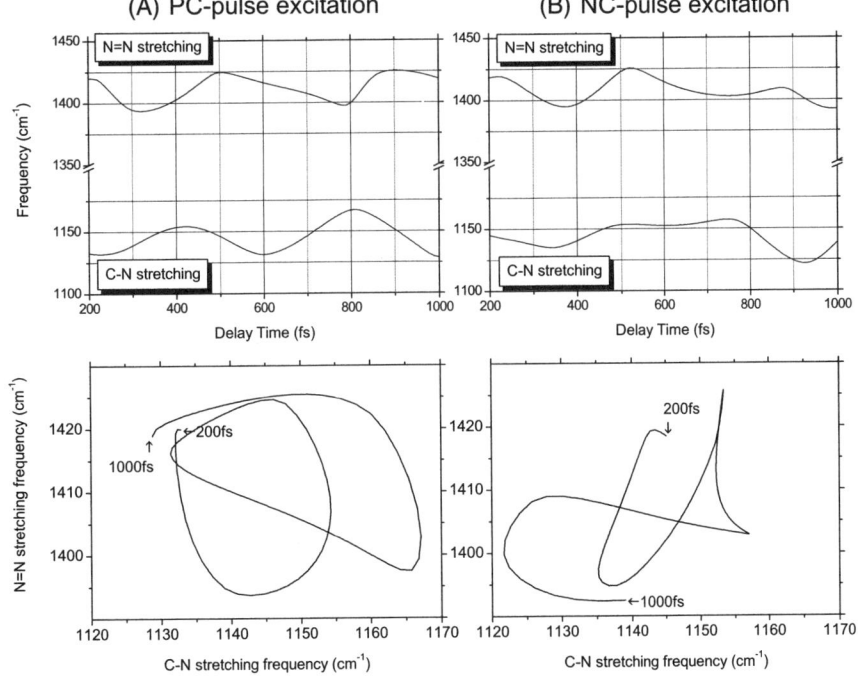

FIG. 2.5 Instantaneous frequencies of C–N and N=N stretching modes calculated from the spectrogram analysis of oscillating components of the $\Delta T/T(t)$ traces by using (A) positively chirped (PC) pulse and (B) negatively chirped (NC) pulse. Delay-tile dependence (upper) and Lissajou's figures (lower).

the Lissajou figures of the modulation frequencies. This difference demonstrates that signals of PC- and NC-pulse excitations correspond to different electronic states—the excited state and the ground state, respectively.[28–30] It is also well known that the excited-state wave packet is much more efficiently generated when the ratio of the vibration period and pulse width is large.[28,29]

The modulation period of 400fs in Figure 2.5A is considered to correspond to the oscillation between structures (B) and (C) as shown in Figure 2.2.[3,37] In this model, C–N and N=N stretching modes are coupled as shown in Figure 2.2B and C through a –N=N–Φ torsion mode, and frequencies of these two vibration modes are modulated by this torsion motion, with a vibration period of about 400fs. These molecular vibrations and their torsion-caused modulations are considered to be related to a *doorway* to the chemical reaction. During a few vibrations of the torsion mode, the isomerization reaction takes place with a stochastic probability, resulting in the macroscopic reaction time constant of about 1ps.[26] The microscopic speed of the configurational change associated with isomerization is fast enough that the N=N and C–N stretching frequencies are not detected to have the extremely modified frequencies of N–N and C=N stretching, respectively. The torsion mode is thus probed by the changes in the N=N and C–N stretching frequencies, even though these do not experience complete quantum mechanical resonance configurations as N–N and C=N do. These frequencies, including intermediate bond orders between 1 and 2 of each bond, are averaged over the torsion time; hence they do not have bond orders that are completely reversed from those of the ground state. It may be better to say that the torsion time is too short to define the bond orders and corresponding stretching frequencies for continuously changing configurations during the torsion motion.

Figure 2.6 shows the results of short-time Fourier transform corresponding to the modulation frequency power spectra of these two instantaneous

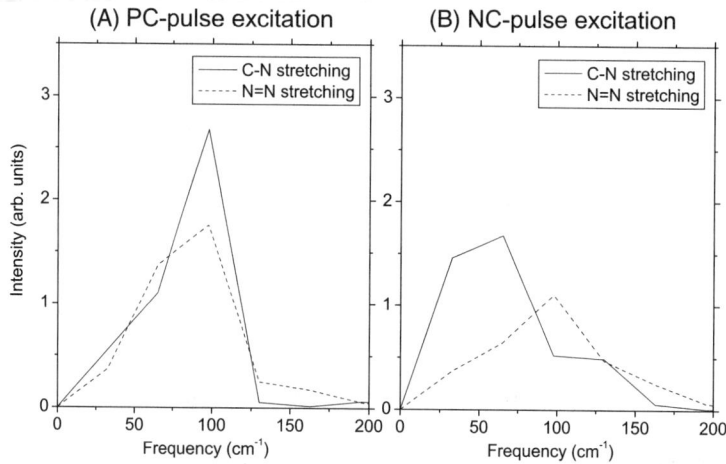

FIG. 2.6 Fourier transform of the frequency modulations of C–N stretching (straight lines) and N=N stretching (dashed lines) modes by using (A) positively chirped (PC) pulse and (B) negatively chirped (NC) pulse.

frequencies of molecular vibration modes in the excited (A) and in the ground (B) states. They were calculated by integrating the TFD along the frequency axis to study the correlation between them. Figure 2.6A, which corresponds to the modulation power spectrum of the instantaneous frequency in the excited state, shows the modulation period of the two vibration modes in the excited state. There exists a relatively good coincidence and strong correlation between the two modes of C–N and N=N stretching. Figure 2.6B which corresponds to the ground states, shows weaker correlation between the modulations of the N=N and C–N stretching frequencies.

It is to be stressed that the sub-5fs real-time spectroscopy enabled by the extremely short pulse allows us to make a detailed analysis of the dynamical mechanism of the chemical reaction. We can also gain information about geometrical change in the excited state by studying the changes of frequency and phases of molecular vibrations relevant to the chemical reaction. This can be done only with the use of a laser with a sufficiently short and stable pulse to determine the change in the molecular vibration period brought about by modulations as small as 5% of the period corresponding to ~1fs.

The first part, Section 2.3.1, was published elsewhere,[38] and the second part, Section 2.3.2, has been submitted for publication.[39]

REFERENCES

1. A. Shirakawa, I. Sakane, M. Takasaka, and T. Kobayashi. *Appl. Phys. Lett.* **74**, 2268 (1999).
2. T. Kobayashi, and A. Shirakawa. *Appl. Phys. B* **70**, S239 (2000).
3. Q. Wang, R. W. Schoenlein, L. A. Peteanu, R. A. Mathies, and C. V. Shank. *Science* **266**, 422 (1994).
4. F. Gai, K. C. Hasson, J. C. McDonald, and P. A. Anfinrud. *Science* **279**, 1886 (1998).
5. Z. F. Liu, K. Hashimoto, and A. Fujishima. *Nature* **347**, 658 (1990).
6. T. Ikeda, and O. Tsutsumi. *Science* **268**, 1873 (1995).
7. A. Burawoy, A. G. Salem, and H. R. Thompson. *J. Chem. Soc.* 4793 (1952).
8. D. Hadzi. *J. Chem. Soc.* 2143 (1956).
9. E. Fischer, and Y. F. Frei. *J. Chem. Soc.* 3159 (1959).
10. E. Lippert, D. Samuel, and E. Fischer. *Ber. Bunsenges. Phys. Chem.* **69**, 155 (1965).
11. H. Rau, in *Photochromism: molecules and systems.* H. Dürr, H. Bouas-Laurent, Eds. (Elsevier, Amsterdam, 1990) Chap. 4, pp. 165–192.
12. I. K. Lednev et al., *Chem. Phys. Lett.* **290**, 68 (1998).
13. T. Fujino, and T. Tahara, *J. Phys. Chem. A* **104**, 4203 (2000).
14. G. Gabor, Y. Frei, D. Gegiou, M. Kaganowitch, and E. Fischer. *Isr. J. Chem.* **5**, 193 (1967).
15. E. Fischer. *Fortsh. Chem. Forsch.* **7**, 605 (1967).
16. D. Huppert, and P. M. Rentzepis. *Appl. Phys. Lett.* **32**, 241 (1978).
17. G. E. Busch, K. S. Greve, E. L. Olson, R. P. Jones, and P. M. Rentzepis. *Chem. Phys. Lett.* **33**, 412 (1975).
18. T. Forster, and G. Hoffman. *Z. Phys. Chem.* (Frankfurt am Main) **75**, 63 (1971).
19. D. Gegiou, K. A. Muszkat, and E. Fischer. *J. Am. Chem. Soc.* **90**, 3907 (1968).
20. D. Magde, and M. W. Windsor. *Chem. Phys. Lett.* **24**, 144 (1974).
21. E. P. Ippen, C. V. Shank, and A. Bergman. *Chem. Phys. Lett.* **38**, 611 (1976).
22. R. Nurmukhametov, D. N. Shigorin, Y. Kozlov, and V. A. Puchkov. *Opt. Spectrosc.* **11**, 327 (1961).
23. G. Gabor, and E. Fischer. *J. Am. Chem. Soc.* **66**, 2478 (1962).
24. T. Kobayashi, and P. M. Rentzepis, *J. Chem. Phys.* **70**, 886 (1979).
25. J. Wachtveitl et al. *J. Photophem. Photobiol. A: Chemistry* **105**, 283 (1997).

26. S. G. Mayer, C. L. Thomsen, M. P. Philpot, and P. J. Reid. *Chem. Phys. Lett.* **314**, 246 (1999).
27. H. Bisle, M. Romer, and H. Rau. *Ber. Bunsenges. Physik. Chem.* **80**, 301 (1976).
28. G. Cerullo, C. J. Bardeen, Q, Wang, and C. V. Shank, *Chem. Phys. Lett.* **262**, 362 (1996).
29. C. J. Bardeen, Q, Wang, and C. V. Shank. *Phys. Rev. Lett.* **75**, 3410 (1995).
30. C. J. Bardeen, Q. Wang, and C. V. Shank. *J. Phys. Chem. A* **102**, 2759 (1998).
31. R. Trebino, D. J. Kane. *J. Opt. Soc. Am. A* **10**, 1101 (1993).
32. J. Hunter, W. E. White. *J. Opt. Soc. Am. B* **11**, 2206 (1994).
33. P. Hamm, S. M. Ohline, W. Zinth. *J. Chem. Phys.* **106**, 519 (1997).
34. H. Okamoto, H. Hamaguchi, and M. Tasumi. *Chem. Phys. Lett.* **130**, 185 (1986).
35. N. Biswas, and S. Umapathy. *Chem. Phys. Lett.* **236**, 24 (1995).
36. N. Biswas, and S. Umapathy. *J. Chem. Phys.* **107**, 7849 (1997).
37. S. Pedersen, L. Bañares, and A. H. Zewail. *J. Chem. Phys.* **97**, 8801 (1992).
38. T. Kobayashi, E. O. Degenkolb, and P. M. Rentzepis. *J. Phys. Chem.* **83**, 2431 (1979)
39. T. Saito and T. Kobayashi, *Phys. Chem.* (submitted).

3

PHOTO-ORIENTATION BY PHOTOISOMERIZATION

ZOUHEIR SEKKAT

Department of Applied Physics, Osaka University, Suita, Osaka 565-0871, Japan
School of Science and Engineering, Al Akhawayn University in Ifrane, 53000 Ifrane, Morocco

ABSTRACT

In this chapter, I discuss the phenomenon of molecular orientation induced by photoisomerization whereby experiment merges with theory to assess molecular movement during isomerization. The theory unifies photochemistry with optics, and it provides rigourous analytical tools for powerful quantification of coupled photoisomerization and photo-orientation. Experiments on spectrally distinguishable isomers detail the mechanisms of chromophore reorientation during photo- and thermal isomerization. In particular, I present contrasting series of photo-orientation experiments on azobenzene derivatives, which undergo isomerization by molecular-shape change, and chromophores, e.g., spiropyrans and diarylethenes, which photoisomerize by reversible photocyclization. In all of these chromophores, both cis and trans isomers are oriented by polarized photoisomerization, and in diarylethenes and spiropyranes, the ultaviolet and the visible transitions of the cis isomer are oriented perpendicularly to each other.

3.1 INTRODUCTION

Induction of photochemical anisotropy in certain materials upon irradiation with polarized light (the Weigert effect) has been known since the beginning of this century.[1] In the 1960s, this effect was widely studied in viscous solutions containing azo-dye molecules, which are known to undergo cis↔trans photoisomerization upon light irradiation.[2] During the 1980s, the Weigert effect was used for polarization holography applications in azo-dye–containing polymer films.[3] Recently, there has arisen a great deal of interest in azo-dye–containing materials, because of their possible application in the areas of optoelectronics, photonics, and optical signal processing.[4–6] In this regard, photoinduced anisotropy in azo-polymer films has been intensively studied, and a tremendous amount of work focusing on applications in rewritable data storage and photon-mode liquid crystal alignment, as well as fundamental aspects of photo-orientation, have been done by a large number of research groups.[6–15]

Photo-orientation of azo-dye molecules within trans↔cis photoisomerization occurs when these photochromic molecules are photo-selected by linearly polarized light of appropriate wavelength. The azo-dye molecules then experience successive cycles of trans↔cis isomerization and eventually align perpendicular to the irradiating light polarization (*vide infra*). In solutions, photo-orientation can be neglected when rotational diffusion has enough time to randomize induced molecular orientation. Small molecules diffuse rotationally in a few picoseconds in low-viscosity solutions. In solid polymer films, however spontaneous molecular mobility can be strongly hindered, depending on temperature and pressure, and photo-orientation effects can be appreciable.

There are three equally important aspects of photo-orientation by photoisomerization that need to be addressed. A first aspect is the bridging of photo-orientation with the pure photochemical point of view of isomerization. In pure photochemistry studies, photo-orientation, or photoinduced birefringence, is generally disregarded, even when it is appreciable (and isotropic absorbance can be properly measured when the polarizations of the analysis and irradiation lights are at the magic angle, i.e., ~ 54.7 degrees, to each other, thereby eliminating the contribution of anisotropy to measured absorbances). However, the analytical expressions used for the extraction of reaction parameters—which are based on absorbance values, such as the photoisomerization quantum yields—were not corrected for anisotropy. In this chapter, I will present the analytical theory that allows for the quantification of coupled photoisomerization and photo-orientation in A↔B photo-isomerizable systems where B is unknown. I will use this theory to study quantitatively both the photoisomerization and the photo-orientation of a series of photoisomerizable chromophores in films of polymer, including azobenzene derivatives and photochromic spectrally distinguishable spiro-pyrans and diarylethene-type chromophores. The way these chromophores move upon isomerization and the symmetry of the isomers' transitions will also be discussed.

A second aspect relates to the mechanisms of photo-orientation. Photo-orientation by photoisomerization occurs through a polarization-sensitive photoexcitation, i.e., photoselection, and the probability of exciting a transition in an isomer is proportional to the cosine square of the angle between that transition and the polarization of the excitation light. Transitions that lie along the polarization of the irradiation light will be excited with the highest probability, and molecules may be isomerized and reoriented and may fade in the direction of the polarization of the irradiation light. Now, and because photoisomerizable chromophores usually have two isomers, cis and trans, that can be interconverted into each other by light or heat, e.g., thermodynamically (*vide infra*), one can ask legitimate questions. Why does the molecule change orientation upon isomerization? Which isomer is oriented during which isomerization reaction? Do isomers orient upon photoselection without isomerization? How may photoisomerization quantum yields influence photo-orientation? Questions are also posed in regard to the systematic choice of the type of the photochromic molecules for photo-orientation studies.

A third aspect of photo-orientation in films of solid polymers is how photo-orientation is influenced by the polymer structure, molecular environment, and film configuration. Polymer structural effects on photo-orientation as well as the effect of intermolecular interaction and free volume are discussed in other chapters of this book.

Section 3.2 of this chapter recalls the pure photochemical point of view of photoisomerization of azobenzene derivatives. Section 3.3 discusses the theory of photo-orientation by photoisomerization and gives analytical expressions for the measurement of coupled photoisomerization and photo-orientation parameters. Sections 3.4 and 3.5 review observations of photo-orientation in azobenzene and push-pull azobenzene derivatives, respectively. Among other things, these sections address photo-orientation in both cis and trans isomers and discuss the effect of trans↔cis cycling, i.e., the photo-chemical quantum yields, on photo-orientation. Section 3.6 discusses the effect of the symmetry of photochemical transitions on photo-orientation in spiropyran and diarylethene-type chromophores. Finally, I make some concluding observations in Section 3.7.

3.2 PHOTOISOMERIZATION OF AZOBENZENES

The isomerization of azobenzenes is discussed in detail by Rau in Chapter 1. In this section, I shall recall the basic features of photoisomerization of azobenzene derivatives, a necessary groundwork for photo-orientation studies. Azobenzene derivatives and other photoisomerizable molecules have two geometric isomers, the trans and the cis forms, and the isomerization reaction is a light- or heat-induced interconversion of the two isomers. (See the top of Figure 3.1 for the trans and cis azobenzenes.) The trans isomer is thermodynamically more stable than the cis isomer—the energy barrier at room temperature is about 50 kJ/mol for the azobenzene—and generally, the thermal isomerization is in the cis→trans direction. Light induces isomerization in both directions. Photoisomerization begins by elevating the isomers to electronically exited states, after which nonradiative decay brings them to the ground state either in the "cis" form or in the "trans" form, the ratio depending upon the quantum yields of the isomerization reaction. From the cis form, molecules come back to the trans form by two mechanisms: spontaneous thermal reaction and reverse cis→trans photoisomerization.

The lower part of Figure 3.1 shows a simplified model of the excited states. Only two excited states are represented, but each represents a set of actual levels. The lifetimes of all these levels are assumed to be very short in comparison of those of the two excited states. σ_t and σ_c form the cross section for absorption of one photon by the trans and the cis isomers, respectively. The cross sections are proportional to the isomers' extinction coefficients. γ is the thermal relaxation rate; it is equal to the reciprocal of the lifetime of the cis isomer (τ_c). ϕ_{tc} and ϕ_{ct} are the quantum yields (QYs) of photoisomerization; they represent the efficiency of the trans→cis and cis→trans photochemical conversion per absorbed photon, respectively. They can be calculated for isotropic media by Rau's method, which was adapted from Fisher (see Appendix A); for anisotropic media, they can be calculated by a method described in this chapter. Two mechanisms may occur during the photoisomerization of azobenzene derivatives—one from the high-energy π–π^* transition, which leads to rotation around the azo group, i.e., –N=N– double bond, and the other from the low-energy n–π^* transition, which induces isomerization by inversion through one of the nitogen nuclei. Both proposed mechanisms lead to the same eventual conformational change, but

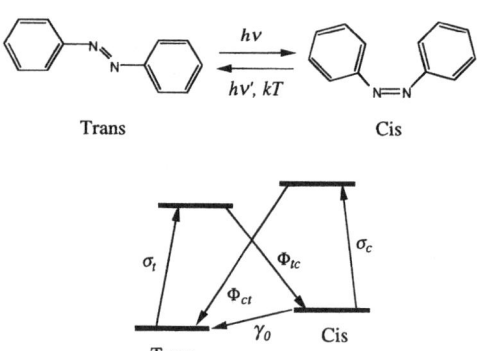

FIGURE 3.1 (Top) Trans↔cis isomerization of azobenzenes. (Bottom) Simplified model of the molecular states.

for each, the process of photoisomerization differs. For the photoisomerization of azobenzenes, the free volume needed for inversion is lower than what is needed for rotation.

Most photo-orientation studies in thin solid films have been perforformed on push-pull azobenzene derivatives such as disperse red one (DR1; see Figure 3.2). DR1 is a pseudo-stilbene–type azobenzene molecule. This means that the π-π^* and n-π^* transitions overlap each other, a feature that leads to a large structureless band in the trans isomer and a strong dependence of the maximum of absorption on the polarity of the host material, which may be a polymer or a solvent. In thin polymeric films, the kinetics of photoisomerization are not of first order, and they represent a complex behavior that is a consequence of the chromophore's local free-volume distribution in the polymer. The kinetics of isomerization of DR1 in films of a poly-methyl-methacrylate (PMMA) polymer are at least biexponential. Although this observation could be explained by the dual nature of the long wavelength λ ($\lambda > 400$ nm) photochemical transition explained above, the multiexponential behavior of the thermal cis→trans isomerization can be rationalized only by the existence of a distribution of local free volumes in the polymer film. In fact, the free volume clearly influences photo-isomerization and photo-orientation (*vide infra*). The thermal cis→trans isomerization of DR1 in PMMA is monoexponential during the first 10 seconds; this reaction is fast in both doped (4 s) and functionalized (5 s) polymers, as it is for stilbene-like azobenzene molecules. For DR1 in PMMA (guest-host), the activation energy is 16 kJ/mol, and the photochemical quantum yields are $\phi_{tc} = 0.11 \pm 0.03$ and $\phi_{ct} = 0.7 \pm 0.1$ at room temperature.[16] The experimental and theoretical methods for QY determination, i.e., the methods of Fisher[17] and Rau,[18] are summarized in Appendix 3A.

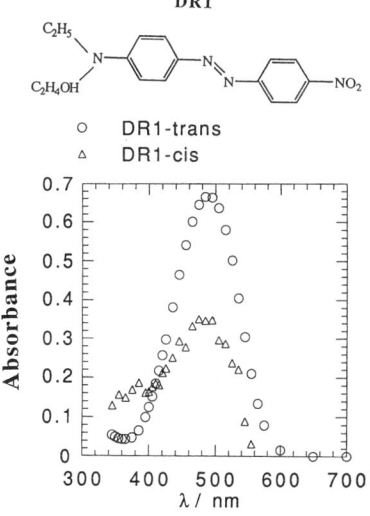

FIGURE 3.2 (Top) Trans-DR1. (Bottom) Absorption spectra of cis- and trans-DR1. The cis-DR1 spectrum was determined by Fisher's method. After reference 16, redrawn by permission.

3.3 PHOTO-ORIENTATION BY PHOTOISOMERIZATION

3.3.1 Base Ground Work

Experiments by Neoport and Stolbova,[2] and Todorov et al.[3] demonstrated that methyl-red and congo-red and methyl-orange chromophores, which are azobenzene and diazo derivatives, are oriented perpendicular to the irradiation light polarization. The understanding of photo-orientation by photoisomerization has been advanced by means of intensity-dependent, real-time anisotropy experiments in DR1-containing poly-methyl-metha-crylates.[19,20] Figure 3.3 shows the observed induced anisotropy, i.e., dichrism, by polarized light irradiation into a guest-host film of PMMA containing DR1 for low- and high-irradiation intensities. For low-irradiation intensity, the cis concentration is small, and $Abs_{//}$ (absorbance of an analysis light polarized parallel to the irradiation light polarization) and $Abs_{\perp}$ (absorbance of an analysis light polarized perpendicular to the irradiation light polarization) evolve in opposite directions, indicative of a near-pure orientation of the trans isomer. For high irradiation intensities, the cis concentration is appreciable, and both $Abs_{//}$ and $Abs_{\perp}$ evolve in the same direction. When the irradiation light is turned off, cis→trans thermal isomerization converts back all cis to trans isomers, as can be seen from the total recovery of the isotropic absorbance in Figure 3.3. A net remnant anisotropy, i.e., orientation of the trans isomer, is also observed. These observations imply that (1) the molecular linear polarizability of the cis form being appreciably smaller than that of the trans form, photoselection resulting from polarized light irradiation burns a hole into the molecular

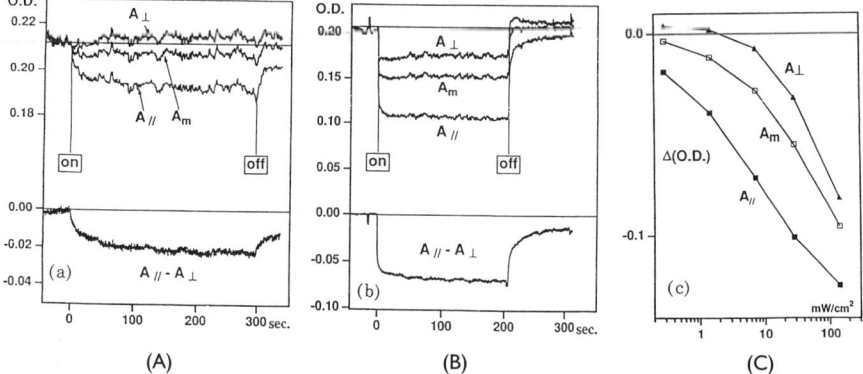

FIGURE 3.3 Variation of the absorbance, O.D., of a film of DR1-doped PMMA (2.5% w/w) with (A) low, 0.28 mW/cm², and (B) high, 28 mW/cm², irradiation (488 nm Ar⁺ laser) intensity. The analysis light was the 514 nm from the same Ar⁺ laser, and both $A_{//}$ and $A_{\perp}$ are measured. A_m represents the isotropic absorbance $(A_{//}+2A_{\perp})/3$. (C) Variation of the absorbance—parallel, perpendicular, and isotropic—with the irradiation light intensity. The data were taken 3 minutes after the irradiation was turned on. At low irradiation intensity, the cis population is small and $A_{\perp}$ increases as a consequence of the orientational distribution. At high irradiation intensity, saturation of the orientational hole-burning process tends to equalize $A_{//}$ and $A_{\perp}$. (After reference 20, by permission.)

orientational distribution in a nonpolar manner (nonpolar orientational hole burning), a feature that causes both $Abs_{//}$ and $Abs_{\perp}$ to decrease upon irradiation, and (2) for unknown reasons, the long molecular axis of the azo chromophore fades from the exciting light polarization, and molecular rotation takes place in the whole photoisomerization process (orientational redistribution).

Additional progress has been made by studying photoisomerization in spectrally individualizable isomers.[21,22] These studies are summarized in this chapter. Briefly, in contrast to trans- and cis-DR1, trans- and cis-azobenzene are spectrally distinguishable. It was also shown that not only the trans isomer but also the cis isomer is oriented by photoisomerization, and the orientation memory of the azobenzene molecule is preserved when it changes shape from cis to trans during the cis→trans thermal isomerization. In azobenzene derivatives, the QYs and the rate of the cis→trans thermal isomerization play a major part in the dynamics and efficiency of photo-orientation. Photochromic spiropyrans (SP) and diarylethenes (DE) isomerize by ring closing/opening, a feature that contrasts with azobenzene derivatives, which isomerize by shape change. The sign of the apparent photo-orientation of SP and DE depends on the photochemical transition band, i.e., on the analysis wavelength. It will be shown that the B isomer of SP and DE exhibit perpendicular transitions in the UV versus the visible transition bands. These experimental findings will be discussed in detail and in succession, after the introduction to the theory of photo-orientation.

3.3.2 Theory of Photo-Orientation

Inasmuch as the optical ordering of photoisomerizable molecules is being intensively studied, its theoretical quantification is needed to bridge independent studies in the areas of optics and photochemistry. Zimmerman et al.[23], and Fisher,[17] and Rau et al.[18] developed an optical pumping population change based theoretical background for photoisomerization within a pure photochemical framework. Michl et al.[24] performed intensive research on infrared vibrations and ultraviolet and visible (UV-vis) electronic transitions of molecules that are already oriented either by introduction into stretched polymer films or after photo-orientation. The theory that coupled photoisomerization and photo-orientation processes was developed a few years ago, and its mathematical foundation is based on the formalism of Legendre polynomials.[20,25] However, further theoretical developments are still needed to quantify coupled photoisomerization and photo-orientation processes. The most important concept within the framework of photo-orientation that needs to be clearly addressed is the polarization nature of the optical transition itself. In this section, I will present a model[26] based on purely polarized optical transitions; I will also present the related rigorous solutions to the general equations of the theory of molecular optical orientation for the full quantification of the coupled isomerization and optical ordering processes. This theory can be used for any type of photoisomerizable chromophore, and I will discuss it for both spectrally overlapping and distinguishable isomers.

3.3.2.1 Purely Polarized Transitions Symmetry

In the following discussion, A and B refer to the trans and cis isomers, respectively. We assume that A→B photoconversion occurs upon excitation of a purely polarized transition with light linearly polarized along the laboratory axis Z, and we define a site-fixed, right-handed orthogonal system of axes for each of the isomers A and B in which the molecule can exist, such that the angle between the Z_A and Z_B axes is χ. In isomers A and B, the electric dipole moments M_A^{UV} and M_B^{UV} responsible for excitation of a photochemical transition, say an ultraviolet (UV) transition, at a given irradiation wavelength are along the Z_A and Z_B axes, respectively. When the chromophore isomerizes, the transition at the irradiation wavelength changes from Z_A to Z_B, or the inverse, depending on whether the isomerization is in the trans→cis or the cis→trans direction. In isomer B, the electric dipole moment M_B^{Vis} responsible for excitation of a different photochemical transition, say a visible transition, at a given irradiation wavelength is at an angle labeled ω with respect to the Z_B axis, and lies in the plane that contains the latter and bisects the angle between X_B and Y_B (see Figure 3.4). For each of the isomers A and B, any polarized transition can be represented in the isomer's fixed molecular coordinates by an inclination angle, say ω, with respect to a reference transition that is fixed rigidly to the molecular coordinates, say the transition that corresponds to the irradiation wavelength, in the same manner as the UV-vis transitions are represented for the B isomer in Figure 3.4.

This model alleviates the concept of the somewhat ambiguous molecular anisometry that is based on an arbitrary choice of fixed molecular axes. So, for each of the A and B isomers, the isotropic absorbance $\overline{A_{A,B}} = (Abs_{//}^{A,B} + 2Abs_{\perp}^{A,B})/3$, the anisotropy $\Delta A_{A,B} = Abs_{//}^{A,B} - Abs_{\perp}^{A,B}$, and the optical order parameter $S_{A,B} = \Delta A_{A,B}/3\overline{A_{A,B}}$ are given by:

$$\overline{A_{A,B}} = \varepsilon_{A,B}C_{A,B}; \quad \Delta A_{A,B} = 3\varepsilon_{A,B}P_2(\cos \omega_{A,B})C_{A,B}A_2^{A,B}$$

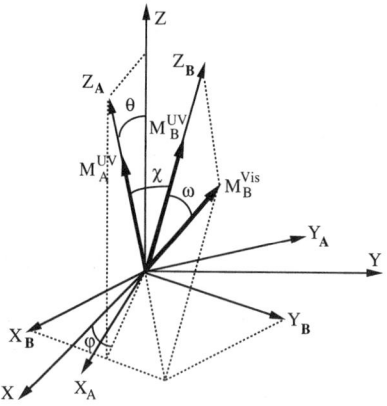

FIGURE 3.4 (X, Y, Z) indicate the laboratory coordinates axes, and $(X_{A,B}, Y_{A,B}, Z_{A,B})$ indicate the isomers' fixed molecular coordinates axes. The angles θ, φ, χ, and ω, and the transition electric dipole moments M_A^{UV}, M_B^{UV}, and M_B^{Vis}, are as defined in the text.

and

$$S_{A,B} = P_2(\cos \omega_{A,B})A_2^{A,B} \tag{3.1}$$

where $A_2^{A,B}$ is the isomer's geometrical order parameter. It is independent of the spectral properties of the chromophore, and $P_2(\cos \omega_{A,B})$ is the second-order Legendre polynomial of $\omega_{A,B}$ given by:

$$P_2(\cos \omega_{A,B}) = (3\cos^2 \omega_{A,B} - 1)/2 \tag{3.2}$$

with $\omega_{A,B}$ the angle that defines the orientation of a transition that corresponds to the analysis wavelength versus the irradiation wavelength transition. In other words, if analysis is done at the irradiation wavelength, $\omega_{A,B} = 0$ and $P_2(\cos \omega_{A,B}) = 1$. $Abs_{/\!/}^{A,B}$ and $Abs_{\perp}^{A,B}$ stand for absorption of light polarized parallel and perpendicular to the polarization of the irradiation light, respectively. We represent by C_A and C_B, and ε_A^λ and ε_B^λ, the concentrations and the isotropic extinction coefficients, i.e., those coefficients that can be measured from the isotropic absorbance spectra, of the isomers A and B, respectively. $\varepsilon_{A,B}^\lambda$ is proportional to $|M_{A,B}^\lambda|^2$. For all the equations, the sub- and superscripts A and B, if any, refer to the isomers A and B, respectively. It is noteworthy that if analysis is performed at a wavelength that is absorbed by either isomer A or B, the case of individualizable isomers, the observed absorbance and anisotropy are directly proportional to the concentration and orientation of only that isomer. Photo-orientation observation in both spectrally individualizable and overlapping isomers will be addressed after the discussion of the phenomenological theory of photo-orientation.

3.3.2.2 Phenomenological Theory and General Equations

The time-dependent expression of photo-orientation is derived by considering the elementary contribution per unit time to the orientation by the fraction of the molecules $dC_{A,B}(\Omega)$, whose representative moment of transition is present in the elementary solid angle $d\Omega$ near the direction $\Omega(\theta,\varphi)$ relative to the fixed laboratory axes (see Figure 3.4). This elementary contribution results from orientational hole burning, orientational redistribution, and rotational diffusion. The transitions are assumed to be purely polarized, and the irradiation light polarization is along the Z axis. The elementary contribution to photo-orientation is given by:

$$\frac{dC_{A,B}(\Omega)}{dt} = -3F'\phi'_{AB}\varepsilon'_A \cos^2 \theta \, C_A(\Omega) + 3F'\phi'_{AB}\varepsilon'_B \int_{\Omega'} C_B(\Omega')\cos^2 \theta' P^{BA}(\Omega' \to \Omega)d\Omega'$$

$$+ \frac{1}{\tau_B}\int_{\Omega'} C_B(\Omega') \, Q(\Omega' \to \Omega)d\Omega' + D_A\Re.\left[\Re C_A(\Omega) + C_A(\Omega)\Re \frac{U_A}{kT}\right],$$

$$\frac{dC_B(\Omega)}{dt} = -3F'\phi'_{BA}\varepsilon'_B \cos^2 \theta \, C_B(\Omega) + 3F'\phi'_{AB}\varepsilon'_A \int_{\Omega'} C_A(\Omega')\cos^2 \theta' P^{AB}(\Omega' \to \Omega)d\Omega'$$

$$+ \frac{1}{\tau_B} \, C_B(\Omega') + D_B\Re.\left[\Re C_B(\Omega) + C_B(\Omega)\Re \frac{U_B}{kT}\right] \tag{3.3}$$

$P^{AB}(\Omega' \to \Omega)$, $P^{BA}(\Omega' \to \Omega)$ and $Q(\Omega' \to \Omega)$ are the probabilities that the electric transition dipole moment of the chromophore will rotate in the A→B and B→A photoisomerizations, and B→A thermal isomerization, respectively. The orientational hole burning is represented by a probability proportional to $\cos^2 \theta$, and the last term in each of the equations in Equaton 3.3 describes the rotational diffusion due to Brownian motion. The latter is a Smoluchowski equation for the rotational diffusion characterized by a constant of diffusion D_A and D_B for the A and B isomers, respectively, where $\mathfrak{R}$ is the rotational operator. k is the Boltzmann constant, T is the absolute temperature, and $U_{A,B}$ is an interaction energy to which the isomers can be subjected. Depending on the type of interaction, $U_{A,B}$ can be polar or nonpolar. It is polar when the chromophores are isomerized in the presence of an electric field (the so called photo-assisted poling, discussed in detail elsewhere[25]); it is nonpolar when intermolecular interactions, such as liquid crystalline-type interactions, are present. I will not discuss these two cases. I will consider the case of $U_{A,B} = 0$, where friction is the only constraint in addition to isomerization. F' is a factor that takes into account that only some part of the totally absorbed amount of light induces photoreaction;[27] it is defined in Appendix 3A. The notations and units used in photochemistry are adopted, because the final theoretical expressions need to be compared to linear dichroism, i.e., polarized absorbance, measurements. In Equation 3.3, as well as in all the equations used in the rest of the chapter, the primed quantities, except for θ' and Ω', refer to an analysis at the irradiation wavelength, and the unprimed ones refer to an arbitrary analysis wavelength. The normalizations are:

$$\int_{\Omega} C_A(\Omega) \, d\Omega = C_A \quad \int_{\Omega} C_B(\Omega) \, d\Omega = C_B$$

$$C_A + C_B = C$$

$$\int_{\Omega'} P^{AB,BA}(\Omega' \to \Omega) \, d\Omega' = 1 \quad \int_{\Omega'} Q(\Omega' \to \Omega) d\Omega' = 1 \tag{3.4}$$

where C is the total concentration of the chromophores. With bulk azimuthal symmetry, the symmetry axis is the Z axis, i.e., the direction of the polarization of the irradiation light. The statistical molecular orientation for each of the photo-oriented A and B isomers is described by an orientational distribution function, $G_{A,B}(\theta)$, that depends only on the polar angle; it can be expressed using the standard basis of Legendre polynomials, $P_n(\cos \theta)$, with $A_n^{A,B}$ as expansion coefficients (order parameters) of order n (integer). $C_{A,B}(\Omega)$ is given by:

$$C_{A,B}(\Omega) = C_{A,B} \, G_{A,B}(\theta)$$

with
$$Q_{A,B}(\theta) = \frac{1}{2\pi} \sum_{n=0}^{\infty} \frac{2n + 1}{2} A_n^{A,B} P_n(\cos \theta)$$

and
$$A_n^{A,B} = \int_0^{\pi} G_{A,B}(\theta) \, P_n(\cos \theta) \sin \theta \, d\theta$$

and
$$A_0^{A,B} = 1 \tag{3.5}$$

C and $C_{A,B}$ correspond to C_0 and $C_{c,t}$ in Fisher's method, respectively. The redistribution processes $P^{AB}(\Omega' \to \Omega)$ and $P^{BA}(\Omega' \to \Omega)$, and $Q(\Omega' \to \Omega)$ depend only on the rotation angle χ between Ω and Ω', and they can also be expressed in terms of Legendre polynomials, with $P_N^{A \to B}$ and $P_n^{B \to A}$, and $Q_n^{B \to A}$ as expansion parameters, respectively. These parameters characterize the molecules' orientational memory after the A→B and B→A photo-isomerization reactions and B→A thermal isomerization.

$$P^{AB}(\chi) = \frac{1}{2\pi} \sum_{q=0}^{\infty} \frac{2q+1}{2} P_q^{A \to B} P_q(\cos \chi),$$

$$P^{BA}(\chi) = \frac{1}{2\pi} \sum_{q=0}^{\infty} \frac{2q+1}{2} P_q^{B \to A} P_q(\cos \chi),$$

$$Q(\chi) = \frac{1}{2\pi} \sum_{m=0}^{\infty} \frac{2m+1}{2} Q_m^{B \to A} P_m(\cos \chi),$$

with
$$P_0^{A \to B} = P_0^{B \to A} = Q_0^{B \to A} = 1 \tag{3.6}$$

When Legendre formalism is used, the variations of the cis and trans orientational distributions are given by the variations of their expansion parameters, i.e., $C_n^{A,B} = C_{A,B} A_n^{A,B}$. Indeed, by substituting Equations 3.4 through 3.6 into Equation 3.3 and using the orthogonality of Legendre polynomials, the following recurrence equations (i.e., Equation 3.7), and the important relation (Equation 3.8), the general rate equations (i.e., Equation 3.3) resume to the system of equations given by Equation 3.9.

$$\frac{d}{dx}\left[(x^2 - 1)\frac{dP_n(x)}{dx}\right] = n(n+1)P_n(x), \tag{3.7}$$

$$x^2 (2n+1)P_n(x) = \frac{(n+1)(n+2)}{2n+3}P_{n+2}(x) + \left[\frac{(n+1)^2}{2n+3} + \frac{n^2}{2n-1}\right]P_n(x)$$

$$+ \frac{n(n-1)}{2n-1} P_{n-2}(x),$$

$$\int_0^{2\pi} P_n(\cos \chi)d\Phi = 2\pi P_n(\cos \theta)P_n(\cos \theta') \tag{3.8}$$

where $\cos\chi = \cos \theta \cos \theta' + \sin s\theta \sin \theta' \cos \Phi$, and $\Phi = \varphi - \varphi'$ (see Figure 3.4). φ' is not shown in that figure, but it is the equivalent of φ for Z_A.

$$\frac{dC_{A,n}}{dt} = -3F'\phi'_{AB}\varepsilon'_A\{C_A\} + 3F'\phi'_{BA}\varepsilon'_B P_n^{B \to A}\{C_B\} + kQ_n^{B \to A}C_{B,n} - n(n+1)D_A C_{A,n}$$

$$\frac{dC_{B,n}}{dt} = 3F'\phi'_{AB}\varepsilon'_A P_n^{A \to B}\{C_A\} - 3F'\phi'_{BA}\varepsilon'_B\{C_B\} - kQ_n^{B \to A}C_{B,n} - n(n+1)D_A C_{B,n}$$

where

$$\{C_A\} = \{\kappa_{n+}C_{A,n+2} + \kappa_n C_{A,n} + \kappa_{n-}C_{A,n-2}\},$$

$$\{C_B\} = \{\kappa_{n+}C_{B,n+2} + \kappa_n C_{B,n} + \kappa_{n-}C_{B,n-2}\}, \tag{3.9}$$

$$\kappa_{n+} = \frac{(n + 1)(n + 2)}{(2n + 1)(2n + 3)}, \; \kappa_n = \frac{(n + 1)(n + 2)}{(2n + 1)(2n + 3)}, \text{ and } \kappa_{n-} = \frac{n(n - 1)}{(2n - 1)(2n + 1)}$$

This system of equations shows, through even orders, that polarized light irradiation creates anisotropy and photo-orientation by photoisomerization. A solution to the time evolution of the cis and trans expansion parameters cannot be found without approximations; this is when physics comes into play. Approximate numerical simulations are possible. I will show that for detailed and precise comparison of experimental data with the photo-orientation theory, it is not necessary to have a solution for the dynamics, even in the most general case where there is not enough room for approximations, i.e., that of push-pull azo dyes, such as DR1, because of the strong overlap of the linear absorption spectra of the cis and trans isomers of such chromophores. Rigorous analytical expressions of the steady-state behavior and the early time evolution provide the necessary tool for a full characterization of photo-orientation by photoisomerization.

3.3.2.3 Dynamical Behavior of Photo-Orientation

3.3.2.3.1 Onset of Photo-Orientation

The purpose of this section is to give an approximate analytical expression that reproduces the dynamics of anisotropy during photo-orientation. The experimentally observed evolution of photo-orientation when only one isomer is photoisomerized as well as when the two isomers are simultaneously photoisomerized presents a characteristic behavior that can be approximated by a double-exponential function. To solve analytically the system of Equation 3.9, I shall introduce approximations that are physically valid for at least the azobenzene molecule in a polymeric environment, and I shall neglect the expansion parameters above the third order. The fourth Legendre polynomial moment is a small correction to the second Legendre polynomial moment, which gives the anisotropy. I have assumed that only the trans isomer significantly absorbs the irradiation light, i.e., the pump light, and that the rates of both the cis→trans thermal isomerization and the diffusion in the cis and trans forms are small. Analytical solutions are found (see Equation 3.10) for the cis and trans populations (αC and $(1 - \alpha)C$ respectively) and the even-order parameters (A_2^B and A_2^A respectively) that characterize the orientation. If the irradiating light is turned on at the time $t = 0$, the solution is given by:

$$\alpha = 1 - \frac{1}{2}\{0.78\exp(-k_2 t) + 1.22\exp(-k_0 t)\},$$

$$A_2^A = \frac{1}{5(1 - \alpha)}\{\exp(-k_2 t) - \exp(-k_0 t)\}, \qquad (3.10)$$

$$A_2^B = \frac{1 - \alpha}{\alpha} P_2^{A \to B} A_2^A,$$

where $k_2 = 2.23 * 1000 I_0' (1 - 10^{-A_0'}) \phi_{AB}' \varepsilon_A'$, and $k_0 = 0.35 * 1000 I_0' (1 - 10^{-A_0'}) \phi_{AB}' \varepsilon_A'$

For photo-orientation analysis of actual data, Equation 3.10 must be combined with Equations 3.1 and 3.2. The time-evolution simulation (see

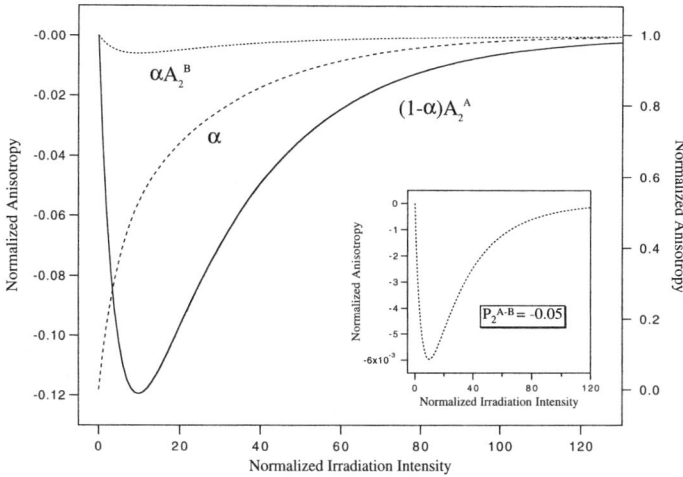

FIGURE 3.5 Theoretical simulations of the effect of polarized irradiation on the variation of the cis population (a) and the cis (aA_2^B) and trans (($1 - a)A_2^A$) orientational anisotropies.

Figure 3.5) of the cis and trans population and orientation under linearly polarized light irradiation shows that for high-irradiation times or high-irradiation intensities, the system should be saturated and the isotropy should be restored, because even molecules that are perpendicular to the irradiating light polarization present an appreciable probability for being photoisomerized. Such a behavior has been observed in a 9Å ultra-thin (optically thin) SAM containing azobenzene molecules,[28] and in an azobenzene-polyglutamate polymer, the dichroism decreased with increasing irradiation times. The behavior illustrated in Figure 3.5 was also observed for DR1 as well as for spiropyran and diarylethene-type molecules in PMMA films (vide infra). Since both the cis and trans isomers of these chromophores present strong spectral overlap at the irradiation wavelength, it is important to note that double-exponential functions in the form of Equation 3.10 can be used to approximate the dynamics of photo-orientation for systems in which all three isomerizations (i.e., photo-induced and thermally activated) can occur. However, care must be taken in comparing fitted values to actual physical quantities outside the approximations framework discussed previously.

3.3.2.3.2. Relaxation of Photo-orientation

When the irradiating light is switched off, the thermal cis→trans isomerization rate, k, governs mainly the population change from cis to trans isomers, and the cis and trans rotational diffusion rates, (i.e., k_D^A and k_D^B, respectively) influence only the relaxation of the isomers' orientation. This orientation diffuses in both cis and trans forms during the cis lifetime, and after that time it diffuses only in the trans form. If the irradiating light is turned off at the time $t' = 0$, the cis and trans population and orientation changes are rigorously described by:

$$\alpha = \alpha_0 \exp(-kt), \quad A_n^B = A_{n,0}^B \exp\left[-(k + k_D^B)t\right],$$

$$\frac{1 - \alpha}{1 - \alpha_0} \frac{A_n^A}{A_{n,0}^A} = \exp(-k_D^A t) \left\{ 1 + \frac{\alpha_0}{1 - \alpha_0} \frac{A_{n,0}^B}{A_{n,0}^A} \frac{Q_n}{\Gamma} \left(1 - \exp(-k\Gamma t)\right) \right\}, \quad (3.11)$$

where $\quad \Gamma = 1 - \dfrac{k_D^A - k_D^B}{k}$, and $k_D^{A,B} = n(n + 1)D_{A,B}$

Equation was derived without approximations. It is noteworthy that these solutions do not couple tensorial components of different orders and that they confirm that rotational diffusion and cis→trans thermal isomerization are isotropic processes that do not favor any spatial direction. In Section 3.4, I discuss, through the example of azobenzene, how Equation 3.11 can be used to study reorientation processes during cis→trans thermal isomerization after the end of irradiation. The next subsection gives analytical expressions at the early-time evolution and steady-state of photo-orientation, for the full quantification of coupled photo-orientation and photoisomerization in A↔B photoisomerizable systems where B is unknown.

3.3.2.4 Early Time Evolution of Photo-Orientation

3.3.2.4.1 A→B Photo-Orientation

At the early time evolution of photoselection, the cis population is negligible compared to the trans population, and the quantification of coupled photo-orientation and photoisomerization can be done for spectrally overlapping isomers as well as for individualizable isomers. For spectrally overlapping isomers, the analysis light is absorbed by both isomers, and the slopes, $p(\Delta)$ and $p(\Delta A)$, of $\bar{A}$ (proportional to the population change) and ΔA (proportional to the orientation) during irradiation, respectively, are rigorously given by:

$$p(\Delta) = 1000I_0' \left(1 - 10^{-A_0'}\right)\phi_{AB}'(\varepsilon_B - \varepsilon_A) \tag{3.12}$$

$$p(\Delta A) = \frac{6}{5}1000I_0' \left(1 - 10^{-A_0'}\right)\phi_{AB}'\left\{P_2^{A\to B} P_2 (\cos \omega_B)\varepsilon_B - P_2 (\cos \omega_A)\varepsilon_A\right\} \tag{3.13}$$

For individualizable isomers, and for an analysis light that can be absorbed only by the B isomer, $p(\Delta)$ and $p(\Delta A)$ are rigorously given by:

$$p(\Delta) = 1000I_0' \left(1 - 10^{-A_0'}\right)\phi_{AB}'\varepsilon_B \tag{3.14}$$

$$p(\Delta A) = \frac{6}{5}1000I_0' \left(1 - 10^{-A_0'}\right)\phi_{AB}'P_2 (\cos \omega_B)P_2^{A\to B}\varepsilon_B \tag{3.15}$$

I will go on to show how Equations 3.12 through 3.15 can be used to determine ϕ_{AB}' and $P_2 (\cos \omega_B)P_2^{A\to B}$ from A→B photo-orientation experiments. The algebra of the derivation of Equations 3.12 through 3.15 is detailed in Appendix 3B.

3.3.2.4.2 B→A Photo-Orientation

B→A photo-orientation is observed only for individualizable isomers, and its evolution is described by a double-exponential behavior in the form of

Equation 3.10. The slopes of the early time evolution of the changes of the normalized isotropic absorbance, $p(\Delta_N)$, and the anisotropy, $p(\Delta A_N)$, are given by Equations 3.16 through 3.18. These equations hold for an analysis light that is absorbed only by the B isomer and describe the orientational distribution of that isomer. Equations 3.16 through 3.18 are rigorously given by:

$$p(\Delta_N) = -k - 1000 I_0' \frac{1 - 10^{-A_0'}}{A_0'} \varepsilon_B' \phi_{BA}' \qquad (3.16)$$

$$p(\Delta A_N) = -\frac{6}{5} 1000 I_0' \frac{1 - 10^{-A_0'}}{A_0'} P_2 (\cos \omega_B) \varepsilon_B' \phi_{BA}' \qquad (3.17)$$

with $\qquad \Delta_N = (Abs - A_0')/A_0'; \text{ and } \Delta A_N = (Abs_{//} - Abs_\perp)/A_0' \qquad (3.18)$

Equations 3.16 through 3.18 were derived by a method similar to that used in deriving Equations 3.12 through 3.15. They allow for the measurement of ϕ_{BA}' and $P_2 (\cos \omega_B)$ from B→A photo-orientation experiments.

3.3.2.5 Steady State of A↔B Photo-Orientation

During the steady state of photo-orientation, the expansion parameters $C_{A,n}$ and $C_{B,n}$ are constants, i.e., $dC_{A,n}/dt = dC_{B,n}/dt = 0$. If the first equation of the system of Equation 3.9 is multiplied by $P_2^{A \to B}$ and added to the second equation of that system, the following relation is obtained after rearrangement.

$$0 = 3\frac{F'}{k} \phi_{BA}' \varepsilon_B' \{C_B\} + \frac{(P_n^{A \to B} Q_n^{B \to A} - 1) - k_D^B/k}{(P_n^{A \to B} Q_n^{B \to A} - 1)} C_{B,n} - \frac{P_n^{A \to B} k_D^A/k}{(P_n^{A \to B} P_n^{B \to A} - 1)} C_{A,n}$$

$$(3.19)$$

Equation 3.19 is valid for $n \neq 0$, and it allows for the derivation of the steady-state order parameters of the isomers' orientational distribution. Indeed, by making the two following assumptions, Equation 3.19 resumes to Equation 3.20 for $n = 2$ after rearrangement.

1. The diffusion rates of both the A and B isomers, k_D^A and k_D^B, respectively, are negligible in comparison to the rate, k, of the B→A thermal recovery. This a good approximation to use when the chromophores are introduced into polymers, because spontaneous molecular movement in polymeric materials is most efficient near the polymer glass transition temperature, Tg, and strongly hindered far below Tg. At room temperature, k_D^B/k is in fact small, i.e., ~ 0.03, for SP in films of PMMA,[29] and in functionalized azopolymers, photo-orientation can be quasi-permanent.[30] Note that $k_D^A/k \ll 1$ is equivalent to $P_n^{A \to B} = 0$, which means that cis orientation loses all memory of trans orientation during the A→B photoisomerization. In other words, the chromophore orientation is thermalized by strong shaking in the excited state upon photon absorption.

2. The process of isomeric-type reorientation is assumed—that is, that process in which the reorientation of the transition dipole is due only to the isomeric change in shape and where the parameters, i.e., $P_n^{A \to B}$ and $P_n^{B \to A}$ and $Q_n^{B \to A}$, that describe the reorientation of the transition during the photo-induced and thermal isomerization reactions are equal, say equal to Q.

$$\frac{36}{35}A_4^B + \left(\frac{11}{7} + \frac{k}{\varepsilon_B' \phi_{BA}'} \frac{1}{F'}\right)A_2^B + \frac{2}{5} = 0 \qquad (3.20)$$

The solution of Equation 3.20 must be of the form:

$$A_2^B = \frac{1}{x + \dfrac{k_2}{\varepsilon_B' \phi_{BA}'} \dfrac{1}{F'}} \quad \text{and} \quad A_4^B = \left(1 + \frac{k_4}{\varepsilon_B' \phi_{BA}'} \frac{1}{F'}\right)A_2^B \qquad (3.21)$$

Because Equation 3.20 is valid for any irradiation intensity, $x = -13/2$, and $2k_2/5 + 36k_4/35 = -k$. Rigorously, k_2 and k_4 are proportional to k, and $k_2 = -k$ is a physically reasonable solution.

$$\frac{1}{A_2^B} = -\frac{13}{2} - \frac{k}{\varepsilon_B' \phi_{BA}'} \frac{1}{F'} \quad \text{and} \quad \frac{1}{S^B} = \frac{1}{P_2(\cos \omega_B)}\left(-\frac{13}{2} - \frac{k}{\varepsilon_B' \phi_{BA}'} \frac{1}{F'}\right) \qquad (3.22)$$

No truncation above any order has been made for the determination of A_2^B and S^B, and the solution given by Equation 3.21 is certainly physical, because it corresponds to actually observed behavior (*vide infra*). Equation 3.22 is useful when isomer B is spectrally distinguishable from isomer A. For spectrally overlapping isomers, Equation 3.23 gives the order parameters, i.e., the geometrical A_2, and spectral, S, that characterize the orientational distribution of the whole—trans and cis—molecular distribution.

$$\frac{1}{S} = \frac{1}{A_2} = -\frac{13}{2} - \frac{k}{\varepsilon_B' \phi_{BA}'} \frac{1}{F'} \qquad (3.23)$$

Equation 3.23 is derived without truncation above any order by assuming that the geometrical order parameters, A_2, of the orientational distribution of the A and B isomers are equal at the photostationary state of irradiation. Although this assumption physically mirrors a uniform molecular orientational distribution, it does simplify considerably the expression of the photostationary-state orientational order and provides a simple law for steady-state photo-orientation characterization. Equation 3.23 holds when analysis is performed at the irradiation wavelength, and fits by Equations 3.22 and 3.23 allow for the measurement of ϕ_{BA}' and $P_2(\cos \omega_B)$ (*vide infra*). Inasmuch as measured values of $P_2(\cos \omega_B)$ can be rigorous because the value of $-13/2$ was derived without compromise, measured values of ϕ_{BA}' depend on the assumption of $k_2 = -k$.

When the analysis of photo-orientation is performed at a wavelength different from the irradiation wavelength, the symmetry of the molecular transitions in both A and B isomers can be found. Indeed, setting n=0 in Equation 3.9, yields the following relation for the photostationary state of irradiation:

$$\frac{\varepsilon_A' \phi_{AB}'}{\varepsilon_B' \phi_{BA}'}(1 + 2A_{2A}) = \frac{\alpha}{1 - \alpha}\left\{(1 + 2A_{2B}) + \frac{k}{\varepsilon_B' \phi_{BA}' F'}\right\} \qquad (3.24)$$

In this equation, the trans and cis photostationary-state order parameters, A_{2A} and A_{2B} respectively, are given by Equation 3.23, and when the irradiation intensity, F', is extrapolated to infinity, $A_{2A}^\infty = A_{2B}^\infty = A_2^\infty = -2/13$, and Equation 3.24 resumes to Fisher's stationary state relation for thermally irreversible systems:

$$\frac{\varepsilon'_A \phi'_{AB}}{\varepsilon'_B \phi'_{BA}} = \frac{\alpha^\infty}{1 - \alpha^\infty} \tag{3.25}$$

So, when photo-orientation analysis is not performed at the irradiation wavelength, the order parameter at infinite irradiation intensity of spectrally overlapping isomers reads:

$$S = -\frac{2}{13} \frac{P_2(\cos \omega_B) + a P_2(\cos \omega_A)}{1 + a} \text{ with } a = \frac{\varepsilon_A}{\varepsilon_B} \frac{\varepsilon'_B \phi'_{BA}}{\varepsilon'_A \phi'_{AB}} \tag{3.26}$$

$P_2(\cos \omega_A)$ and $P_2(\cos \omega_B)$ are the second-order Legendre polynomials of ω_A and ω_B, which are the angles between the irradiation and analysis transitions of isomers A and B, respectively. ω_A and ω_B can be experimentally determined by a two-step photo-orientation experiment: irradiate at a wavelength λ_1 and analyze at λ_2; irradiate at λ_2 and analyze at λ_1. Although S changes for each step, $P_2(\cos \omega_B)$ and $P_2(\cos \omega_A)$ remain unchanged, and their determination is straightforward. a can be determined beforehand by determining the cis spectrum by Fisher's method and the quantum yields by photo-orientation analysis at the irradiation wavelength.

3.4 PHOTO-ORIENTATION OF AZOBENZENES: INDIVIDUALIZABLE ISOMERS

In spectrally overlapping isomers, such as cis- and trans-DR1, both forms may be simultaneously isomerized and the molecule may rotate in each isomerization reaction, i.e., the trans→cis photoisomerization, and the photo-induced and thermally activated back cis→trans reactions. All three reactions may happen simultaneously given that both the trans and cis isomers exhibit their absorption maximum in the same visible region (see Figure 3.2 for trans- and cis-DR1), and the lifetime of the cis isomer is relatively short (4 to 5 seconds in polymethyl-methacrylates).[16] In contrast, azobenzene-containing polymeric films are more appropriate for closely probing this reorientation process in each of the three isomerization reactions. The trans and cis isomers of azobenzene present different absorption bands in the UV (around 360 nm) and the visible (around 450 nm) regions (see Figure 3.6). The lifetime of the azobenzene cis isomer is generally on the order of hours, depending on the polarity of the host material, which may be a polymer or a solvent; therefore, the azobenzene cis isomer can be considered stable on a time scale of minutes, i.e., the time scale of the experiment. It will be shown that azobenzene molecules are oriented within both the direct trans→cis photoisomerization and the thermal cis→trans back isomerization.

3.4.1 Reorientation within the trans→cis Photoisomerization

Figure 3.7 shows the dichroism observed in spin-cast films of an azo-polyglutamate (see the formula in the caption for Figure 3.6). These UV-vis spectra were obtained after irradiation with linearly polarized UV light

(A)

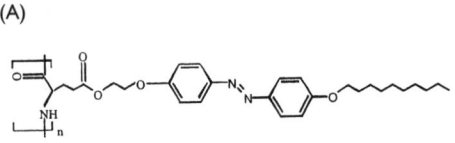

(B)

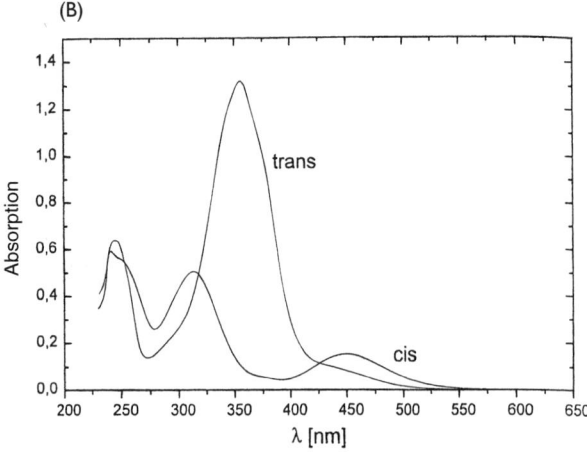

FIGURE 3.6 (A) Structure formula and (B) UV-vis spectra of a poly-(5-(2-(4-(4-decyloxyphenylazo) phenoxy) ethyl)-L-glutamate), denoted by $P_{2,10}$, solution in $CHCl_3$ in the dark (trans) and at the photostationary state (cis) of a 360 nm UV light irradiation. After reference 21, redrawn by permission.

(360nm; 0.2 mW/cm^2) for 35 minutes, and recording both $Abs_{//}$ and $Abs_{\perp}$. It is clear that the absorption $Abs_{\perp}$ is higher than the absorption $Abs_{//}$. Identical spectra were recorded for both $Abs_{//}$ and $Abs_{\perp}$ before UV irradiation (only $Abs_{\perp}$ shown), demonstrating that the sample was in-plane optically isotropic at that time. These findings are true for the trans absorption band in the UV region around 360 nm as well as for the cis absorption band in the visible region around 450 nm (see the inset in Figure 3.7). This shows clearly that both the trans and cis azobenzene molecules are preferentially distributed perpendicular to the initial UV polarization and that the cis isomer aligns perpendicular to the initial UV polarization within the trans→cis isomerization.

Irradiation with unpolarized blue light does not erase the induced dichroism, because only the cis isomer has significant absorption in the blue region around 450 nm; consequently, the trans molecules cannot be excited and reoriented. The in-plane isotropy in both the trans and the cis molecular distributions can be restored only after successive unpolarized UV and blue light irradiations. The initial spectra of a freshly prepared sample prior to irradiation is not restored by this procedure, however, because a net out-of-plane orientation of the azomolecules remains. It is noteworthy that spectra recorded with different analysis light polarizations do not intersect at isosbestic points because of the anisotropy that exists in the sample. Heating the azo-polyglutamate sample at 80°C for 30 minutes and 14.5 hours failed

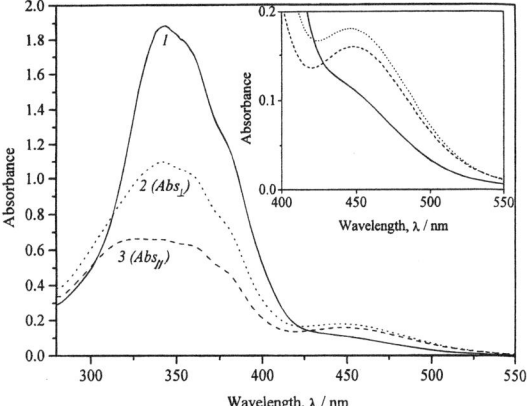

FIGURE 3.7 UV-vis absorption spectra of the azo-polyglutamate film before (1) and after 35 min (2, 3) of linearly polarized UV (360 nm) irradiation. The probe light was also linearly polarized and spectra were obtained for both parallel (3) and perpendicular (2) orientations. The inset is an expanded view of the cis absorption. After reference 22, redrawn by permission of ACS.

to erase the UV-irradiation–induced dichroism in the sample. At this temperature, the sample is still at least 120°C below the sidechain isotropization temperature (T > 200°C) of the polymer.[31] The absorbance of linearly polarized probe light (at 360 and 450 nm) at various angles, Ψ, between the polarizations of probe and UV lights, shows sinusoidal behavior and confirms the orientation of both trans and cis molecular distributions (Figure 3.8). As in Figure 3.7, the highest absorption for both trans and cis distributions is observed when the probe and irradiation beams have perpendicular polarizations.

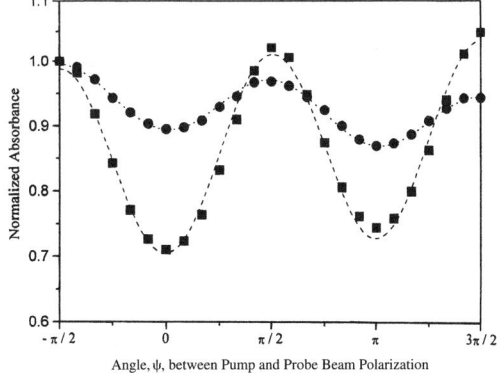

FIGURE 3.8 Dependence of the absorbance of linearly polarized probe light at 360 nm (squares) and 450 nm (circles) on the angle, Ψ, between the probe and the UV light polarization. This behavior is fitted by a $\cos^2 \Psi$ with an amplitude that decays with the cis→trans thermal isomerization rate. After reference 22, redrawn by permission of ACS.

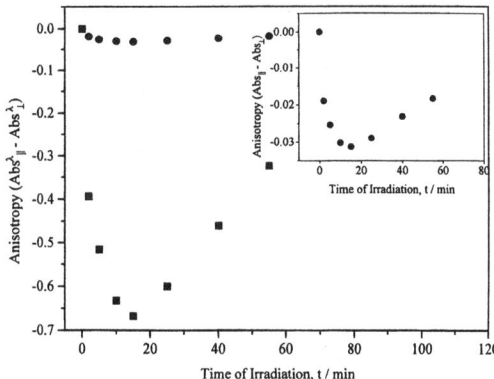

FIGURE 3.9 Evolution of the anisotropy ΔA^λ for the wavelengths 360 nm (squares) and 450 nm (circles), also shown in the inset, with the time of UV (360 nm) irradiation. After reference 22, redrawn by permission of ACS.

Figure 3.9 shows the UV-irradiation–induced dichroism observed in the azo-polyglutamate film sample for a constant UV irradiation intensity and different irradiation times for both cis and trans molecular distributions. This was performed by irradiating the sample with linearly polarized UV light for a defined period of time, recording the absorption spectrum, and subsequently calculating the anisotropy, $\Delta A^\lambda = Abs_{/\!/}^\lambda - Abs_\perp^\lambda$, and the spectral order parameter, $S^\lambda = \Delta A^\lambda/(Abs_{/\!/}^\lambda + 2Abs_\perp^\lambda)$, at the wavelengths, λ, equal to 360 and 450 nm for the trans and cis molecular distributions, respectively. Figure 3.9 shows that the dichroism increases with increased irradiation times until a maximum value is reached; further irradiation progressively produces less dichroism. This behavior is described theoretically by Equation 3.10, because the assumptions under which those equations were derived hold true for the azobenzene molecule in the polyglutamate polymer. Indeed, the absorption of cis azobenzene at 360 nm is much smaller—at least 14 times lower—than that of trans-azobenzene at this same wavelength, and the cis state is quasi-stable at the minutes time scale (the cis lifetime is 3.5 hours), i.e., the duration of the experiment, and the diffusion time is considerably higher than minutes because the UV-induced dichroism is quasi-permanent.

Table 3.1 gives the ratio of the photoinduced anisotropy of the cis relative to that of trans isomers, i.e., $\Delta A^{450nm}/\Delta A^{360nm}$. For these calculations, $\Delta A^{450nm}/\Delta A^{360nm}$ was experimentally measured for each UV irradiation time. The theoretical expression $\Delta A^{450nm}/\Delta A^{360nm}$ is obtained by combining Equations 3.1 and 3.10. It reads:

TABLE 3.1   Time of UV (360 nm) Irradiation (0.2 mW/cm²), Cis Anisotropy Relative to Trans Anisotropy

Time of UV irradiation (min)	2	5	10	15	25	40	55
$-10^2 \, (\Delta A^{450nm} / \Delta A^{360nm})$	4.83	4.93	4.78	4.68	4.82	5.03	5.68

$$\frac{\Delta A_B^{450}}{\Delta A_A^{360}} = -\frac{\varepsilon_B^{450}}{\varepsilon_A^{360}} P_2 \left(\cos \omega_B^{360 \to 450}\right) P_2^{A(360) \to B(360)} \qquad (3.27)$$

Equation 3.27 dictates that $\Delta A^{450nm}/\Delta A^{360nm}$ is constant regardless of the irradiation dose, a result confirmed by the data in Table 3.1. It is noteworthy that the parameters $P_2(\cos \omega_B^{360 \to 450})$ and $P_2^{A(360) \to B(350)}$, which characterize the symmetry of the 360 and 450 nm transitions in the cis isomer and the reorientation of the UV transition of the azobenzene chromophore during the trans→cis photoisomerization, respectively, can be determined when $P_2(\cos \omega_B^{360 \to 450})$ is measured at the steady state of photo-orientation using Equation 3.22. This type of experiment will be discussed eventually for spiropyran and diarylethene chromophores in films of PMMA. Next, I compare reorientation observations after cis→trans thermal isomerization of azobenzene to the theoretical developments in Section 3.2.3.2.

3.4.2 Reorientation within the cis→trans Thermal Isomerization

The process of reorientation during cis→trans thermal isomerization can be seen at the value of Q_2 in Equation 3.11, which shows that the cis anisotropy does not contribute to the trans anisotropy if the trans isomer loses total memory of the orientation in the cis isomer ($Q_2 = 0$). It is informative to note that in the realistic physical case—i.e., the case of the azobenzene molecule chemically attached to a polymer, where the cis and trans diffusion rates are negligible in comparison to the cis→trans isomerization rate—the relaxation of the cis and trans anisotropy, ΔA^B and ΔA^A, can be written respectively in the form:

$$\frac{\Delta A^B}{\Delta A_0^B} = \exp(-kt) \text{ and } \frac{\Delta A^A}{\Delta A_0^A} = 1 + Q_2 \frac{\Delta A_0^B}{\Delta A_0^A} \left(1 - \exp(-kt)\right) \qquad (3.28)$$

where $\Delta A_0^{A,B}$ is the anisotropy at the moment the irradiation is stopped. Figure 3.10 shows that trans-azobenzene anisotropy increases during cis→trans thermal isomerization in the $P_{2,10}$ azo-polyglutamate. Thus, it can be concluded that the azobenzene molecule has retained memory of its orientation when returning from the cis to the trans form ($Q_2 \neq 0$ in Equation 3.28). A similar behavior (not shown) was found for an azobenzene self-assembled monolayer.[28] The value of Q_2 can be estimated by comparing reorientation measurements to Equation 3.28. In the next section, I discuss the photo-orientation of push-pull azo dyes.

3.5 PHOTO-ORIENTATION OF AZO DYES: SPECTRALLY OVERLAPPING ISOMERS

This section describes how coupled photo-orientation and photo-orientation can be quantified in spectrally overlapping isomers. Four azo-polyurathanes (Azo-PURs), PUR-1, PUR-2, PUR-3, and PUR-4 (see Figure 3.11) were photo-oriented by polarized 488 nm blue light from an Argon-ion laser, and real-time dichroism analysis at the irradiation wavelength was utilized to

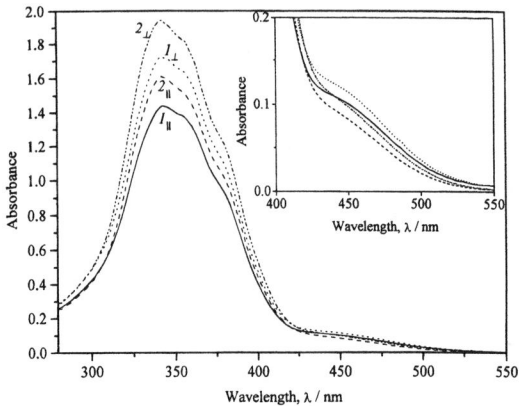

FIGURE 3.10   Relaxation in the dark of the dichroism induced by linearly polarized 360 nm light irradiation in the $P_{2,10}$ azo-polyglutamate. The spectra were obtained after 2 minutes' unpolarized 450 nm irradiation ($1_{//}$ and $1_{\perp}$) and after 6 days' relaxation in the dark ($2_{//}$ and $2_{\perp}$). Note that the absorbances have shifted to higher values due to the cis→trans isomerization, while the initial dichroism was increased. The inset is an expanded view of of the region around 450 nm. After reference 22, redrawn by permission of ACS.

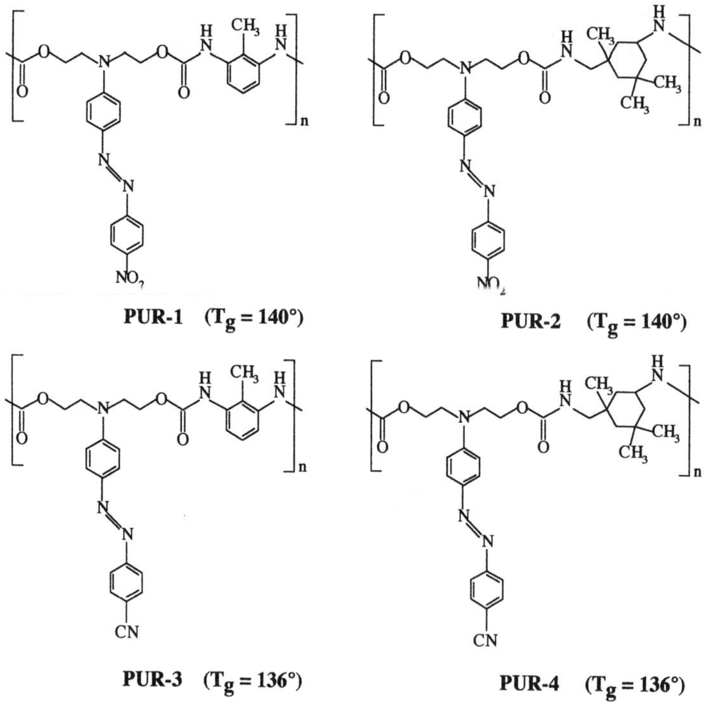

PUR-1   ($T_g = 140°$)

PUR-2   ($T_g = 140°$)

PUR-3   ($T_g = 136°$)

PUR-4   ($T_g = 136°$)

FIGURE 3.11   Chemical structures of the Azo-PUR polymers. The polymers' Tgs are indicated.

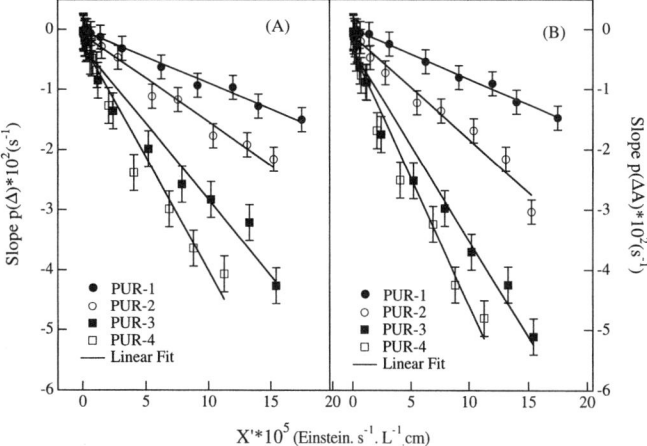

FIGURE 3.12 The fitted slopes $P(\Delta)$ and $P(\Delta A)$ of the observed early time evolution of the isotropic absorbance (A) and anisotropy (B), respectively. The full lines are linear theoretical fits to Equations 3.12 and 3.13. $X' = F' * A_0'$.

record the dynamics of photo-orientation. In these polymers, the chromophores are push-pull derivatives of the azobenzene, a feature that leads to a strong overlap in the absorption spectra of the cis and trans isomers and increases the rate of the cis→trans thermal isomerization. Details about the polymers' structure and glass transition temperature (Tg), the films' preparation, and the polymers' structural effects on photo-orientation are discussed in Chapter 4. For all of the Azo-PURs, the fitted slopes of the early time evolution of the isotropic absorbance and the anisotropy, $p(\Delta)$ and $p(\Delta A)$, respectively, exhibit a linear dependence on the irradiation light intensity during photo-orientation (see Figure 3.12), in agreement with Equations 3.12 and 3.13. In addition, the photostationary-state order parameter behaves according to Equation 3.22, i.e., $1/S$ obtained at the steady state of irradiation-induced A↔B orientation showed a linear dependence on the reciprocal of the irradiation light intensity (see Figure 3.13). The fits by Equations 3.12, 3.13, and 3.23 yielded ϕ_{AB}^{488}, ϕ_{BA}^{488}, and Q for all polymers (see Table 3.2).

When the irradiation intensity is extrapolated to infinity, the experimentally observed order parameter for all Azo-PURs is near −2/13, a value

TABLE 3.2 Coupled photoisomerization and photo-orientation parameters for Azo-PURs. The extinction coefficients are in units of L.mol⁻¹.cm⁻¹

Azo-PUR	ε_A^{488}	ε_B^{488}	ϕ_{AB}^{488}	ϕ_{BA}^{488}	Q^{488}
PUR-1	41300	21200	0.004±0.002	0.57±0.20	1.19±0.02
PUR-2	41300	21200	0.007±0.002	0.29±0.05	1.01±0.04
PUR-3	28000	5143	0.011±0.001	0.21±0.17	0.81±0.05
PUR-4	28000	5143	0.017±0.001	0.29±0.20	1.37±0.07

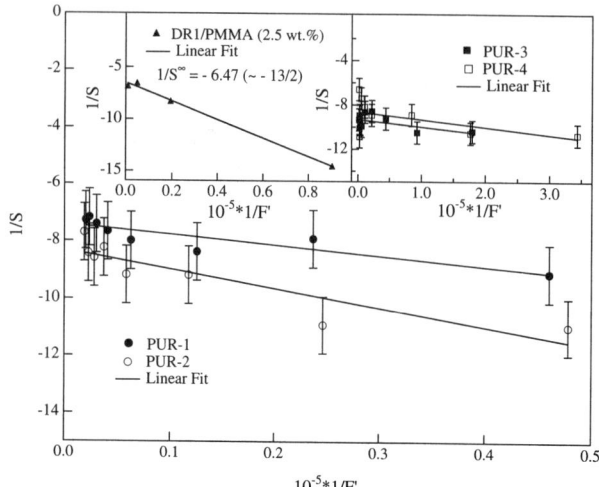

FIGURE 3.13 The reciprocal of the order parameter versus the reciprocal of the irradiation intensity for PUR-1 and PUR-2. The insets show this dependence for PUR-3 and PUR-4, and DR1/PMMA, as adapted from reference 20.

predicted by Equation 3.23 and that is referred to as the *photo-orientation constant*, or the constant of photo-orientation by photoisomerization. This value, −2/13, sets the maximum orientation that can be achieved by photoisomerization and rationalizes, at least for high irradiation intensity, the concept of a uniform stationary-state molecular order for both the A and B isomers. Together with Azo-PURs, amorphous azo-polymers should exhibit a photostationary-state order parameter at infinite irradiation flux near the photo-orientation constant. A good example from the literature is provided by DR1 molecules introduced as guests into films of PMMA (Figure 3.3). In this figure, both the anisotropy and the isotropic absorbance were measured as functions of the irradiating intensity, and the order parameter adapted from those measurements (shown as an inset to Figure 3.13) also exhibits an infinite flux value of S near −2/13. Even though the analysis wavelength, 514 nm, was different from the irradiation wavelength, 488 nm, these wavelengths are close enough within the same absorption band of the DR1 chromohopre.

To evaluate photoisomerization and photo-orientation parameters, ε_A and ε_B should be known. ε_A was calculated from the absorption spectrum of the polymer solution before irradiation, assuming the same extinction coefficient in the film and in solution; ε_B was determined by the Fisher's method, modified by Rau, which holds not only for isotropic but also for anisotropic samples when the isotropic absorbance is considered (*vide infra*). For this determination, the isotropic absorbance change was recorded versus the irradiating light intensity, and the sample absorbance change was extracted for an irradiation flux extrapolated to infinity for three different combinations of irradiation and analysis wavelengths: 488-488, 532-488, and 532-532 nm, irradiation and analysis, respectively. These experiments were done by recording the transmitted light of a probe propagating perpendicular to the sample and polarized at the magic angle, ~ 54.7 degrees, from the vertically polarized pump, so as to eliminate anisotropy contributions to absorption changes. The extent of cis concentration, i.e., the α value, found at 488 nm for PUR-1 and PUR-3 are 0.158 and 0.077, respectively. The extinction coefficients are also given in Table 3.2.

For all Azo-PURs, the quantum yields of the forth, i.e., trans→cis, are small compared to those of the back, i.e., cis→trans, isomerization—a feature that shows that the azo-chromophore is often in the trans form during trans↔cis cycling. For PUR-1, trans isomerizes to cis about 4 times for every 1000 photons absorbed, and once in the cis, it isomerizes back to the trans for about 2 absorbed photons. In addition, the rate of cis→trans thermal isomerization is quite high: 0.45 s^{-1}. Q~1 shows that upon isomerization, the azo-chromophore rotates in a manner that maximizes molecular nonpolar orientation during isomerization; in other words, it maximizes the second-order Legendre polynomial, i.e., the second moment, of the distribution of the isomeric reorientation. Q~1 also shows that the chromophore retains full memory of its orientation before isomerization and does not shake indiscriminately before it relaxes; otherwise, it would be Q~0. The fact that the azo-chromophore moves, i.e., rotates, and retains full orientational memory after isomerization dictates that it reorients only by a well-defined, discrete angle upon isomerization. Next, I discuss photo-orientation processes in chromophores that isomerize by cyclization, a process that differs from the isomeric shape change of azobenzene derivatives.

3.6 PHOTO-ORIENTATION OF PHOTOCHROMIC SPIROPYRANS AND DIARYLETHENES

Most optically induced reorientation studies have focused on azo-dye–containing materials; studies of the light-induced orientation of chromophores other than azobenzene derivatives and azobenzene-type molecules have rarely been reported. It will be shown that it is possible to individualize photo-oriented isomers of diarylethene and spiropyran derivatives in thin films of PMMA. Such photochromic chromophores have been extensively studied, not only from a photochemical point of view,[32–34] but also for use in near-field and three-dimensional optical data storage,[35–37] and for photo-assisted poling and hyperpolarizability switching.[38] Photoisomerization of diarylethenes and spiropyrans is different from that of azobenzene derivatives. The photoisomerization of the former two leads to reversible ring closing (photo-cyclization) and opening, and their apparent photo-orientation changes sign for the UV versus the visible photochemical transition bands, a feature that is due to the perpendicular UV and visible transitions in the B isomer. The photo-orientation processes of both isomers of each of the diarylethene and spiropyran chromophores studied will be separated by using polarized UV-vis and real time dichroism spectroscopies and by taking advantage of the natural spectral differences exhibited by the photoisomers in the UV-vis region.

The structural formula of 1,2-dicyano-1,2-bis-(2,4,5-trimethyl-3thienyl)ethene, and 6-nitro-1′,3′,3′-trimethylspiro[2H-1-benzopyrane-2,2′-indoline], referred to here as DE and SP, respectively, and their photochemical isomers are shown in Figure 3.14. The DE and SP chromophores have two photochemical isomers, a stable isomer and a thermally unstable isomer, namely the open-ring and close-ring forms for DE and the spiropyran and photomerocyanine for SP. The stable and thermally unstable isomers are henceforth referred to here as the A and B isomers, respectively. Light irradiation produces photoreaction in both the A→B and A←B directions, and the thermal reaction proceeds in the A←B direction. In contrast to the colored photomerocyanine form, which usually fades after several minutes at room temperature,[32,37] the colored close-ring form of DE is stable for more than three months at 80°C.[34] Both the A and B isomers of DE and SP can absorb UV light and simultaneously induce the A→B and A←B photoisomer-izations. When either DE or SP is irradiated with visible light, only the B isomer can appreciably absorb light and induce the A←B photoisomerization. Irradiation was performed by linearly polarized UV (365 ± 15 nm) and green (546 ± 5 nm, and 532 nm) lights from a high-pressure mercury lamp and a diode-pumped, frequency-doubled Nd:YVO$_4$ laser to induce the A↔B and A←B photoreactions, respectively. A UV-vis spectrometer was used to record linearly polarized spectra, i.e., $Abs_{//}$ and $Abs_{\perp}$, of the film samples, and the dynamics of photo-orientation were recorded by real-time dichroism with probe-light wavelengths at 633 (He-Ne laser) and 532 nm for SP and DE, respectively.

3.6.1. Photoisomerization of Spiropyrans and Diarylethenes

Photoisomerization of both DE and SP clearly occurs in films of PMMA. Spectra (not shown) taken before and after several amounts of UV and green irradiation exhibit shape changes and isosbestic points at about 329, 377, and 429 nm for DE and 317, 339, and (less pronounced) 440 nm for SP. These spectra clearly demonstrate the forth A→B and back B→A photoiso-

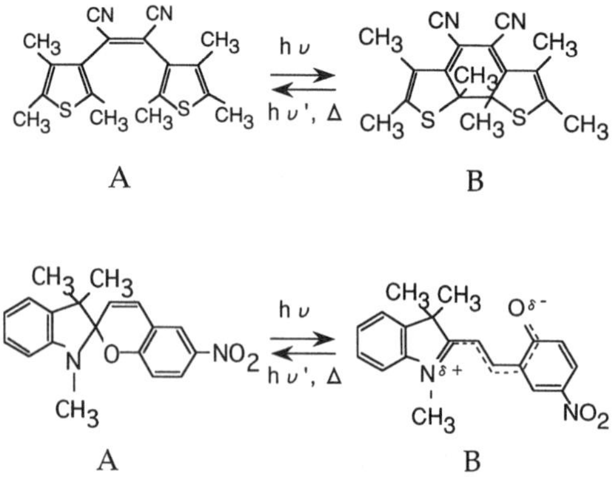

FIGURE 3.14 Chemical structures and isomerization of (top) diarylethene and (bottom) spiropyran isomers.

merization reactions. The isosbestic points shifted slightly when film samples were irradiated with green versus UV lights. No spectral change was observed during the B→A thermal back reaction of the DE chromophore over 24 hours, meaning that the closed-ring form (the B isomer) of the DE chromophore is stable at room temperature for this time period. The colored photomerocyanine isomer (B isomer) of the SP chromophore persists for several minutes at room temperature. A biexponential fading (B→A thermal recovery) of this form at room temperature was observed with rate constants (k) of $k_1 = 0.00125$ s^{-1} and $k_2 = 0.00009$ s^{-1} and weighting factors (amplitudes) 0.5157 for k_1 and 0.4017 for k_2 in reasonable agreement with previous results.[38,39] Although the polymer free volume, the free-volume distribution, or both, could influence the fading-rate constants, in the absence of aggregation, the biexponential behavior is usually attributed to the dual form of the photomerocyanine (i.e., the quinonic and zwitterionic forms). The assessment of the photo- and thermal isomerization features is necessary for photo-orientation studies.

3.6.2 Spectral Features of Photo-orientation

Figure 3.14 shows the dichroic spectra observed in films of DE/PMMA and SP/PMMA. The insets in Figure 3.15A are expanded views of both the UV and the visible absorptions of the DE chromophore. These spectra were obtained 30 s after polarized UV irradiation (irradiation dose: 78 mJ/cm^2). It is clear that Abs$_{//}$ and Abs$_{\perp}$ are different; in other words, the irradiated samples show anisotropic absorbance upon polarized UV irradiation. Identical spectra were recorded for Abs$_{//}$ and Abs$_{\perp}$ before UV irradiation, demonstrating that the samples were in-plane isotropic at that time.

It is particularly remarkable in Figure 3.15 that Abs$_{//}$ is higher than Abs$_{\perp}$ in the visible band, where the absorption of the A form is negligible and only the B form exhibits an appreciable absorption. This result is confirmed by real-time dichroism experiments *(vide infra)*. Figure 3.16 shows the absorbance of linearly polarized probe light (at 360 and 520 nm for DE, and 570 nm for SP) at various angles, Ψ, between the polarizations of the probe and UV lights. The UV irradiation dose was 401 mJ/cm^2. Sinusoidal behavior is clearly shown, which demonstrates the nonpolar orientational distribution of the isomers' transitions in both the UV and visible bands. For DE, perpendicularly oriented transitions are clearly shown for the UV and visible bands. The small drift (smearing-out of the modulation) in the absorption data is due to rotational diffusion of the chromophores. Photo-orientation also occurs during B→A photoisomerization for both DE and SP in PMMA (not shown), and the spectral features of photo-orientation are similar to those observed after A→B photo-orientation.

3.6.3 Photo-Orientation Dynamics and Transitions Symmetry

Real-time dichroism experiments were used to investigate the dynamics of photo-orientation of SP and DE in films of PMMA. The samples were irradiated with linearly polarized light while *in situ* transmittance

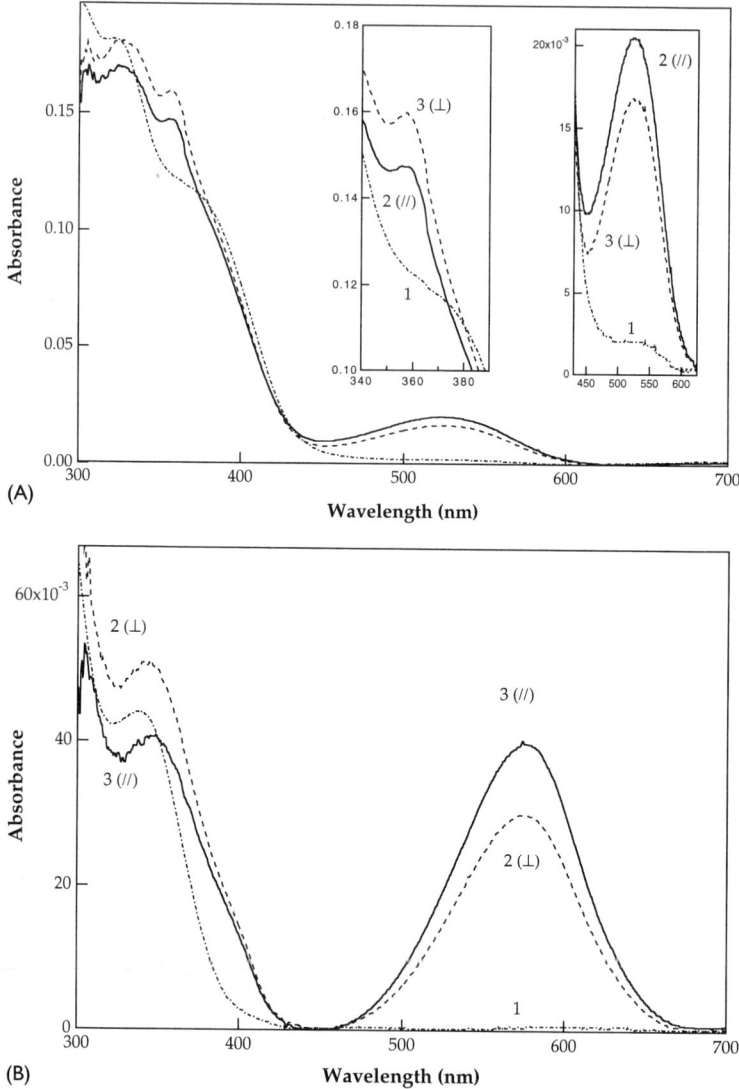

FIGURE 3.15 UV-vis absorption spectra of PMMA films containing (A) diarylethene and (B) spiropyran before (1) and after (2, 3) linearly polarized UV irradiation. The probe light was also linearly polarized, and spectra were obtained for both parallel, Abs$_{//}$, and perpendicular, Abs$_{\perp}$, orientations. The insets in (A) are expanded views of both the UV and the visible absorptions of the diarylethene chromophore. Identical spectra were obtained for both Abs$_{//}$ and Abs$_{\perp}$ before UV irradiation. For both the diarylethene and spiropyran chromophores, inasmuch as the orientational hole burning and orientational redistribution processes occur with Abs$_{\perp}$ > Abs$_{//}$ within the irradiation wavelength band, note the inversion of the sign of the dichroism between the UV and visible bands of the spectra. After reference 29, redrawn by permission.

measurements were performed with a probe light polarized either parallel or perpendicular to the initial irradiating light polarization. The probe beam was propagating perpendicular to the plane of the sample and linearly polarized at 45 degrees with respect to the plane of incidence of the irradiating beam. The transmitted parallel and perpendicular components

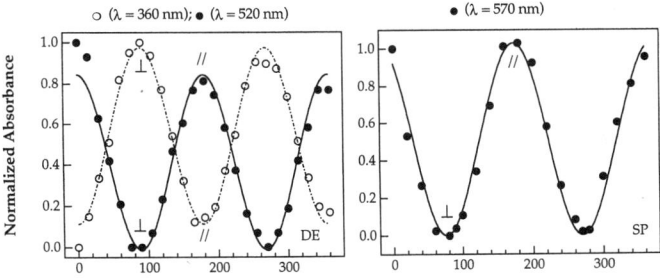

Angle, Ψ (degrees), between Pump and Probe Polarizations

FIGURE 3.16 Dependence of the absorbance of PMMA films containing (left) diarylethene and (right) spiropyran on the angle, Ψ, between UV irradiation and probe beam polarizations. The normalized absorbance is defined as (Abs - Abs$_{min}$)/(Abs - Abs$_{max}$), where Abs$_{min}$ and Abs$_{max}$ are the maximum and minimum absorbances, respectively. The // and ⊥ signs stand for the directions parallel and perpendicular to the UV polarization, respectively. The markers are experimental data points, and the full and dashed lines are $\cos^2 \Psi$ theoretical fits. The analysis wavelengths are indicated. Note the apparent π/2 angle-shift between the UV and visible orientational distributions of the diarylethene chromophore. After reference 29, redrawn by permission.

were separated by a Wollaston prism and detected separately. The probe beam was the 633 nm red and 532 nm green light for SP and DE, respectively, so that photo-orientation processes of the isomer B were probed independently from those of the isomer A. $Abs_{//}^B$ and $Abs_{\perp}^B$ were calculated from the amount of absorbed light, and the isotropic absorbance $\overline{A_B}$, the anisotropy ΔA_B, and the order parameter S_B were deduced. S_B was calculated at the steady state of photo-orientation. Photo-orientation of SP in PMMA is discussed first.

Figure 3.17 shows the time evolution of $\overline{A_B}$ and ΔA_B of SP/PMMA during and after linearly polarized UV irradiation for different irradiation power

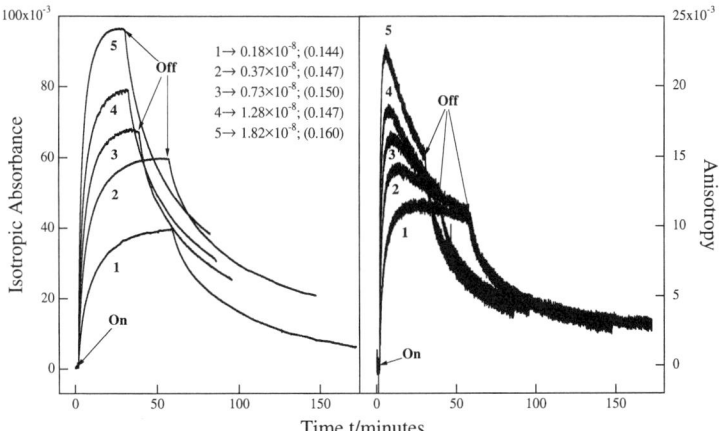

Time t/minutes

FIGURE 3.17 Real-time evolution of the isotropic absorbance (left) and the anisotropy (right) of SP in PMMA upon linearly polarized UV irradiation for several irradiation intensities. The numbers from 1 to 5 indicate the value of the irradiation intensity in units of Einstein.s^{-1}.cm^{-2} with the corresponding sample absorbance (value in parentheses) at the irradiation wavelength (365 nm). The moments of turning the irradiation light on and off are indicated. After reference 26, redrawn by permission.

values. The occurrence of anisotropy is indicative of photo-orientation of the chromophores. The green-light–induced orientation of the chromophores showed a dynamical behavior (not shown) similar to that of Figure 3.17. In this experiment, the SP/PMMA samples were irradiated by unpolarized UV light to the photo-stationary state, and linearly polarized green irradiation followed. Spiropyran molecules degrade after successive irradiation cycles; therefore, each photo-orientation experiment has been done on a different previously nonirradiated sample so as to avoid degradation complications. For high irradiation doses, higher than those reported in Figures 3.17 and 3.18, the evolution of the isotropic absorbance exhibits a reversal at the photostationary state due to the degradation of the chromophores.

The fitted slopes of the early time evolution of $\overline{A}_B$ and ΔA_B showed a linear dependence on the irradiation light intensity for both UV- and green-light–induced orientation, as predicted by the theory (see Figure 3.18). $1/S_B$ obtained at the steady state of the UV-light–induced A↔B photo-orientation showed a linear dependence on the inverse of the irradiation light intensity, also predicted by the theory. The solid lines in Figure 3.18 are linear fits by Equations 3.14 through 3.18 and 3.22, which yielded $\phi_{AB}^{365} = 0.053$, $P_2\left(\cos\omega_{633}^{365}\right)P_2^{A_{365}\rightarrow B_{365}} = 0.493$, $\phi_{BA}^{365} = 0.030$, $P_2\left(\cos\omega_{633}^{365}\right) = -0.345$, $\phi_{BA}^{546} = 0.003$, and $P_2\left(\cos\omega_{633}^{546}\right) = 0.642$. $\omega_{633}^{365} = 71.25$ and $\omega_{633}^{546} = 29.24$ degrees are the angles, calculated using Equation 3.2, between the B isomer's transition moments at 365 and 633, and 546 and 633 nm, respectively. $\varepsilon_B^{546} = 11380$ and $\varepsilon_B^{633} = 3460$ L mol^{-1}cm^{-1} were adapted from the literature,[38] and the fastest component of the thermal isomerization rate, k = 0.00125 s^{-1}, was used in Equation 3.20 for the determination of ϕ_{BA}^{546}.

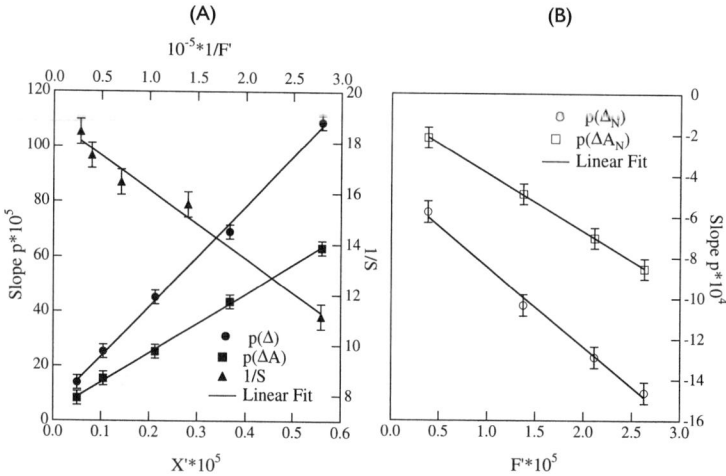

FIGURE 3.18 Experimentally observed dependence of the inverse of (A) the order parameter, 1/S, and the fitted slopes, $p(\Delta)$ and $p(\Delta A)$, and (B) $p(\Delta_N)$ and $p(\Delta A_N)$, of the observed change of the early time evolution of the isotropic absorbance and the anisotropy, respectively, on the irradiation intensity (UV for Figure A and green for Figure B). In Figure A, the arrows indicate that 1/S, and $p(\Delta)$ and $p(\Delta A)$, are plotted versus the top and bottom axis, respectively. The full lines are linear theoretical fits by Equations 3.14 through 3.18, and 3.22; F' is defined in Appendix 3.1; and $X' = F'^*A_0'$. After reference 26, redrawn by permission.

The quantum yields are reasonably small for photoisomerization processes in polymeric environments, meaning that molecular movement can be hindered far below the polymer Tg. These quantum yields are in agreement with those reported in the literature.[38] $P_2 (\cos \omega_{633}^{365})\ P_2^{A_{365} \rightarrow B_{365}} = 0.493$ shows that the orientation of the chromophore is partially retained, i.e., not thermalized, after the UV-light–induced A→B photoisomerization. $\omega_{633}^{365} = 71.25$ degrees demonstrates that the direction of the 365 and 633 nm transitions of the B isomer are nearly perpendicular to each other, a feature that explains the observed inversion of the sign of the anisotropy of photo-oriented SP in PMMA for the UV versus the visible transition band. Next, I discuss the photo-orientation features of DE in PMMA.

Diarylethenes contrast with spiropyrans by the thermal stability of the B isomer, a feature that brings about interesting photo-orientation effects in spectrally distinguishable photoisomers. The order parameter is independent of the irradiation light intensity at the photostationary state. Quantified photo-orientation of diarylethenes reveals that the closed form of such chromophores also exhibits perpendicular UV and visible transition dipole moments. Figure 3.19 shows the time evolution of $\overline{A}_B$ and ΔA_B of DE/PMMA during and after linearly polarized UV irradiation for different irradiation power values. The green-light–induced orientation of the chromophores shows a dynamical behavior that reverses for long irradiation times (see Figure 3.20). Even though diarylethenes are robust chromophores, the reversal of the mean absorbance observed at high UV irradiation intensities is due to the isomer's degradation. In fact, the colored intensity of DE decreases to 80% after 10 UV-vis irradiation cycles.[34] Here too, to avoid possible

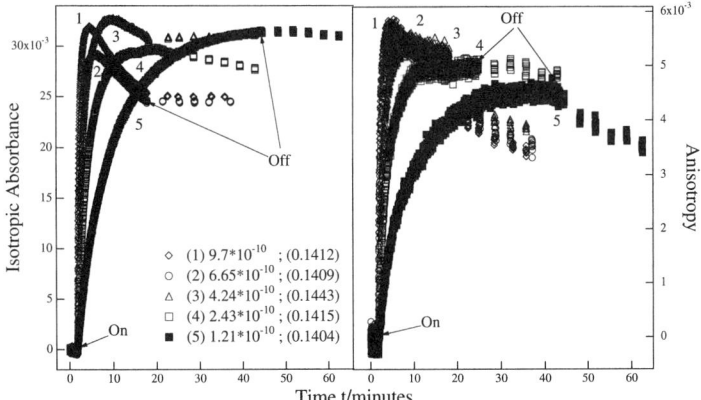

FIGURE 3.19 Real-time evolution of the isotropic absorbance (left) and the anisotropy (right) of DE in PMMA upon linearly polarized UV irradiation for several irradiation intensities. Only the 365 nm UV photo-orientation is shown; 405 nm photo-orientation showed similar dynamical behavior. The numbers from 1 to 5 indicate the value of the irradiation intensity in units of Einstein.s^{-1}.cm^{-2} with the corresponding sample absorbance A$_0'$ (value in parentheses) at the irradiation wavelength (365 nm). The moments of turning the irradiation light on and off are indicated. Note that after irradiation, the isotropic absorbance is stable because the B isomer is thermally stable, and the anisotropy relaxes due to molecular rotational diffusion. After reference 42, redrawn by permission of ACS.

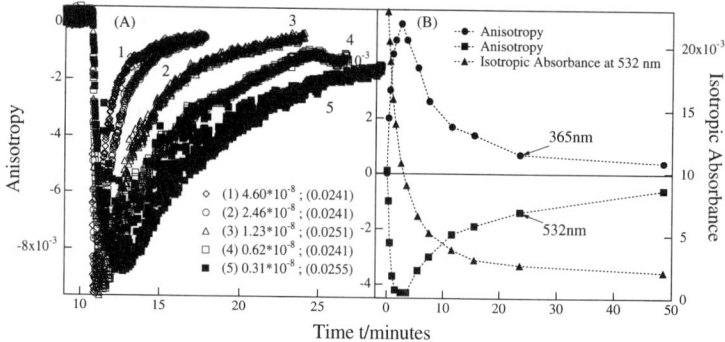

FIGURE 3.20 (A) Same as Figure 19, but for 532 nm analysis and irradiation. (B) Anisotropy observed by a UV-vis spectrophotometer after 546 nm irradiation (~1 mW/cm²). In Figure B, the time refers to the duration of irradiation. After reference 42, redrawn by permission of ACS.

photodegradation complications after successive irradiation cycles, each photo-orientation experiment was done on a different, previously nonirradiated sample, and for data analysis, only the slopes and the maximum absorbances at high irradiation intensities were considered.

As was the case for SP and azo dyes, the fitted slopes, $p(\Delta)$ and $p(\Delta A)$, of the early time evolution of $\overline{A}_B$ and ΔA_B, respectively, showed a linear dependence on the irradiation light intensity for both UV- and green-light–induced orientation, and the solid lines in Figure 3.21(A) are linear theoretical fits by Equations 3.14 and 3.15. The value of S_B of DE in PMMA, obtained at the photostationary state of UV irradiation for different irradiation intensities and wavelengths, is fairly constant, i.e., ~0.060 and ~0.024 for 365 and 405 nm irradiation, respectively, a feature that is theoretically rationalized by Equation 3.27. Indeed, Equation 3.20 dictates that for a system without B→A themal isomerization, $P_2(\cos \omega)$ is given at the photostationnary state by:

$$S_B = -\frac{2}{13} P_2(\cos \omega_B).$$
(3.27)

ε_B was determined by Fisher's method for photo-oriented samples. The obtained values, together with the photo-chemical quantum yields and the parameters obtained from the photo-orientation experiments, are summarized in Table 3.3. The fitted slope of the isotropic absorbance of the green

TABLE 3.3 Data of coupled photoisomerization and photo-orientation for DE. The extinction coefficients are expressed in units of L.mol⁻¹.cm⁻¹

λ/nm	ε_A	ε_B	ϕ_{AB}	ϕ_{BA}	$P_2(\cos \omega_{532}^{UV})$	ω_{532}^{UV}	$P_2^{A_{UV} \to B_{UV}}$
365	4436	8332	1.10	—	−0.39	74.3	−0.43
405	2402	423	0.32	—	−0.15	61.3	−1.04
532	0	4574	—	0.16	—	—	—

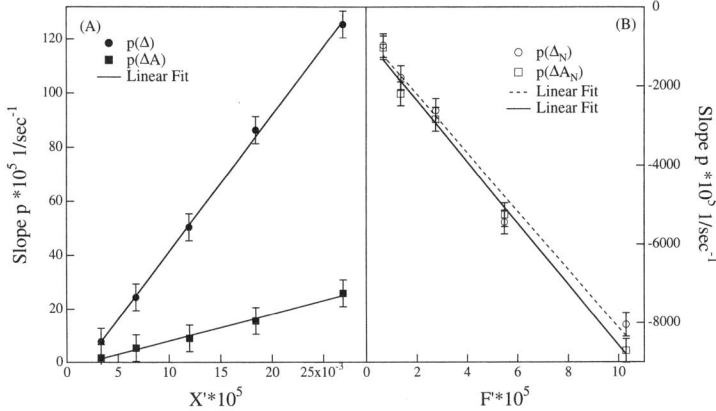

FIGURE 3.21 Slopes, (A) $p(\Delta)$ and $p(\Delta A)$, and (B) $p(\Delta_N)$ and $p(\Delta A_N)$, of the observed change of the early time evolution of the isotropic absorbance and the anisotropy, respectively, on the irradiation intensity (UV for Figure A and green for Figure B). The full lines are linear fits. F' and A_0' are as defined earlier. After reference 42, redrawn by permission of ACS.

photo-orientation experiments in Figure 3.21B yielded ϕ_{BA}^{532}. It is noteworthy that for systems without B→A thermal isomerization, if analysis is performed at the irradiation wavelength, predictions of B→A photo-orientation suggest a slope ratio $p(\Delta A)/p(\Delta)$ equal to 1.2 (see Equations 3.16 and 3.17), and the slopes of the green B→A photo-orientation of DE, calculated from Figure 3.21B, gave a ratio of $p(\Delta A)/p(\Delta)\sim1.1$.

The values of the quantum yields found in PMMA by irradiation at 532 and 405 nm, i.e., $\phi_{BA}^{532} \sim 0.16$ and $\phi_{AB}^{405} \sim 0.32$, respectively, are close to those, i.e., $\phi_{BA}^{546} \sim 0.14$ and $\phi_{AB}^{334} \sim 0.27$ to 0.33, found in CCl_4 and benzene solutions for other diarylethene derivatives that are structurally related to DE.[35] Even though little free-volume change is needed for DE ring opening and closing, the PMMA matrix does not seem to hinder the isomerization movement of the chromophore. It is noteworthy that polymer thin films present a distribution of free volumes to solutes and that the films' properties are averaged for chromophores in different sites. Quantum yields depend on the excitation wavelength, and when side reactions are present, they can be larger than 1. Indeed, a single photon may lead to the isomerization of more than one chromophore by side-reaction isomerization. The value of 1.1 found for ϕ_{AB}^{365} at 365 nm might reflect the existence of a possible side reaction. Other authors have found a quantum yield of 2 for the isomerization of another diarylethene derivative in films of poly(vinyl butyral).[40]

$P_2^{A_{365}\rightarrow B_{365}} \sim -0.4$ and $P_2^{A_{405}\rightarrow B_{405}} \sim -1$ show that the orientation of the UV transition dipole of the chromophore is partially retained—not thermalized—upon isomerization from A to B after UV irradiation, a feature that suggests that the chromophore does not tumble indiscriminately before it cools off as it does when isomerized from B to A by green irradiation. Indeed, Figure 3.20B shows that the green-light–induced orientation observed at both 532 nm and 365 nm disappears after all B forms are isomerized to A forms, which demonstrates that isomer A is not oriented by green-light–induced

A←B isomerization and that the observed anisotropy at both 532 and 365 nm is due to the orientation of isomer B only. If orientation occurs in A at any time by green irradiation, some anisotropy should remain at 365 nm after all B are isomerized to A. This behavior is theoretically rationalized by $P_2^{A_{365}\rightarrow B_{532}} \sim 0$, because the orientation of A is proportional to that of B through $P_2^{A_{365}\rightarrow B_{532}}$ (*vide infra*).

The lack of orientation in A may be due to the large amount of energy that needs to be dissipated during the photochemical process induced by the 532 or 546 nm photon; perhaps when the molecule is excited, it shakes strongly before it relaxes. $\omega_{532}^{365} = 74.3$ and $\omega_{532}^{405} = 61.3$ degrees demonstrate that the direction of the UV, i.e., 365 and 405 nm, and visible, i.e., 532 nm, transitions of the B isomer (the closed form) are oriented toward perpendicular directions and rationalize the result of Figure 3.20B. This finding is reinforced by the result of Figure 3.22, which shows that the calculated 365 and 532 nm transitions of the closed form of DE, with oscillator strengths of 0.960878 and 0.398422 respectively, are indeed perpendicular to each other. The UV and visible transitions of the closed form of DE were calculated using the CNDO/S (completely neglected differential overlap/spectroscopy) with the associated AM1 parametrization for geometry optimization, which are available with the MOPAC molecular orbital software.[41]

3.7 CONCLUSION

Polarized light absorption orients both isomers of photisomerizable chromophores, and quantified photo-orientation both reveals the symmetrical nature of the isomers' photochemical transitions and shows how chromophores move upon isomerization. Photo-orientation theory has matured by merging optics and photochemistry, and it now provides analytical means for powerful characterization of photo-orientation by photoisomerization. In azobenzenes, it was found that the photochemical quantum yields and the rate of the cis→trans thermal isomerization strongly influence photo-

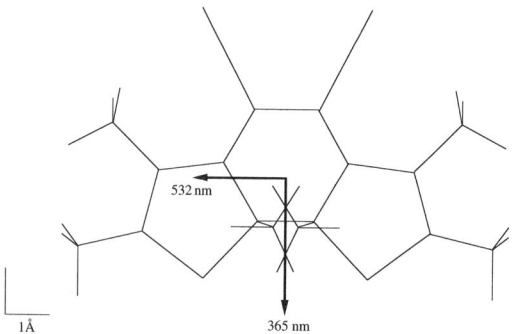

FIGURE 3.22 Drawing of the top view and transition-moment vectors at 365 and 532 nm of the closed form of DE obtained by CNDO/S AM1 MOPAC molecular orbital calculations. The 532 nm is parallel to the long axis of the molecule. After reference 42, redrawn by permission of ACS.

orientation, and chromophores' orientation is not thermalized during cis↔trans isomerization shape change. In photochromic spiropyrans and diarylethenes, the sign of apparent photo-orientation depends on the photo-chemical transition band, and the B isomer exhibits perpendicular transitions in the UV versus the visible transition bands.

ACKNOWLEDGMENTS

It is my pleasure to thank many friends and colleagues for their support and many helpful discussions. In particular, the work on DR1 was done with J. A. Delaire and M. Dumont at the University of Orsay, and the work on azobenzene was done with W. Knoll and J. Wood at the Max-Planck Institute for Polymer Research in Mainz. H. Menzel from Hannover University provided the azo-polyglutamate polymer. The work on azo-polyurethanes and spiropyrans and diarylethenes was done at Osaka University, Handai, with D. Yasumatsu, H. Ishitobi, and S. Kawata, and M. Irie from Kyushu University. I would like to thank the Japan Society for the Promotion of Science for research support under the Research for the Future Program and to thank the Max-Planck Society for financial support during my stay at the Max-Planck Institute for Polymer Research in Mainz.

APPENDIX 3A
QUANTUM YIELDS DETERMINATION

The procedure for determining QYs is summarized as follows. On the one hand, both trans and cis isomers may be excited by the same irradiation and interconverted, and an equilibrium of the two isomers, called the *photostationary state*, is reached. On the other hand, thermal cis→trans isomerization moves this equilibrium in favor of the trans isomer. So, the first part of the experiment consists of eliminating the effect of thermal isomerization on the equilibrium by determining the absorbance of hypothetical photostationary states of the photoisomerization reactions only, by extrapolating the irradiating intensity to infinity for combinations of two irradiation and analysis wavelengths (Rau's method). In the second part of the experiment, the obtained extrapolated values of the absorbance are used (in Fisher's method) to determine the extinction coefficients ε_c, i.e., the absorbance spectrum, of the cis isomer, and the determination of the quantum yields is straightforward.

3A.1 RAU'S METHOD

During irradiation, the concentration C_c of the cis form is given by:

$$\frac{dc_c}{dt} = 1000 I_0'(1 - 10^{-A'})(\varepsilon_t' \phi_{tc}' c_t - \varepsilon_t' \phi_{ct}' c_c)/A' - kc_c \tag{3A.1}$$

The primed quantities refer to a measurement at the irradiation wavelength; the unprimed quantities refer to a measurement at the analysis wavelength. I_0' is the incident photon flux, A' the total absorbance of the sample, k the first-order thermal relaxation rate of the cis isomer, ε_t' (respectively ε_c') the extinction coefficient of the trans (respectively cis) isomer, ϕ'_{tc} (respectively ϕ'_{ct}) the quantum yield of the trans→cis (respectively cis→trans) photoisomerization, and c_t the concentration of the trans isomers. I'_0, is the intensity of the irradiating light (flux of photons per square centimeter), and the factor 1000 occurs when I'_0 is expressed in mol of photon/cm^2. The extinction coefficients (proportional to the cross section) and the concentrations are expressed in L.mol^{-1}.cm^{-1} and mol.L^{-1}, respectively. Equation 3A.1 can be rewritten as:

$$dy / dt = F'(t)\varepsilon'_t \phi'_{tc} - (F'(t)Q' + k)y \tag{3A.2}$$

We denote by c_0 the total concentration of the isomers ($c_0 = c_t - c_c$), y the molar fraction of cis form ($y = c_c / c_0$), $F'(t)$ the following dependent time function ($F'(t) = 1000I'_0(1 - 10^{-A'})/A'$), and Q' the following factor ($Q' = \varepsilon'_t \phi'_{tc} + \varepsilon'_c \phi'_{ct}$). For the photostationary state, denoted by the index ∞, dy/dt is equal to zero and:

$$\varepsilon'_t \phi'_{tc} = (F'_\infty Q' + k)y_\infty / F'_\infty \tag{3A.3}$$

The total absorbance $A(t)$ can be expressed as a function of y:

$$A(t) = \varepsilon_t c_t L + \varepsilon_c c_c L = [(\varepsilon_c - \varepsilon_t)y + \varepsilon_t]c_0 L \tag{3A.4}$$

In this equation, L is the thickness of the sample along the analysis beam. From Equations 3A.3 and 3A.4 written for the photostationary state, the equation is as follows:

$$y_\infty = \frac{F'_\infty \varepsilon'_t \phi'_{tc}}{F'_\infty Q' + k} = \frac{A_\infty - \varepsilon_t c_0 L}{(\varepsilon_c - \varepsilon_t)c_0 L} = \frac{A_\infty - A_t}{(\varepsilon_c - \varepsilon_t)c_0 L} = \frac{\Delta}{(\varepsilon_c - \varepsilon_t)c_0 L} \tag{3A.5}$$

In this equation, both A_∞ and F'_∞ depend on the irradiation intensity I'_0. A_t stands for the optical density of a similar sample containing only the trans isomer. Δ is the optical density's variation when a sample (initially containing only the trans isomer) is irradiated to the photostationary state. The second and the last terms of this equation can be arranged to give:

$$\frac{1}{\Delta} = \frac{F'_\infty Q' + k}{F'_\infty \varepsilon'_t \phi'_{tc} c_0 L(\varepsilon_c - \varepsilon_t)} = \frac{F'_\infty Q' + k}{F'_\infty A'_t \phi'_{tc}(\varepsilon_c - \varepsilon_t)} =$$

$$\frac{Q'}{(\varepsilon_c - \varepsilon_t)A'_t \phi'_{tc}} + \frac{k}{(\varepsilon_c - \varepsilon_t)A'_t \phi'_{tc} F'_\infty} = \frac{\varepsilon'_t \phi'_{tc} + \varepsilon'_c \phi'_{ct}}{(\varepsilon_c - \varepsilon_t)A'_t \phi'_{tc}} + \frac{k}{1000(\varepsilon_c - \varepsilon_t)A'_t \phi'_{tc}} X \tag{3A.6}$$

where $X = A'_\infty/(1 - 10^{-A'})I'_0$. By plotting the left-hand side of Equation 3A.6 versus X at different irradiation intensities I'_0, we may obtain ϕ'_{tc} from the slope and ϕ'_{ct} from the intercept, provided that the extinction coefficients are known. ε_t and ε'_t can be experimentally measured, whereas Fischer's method is needed to determine each ε_c and ε'_c.

3A.2 FISCHER'S METHOD

This method is valid for systems without thermal relaxation. Therefore, all data concerning the photostationary state were extrapolated to infinite flux.

These extrapolated data are denoted by the exponent ∞. Under these conditions, the ratio of the equilibrium concentrations c_∞^∞ are given by:

$$\frac{c_{t\infty}'^\infty}{c_{c\infty}'^\infty} = \frac{\phi_{ct}' \varepsilon_c'}{\phi_{tc}' \varepsilon_t'} = \frac{\phi_{ct}' A_c'}{\phi_{tc}' A_t'} \quad (3A.7)$$

A_c' is the equivalent of A_t' for the cis isomer. When comparing the results of irradiation at any two wavelengths λ' and λ'', we have two equations of type 3A.7. By taking the ratio between these two equations, ϕ_{ct}'/ϕ_{tc}' and ϕ_{ct}''/ϕ_{tc}'' will cancel (assuming the ratio does not depend on the irradiation wavelength). We then get Equation 3A.8:

$$\left(\frac{c_{t\infty}'^\infty}{c_{c\infty}'^\infty}\right) / \left(\frac{c_{t\infty}''^\infty}{c_{c\infty}''^\infty}\right) = \left(\frac{A_c'}{A_t'}\right) / \left(\frac{A_c''}{A_t''}\right) \quad (3A.8)$$

If we introduce the extent α^∞ of trans $\rightarrow$ cis conversion at infinite flux, then:

$$c_{t\infty}^\infty / c_{c\infty}^\infty = (1 - \alpha^\infty) / \alpha^\infty \quad (3A.9)$$

α is the equivalent of y in Rau's method. Rewriting Equation 3A.9 for irradiation wavelengths λ' and λ'' and inserting them in the left-hand side of Equation 3A.8 leads to:

$$\left(\frac{1 - \alpha'^\infty}{\alpha'^\infty}\right) / \left(\frac{1 - \alpha''^\infty}{\alpha''^\infty}\right) = \left(\frac{A_c'}{A_t'}\right) / \left(\frac{A_c''}{A_t''}\right) \quad (3A.10)$$

Next, A_c and α are expressed in terms of experimentally measurable data. The optical density of a mixture of cis and trans, where the overall concentration $c_c + c_t$ is constant (c_0), is given by:

$$A = A_t(1 - \alpha) + A_c \alpha \quad (3A.11)$$

This equation is also valid when at the infinite flux photostationary state:

$$A_c = A_t + \Delta^\infty / \alpha^\infty \quad (3A.12)$$

Recall that Δ was introduced in Equation 3A.5 and that it is measured at the same wavelength as the irradiation. The infinite flux extrapolated value Δ^∞ is the intercept of the curve corresponding to Equation 3A.6. Introducing Equation 3A.12 for λ' and λ'' into Equation 3A.10, we have:

$$\left(\frac{1 - \alpha'^\infty}{\alpha'^\infty}\right) / \left(\frac{1 - \alpha''^\infty}{\alpha''^\infty}\right) = \left(1 + \frac{\Delta'^\infty}{A_t' \alpha'^\infty}\right) = \left(1 + \frac{\Delta''^\infty}{A_t'' \alpha''^\infty}\right) = \left(1 + \frac{\delta'^\infty}{\alpha'^\infty}\right) / \left(1 + \frac{\delta''^\infty}{\alpha''^\infty}\right)$$

$$(3A.13)$$

In this equation, δ'^∞ and δ''^∞ denote the relative change of absorbance observed at wavelengths λ' and λ'', respectively, when a solution of *trans*-isomers is photoequilibrated with an infinite-flux light at the respective wavelength. Furthermore, the ratio ρ ($\rho = \alpha'^\infty / \alpha''^\infty$) of α^∞ at two different excitation wavelengths λ' and λ'' is equal to the ratio of the Δ's measured at the maximum Δ wavelength when irradiating with wavelengths λ' and λ''. Finally, one gets:

$$\alpha''^\infty = (\delta'^\infty - \delta''^\infty) / ([1 + \delta'^\infty - \rho(1 + \delta''^\infty)]) \quad (3A.14)$$

All these parameters can be measured experimentally, and the numerical value of α''^∞ determined by this equation can then be used to calculate the absorption spectrum of pure cis by means of Equation 3A.12. So, ε_c is known for any wavelength, and its value can be introduced in Equation 3A.6, which allows the determination of ϕ'_{tc} and ϕ'_{ct}. It is shown in the text that Fisher's method is still valid for the determination of the cis absorption spectrum in photo-oriented films when the irradiation light intensity is extrapolated to infinity.

APPENDIX 3B
DEMONSTRATION OF EQUATIONS 3.12 THROUGH 3.15

In this appendix, I show how Equation 3.12 and 3.13 were derived. At the early time evolution, the cis concentration is negligible. If we introduce the extent, α, of the concentration of isomer B, i.e., $C_B = \alpha C$ and $C_A = (1 - \alpha)C$, with C the total concentration, and C_A and C_B the concentrations of the A and B isomers, respectively, we obtain the following equation by setting $n = 0$ in Equation 3.9:

$$\frac{d\alpha}{dt} = F'\varepsilon_A'\phi_{AB}'(1 + 2A_{2A})$$ (3B.1)

A_{2A} must be determined to find the expression of α. The following two equations, derived from Equation 3.9 for the early time evolution, i.e., $\alpha \ll 1$, allow us to do that.

$$\frac{dC_{A,n}}{dt} = -3F'\phi_{AB}'\varepsilon_A'\{C_A\} - n(n + 1)D_A C_{A,n}$$ (3B.2)

$$\frac{dC_{A,n}}{dt} = -\frac{1}{P_n^{A \to B}}\frac{dC_{B,n}}{dt} - n(n + 1)D_A C_{A,n}$$ (3B.3)

Setting $n = 2$ in Equation 3B.2 yields the second- and fourth-order parameter of the trans distribution. These are given by:

$$A_{2A} = -\frac{2}{5}F'\varepsilon_A'\phi_{AB}'.t \text{ and } A_{4A} = (\frac{11}{18}F'\varepsilon_A'\phi_{AB}' + \frac{7}{3}D_A).t$$ (3B.4)

where t represents the time. Now, substituting A_{2A} into Equation 3B.1, and noting that at the early time evolution $t^2 \ll t$, Equation 3B.1 yields:

$$\alpha = F'\varepsilon_A'\phi_{AB}'.t$$ (3B.5)

At a given analysis wavelength, the absorption change, $\Delta = A - A_A$, is given by:

$$\Delta = \alpha \, C \, (\varepsilon_B - \varepsilon_A) \tag{3B.6}$$

where $A_A = \varepsilon_A$. C and A are the sample absorbances before and during irradiation, respectively. Note that the sample is all-trans prior to irradiation. Substituting 3B.5 into 3B.6 yields the following equation, the slope of which is given by Equation 3.1.

$$\Delta = 1000 I_0'(1 - 10^{-A_0'})\phi_{AB}'(\varepsilon_B - \varepsilon_A).t \tag{3B.7}$$

For the demonstration of Equation 3.2, we set $n = 2$ in Equation 3B.3 and solve for αA_{2B} with $t^2 \ll t$ in mind. This gives:

$$\alpha A_{2B} = \frac{2}{5}P_2^{A\to B}F'\varepsilon_A'\phi_{AB}'.t \quad \text{and} \quad A_{2B} = \frac{2}{5}P_2^{A\to B} \tag{3B.8}$$

The total anisotropy, $\Delta A = \Delta A_A + \Delta A_B$, is derived by summing the anisotropies due to the orientation of both isomers, i.e., $\Delta A_{A,B} = 3 \, C \, \varepsilon_{A,B} P_2(\cos \omega_{A,B})A_{2A,2B}$, and by using Equation 3B.8. ΔA reads:

$$\Delta A = \frac{6}{5}1000 I_0'(1 - 10^{-A_0'})\phi_{AB}'\{P_2^{A\to B}P_2(\cos \omega_B)\varepsilon_B - P_2(\cos \omega_A)\varepsilon_A\}.t \tag{3B.9}$$

When isomeric reorientation is assumed (*vide infra*) and irradiation is performed at the irradiation wavelength, the slope of ΔA resumes to:

$$p(\Delta A) = \frac{6}{5}1000 I_0'(1 - 10^{-A_0'})\phi_{AB}' \, (Q\varepsilon_B - \varepsilon_A) \tag{3B.10}$$

REFERENCES

1. Weigert, F. *Verh. Phys. Ges.* **1919**, 21, 485.
2. Neoport, B., S., and Stolbova, O. V. *Opt. Spectrosc.* **1961**, 10, 146.
3. Todorov, T., Nocolova, L., and Tomova, T. *Appl. Opt.* **1984**, 23, 4309.
4. Sekkat, Z., and Knoll, W. In *Advances in Photochemistry*. Neckers, D. C., Volman, D. H., and Bunau, G. Von, Eds. (Wiley and Sons) **1997**, 22, 117; *SPIE Proceeding* **1997**, 2998, 164; Sekkat, Z., Knoesen, A., Knoll, W., and Miller, R. D. *SPIE Critical Reviews*. Najafi, I., and Andrews, M. P., Eds. **1997**, CR68, 374.
5. Delaire, J.A., and Nakatani, K. *Chem. Rev.* **2000**, 5, 1817; and references therein.
6. Eich, M., and Wendorff, J. H. *J. Opt. Soc. Am. B.* **1990**, 7, 1428.
7. Sekkat, Z., and Dumont, M. *Appl. Phys. B.* **1991**.
8. Shi, Y., H. Steier, W., Yu, L., Shen, M., and R. Dalton, L. *Appl. Phys. Lett.* **1991**, 58, 1131.
9. Rochon, P., Gosselin, J., Natansohn, A., and Xie, S. *Appl. Phys. Lett.* **1992**, 60, 4.
10. Hvilsed, S., Andruzzi, F., and Ramanujam, R. *Opt. Lett.* **1992**, 17, 1234.
11. Seki, T., Sakuragi, M., Kawanishi, Y., Suzuki, Y., Tamaki, T., Fukuda, R., and Ichimura, K. *Langmuir* **1993**, 9, 211.
12. Gibbons, W. M., Shannon, P. J., Sun, S. T., and Sweltin, B. J. *Nature* **1991**, 95, 509.
13. Rochon, P.; Batalla, E.; Natansohn, A. *Appl. Phys. Lett.* **1995**, 66, 136.
14. Kim, D. Y., Tripathy, S. K., Li, L., and Kumar, J. *Appl. Phys. Lett.* **1995**, 66, 1166; Viswanathan, N. K., Kim, D. Y., Bian, S., Williams, J., Liu, W., Li, L., Samuelson, L., Kumar, J., and Tripathy, S. K. *J. Mater. Chem.* **1999**, 9, 1941.
15. Sekkat, Z., Wood, J., Aust, E. F., Knoll, W., Volksen, W., and Miller, R. D. *J. Opt. Soc. Am. B.* **1996**, 13, 1713; 15.
16. Loucif-Saibi, R., Nakatani, K., Delaire, J. A., Dumont, M., and Sekkat, Z. *Chem. Mater.* **1993**, 5, 229.

17. Fisher, E. *J. Phys. Chem.* **1967**, 71, 3704.
18. Rau, H., Greiner, G., Gauglitz, G., and Meier, H. *J. Phys. Chem.* **1990**, 94, 6523.
19. Sekkat, Z., and Dumont, M. *Appl. Phys. B.* **1992**, 54, 486.
20. Sekkat, Z., and Dumont, M. *Synth. Metals* **1993**, 54, 373, 53, 121.
21. Sekkat, Z., Büchel, M., Orendi, H., Menzel, H., and Knoll, W. *Chem. Phys. Lett.* **1994**, 220, 497.
22. Sekkat, Z., Wood, J., and Knoll, W. *J. Chem. Phys.* **1995**, 99, 17226.
23. Zimmerman, G., Chow, L. Y., and Paik, U. Y. *J. Am. Chem. Soc.* **1958**, 80, 3528.
24. Thulstrup, E. W., and Michl, J. *J. Am. Chem. Soc.* **1982**, 104, 5594.
25. Sekkat, Z., and Knoll, W. *J. Opt. Soc. Am. B.* **1995**, 12, 1855.
26. Ishitobi, H., Sekkat, Z., and Kawata, S. *Chem. Phys. Lett.* **2000**, 316, 578.
27. G. Gauglitz, in *Photochromism Molecules and Systems.* H. Dürr and H. Bouas-Laurent, Eds. Elsevier, Amsterdam, **1990**, Chap. 2
28. Sekkat, Z., Wood, J., Geerts, Y., and Knoll, W. *Langmuir* **1995**, 11, 2856.
29. Ishitobi, H., Sekkat, Z., and Kawata, S. *Chem. Phys. Lett.* **1999**, 300, 421.
30. Sekkat, Z., Wood, J., Knoll, W., Volksen, W., Miller, R. D., and Knoesen, A. *J. Opt. Soc. Am. B.* **1997**, 14, 829.
31. Menzel, H. *Macromolecules*, **1993**, 26, 6226.
32. *Photochromism Molecules and Systems.* Dürr, H., and Bouas-Laurent, H., Eds. Elsevier, Amsterdam, **1990**.
33. Eckhardt, H., Bose, A., and Krongauz, V. A. *Polymer* **1987**, 28, 1959.
34. Irie, M., and Mohri, M. *J. Org. Chem.* **1988**, 53, 803.
35. Hamano, M., and Irie, M. *Jpn, J. Appl. Phys.* **1996**, 35, 1764.
36. Toriumi, A., Herrmann, J. M., and Kawata, S. *Opt. Lett.* **1997**, 22, 555.
37. Parthenopoulos, D. A., and Rentzepis, P. M. *Science* **1989**, 245, 843.
38. Atassi, Y., Delaire, J. A., and Nakatani, K. *J. Phys. Chem.* **1995**, 99, 16320.
39. Arsenov, V. D., Mal'tsev, S. D., Marevtsev, V. S., Cherkashin, M. I., Freidson, Y. S., Shibayev, V. P., and Plate, N. A. *Vysokomol. Soedin.* **1974**, A 16, 390; and **1982**, A 24, 2298. In these papers, the rate constants in PMMA of the thermal back reaction of the SP molecule are $k_1 = 0.0012$ s^{-1} and $k_2 = 0.0001$ s^{-1} versus $k_1 = 0.00125$ s^{-1} and $k_2 = 0.00009$ s^{-1} in reference 29.
40. Tsujioka, T., Kume, M., and Irie, M. *J. Photochem. Photobio. A: Chem.* **104**, 203 (1997).
41. Kurtz, H. A., Stewart, J. J. P., and Dieter, K. M. *J. Comput. Chem.* **11**, 82 (1990).
42. Ishitobi, H., Sekkat, Z., and Kawata, S. *J. Am. Chem. Soc.* **2000**, 122, 12802.

II

PHOTOISOMERIZATION IN ORGANIC THIN FILMS

4

PHOTOISOMERIZATION AND PHOTO-ORIENTATION OF AZO DYE IN FILMS OF POLYMER: MOLECULAR INTERACTION, FREE VOLUME, AND POLYMER STRUCTURAL EFFECTS

ZOUHEIR SEKKAT[*,†]
WOLFGANG KNOLL[‡]
Department of Applied Physics, Osaka University, Suita, Osaka 565-0871, Japan
†*School of Science and Engineering, Al Akhawayn University in Ifrane, 53000 Ifrane, Morocco*
‡*Max-Planck Institut für Polymerforschung, Ackermannweg 10, 55128 Mainz, Germany*

ABSTRACT

We review our work on photoisomerization and photo-orientation in films of polymer by focusing on the influence of the chromophores' environment on induced molecular movement. We compare photoisomerization of azobenzene derivatives in supramolecular assemblies, i.e., Langmuir-Blodgett-Kuhn (LBK) multilayers, molecularly thin self-assembled monolayers (SAMs), and amorphous spin-cast films. In azo-silane SAMs, photoisomerization and photo-orientation occur in molecularly thin layers much as they do in bulk spin-cast films, and photoisomerization modulates reversibly the optical thickness of 9 Å thin layers. In LBK multilayers of azo-polyglutamates, the polymers' side-chain structure influences the stability of the layers' stacking, and the azobenzene molecule is orientationally trapped and isomerizes between a highly oriented and a bend configuration, thereby controlling the film's optical order. Highly organized LBK structures impede the orientational freedom required for the chromophores' photo-(re)orientation. In amorphous spin-cast azo-polymer films, molecules are initially randomly distributed without intermolecular interaction, and photoisomerization and photo-orientation depend on the polymer structure and the free volume. Near-pure photo-orientation occurs in an azo-polyurethane polymer, and photo-orientation is observed 325°C below Tg of a rigid azo-polyimide polymer containing no flexible connector or tether. Isomerization is slowed by rigid embedding into rigid backbones, and high-pressure application reduces the polymers' free volume and suppresses the chromophores' photoisomerization.

4.1 INTRODUCTION

Organic materials that incorporate photosensitive molecular units are macro-scopically photoresponsive, and their structural and/or optical properties can be manipulated by light.[1,2] Photo-induced molecular structural change of the photochromic units into polymers leads to interesting macroscopic properties, such as changes in phase transition, viscosity, solubility, wettability, elasticity, and so on.[3] The molecular geometrical change that occurs in the trans↔cis photoisomerization process may lead to a loss of the initial orientation of the molecules after an isomerization cycle, and anisotropy can be induced.[4–9] There are numerous photochemical reactions that can lead to photochromism,[10] among which the trans↔cis photoisomerization of azobenzenes is the cleanest photo-reaction known to date.[11] In azo dye doped polymeric films, where the mobility of the guest molecules is still appreciable, photoisomerization leads to reversible polarization holography.[4,5] In azo dye functionalized polymeric films, where the mobility of the azo chromophores is greatly reduced, photo-isomerization creates a permanent alignment, which leads to writing erasing optical memory[6–9] or to permanent second-order nonlinear-optical effects[12–19] or cubic optical nonlinearities.[20] This chapter addresses the effect of the chromophore microcogent environment, including intermolecular interaction, free volume, and polymer structural effects on the photoisomerization and photo-orientation of azobenzene derivatives in polymeric thin films.

Several photoisomerization and photo-orientation studies have been reported in amorphous and liquid crystalline and hybrid azo-polymers,[4–28] Langmuir-Blodgett-Kuhn (LBK) multilayers,[29–33] alignment layers for liquid crystal molecules,[34–36] self-assembled monolayers (SAMs),[37–39] dendrimers,[40] phospholipids,[41] polypeptides,[42] peptide oligomers,[43] zeolites,[44] and so on. Photoinduced mass movement of azo-polymer chains has been reported and polarization-sensitive surface relief gratings have been fabricated.[45,46] Recent studies on optical ordering processes in amorphous polymers have addressed the role of Tg and polymer structural effects, including the main chain rigidity; the nature of the connection of the chromophore to a rigid, semirigid, or flexible mainchain; and the free volume, the free-volume distribution, or both.[47]

The polymer structure and Tg are not the only important parameters for polymers; the molecular weight and its distribution are also important. In fact, the glass relaxation is characterized by Tg, which is affected by several factors including the molecular weight, swelling, cross-linking, and hydrostatic pressure. Pressure effects on photoisomerization-induced molecular movement processes in NLO azo-polymers substantially below Tg have also been reported, and it has been shown that both photoisomerization and photo-orientation in a poly(methy-methacrylate) polymer is strongly hindered under hydrostatic pressure.[48] Photoisomerization of azobenzenes depends on the free volume.[11] Even though azo dyes can sometimes trigger polymer segmental motion and swelling by photoisomerization,[1,47,49] applied pressure can bury the chromophores into the polymer by compression and free-volume reduction.

Many of the light-induced nonpolar orientation studies have been performed in liquid crystalline polymers and poly(methyl-methacrylate) (PMMA) polymers containing azo dye with Tgs around 130°C. Recently, photo-orientation in higher-Tg polymers has been of interest, and the correlation of optical ordering (nonpolar and polar) to the polymer structure in a series of very high Tg (up to 350°C) rigid or semirigid NLO polyimides has been reported.[47] In particular, it has been shown that sub-Tg molecular movement, which generally is believed to be governed by the difference between Tg and the operating temperature T,[50] strongly depends on the molecular structure of the unit building blocks of the polymer, and that polymers with similar Tgs can exhibit significantly different photoinduced properties. The long-term stability of induced molecular order should, in principle, improve with the increased difference between the use temperature and the glass transition temperature of the polymer.

In this chapter, we will show that the occurrence of near-pure photo-orientation can also be strongly influenced by the polymer molecular structure by using a series of polyurethane polymers containing azo dye, and we review the work on high temperature azo-polyimides. We show in an NLO polyimide with a 350°C Tg containing no flexible connectors or tethers to an NLO azo chromophore that is connected through the donor substituent as a part of the polymer backbone, that photoisomerization is also capable of moving molecular units at room temperature, whereas, in the absence of photoisomerization, appreciable molecular movement is induced only by heating the polymer above its Tg.[47] We will present evidence that the process of isomerization itself depends on the polymer molecular structure of these high-Tg polyimides. The evidence clearly shows that molecular movement depends on the structure of the unit building blocks of the polymer.

Organized azo-molecular assemblies allow for the study of photoisomerization and photo-orientation in sterically and orientationally well-defined media, such as LBK multilayers, the molecular-interactions–based order of which can be altered by photoisomerization of the azo units.[30] In this chapter, we also discuss the photoisomerization-induced changes in the structural and optical properties of highly organized organic films containing azobenzenes. Photoisomerization and photo-orientation of azobenzenes is compared in amorphous spin-cast films, in LBK supramolecular assemblies, and in self-assembled monolayers.

This chapter is organized as follows. Section 4.2 addresses the study of photoisomerization and photoinduced orientation of azobenzene molecules at the molecular level in SAMs of azo-silane molecules. Section 4.3 discusses photoinduced effects in supramolecular assemblies, i.e., LBK multilayer structures containing azobenzene molecules, and compares the photoinduced movement of azobenzenes in these structures to that observed in spin-cast films. Section 4.4 focuses on the isomerization and sub-Tg photoinduced orientation in a series of very high Tg (up to 350°C) nonlinear optical polyimide and thermoplastic donor-embedded polyurethane polymers containing azo dye, especially focusing on polymer structure-Tg-photoinduced molecular movement relationships. Section 4.5 describes pressure effects on photoisomerization and photo-orientation in films of a PMMA polymer containing azo dye. Finally, we make some concluding remarks in Section 4.6.

4.2 PHOTOISOMERIZATION OF AZOBENZENES IN MOLECULARLY THIN SELF-ASSEMBLED MONOLAYERS: PHOTO-ORIENTATION AND PHOTO-MODULATION OF THE OPTICAL THICKNESS

In this section, we discuss the photo-orientation of azobenzenes in molecularly thin SAMs by means of UV-vis spectroscopy and surface plasmons (SPs). The structural formula of 4-(6-carboxy-(3-amidopropyl)triethoxysilane)-4'-pentylazobenzene, referred to as azo-silane, which leads to a self-assembled monolayer, is shown in Figure 4.1 (top). Azo-silane SAMs (see schematic in Figure 4.1, bottom) for the UV-vis spectroscopy and surface plasmons experiments were prepared as reported elswhere.[39]

4.2.1 Photoisomerization of Azo-SAMs

UV (360 nm) and blue (450 nm) light irradiations of the ultrathin azo-silane SAMs clearly induce the forth, i.e., trans→cis, and back, i.e., cis→trans, photoisomerization of azobenzene molecules (see Figure 4.2A). The real-time dependence of the absorbance of the sample during the thermal cis→trans back reaction is not a monoexponential decay (see Figure 4.2B). This decay shows a complex multiexponential relaxation behavior that could be fit neither by a monoexponential decay nor by a biexponential relaxation. Nevertheless, a monoexponential decay could be fit to the data acquired over

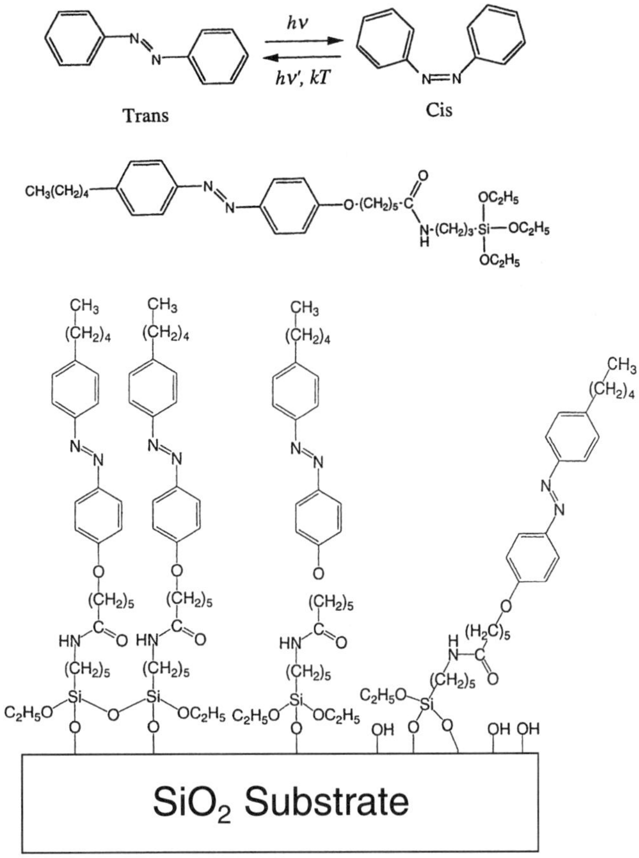

FIG. 4.1 (Top) Trans-cis isomerization of azobenzene, and structural formula of 4-(6-carboxy-(3-amidopropyl)triethoxysilane)-4'-pentylazobenzene, referred to in the text as azo-silane, which leads to a self-assembled monolayer, (Bottom) Idealized schematic drawing of a SAM on a SiO$_x$ substrate.

the first few hours of the relaxation, with a rate constant of 1/(10 hour), showing that the thermal cis→trans back reaction is relatively slow for the azobenzene in the azo-silane SAM. This is typical of azobenzene-type molecules. The nonmonoexponential thermal back-reaction kinetics reflect a distribution of mobilities for the azobenzene molecules in the azo-silane SAM that is the consequence of steric hindrance at the molecular level.

4.2.2 Photo-Orientation in Molecularly Thin Layers (Smart Monolayers)

Figure 4.3 shows the photo-orientation, i.e., the dichroism, observed in the azo-silane SAMs. These spectra were obtained after 3 minutes of irradiation with linearly polarized UV light. It is clear that the absorption, Abs$_\perp$, recorded with the probe light linearly polarized perpendicular to the initial UV-light polarization, is higher than the absorption, Abs$_{//}$, recorded with the probe light linearly polarized parallel to the initial UV-light polarization. Identical spectra were recorded for both Abs$_{//}$ and Abs$_\perp$ prior to UV irradiation (only

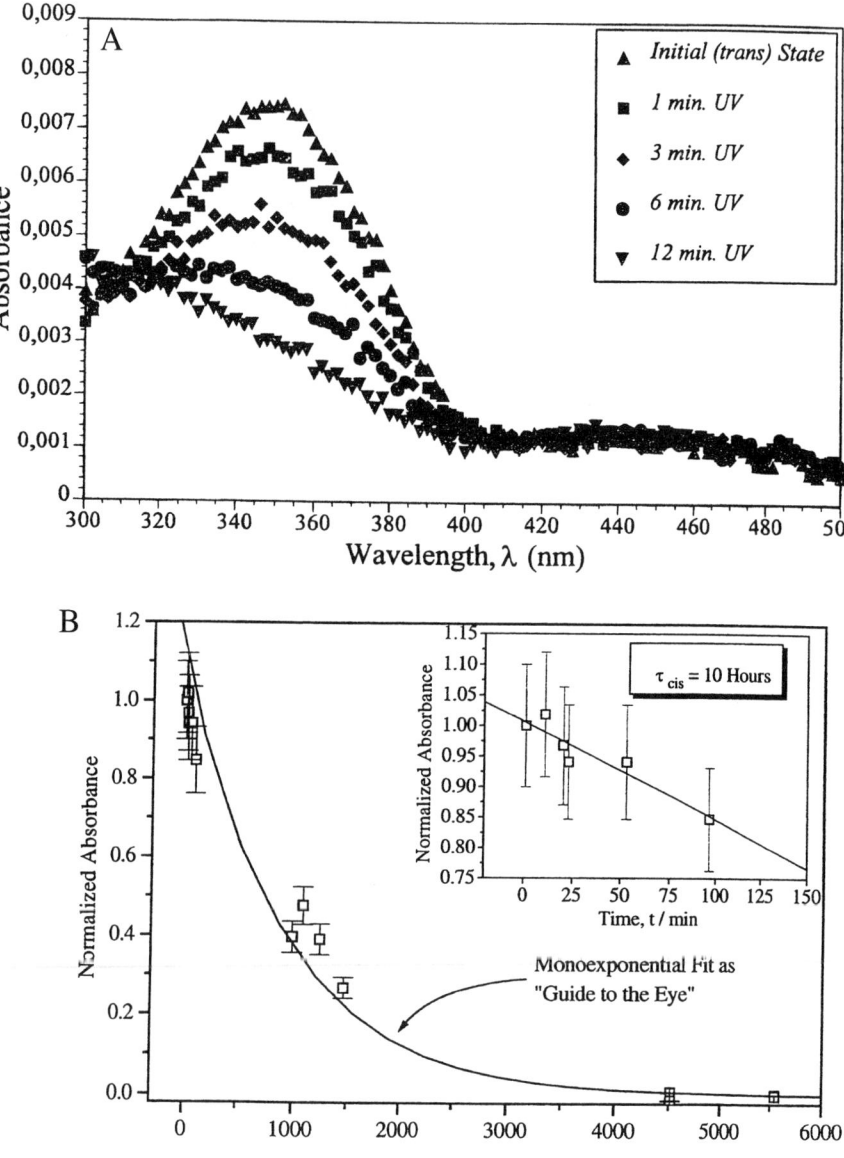

FIG. 4.2 (a) UV-vis absorption spectra of the azo-silane layer before and after various timed doses of UV (360 nm, ~ 2mW/cm^2) irradiation. Cis→trans photoisomerization by blue (450 nm) irradiation produced a reverse effect and restored the initial spectra. (B) Real-time kinetics of the thermally activated cis→trans isomerization, with an inset showing an expanded view of the first points of the figure. After reference 39, redrawn by permission of ACS.

Abs$_\perp$ is shown), demonstrating that the monolayer was in-plane optically isotropic at that time. The UV-induced dichroism could be erased on irradiation with unpolarized blue light upon which the initial spectra for both Abs$_{//}$ and Abs$_\perp$ were recovered. Further irradiation with linearly polarized UV light restores the initial dichroism, and successive cycles of UV (linearly polarized)/blue (unpolarized) irradiation have shown that this dichroism can be written/erased and rewritten without any fatigue over 20 cycles. When the sample was kept in the dark, the spectra from the relaxed sample was shifted to higher absorbance due to the thermal back reaction, but the initial dichroism was retained. It was interesting to see that the absorption of probe light vertically polarized and that of probe light horizontally polarized was

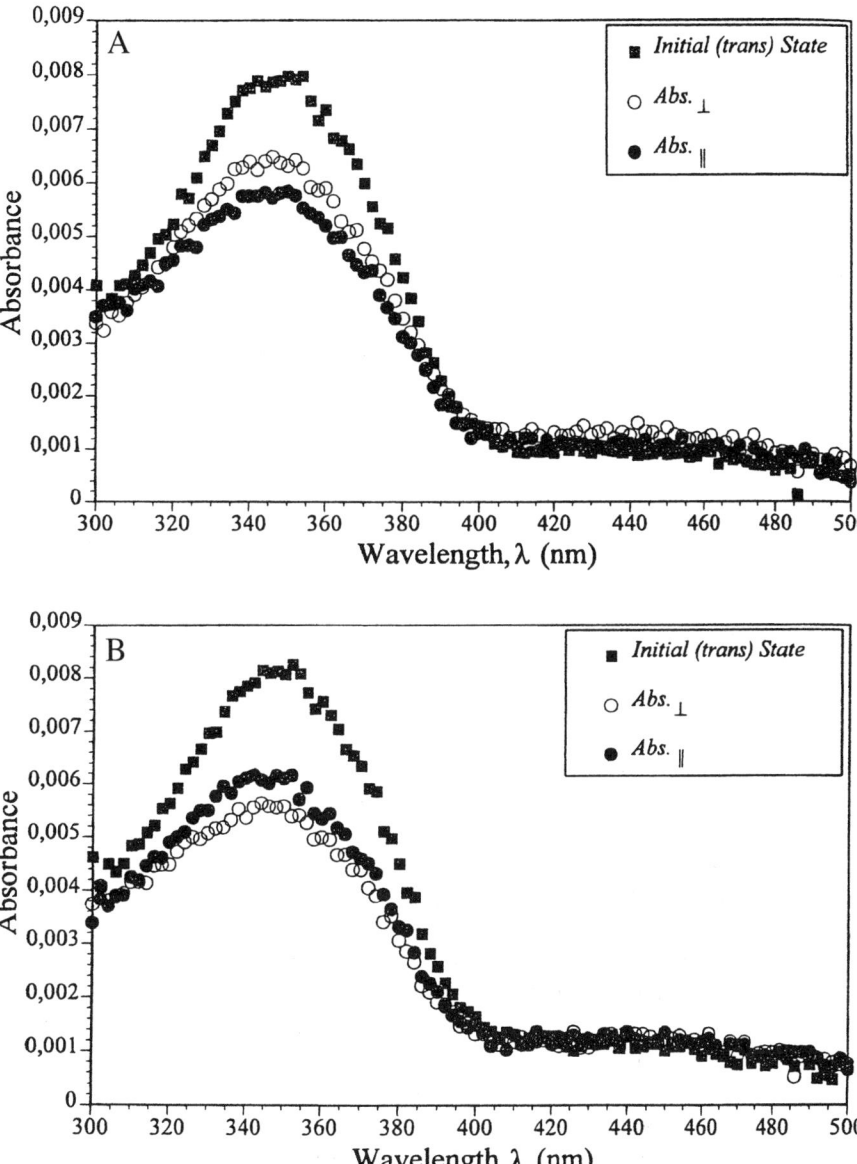

FIG. 4.3 UV-vis absorption spectra of the azo-silane layer before (Initial) and after linearly polarized UV (360 nm, ~ 2mW/cm²) irradiation. The probe light was also linearly polarized and spectra were obtained for both parallel and perpendicular orientations. The polarization of the irradiation UV light was rotated by 90 degrees from (A) for (B). After reference 39, redrawn by permission of ACS.

interchanged when the initial UV polarization was rotated from vertical to horizontal (see Figure 4.3B). The same amount of dichroism was observed for the same amount of UV irradiation, independent of the direction of UV polarization. This crossed dichroism could then be erased and rewritten as described previously.

The light-polarization sensitivity of this photoisomerization reaction is discussed earlier in this book. It is worth recalling that, in principle, for high intensities of irradiating light, or for long irradiation times, even the molecules aligned perpendicular to the polarization of the irradiating light will be excited. The system will then be saturated and isotropy will be restored. We have accounted for this phenomenon in our experiments by recording that the amount of photoinduced dichroism increases with increasing irradiation times until a maximum value is reached. Longer irradiation times progressively produce less dichroism until isotropy is finally regained.

4.2.3 Photo-Modulation of the Optical Thickness of Molecularly Thin Layers

We used surface plasmons (SPs) to estimate the thickness of the azo-silane SAMs.[39] Details about surface plasmons and guided waves can be found elsewhere.[1,51] Briefly, surface plasmons are transverse magnetic waves that propagate along a metal dielectric interface, their field amplitude decaying exponentially perpendicular to the interface. SPs and guided waves can be introduced by the Kretschmann configuration setup, wherein a thin metal film (~ 50 nm) is evaporated on the base of a glass prism. The metal film acts as an oscillator that can be driven by the electromagnetic wave impinging upon that interface, and a resonance phenomenon that depends on the incidence angle made by the wave with the interface can occur in the attenuated total reflection (ATR) scan.

In the ATR scan, the incident light is totally reflected above a critical angle θ_c. Dips in the reflectivity curve above θ_c indicate the resonant excitation of SPs at the metal/air interface, or guided waves in films deposited on top of the metal layer (see Figure 4.4). The coupling angle depends on the resonance condition for SPs and guided waves; it is possible to excite two sets of guided modes in waveguide films. The transverse electric (TE) modes are sensitive to the in-plane refractive index, n_y, of the waveguide, and the transverse magnetic (TM) modes are sensitive to both the in-plane refractive index in the guided wave propagation direction (n_x, x perpendicular to y) and the out-of-plane refractive index, n_z, in the direction normal to the waveguide plane. Study of the resonance angles enables accurate determination of the optical thickness (refractive index × thickness) of thin coatings by SPs, and the anisotropic refractive indices, n_x, n_y, and n_z, and thickness, d, of thicker coatings by waveguide modes. Waveguide spectroscopy is used later in this chapter to study thick layers of azo-polyglutamates and azo-polyimides.

For the azo-silane layers, assuming a value equal to 1.45, at 632.8 nm, for the refractive index, n_z, normal to the plane of the layer, our SAM could best be described by a layer thickness of 9 Å (i.e., an optical thickness of 13.1 Å). This is considerably thinner than would be expected for a fully extended azo-silane molecule (ca. 30 Å). This may be better understood by comparing the

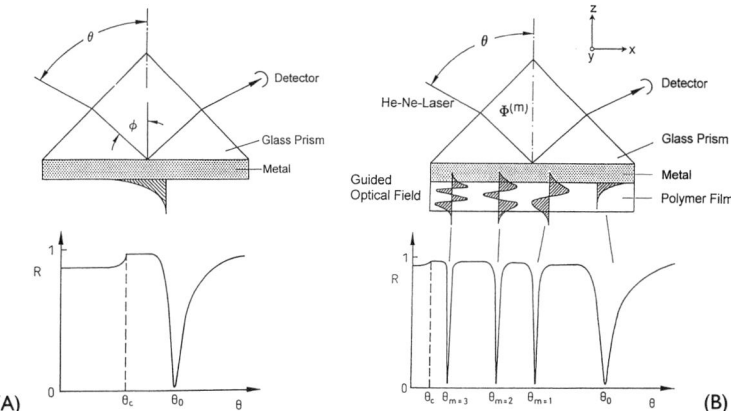

FIG. 4.4 ATR setup for the excitation of surface plasmons in Kretschmann geometry. (As top) A thin metal film (thickness ~ 50 nm) is evaporated into the base of the prism and acts as a resonator driven by the photon field. (As bottom) The resonant excitation of the surface plasmon wave is seen in the reflectivity curve as a sharp dip at coupling angle θ_0. (B) Setup for the excitation of guided waves. Modes are excited at the (external) angle θ_m.

area per azo unit (60 Å^2) obtained from the absorbance spectra—we estimate this area per azo unit from the extinction coefficient of the azo-silane in ethanol ($\varepsilon = 3 \times 10^4$ L mol^{-1} cm^{-1}) and the absorbance at λ_{max} for the azo-silane film (assuming that ε on the surface is approximately the same as ε in solution)—with that of closely packed azo units at the air/water interface (25 Å^2),[52] from which it is clear that we do not have a densely packed film. Additionally, the presence of the much smaller (ca. 6 Å-long) (3-aminopropyl)triethoxysilane as an impurity in the silanizing solution[39] probably acts as a diluent of the azo-silane on the surface. SPs were recorded at several different points of this SAM; they always gave the same resonance angle, and consequently the same SAM optical thickness. This shows that the SAM covers the sample surface homogeneously.

Irradiating the sample with UV light (360 nm) shifts all the dark-adapted trans-chromophores through trans→cis isomerization to a photostationary equilibrium with a high cis isomer content. As a result, the optical refractive index anisotropy is changed, and the refractive index n_z (more accurately, the optical thickness) is reduced. This shifts the surface plasmon resonance to smaller angles corresponding to an optically thinner SAM. This shift is too small to be seen by comparing the plasmon resonance curves before and after the irradiation. However, this photoinduced change in the optical thickness can be followed on line (during irradiation) by recording the reflected intensity at a fixed angle of incidence; and a kinetic analysis of the optical thickness change can give information about the reaction rates, the equilibrium changes, and the reversibility. The latter important aspect is shown in Figure 4.5; more important yet is that this optical switching is observed for a 9 Å monomolecular layer. The first rapid decrease of reflectivity upon UV irradiation

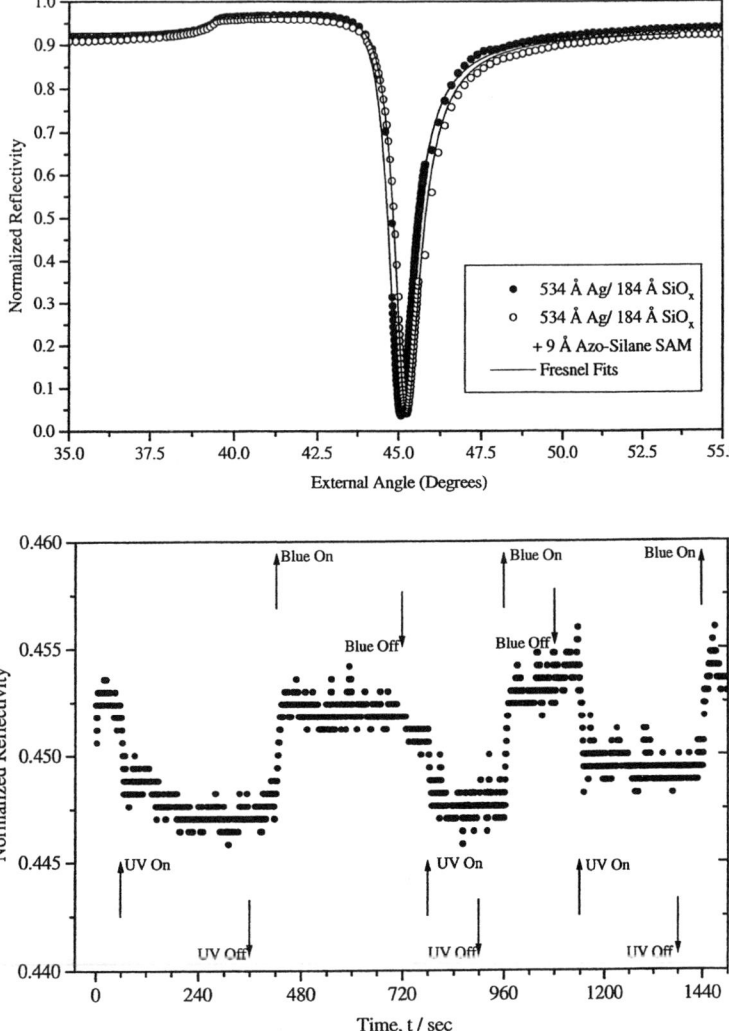

FIG. 4.5 (Top) Surface plasmon resonance of the Bare Ag/SiO$_x$ substrate before (black circles) and after (white circles) coating with a SAM of azo-silane; the inset shows the fitted geometrical thickness of the Ag, SiO$_x$, and azo-silane SAM. The full lines are theoretical fits from Fresnel calculations. (Bottom) Optical thickness change as obtained by recording the reflected intensity of the azo-silane SAM sample at a fixed angle of incidence (θ = 45 degrees) during irradiation; the moments of turning the irradiation light on and off are indicated by arrows. After reference 39, redrawn by permission of ACS.

can be restored by switching to visible light (450 nm), which isomerizes the azobenzenes back into the trans form. These reaction cycles can be conducted many times. It is important to note that, given the signal-to-noise level of these data, index changes smaller than 0.001 can be monitored in refractive layers that are only 9 Å thick.

4.3 PHOTOISOMERIZATION AND PHOTO-ORIENTATION OF AZOBENZENES IN SUPRAMOLECULAR ASSEMBLIES: PHOTO-CONTROL OF THE STRUCTURAL AND OPTICAL PROPERTIES OF LANGMUIR-BLODGETT-KUHN MULTILAYERS OF HAIRY-ROD AZO-POLYGLUTAMATES

Hairy-rod molecules, consisting of rod-like molecules with covalently attached flexible side chains[53–56] have been designed and used for forming nano-structured systems by using the LBK deposition technique. The flexible side chains enable the stiff molecules to be soluble in organic solvents and improve their transfer through LBK deposition. At the air/water interface, the side chains are oriented away from the water subphase, and the rods lay flat on the interface. For small substrates (in comparison to the width of the trough), the flow process during LBK deposition orients the rods parallel to the dipping direction, and the side chains form a liquid-like matrix for the rods in the reinforced liquid model.[54]

Hairy-rod molecules containing azobenzene units in the side chains can be assembled into LBK structures.[57] The azobenzene molecules undergo trans↔cis photoisomerization under light irradiation of appropriate wavelength and alter the structural properties of LBK films. We will discuss the photoinduced movement of the azobenzene molecule in a series of LBK structures of azo-polyglutamates in light of their optical and structural properties. The polymers we used are poly-(5-(2-(4-(4-((decyloxy)phenyl)azo)phenoxy)ethyl)-L-glutamate), denoted by $P_{2,10}$, and poly-(5-(2-(4-(4-((hexylphenylazo)phenoxy)ethyl)-L-glutamate) and poly-(5-(6-(4-(4-((hexylphenylazo)phenoxy)ethyl)-L-glutamate), denoted by $P_{2,6}$ and $P_{6,6}$, respectively. The structures of these polymers are shown in Figure 4.6. In these molecules, the main chain forms

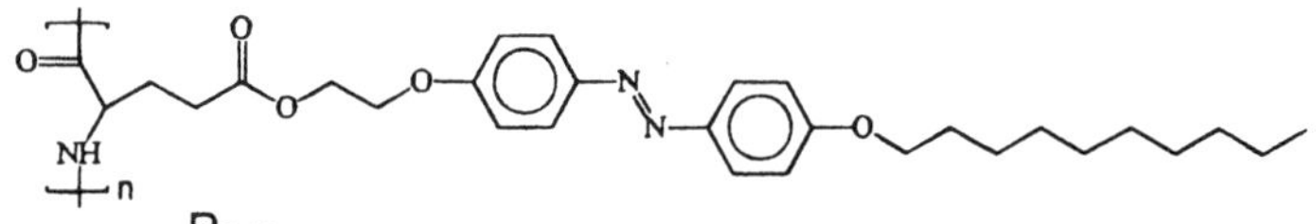

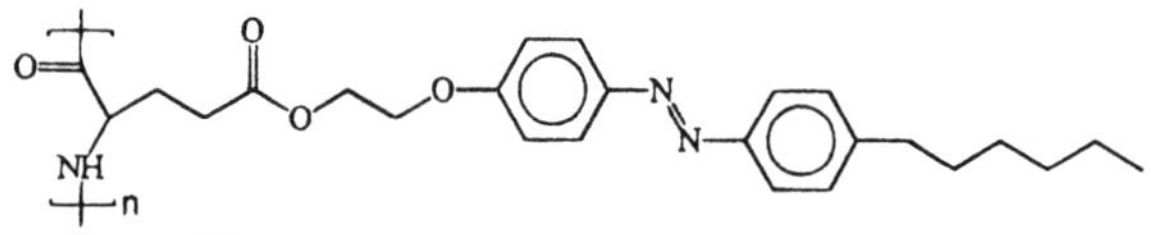

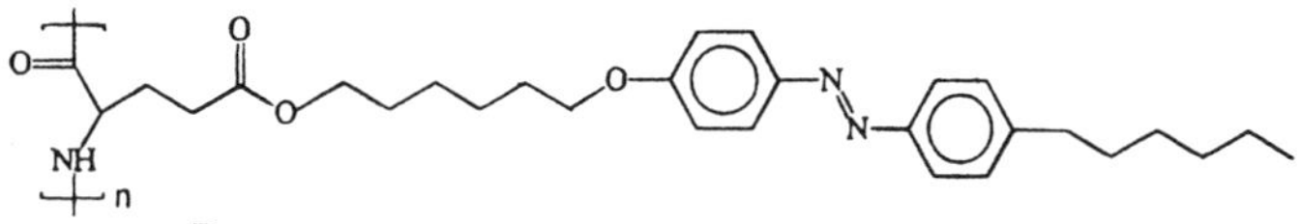

FIG. 4.6 Chemical structures of poly-(5-(2-(4-(4-((decyloxy)phenyl)azo)phenoxy)ethyl)-L-glutamate), denoted by $P_{2,10}$, and poly-(5-(2-(4-(4-((hexylphenylazo)phenoxy)ethyl)-L-glutamate) and poly-(5-(6-(4-(4-((hexylphenylazo)phenoxy)ethyl)-L-glutamate), denoted by $P_{2,6}$ and $P_{6,6}$, respectively.

an α-helix that is stabilized by hydrogen bonds, with the side chains pointing to the outside. These rigid-rod polyglutamates with flexible side chains constitute a type of the so-called hairy rod polymers designed by Wegner.[53] The synthesis is described elsewhere.[57]

We used the waveguide spectroscopy technique to study the optical properties of the LBK azo-polyglutamate polymers. Waveguide films could be prepared by transferring up to 156 monolayers of azo-polyglutamates onto the silver-coated glass substrate using the vertical dipping method.[30] Both TE- and TM-guided light modes could be coupled into these LBK structures by means of the ATR method. When an LBK azo-polyglutamate sample is irradiated, as shown in Figure 4.7, by unpolarized UV light (360 nm), trans→cis photo-isomerization takes place, and a waveguide mode shifts its angular position to lower incidence angles (see Figure 4.8). After blue light (450 nm) irradiation and the subsequent cis→trans back photo-reaction, the mode recovers exactly its initial angular position before UV irradiation. The transient behavior of the photoinduced switching process is reported in Figure 4.8 which shows the time evolution of a TM mode guided in an LBK structure made out of 156 monolayers of $P_{2,10}$. A similar modulation behavior (not shown) of a surface plasmon mode was observed when 20 LBK monolayers of $P_{2,10}$ were transferred, as a thin coating, to the silver-evaporated glass substrate. The transients were accomplished by alternating the irradiation between UV and blue light. Figure 4.8 exhibits the efficiency of the optical switching process and shows that it can be repeated for several cycles without fatigue of the LBK structures.

Figure 4.9 shows the evolution of in-plane (n_x, n_y) and out-of-plane (n_z) refractive indices of the 0.37 μm (156 monolayers)-thick $P_{2,10}$ LBK structure, under successive UV and blue unpolarized light irradiation cycles. The mean refractive index ($n = (n_x + n_y + n_z)/3$) is also reported in this figure. In all the columns in Figure 4.9 (labeled New, UV, and B, and corresponding, respectively, to the LBK structure before any irradiation, and after UV and blue-light irradiations), a small and persistent in-plane anisotropy (n_y–n_x) can be noted between the dipping direction (y), and the direction x perpendicular

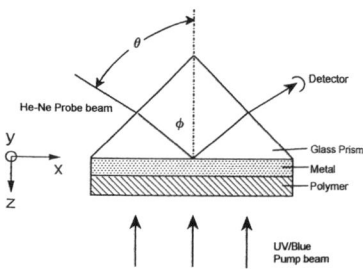

FIG. 4.7 Waveguide spectroscopy experimental arrangement in the ATR-Kretschmann setup. The probe is a 632.8 nm He-Ne laser beam, and the reflectivity of the sample is recorded as a function of the incidence angle. The irradiation (pump) beam direction of propagation is perpendicular to the plane of the sample.

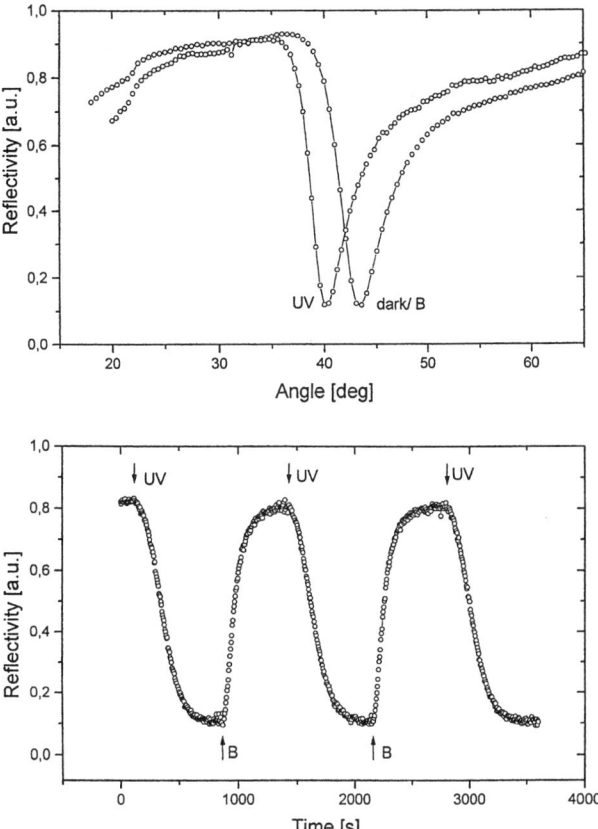

FIG. 4.8 (Top) TM light mode guided into a LBK polymer film consisting of 156 $P_{2,10}$ monolayers. Dark, UV, and B refer, to the angular position of the mode before any irradiation, after 360-nm UV irradiation, and after 450-nm blue light irradiation, respectively. (Bottom) Time evolution of trans↔cis photoisomerization induced forth and back switching in LBK structures of $P_{2,10}$ films consisting of 156 monolayers. The moments for turning the photoactive light (UV or B) on and off are indicated by arrows. After reference 30, redrawn by permission of ACS.

to it. This is due to the LBK film deposition process, in which the flow orients the rods parallel to the transfer direction.

Figure 4.9 also shows, in the columns labeled New and B, that the out-of-plane refractive index is much higher than the in-plane refractive indices ($n_z-n_{x,y} \approx 0.14$, where $n_{x,y}$ is the in-plane mean refractive index). This means that the side chains with the azobenzene in the trans form are highly oriented, and point outward perpendicular to the plane of the substrate. Figure 4.10 schematically represents the double-layer structure of hairy-rod polymers. In this structure, the highly anisometric trans isomer can be represented by a long molecular axis oriented along the alkyl side chains in the sample's out-of-plane direction. In the New film, this orientation is due to the structure of the monolayer at the air/water interface on the through, which is conserved by transfer and is common to all azo-polyglutamates. When the LBK film is exposed to the UV light, n_z decreases significantly ($\Delta n_z \sim 0.1$), and n_x and n_y

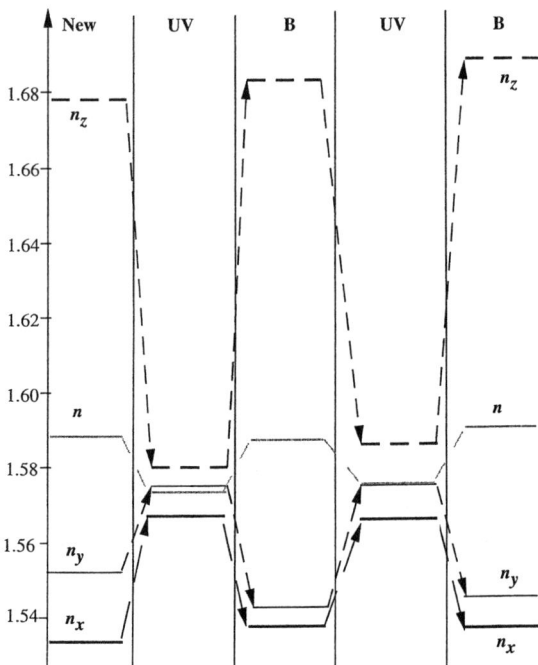

FIG. 4.9   Evolution of the 633-nm indices of refraction of the LBK ($P_{2,10}$, 156 monolayers) structure, under successive UV (360 nm) and blue (450 nm) light irradiations, indicated by the UV and B column heads. After reference 30, redrawn by permission of ACS.

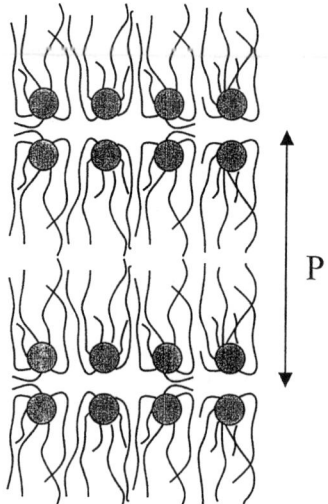

FIG. 4.10   Schematic drawing of the hairy-rod polyglutamate double-layer structure. P indicates the structure's periodicity. The trans isomer can be represented by a highly anisometric, i.e., nondegenerate, transition dipole oriented along the alkyl side chains.

both increase by nearly the same amount (see the UV columns). This shows that the polarizability of the azobenzene molecules breaks down because of the change in their electronic and structural properties induced by the photoisomerization from a planar to a bend structure. This also can be seen at the value of n (the mean refractive index), which is not conserved after the UV irradiation (see the UV columns in Figure 4.9).

The thickness of the LBK film does not change under UV and blue light irradiation. This was confirmed by the X-ray experiments and has also been shown, by means of X-ray reflectometry, for $P_{2,6}$ and $P_{6,6}$. When the sample is exposed to blue light, n_z increases, and n_x and n_y decrease to nearly their initial values before UV irradiation. This means that the packing of the end chains cannot be broken by the isomerization. The film structure is stable and the chromophores retain full memory of their initial orientation in the dark state before UV irradiation. This order can be explained by the crystallinity of the tails; their packing is the ordering force that stabilizes the film structure. The isomerization in this film can be repeated without fatigue under successive UV and blue light irradiations (Figure 4.9), and it always achieves the same refractive indices. The stability of the structure was also confirmed by X-ray reflectometry experiments. As a result of the dramatic photoinduced change in the molecular orientational order, the three-dimensional optical order parameter of the sample changes forth and back by about one order of magnitude under UV and blue light irradiation.

For comparison, we also prepared LBK structures with the $P_{2,6}$ and $P_{6,6}$ polymers. Table 4.1 summarizes the thickness and the refractive indices in the three principal directions of $P_{2,6}$ and $P_{6,6}$, together with the data for $P_{2,10}$. In this case, the dipping direction is represented by the x-axis. All three materials show an in-plane $(n_x - n_y)$ anisotropy directed in the dipping direction, but the main contribution is found perpendicular to the surface, due to the monolayer structure at the air/water interface, which is conserved by the transfer. The nearly complete loss of anisotropy in the cis isomer (labeled

TABLE 4.1 Thickness (d) and the 633-nm Indices of Refraction (n_x, n_y, n_z) in the Three Principal Directions of Different Azo-Polyglutamates' ($P_{2,10}$, $P_{2,6}$, $P_{6,6}$) LBK Structures[a]

		n_x	n_y	n_z	d, Å
$P_{2,10}$	new	1.552	1.533	1.678	23.7
	UV	1.574	1.566	1.586	23.8
	B	1.545	1.537	1.690	23.8
$P_{2,6}$	new	1.572	1.538	1.689	24.2
	UV	1.598	1.589	1.603	24.3
	B	1.583	1.592	1.636	24.3
$P_{6,6}$	new	1.589	1.547	1.637	24.4
	UV	1.581	1.573	1.568	24.8
	B	1.611	1.562	1.606	24.8

[a]Here, x refers to the dipping direction. New refers to freshly prepared samples, and UV and B refer to samples irradiated with UV and blue light, respectively.

"UV" in the table) is also seen for the three polymers. It can be seen that $P_{2,6}$ and $P_{2,10}$, which have the same spacer-length, show similar behaviors when the LBK structures are freshly prepared (labeled "New" in the table). In the case of the New $P_{6,6}$ film, the anisotropy in the dipping direction is more pronounced. After the first UV-blue irradiation cycle, the optical anisotropy of the $P_{2,6}$ LBK structure is partially lost compared to the $P_{2,10}$ LBK structure. There are two reasons for this: (1) the molecular tail-length is longer for $P_{2,10}$ than for $P_{2,6}$, leading to a higher Van der Waals interaction; and (2) there is additional electrostatic interaction of the oxygen in the alkoxy-chains[58] of $P_{2,10}$ compared to the alkyl-chains of $P_{2,6}$. These interactions in the side chains of $P_{2,10}$ conserve the high order and produce the stability. Table 4.1 also shows that the refractive index of the New $P_{6,6}$ LBK structure in the z direction is smaller than for the $P_{2,6}$ and $P_{2,10}$ LBK structures, but the in-plane anisotropy is larger. After the first UV-blue irradiation cycle, this in-plane anisotropy is even more pronounced compared to $P_{2,6}$ and $P_{2,10}$, which become nearly isotropic, whereas the anisotropy of $P_{6,6}$ observed between the out-of-plane and in-plane directions becomes smaller. This can be explained by the longer spacer in the case of $P_{6,6}$; this would allow more mobility to the side chains, which prefer orientation in the direction of the main chains.

For all the azo-polyglutamate LBK-polymer films, all the experiments mentioned previously were repeated setting the UV and blue lights to be linearly polarized, in the plane of the sample, successively parallel and perpendicular to the dipping direction; the same results were obtained for both UV and blue-light directions of polarization. In other words, no photo-selection effect was detected in these LBK structures. This contrasts with the results obtained with spin-cast films from the $P_{2,10}$ polymer (*vide infra*). This may be due to the subtle nature of the reorientation of the azobenzenes imposed by the LBK deposition technique, because efficient photoselection, following photoisomerization-induced reorientation, can be achieved in spin-cast azo-polymer films.[4–9] To isomerize, azobenzene molecules must feel highly organized supramolecular assemblies constructed by means of strong molecular physico-chemical interactions. In contrast to spin-cast films, for which the molecular units are initially randomly distributed, the degree of freedom for the azo units is considerably reduced in such supramolecular assemblies. This can be seen from the behavior of the azobenzene molecules in the $P_{2,10}$ LBK structures, which have to recover their initially highly oriented arrangement after a complete trans↔cis photoisomerization cycle. The azo units are constrained by the LBK structure to a highly oriented trans configuration or to a bend cis configuration. Next, we discuss polymer structural effects on photoisomerization and photo-orientation in spin-cast polymer films, where the azo dyes are initially randomly distributed without intermolecular interaction imposed on the molecular order.

4.4 POLYMER STRUCTURAL EFFECTS ON PHOTO-ORIENTATION

Photo-orientation results in anisotropy and depends on the environment of the photoisomerizable chromophore. To increase the stability of photo-orientation, azo dyes have been attached to polymer main chains as pendant side groups. In this section, we discuss the correlation of the polymer architecture to sub-Tg, light-induced molecular movement in high-temperature NLO azo-polyimides and in donor embedded azo-polyurethanes, and we show that light can still create orientation of azo dyes up to 325°C below the Tg of rigid azo-polyimides. The chromophores were embedded rigidly into rigid polyimide backbones without any flexible connector or tether. This rigid embedding of the chromophores into a rigid polyimide backbone decreases appreciably the rate of the cis→trans thermal isomerization. It will be shown that the isomerization process itself depends on the molecular structure of the unit building blocks of the polymer. The effect of the cis→trans thermal isomerization rate on photo-orientation will be discussed for azo-polyurethanes in which the azo-chromophore is substituted with groups of different electron-withdrawing strengths, i.e., Cyano versus Nitro groups. It will also be shown that the photo-orientation dynamics and efficiency are strongly influenced by a seemingly small difference into the polyurethane backbone.

4.4.1 Photoisomerization and Photo-Orientation of High-Temperature Azo-Polyimides

We used spin-coated films of high-Tg, nonlinear optical (NLO) azo-polyimides, i.e., PI-1, PI-2, PI-3a and PI-3b (see Figure 4.11), the unit building blocks of which have distinct molecular structures. PI-1 and PI-2 are both donor-embedded systems in which the NLO chromophore is incorporated rigidly into the backbone of the polymer without any flexible connector or tether. PI-3a and PI3b, on the other hand, are true side-chain systems in which the NLO azo dye is attached to the main chain via a flexible tether. Compared with the donor-embedded system, the flexible side chain system allows freer movement of the azo-chromophore. Details of the polymer synthesis and characterization and of the sample preparation can be found in references 59 and 60. The Tg values of the polyimides were determined by differential scanning calorimetery at a heating rate of 20°C per minute. The Tg values for PI-1, PI-2, PI-3a, and PI-3b were 350, 252, 228, and 210°C, respectively. Note that introducing a flexible unit into the polyimide backbone via the precursor dianhydride lowers the Tg of the donor-embedded polymers (PI-1 and PI-2) by ~ 100°C (350 versus 252°C).

Figure 4.12 shows 633-nm ATR modes in PI-1 before and after TE-polarized 532-nm (30 mW/cm^2) irradiation. The mode shifts after irradiation showed that birefringence is achieved in PI-1 at room temperature. The ATR accurate measurement of the refractive index components n_x and n_y (in-plane) and n_z (normal to the plane) yielded $n_x=n_y=1.649$ and $n_z=1.628$ before irradiation, and $n_x=1.652$, $n_y=1.617$, and $n_z=1.635$ 35 minutes after irradiation, where the y direction corresponds to the polarization direction of the irradiation light. The same observation was made for PI-2. Irradiation of the samples at room temperature, therefore, induces considerable birefringence in the samples.

For both PI-1 and PI-2, assuming that $n^2 \sim \varepsilon$ (n, mean refractive index) for optical frequencies, the mean dielectric constant, (ε) decreases upon irradiation ($\Delta\varepsilon_{PI-1} = - 0.023$, $\Delta\varepsilon_{PI-2} = - 0.016$). The dielectric constant is proportional to the chromophore density and the decrease upon irradiation suggests a quasi-

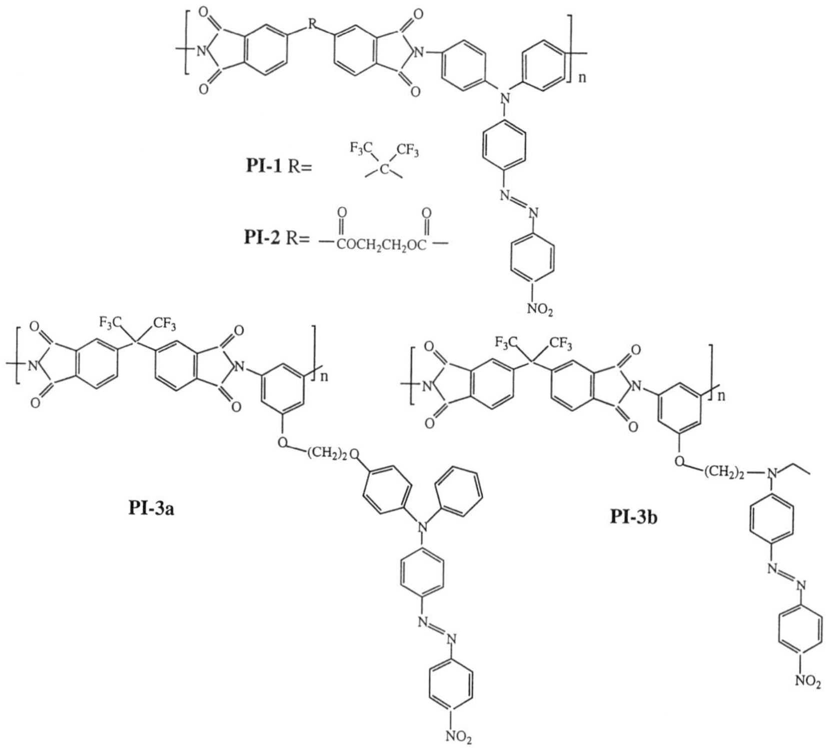

FIG. 4.11   Chemical structures of the azo-containing polyimide polymers.

stable population of the cis isomer. The stability of the cis isomer in PI-1 and PI-2 is unusual, because donor-acceptor substituted azo chromophores usually undergo rapid cis→trans isomerization on a time scale of seconds to minutes. This effect is shown in Figure 4.13. In this case, the mean absorbance at λ_{max} $(A_y + 2A_x)/3$ first drops substantially upon irradiation then finally recovers to near its initial value after ~ 25 h. The order parameter, $S = (A_y - A_x)/(A_y + 2A_x)$, however, does not return to its original value even after 350 h, suggesting the orientational birefringence is retained even after the cis→trans thermal isomerization is completed. A similar effect is observed for the irradiated sample of PI-2. In this case, $\Delta\varepsilon \sim 0$ after 26 h at room temperature although the film still remained quite birefringent, $\Delta n_x = 0.014$, $\Delta n_y = -0.028$, $\Delta n_z = 0.013$. The kinetics of the return of the trans isomer from the cis isomer in both PI-1 and PI-2 exhibit a fast component of few seconds (not shown) and a longer one on the order of minutes to hours, which can be fitted with a triexponential decay with a long time constant of 6 to 8 h (see Figure 4.14 for PI-1).

By means of comparison, in related experiments on PI-3a and PI-3b, the mean absorbance returned to its initial value within two minutes after removal of the irradiation light. Clearly, the polymer rigidity intrinsic to the donor-embedded samples must affect the rate of isomerization of the cis azobenzene derivative, suggesting that motion of the chromophore and the polymer backbone are somehow coupled.

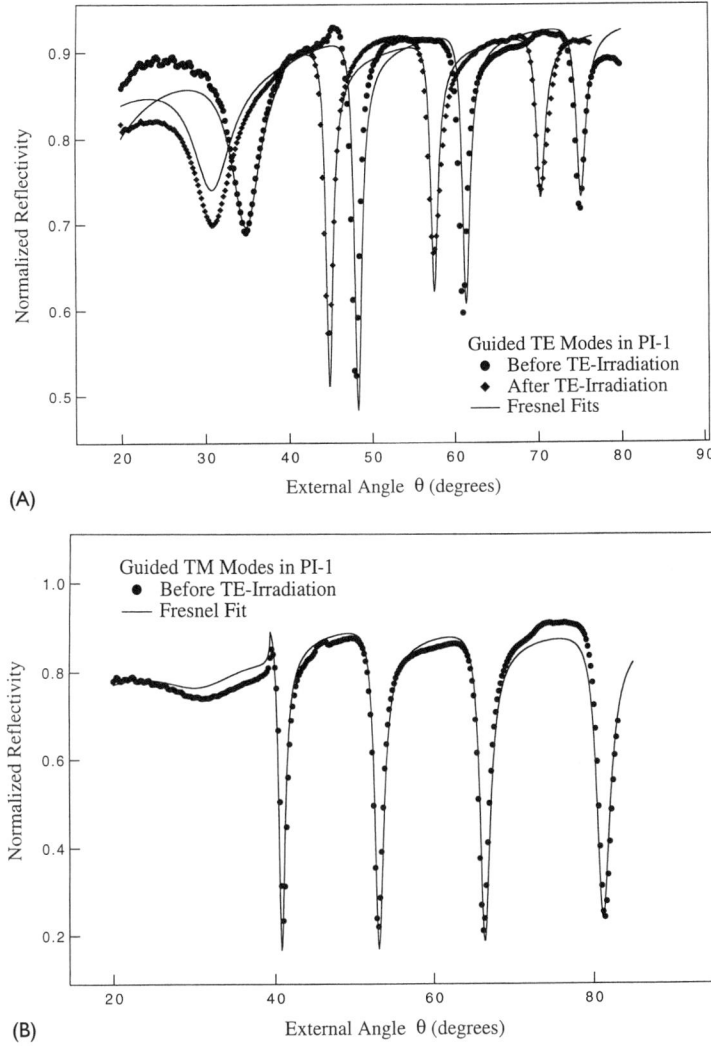

FIG. 4.12 TE (A) and TM (B) waveguide modes coupled into PI-1. Because the angular position of the TM modes didn't change appreciably after irradiation, only TM modes before irradiation are shown. After reference 47, redrawn by permission of OSA.

PI-1 and PI-2 exhibits a stronger coupling between the chromophore and the polymer backbone than exists in the PI-3a and PI-3b tethered side-chain systems. In PI-1 and PI-2, isomerization of the azo chromophore requires some correlated motion of a substantial region of the polymer backbone. The slowdown of the thermal back isomerization is caused by additional activation energy required for the backbone rearrangement. This clearly demonstrates that embedding the azo chromophore through the donor substituents into a fairly rigid polymer backbone hinders substantially the movement of the chromophore. The exceptional thermal stability of dc-field-induced polar order in this class of polyimides gives further proof of this.[60]

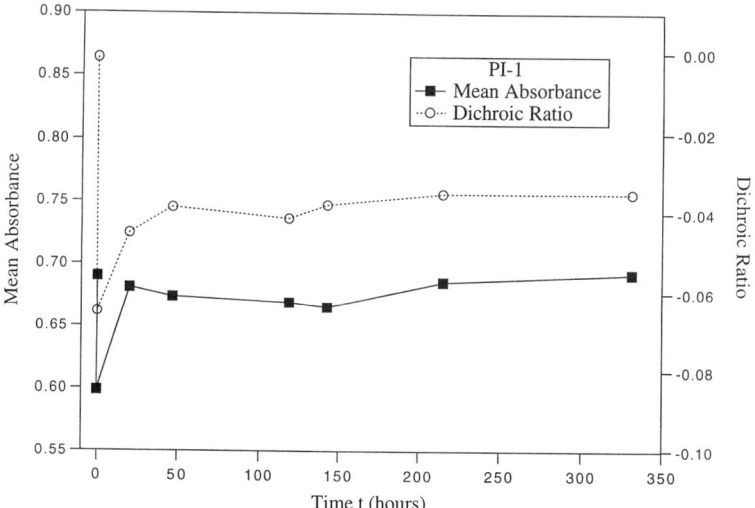

FIG. 4.13 Mean absorbance and order parameter, referred to in the figure as dichroic ratio, of PI-1 versus time after end of irradiation. After reference 47, redrawn by permission of OSA.

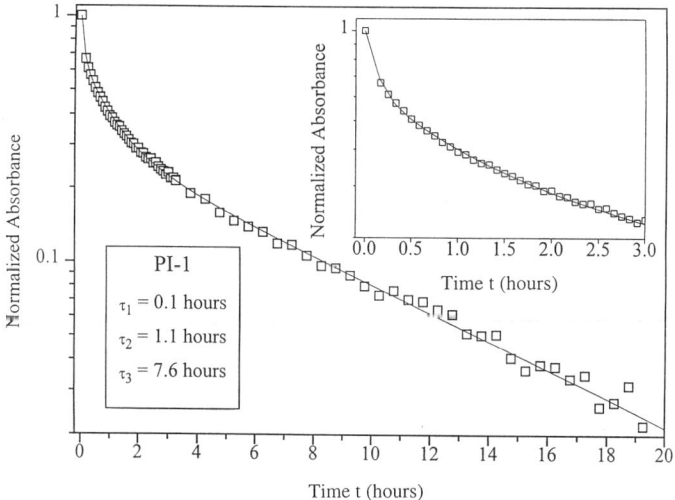

FIG. 4.14 A logarithmic plot of the long-time cis→trans thermal back reaction of PI-1. The inset shows an expanded view of the first points of the figure. Squares indicate the experimental absorbance, and the solid line indicates a triexponential theoretical fit with the time constants given in the figure. The time t = 0 corresponds to the state of the system approximately 2 minutes after the irradiation is turned off. After reference 47, redrawn by permission of OSA.

PI-1 and PI-2 films exhibit dichroic absorbance after polarized irradiation. Figure 4.15 shows polar plots of the absorbance of linearly polarized probe light (at 488 nm) as a function of the angle, ψ, between the polarization of the probe and irradiation light (532 nm; 30 mW/cm^2). Nonpolar orientation is clearly shown for both PI-1 (left) and PI-2 (right). The highest absorption is observed when the probe and irradiation beams have perpendicular polarizations,

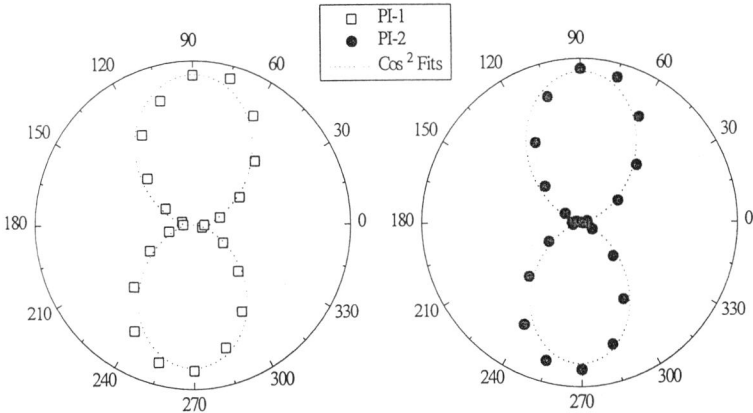

FIG. 4.15 Polar plots depicting the absorbance of PI-1 (left) and PI-2 (right) versus the angle between irradiation and probe-beam polarizations. The markers are experimental data points and the dashed curves are second-order Legendre polynomial theoretical fits. After reference 47, redrawn by permission of OSA.

confirming that the azo molecules are preferentially distributed along the perpendicular to the polarization of the green light. Before irradiation, identical values were recorded for the absorption of light linearly polarized both parallel and perpendicular to the irradiating light polarization, indicating that in-plane the sample was optically isotropic. In contrast to the quasi-stable photo-orientation observed in PI-1 and PI-2, PI-3 films show some relaxation of the light-induced orientation, confirming the previous suggestions regarding the coupling of the chromophore motion with that of the polymer backbone.

Photo-orientation in PI-1 and PI-2 is not erased even after heating at 170°C for one hour. This orientation is, however, completely randomized upon heating the samples above their Tgs for 10 minutes. At 170°C, PI-1 and PI-2 are still about 180 and 80°C below their respective Tgs. Molecular movement in polymeric materials is governed primarily by the difference between the operating temperature T and the Tg of the polymer (*vide infra*); in other words, the smaller this difference (Tg-T), the greater the molecular mobility. The sign of the dichroism is inverted from negative to positive when the irradiation light polarization is rotated through 90°. The horizontal and vertical absorbances were exactly interchanged by this procedure (spectra not shown). This inversion of the sign of the dichroism shows that the photo-isomerization reaction can easily reorient the chromophores at temperatures at least 325°C and 225°C below the Tg of PI-1 and PI-2 respectively, whereas heating both polymers even at 170°C failed to do so. This strongly suggests that the photoisomerization process is capable to some extent of moving the polyimide backbone via a coupling to the photoinduced movement of the azo chromophores. The contrasting behavior of the donor-embedded systems (PI-1 and PI-2) versus the flexibly tethered true side-chain systems (PI-3a and PI-3b) observed by the ATR technique and UV-vis dichroism is also confirmed by photo-orientation dynamics with real-time dichroism and dynamical ATR-birefringence,[47] and photo-assisted poling experiments (*vide infra*). Sub-Tg photoisomerization-induced molecular movement depends strongly on the molecular structure of the unit building blocks of the polymer, a feature confirmed in azo-polyurethanes.

4.4.2 Photoisomerization and Photo-Orientation of Flexible Azo-Polyurethanes

Four azo-polyurethane derivatives used in our study, PUR-1, PUR-2, PUR-3, and PUR-4, each with distinct molecular structures of the unit building blocks, are shown in Figure 4.16. The Tg values for PUR-1 and PUR-2 were 140°C, and for PUR-3 and PUR-4 were 136°C, as measured by differential scanning calorimetry. These azo-PUR polymers were commercial samples (Chromophore Inc.). All of these polymers are donor-embedded systems, where the chromophore is incorporated flexibly into the backbone of the polymer through the electron-donating substituent. The azo chromophore in PUR-1 and PUR-2 has a nitro, NO_2, versus a Cyano, CN, electron-withdrawing group for PUR-3 and PUR-4, a feature that slightly decreases the Tg and blue-shifts the maximum of absorption (see Figure 4.17 for the polymers' spectra). This feature also slows the cis→trans thermal back isomerization of the azo dye in PUR-3 and PUR-4. The difference between PUR-1 and PUR-3, and PUR-2 and PUR-4, is the aromatic ring into the polymer backbone, a seemingly small structural difference that does not affect the polymer Tg but

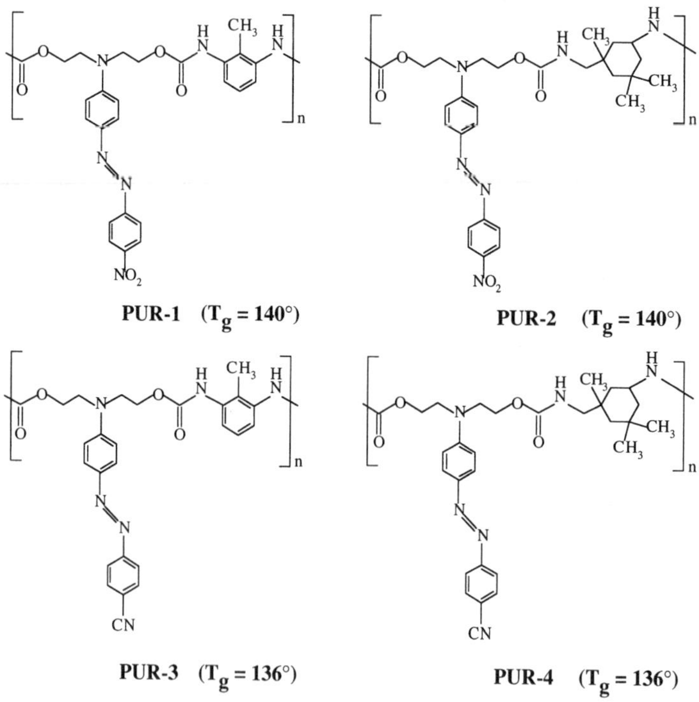

FIG. 4.16 Chemical structures of the Azo-PUR polymers.

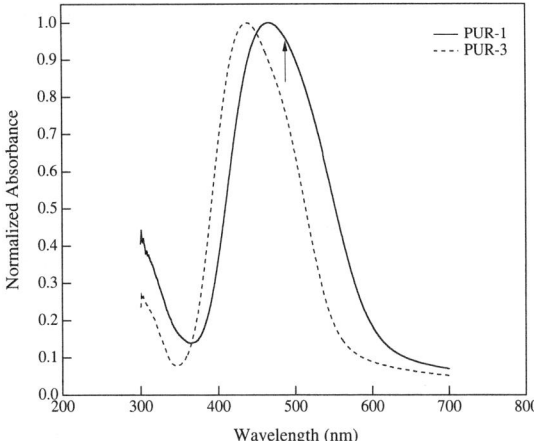

FIG. 4.17 Normalized spectra of PUR-1 and PUR-3. The spectra are normalized by the value of the maximum absorbance, and the arrow indicates the irradiation and analysis wavelength.

does noticeably affect the orientation dynamics of the polymers (*vide infra*). It will be shown that the photo-orientation dynamics of Azo-PURs is also influenced by the rate of the cis→trans thermal isomerization.

Azo-PURs represent a good example of donor-embedded polymers. The structure of these polymers contrasts with that of high temperature azo-polyimides in that both the ethylene spacers and the donor portion of the chromophore are incorporated into the polymer backbone. The particular structure of the azo-PUR systems studied enables a very high chromophore concentration per weight relative to the polymer backbone (~ 80 wt. % in azo-PURs versus 40 and 15% in azo-polyimides and azo-PMMA co-polymers, respectively), a feature that makes it easier for the backbone to respond to the photoinduced movement of the chromophores. Such structural features should improve the efficiency of polar and nonpolar photo-orientation. We performed nonpolar photo-orientation studies on the azo-PUR polymer series shown in Figure 4.1, and we used real-time dichroism experiments to investigate the dynamics of photo-orientation of the azo chromophores in films of PURs with the blue light (λ = 488 nm) from an Argon-ion laser as the irradiation and analysis light. Polymer films were spin-cast from solution onto glass substrates, heated above Tg to 150°C for one hour to remove residual solvent, and allowed to cool slowly to room temperature. Film samples were irradiated by linearly polarized light; $Abs_{//}$ and $Abs_{//}$ were calculated from the amount of absorbed light polarized parallel and perpendicular to the irradiation light polarization, respectively; and the anisotropy, $\Delta A = Abs_{//} - Abs_{\perp}$, was deduced.

Figures 4.18 and 4.19 show the time evolution of $Abs_{//}$ and $Abs_{\perp}$ of PUR-1 and PUR-3, respectively, during and after linearly polarized irradiation for different irradiation power values. The dynamics of photo-orientation of PUR-2 (not shown) resemble those of PUR-1, and PUR-4 shows a photo-orientation dynamical behavior (not shown) similar to that of PUR-3 (*vide infra*). When irradiation starts at time t = 5 minutes, anisotropy occurs and

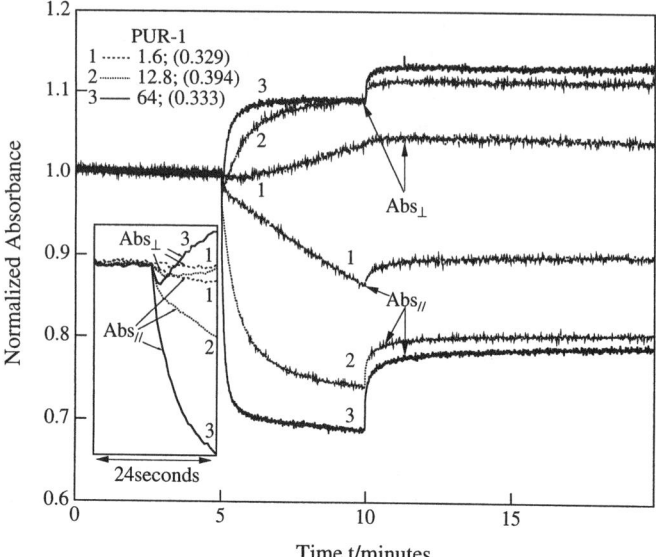

FIG. 4.18 Photo-orientation dynamics of PUR-1. Absorbance, normalized by the absorbance value before irradiation. The irradiation light is turned on and off at 5 and 10 minutes. The numbers 1–3 indicate and increasing irradiation intensity the value of which is given in units of mW/cm² with the corresponding sample's absorbance prior to irradiation in units of OD (value between brackets). The inset shows and expanded view of the first few seconds of photo-orientation.

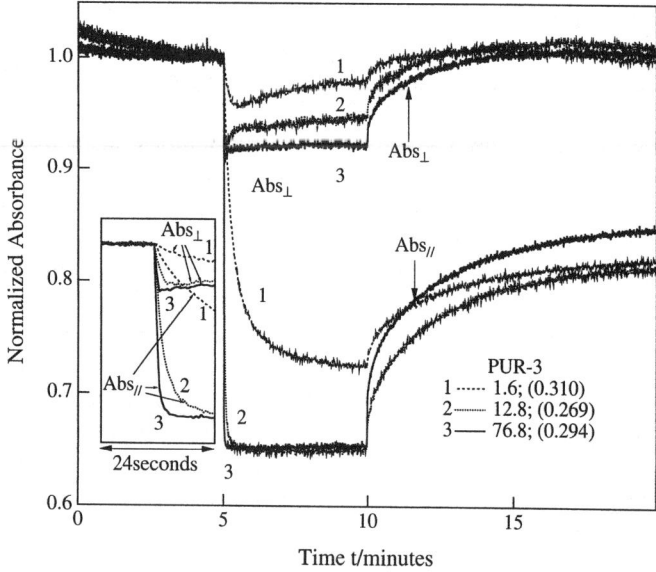

FIG. 4.19 Photo-orientation dynamics of PUR-3. The samples' OD prior to irradiation was ~ 0.3.

demonstrates chromophore photo-orientation. When irradiation is turned off at time t = 10 minutes, the observed relaxation indicates that cis→trans thermal isomerization, which is completed after a few seconds in PUR-1 and PUR-2, and takes several minutes to more than an hour in PUR-3 and PUR-4 (*vide infra*), converts cis to trans isomers, and the remnant anisotropy demonstrates that the trans molecules are oriented after isomerization.

Photo-orientation by photoisomerization occurs through a polarization-sensitive photoexcitation, i.e., photoselection. Two competing limiting cases of photoselection are worth discussing. If the chromophores are photoisomerized only through photoselection and are not rotated, a large cis population is anisotropically generated, and a hole is burned into the trans isomer's orientational distribution (orientational hole burning, OHB, cosine square probability of photoexcitation). In this case, both $Abs_{//}$ and $Abs_{\perp}$ change in the same direction with $(Abs_{//} - Abs_0) = + 3 (Abs_{\perp} - Abs_0)$. Abs_0 is the sample's absorbance before irradiation. Pure photo-reorientation occurs when only the trans isomer is rotated by a discrete angle for each absorbed photon, a feature that implies high reorientation rates for high-irradiation intensities. Pure photo-reorientation can involve the cis isomer, but only when it returns immediately to the trans isomer; therefore, the concentration of cis isomers is negligible during pure photo-orientation, and the chromophore is in the trans form most of the time during cis↔trans isomerization cycling. Pure photo-reorientation is theoretically characterized by high anisotropy values for high-irradiation intensities and by a dynamic behavior in which $Abs_{//}$ and $Abs_{\perp}$ evolve in opposite directions starting from the moment when polarized light impinges the sample with $(Abs_{//} - Abs_0) = - 2 (Abs_{\perp} - Abs_0)$. The factors + 3 and − 2 originate from the orientational averaging of the chromophores' polarizability after isomerization and the orientation by photoselection, respectively. Upon polarized irradiation, both OHB and pure photo-reorientation decrease $Abs_{//}$, whereas pure photo-reorientation increases $Abs_{\perp}$ and OHB decreases it in a competing manner. The trends of Figures 4.18 and 4.19 can be explained by the competitive scheme of OHB versus pure photo-reorientation. Upon polarized irradiation, $Abs_{//}$ decreases in all four Azo-PURs, and $Abs_{\perp}$ increases for PUR-1 and PUR-2 and decreases for PUR-3 and PUR-4. OHB is dominant in PUR-3 and PUR-4 because of a long-living cis isomer; in addition, the increase, after some time, of $Abs_{\perp}$ in PUR-3 is indicative of molecular reorientation following OHB.

Near-pure photo-orientation of PUR-1 by polarized irradiation is shown in Figure 4.16. Indeed, when the irradiating light is turned on, $Abs_{\perp}$ starts nearly immediately exceeding the absorbance before irradiation, i.e., Abs_0. At this same time, $Abs_{//}$ and $Abs_{\perp}$ change in opposite directions, and the higher the pump intensity the faster and the larger the increase of $Abs_{\perp} - Abs_0$ and of the anisotropy. The near-pure photo-orientation dynamics observed for PUR-1 fits the very first model developed for photo-orientation by photo-isomerization, which assumed that the chromophore is constantly in the trans state, or in other words, returns immediately to the trans state upon excitation, and rotates during the excitation cycle by a discrete angle. For all Azo-PURs, the quantum yields of the forth isomerization (trans→cis,) are small compared to those of the back (cis→trans) isomerization (cf. Chapter 3); in addition, the rate of cis→trans thermal isomerization is quite high, a feature that shows that the azo-chromophore is in the trans form most of the time during trans↔cis cycling. The near-pure photo-orientation observed in PUR-1 is in clear contrast to photo-orientation observations in all of the azo-polymers studied to date, including PUR-2, PUR-3, and PUR-4, wherein both $Abs_{//}$ and $Abs_{\perp}$ change in the same direction upon photo-orientation (OHB), as shown in Figure 4.19 for PUR-3, and $Abs_{\perp}$ can exceed Abs_0 only for weak pump intensities to minimize the concentration, of the cis population. Indeed, weak pump intensities minimize the cis concentration, thereby favoring orientational redistribution over OHB.

Near-pure photo-orientation occurs in PUR-1 and not in PUR-2, PUR-3 and PUR-4 because of a fast cis→trans isomerization of PUR-1 versus PUR-3 and PUR-4 and a seemingly small difference into the polymer backbone of PUR-1 versus PUR-2. Although PUR-1 and PUR-3 have the same polymer backbone, the CN electron-withdrawing group of the azo chromophore in PUR-3 and PUR-4 slows the cis→trans thermal isomerization. The isotropic absorbance of PUR-1 is recovered more quickly than that of PUR-3 upon cis→trans thermal isomerization after the end of irradiation, with 0.45 and 0.14 s-1 as the fastest isomerization rates for PUR-1 and PUR-3, respectively (see Table 4.2). This table also shows that cis→trans thermal isomerization proceeds with faster rates for PUR-1 and PUR-2 versus PUR-3 and PUR-4, and that the rates of polymers with the same electron-withdrawing groups are similar. The cis→trans thermal isomerization is slowed down in PUR-3 and PUR-4 because of a higher energy barrier that needs to be crossed by the cis form to isomerize back to the trans form, a feature that decreases the number of cycles per unit time for the azo-chromophore in PUR-3 and PUR-4 versus PUR-1 and PUR-2.

The effect of the structure of the polymer backbone on photo-orientation can be seen from the dynamic behavior as well as from the steady-state values of the photoinduced anisotropy in all azo-PURs. The photo-orientation dynamics of PUR-2 resemble but also contrast with those of PUR-1. In PUR-2, $Abs_{\perp}$ exceeds Abs_0, but not quite, as is the case for PUR-1, and the photostationary-state anisotropy is smaller than that of PUR-1, as can be seen in Figure 4.20. PUR-1 and PUR-2 exhibit exactly the same extinction coefficient at the analysis wavelength because they have the same azo chromophore; furthermore, the rate of the cis→trans thermal isomerization is nearly the same in both polymers. The seemingly small difference into the

TABLE 4.2 Rate Constants, k_i, and Weighting Coefficients, a_i, i=1, 2, 3, of the Cis→Trans Thermal Isomerization of the Chromophore in Azo-PURs

	PUR-1	PUR-2	PUR-3	PUR-4
$k_1(s^{-1})$; a_1	0.450; 0.75	0.410; 0.67	0.140; 0.18	0.27; 0.16
$k_2(s^{-1})$; a_2	0.031; 0.31	0.024; 0.35	0.007; 0.47	0.01; 0.46
$k_3(s^{-1})$; a_3	—	—	0.0003; 0.3	0.0002; 0.40

polymer backbone, i.e., the presence of the aromatic ring into the backbone of PUR-1, clearly influences the photo-orientation dynamics and efficiency for PUR-1 versus PUR-2. This influence of the polymer backbone was also observed for PUR-3 versus PUR-4, confirming the influence of polymer structural effects on photo-orientation.

The levels of photo-induced anisotropy in all four Azo-PURs are also correlated to the polymer structure. Figure 4.20 shows real-time dichroism values of the photoinduced anisotropy with 488 nm irradiation and analysis in all four Azo-PURs obtained at near the steady state of irradiation versus the irradiation intensity. The observed polymer structural effects of Figure 4.20 are confirmed by other data (not shown) independently obtained after photoisomerization by linearly polarized 488 nm irradiation of all four polymers and measurement of $Abs_{//}$ and $Abs_\perp$ by a UV-vis spectrophotometer at the maximum wavelength, i.e., 470 nm for PUR-1 and PUR-2, and 440 nm for PUR-3 and PUR-4. Figure 4.20 clearly shows a higher photostationary-state anisotropy for PUR-1 than for PUR-2, PUR-3, and PUR-4. Even though the observed anisotropy depends on the cis and trans balance in concentration, on the isomers' extinction coefficients, and on the isomers' orientation, polymers with the same chromophore but different backbones exhibit different levels of induced anisotropy under the same irradiation conditions. The series of data taken by both real-time dichroism and UV-vis steady-state values demonstrate that the photo-orientation efficiency in the azo-polyurethanes studied decreases according to the series PUR-1>PUR-2>PUR-3>PUR-4. The fast cis→trans thermal isomerization and the addition of the aromatic ring into the polymer backbone facilitate the movement of the chromophore.

Next, we discuss whether the chromophores' spontaneous, i.e., thermally activated, relaxation is primarily influenced by the movement of the polymer backbone (α-relaxation), by that of the chromophore (β-relaxation), or by

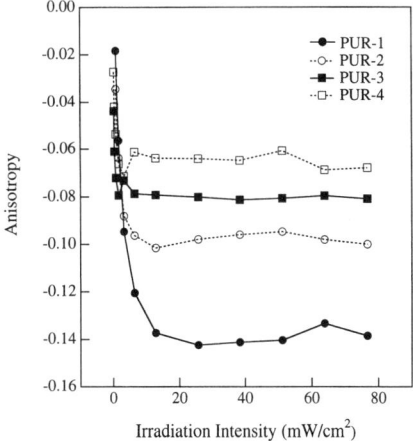

FIG. 4.20 Photostationary-state anisotropy in Azo-PURs versus irradiation intensity. Each data point was taken after 5 minutes of irradiation. The data are those of real-time dichroism with 488 nm irradiation and analysis.

both. To do so, we have recorded the erasure of the anisotropy, which is indicative of the chromophores' disorientation, in each polymer versus temperature (see Figure 4.21). The data in Figure 4.21 were obtained by heating an all-trans photo-oriented sample, that is a sample that was photo-oriented to the photostationary state and relaxed in the dark for 5 hours to complete the cis→trans thermal isomerization and in the oven for 15 minutes at a given temperature. Then we immediately recorded $Abs_{//}$ and $Abs_\perp$, from which we computed the anisotropy, and the measurement at the next temperature value followed. The sample's initial isotropic absorbance remained unchanged on heating. It is clear from Figure 4.21 that polymers with the same backbone follow the same path independent of the chromophore, which shows that the chromophores' thermally activated orientational relaxation is primarily governed by the polymer backbone rather than by the chromophore itself. This α-relaxation–triggered β-relaxation molecular movement is especially pronounced near the polymers' Tg, where substantial spontaneous molecular movement occurs. Note that the Tgs of the Azo-PURs that have the same backbone are slightly but noticeably different (140 versus 136°C). The effect of the polymer free volume on isomerization movement can be studied at pressure, as we discuss in the next section.

4.5 PRESSURE EFFECTS ON PHOTOISOMERIZATION AND PHOTO-ORIENTATION

Because chromophores' orientation is important for creating anisotropy and optical nonlinearities, intensive studies have been performed to understand induced molecular orientation and relaxation processes in polymers.[61] To gain further insight into the physics of thin polymer films and the effects of molecular orientation in solid polymers, studies at high pressure could be beneficial.[62] Pressure as a thermodynamic parameter is widely used to study

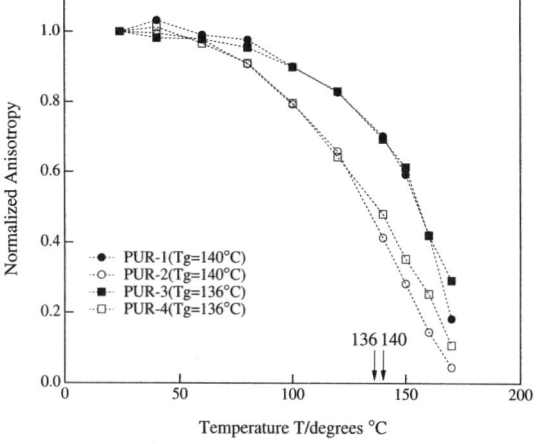

FIG. 4.21 Erasure of the anisotropy versus temperature for Azo-PURs. The data points are normalized by the value of the anisotropy at 20°C. The polymers' Tgs are indicated.

activation and reaction volumes in solution.[63] For photo and thermal isomerization reactions, kinetics of both ground- and excited-state isomerization are used to assess solvent and high-viscosity solution effects on isomerization rates and isomerization movement.[64-66] Regarding the effects of pressure on photoisomerization and photo-orientation in polymeric thin films, we show that both photoisomerization and photo-orientation of disperse red one (DR1), an azo dye, flexibly tethered to a poly(methyl-methacrylate) (PMMA) polymer, referred to in the text as PMMA-DR1, are hindered by the application of hydrostatic pressure.[48]

We used ATR prism coupling as well as a Kerr gate optical setup to probe the influence of pressure on photoisomerization and photo-orientation in PMMA-DR1. The polymer was obtained from IBM Almaden; its chemical structure is shown in Figure 4.22. The dye is tethered to the PMMA copolymer with 10% DR1 per monomer unit. The molecular weight (M_ω) and glass transition temperature (Tg) of this polymer are M_n = 163000 g/mol and Tg = 123°C, respectively. PMMA-DR1 films (thickness ~ 1 μm) were prepared by spin-coating from a diethyleneglycol dimethyl ether solution either directly on top of a sapphire slide for the Kerr gate experiments, or on top of a 50 nm-thick gold layer that was then evaporated directly on top of a sapphire slide for the ATR prism coupling experiments. The films were dried at 90°C for 12 h and heated at 130°C for several minutes in a vacuum to remove traces of the remaining solvent. After heating, the films were allowed to cool slowly to room temperature. Details about the characteristics of the pressure chamber can be found elsewhere.[48] The temperature inside the chamber was controlled by a thermostat, and all the experiments discussed in this paper were performed at 25°C. Water was used as a pressure medium. For the Kerr gate optical experiments, the pressure cell containing the film sample was placed between crossed polarizers oriented at 45 and -45 degrees with respect to the vertical, and a probe beam propagated in succession through the polarizer,

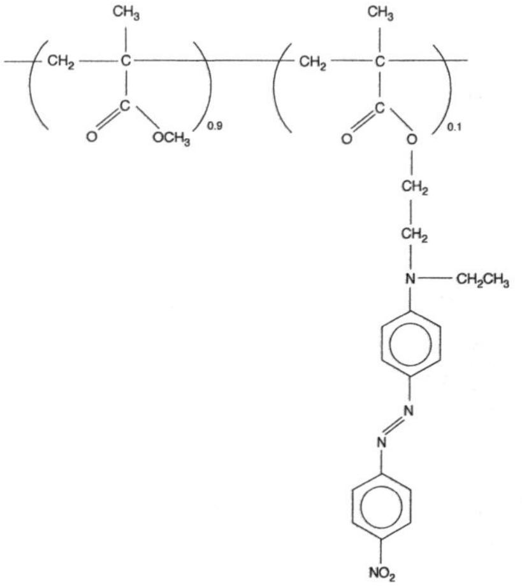

FIG. 4.22 Chemical structure of PMMA-DR1.

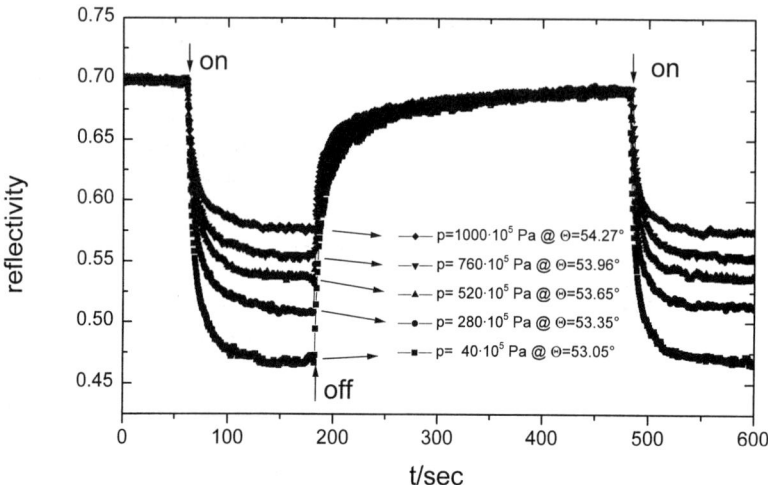

FIG. 4.23 TM mode reflectivity cycling in PMMA-DR1 by photoisomerization at high pressure. The moment of turning the irradiation light on and off are indicated. After reference 48, redrawn by permission of OSA.

the cell i.e., the input sapphire window/water/PMMA-DR1 film/sapphire slide/water/output sapphire window, and the analyzer. Experimental details can be found elsewhere.[48] Briefly, the probe beam was $\lambda = 633$ nm light from a He-Ne laser propagating perpendicular to the windows and the sample, and the irradiation beam was the green light of a laser ($\lambda = 543$ nm at 3 mW power with a 2-mm-diameter spot). The irradiating beam was also linearly polarized and propagating at normal incidence, exposing the PMMA-DR1 film through the input sapphire slide window, but blocked in front of the detector by a red filter.

Figure 4.23 shows cycles of reversible change of reflectivity of a 633-nm TM mode at a fixed large-incidence angle coupled by an ATR prism into PMMA-DR1 during alternating photoinduced and thermal back isomerization of DR1 at five different pressure values: 40, 280, 520, 760, and 1000×10^5 Pa. The angle was fixed in the left wing of the mode so that the photo-isomerization-induced decrease of reflectivity would indicate a shift of the resonance to smaller incidence angles. The curves represent the evolution of the refractive index in the plane of incidence. The irradiating light (543.5 nm; 6 mW/cm^2) was TM-polarized, and the moments when it was turned on and off are indicated. The angle corresponding to the minimum reflectivity of the mode is indicated for each pressure value. Clearly, it is increasingly difficult for isomerization to proceed in PMMA-DR1 with increasing pressure.

Figure 4.24 shows the influence of pressure on the photoinduced anisotropy in PMMA-DR1 observed by the Kerr gate experiment for several applied hydrostatic pressures up to 150 MPa, as indicated. The moments when the irradiating light was turned on and off are indicated. After the thermal isomerization is completed after the end of the irradiation, circularly polarized irradiation randomized the in-plane orientation, and photo-orientation at the next higher pressure value followed. Figure 4.24 clearly shows the time

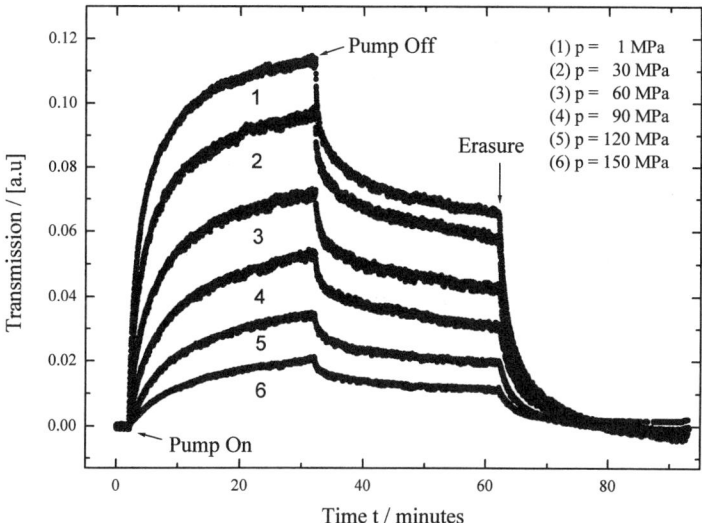

FIG. 4.24 Influence of pressure on photo-orientation of DR1 observed by the Kerr gate setup for applied hydrostatic pressures up to 150 MPa. The numbers 1 to 6 refer to the applied pressure, and the moments when the irradiating light is turned on and off are indicated. After the thermal isomerization is completed, circularly polarized irradiation, indicated by "Erasure," randomized the in-plane orientation to erase the anisotropy, and photo-orientation at the next-higher pressure followed. Note that the level of the observed anisotropy decreases with the increased pressure, a feature that indicates decreasing capability of photo-orientation with increasing pressure. After reference 48, redrawn by permission of OSA.

course of the photoinduced anisotropy to be slowed down and the efficiency to decrease with increasing pressure. This feature is indicative of the increasing difficulty of trans-DR1 to move at the increased pressure, in that the cis concentration is reduced by pressure application. Pressure increases the friction of the chromophore by changing the shape and reducing the volume of the cavity that surrounds it. Photoisomerization and photo-orientation occur for those trans isomers that have enough free volume to undergo reorientation, isomeric change in shape, or both. At high pressure, some trans isomers lack the freedom, i.e., the local free volume, necessary for photoisomerization and reorientation. It is well known that the photoisomerization of azobenzenes depends on the free volume,[11,49,67] and even though azo dyes can sometimes trigger polymer segmental motion and swelling by photoisomerization, pressure quenches the movement of some of the trans-DR1 into the polymer by compression and free-volume reduction. Pressure also freezes high-energy conformations of the host, which contributes to reduced mobility of the chromophore,[63] an effect that may occur in the PMMA chains and add to the hindrance of DR1 movement. The suppression of some free volume by pressure is supported by independent waveguide spectroscopy experiments that show that the thickness of μm thick films of PMMA-DR1 decreases by ~ 16 nm and the 633-nm refractive index increases linearly by ~ 0.012 for each 100 MPa of pressure applied. Pressure increases the films' density.

Pressure-induced changes in the refractive index and the thickness are theoretically rationalized by Tait's and Lorentz-Lorenze's equations. Studies of the effect of pressure on water as a pressure medium were performed without polymer film, and the refractive index of water, n_{water}, increased at pressure. n_{water} is known for each pressure value and taken into account for the determination of the optical constants of the polymer film. The refractive index of water is considerably smaller than that of PMMA-DR1 at all pressures (for example at 100 MPa, $n_{water}^{633nm} = 1.345\pm0.003$ and $n_{PMMA-DR1}^{633nm} = 1.630\pm0.003$). The increase of $n_{PMMA-DR1}$ up to 1.636 at 150 MPa (*vide infra*) cannot be due to the absorption of water by the polymer, otherwise $n_{PMMA-DR}$ should decrease rather than increase. Even though water molecules are small, they do not penetrate into the polymer film studied. Figure 4.24 shows that most photo-orientation is suppressed at 150 MPa. Assuming that pressure does not affect noticeably the lateral dimensions of the films, the thickness variation mentioned previously implies a ~ 2.4% volume change at that pressure value. A volume fraction, i.e., free volume, is necessary for the isomeric and reorientational movement of most of the azo chromophores in PMMA-DR1. This near 2.4% volume change is due to a change in density that couples to a change in the refractive index of the material. It can be rationalized by the following Clausius-Mosotti equation:[68]

$$\Delta n = \frac{(n^2 - 1)(n^2 + 2)}{6n} \frac{\Delta \rho}{\rho} \qquad (4.1)$$

where n, and ρ are the isotropic refractive index and the density of the material, respectively, and Δn and $\Delta \rho$ are the corresponding changes induced by pressure. We found experimentally that n, i.e., $n_{PMMA-DR1}$, is equal to 1.636 at 150 MPa for a 633-nm probe light and that consequently Equation 4.1 predicts a Δn value of 0.019 for a 2.4% change in density, whereas the experimentally measured value for Δn at 150 MPa is ~ 0.018. This value is obtained by multiplying the slope, i.e., ~ $0.012*10^{-2}$/MPa, of Δn versus pressure by 150 MPa. The Clausius-Mosotti equation well supports the claimed ~ 2.4% change in the sample's density at 150 MPa; it also supports the arguments we put forth concerning free-volume reduction by pressure. The free volume size necessary for one DR1 molecule isomerization in PMMA-DR1 is discussed next.

The gradual reduction of DR1 photo-orientation with increased pressure implies a distribution of local free-volume elements of different sizes available to the trans isomers in PMMA-DR1, a concept that has theoretical support,[49,69,70] and that is experimentally observed and discussed in the literature for photoisomerization in polymers[49] and poled PMMA-DR1.[13,19] Photo-orientation in the glassy state requires a minimum, critical size of local free volume in the vicinity of the chromophore—the photo-orientation activation volume, i.e., the volume swept by the chromophore during photoisomerization and photo-orientation geometrical rearrangement (*vide infra*).[71] During the early time evolution, the cis concentration is negligible, and the observed anisotropy is dictated by the trans orientation. In fact, the rate k, i.e., the slope, of the early time photo-orientation is proportional to the quantum yield of the trans→cis photoisomerization, which, as far as friction effects are concerned, is the only material parameter that can be pressure dependent in k.[72] Reaction rates at high pressure are theoretically and experimentally rationalized by:[63–66]

$$\ln k = (\ln k_0) - \Delta V^{\ddagger} \frac{P}{RT} \text{ with } \left(\frac{\partial \ln k}{\partial P} \right)_T = - \Delta V^{\ddagger} \frac{1}{RT} \qquad (4.2)$$

where $\Delta V^{\ddagger}$ is the activation volume, that is, the minimum volume needed for the reaction to proceed, P is the pressure, T is the absolute temperature, R is the universal gas constant, and k_0 is the reaction rate that corresponds to the zero pressure. For PMMA-DR1, we found that *ln k* decreases linearly with the applied pressure (see Figure 4.25), a finding that suggests that Φ_{TC} may be described at pressure by:

$$\ln \Phi_{TC} = (\ln \Phi_{TC}^0) - \Delta V^{\ddagger} \frac{P}{RT} \qquad (4.3)$$

where Φ_{TC}^0 is the analog of k_0. From the slope of Figure 4.23, we measured an activation volume for one chromophore photo-orientation of (111 ± 7) Å^3 by using Equation 4.2—a value quite close to the theoretical 101 Å^3 swept volume and the 127 Å^3 needed for azobenzene as an extra volume to isomerize.[73] In light of this finding, we have added to the debate about the relationship between the activation and swept volumes and infer, as suggested earlier in this chapter, that these two volumes represent the same physical volume.

A comment must be made about the cis→trans thermal isomerization rate at pressure. At room temperature, the thermal back reaction of DR1-PMMA follows a complex, nonexponential recovery, most of which is completed after a few seconds with a rate of 0.25 s^{-1}, and deviates from a single exponential decay after the first 10 seconds.[13] Larger relaxation times at Tg −98°C include slow polymer motion coupled with the chromophores' rotational diffusion. We confirmed that this behavior is true in the polymer

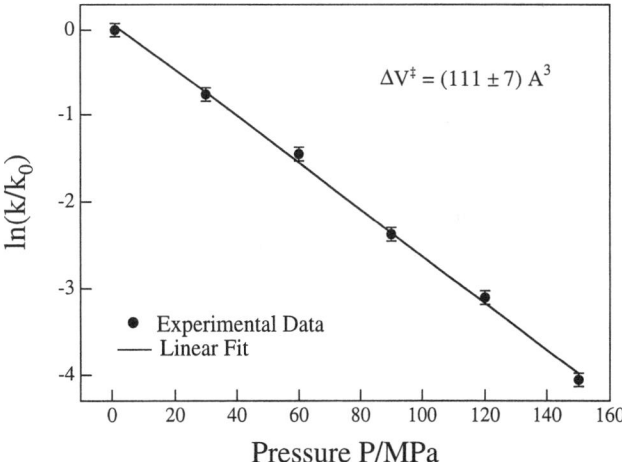

FIG. 4.25 ln(k/k$_0$) versus the applied pressure. k and k$_0$ are the rates given by the slope of the early time evolution of photo-orientation at a given pressure and at 1 MPa, respectively. The markers are experimental data points and the line is a theoretical fit by Equation 4.2. The number in parentheses corresponds to the activation volume of the trans→cis photo-orientation. After reference 48, redrawn by permission of OSA.

studied in this chapter. In particular, we adjusted an exponential decay to the first 10 seconds of the recovery at all 6 pressures, and we found a thermal isomerization rate for PMMA-DR1 in the range $0.17 - 0.23$ s^{-1}, with no particular dependence on pressure. This result rules out pressure-induced static effects and reinforces the friction effects discussed; it also shows that if trans has enough sweep volume to isomerize to cis, cis will also have enough sweep volume to isomerize back, a feature supported by the more compact and globular, i.e., twisted, conformation of the cis- versus the trans-DR1.

4.6   CONCLUSION

Photoisomerization and photo-orientation of azobenzenes and their derivatives in films of polymer are influenced by the environment of the chromophore. In highly organized supramolecular structures constructed by LBK multilayers, the azobenzene molecules can be trapped in a well-defined manner; in amorphous spin-cast films, molecular movement is strongly influenced by the structure of the unit building blocks of the polymer. Besides the thermodynamical parameters, temperature and pressure, the factors that strongly influence photoisomerization and photo-orientation in films of polymer are the free volume, the nature of the connection of the chromophore to the polymer main chain, the rigidity of the backbone, the molecular weight, and the glass transition temperature. This chapter summarizes a number of examples and proofs of environmental and polymeric structural effects on photoisomerization and photo-orientation. In LBK azo-polyglutamates, the lengths of the spacer and the tail, i.e., the alkyl chain, attached to the chromophore clearly influence the movement of the azobenzene side chains, and studies in loosely packed, molecularly thin SAMs reveal a similar isomerization and reorientation process in the bulk and at the molecular level. In clear contrast to spin-cast films, photo-(re)orientation does not appear in LBK structures due to strong intermolecular interactions. In high-Tg, nonlinear optical polyimides, the isomerization is slowed down when the chromophore is firmly embedded into a rigid backbone, and photoisomerization is capable of inducing molecular orientation as much as 325°C below Tg of a polyimide containing no flexible connector or tether. In azo-polyurethanes, the azo dye photo-orientation efficiency is influenced by the polymer structure as well as by the isomerization rate and the photochemical quantum yields, and pressure studies in films of PMMA-DR1 show that the chromophore's local free volume plays a major role in photoisomerization and photo-orientation.

ACKNOWLEDGMENTS

This work is based on collaborations with many friends and colleagues, and it is our pleasure to thank them for their support and for many helpful discussions. In particular, J. Wood participated in most of the work, Y. Geerts and K. Müllen provided the azo-silane compound, M. Büchel and S. Paul contributed to the azo-polyglutamate LBK experiments, and B. Weichart and H. Menzel from Hannover Universität provided the azo-polyglutamate polymers. The work on azo-polyimides was done at the University of California, Davis and IBM Almaden (San Jose) with

R. D. Miller and A. Knoesen. We would like to thank W. Volksen, V. Y. Lee, P. Prêtre, L. M. Wu, and D. Yankelevich for helpful discussions. The work on azo-polyurethanes was done at Osaka University, Handai, with D. Yasumatsu and S. Kawata. We thank C. Hawker at IBM Almaden for the PMMA-DR1 material, and the pressure experiments were done with G. Kleideiter. Zouheir Sekkat acknowledges research support from the Japan Society for the Promotion of Science under the Research for the Future Program and thanks the Office of Naval Research for research support during his stay at UC Davis, and the Max-Planck Society for financial support during his stay at the Max-Planck Institut für Polymerforschung in Mainz. Financial support also came from the National Science Foundation (NSF) through the Center on Polymer Interfaces and Macromolecular Assemblies (CPIMA).

REFERENCES

1. Sekkat, Z. and Knoll, W. In *Advances in Photochemistry*. Neckers, D. C., Volman, D. H., and Bunau, G. Von, Eds. (Wiley and Sons) **1997**, 22, 117; *SPIE Proceeding* **1997**, 2998, 164; Sekkat, Z., Knoesen, A., Knoll, W., and Miller, R. D. *SPIE Critical Reviews*, Najafi, I. and Andrews, M. P., Eds. **1997**, CR68, 374.
2. Delaire, J.A. and Nakatani, K. *Chem. Rev.* **2000**, 5, 1817; and references therein.
3. Ichimura, K. Ch. 26 in *Photochromism Molecules and Systems*, Dürr, H. and Bouas-Laurent, H., Ed. Elsevier, Amsterdam, **1990**, pp. 903–917.
4. Todorov, T., Nicolova, L., and Tomova, T. *Appl. Opt.* **1984**, 23, 4309.
5. Sekkat, Z. and Dumont, M. *Appl. Phys.* B. **1991**.
6. Shi, Y., H. Steier, W., Yu, L., Shen, M., and R. Dalton, L. *Appl. Phys. Lett.* **1991**, 58, 1131.
7. Rochon, P., Gosselin, J., Natansohn, A., and Xie, S. *Appl. Phys. Lett.* **1992**, 60, 4.
8. Sekkat, Z. and Dumont, M. *Synth. Metals* **1993**, 54, 373, 53, 121.
9. Sekkat, Z., Büchel, M., Orendi, H., Menzel, H., and Knoll, W. *Chem. Phys. Lett.* **1994**, 220, 497.
10. *Photochromism Molecules and Systems*, Dürr, H. and Bouas-Laurent, H., Eds. Elsevier, Amsterdam, **1990**.
11. Rau, H. In *Photochemistry and Photophysics*. Rabeck, J. F., Ed. CRC Press (Boca Raton, F., **1990**), 2, 119; and references therein.
12. Sekkat, Z. and Dumont, M. *Appl. Phys.* B. **1992**, 54, 486; *Mol. Cryst. Liq. Cyrs. Sci. Technol., SecB: Nonlinear Optics* **1992**, 2, 359; SPIE Proceeding **1992**, 1774, 188.
13. Loucif-Saibi, R., Nakatani, K., Delaire, J. A., Dumont, M., and Sekkat, Z. *Chem. Mater.* **1993**, 5, 229.
14. Palto, S. P., Blinov, L. M., Yudin, S. G., Grewer, G., Schönhoff, M., and Lösche, M. *Chem. Phys. Lett.* **1993**, 202, 308.
15. Blanchard, P. M. and Mitchell, G. R. *Appl. Phys. Lett.* **1993**, 63, 2038, 941.
16. Barnik, M. I., Blinov, L. M., Weyrauch, T., Palto, S. P., Tevosov, A. A., and Haase, W. In: *Polymers for Second Order Nonlinear Optics*, Lindsay and Singer, Eds. **1995**, ACS Symposium Series 601, Ch. 21, pp. 288–303.
17. Bauer, S., Bauer-Gogonea, S., Ylmaz, S., Wirges, W., and Gerhard-Multhaupt, R. In: *Polymers for Second Order Nonlinear Optics*, Lindsay and Singer, Eds. **1995**, ACS Symposium Series 601, Ch. 22, pp. 304–316.
18. Hill, R. A., Dreher, S., Knoesen, A., and Yankelevich, D. *Appl. Phys. Lett.* **1995**, 66, 2156.
19. Charra, F., Kajzar, F., Nunzi, J. M., Raimond, P., and Idiart, E. *Opt. Lett.* **1993**, 12.
20. Sekkat, Z., Knoesen, A., Lee, V. Y., and Miller, R. D. J. *Phys. Chem.* B **1997**, 101, 4733.
21. Sekkat, Z., Kang, C. S., Aust, E. F., Wegner, G., and Knoll, W. *Chem. Mater.* **1995**, 7, 142, Sekkat, Z., Wood, J., and Knoll, W. J. *Chem. Phys.* **1995**, 99, 17226.
22. Buffeteau, T. and Pézolet, M. *Appl. Spectro.* **1996**, 50, 948.
23. Böhm, N., Materny, A., Kiefer, W., Steins, H., Müller, M. M., and Schottner, G. *Macromolecules* **1996**, 29, 2599.
24. Eich, M. and Wendorff, J. H. J. *Opt. Soc. Am.* B. **1990**, 7, 1428.
25. Anderle, K., Birenheide, R., Werner, M. J. A., and Wendorff, J. H. *Liq. Cryst.* **1991**, 9, 691.
26. Wiesner, U., Reynolds, N., Boeffel, C., and Spiess, W. H. *Liq. Cryst.* **1992**, 11, 251.
27. Hvilsed, S., Andruzzi, F., and Ramanujam, R. *Opt. Lett.* **1992**, 17, 1234.

28. Lagugné Labarthet, F., Sourisseau, C. J. *Ram. Spectro.* **1996**, 27, 491.
29. Sawodny, M., Schmidt, A., Stamm, M., Knoll, W., Urban, C., and Ringsdorf, H. *Polym. Adv. Technol.* **1991**, 2, 127.
30. Büchel, M., Sekkat, Z., Paul, S., Weichart, B., Menzel, H., and Knoll, W. *Langmuir* **1995**, 11, 4460.
31. Stumpe, J., Fischer, Th., and Menzel, H. *Macromolecules* **1996**, 29, 2831.
32. Wang, R., Iyoda, T., Hashimoto, K., and Fujishima, A. *J. Phys. Chem.* **1995**, 99, 3352.
33. Schönhoff, M., Mertesdorf, M., and Lösche, M. J. *Phys. Chem.* **1996**, 100, 7558.
34. Seki, T., Sakuragi, M., Kawanishi, Y., Suzuki, Y., Tamaki, T., Fukuda, R., and Ichimura, K. *Langmuir* **1993**, 9, 211.
35. Gibbons, W. M., Shannon, P. J., Sun, S. T., and Sweltin, B. J. *Nature* **1991**, 95, 509.
36. Sekkat, Z., Büchel, M., Orendi, H., Knobloch, H., Seki, T., Ito, S., Koberstein, J., and Knoll, W. *Opt. Commun.* **1994**, 11, 324.
37. Ichimura, K., Hayashi, Y., and Akiyama, H. *Langmuir* **1993**, 9, 3298.
38. Wolf, M. O., and Fox, M. A. *Langmuir* **1996**, 12, 955.
39. Sekkat, Z., Wood, J., Geerts, Y., and Knoll, W. *Langmuir* **1995**, 11, 2856, *Langmuir* **1996**, 12, 2976.
40. Junge, M., and McGrath, D. V. *Chem. Commun.* **1997**, 9, 857.
41. Song, X., Geiger, C., Vaday, S., Perlstein, J., and Whitten, D. G. *J. Photochem. Photobiol. A: Chem.* **1996**, 102, 39.
42. Fissi, A., Pieroni, O., Balestreri, E., and Amato, C. *Macromolecules* **1992**, 29, 4680.
43. Berg, R. H., Hvilsed, S., and Ramanujamm, P. S. *Nature* **1996**, 383, 505.
44. Hoffmann, K., Marlow, F., and Caro, J. *Adv. Mater.* **1997**, 9, 567.
45. Rochon, P., Batalla, E., and Natansohn, A. *Appl. Phys. Lett.* **1995**, 66, 136.
46. Kim, D. Y., Tripathy, S. K., Li, L., and Kumar, J. *Appl. Phys. Lett.* **1995**, 66, 1166, Viswanathan, N. K., Kim, D. Y., Bian, S., Williams, J., Liu, W., Li, L., Samuelson, L., Kumar, J., and Tripathy, S. K. *J. Mater. Chem.* **1999**, 9, 1941.
47. Sekkat, Z., Wood, J., Aust, E. F., Knoll, W., Volksen, W., and Miller, R. D. *J. Opt. Soc. Am. B.* **1996**, 13, 1713, Sekkat, Z., Wood, J., Knoll, W., Volksen, W., Miller, R. D., and Knoesen, A. *J. Opt. Soc. Am. B.* **1997**, 14, 829, Sekkat, Z., Wood, J., Knoll, W., Volksen, W., Lee, V. Y., Miller, R. D., and Knoesen, A. *Polymer Preprints, Am. Chem. Soc. Div. Polym. Chem.* **1997**, 38, 977, Sekkat, Z., Knoesen, A., Lee, V. Y., and Miller, R. D. *J. Polym. Science. B. Polym. Phys.* **1998**, 36, 1669, Sekkat, Z., Prêtre, P., Knoesen, A., Volksen, W., Lee, V. Y., Miller, R. D., Wood, J., and Knoll, W. *J. Opt. Soc. Am. B.* **1998**, 15, 401.
48. Kleideiter, G., Sekkat, Z., Kreitre, M., Dieter Lechner, M., and Knoll, W. *J. Mol. Struc.* **2000**, 521, 167, Sekkat, Z., Kleideiter, G., and Knoll, W. *J. Opt. Soc. Am. B.* **2001**, 18, 1854.
49. Isenbach, C. D. *Bunsenges. Ber. Phys. Chem.* **1980**, 84, 680.
50. Williams, M. L., Landel, R. F., and Ferry, J. D. *J. Am. Chem. Soc.* **1955**, 77, 3701.
51. Knoll, W. *Ann. Rev. Phys. Chem.* **1998**, 49, 569.
52. Seki, T., and Ichimura, K. *Polym. Commun.* **1989**, 30, 109.
53. Wegner, G. *Thin Solid Films* **1992**, 216, 105.
54. Vierheller, T. R., Foster, M. D., Schmidt, A., Mathauer, K., Knoll, W, Wegner, G., Satija, S, and Majkrzak, C. F. *Macromolecules*, **1994**, 27, 6893.
55. Schmidt, A., Mathauer, K, Reiter, G., Foster, M. D., Stamm, M., Wegner, G., and Knoll, W. *Langmuir*, **1994**, 10, 3820.
56. Orthmann, E., and Wegner, G. *Angew. Chem*, **1986**, 98, 1114.
57. Menzel, H., Weichart, B., Schmidt, A., Paul, S., Knoll, W., Stumpe, J., and Fischer, T. *Langmuir*, **1994** 10, 1926, Menzel, H. *Macromolecules* **1993**, 26, 6226.
58. Sato, T., Ozaki, Y., and Iriyama, K. *Langmuir* **1994**, 10, 2363.
59. Miller, R. D., Burland, D. M., Jurich, M. C., Lee, V. Y., Moylan, C. R., Twieg, R. J., Thackara, J., Verbiest, T., and Volksen, W. *Macromolecules* **1995**, 28, 4974.
60. Verbiest, T., Burland, D. M., Jurich, M. C., Lee, V. Y., Miller, R. D., and Volksen, W. *Science* **1995**, 268, 1604.
61. Kaatz, P., Prêtre, P., Meier, U. Stalder, U., Bosshard, C., Gunter, P., Zysset, B., Stahelin, M., Ahlheim, M., Lehr, F. *Macromolecules* **1996**, 29, 1666–1678.
62. Strutz, S. J. and Hayden, L. M. *Journal of Polymer Science: Part B: Polymer Physics* **1998**, 36, 2793–2803.

63. Draljaca, A., Hubbard, C. D., van Eldik, R., Asano, T., Bsilevsky, M. V., and le Noble, W. J. *Chem. Rev.* **1998** 98, 2167–2289.
64. Flom, S. R., Nagarajan, V. and Barbara, P. F. *J. Phys. Chem.* **1986**, 90, 2085–2092.
65. Asano, T., Furuta, H., and Sumi, H. *J. Am. Chem. Soc.* **1994**, 116, 5545–5550.
66. Hara, K., Ito, N., and Kajimoto, O. *J. Phys. Chem.* **1997**, A 101, 2240–2244.
67. Chen, D. T.-L., and Morawetz, H. *Macromolecules*, **1976**, 9, 463.
68. Newell, A. C., and Moloney, J.V. *Nonlinear Optics*, Adison-Wesley (Cal., **1992**).
69. Cohen, M.H., and Turnbull, D. *J. Chem. Phys.* **1956**, 31, 1164.
70. Robertson, R. E. *Macromolecules* **1985**, 18, 953.
71. Paik, C. S., and Morawetz, H. *Macromolecules* **1972**, 5, 171
72. Ishitobi, H., Sekkat, Z., and Kawata, S. *Chem. Phys. Lett.* **2000**, 316, 578, Ishitobi, H., Sekkat, Z., and Kawata, S. *J. Am. Chem. Soc.* **2000**, 122, 12802.
73. Victor, J., and Torkelson, M. *Macromolecules* **1987**, 20, 2241–2250.

5

CHIRAL POLYMERS WITH PHOTOAFFECTED PHASE BEHAVIOR FOR OPTICAL DATA STORAGE

MIKHAIL V. KOZLOVSKY*,[†]
LEV M. BLINOV[†]
WOLFGANG HAASE*
*Institute of Physical Chemistry, Darmstadt University of Technology, D-64287, Darmstadt, Germany
[†]Institute of Crystallography, Russian Academy of Sciences 117333, Moscow, Russia

5.1 INTRODUCTION

During the decade of the 1990s, substantial interest has grown in photoaddressed polymer materials. These systems are characterized by their capability of reversible reorientation and/or photochemical transformations under the action of UV or visible light, followed by corresponding change in such optical properties as birefringence, transparency, or color. Therefore, they are widely investigated as optical data storage and photorecording media, coatings for contactless surface controlling, NLO active materials, and so on.[1,2,3]

Among these photosensitive polymer materials, foremost are the systems containing dichroic azobenzene moieties either as low molar mass components or as fragments of the macromolecule itself. The key physical process involved is the repetitive *E/Z*-isomerization of azobenzene groups, as shown next.

E-form (or *trans*-isomer)　　　　　　　　　　　　Z-form (or *cis*-isomer)

In this process, the *trans*-to-*cis* transformation is induced by UV light illumination, whereas the backward conversion at ambient temperature occurs spontaneously but can be accelerated by increasing the temperature or by illumination with yellow light.

The *E/Z*-isomerization process is characterized by angular-dependent excitation and leads, therefore, to the photoselection of a preferred azobezene dye orientation. In other words, the dichroic dye units choose an orientation where the electronic transition moment is perpendicular to the light electric vector. It promotes, in turn, the cooperative reorientation of neighboring moieties, which include other fragments of the macromolecule, such as the main chain or photochemically inactive comonomer units, and low molar mass additives. Thus, a macroscopic orientation of the sample arises, and it remains long after the illumination is stopped and all the dye moieties return to the thermodynamically equilibratory *E*-state.

To date, various possibilities for incorporating an azobenzene dye into a polymer system have been suggested and tested. First, solid dispersions of low molar mass azo dyes in amorphous polymers were studied.[4–8] Then, the amorphous copolymers containing chemically bound azobenzene fragments were thoroughly investigated.[9–16] Many attempts at combining the photochromism of azobenzene systems with liquid crystallinity have also been performed, probably because of the prominent role of liquid crystals in modern display technology. In the simplest way, azobenzene dye can be dissolved in an LC polymer matrix.[17–19] In addition, with polymer-stabilized liquid crystals (PSLCs), one can introduce orientation into a low molar mass liquid crystal through azobenzene-containing polymer networks,[20,21] while other authors, in an opposite manner, use LC polymer networks and low molar mass azo dyes.[22,23] Optical and electro-optical switching in polymer-dispersed liquid crystals (PDLCs) with azobenzene molecules has been reported as well.[24,25] It is worth noting as well of that Wu *et al.* have performed systematic studies on contactless photoalignment of liquid crystals by thin surface layers of azodye polymers.[26–28]

The most intensively developed materials, however, are the azobenzene-containing side chain LC polymers, which show a unique combination of liquid crystallinity and photochromic behavior in a single macromolecule, whereas copolymerization of comonomers with different functionality allows fine tuning of phase behavior, photo-optical properties, and other parameters to the requirements of a particular application. Starting from the first publications of Eich *et al.*,[29–31] work in that area has been done by several research groups.[32–42] As compared with amorphous polymers, the liquid crystalline systems possess an initial preorientation, which should be overcome during the photo-orientation process. Generally, the higher the degree of mesogenic group ordering, the lower the values of the photoinduced birefringence, because the orientational order in the LC state restricts the reorientation of the mesogenic side chains. That disadvantage can be easily overridden, however, by preliminary misalignment of the LC order using irradiation with non-polarized or circularly polarized UV light. On the other hand, higher ordering of the mesogenic and dye groups in the mesophase leads to higher glass temperatures and, potentially, to enhanced stability of the recorded data or images. Moreover, the competition between the thermodynamic equilibrium LC structure and the photo-optically induced one might result in a number of advanced and sophisticated photorecording techniques, as we will show later in this chapter.

There have also been reports on the preparation of polar materials by a photo-electro-poling technique that combines the optically induced quadrupolar depletion of chromophores in the direction of the light electric vector with an additional field-induced orientation of dipolar chromophores.[43–45] The latter allows the preparation of "cold electrets," which are interesting for nonlinear optical applications, such as optical harmonic generation, wave mixing, etc.[3]

Among the known LC phases used for photoaddressed polymer applications, chiral systems attract particular interest due to such features as selective reflection of light, circular polarization of the propagating/reflected light, and ferroelectricity. Thus, for example, Bobrovsky *et al.* have reported recently on the controlled change of a helical pitch and, hence, of a color of cholesteric photochromic copolymers,[46] and dual photoaddressing using light of different wavelengths in terpolymers containing both azobenzene and benzylidene-*p*-menthanone photochromic moieties.[47] Moreover, the photo-induced chromophore reorientation was shown to unwind totally the helical structure of cholesteric oligomers and polymers.[48]

We should note here that most of the publications consider the photo-orientation process and the birefringence it introduces to be the only effect of the light illumination onto dichroic LC polymers. Nevertheless, the elongated *E*-form and the bent *Z*-form of the same azo dye molecule generally should reveal quite different phase behaviors: The former favors formation of LC phases, but the latter can hardly accommodate to the mesophase. Hence, the illumination should affect substantially mesophase properties and phase transitions of dichroic LC polymers. Eich, Wendorff, Reck and Ringsdoff's early paper[29] discusses a large shift—up to 10 K—of the clearing point in a nematic azo dye copolymer under illumination. It is no surprise, therefore, that the UV illumination within the temperature gap between the clearing points of completely-*trans* and completely-*cis* isomers, i.e., isotropization temperatures in darkness and under illumination, causes the (isothermal) clearing transition, as reported by Hayashi *et al.*[49] for a high-ordered smectic phase of an azo dye polyacrylate. In that case, the temperature variation of the photo-induced phase transition does not exceed 6 K. However, Ikeda *et al.*[50] report an isothermal photoinduced N-Iso phase transition of photochromic side chain poly(meth)acrylates within a much broader temperature range (from 130°C down to room temperature), including the region below T_g. The backward transition to the nematic state occurs thermally in darkness, taking some seconds close to the clearing point but much longer at lower temperatures, with an activation energy of about 105 kJ/mol. That I–N transition can be accelerated, however, by illumination with visible light ($\lambda > 420$ nm), *cis*-to-*trans* transformation being the rate-determining step of the transition process. According to Ikeda *et al*,[50] the presence of electron-donating and -accepting groups in *p,p'*-positions of the azobenzene core increases the rate of the I–N transition, thus shortening the response time (good for switching applications) but reducing the stability (a disadvantage for photorecording applications).

It should be also mentioned that the conventional appearance of liquid crystals as a polydomain texture results in remarkable light scattering by the domain boundaries. Special efforts should be taken to produce large transparent (monodomain) samples of LC polymers, including shear flow, film stretching, thermal training, external electric or magnetic fields, etc. Such orientation procedures show poor reproducibility, however; they require time and sophisticated equipment. For that reason, it is usual that only thin films of azo dye LC polymers, in which the scattering can be neglected, are investigated as photochromic polymer materials. Among those are Langmuir-Blodgett films consisting of several molecular layers,[38–42,53–55] 0.05 to 0.5-μm-thick spin-coated films,[56–58] and 1 to 5-μm-thick cast films.[26–28,59]

For all these reasons, our recent discovery of chiral side chain polymers, which appear visually amorphous and optically isotropic but possess some hidden LC ordering, is of particular interest. The structure of that "isotropic smectic" (IsoSm*) phase is not completely determined yet; nevertheless, it was successfully used for the creation of novel azo dye copolymer materials that combine spontaneous transparency with mesomorphic (micro)structure, which incorporates photosensitive moieties. In the next section, we consider the structure and properties of the IsoSm* phase of chiral dichroic copolymers and discuss its application to two types of photorecording. One is related to the photoinduced birefringence in polymer films (holographic grating recording), whereas the other is based on the possibility of governing mesomorphic phase transitions in such polymers by light illumination (the light-controlled phase transition—LCPT—recording). Also, we will report an application of such films for contactless orientation of conventional (nematic) liquid crystals.

5.2 AMORPHOUS, OPTICALLY ISOTROPIC MESOPHASE OF CHIRAL SIDE-CHAIN POLYMERS WITH A HIDDEN LAYER STRUCTURE—THE "ISOTROPIC SMECTIC" PHASE

In 1991, Bata *et al*. reported an unusual phase behavior of a chiral side chain polymethacrylate, P8*M.[60]

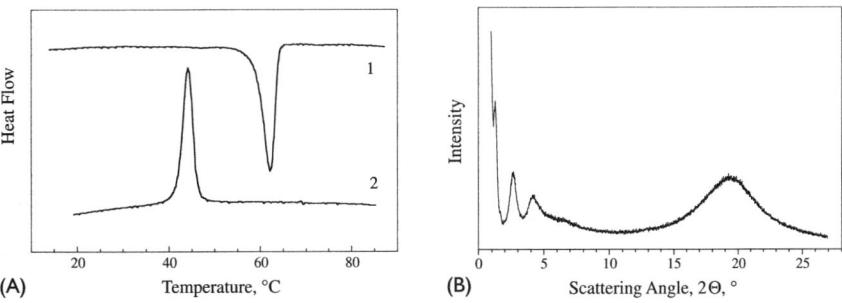

FIG. 5.1 Structural data for P8*M: (A) DSC traces on heating (*1*) and cooling (2); (B) X ray scattering profile below the phase transition (40°C).

This showed a single-phase transition with remarkable hysteresis: at 64°C on heating but at 44°C on cooling. The transition was detected by various experimental techniques including DSC, (see Figure 5.1A), X ray, and broadline NMR.[60,61]

As seen in Figure 5.1B, the polymer forms a layered structure below the transition point, showing a well-developed system of SAXS peaks. Moreover, the ORD measurements (see Figure 5.2) exhibit drastic growth of the optic rotation in the IsoSm* phase towards shorter wavelengths, in contrast to the proper isotropic melt of the same polymer above the transition point, suggesting some helix-like superstructural ordering in the mesophase with a pitch below 300 nm.[62] That hypothesis was further supported with CD, dielectric spectroscopy, and UV-vis spectroscopy data.[63,64] The critical importance of chirality for the formation of the IsoSm* phase is also confirmed by

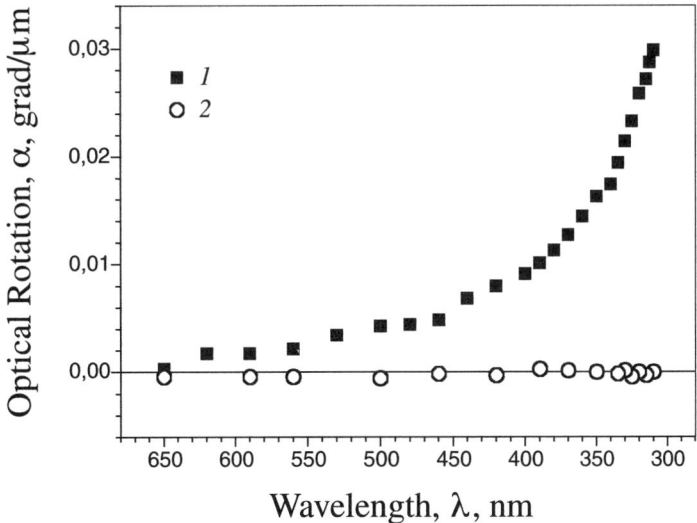

FIG. 5.2 ORD data from P8*M (an 18-μm film) below the phase transition at (*1*) 20°C and above the phase transition at (2) 80°C.

(A)   (B)

FIG. 5.3 Optical appearance of polymer films: (A) racemic polymer, P8$^\pm$M; (B) chiral polymer, P8*M.

the fact that the racemic isomer, P8$^\pm$M, forms only in the conventional Sm A phase.[62] On the other hand, the absence of a pyroelectric effect and spontaneous polarization in the chiral polymer excludes completely any tilted smectic structure.

At the same time, to our surprise, P8*M appears visually transparent and nonbirefringent, both in bulk samples and in thin films. This is illustrated by Figure 5.3, where the spot of chiral polymer can hardly be recognized between two glasses in contrast to its racemic isomer, P8$^\pm$M. We should stress here that it is impossible to induce any birefringence in P8*M using shear flow, electric field up to 15 V/μm, or magnetic field up to 2.5 T.

P8*M is not the only polymer forming the isotropic smectic phase. To date, we have observed formation of that phase for a half-dozen chiral polymethacrylates and polysiloxanes. Table 5.1 summarizes the chemical structure and phase behavior of synthesized side-chain homopolymers, which carry chirally substituted side chains derived from asymmetric esters of terephthalic acid and hydroquinone. Such a structure with alternating orientation of carboxylic link groups seems to favour the formation of the IsoSm* phase, whereas isomeric derivatives of p-hydroxybenzoic acid, where all carboxylic links have the same orientation, form only conventional Sm A and Sm C* phases.[65] Molar mass of all the synthesized homo- and copoly(meth)acrylates is within the range of 1 to 2·10^5 g·mol^{-1}; the polysiloxanes have the average degree of polymerization, $p \sim 35$.

It is worth noticing that the combination of properties of the IsoSm* phase can hardly be explained in terms of known LC phases. The ultrashort pitch TGB-like structure suggested in reference 64 still remains the only structural model that can explain the observed lack of birefringence in the IsoSm* phase.

The twist grain boundary (TGB) phases predicted by Renn and Lubensky[66,67] have been intensively studied in the few last years.[68–71] The general structure of the TGB phase is shown schematically in Figure 5.4. Because the symmetry of the Sm A phase does not allow continuous helical twisting, the chiral superstructure is realized in a stepwise manner: Small smectic grains rotate around a helical axis, while screw dislocations build the

TABLE 5.1   Chiral Homopolymers with Side Chains Based on Asymmetric Esters of Terephthalic Acid and Hydroquinone

Polymer[#]	Main chain	Side-chain structure	n	Phase transitions[†]
P5*A	Acrylate			Sm F* 77 Sm C* 97 Iso
P5*M	Methacrylate			gl 40 Sm C* 74 Sm A 85 Iso
P4*A	Acrylate		2	Sm B 50 Sm C* 88 Iso
P4*M	Methacrylate		2	gl 40 Sm C* 78 Iso
P6*M	Methacrylate		4	gl 30 IsoSm* 53 Iso
P6*ST	Siloxane		4	Sm B 33 Sm C* 51 Iso‡
P7*M	Methacrylate		5	gl 30 IsoSm* 57 Iso
P7*ST	Siloxane		5	gl 18 IsoSm* 54 Iso
P8*A	Acrylate		6	Sm B 55 Sm A 61 Iso
P8*M	Methacrylate		6	gl 30 IsoSm* 64 Iso
P8*ST	Siloxane		6	gl 25 IsoSm* 61 Iso
P8*S	Siloxane		6	gl 24 IsoSm* 47 Iso (metastable) or gl 24 Sm C* 43 Sm A 47 Iso (equilibrium)
PL4*A	Acrylate		4	gl 10 Iso
PL4*M	Methacrylate		4	gl 15 Iso
PL4*S	Siloxane		4	gl 10 Iso
PL6*S	Siloxane		6	gl 15 Iso

[#]The number in the polymer abbreviations corresponds to the total number of carbon atoms in the chiral terminal group, P stands for polymer, and the last letter relates to the type of polymer main chain. [†]In heating, [‡]For the sample annealed 24 h at 20°C.

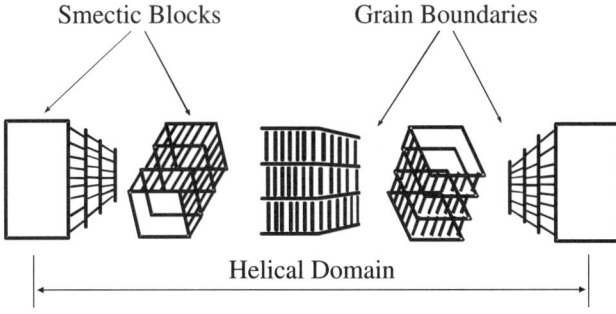

FIG. 5.4   Sketch of the TGB A* structure.

grain boundaries. Within the model, which we suggest for the particular case of the isotropic smectic phase, the small Sm A grains are ~10 nm in diameter, and the helical domains are about 250 to 300 nm long.[64] Thus, all the structural elements are smaller in dimension than the visible light wavelength, and the mesophase appears visually as optically isotropic, in contrast to known TGB A* phases.[72] In the near-UV range, the selective reflection from the TGB lattice cannot be observed, because of the strong absorption of phenyl benzoate groups. The phase still keeps the smectic ordering at the molecular dimensions.

We should state that our suggested model is not completely justified yet, but it remains the only one that can explain the whole ensemble of properties observed for the isotropic smectic phase of P8*M and related polymers. Keeping that in mind, we will refer hereafter to the mesophase as the IsoSm* phase, which most probably has the ultrashort pitch TGB A* structure.

5.3 PHOTOINDUCED BIREFRINGENCE IN PHOTOCHROMIC ISOSM* COPOLYMERS

Because of the combination of short-scale smectic ordering, transparency, and optical isotropy, the IsoSm* phase is promising for numerous optical and photo-optical applications. We therefore prepared three series of copoly-methacrylates combining the monomer matrix of P8*M with dichroic comonomers based on different azobenzene dyes, namely p-cyanoazobenzene (SK series), p-mehoxyazobenzene (KW series), and p-trifluoromethoxy-azobenzene (KM series).[73,74] Figure 5.5 presents the chemical stuctures of the comonomer units and the corresponding range of compositions, where the formation of the IsoSm* phase is observed. The copolymers are referred to as SKn, KWn, and KMn, respectively, where n is the concentration of azo dye comonomer, mol. %. For all three copolymer series, the corresponding phase diagram can be divided in three sections. At lower n values, the copolymers form an equilibrium IsoSm* phase, whereas at higher dye concentration they form a conventional Sm A phase. The intermediate range of concentrations corresponds to the metastable formation of the IsoSm* phase, which is formed at fast cooling ($\geq$ 15 K/min), whereas slow cooling results in formation of an Sm A phase. Thus, a copolymer can appear at ambient temperature

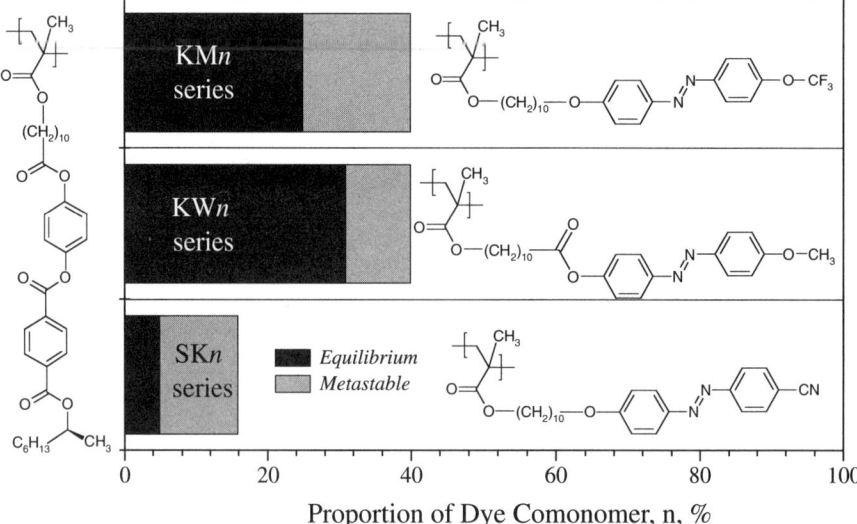

FIG. 5.5 Range of compositions favoring the formation of the IsoSm* phase for chiral azo dye side-chain copolymers.

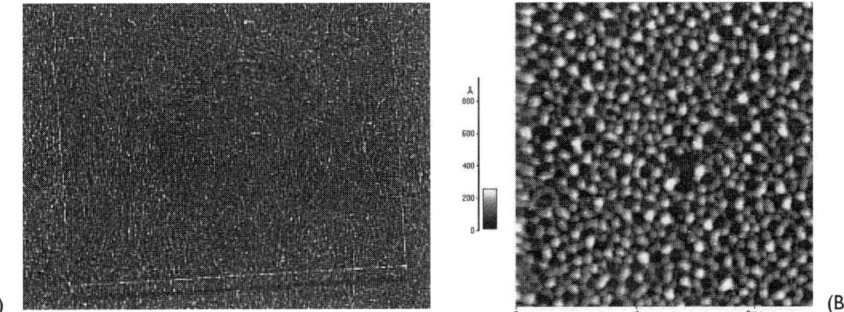

FIG. 5.6 (A) Optical appearance of an azo dye IsoSm* film (KW11); (B) surface relief of an azo dye IsoSm* copolymer (KW40, cast film).

either as high-transparent, optically isotropic IsoSm* glass, or as turbid, scattering, and birefringent Sm A glass, depending on sample prehistory. We should underline here that particularly in that intermediate range of copolymer proportions, the phase transformations can be controlled by light illumination during cooling, thus allowing the LCPT recording (see Section 5.6).

As seen from Figure 5.5, the chiral dye copolymers form the IsoSm* phase in a broad range of compositions. The phase appears orange in colour but transparent, amorphous, and elastic (Figure 5.6A), in contrast to the turbid, birefringent, and fragile polydomain texture of Sm A copolymers with a higher proportion of dye moieties. At the same time, the AFM profile of a free surface of a photochromic IsoSm* copolymer[75] shows a grain texture with the periodicity of 250 nm (Figure 5.6B), thus giving more evidence for the suggested short-pitch TGB structure of the mesophase.

It should be noted that the Iso → IsoSm* phase transition on cooling is thermodynamically controlled for the former two series but kinetically controlled for the KM series.[74] As an example, Figure 5.7A presents DSC data from KM11. As seen from the figure, the second DSC scan from the polymer sample shows no transition peaks, but glass transition inflexion only; the

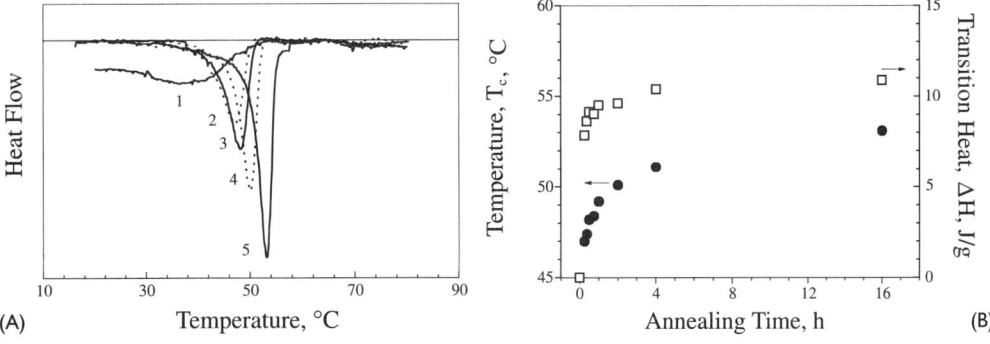

FIG. 5.7 (A) DSC curves from KM11: (1) second scan (no annealing), (2–5) sample annealed at 32°C for 0.25 h, 0.5 h, 2 h, and 16 h, respectively; (B) corresponding growth of the transition temperatures (●) and enthalpies (□) with the annealing time.

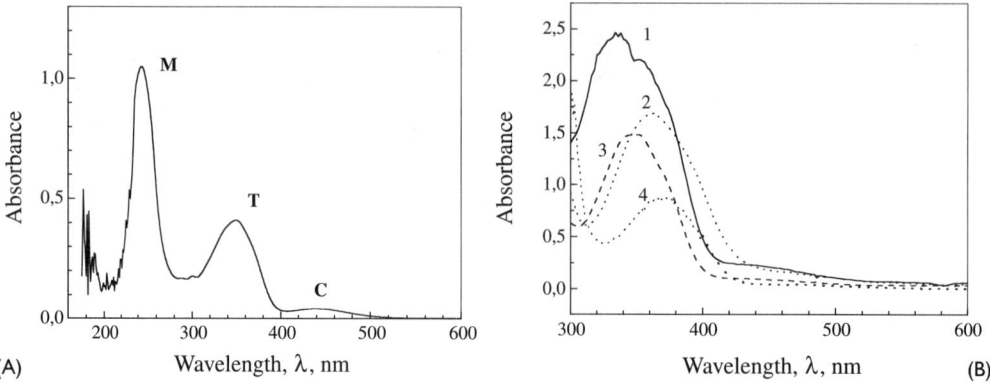

FIG. 5.8 Absorption spectra of the photochromic IsoSm* copolymers: (A) KW40 (chloroform solution, 12 mg/l); (B) spin-coated films of (*1*) KM55, (2) SK28, (3) KW40, and (4) SK5.

transition peak can, however, be recovered with sample annealing. The corresponding growth in the transition temperatures and enthalpies with annealing time is shown in Figure 5.7B. It is clearly seen that the transition is completed within about one day.

The copolymers manifest three characteristic spectral bands in UV-vis absorption spectra (see Figure 5.8A): Band **M** at about 240 nm corresponds to the absorption of mesogenic phenyl benzoate groups; band **T** at about 330 to 370 nm (for different copolymer series, see Figure 5.8B) relates to the π-π* transition in the *trans*-isomer of the azo chromophore; and less developed band **C** at 440 to 460 nm to the n-π* transition.[76–78] We should comment here that for individual molecules, band C is allowed only for the bent *cis*-isomers, the number of which is very small under ambient conditions. On the other hand, the selection rule might be cancelled by the interaction of the chromophores with the surrounding medium, and the elongated *trans*-isomers might be allowed to participate in the band C absorption. Our study shows, for instance, that for SK8, the n-π*oscillator strength of a *cis*-isomer (allowed by symmetry) in a thin film is only two times higher than that of the *trans*-isomer (forbidden in the simplest model).[77]

Furthermore, the copolymers show pronounced photochromism, similar to other azobenzene derivatives: UV irradiation transforms most of the azobenzene units from the *trans*-isomer to the *cis*-isomer, so that the band **T** disappears but the band **C** increases considerably (see Figure 5.9A), whereas the backward isomerization occurs thermally in darkness but can be accelerated by yellow light irradiation (see Figure 5.9B). Once again, we should note that a strong illumination by green light (within the band C) can result in the *trans-cis* transformation because of the n-π* transition, as considered in the previous paragraph. The process is not as efficient as the π-π* transition under UV irradiation but, nevertheless, it shifts slightly the equilibrium ratio between the isomers in favour of *cis*-isomers.

We should underline here one peculiarity of the chiral photochromic copolymers under discussion: The *cis*-isomers have an extremely long lifetime

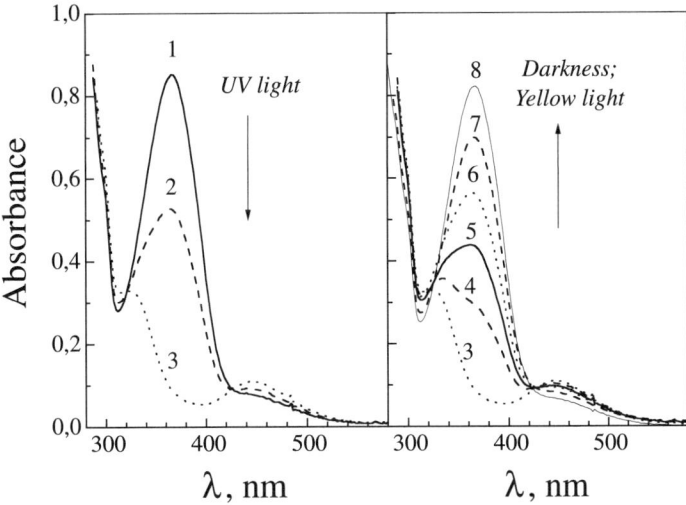

FIG. 5.9 (A) Changes in the UV absorption spectra of copolymer films under UV irradiation. (B) Dark relaxation and yellow light illumination afterwards (1) virgin film; (2) UV illumination for 5 s; (3) the same for 30 and 60 s; (4–7) dark relaxation for 3 min, 6 min, 9 min, and 24 h, respectively; (8) illumination with yellow light, 30 s.

as compared with other photochromic materials based on azobenzene derivatives, including low molar mass chiral dyes and achiral LC polymers (see Table 5.2). This opens up the possibility of preparing thin polymer films (both cast and spin-coated) in their *cis*-form, from UV-irradiated solutions. After evaporation of the solvent, the films keep the acquired *cis*-form for hours and can be further used for photorecording purposes.[77]

As mentioned in Section 5.1, the reorientation of azobenzene photochromic moieties under illumination with polarized light and the birefringence that is induced from it are well known and covered in many publications.

TABLE 5.2   Lifetime of cis-Forms for Various Azobenzene Derivatives

Azo-dye compound	Temperature, °C	Relaxation time, τ	Reference
Chiral low molar mass dye			
	25	~ 3 min	79
Achiral LC polymer			
	50	~ 2. 5 min	52
Chiral LC copolymer			
SK8	20	2.5 h	76, 77
KW40	20	10 h	76, 78, 80

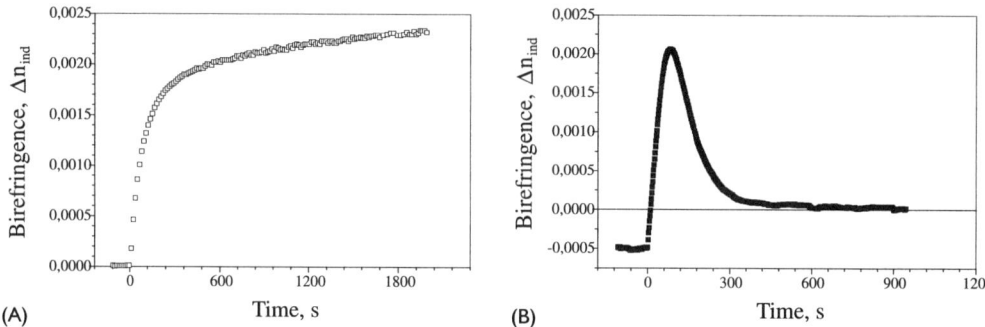

FIG. 5.10 (A) Evolution of the birefringence, Δn_{ind}, induced by polarized blue-green light (400 < λ < 500 nm) in a 19- μm-thick KW40 film; (B) the same, in a 3-μm-thick film, induced by UV light (initial birefringence of the film induced by preliminary illumination with polarized blue-green light).

The value of the photoinduced birefringence, Δn_{ind}, the kinetics of its growth under illumination, and its stability thereafter depend on many factors, including the chemical structure of the azobenzene dye and of the polymer matrix, light intensity, wavelength, temperature, film thickness, etc. As an example, Figure 5.10A presents the growth of the Δn_{ind} values with illumination time in a 19 μm thick pressed film of KW40.[81]

The plot presented in Figure 5.10B gives evidence that polarized UV light, in contrast to visible light, induces birefringence in azo dye copolymer films only at the initial step of the irradiation, whereas further illumination results in disappearance of the phase retardation. This effect can be explained by the complete and reversible reorientation of both chromophores and mesogenic units along the light wave vector, as discussed by Han *et al.*[78,82]

The effect of illumination wavelength (within the visible range) on the photoinduced birefringence is shown in more detail in Figure 5.11. As seen in the figure, when cutting off the wavelength range below λ ~ 500 nm, one cannot induce any further birefringence. On the other hand, the blue-violet light (λ < 420 nm) seems to have a negligible effect on the Δn value (see Figure 5.11B). We can conclude, therefore, that the irradiation within absorption band C (Figure 5.8) is responsible for the long-term photoinduced birefringence, as shown in Figure 5.10A.

The dependence of photoinduced birefringence on light intensity, as presented in Figure 5.12, shows saturation at 150 to 200 mW/cm². It should be noted, however, that the Δn_{ind} values saturate versus the irradiation power but not versus the illumination time (Figure 5.10A).

Of considerable importance is also the fact that nonpolarized UV irradiation can erase completely the previously induced birefringence, newly writing another pattern. As seen in Figure 5.13, irradiation of a UV-treated copolymer film by *p*-polarized visible light results in a negative phase delay, a subsequent irradiation with nonpolarized UV light eliminates birefringence, and subsequent irradiation by *s*-polarized visible light induces positive birefringence.

When comparing the data of Figure 5.13 with those from Figure 5.10A for the same film thickness (curve 2), it is clearly seen that the preliminary

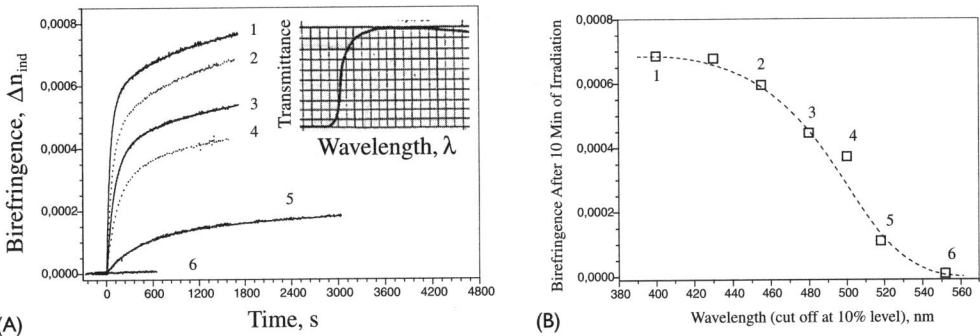

FIG. 5.11 (A) Evolution of the birefringence, Δn_{ind}, induced by polarized light in a copolymer film (KW07, 23 μm) through color filters with different transmission ranges. Typical transmission curve of a filter shown as insert; (B) The Δn value induced for 10 min, versus the wavelength corresponding to the filter transmission 10%. Point numbers correspond to the curves in Figure 5.11A.

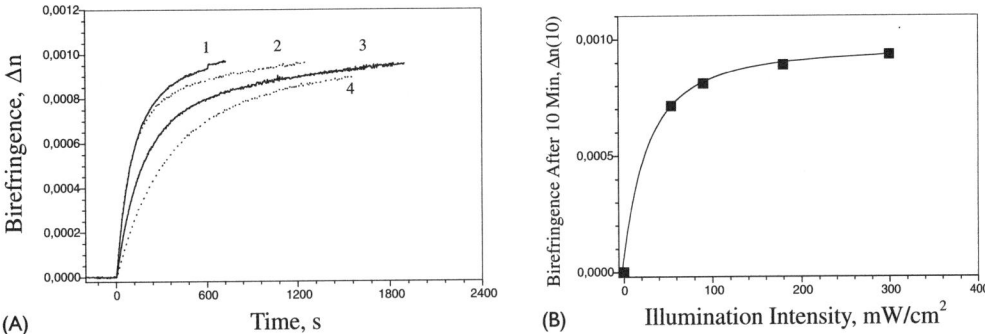

FIG. 5.12 (A) Evolution of the birefringence, Δn_{ind}, induced by polarized white light in a copolymer film (KW19, 22 μm) without filters (*1*) or through grey filters of different optical density (*2–4*); (B) the Δn value induced for 10 min, versus the light intensity.

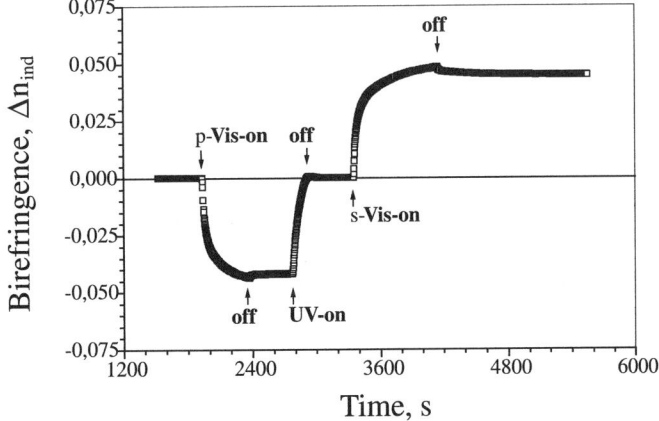

FIG. 5.13 Evolution of the induced birefringence in a dye copolymer film (KW40, 3 μm) under consequent irradiation with *s*-polarized blue-green light, UV non-polarized light, and *p*-polarized blue-green light.

irradiation with nonpolarized UV light increases drastically the birefringence induced by polarized visible light: $\Delta n_{ind} \approx \pm 0.045$ for the illuminated film versus $\Delta n_{ind} \approx \pm 0.001$ for that reached without UV treatment. This gives rise to the possibility of amplifying the sensitivity of the material to visible light by previous exposure to UV light. To some extent, the phenomenon is similar to the "nonvolatile holographic storage effect" observed recently in a doubly doped inorganic ferroelectric.[83] Our data show, however, that the UV sensibilization is much more pronounced for the copolymers of the KW series[78–80,84] than for the SK and KM series.[77,81] That can be explained by the stronger electron-accepting substituents at the tail of the azobenzene chromophores for the latter two series, –CN for SK copolymers and $-OCF_3$ for KM copolymers as compared with $-OCH_3$ for KW copolymers, which lead to reduced stability and shorter lifetimes for the *cis*-isomers, as reported by Ikeda et al.[50–52]

There is one more possibility for substantially increasing the stability of photoinduced birefringence in chiral LC copolymers. If the light-induced reorientation of photochromic azo groups occurs not in a well-formed (mature) mesophase, but during phase formation, it is less restricted by main-chain conformation, so that higher Δn values can be achieved. Moreover, corresponding structural reorganization should involve the polymer main chain itself. As a result, its conformation "memorizes" the preferred orientation of chromophores and the thereby-induced cooperative reorientation of colorless mesogenic moieties in glass, and the copolymer film keeps the macroscopic orientation for much longer times.[85]

Table 5.3 compares Δn values recorded in a 22-µm film of KW19 after 20 min of irradiation with polarized white light from a polarizing microscope for isothermal recording at 26°C and recording during sample cooling from 66 to 26°C. It is clearly seen that the thermally assisted recording is twice as efficient and much more stable. To illustrate the stability of such an "imprinted" birefringence (i.e., photoinduced during film cooling), Figure 5.14 presents a 2.5-year-old birefringent spot, induced in the copolymer film during microscopic observations.

The kinetically determined Iso–TGB A* phase transition in copolymers of the KM series (shown in Figure 5.7) allows also for isothermal nonequilibrium photorecording in the copolymers. Figure 5.15 shows the evolution of photoinduced birefringence in a film just cooled from 100°C, where the

TABLE 5.3 Comparison of the Isothermally Recorded and "Imprinted" Birefringence in Dye Copolymer Film (KW19, 22-µm Film)

Birefringence, Δn	Isothermal recording (20 min at 26°C)	Thermally assisted recording (20 min as cooled from 66°C to 26°C)
Immediately after the recording	0.0008	0.0020
18 h later	0.0003	0.0019

FIG. 5.14 Microphotograph of a 2.5-year-old photorecorded birefringent spot (left) in ~20-μm-thick film of SK8.

mesophase is still being formed with time, as compared with the birefringence induced under the same conditions in a film kept for 24 h at ambient temperature, where the short-pitch TGB A* phase is already organized completely. As seen from the figure, the illumination during the mesophase formation results in Δn_{ind} values that are three times higher.

We should emphasize that by using thick but still transparent IsoSm* films, we can obtain huge phase delay even for moderate Δn_{ind} values. For instance, the maximum value of the photoinduced birefringence shown in Figure 5.15 (curve 1) was measured for a 24-μm-thick film and corresponds to the phase delay value $\Delta\Phi = -245°$, i.e., close to the three-quarter-wave plate.

To summarize, photochromic chiral IsoSm* copolymers are capable of photoreorientation under illumination with polarized light. If irradiated within the UV wavelength range (band T), the Δn_{ind} value achieves maximum

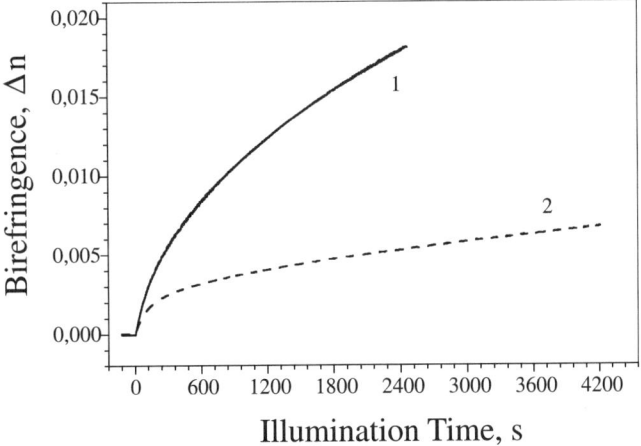

FIG. 5.15 Growth of the photoinduced birefringence in (1) just-cooled and (2) 1-day-old copolymer films (KM25, 23 μm) under light illumination (nonfiltered white light, 0.3 W/cm²).

within 1 to 2 minutes but disappears if the irradiation is continued further. On the other hand, the birefringence induced by blue-green light (band C) shows no saturation, due to repeated *cis-trans-cis* transformations. The sensitivity of films to visible light can be substantially increased by preillumination with nonpolarized UV light. Photorecording during the phase formation (on cooling, or also during the kinetically determined isothermal mesophase formation for copolymers of the KM series) results in much higher Δn_{ind} values as compared with photorecording in a completely organized mesophase.

The combination of these features presents photochromic IsoSm* films as interesting and diverse media for optical data storage. In Section 5.4, we consider holographic grating recording in copolymer films and related (combined) photorecording techniques.

5.4 HOLOGRAPHIC GRATING RECORDING

Optical holography provides unique opportunities for information storage and visualization. To date, most investigations have focused on doped inorganic crystals,[86] the polymer organic materials attract more and more attention as holographic media, however, because of easy processing and "custom-tailored" adjustment of properties with well-developed polymer technology techniques. Among those, amorphous and liquid crystalline azobenzene side-chain polymers[9–42] are distinguished for high diffraction efficiency, easy erasing/rewriting of recorded data, and reasonably short response time.

Generally, there are three mechanisms for writing holograms in azo-dye polymers: (1) (rotational) molecular reorientation; (2) redistribution of *cis*- and *trans*-isomer population; and (3) formation of surface relief grating.[56,87] The first two mechanisms represent bulk effects, and the latter is related to the surface. The contribution of these mechanisms into the total diffraction efficiency depends on many parameters, including the extinction coefficient at a particular wavelength of a laser beam, the quantum yield of the photochemical isomerization, the temperature-dependent relaxation time of the *cis*-isomers, the angular diffusion, or even the film confinement conditions. To elucidate the mechanism of photorecording, and to optimize the conditions for readout of recorded data, different polarization geometries of the interfering pump beams can be chosen.[84,87]

Holographic grating recording in the chiral photochromic copolymers of the SK-, KW-, and KM-series has been studied for several polarization configurations. Figure 5.16A shows the experimental setup for the grating recording, and the light-field interference patterns for *s,s*-, *p,p*-, *s,p*-, and *R,L*-configurations of laser beams are presented in Figure 5.16B. Here, *s*- and *p*-polarizations correspond to linear polarization with the electric vector of the incident beam perpendicular and parallel to the plane of light incidence, whereas the *R*- and *L*-polarizations correspond to right- and left-handed circularly polarized light. The experimental scheme of Figure 5.16A allows for all four types of recording with the proper choice of waveplates, WP. Some examples of holographic gratings recorded in an IsoSm* copolymer film are shown in Figure 5.17.

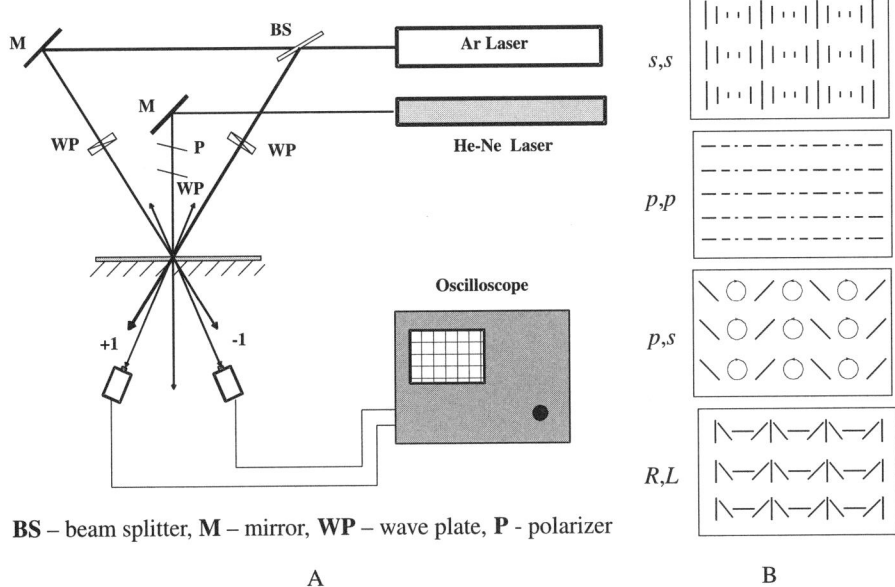

BS – beam splitter, M – mirror, WP – wave plate, P - polarizer

FIG. 5.16 (A) Experimental setup for holographic grating recording; (B) schematic representation of electric field patterns on a vertically installed sample.

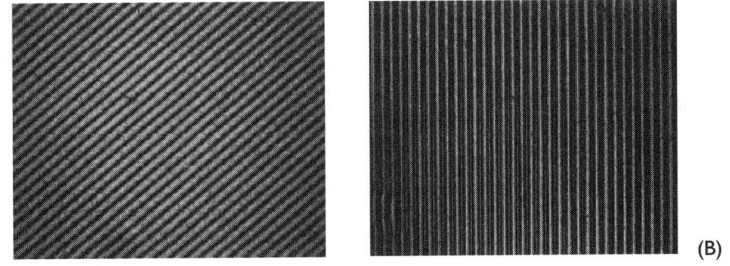

FIG. 5.17 Microphotographs of an (A) s,s-grating and a (B) s,p-grating from SK8 (22-µm pressed film) in crossed polarizers.

The quality of holographic grating is estimated by the efficiency of the probe beam scattering from the grating, calculated as the ratio of light intensities for the first diffraction order and for the incident beam, $\eta = I_1/I_0$. The He-Ne laser source is usually selected for the probe beam, because it is inactive in the photochemical processes involved. The η value depends substantially on the wavelength and intensity of the recording beam, the recording time, and the prehistory of the sample.

It has been shown that the scattering efficiency of the grating recorded in a 23 µm thick virgin film of SK8 grows linearly with the exposure time at constant beam power but falls drastically with the wavelength of the recording beam.[77] In contrast, the η value shows a superlinear growth upon

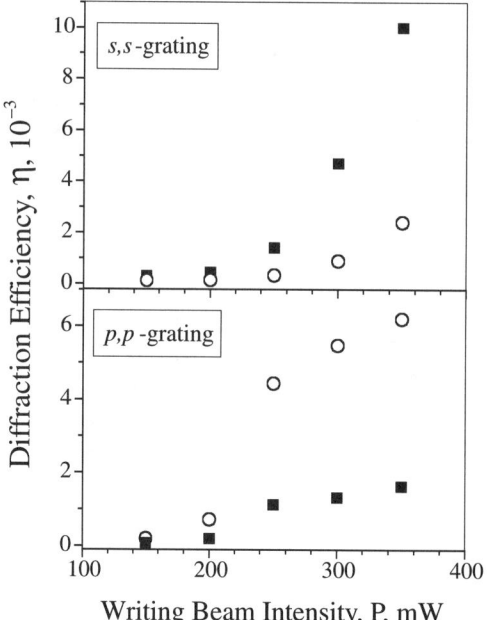

FIG. 5.18 First-order s- and p-diffraction efficiencies (■ and ○, correspondingly) of s,s- and p,p-gratings recorded with a fixed exposure time, t_w = 15 s, as a function of the power of the initial s-polarized (top) and p-polarized (bottom) Ar laser beam (for the ∅ 1.6-mm spot).

increase in the beam intensity, for a fixed exposure time, as shown in Figure 5.18. The superlinearity is especially pronounced at higher writing powers. The data presented in the figure show unambiguously that in the cases of both s,s- and p,p-gratings, the diffraction efficiency of the probe beam is higher for polarization coinciding with that of the writing beams. On first sight this looks strange because, in uniaxially symmetric materials, the difference $n_{\parallel} - n_{iso}$ is always larger than $n_{\perp} - n_{iso}$ (n_{iso}, $n_{\parallel}$, and $n_{\perp}$ are refraction indices for the nonoriented phase, and parallel and perpendicular to the director, respectively), and if the chromophores are deviated outwards of the light electric vector into the perpendicular position, the difference δn between illuminated and dark parts would be greater for the perpendicular polarization. However, a more precise analysis of a simple three-dimensional model[88] shows that what we observe here is a typical situation when the angular distribution of chromophores in darkness is close to isotropic (the case of the chiral optically isotropic TGB A* phase, IsoSm*).

At the same time, only low-contrast s,s- and p,p-gratings could be written at λ = 514 nm in a virgin 3-µm-thick film of another copolymer, KW40, using a power of the output Ar laser beam as high as 25 W/cm^2 (writing time t_w=2s).[78] However, as was mentioned in Section 5.3, pretreatment with unpolarized UV light increases the proportion of long-living cis-isomers and amplifies therewith the photo-optical response, the effect being especially pronounced for the copolymers of KW series with lower longitudinal dipole moment of the photochromic groups. Thus, film sensibilization with UV pre-

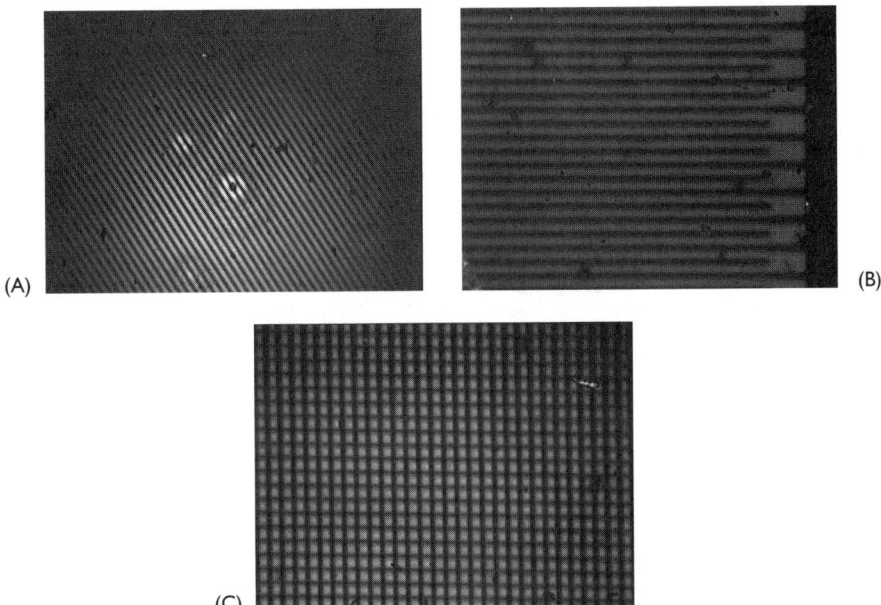

FIG. 5.19 (A) Image of a holographic grating recorded on a UV-sensibilized KW40 film (grating period is 38 μm). (B) Image of a mask recorded by unpolarized UV light and developed by polarized visible light (grating period is 40 μm). (C) Holographic development (grating period is 38 μm) of an image of mask with grating period 40 μm.

treatment (2 mW/cm^2 at $300 < \lambda < 400$ nm) gives a 100-fold gain in the exposure for holographic gratings, which can easily be recorded with a power density of less than 0.5 W/cm^2 and t_w=1s. An image of such a grating is shown in Figure 5.19, and Figure 5.20 (left) shows the diffraction efficiency, η, from the s,s- gratings versus writing time. The maximum diffraction efficiency is reached within 3s, $\eta_{max} = 0.27\%$. In contrast to the efficiency reported previously for SK8 films, the diffraction efficiency is higher for the orthogonal polarization of the recording and probe beams. This might be explained by a different mechanism responsible for the modulation of the birefringence index in the gratings for the two copolymers—namely, enrichment and depletion of the material with cis-isomers should play the key role for KW40, but the photoinduced reorientation of chromophores, similar to the mechanism reported for Langmuir-Blodgett films of low molar mass azo dyes, plays the key role for SK8.[88]

As the proportion of cis-isomers relaxes thermally to the equilibrium state, the enhanced sensitivity of the film to the writing beam degrades completely, for about 10 hours at ambient temperature (the lifetime of cis-isomers). The desensibilization process may be accelerated by additional exposure of the film to the white light of an incandescent lamp, as shown in Figure 5.20, right (power density 2 mW/cm^2 in the range 450 to 650nm): A 30 min irradiation is sufficient to suppress the sensitivity by converting all cis-isomers into their $trans$-counterparts. It is of great importance, nevertheless,

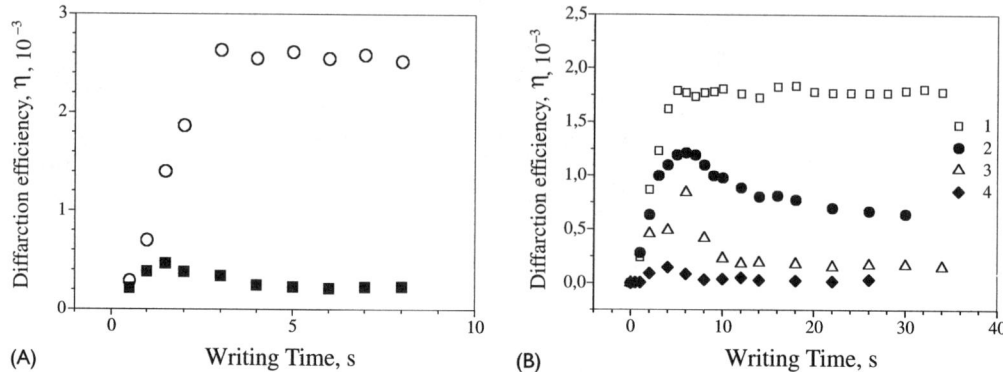

FIG. 5.20 (A) Diffraction efficiencies, η, of *s,s*-gratings in the 10 μm thick KW40 film, as read out by an *s*- (■) or *p*- (○) polarized probe beam, versus writing time. (B) Grating diffraction efficiencies, η_p, for the *p*-polarized reading beam, as functions of the writing time. After UV irradiation (t_{exp} =0), the film was exposed additionally to the light of an incandescent lamp, and gratings were recorded at different exposition times, t_{exp}, shown at the curves.

that a grating, once recorded, remain stable for a much longer time (several weeks under ambient conditions) even when all the *cis*-isomers have already converted into the *trans*-form. And moreover, under the strong additional illumination by visible light that eliminates *cis*-isomers, the grating remains stable.

The UV-sensibilization of copolymer films also allows the possibility of writing hidden images as sensibilized patterns in an "inert" (much less sensitive) area by UV illumination through a mask. The film remains visually isotropic and uniform, but the image can be developed later by illumination with polarized visible light. As an example, Figure 5.19B presents the pattern recorded with unpolarized UV light (300 < λ < 400 nm, exposition 10 min, power density 3.3 mW/cm²) and developed with polarized blue-green light (400 < λ < 500 nm, power density 5 mW/cm²). Moreover, if the developing illumination with visible light is not uniform, the resulting image represents the overlap of recording and developing patterns, as shown in Figure 5.19C.

We should point out that photochromic IsoSm* films are even more sensitive to holographic grating recording with circularly polarized light (*R,L*-gratings), as compared with *s,s*-, *p,p*-, or *s,p*-gratings recorded with linearly polarized light (see Figure 5.16B for the corresponding interference patterns). The interference of two coherent beams having opposite circular polarizations creates no intensity modulation, but only a rotation of the linear polarization in the interference plane;[87] a comprehensive theoretical consideration is given by Cipparone *et al.*[84] Thus, virgin spin-coated ~0.25-μm-thick films of KW40 allow for recording a phase grating at a beam power density as low as 0.5 W/cm², as compared with the 25 W/cm² required for a *p,p*-grating in the same polymer (without the sensibilization). The gratings are unstable, however, and degrade within several minutes when recorded under low beam intensity. As an example, Figure 5.21 illustrates the decay of diffraction intensity from an *R,L*-grating recorded in a spin-coated film of

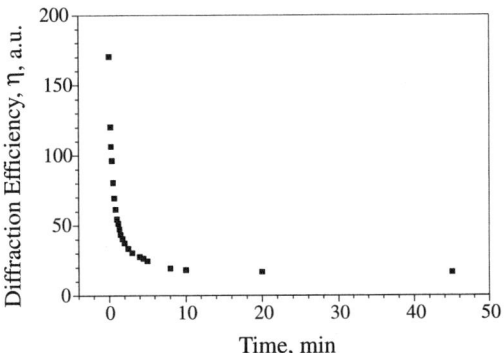

FIG. 5.21 Relaxation of the phase grating (SK8; ~ 0.25-μm spin-coated virgin film; writing beam 0.5 W/cm² at λ = 514 nm for 5 min) after switching off the pump beam.

SK5. Two relaxation domains, the fast one with $\tau_1 \sim 10^1$ min and the slow one with $\tau_2 \sim 10^3$ min, can be recognized in the figure.

On the other hand, images recorded with higher power density or in UV-sensibilized IsoSm* films (see Figure 5.17 and 5.19) are stable under ambient conditions for at least months. It is possible to erase them, however. Heating the polymer to above the phase transition point (~60 to 65°C) and illumination with unpolarized UV light both erase any recording and make the same spot ready for new recording (as shown in Figure 5.10B). We made several such cycles on the same spot without any trace of material degradation. This is consistent with observations that polymer films with azo compounds can survive thousands of *trans-cis-trans* photoisomerisation cycles.[89] On the other hand, a grating may be erased using a single Ar laser beam providing a spatially uniform illumination. An example is shown in Figure 5.22. The grating in SK8 film was recorded with two beams of initial nonsplit beam power at 15.4 W/cm² and for t_w = 15 s and erased with a single beam of 6.2 W/cm². Almost complete erasure was achieved at ~ 50 s.

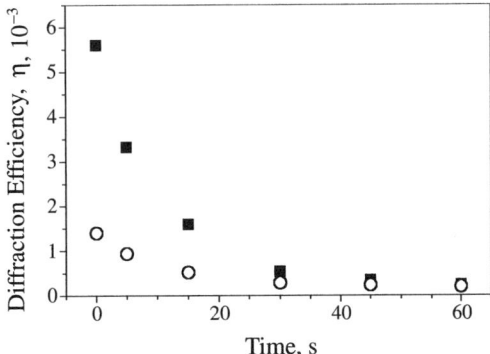

FIG. 5.22 First-order s- and p-diffraction efficiencies (□ and ○, correspondingly) of the s,s- grating written in SK8 (23-μm film), as functions of the "erasing time."

We conclude this section with the remark that high-transparent, optically isotropic, chiral and photosensitive IsoSm* copolymers represent novel and versatile media for holographic grating recording, allowing UV sensibilization, recording of both intensity and phase gratings, and holographic development of hidden sensibilization patterns.

5.5 PHOTOINDUCED ALIGNMENT OF LOW MOLAR MASS LIQUID CRYSTALS

In recent years, many studies have been devoted to the photoinduced alignment of nematic liquid crystals by photosensitive polymers. Advantages of this technique for optoelectronic technology are quite evident in comparison with such traditional methods as mechanical buffing of an orienting polymer layer: It is a "clean" technique that may be applied not only to flat substrates, but also to curved ones or even those not accessible to mechanical contact at all. In addition, buffing is irreversible, but photoinduced liquid crystal alignment may be changed optically. It may also be combined with buffing.

The most popular photoalignment technique is based on photo-crosslinking various poly(vinylcinnamates), PVC.[90,91] The initially optically isotropic PVC layer, which provides a degenerate planar alignment of a nematic liquid crystal (NLC), becomes optically anisotropic due to a lack of chromophores in a direction selected by polarized UV light and orients the NLC *perpendicular to the light electric vector*. The same alignment occurs when rod-like azo dyes are reoriented by light action.[92] On the other hand, NLC alignment *parallel to the light electric vector* may be achieved using coumarin dye layers.[93,94] This case is more interesting because, for reasons of symmetry, it is possible to produce an orientation of nematics with an arbitrary tilt angle using an oblique incidence of the exciting light. In both cases mentioned in this paragraph, only two different orientations of a liquid crystal are possible: degenerate planar (initial), and uniaxial either along or perpendicular to the light vector.

We tested the azo dye IsoSm* copolymer, KW40, as a photo-orienting layer for low molar mass nematic liquid crystals MBBA, 5CB, and E7. The studies were done using hybrid liquid crystal cells made of 1 mm thick ITO covered glass plates transparent in the range of $\lambda > 350$nm. Hybrid cells are the most convenient cells for anchoring energy determination because, in the high field limit, the field-induced distortion is easily modelled by the semi-infinite medium.[95] The homeotropic orientation was made using a surfactant chromium distearyl chloride (chromolane). On the opposite plates, spin-coated films of KW40 of different thicknesses (100 to 600 nm) were deposited from a 1 to 2% cyclohexanon solution of the copolymer. The films were irradiated by normally incident polarized UV or visible light before filling the cell with a liquid crystal.

The initial texture of a hybrid cell filled with E7 is seen outside the dark spot in Figure 5.23. It is a typical texture of a hybrid cell with *degenerate planar anchoring*. Such texture was also observed for MBBA and 5CB. After strong irradiation with UV light (either polarized or unpolarized; it does not matter) corresponding to the end of the curve shown in Figure 5.10B, a

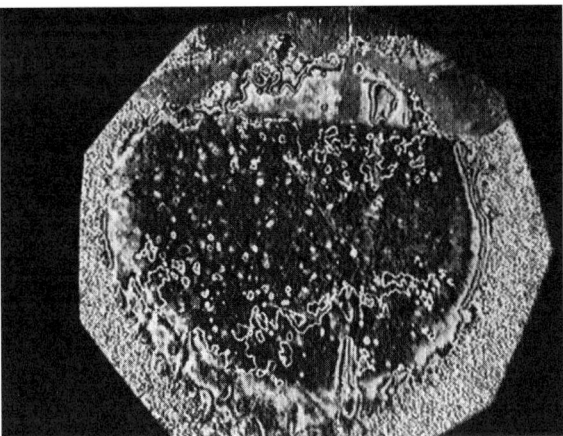

FIG. 5.23 Microphotograph of a 3.3-mm-diameter spot (dark on the bright background) of NLC E7 oriented with the director along the electric vector of exciting UV light. Hybrid cell, 11.5 μm, the KW40 layer was irradiated for 7 s with polarized UV light (160 mW/cm^2, λ = 300-400 nm) before filling the cell.

homeotropic texture is observed. If however, the irradiation of the copolymer film by polarized UV light is stopped in a few seconds (around the maximum shown in Figure 5.10B), after filling the cell a *homogeneous* orientation is observed with the *director oriented parallel to the UV light electric vector*. It is seen between crossed polarizers as a dark spot on the bright background of the degenerate planar texture (see Figure 5.23). There are two types of defects observed for E7, 5CB, and MBBA. One of them is typical of hybrid cells without a pretilt angle at the planar interface: Different domains having opposite splay-bend curvatures are separated by disclination lines. The other type, similar to long brushes, originates from the considerable solubility of our copolymer in liquid crystals: Some material is washed out while the cell is being filled.

Finally, a *homogeneous* orientation is observed with the *director oriented perpendicular to the light electric vector* if the layer was pretreated with strong UV light and additionally irradiated by *s*- or *p*-polarized visible light. This case corresponds to the plateau regions (either positive or negative) visible in Figure 5.13. The textures are similar to those shown in Figure 5.6.

Because the homogeneous orientation with the director oriented parallel to the light electric vector is the more rare and interesting, the anchoring energy measurements were carried out for this latter case. Figure 5.24 shows the experimental optical path difference, $R = <\Delta n>d$, as a function of the following electric coherence length:

$$\xi = \frac{1}{E}\left(\frac{4\pi K}{\varepsilon_a}\right)^{1/2} \quad (5.1)$$

In this equation, ε_a is the dielectric anisotropy, $K = (K_{11}+K_{33})/2$ is an average splay-bend elastic modulus, and E is the applied electric field. In this presentation, the curve must be linear when the Rapini approximation for the

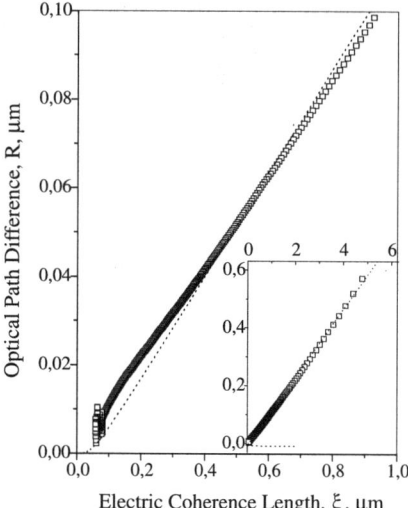

FIG. 5.24 Optical path difference, R, versus the coherence length, ξ, for a 23-μm-thick hybrid cell filled with E7 [homeotropic boundary: chromolane; planar boundary: KW40 irradiated for 3 s by polarized UV light of intensity 160 mW/cm^2 (λ = 300 – 400 nm) before filling cell.] Insert : Large scale for ξ, ξ = 0 – 6 μm. Main plot: Strong field region, ξ = 0 – 1 μm.

anchoring energy is used and the flexoelectricity is not taken into account.[96] Actually, the curve for a rather large range of ξ seems to be linear (see the insert in Figure 5.24).

The more careful fitting of the $R(\xi)$ curve to the two-dimensional theoretical model[95] is shown by the dotted curve in the insert. All measurements were made at temperature 29 to 30°C and the following parameters of E7 were used for calculations: refraction indices $n_{\parallel}$ = 1.7238, $n_{\perp}$ = 1.5185; dielectric susceptibilities $\varepsilon_{\parallel}$ = 18.6, $\varepsilon_{\perp}$ = 5.1; elastic moduli K_{11} = 1.04·10^{-6} dyn, K_{33} = 1.69·10^{-6} dyn. The curve corresponds to anchoring energy $W \approx 1$ erg/cm^2 and the apparent pretilt angle ϑ_{app} = 7°. For the normal incidence of UV irradiation, the pretilt angle should not appear. In fact, a finite ϑ_{app} reflects the inhomogeneous (domain) structure with out-of-plane (three-dimensional) distortion of the hybrid cell. In the main plot, the experimental and theoretical curves are given in the strong field limit. A remarkable deviation from the simple theory is observed, as was the case with buffed polyimide orienting layers.[95] This is probably related to some flow phenomena induced by a strong electric field.

5.6 PHOTOAFFECTED PHASE BEHAVIOUR AND THE LCPT PHOTORECORDING

As mentioned earlier in this chapter, the elongated *trans*-form and the bent, banana-shaped *cis*-form of the same azobenzene chromophore differ substantially in molecular dimensions and therefore can reveal quite different phase behaviour. We have found that, in the case of chiral photochromic copoly-

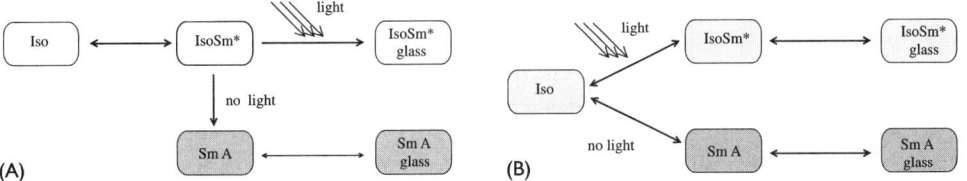

FIG. 5.25 Scheme of phase transitions in darkness and under illumination in chiral photochromic copolymers (A) SK8 and SK16 and (B) KW40 and KM45.

mers under discussion (Figure 5.5), the difference in mesomorphic behaviour is especially pronounced within the intermediate range of copolymer compositions, corresponding to the metastable formation of the IsoSm* phase. In that range, the unpolarized light illumination can even switch the sequence of mesophases for the copolymer towards the formation of the IsoSm* phase or prevent some phase transitions.[97] The phenomenon of the light-controlled phase transitions (the LCPT effect) has been observed to date for at least four copolymers (the corresponding schemes of phase transformations in darkness and under light illumination are shown in Figure 5.25). To be specific, for SK8 and SK16, light illumination forbids the TGB A* → Sm A transition at 29°C on cooling, and for KW40 and KM45, it switches the phase sequence from Iso → Sm A to Iso → TGB A*. Once the phase is formed and frozen in glass at ambient temperature, it keeps the acquired structure (either transparent, IsoSm*, or turbid, Sm A) for years with no visible changes. Figure 5.26 illustrates the optical appearance of an LCPT copolymer film cooled down to room temperature either in darkness or under illumination with visible light. As seen from the scan, a turbid, birefringent multidomain Sm A texture is formed in the shadow area, but the illuminated spot acquires a transparent, optically isotropic IsoSm* structure (the short-pitch TGB A* phase).

FIG. 5.26 Polymer film (SK16, ~23 μm) cooled from 70°C to ambient temperature at 0.5 K/min. The central spot was illuminated by a microscope lamp (nonpolarized), whereas the outside ring was not illuminated.

The two phase states of the same copolymer at the same temperature can be distinguished not only by their optical properties but also by X ray scattering[97] and dielectric relaxation measurements.[62] It is also worth noticing that the phase behavior of the polymers can be switched not only by nonfiltered white light of low intensity (e.g., from a desk lamp with power density 11 W/cm^2 within the wavelength range 350 to 2000 nm), but also selectively by blue-green light (400 < λ < 500 nm) with power density as low as 0.3 W/cm^2. That wavelength range corresponds to band C in the absorption spectra (Figure 5.8), and the same light causes the light-induced birefringence (Figure 5.11).

The LCPT effect originates undoubtedly from a change in the proportion of *cis*- and *trans*-isomers of azobenzene chromophores in the polymer film. It is still not clear, however, which of those isomers is more favorable for the formation of the IsoSm* structure and which prefers the straight Sm A type of packing. We found it reasonable to first suggest, that illumination within the n-π* band suppresses *cis*-isomers completely and that the TGB A* structure cannot transform to the Sm A phase under illumination because of the excess of *trans*-configured azobenzene fragments, which would possess higher helical twisting power and "frustrate" the uniform smectic structure. That suggestion can be supported with our early observations[98] that the rod-like *trans*-isomer of a chiral low molar mass dye induces higher spontaneous polarization in the Sm C matrix than the banana-shaped *cis*-form does. Also, a ~3-µm-thick SK8 film cast from UV-irradiated solution (the *cis*-film) evidently forms Sm A phase, in contrast to the mostly *trans*-film cast from virgin solution.

On the other hand, as mentioned in Section 5.3, the *trans*-isomer also absorbs within band C, especially in the condensed state of the copolymer, so that illumination with blue-green light creates not a pure *trans*-film but a new steady-state proportion of both isomers. Moreover, UV irradiation of the Sm A film of SK8 has been reported to transform it back to the IsoSm* phase.[77] Therefore, the *cis*-isomers seem to be preferred for the formation of the TGB A* phase.

Whatever the explanation, the LCPT effect has been used successfully for photorecording purposes (a primitive setup for LCPT recording is shown in Figure 5.27). Actually, if a copolymer film is being cooled slowly from above the transition point and illuminated simultaneously through a mask, the Sm A phase forms in a shadow area under the mask. It scatters incident light strongly and appears white. At the same time, the cooling under illumination ends up with the short-pitch TGB A* (IsoSm*) phase. In other words, the written image is realized as a negative Sm A image on the IsoSm* background.

There are, then, three possibilities for reading out the recorded image. First, the image can be viewed in the scattering mode, if observed in the daylight or in standard illumination; it appears then as a white graph on the colorless background. This is illustrated in Figure 5.28A, where the name of one of this chapter's authors is recorded. The image has already been kept for three years at ambient temperature and shows no visible changes. Second, the image can be read out in the transmission (dia) mode when looking through

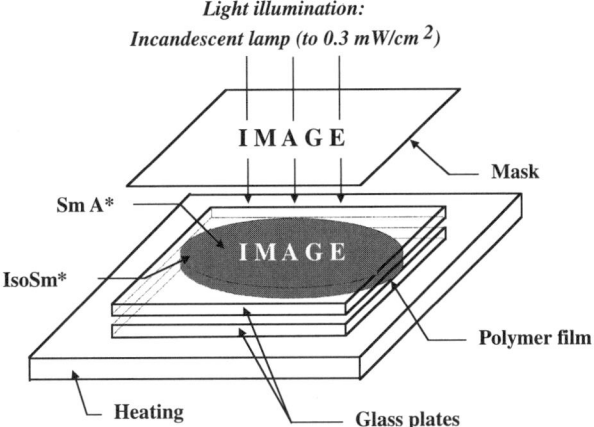

FIG. 5.27 Sketch of the photorecording setup.

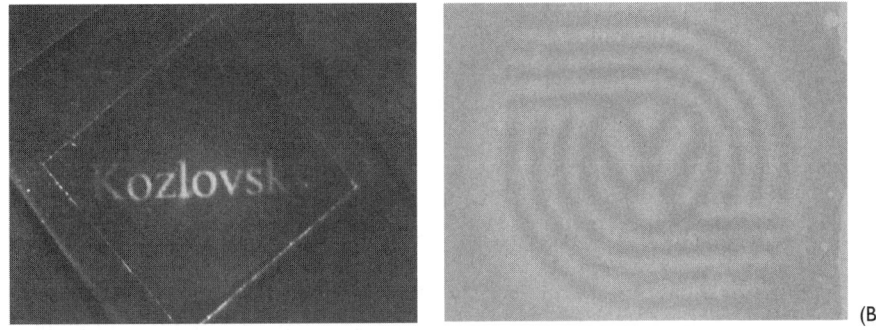

(A) (B)

FIG. 5.28 Examples of LCPT-recorded images in a ~ 10-μm-thick SK8 film, observed in (A) the scattering mode and (B) the transmission mode.

the copolymer film; and appears then as a dark graph on the transparent background, as illustrated in Figure 5.28B. We should emphasize here that no polarized light is necessary for either LCPT recording or reading of the image. Third, if crossed polarizers are used, the image can be read out as a white graph on the black background.

In contrast to photoinduced birefringence, the mesophase patterns in Figure 5.28 cannot be erased by light irradiation with UV light or with circularly polarized visible light, and only heating above the transition temperature erases the pattern. After such a thermal treatment, a new image can be written in the same film, up to 10 cycles of LCPT recording have been tested with a 10-μm-thick SK8 film without any loss in image quality.

To summarize, the LCPT effect allows a novel recording technique to be implemented. The technique is very simple to realize and allows for recording large-area images using nonpolarized light of low intensity.

5.7 CONCLUSIONS

The chiral side chain polymers derived from asymmetric esters of terephthalic acid and hydroquinone can form (in a broad temperature range, including ambient temperature) an unusual mesophase (the "isotropic smectic" phase, IsoSm*) characterized by high transparency and optical isotropy within the visible wavelength range, combined with a hidden layered smectic ordering and some elements of helical superstructure at shorter dimensions of 10 to 250 nm. The short-pitch TGB A* model seems to be the most adequate for the mesophase structure.

If functionalized with an azobenzene chromophore, the resulting copolymers also form the IsoSm* phase in a broad range of compositions. They show photochromism and photoinduced birefringence typical of azo dye LC copolymers, but they differ from the latter by their spontaneous formation of transparent, optically isotropic films, thus allowing for easy production of thick (up to 30 μm) photoactive layers. On the other hand, as compared with amorphous azobenzene-containing polymers, our copolymers possess some hidden liquid-crystalline ordering, which allows "imprinting" of photo-induced orientation of chromophores into the mesogenic matrix and/or main chain conformation during phase formation. Such imprinting can occur either on cooling or isothermally (in the case of kinetically governed transitions) and results in larger values of photoinduced birefringence and its higher stability.

Birefringence can be induced in optically isotropic photochromic IsoSm* copolymers by UV illumination and/or by visible light within the blue-green wavelength range of 400 to 500 nm. The latter is rather stable and shows no saturation up to several hours of illumination, whereas the birefringence induced by polarized UV light (300 to 400 nm) within the first few seconds disappears with further illumination. On the other hand, pretreatment with UV light increases substantially the sensitivity of the copolymer to the next photorecording within the blue-green range. Moreover, hidden UV images can be written by UV irradiation through a mask, and the visible images can be further developed by polarized uniform visible light.

Generally, photochromic IsoSm* films are considerably more sensitive to phase grating recording (created, e.g., by two laser beams of opposite circular polarization) than to intensity grating recording (with linearly polarized light). The gratings can be stable (for several months) or unstable (disappearing in few minutes) depending on the laser beam intensity, film pretreatment, and type of grating. Stable gratings can be erased with UV or circular light irradiation, or by heating the film above the phase transition point.

The most remarkable feature of the reported chiral azo dye copolymers, however, is the possibility of controlling their phase transitions by light illumination. This capability is observed for the copolymer compositions within the range of metastable IsoSm* phase formation, and it has been applied for the novel photorecording technique, LCPT (light-controlled phase transition) recording. The recording is carried out with unpolarized visible light of low intensity, and the images are recorded as Sm A scattering patterns on the IsoSm* background. They can be easily read out with nonpolarized light, both in the scattering mode and in the transmission (dia) mode.

ACKNOWLEDGMENTS

This work was supported by the Volkswagen Foundation (Project I 47 741), Deutsche Forschungsgemeinschaft (Project Re 923/8-1), and Russian Funds for Basic Research (Grant No. 01-02-16287). We are grateful to K. Yoshino and M. Ozaki (Department of Electronic Engineering, Osaka University, Japan), V. Lazarev (Institute of Crystallography, Russian Academy of Sciences, Moscow, Russia), and G. Gipparrone, N. Scaramuzza, and M. de Santo (Dipartimento di Fisica, Universitá della Calabria, Rende (Cs), Italy) for their participation in our experiments.

REFERENCES

1. McArdle, C. B. Ed. *Applied Photochromic Polymer Systems*. New York: Blackie & Son, Ltd. (1992).
2. Shibaev, V. P. Ed. *Polymers as Electro-optical and Photo-optical Media*. Berlin: Springer Verlag (1996).
3. Blinov, L. M. Photoinduced molecular reorientation in polymers, Langmuir-Blodgett films and liquid crystals. *J. Nonlin. Opt. Phys. Mater.* **5**, 165 (1996).
4. Todorov, T., Nikolova, L., and Tomova, N. Polarization holography. 1: A new high-efficiency organic material with reversible photoinduced birefringence. *Appl. Opt.* **23**, 4309 (1984).
5. Ivanov, M., Todorov, T., Nikolova, L., Tomova, N., and Dragostinova, V. Photoinduced changes in the refractive index of azo-dye/polyme systems. *Appl. Phys. Lett.* **66**, 2174 (1995).
6. Havinga, E. E., and van Pelt, P. Electrochromism of substituted polyalkenes in polymer matrixes; influence of chain length on charge transfer. *Ber. Bunsenges. Phys. Chem.* **83**, 816 (1979).
7. Blinov, L. M., Barnik, M. I., Weyrauch, T., Palto, S. A., Tevosov, A. A., and Haase, W. Photoassisted poling of polymer materials studied by stark spectroscopy (electroabsorption) technique. *Chem. Phys. Lett.* **231**, 246 (1994).
8. Barnik, M. I., Blinov, L. M., Weyrauch, T., Palto, S. A., Tevosov, A. A., and Haase, W. Stark spectroscopy as a tool for the characterization of poled polymers for nonlinear optics. In: G. A. Lindsay, K. O. Singer, Eds. *Polymers for Second-Order Nonlinear Optics*, 288 (1995).
9. Natansohn, A., Rochon, P., Gosselin, J., and Xie, S. Azo polymers for reversible optical storage. 1. Poly[4′-[[2-(acryloyloxy)ethyl]ethylamino]-4-nitroazobenzene]. *Macromolecules* **25**, 2268 (1992).
10. Ho, M.-S., Natansohn, A., and Rochon, P. Azo polymers for reversible optical storage. 9. Copolymers containing two types of azobenzene side groups. *Macromolecules* **29**, 44 (1996).
11. Natansohn, A., Rochon, P., Meng, X., Barrett, C., Buffeteau, T., Bonenfant, S., and Pézolet, M. Molecular addressing? Selective photoinduced cooperative motion of polar ester groups in copolymers containing azobenzene groups. *Macromolecules* **31**, 1155 (1998).
12. Hvilsted, S., Andruzzi, F., Kulinna, C., Siesler, H. W., and Ramanujam, P. S. Novel side-chain liquid crystalline polyester architecture for reversible optical storage. *Macromolecules* **28**, 2172 (1995).
13. Rasmussen, P. H., Ramanujam, P. S., Hvilsted, S., and Berg, R. H. A remarkably efficient azobenzene peptide for holographic information storage. *J. Am. Chem. Soc.* **121**, 4738 (1999).
14. Rasmussen, P. H., Ramanujam, P. S., Hvilsted, S., and Berg, R. H. Accelerated Optical Holographic Recording Using Bis-DNO. *Tetrahedron Lett.* **40**, 5953 (1999).
15. Si, J., Mitsui, T., Ye, P., Li, Z., Shen, Y., and Hirao, K. Optical storage in an azobenzene-polyimide film with high glass transition temperature. *Opt. Commun.* **147**, 313 (1998).
16. Fukuda, T., Matsuda, H., Shiraga, T., Kimura, T., Kato, M., Wiswanathan, N. K., Kumar, J., and Tripathy, S. K. Photofabrication of surface relief grating on films of azobenzene polymer with different dye functionalization. *Macromolecules* **33**, 4220 (2000).
17. Ringsdorf, H., Schmidt, H.-W., Baur, G., Kiefer, R., and Windscheid, F. Orientational order-

ing of dyes in the glassy state of liquid-crystalline side group polymers. *Liquid Crystals* **4**, 319 (1986).

18. Okazaki, S., Uto, S., Ozaki, M., and Yoshino, K. Guest-host electro-optic switching in spin-coated polymer ferroelectric liquid crystal film. *Appl. Phys. Lett.* **71**, 3373 (1997).

19. Jakob, E., Wolarz, E., Bialecka-Florjanczyk, E., Bauman, D., and Haase, W. Dielectric relaxation in a mixture of a side-chain liquid crystalline polymer and a low molecular mass azo dye. *Mol. Cryst. Liq. Cryst.* **258**, 253 (1995).

20. Corvazier, L., and Zhao, Y. Induction of liquid crystal orientation through azobenzene-containing polymer networks. *Macromolecules* **32**, 3195 (1999).

21. Zhao, Y., Chénard, Y., and Paiement, N. Liquid crystalline anisotropic gels based on azobenzene-containing networks. *Macromolecules* **33**, 1049 (2000).

22. Kurihara, S., Sakamoto, A., and Nonaka, T. Fast photochemical switching of a liquid-crystalline polymer network containing azobenzene molecules. *Macromolecules* **31**, 4648 (1998).

23. Kurihara, S., Sakamoto, A., Yoneyama, D., and Nonaka, T. Photochemical switching behavior of liquid crystalline polymer networks containing azobenzene molecules. *Macromolecules* **32**, 6493 (1999).

24. Kurihara, S., Matsumoto, K., and Nonaka, T. Optical shutter driven photochemically from anisotropic polymer network containing liquid crystalline and azobenzene molecules. *J. Appl. Phys. Lett.* **73**, 160 (1998).

25. Yamane, H., Kikuchi, H., and Kajiyama, T. Laser-addressing rewritable optical information storage of (liquid crystalline side chain copylymer/liquid crystals/photo-responsive molecule) ternary composite systems. *Polymer* **40**, 4777 (1999).

26. Wu, Y., Demachi, Y., Tsutsumi, O., Kanazawa, A., Shiono, T., and Ikeda, T. Photoinduced alignment of polymer liquid crystals containing azobenzene moieties in the side chain. 1. Effect of light intensity on alignment behavior. *Macromolecules* **31**, 349 (1998).

27. Wu, Y., Kanazawa, A., Shiono, T., Ikeda, T., and Zhang, Q. Photoinduced alignment of polymer liquid crystals containing azobenzene moieties in the side chain. 4. Dynamic study of the alignment process. *Polymer* **40**, 4787 (1999).

28. Wu, Y., Mamiya, J.-I., Kanazawa, A., Shiono, T., Ikeda, T., Nagase, Y., and Zhang, Q. Photoinduced alignment of polymer liquid crystals containing azobenzene moieties in the side chain. 6. Biaxiality and three-dimensional reorientation. *Macromolecules* **32**, 8829 (1999).

29. Eich, M., Wendorff, J. H., Reck, B., and Ringsdorf, H. Reversible digital and holographic optical storage in polymeric liquid crystals. *Makromol. Chem., Rapid Commun.* **8**, 59 (1987).

30. Eich, M. and Wendorff, J. H. Erasible holograms in polymeric liquid crystals. *Makromol. Chem. Rapid Commun.* **8**, 574 (1987).

31. Wendorff, J. H., and Eich, M. Nonlinear optical phenomena in liquid crystalline side chain polymers. *Mol. Cryst. Liq. Cryst.* **169**, 133 (1989).

32. Anderle, K. and Wendorff, J. H. Holographic recording using liquid crystalline side chain polymers. *Mol. Cryst. Liq. Cryst.* **243**, 51 (1994).

33. Ringsdorf, H., Urban, C., Knoll, W. and Sawodny, P. Photoreactive chiral liquid-crystalline side-group copolymers containing azobenzene mesogens. *Makromol. Chem.* **193**, 1235 (1992).

34. Sawodny, M., Schmidt, A., Stamm, M., Knoll, W., Urban, C., and Ringsdorf, H. Langmuir-Blodgett-Luhn multilayer assemblies of liquid-crystalline azo-dye side chain homo- and copolymers. *Thin Solid Films* **210/211**, 500 (1992).

35. Shibaev, V. P., Kostromin, S. G., and Ivanov, S. A. Photoregulation of the optical properties of comb-shaped polymers with mesogenic side groups and the problems of data recording. *Polym. Sci.* **39**, 118 (1997).

36. Kostromin, S. G., Shibaev, V. P., Geßner, U., Cackovic, H., and Springer, J. Oligoacrylates with cyanobiphenyl mesogenic groups. *Vysokomol. Soed. A* **38**, (1996).

37. Shibaev, V. P. Some new physico-chemical aspects of side-chain liquid crystal polymers. *Mol. Cryst. Liq. Cryst.* **243**, 201 (1994).

38. Stumpe, J., Fischer, T., Ziegler, A., Geue, T., Menzel, H. Photoorientation in LB multilayers of thermotropic polymers. *Mol. Cryst. Liq. Cryst.* **299**, 245 (1997).

39. Fisher, T., Läsker, L., Czapla, S., Rübner, J., and Stumpe, J. Interdependence of photoorientation and thermotropic self-organization in photochromic liquid crystalline polymers. *Mol. Cryst. Liq. Cryst.* **298**, 213 (1997).

40. Fischer, T., Läsker, L., Rutloh, M., Czapla, S., and Stumpe, J. Competition of self-organization and photo-orientation in liquid crystalline polymers. *Mol. Cryst. Liq. Cryst.* **299**, 293 (1997).

41. Sekkat, Z., Wood, J., Knoll, W., Volksen, W., Miller, R. D., and Knoesen, A. Light-induced orientation in azo-polyimide polymers 325°C below the glass transition temperature. *J. Opt. Soc. Am. B* **14**, 829 (1997).

42. Sekkat, Z., and Knoll, W. Photoreactive organic thin films in the light of bound electromagnetic waves. In: D. Neckers, D. Volman, and G. Von Buenau, Eds. *Adv. in Photochemistry* **22**, 117 (1997).

43. Sekkat, Z., and Dumont, M. Poling of azo dye doped polymeric films at room temperature. *Appl. Phys. B* **54**, 486 (1992).

44. Sekkat, Z., Aust, E. F., and Knoll, W. Photo-induced poling of polar azo dyes in polymer films. In: G. Lindsay and K. Singer, Eds. *Polymers for Second-Order Nonlinear Optics*, ACS *Symp. Ser.* **601**, Ch. 19, 1995.

45. Blinov, L. M., Palto, S. P., Tevosov, A. A., Barnik, M. I., Weyrauch, Th., and Haase, W. Electrically and photoelectrically poled polymers for nonlinear optics: Chromophores' polar order and its relaxation studied by electroabsorption. *Mol. Mater.* **5**, 311 (1995).

46. Bobrovsky, A., Boiko, N., and Shibaev, V. Photo-optical properties of new combined chiral photochromic liquid crystalline copolymers. *Liquid Crystals* **25**, 393 (1998).

47. Bobrovsky, A., Boiko, N., and Shibaev, V., A new type of multifunctional materials based on dual photochromism of ternary photochromic liquid crystalline copolymers for optical data recording and storage. *J. Mater. Chem.* **10**, 1075 (2000).

48. Petri, A., Kummer, S., Anneser, H., Feiner, F., and Bräuchle, C. Photoinduced reorientation of cholesteric liquid crystalline polysiloxanes and applications in optical information storge and second harmonic generation. *Ber. Bunsenges. Phys. Chem.* **97**, 1281 (1993).

49. Hayashi, T., Kawakami, H., Doke, Y., Tsuchida, A., Onogi, Y., and Yamamoto, M. Photoinduced phase transition of side-chain liquid crystalline copolymers with photochromic group. *Eur. Polym. J.* **31**, 23 (1995).

50. Ikeda, T., Horiuchi, S., Karanjit, D. B., Kurihara, S., and Tazuke, S. Photochemically induced isothermal phase transition in polymer liquid crystals with mesogenic phenyl benzoate side chains. 1. Calorimetric studies and order parameters. *Macromolecules* **23**, 36 (1990).

51. Kanazawa, A., Hirano, S., Shushido, A., Hasegawa, M., Tsutsumi, O., Shiono, T., Ikeda, T., Nagase, Y., Akiyama, E., and Takamura, Y. Photochemical phase transition behavior of polymer azobenzene liquid crystals with flexible siloxane units as a side-chain spacer. *Liquid Crystals* **23**, 293 (1997).

52. Tsutsumi, O., Kitsunai, T., Kanazawa, A., Shiono, T., and Ikeda, T. Photochemical phase transition behavior of polymer azobenzene liquid crystals with electron-donating and -accepting substituents at the 4,4′-positions. *Macromolecules* **31**, 355 (1998).

53. Gu, J., Liang, B., Liu, L., Tian, Y., Chen, Y., Lu, B., and Lu, Z. Photoinduced properties of liquid crystalline azobenzene polymer in Langmuir-Blodgett films investigated by surface plasmon resonance. *Thin Solid Films* **327–329**, 427 (1998).

54. Tian, Y., Ren, Y., Sun, R., Zhao, Y., Tang, X., Huang, X., and Xi, S. Synthesis of a series of chiral copolymers with azo groups and investigations of reversible liquid crystalline alignment induced by the LB films of these materials. *Liquid Crystals* **22**, 177 (1997).

55. Fischer, T., Menzel, H., and Stumpe, J. Photo-reorientation of azobenzene side groups of thermotropic 'hairy rod' polyglutamate in LB multilayers. *Supramol. Sci.* **4**, 543 (1997).

56. Andruzzi, L., Altomare, A., Ciardelli, F., Solaro, R., Hvilsted, S., and Ramanujam, P. S. Holographic gratings in azobenzene side-chain polymethacrylates. *Macromolecules* **32**, 448 (1999).

57. Labarthet, F. L., Freiberg, S., Pellerin, C., Pézolet, M., Natansohn, A., and Rochon, P. Spectroscopic and optical characterization of a series of azobenzene-containing side-chain liquid crystalline polymers. *Macromolecules* **33**, 6815 (2000).

58. Si, J., Mitsui, T., Ye, P., Li, Z., Shen, Y., and Hirao, K. Optical storage in an azobenzene-

polyimide film with high glass transition temperature. *Opt. Commun.* **147**, 313 (1998).

59. Kulinna, C., Hvilsted, S., Hendann, C., Siesler, W. H., Ramanujam, P. S. Selectively deuterated liquid crystalline cyanobenzene side-chain polyesters. 3. Investigations of laser-induced segmental mobility by Fourier Transform Infrared Spectroscopy. *Macromolecules* **31**, 2141 (1998).

60. Bata, L., Fodor-Csorba, K., Szabon, J., Kozlovsky, M. V., and Holly, S. A chiral side-chain polymer with layered structure. *Ferroelectrics* **122**, 149 (1991).

61. Kozlovsky, M. V., Fodor-Csorba, K., Bata, L., Shibaev, V. P. Chiral smectic side-chain copolymers. 1. Synthesis and phase behavior. *Eur. Polym. J.* **28**, 901 (1992).

62. Kilian, D., Kozlovsky, M. V., and Haase, W. Dielectric measurements on the 'isotropic smectic phase' of dyed side-chain polymers. *Liquid Crystals* **26**, 705 (1999).

63. Kozlovsky, M. V., Haase, W., Kuball, H.-G. Influence of chirality on mesophase state of side-chain polymers with phenyl benzoate mesogenic groups. In: *6th European Conference on Liquid Crystals*; March 25–30, 2001, Halle (Saale), Germany. Abstracts, 5-P14.

64. Demikhov, E., Kozlovsky, M. V. Amorphous chiral smectic A phase of side-chain copolymers. *Liquid Crystals* **18**, 911 (1995).

65. Vill, V. *LiqCryst 3.3. Database of liquid crystalline polymers.* Hamburg: LCI Publischer GmbH (1999).

66. Renn, S. R., Lubensky, T. C. Abrikosov dislocation lattice in a model of the cholesteric—to—smectic-A transition. *Phys. Rev. A* **38**, 2132 (1988)

67. Renn, S. R., Lubensky, T. C. Existence of a Sm-C grain boundary phase at the chiral NAC point. *Mol. Cryst. Liq. Cryst.* **209**, 349 (1991).

68. Goodby, J. W., Waugh, M. A., Stein, S. M., Chin, E., Pindak, R., and Patel, J. S. A new molecular ordering in helical liquid crystals. *J. Am. Chem. Soc.* **111**, 8119 (1989)

69. Navaliles, L., Barois, P., Nguyen, H. T. X-ray measurement of the twist grain boundary angle in the liquid crystal analog of the Abrikosov phase. *Phys. Rev. Lett.* **71**, 545 (1993).

70. Goodby, J. W. Twist grain boundary (TGB) phases. In: D. M. P. Mingos Ed. *Liquid crystals II. Structure and bonding*, Vol. 95, New York: Springer Verlag, 83 (1999).

71. Kitzerow, H.-G. Twist grain boundary phases. In: C. Bahr, H. Kitzerow Eds. *Chirality in liquid crystals*. New York: Springer Verlag, 297 (2000).

72. Dierking, I., Lagerwall, S. T. A review of textures of the TGBA* phase under different anchoring geometries. *Liquid Crystals*, **26**, 83 (1999)

73. Kozlovsky, M. V., Haase, W., Stakhanov, A., and Shibaev, V. P. Chiral smectic side-chain copolymers. 3. Copolymers containing a diazo chromophore. *Mol. Cryst. Liq. Cryst.* **321**, 177 (1998).

74. Kozlovsky, M. V., Meier, J. G., and Stumpe, J. Chiral side-chain copolymers. 4. Kinetics of the phase transition from conventional isotropic liquid to the TGB-like "isotropic smectic" state. *Macromol. Chem. Phys.* **201**, 2377 (2000).

75. Blinov, L. M., Barberi, R., Kozlovsky, M. V., Lazarev, V. V., and de Santo, M. P. Optical anisotropy and four possible orientations of a nematic liquid crystal on the same film of a photochromic chiral smectic polymer. *J. Nonlinear Opt. Phys. Mat.* **9**, 1 (2000).

76. Blinov, L. M., Kozlovsky, M. V., Ozaki, M., Skarp, K., and Yoshino, K. Photoinduced dichroism and optical anisotropy in a liquid crystalline azobenzene side-chain polymer caused by anisotropic angular distribution of trans and cis isomers. *J. Appl. Phys.* **84**, 3860 (1998).

77. Blinov, L. M., Kozlovsky, M. V., and Gipparone, G. Photochromism and holographic grating recording on a chiral side-chain polymer containing azobenzene chromophores. *Chem. Phys.*, **245**, 473 (1999).

78. Blinov, L. M., Gipparone, G., Kozlovsky, M. V., Lazarev, V. V., Scaramuzza, N. Holographic "development" of a hidden UV image recorded on a liquid crystalline polymer. *Optics Commun.* **73**, 137 (2000).

79. Blinov, L. M., Kozlovsky, M. V., Ozaki, M., Yoshino, K. Effect of the trans-cis isomerization of a chiral azo-dye on the dye induced ferroelectricity in an achiral liquid crystal. *Mol. Mat.* **6**, 235 (1996)

80. Blinov, L. M., Barberi, R., Cipparrone, G., Kozlovsky, M. V., Lazarev, V. V., Ozaki, M., de Santo, M. P., Scaramuzza, N., and Yoshino, K. Reversible UV Image Recording on a photochromic side-chain liquid crystalline polymer. *Mol. Cryst. Liq. Cryst.* **355**, 359–380 (2001).

81. Kozlovsky, M. V., and Lazarev, V. V. The photoinduced birefringence in thick films of chiral azodye copolymers (submitted to *Macromol. Chem. Phys.*).
82. Han, M., Morino, S., and Ichimura, K. Factors affecting in-plane and out-of-plane photo-orientation of azobenzene side chains attached to liquid crystalline polymers induced by irradiation with linearly polarized light. *Macromolecules* **33**, 6360 (2000).
83. Buse, K., Adibili, A., and Psaltis, D. Non-volatile holographic storage in doubly doped lithium niobate crystals. *Nature* **393**, 665 (1998).
84. Gipparone, G., Mazzulla, A., Kozlovsky, M. V., Palto, S. P., Yudin, S. G., and Blinov, L. M. Polarization gratings in photosensitive Langmuir-Blodgett films and chiral liquid crystalline polymers. *Mol. Mat.* **12**, 359 (2000).
85. Kozlovsky, M. V., and Haase, W. Photorecording in the "isotropic smectic phase" of dyed side-chain polymers. *Macromol. Symp.* **137**, 47 (1999).
86. Petrov, M. P., Stepanov, S. I., Khomenko, A. V. *Photorefractive Crystals in Coherent Optical Systems*. Berlin: Springer, (1991).
87. Blanche, P.-A., Lemaire, Ph. C., Maertens, C., Dubois, P., and Jérôme, R. Polarization holography reveals the nature of the grating in polymers containing azo-dye. *Opt. Commun.*, **185**, 1 (2000).
88. Blinov, L. M., Cipparrone, G., and Palto, S. P. Phase grating recording on photosensitive Langmuir-Blodgett films. *Int. J. Nonlin. Opt. Phys.* **7**, 369 (1998).
89. Couture, J. J. A. Polarization holographic characterization of organic azo dyes/poly(vinyl alcohol) films for real-time applications. *Appl. Opt.* **33**, 2858 (1991).
90. Dyadyusha, A. G., Kozenkov, V. M., Marusii, T. Ya., Reznikov, Yu. A., Reshetnyak, V. Yu., and Khizhnyak, A. I. Light induced planar orientation of nematic liquid crystals on photo-anisotropic surface without a microrelief. *Ukr. Fiz. Zh.* **36**, 1059 (1991).
91 Schadt, M., Schmit, K., Kozenkov, V., and Chigrinov, V. Surface-induced parallel alignment of liquid crystals by linearly polymerized photopolymers. *Jpn. J. Appl. Phys.* **31**, 2155 (1992).
92. Gibbons, W., Shannon, P., Shao-Tang Sun, and Swetlin, B., Surface mediated alignment of nematic liquid crystals with polarized laser light. *Nature* **351**, 49 (1991).
93. Schadt, M., Seiberle, and Schuster, A. Optical patterning of multidomain liquid-crystal displays with wide-viewing angles. *Nature* **381**, 212 (1996).
94. Perny, S., Le Barny, P., Delaire, J., Buffeteau, T., Sourisseau, C., Dozov, I., Forget, S., and Martinot-Lagarde, P. Holoinduced orientation in poly(vinylcinnamate) and poly(7-methacryloyloxycoumarin) thin films and the consequences on liquid crystal alignment. *Liquid Crystals* **27**, 329 (2000).
95. Palto, S., Barberi, R., Iovane, M., Lazarev, V. V., and Blinov, L. M. Measurements of zenithal anchoring energy of nematics at the planar interface in hybrid cells. *Mol. Mat.* **12**, 277 (2000).
96. Yokoyama, H., and Van Sprang, H. A. A novel method for determining the anchoring energy function at a nematic liquid crystal-wall interface from director distortions at high fields. *J. Appl. Phys.* **57**, 4520 (1985).
97. Kozlovsky, M. V., Shibaev, V. P., Stakhanov, A. I., Weyrauch, T., Haase, W., and Blinov, L. M. A new approach to photorecording based on hindering the TGB A* – Sm A* phase transition in photochromic chiral liquid crystalline polymers. *Liquid Crystals* **24**, 759 (1998).
98. Blinov, L. M., Kozlovsky, M. V., Nakayama, K., Ozaki, M., and Yoshino, K. Photochromism of azo-dyes and effect of liquid crystalline ordering on its efficiency and kinetics. *Japan. J. Appl. Phys.* **35**, 5405 (1996).

6
PHOTOISOMERIZATION IN LANGMUIR-BLODGETT-KUHN STRUCTURES

HENNING MENZEL
Institut für Technische Chemie, Technische Universität, Braunschweig

6.1 INTRODUCTION

Fabrication of photoreactive organic thin films (POTF) requires the use of special chromophores that change their molecular structure and/or properties upon irradiation and thereby constitute the basic photoactive function. The changes in the chromophores' properties should be adequately large to obtain a satisfactory response toward irradiation. It is not sufficient to have an appropriate chromophore, however; the structure of the film-forming matrix must also be tailored to meet the requirements to keep the chromophores photoactive. Moreover, it is possible to design the matrix in a way that boosts the photoinduced changes caused by the photoreaction and enhances the system's response. Photoisomerization, a very common mechanism used in photoactive films, is a process requiring some free volume and is, therefore, strongly influenced by the matrix. But reciprocally, it also affects the matrix in its structure and order.

Langmuir-Blodgett-Kuhn (LBK) films are prepared by a dipping process in which monolayers floating at the air/water interface are deposited on a solid substrate. LBK films are highly ordered supramolecular assemblies. By selection of the amphiphilic molecules and adjustment of the transfer conditions, the LBK technique offers extensive structural control and opens a way to optimize matrices for photoactive dyes.

6.2 OTHER DYES USED IN PHOTOACTIVE LBK FILMS

Several photoactive dyes have been incorporated into LBK films. Among these, the class of photoisomerizing dyes, namely azobenzene derivatives, are particularly versatile and have been investigated in great detail.

Azobenzene derivatives undergo *trans* to *cis* isomerization when irradiated (see Figure 6.1). The isomerization is thermally and photochemically reversible. Upon isomerization, not only the spectral properties, but also the molecular dimensions of the molecule, change considerably (see Chapter 1). *Trans*-azobenzene is an elongated mesogenic molecule, whereas the *cis*-form is crooked and nonmesogenic. Azobenzene derivatives have been shown to be highly versatile, easily accessible, and relatively stable against photobleaching. Therefore, I will focus mainly on LBK films having azobenzene moieties as the photoactive dye. However, several other dyes have been incorporated in LBK films, which I shall introduce briefly.

6.2.1 Stilbenes

Stilbenes are chromophores that can undergo photoisomerization as do azobenzenes, but they also show photodimerization (see Figure 6.2). Besides other parameters, the packing of the chromophores influences the ratio of

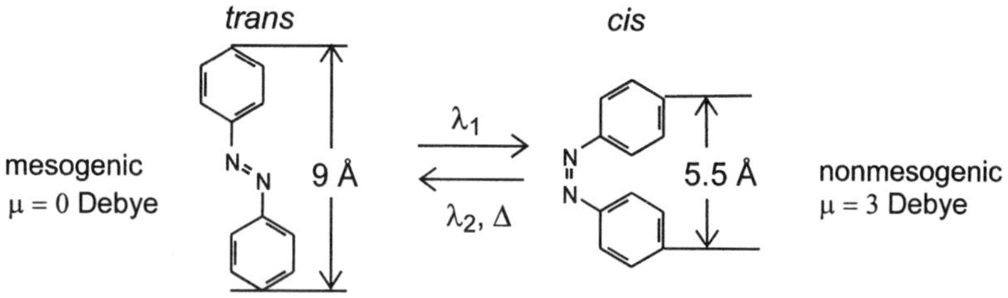

FIG. 6.1 Photoisomerization of azobenzene and the concomitant changes in molecular structure and properties.

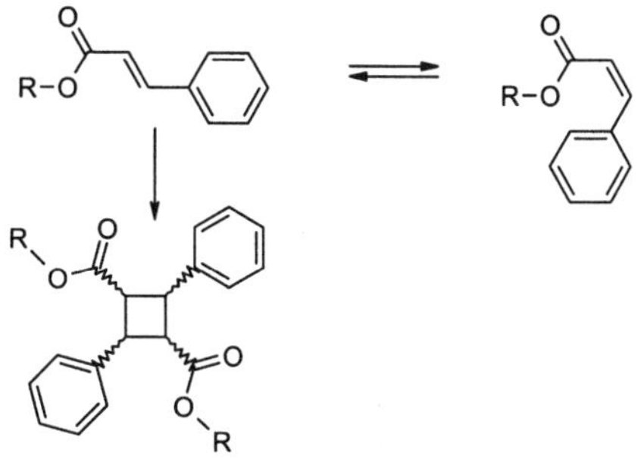

FIG. 6.2 Reversible photoisomerization and photodimerization of cinnamoyl moieties.

these two reactions.[1] Stilbenes have been incorporated in a fatty acid derivative **1**, which could readily be deposited to form multilayer assemblies on rigid supports. The isomerization of **1** was found to be rather inefficient, with a very low quantum yield. In addition to the isomerization, dimerization occurred.[2,3]

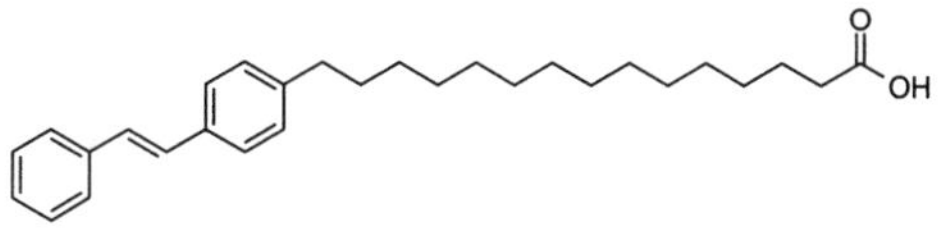

Furthermore, the photoisomerization is not completely reversible, because the phenanthrene derivative **2** is formed in the presence of an oxidizing agent with photogenerated dihydrophenanthrenes as intermediates. Stilbene derivatives, therefore, are not very promising for the fabrication of reversible POTFs. For cinnamoyl groups, *trans* to *cis* isomerization is the main reaction only in dilute systems.[4] In more concentrated solutions, however, isomerization and dimerization compete with each other. In crystals and other systems in which the isomerization is restricted due to lack of free volume, the dimerization can be the main photoreaction.[5] Although the equilibrium between *trans*- and *cis*-isomers is established very quickly in LBK films with appropriate fluidity,[6] the cinnamoyl groups mostly have been used for crosslinking LBK films.[6,7,8]

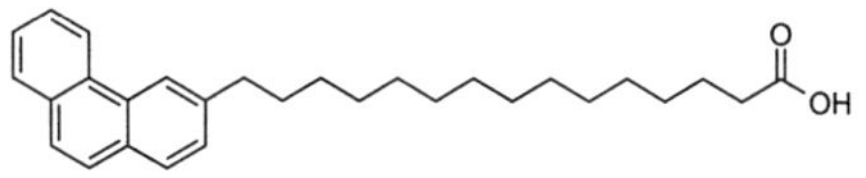

6.2.2 Salicylidene Aniline

Reversible photoreactions have been reported for the salicylidene aniline derivative **3**, which is isomerized at $\lambda = 308$ nm, as shown in Figure 6.3. This reaction is reversible because of a thermal back reaction. Derivative **3** is highly reactive in LBK films, because only small-volume changes occur, involving a proton transfer and a rotation of the phenyl ring.[9] The thermal back reaction is retarded in the LBK film as is the case with crystals, however, and the isomerized form is much more stable than in solution.

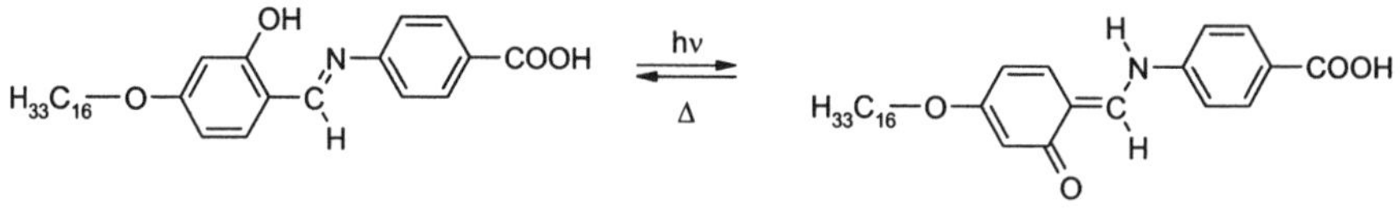

FIG. 6.3 Reversible photochroism of amphiphilic salicylidene aniline **3** in LBK films.

6.2.3 Spiropyrans

The photoreaction of spiropyrans **4** comprises a bond cleavage resulting in a zwitterionic species, the merocyanine, and subsequent *cis* to *trans* isomerization (see Figure 6.4). There are several species in equilibrium, so the chemistry and photochemistry of the spiropyrans are rather complex.[10] The photoreaction of spiropyrans is sterically quite demanding and has been reported to be restricted in LBK films.[11]

There are also reports describing unhindered photoreaction of spiropyrans in LBK films.[12] Nevertheless, the matrix formed by the LBK films influences the reaction, and the optical properties of the material can be fine-tuned by the choice of the film. For example, it is possible to selectively aggregate the chromophores and to build up a layered multifrequency recording medium for optical data storage as has been shown by Hibino *et al.*[12] The authors used amphiphiles with differing polarities and concentrations to generate a medium of tailored polarity that favors the formation of aggregates. Furthermore, the colored ionic form was stabilized by adsorption to a monolayer of a cationic amphiphile, or if it forms a monolayer, by adsorption of an polyelectrolyte.[12]

6.2.4 Other Chromophores

Besides azobenzene, spirobenzopyran, and salicylidene derivatives, other photochromic dyes have been used in LBK films. For example, anthocyanine dyes **5** (see Figure 6.5) have been used to obtain monolayers that change the area at constant pressure upon irradiation.[13] Diphenyldiacetylene chromophores **6** (Figure 6.5) have shown photoinduced anisotropy upon polar-

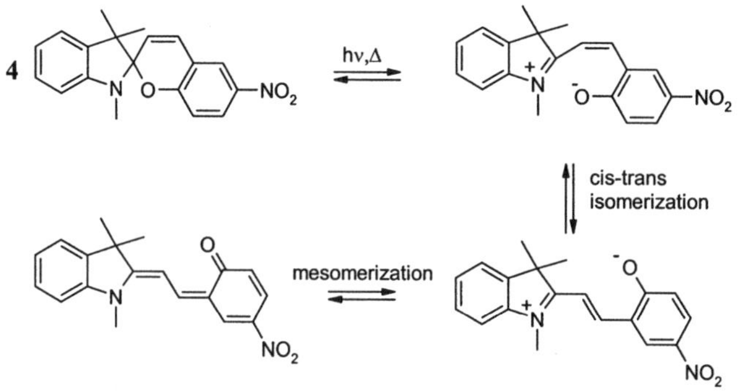

FIG. 6.4 Photochemical reactions for spirobenzopyrans **4**.

FIG. 6.5 Anthoyanine dye **5** and its simplified photoreaction (top) and diphenyldiacetylene amphiphile **6** (bottom).

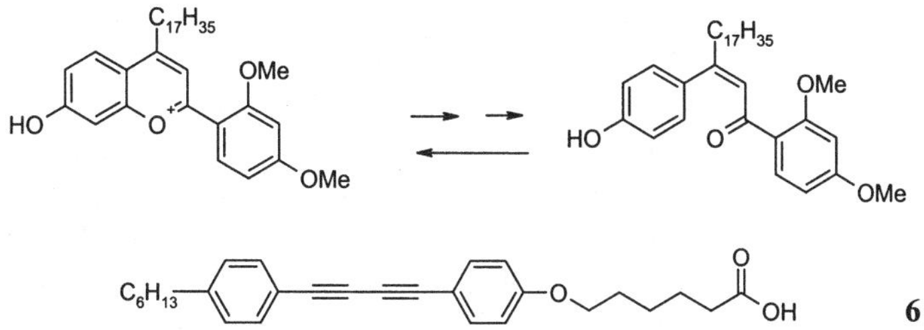

ized irradiation, as observed in systems having azobenzene moieties[14] (see Section 6.5.2). By employing structurally better-defined LBK films of poly-electrolyte complexes of diphenyldiacetylene amphiphiles, however, it has been established that the photoinduced anisotropy is not due to a photo-reorientation process.[15,16]

6.3 UV-VIS SPECTROSCOPY AS AN ANALYTICAL TOOL FOR THE INVESTIGATION OF AZOBENZENE LBK FILM STRUCTURE

6.3.1 Trans-Cis Interconversion

The *cis*-isomer of azobenzene derivatives is difficult to isolate in pure form because of the thermal back reaction, and its extinction coefficient cannot be determined easily. So the determination of the *cis/trans* ratio by UV-vis spectroscopy is not straightforward. Some approximation methods have been proposed, however. Despite being rather simple and crude, the difference spectra method of Brode *et al.*[17] has been used widely. This method can be used only for an initial approximation, because the error is rather large. For example, the *cis*-isomer percentage for an irradiated LBK film of azobenzene fatty acid **7** was calculated by this method to be 30%, whereas more accurate electrochemical methods indicate a conversion of only 19%.[18]

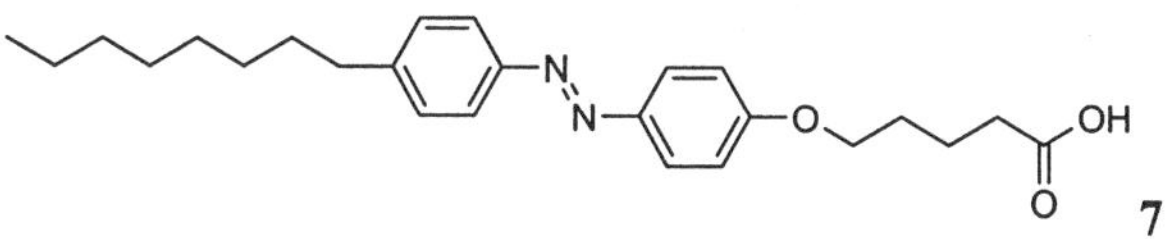

A much more sophisticated spectroscopical method was proposed by Fischer.[19] This method allows the calculation of photostationary states in systems containing substances A and B when only A is known. The only assumption made in this method is that the ratio of the quantum yields $\phi A/\phi B$ does not change with the wavelength. The approximate validity of this assumption has been shown.[20] Further requirements of this method are that two photostationary states must be able to be achieved by irradiation at different wavelengths and that the changes in the optical spectrum must be due only to the changes in the *cis/trans* ratio. In solutions of azobenzene derivatives, these requirements are fulfilled in most cases. In the solid state, such as in densely packed LBK films, they are not fulfilled. The optical properties of the chromophore depend on the dense packing in the LBK film (see Section 6.3.2) and are changed when increasing amounts of *cis*-isomer alter the packing of the chromophores. In this case, all spectroscopic methods for determining the *cis/trans* ratio will fail. Besides spectroscopy, electrochemical methods can be employed to determine the *cis/trans* ratio in LBK films. The *cis*-isomer can be reduced at significantly greater anodic potential than the *trans*-isomer; this gives rise to a corresponding peak in the cyclic voltammetry experiment, and the amount of *cis*-isomer then can be calculated from either the cathodic or the anodic charge. The total number of chromophores can be calculated from the transfer conditions in the LBK experiment. It is even possible to use this method for actinometry.[21] It was by employing this method that it was found that only 19% of chromophore 7 can be isomerized in LBK films.[18]

6.3.2 Aggregation

As mentioned previously, the interactions of the chromophores with extended π-systems in densely packed solids like LBK films significantly influence the optical properties. In particular, the aggregation causes a significant peak shift, the direction and extent of which depends on the number of aggregated chromophores and their distance from each other in the aggregate. Furthermore, a particular influence on the extent and direction of the peak shift is found for the orientation of the chromophores to each other i.e., the angle of inclination θ as defined in Figure 6.6. This phenomenon can be described by Mc Rae and Kasha's[22,23] molecular exciton model or the more elaborated extended dipole model by Kuhn *et al.*[24]

Both the molecular exciton model and the extended dipole model can be used to calculate one of the parameters—i.e., the distance of the chromophores, the inclination angle, or the number of aggregated chromophores—if the other parameters are known. However, in most LBK systems under consideration, it is the case that none of the parameters is known exactly. Nevertheless, in some cases well-justified assumptions can be made, and an estimation of, e.g., the inclination angle is possible.[25,26] In any case, the spectral shift observed in ordered arrangements of extended π-systems can be related to the aggregation and orientation of the chromophores. A very

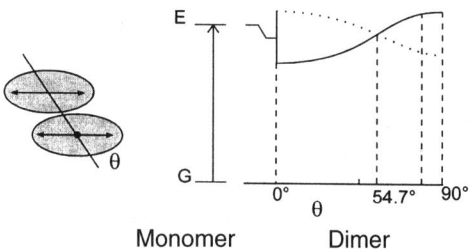

FIG. 6.6 Exciton band energy diagram for a molecular dimer with coplanar transition dipoles inclined to the interconnecting axis by an angle θ (adapted from reference 23 with permission).

H-aggregates	Dimers	Isolated	J-aggregates
$\Delta\lambda = -50$ nm	-15 to -25 nm	0 nm	15 to 25 nm

FIG. 6.7 Spectral characteristics and modes of aggregation of azobenzene amphiphiles (adapted from reference 27 with permission from Elsevier Science).

extensive investigation of this correlation was done by Shimomura and Kunitake.[26,27] They examined azobenzene amphiphiles in various states of organization and correlated the observed peak shift with the organization in the ordered arrangement. Large blue shifts up to $\Delta\lambda = -50$ nm were observed for a crystalline ordering with a parallel packing of the chromophores, that is θ close to 90° (see Figure 6.7), whereas a less-ordered liquid crystalline arrangement resulted in a peak shift of approximately $\Delta\lambda = -15$ to -25 nm. The tilted arrangement of the chromophores in so-called J-aggregates, that is $\theta < 54°$, resulted in a red shift of $\Delta\lambda = 15$ to 25 nm.[26] Using these correlations between the peak shift and the structural order of the film, it is possible to determine the temperature for a crystalline-to-liquid crystalline transition by monitoring the peak position in the spectra.[28]

In azobenzene polymers, the aggregation behavior is even more complex than for low molecular weight azobenzene amphiphiles. Already in solution, the chromophores are aggregated along the polymer backbone to some extent. In the solid state, very broad peaks in the UV-vis spectra of these materials are normally observed. These broad peaks represent the distribution of chromophores in environments with slightly differing degrees of order. A semiquantitative procedure for describing changes in the distribution of order within the polymer assemblies was developed by Menzel et al.[29] In this method, the relative amounts of chromophores in the aggregated and the nonaggregated states were determined to be the areas of the corresponding peaks in the UV-vis spectrum, as resolved by a peak-fitting procedure (see Figure 6.8).

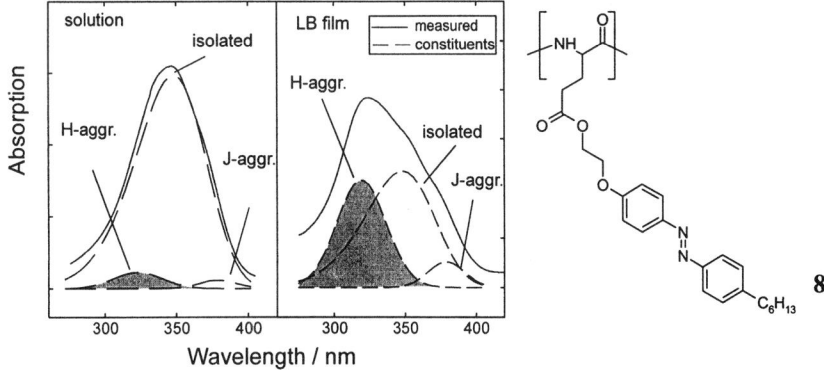

FIG. 6.8 Observed π-π^* band and the constituents, as determined by a peak-fitting procedure for the solution and as a prepared LBK film of polyglutamate **8**.

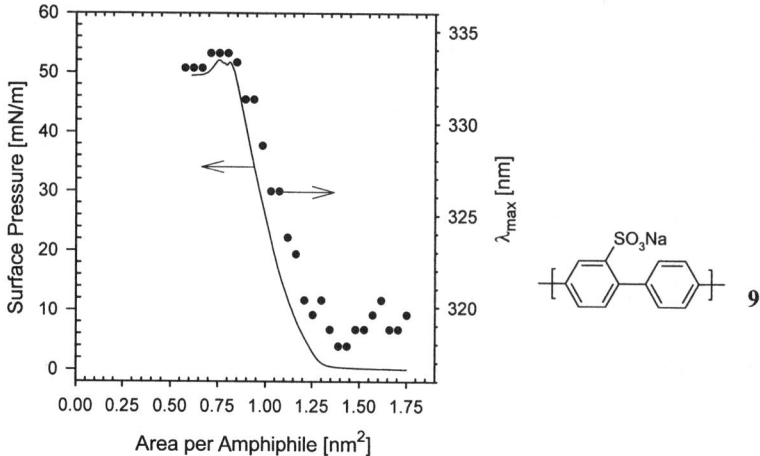

FIG. 6.9 Wavelength at the peak maximum of the poly(phenylene sulfonate) **9** A band upon compression of a monolayer (complex with dioctdecyldimethylammonium bromide) and the corresponding isotherm (reproduced with permission from reference 32).

The peak shift due to aggregation is observed not only in LBK films containing azobenzene chromophores, but also for other chromophores with extended π-systems, such as viologen polymers.[30,31] For monolayers of the poly(p-phenylene sulfonate) **9**/ dioctadecyldimethylammonium bromide (DODA) complex, the peak shift due to aggregation results in a "piezochromic" effect—that is, upon compression of the monolayer, a significant shift of the poly(p-phenylene sulfate) A band is observed (see Figure 6.9). This "photochromic" effect has been shown to be based on the improved π-π interaction upon compression of the monolayer.[32]

6.4 EXAMPLES OF THE INFLUENCE OF STRUCTURE ON PHOTOISOMERIZATION

The rigid, highly ordered structure of LBK assemblies provides an environment that influences the reactivity of any photoreactive functional group in the film. As a consequence, only solid-state reactions resulting in minor changes of the lattice parameters are straightforward in LBK films. For only a few systems is this rigidity advantageous. So the photoreaction of salicylidene anilines (see Figure 6.3) is not restricted in LBK films, but the thermal back reaction is retarded considerably, so the photoproduct is stabilized.[9] However, more space-consuming reactions, such as *trans* to *cis* isomerization, can be severely restricted.[1,2,3,11,33] For the stilbene derivatives, e.g., **1**, **10**, or a thioindigo **11**, very low quantum yields for the *trans* to *cis* isomerization were reported in the LBK films compared to the solution, whereas the *cis* to *trans* isomerization was not hindered. This is due to the fact that *cis*-isomers have higher areal requirements than *trans*-isomers.[2,3,34]

LBK films of the azobenzene containing fatty acid **7** show a similar behavior. When compressed and transferred in the *trans*-form, a peak shift due to aggregation is observed and the photoisomerization is hindered. The chromophores are very densely packed, as established by STM measurements.[35] When compressed and deposited in the *cis*-form, i.e., under illumination with UV light, there is no aggregation, and the *cis* to *trans* isomerization is unrestricted. Alternate irradiation under constant surface area causes changes in the surface pressure.[36]

The time necessary to reach the photostationary *cis*-state for the first time in LBK films of this azobenzene amphiphile transferred in the *trans*-state is considerably longer than for the subsequent cycles.[37] This is explained by the time it takes to break up some aggregates in the dense packing of the LBK film.[35] The extent of isomerization, determined by Brode's method,[17] is given to be 30%. Reestimation employing electrochemical methods showed that the irradiation yielded a *cis*-isomer count of only 19%.[18]

The extent of the isomerization hindrance depends on the structure of the azobenzene amphiphile. When the azobenzene is located in the middle of the alkyl chain (i.e., **7**), the *trans* to *cis* photoisomerization in LBK films is severely hindered but not completely suppressed.[17,18] However, when the azobenzene moiety is located directly at the head group, as in **12**, the isomerization in the monolayer at the air/water interface is completely blocked.[38]

12   **13**

Furthermore, the means of binding the alkyl tail to the azobenzene influences the rigidity of the monolayers and the *trans* to *cis* photoisomerization. If the alkyl chain is bound to the azobenzene unit via an ether linkage (as in **13**), the alkyl chain can adopt a better-ordered arrangement in which the chromophores are aggregated to an higher extent. So, in LBK films of **13**, the hindrance of the isomerization is more pronounced than in LBK films of azobenzene amphiphiles having the alkyl chain directly bound to the chromophore (as in **7**).[33,39]

The structure of aggregates was investigated by Whitten and coworkers. It was proposed that the aggregates have a "herringbone" rather than a simple "cardboard" arrangement. This conjecture was supported by measurements of the induced circular dichroism for aggregates of chiral amphiphiles and by comparison of the fluorescence spectra for aggregates and for the corresponding crystal, the structure of which was known.[40]

The influence of chain packing on the photoisomerization of azobenzene moieties was underpinned by simulations, which also show that sufficient free volume has to be present around the azobenzene moiety for the *trans* to *cis* isomerization to occur.[41]

The packing influences not only the photochemical *trans* to *cis* isomerization but also the thermal back reaction. The kinetics of the back reaction in

LBK films of **7**, measured by electrochemical methods, indicated activation energies similar to those measured for solutions. This was interpreted to be the result of the sterically demanding *cis*-isomer having a rather unfavorable state.[42] Furthermore, the kinetics may be used as an indication that, in the LBK film, the back isomerization proceeds via the inversion mechanism (see Chapter 1). The thermal back reaction of salicylidene anilines **3** in LBK films was also found to deviate significantly from a first-order kinetics.[9]

These examples and investigations on azobenzene moieties in polymers[43] show that the photochromic behavior is mainly controlled by the free volume distribution around the chromophore. To obtain LBK films in which azobenzene moieties can undergo photoisomerization, therefore, the free volume around the azobenzene chromophore must be controlled precisely to allow for the molecular rearrangement inherent in the reversible *trans* to *cis* photoisomerization. This is possible by (1) mixing with other amphiphiles, (2) adjusting the architecture of the amphiphile, or (3) attaching the chromophore to a polymer either by coulomb interaction or covalently.

6.4.1 Mixing with Other Amphiphiles

The molecular aggregation of the azobenzene amphiphile **7** can be reduced significantly by mixing it with phospholipids having saturated or unsaturated alkyl chains. The *trans* to *cis* photoisomerization in a mixed monolayer is accelerated. However, the contribution of the saturated and unsaturated lipids in the mixed monolayers was found to be different.[44] Like low molecular weight amphiphiles, surface active polymers can also be used as a matrix for azobenzene moieties. For example, helical copolyglutamates, which form very good monolayers, have been used as a matrix for oleophilic azodyes in Langmuir-Blodgett films. The dye is "dissolved" in the side chain region of the polymer.[45] Isomerization of the azobenzene dyes was not investigated specifically, but the resulting LBK films were successfully tested in optical data storage experiments. Irradiation resulted in slightly different optical properties, which could be detected by surface plasmon spectroscopy. The origin of the changes to the optical properties was not investigated in detail.[46]

A problem often associated with the approach of mixing a photoactive amphiphile with a nonphotoactive amphiphile is that not necessarily molecular mixing occurs, but domains of the azobenzene amphiphile are formed.[47,48] To overcome this problem, the molecular mixing has been enforced by tailored interactions between the different amphiphiles. For example, low molecular weight azobenzene dyes can be dissolved in polyglutamates having long alkyl chains and some azobenzene groups in the side chain. The molecular mixing is supported by the interaction of the polymer-bound azobenzene with the admixed chromophore. Because the azobenzene is molecularily dissolved in the side chain region, it can be readily isomerized by irradiation. Irradiation causes demixing, however, because the *cis*-azobenzene units do not have the interactions that are present in the all-*trans* film and are expelled from the side chain region.[49]

A mixture that is stable in the *cis*-form was achieved by mixing an ionic azobenzene amphiphile having a long spacer between the head group and the

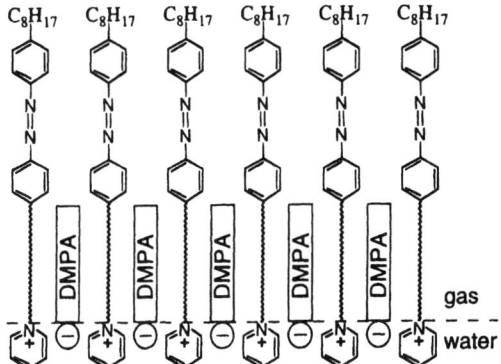

FIG. 6.10 Schematic representation of the molecular packing arrangement in the mixed monolayer of an ionic azobenzene amphiphile and an oppositely charged "spacer" amphiphile (reproduced with permission from reference 50 Copyright (1995) American Chemical Society).

azobenzene unit with an amphiphile having an oppositely charged head group and alkyl chains shorter than the spacer (see Figure 6.10). Due to the design of the system and the coulomb interaction of the head groups, which ensure molecular mixing, the azobenzene moiety is located atop a layer of closely packed alkyl chains. The photoisomerization of the azobenzene is unrestricted and a first-order kinetics can be observed. The structure of the LBK film is not changed by the photoisomerization, and no change in surface pressure is observed upon irradiation.[50]

Azobenzene amphiphiles can be isolated from each other and equipped with sufficient free volume for reversible photoisomerization by inclusion into amphiphilic β-cyclodextrin. The inclusion complex can be transferred to give LBK films in which the azobenzene moieties photoisomerize as they would in solution (see Figure 6.11).[51]

6.4.2 Adjusting the Structure of the Azobenzene Amphiphile

The free volume around the azobenzene that is necessary for an unrestricted photoisomerization can be established by the introduction of bulky head groups or additional alkyl chains in the azobenzene amphiphile. An example of this approach has been published by Markava *et al*. The authors compared azobenzene amphiphiles with one (**14**) and two (**15**) alkyl chains as well as amphiphiles with small (**16**) and large (**17**) head groups. They found that photoisomerization takes place in LBK films of **15** and **17**, where the areal requirements of the amphiphile are high, but not in the LBK films of **14** or **16**, which can be packed more densely.[52]

14

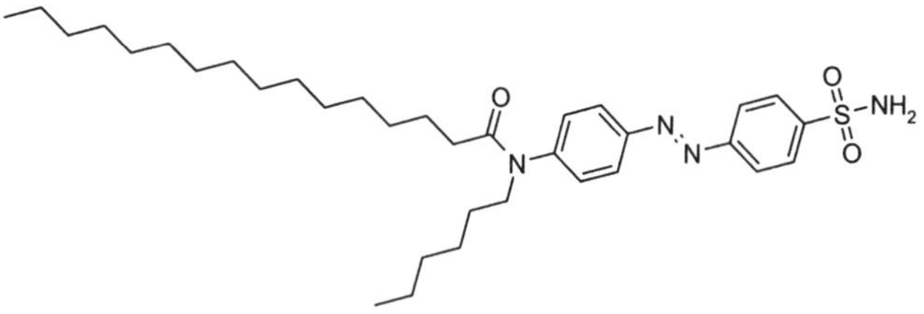

15

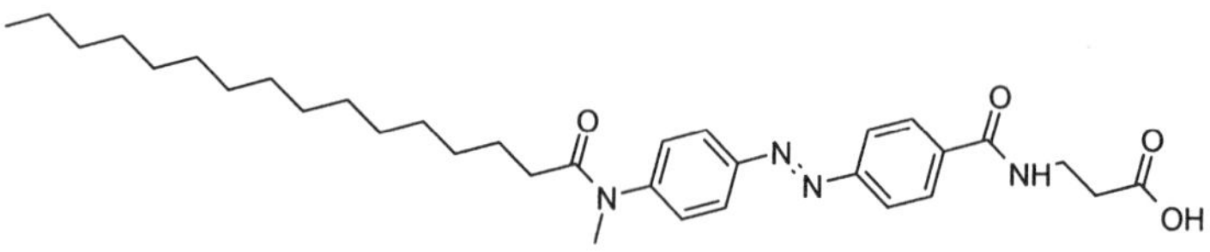

16

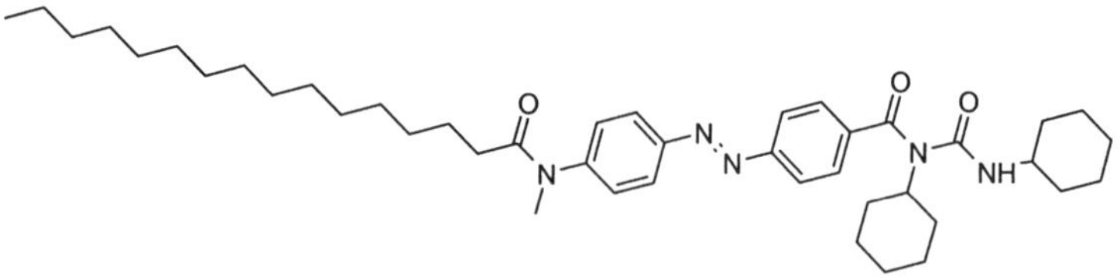

17

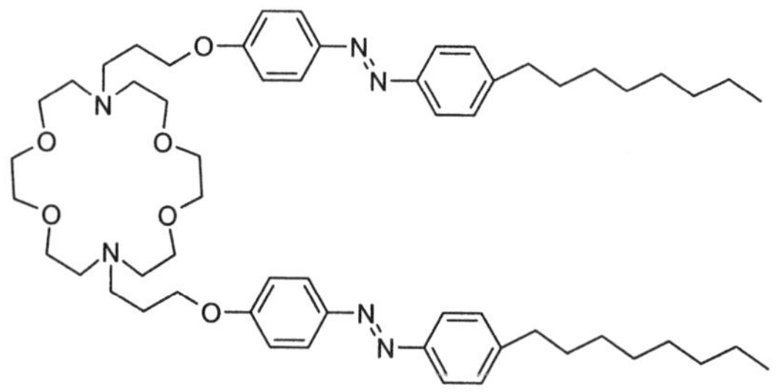

18

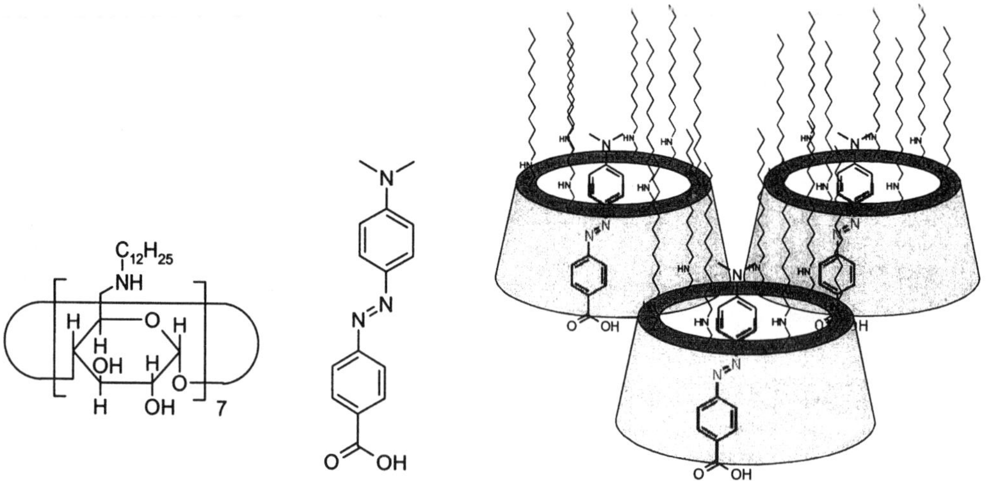

FIG. 6.11 Amphiphilic β-cyclodextrin, amphiphilic azobenzene, and a schematic representation of the molecular arrangement of their inclusion complex in LBK films (adapted with permission from reference 51).

Tethering azobenzene moieties to a crown-ether as polar head group (**18**) gives an amphiphile that forms monolayers at the air/water interface. The large crown-ether functionality provides enough free volume for azobenzene *trans* to *cis* isomerization. The photochemical conversion in this azobenzene monolayer reached 76%,[53] showing a great increase compared with that in a simple long-chain fatty acid azobenzene (**7**) LBK film (30% determined by spectroscopy, 19% determined by electrochemical methods.[18] The percentage is only a little smaller than the value obtained in solution (86%).

In addition to crown-ethers, calix[4]resorcinarenes **19** have also been used as head groups to obtain azobenzene amphiphiles with sufficient photo-isomerization in LBK films. For this purpose, azobenzene moieties have been tethered to the lower rim of the crown conformer of the calixarene. O-octacarboxymethoxylated calix[4]resorcinarenes **19** (X = CH$_2$-COOH) display efficient *trans* to *cis* photoisomerizability in densely packed mono-layers on a water surface, in LBK films, and in surface-adsorbed monolayers, whereas the noncarboxymetylated **19** (X = H) derivative gives films that are too densely packed. However, the aggregation is already suppressed efficiently compared to the azobenzene derivative without calixaren.[54,55]

19

In contrast to this, it is possible to suppress the photoisomerization completely by employing a head group that generates a highly ordered rigid structure. This is the case when urea is used as the head group (**20**) (see Figure 6.12). Due to the hydrogen bonds that form at the air/water interface or when the monolayer is deposited on a hydrated mica surface, there is a characteristic, tilted packing of the chromophores, and photoisomerization is completely hindered. However, H-aggregation of the chromophores was found in the dry state, but photoisomerization proceeded moderately.[56]

6.4.3 Coupling of Azobenzene Moieties to Polymers

6.4.3.1 Coupling between Ionic Azobenzene Amphiphiles and Polyelectrolytes

The area per amphiphile can be adjusted by complexation of ionic amphiphiles with polyelectrolytes having the opposite charge. The areal requirements of the complex are determined by the distance of the ionic

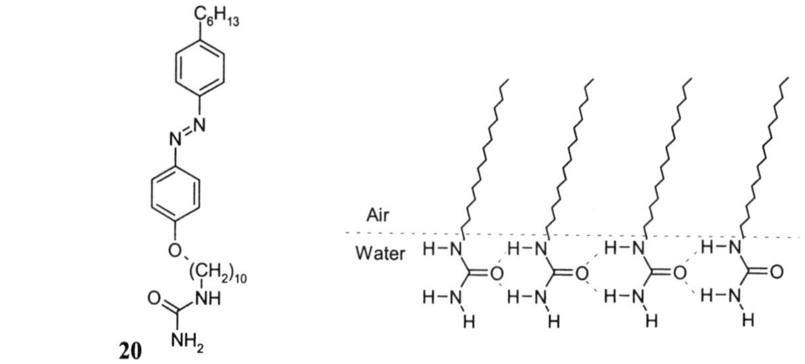

FIG. 6.12 Schematic representation of hydrogen bond formation in monolayers of urea containing amphiphiles **20** at the air/water interface.

groups from each other along the polymer chain. The complexation of ionic azobenzene amphiphiles with suitable polyelectrolytes, therefore, allows fine-tuning of the free volume around the azobenzene moiety so that photoisomerization is possible.

The first example of an azobenzene amphiphile polyelectrolyte complex was reported by Shimomura and Kunitake.[57] They used poly(vinylsulfate) **23** to stabilize an ammonium amphiphile (**21**, n = 5). Because the tertiary ammonium head group is rather large and the distance of the ionic sites in poly(vinylsulfate) is rather small, the packing of the amphiphiles is not significantly loosened by the complexation. In this case, the influence of the poylelectrolyte on the spectral properties and the photoisomerization is small.

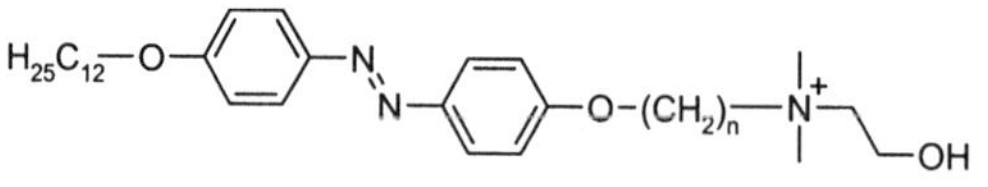

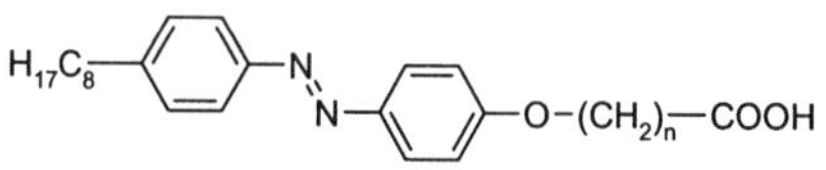

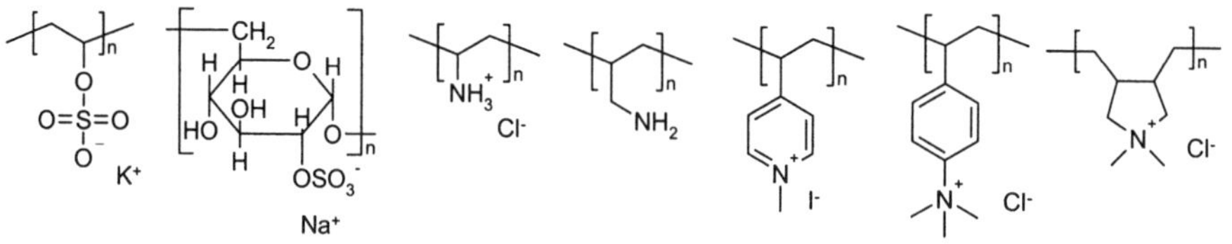

A small but significant influence on the packing in monolayers at the air/water interface was found when dextran sulfate **24**, which has a much larger distance between the ionic sites, is used with azobenzene amphiphile **21** (n = 10).[58] A large effect was found for complexes of an azobenzene fatty acid **22** (n = 3) complexed with poly(allylamine) **26**, however. In this case, the area per amphiphile increased from 0.28 nm^2 to 0.39 nm^2. This increase

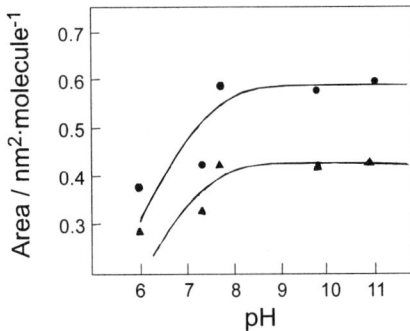

FIG. 6.13 pH-dependence of the area per molecule **22** (n = 5) on a subphase containing polyelectrolyte **28** obtained from surface-pressure isotherms at 30 mN m^{-1} (▲) and extrapolated to Π = 0 (●) (reproduced from reference 61 with permission from Elsevier Science).

results in an increased free volume in the LBK film, and the *trans* to *cis* photoisomerization proceeds to a higher extent than in LBK films of the pure azobenzene amphiphile.[59] At first glance, this result is somewhat astonishing, since the distance of the potential ionic sites in poly(allylamine) **26** is smaller than in dextran sulfate **24** but the effect on the photoisomerization is larger. It can be understood, however, by taking into account that both the azobenzene containing fatty acid **22** and the poly(allylamine) **26** are weak electrolytes. In this case, depending on the pH, some of the polymer is incorporated in the monolayer and the distance between the amphiphiles is increased.[60]

A detailed investigation into the influence of polyelectrolytes was conducted by Nishiyama *et al.*[61] The authors spread an azobenzene amphiphile **22** (n = 5) on subphases containing the polyelectrolyte **28** at different pH values. The degree of ionization does not change for this polyelectrolyte, but it does for the amphiphile. They found that the area per amphiphile measured at 30 mN/m depends very much on the pH (see Figure 6.13). It was rather low at pH = 6.0, where not all of the carboxyl group are deprotonated, but it reached a plateau at approximately pH = 8.

The authors concluded that at lower pH, the protonated azobenzene amphiphiles are packed in between the deprotonated amphiphiles. At pH = 8.0, however, all amphiphiles are deprotonated, and their distance is given by the distance of the ionic sites of the polyelectrolyte (see Figure 6.14). Similar results were obtained with other polyelectrolytes (**26** through **29**). According to this study, the area per amphiphile extrapolated to Π = 0 can be adjusted by choosing the size of the polymeric counterion, which can be calculated according the Corey-Pauling-Koltun (CPK) model (see Table 6.1). Because of the different possible conformations, the calculation of the size of the polymer counterions is somewhat ambiguous. However, the results summarized in Table 6.1 clearly indicate that the area per amphiphile can be controlled by employing polymeric counterions.[61]

The increased area per amphiphile in LBK films in which azobenzene amphiphiles have been complexed with a polymeric counterion results in

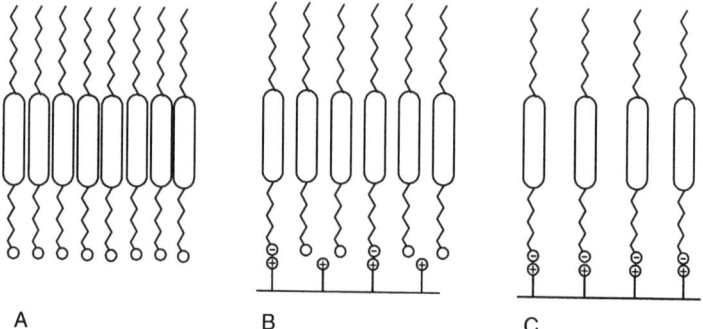

FIG. 6.14 Schematic representation of the pH-dependent change in the monolayer structure of azobenzene amphiphile **22** spread on (A) a pure water subphase, (B) a subphase containing **28** at pH = 6.0, and (C) at pH = 8.0 (adapted from reference 61 with permission from Elsevier Science).

TABLE 6.1   Dependence of the Area per Amphiphile 22 (n = 5) on the Polymeric Counterion[61]

Counterion	Area per molecule 22		CPK model
	$\Pi = 30$ mNm^{-1}	$\Pi = 0$ mN m^{-1}	
Ba^{2+}	0.25	0.32	–
poly(vinylammonium) chloride **25**	0.37	0.54	0.35–0.41
poly(allylamine) **26**	0.39	0.50	0.36–0.43
poly(N-methyl vinylpyridium) iodide **27**	0.41	0.61	0.41–0.63
poly(N-benzylvinyl,N,N,N-trimethyl ammonium) chloride **28**	0.42	0.59	0.56–0.70
poly(N,N-diallyl-N,N-dimethyl ammonium) chloride **29**	0.45	0.65	0.64–0.76

improved photoisomerization. Although photoisomerization is completely suppressed in the case of the densely packed LBK films with Ba^{2+} counterions, it is almost unrestricted in the case of the poly(N-benzylvinyl,N,N,N-trimethyl ammonium) chloride **28** counterion.[61] Similar observations have been made for the complexes of a two-azobenzene-chain amphiphile **30** with polycations, and an almost linear relation was found between the area per molecule and the fraction of *cis*-isomer in the irradiated LBK-film[62] (see Figure 6.15).

Recent investigations of a polyelectrolyte complex of **31** and poly(diallyl-dimethylammonium) chloride **29** have shown that although photoisomerization is unrestricted in the LBK films, some mechanical stress is built up in the films upon *trans* to *cis* isomerization. This was established by AFM, which shows reversible morphological changes. On irradiation, a number of protrusions with a height of ca. 5 nm appeared on the surface. These structures

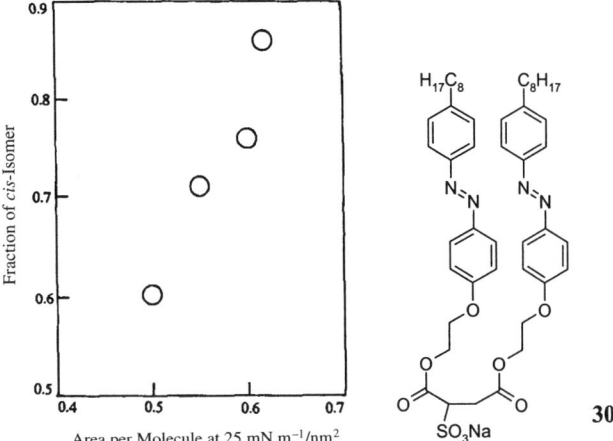

FIG. 6.15 Fraction of the *cis*-isomer at the photostationary state in LBK films as a function of the area per molecule at 25 mN/m for azobenzene amphiphile **30** (reproduced from reference 62 with permission from Elsevier Science).

disappeared upon irradiation with visible light, which causes *cis* to *trans* isomerization.[63]

$$H_{17}C_8 - \text{(ring)} - N\!\!=\!\!N - \text{(ring)} - O-(CH_2)_n-SO_3^- \ Na^+$$

31

6.4.3.2 Covalent Attachment to a Polymer Chain

Several problems occurring in polyelectrolyte complexes can be circumvented by attaching the azobenzene moiety covalently to a polymer chain. Seki and Ichimura used poly(vinylalcohol) as a hydrophilic backbone and attached azobenzene fatty acids **32** (n = 5, 10) to this backbone.

32

The corresponding fatty acid could not be photoisomerized in the LBK film. By attaching the azobenzene chromophore to the hydrophilic backbone, however, the free volume in LBK films was increased and photoisomerization was possible (i.e., 50 to 70% *cis*-isomer compared to 0% for the nontethered azobenzene amphiphile and 90% *cis*-isomer in solution). However, concomitant with the increased free volume, there is a decrease in the orientational order of the chromophores.[64,65] These polymers have been widely used as command surfaces to control the orientation of liquid crystals[66,67] and to investigate the photomechanical effect.[68]

In polyurethane **33**, the azobenzene moieties are separated from each other along the polymer backbone by isophorone units and have the free volume necessary for isomerization and molecular reorientation. This polymer's azobenzene moieties can be photoisomerized readily in mixed LBK films, as demonstrated by measuring the optically induced birefringence that originates from the photoreorientation of the chromophore upon polarized irradiation[69] (see Section 6.5.2).

33

A system similar to the partially substituted poly(vinylalcohol) was proposed by Penner *et al.*[28] They prepared copolymers of a mesogenic azobenzene monomer and hydroxyethylacrylate (HEA) **34** with different compositions.

34   **35**

Copolymers **34** with a 1:5 or 1:10 composition transferred readily as multilayers. UV-vis spectroscopy revealed through a strong blue shift of the π-π*-band compared to the solution that the azobenzene moieties are highly aggregated in the LBK films at room temperature. The extent of the blue shift suggests a crystalline order in the LBK film. The situation changed when the LBK film was heated into the liquid crystalline phase (around 50 to 60°C for the LBK film). The phase transition was accompanied by shift in the peak position from 335 nm (H-aggregates) to 355 nm (dimers/non aggregated).[28] That is, by heating into the liquid crystalline phase, the packing of the chromophores was significantly loosened; there was still sufficient interaction to stabilize the LBK film, however.[28]

Illumination of LBK films of copolymers **34** (1:2.4) in the liquid crystalline state at 63°C, with polarized light at 457 nm, results in a photoinduced optical anisotropy. That is, the chromophores can be reoriented, which indicates that efficient photoisomerization is possible.[70] Photoreorientation is also possible at room temperature after the original structure of the LBK film has been broken up by irradiation with UV light (that is, after a photostationary state with a high content of *cis*-isomer has been established).[71]

The copolymers **35** and the corresponding homopolymer (n = 0) are liquid crystalline, too, and have been deposited as LBK films.[72–75] In the homopolymer's LBK film, the chromophores are aggregated, as evidenced by the shift in the UV-vis spectrum, but isomerization is possible. This was attributed to the higher degree of flexibility in a liquid crystalline system as compared to a crystalline system. Photoisomerization is accompanied by a major change in the structure of the LBK film. A compressed monolayer at the air/water interface with tightly packed chromophores in a liquid crystalline order is deposited on a solid support, giving a smectic-like multilayer assembly. This structure is metastable and photoisomerization from *trans* to *cis* is enough to destabilize the structure. So, upon irradiation with UV light, the double layers are irreversibly destroyed. After a photoinduced *trans-cis-trans* isomerization cycle, a largely disordered structure is obtained, with a complete loss of the layered structure (see Figure 6.19).[73] The structural changes are accompanied by irreversible changes in the optical properties, which can be monitored by surface plasmon microscopy.

On the other hand, in the LBK films of the copolymers **35**, aggregation does not occur because of the spatial decoupling of the chromophores. Photoisomerization takes places readily in these films upon irradiation and causes changes in the optical properties, which again can be probed by surface plasmon spectroscopy. In this case, however, the spectra and the changes in the optical properties are fully reversible upon alternate irradiation.[73,75]

LBK films of photochromic liquid crystalline copolymers containing mesogenic cholesterol and azobenzene side chains attached to a polysiloxane main chain are another example of the approach of using order and flexibility of liquid crystalline phases. In the LBK films, no aggregation of the chromophores is observed, and photoisomerization is possible.[76]

Another class of polymers equipped with azobenzene moieties comprises α-helical polypeptides, in particular poly(L-glutamate)s and poly(L-lysine)s. In solution, these azobenzene-modified polypeptides can undergo photoinduced helix-coil transitions.[77–80] Polypeptides partially (30 to 50%) substituted with azobenzene moieties are surface active and form stable monolayers.[81,82] Because of the partial substitution, there is sufficient free volume, and the azobenzene moieties can be isomerized in the monolayer. The photoisomerization changes the area per molecule, and the monolayer shows a photomechanical effect.[81] LBK films of a photosensitive poly(L-lysine) with 31 mol

36   37   38   39

% azobenzene moieties in the side chains have been deposited and studied using circular dichroism, IR spectroscopy, X ray reflection, and surface plasmon resonance. The macroscopic structure of freshly prepared LBK films consists of deformed polypeptide helixes arranged in layers about 2.0 nm thick.[82]

α-helical polypeptides have rigid rodlike backbones. Equipped with flexible side chains, the then-called "hairy rod"-like polymers can be fabricated into LBK films, having exceptional quality and stability.[83–85] Azobenzene chromophores can be dissolved in the side chain region of such poly(L-glutamate)s having long alkyl chains.[45,49] Miscibility between the azobenzene dye and the side chain region is an issue, however, and this approach works well only for very oleophilic dyes.

More versatile and stable systems are prepared by covalent attachment of flexible alkyl chains and of the azobenzene dye to the polypeptide main chain. Such copolymers **37** form stable monolayers, which can readily be deposited on solid substrates to give LBK films of excellent optical quality.[86] The azobenzene moieties are not aggregated, can be isomerized by irradiation, and show spectral properties very similar to those of a solution. These characteristics indicate that the chromophores are embedded in a liquid-like matrix. Being embedded in a very fluid matrix, however, the isomerization does not produce significant changes in the structure of the LBK film. There is a very slow structural relaxation upon irradiation that resembles the relaxation observed upon heating the LBK film over the "melting temperature" of the side chains.[87] It was concluded that the molecular motions resulting from the repeated isomerization of the azobenzene moieties have the same effect on the LBK film structure as the increased chain mobility does on heating.

The number of chromophores in the LBK film can be enhanced by mixing the poly(glutamate) (**37**) with a low molecular weight azobenzene dye. Both chromophores, the covalently bound and the admixed dye, can be isomerized in LBK films by irradiation with UV light. The *trans* to *cis* isomerization reduces significantly the interaction between the bound and the admixed chromophores, and phase separation occurs. Upon repeated isomerization cycles, the low molecular weight dye is expelled from the LBK film and forms crystals on top of the film. This can be easily detected by polarized light microscopy.[49]

To enhance the chromophore density and to obtain stable films, polyglutamates that have azobenzene moieties in every side chain have been synthesized (**38, 39**).[88,89] In this case, the side chains have to be tailored to keep the "hairy rod" character of the polymer. The length of the alkyl spacer between the chromophore and the polymer backbone (n = 2, 3, 4, 6), as well as the length and means of tethering of the alkyl tail to the chromophore, have to be adjusted ($-C_6H_{13}$ or $-OC_{10}H_{21}$) so that the side chain region is in a liquid crystalline state.[39] The flexibility in the side chain region is necessary to ensure the "hairy rod" character and thereby to gain high monolayer stability and ensure that the monolayer can be transferred to solid substrates. Furthermore, the flexibility in the side chain region facilitates photoinduced *trans* to *cis* isomerization in LBK films with a high content of *cis*-isomer in the photostationary state. Particularly good results were obtained with

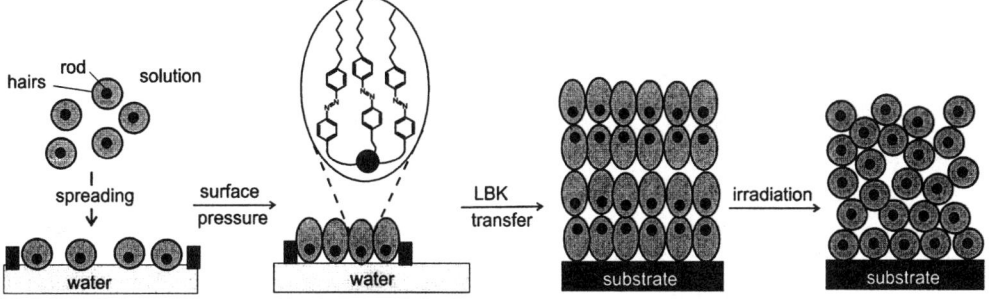

FIG. 6.16 Schematic representation of the deformation of photochromic "hairy rod"-like poly(L-glutamate)s **38** at the air/water interface through interaction with the water surface and the applied surface pressure, and the resulting bilayered structure of LBK films.

poly(L-glutamate)s **38** having an alkyl tail, whereas poly(L-glutamate)s with an alkoxy tail group **39** did not have the required flexibility of the side chains. Most likely, the ether bridge enables the tail to form a higher ordered state that lacks the flexibility of a slightly lower ordered liquid crystalline state.[33,39] However, polymers **39** can be deposited if the *cis*-form is spread, because the crooked *cis*-isomer cannot form highly ordered assemblies.[90]

The structure of mono- and multilayers of polymers **38** was investigated in detail.[29,91] The polymers form domains when spread at the air/water interface. From the peak position of the azobenzene π-π^*-band, it can be concluded that the chromophores are strongly aggregated in the domains even if no surface pressure is applied.[91] Furthermore, it was found that the "hairy rods," which should have a symmetrical distribution of the side chains around the main-chain helix, are deformed by contact with the water surface and the applied surface pressure (see Figure 6.16).

This deformation can be followed by UV-vis spectroscopy directly at the air/water interface. Upon compression of the monolayer, the absorbance changes not very much around the point where the first increase in surface pressure is recorded (see Figure 6.17). This is caused by the disappearance of the voids between the domains. Above the kink in the isotherm, however, the π-π^* band (~323 nm) decreases while the band at 248 nm increases. This increase is due to the increase in chromophore concentration upon compression of the monolayer. The decrease of the π-π^* band is caused by a preferential orientation of the chromophores perpendicular to the surface. Chromophores oriented perpendicularly are not detected because the transition dipole of the π-π^* band is parallel to the probing light. The band at 248 nm, however, is not sensitive to the orientation of the chromophore, because the transition dipole moment has a component perpendicular to the long axis of the molecule[89] (see Figure 6.17).

The structure with the deformed "hairy rods" is preserved when multilayers are deposited on solid supports, as evidenced by X ray reflectometry revealing a marked bilayered structure[29] (see Figure 6.16). Photoisomerization of the aggregated azobenzene chromophores in LBK films is possible and leads to significant structural changes. The bilayered structure is completely

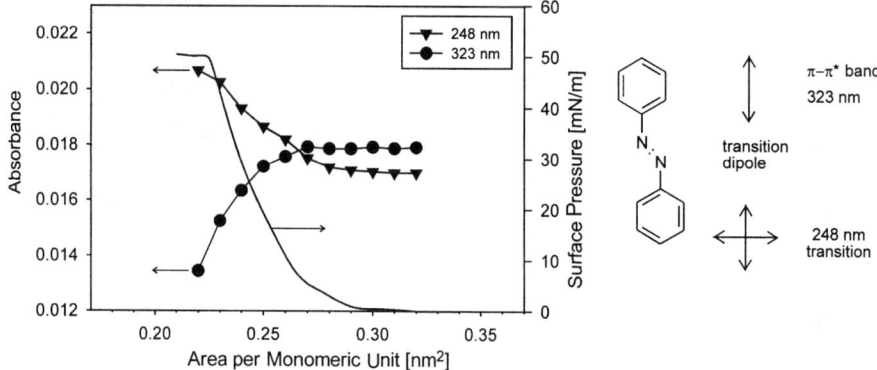

FIG. 6.17 Absorbance at 243 nm and 323 nm for monolayers of polymer **38** (n = 2) at the air/water interface as a function of the surface pressure and the corresponding surface pressure area/isotherm.

lost, indicated by the loss of the Bragg reflexes in the X ray reflectometry[29] (details of the structural changes will be discussed later; see Section 6.5.1) Furthermore, the optical properties, namely the refractive indices of the LBK films, change, as can be established by surface plasmon spectroscopy.[92] The structural changes are accompanied by a significant deaggregation of the azobenzene chromophores. After the structural changes have taken place, the azobenzene moieties can be photoreoriented.[69]

40

Poly(L-glutamate)s having both azobenzene and mesogenic, nonphotochromic biphenyl side chains (**40**) have been prepared to investigate the effect of aggregation on photoisomerization, and in particular on the possibility of reorienting the chromophores by polarized irradiation.[93] The polymers form mono- and multilayers that have the same densely packed, bilayered structure as the homopolymers, but the azobenzene moieties are spatially decoupled and are not aggregated in the LBK films. Nevertheless, photoreorientation by polarized irradiation is hindered, as is the case in LBK films of the homopolymer **38** (n = 2).[93]

Azobenzene chromophores have been incorporated not only in side chain polymers, but also in other polymer structures. There are several examples of incorporation of azobenzene into the polymer backbone[94–97] or in the backbone and in the side chain.[98] In these works, the main interest was to realize photoinduced changes of the polymer coil in solution.[43,94,95] Some of the main-chain azobenzene polymers were investigated in monolayers at the air/water interface. Blair *et al.* prepared rigid polyamides having azobenzene moieties in the main chain linked in 4,4 (**41**) or in 3,3 position (**42**). In both cases, the *trans* to *cis* photoisomerization is a favorable process in the mono-layer, because the *cis*-form of the polymer has the lower area per repeat unit. Both polymers show a contraction of the monolayer upon irradiation.[96,97]

41

42

Azobenzene was also incorporated in the shell of a dendrimer. In solution, the photoisomerization is unrestricted and a first-order thermal back reaction is found, although the packing of any functional group at the outside of a higher generation dendrimer should be rather dense. In LBK films prepared from azobenzene-functionalized dendrimers, aggregation has been established by UV-vis spectroscopy. The aggregation is suppressed when the shell of the dendrimer is functionalized by a 1:1 mixture of azobenzene moieties and long alkyl chains, however. The thusly modified dendrimers show reversible monolayer expansion upon irradiation with UV light.[99] This photomechanical effect can be found for other polymers, too (see for example reference 68), and is not specific to the dendritic character of the polymer.

Azobenzene has been incorporated in the shell of hydrophilic micro-particles, too. Here, the azobenzene moieties have sufficient free volume for isomerization, as shown by the fact that they can be efficiently reoriented by polarized irradiation[100] (see Section 6.5.2).

In conclusion regarding the results obtained with polymer-bound azoben-zene chromophores, establishing liquid crystalline phases that combine order and flexibility in the side chain region of the polymer seems to be the best way to obtain materials with a strong photoresponse. This is because a high density of chromophores can be combined with sufficient flexibility that the chromophores still can be photoisomerized in the dense packing of an LBK film. When a low density of chromophores is acceptable, however, the chro-mophores can be diluted along the polymer chain to reduce their interaction and secure sufficient free volume.

6.5 EXAMPLES OF THE MANIPULATION OF LBK FILM STRUCTURE BY PHOTOISOMERIZATION

6.5.1 Structure Change

The photoisomerization of the azobenzene moieties in LBK films can cause structural changes, as has already been mentioned briefly. Tachibana *et al.*

reported an example of a structural change that takes place in an LBK film of a double-chain azobenzene amphiphile complexed with polyviologen.[101] With repeated isomerization cycles, the tilt angle of the azobenzene moieties increases significantly. Upon storage in the dark, however, the original chromophore orientation—almost perpendicular to the substrate—was restored.

Another interesting example of photoinduced structural changes involves molecules made up of two flat, ring-shaped cyclic octapeptides connected by an azobenzene moiety (see Figure 6.18). The peptide rings were designed in such a way that they form hydrogen bonds on only one side, resulting in dimerization only. In solution, *trans*-isomers have been shown to form intermolecular assemblies (Figure 6.18), whereas *cis*-isomers form discrete dimers

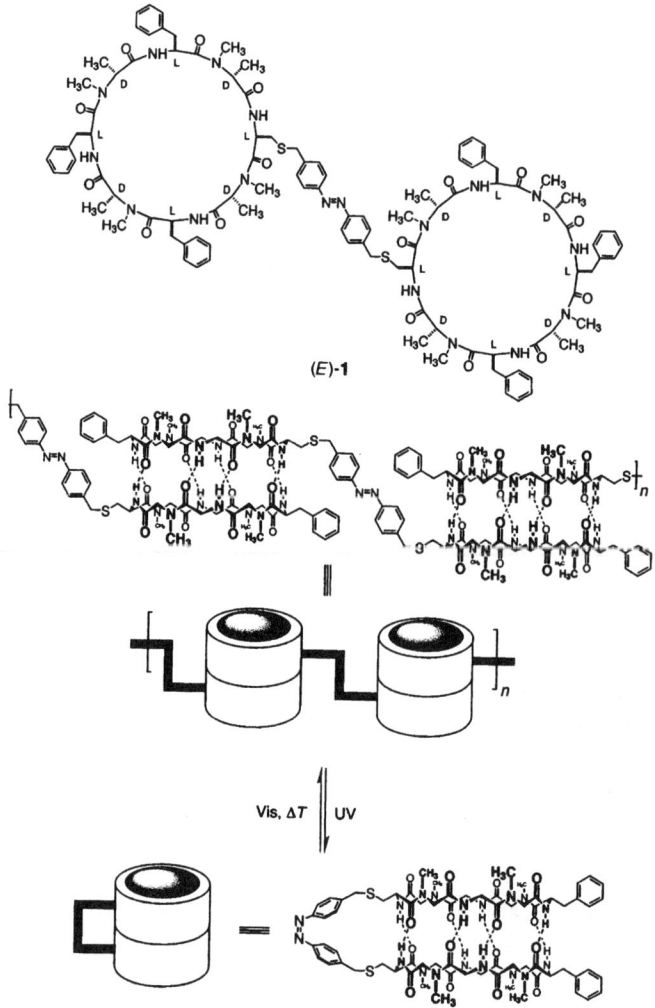

FIG. 6.18 Photoactive peptide made up of two flat, ring-shaped cyclic octapeptides connected by azobenzene, and their cylindrical aggregates in the *trans* state (intermolecular aggregate) and in the *cis* state (intramolecular aggregate). (Reproduced from reference 102 with permission from Wiley-VCH).

(intramolecular aggregates).[102] Due to the stronger interactions of the two peptide rings in the *cis*-form, the thermal back reaction is slower than in model azobenzene. So, irradiation with UV light results in nearly 100% *cis*-isomer, as has been shown by NMR.[102]

These photoresponsive peptides form stable monolayers at the air/water interface,[103] the structure of which depends on the isomerization of the azobenzene. Photoisomerization, therefore, induces changes in the structure, and with that an increase in the areal requirements at constant surface pressure. As confirmed by UV-vis spectroscopy, LBK films of these peptide monolayers deposited on quartz glass also retain the ability to isomerize.

There are several examples of photoisomerization of azobenzene moieties in LBK films that cause order-disorder transitions,[29,68,70,73,104,105] and similar results and interpretations were found for different polymers (34, 35, 38). The common aspects are that (1) the compressed monolayer at the air/water interface consists of chromophores oriented into the gas phase, and (2) the hydrophilic polymer backbone is oriented toward the air/water interface (see Figure 6.19). A very similar situation is found in "hairy rod"-like polymers (see Figure 6.16). This monolayer structure is preserved in the LBK transfer and gives rise to a bilayered structure with a long range period, as can be evidenced by X ray reflectometry. Upon photoinduced *trans* to *cis* isomerization, the interaction between the layers is weakened and the layers become more disordered. The resultant structure can be completely disordered[70,72] (Figure 6.19), a different liquid crystalline structure,[75,105] or a structure in which only the main chains remain ordered[29] (Figure 6.16), or it can lead to a recrystallization in the case of low molecular weight azobenzene amphiphiles.[105]

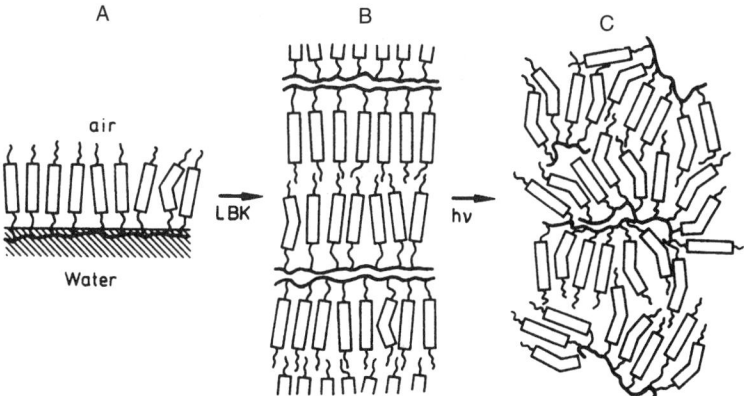

FIG. 6.19 Schematic representation of the structural formation and order-disorder transition for photoactive LBK films, showing (A) the compressed monolayer on the water surface with densely packed chromophore side chains oriented into the gas phase and the polymer backbone facing the water surface, and (B) LBK transfer from the water to a solid support, resulting in well-ordered smetic-like (bilayered) multilayer assemblies. (C) After photoinduced *trans* to *cis* isomerization, a largely disordered structure is obtained and the layered structure is completely lost (reproduced from reference 72 with permission from Wiley-VCH).

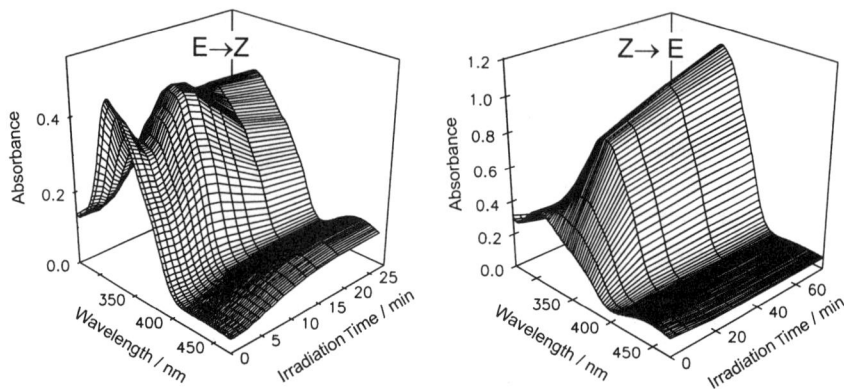

FIG. 6.20 UV-vis spectra of LBK film of polymer **38** (80 layers, 20 mN/m), as a function of irradiation time for irradiation with UV light ($\lambda \approx 360$ nm) (left) and for irradiation with visible light ($\lambda \approx 440$ nm) (right). (Reproduced with permission from reference 93).

UV-vis spectroscopy can be used to monitor these structural changes. The spectra of a LBK film of polymer **38** (n = 2), recorded after different irradiation times, are depicted in Figure 6.20. The decrease of the π-π^* band at 348 nm upon irradiation is not monotone, as would be expected for a unimolecular reaction, but rather there is a transient increase of the π-π^* band absorption. The absorbance at the n-π^* band (at 445 nm), which is characteristic for the *cis*-isomer, increases monotonically. This indicates that the irradiation continuously produces *cis*-isomer, even when the π-π^* band increases. Therefore, within the LBK film there must be additional processes, apart from photoisomerization, that increase the number of detectable chromophores faster than they are consumed by photoisomerization. From these results, in combination with X ray investigations, it was concluded that there is a structural change within the LBK film upon irradiation with UV light.[29,93] Furthermore, the development of the absorption at the π-π^* band indicates that the restructuring of the LBK film does not start immediately upon irradiation but rather that there is an induction period. Obviously, the restructuring starts only after a critical number of chromophores has been photoisomerized to the *cis*-isomer. After completion of the rearrangement within the LBK film, "normal" decrease of the π-π^* band occurs.[93]

Figure 6.21 allows a closer examination of the rearrangement by depicting the absorbance at wavelengths that are characteristic for the aggregated (323 nm) and the nonaggregated *trans*-isomers (348 nm), and for the *cis*-isomer (445 nm). Furthermore, the wavelength of the π-π^* band maximum has been depicted as a function of irradiation time. The irradiation process can be divided in three periods (see the dotted lines in Figure 6.21). The absorbance for both the aggregated and nonaggregated chromophores decreases in the first period, but for the aggregated chromophores, it decreases faster. In the second period, the absorbance increases. The uneven decrease of the absorbance at 348 nm and 323 nm results in a shift of the peak maximum for the π-π^* band from 323 nm to 340 nm in the "decrease" period and indicates

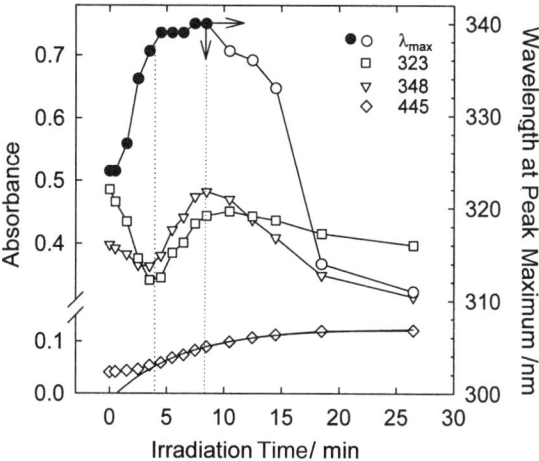

FIG. 6.21 Observed π-π^* band maximum and absorbance at the maximum for the π-π^* band of aggregated (323 nm) and nonaggregated (348 nm) *trans*-chromophores, as well as the absorbance at the n-π^* band (characteristic for the *cis*-isomer) as a function of the irradiation time (at $\lambda \approx 360$ nm) of a LBK film of polymer **38** (reproduced with permission from reference 93).

that the H-aggregates have been destroyed. The new peak position is not changed further until the decreasing π-π^* band of the *trans*-isomer is overwhelmed by that of the *cis*-isomer (the corresponding data are marked by open symbols in Figure 6.21).

From the continuous increase of the n-π^* band absorption (ca. 445 nm), it can be seen that photoisomerization takes place during the irradiation. This increase and the photoisomerization do not proceed by simple first-order kinetics as expected (see Figure 6.21), but rather there is a period during which the photoisomerization is slower. Because this period coincides with the period in which the aggregation is destroyed and the induction for rearrangement occurs, the following model can be suggested. Due to the closely packed structure within the aggregates, the azobenzene photoisomerization is slowed down, but beyond a critical amount of *cis*-isomer, the aggregates are destroyed and there is sufficient free volume for the photoisomerization process and a rearrangement toward a more symmetrical distribution of the side chains around the helical main chains. Because of this rearrangement, chromophores change from being out-of-plane, an orientation for which the π-π^* band cannot be detected, toward being in-plane, resulting in an increase of the π-π^* band. The n-π^* band does not show any increase as a result of the rearrangement of the chromophores. This is because the n-π^* band of the *trans*-isomer is much weaker than that of the *cis*-isomer, and the latter is not directed along the long axis of the azobenzene but has an angle of approximately 35°.[106]

The structural changes in the LBK films' photoisomerization of the azobenzene moieties are accompanied with changes in the optical properties. This can be proven employing surface plasmon resonance spectroscopy.[107,108] If a glass/metal interface is irradiated, the light is not reflected for all incident angles, but there is a certain angle at which the resonance conditions for the excitation of surface plasmons are fulfilled. If the metal layer is very thin, the resonance conditions are influenced by the material present at the back side of the metal layer; in particular, the resonance conditions are sensitive to the refractive index n_z (perpendicular to the surface) of the surrounding medium. The *trans* to *cis* isomerization in photoactive LBK films results in a change of the refractive index, and therefore, the photoisomerization results in a change of the resonance conditions. Irradiated and nonirradiated areas can be distinguished with high lateral resolution with this method. The use of photoactive LBK films as material and surface plasmon microscopy as read-out technique have been suggested for application in optical data storage.[46]

Observation of the reflected intensity at a fixed angle allows the monitoring of changes in the refractive index with time, and with that the structural changes. In such an experiment, the reflectivity of a glass/metal interface covered with an LBK film of a photochromic poly(L-glutamate) **38** (n = 6) shows an increase in the very beginning of UV irradiation (see Figure 6.22). As the irradiation proceeds, however, the reflectivity sharply drops.

The increase in the beginning is due to the *trans* to *cis* isomerization and the concomitant change in the refractive index. As long as the content of the *cis*-isomer is not too high, the structure is not changed; but as soon as the *cis*-content exceeds a critical value, the structural changes set in. At this point, the optical properties of the LBK film change in such a way that the reflectivity of the glass/metal interface drops. At the photostationary state, the reflectivity reaches a plateau. Upon *cis* to *trans* isomerization by irradiation with visible light, the reflectivity is restored partially, but it does not reach the original state. In subsequent irradiation cycles, it can be switched between the plateaus.[92] These results indicate that the change of the LBK film

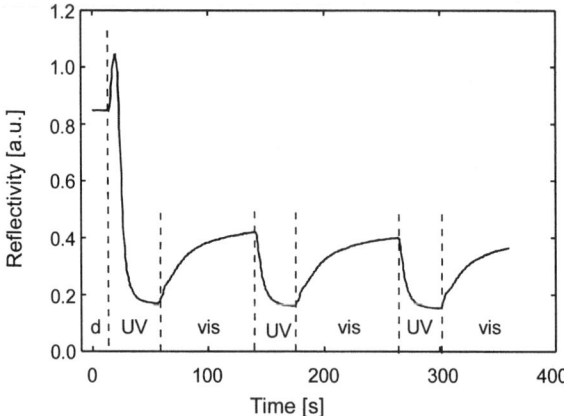

FIG. 6.22 Reflectivity of a glass/metal interface covered with 12 layers of poly(L-glutamate) **38** (n = 6) in a surface plasmon experiment with fixed angles corresponding to the resonance condition for the nonirradiated films as a function of the irradiation time at the first and subsequent irradiation cycles (UV = irradiation at 365 ± 50 nm; vis = irradiation at 440 ± 50 nm). (Reproduced with permission from reference 92).

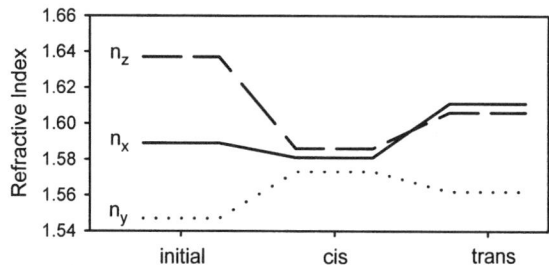

FIG. 6.23 Refractive indices for poly(L-glutamate) **38** (n = 6) LBK films in the initial state before any irradiation, in the *cis*-photostationary state after irradiation at 365 nm, and in the *trans*-photostationary state after irradiation at 440 nm, determined by waveguide spectroscopy (data from reference 90).

structure upon the first irradiation with UV light enhances the effect of the photoisomerization on the refractive index,[72,92] (see Figure 6.22). Employing waveguide spectroscopy in thicker LBK films, the refractive indices can be determined.[90] The values for LBK films of poly(L-glutamate) **38** are shown in Figure 6.23.

The problem with the application of these changes in optical properties for optical data storage is that the significant drop in reflectivity upon the first UV irradiation is related to an irreversible structural change. So the film can be used a as write-once medium only. Furthermore, the sensitivity of the system is low, so writing the information takes a rather long time. The changes in the optical properties in subsequent irradiation cycles are reversible, but they are connected with the unstable *cis*-isomer, which renders the inscribed information volatile.

To circumvent this problem of the unstable *cis*-isomer as information carrier, one can take an approach that combines photochemistry and electrochemistry. The *cis*-isomer can be reduced to a hydrazobenzene species with substantially more anodic potential than the *trans*-form, and there is a large difference in the adsorption spectra of *trans*-azobenzene and the hydrazobenzene. The photochemically inscribed and electrochemically developed information can be read out by monitoring this spectral change. Because both optical writing and electrochemical reduction are necessary for the formation of the hydrazobenzene, optical reading will not destroy the stored information. Electrochemical oxidation will erase the information.[109] Another approach for circumventing the use of the unstable *cis*-isomer in optical data storage applications involves using photoinduced birefringence (see Section 6.5.2).

The structural changes within LBK films upon irradiation can cause morphological changes, too. Irradiation of LBK films of azobenzene amphiphile **43** results in an increase of the surface roughness, as shown by AFM measurements.[105] The roughness is most likely caused by a recrystallization of the azobenzene amphiphiles in the irradiated area.

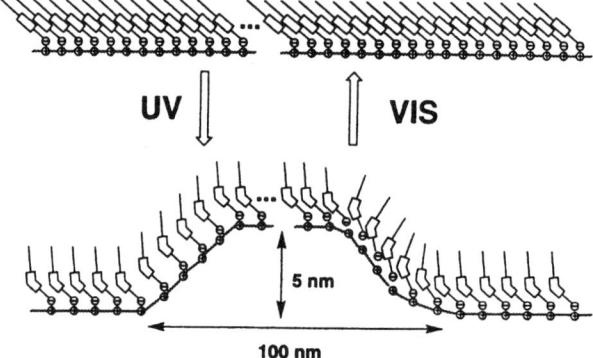

FIG. 6.24 Schematic representation of the structural change of a single-layer LBK film of the **31/29** complex upon photoisomerization of the azobenzene moieties (reproduced from reference 63 with permission. Copyright (1998) American Chemical Society).

Morphological changes in LBK films upon irradiation were observed for a complex of an azobenzene amphiphile with an anionic head group **31** and a cationic polyelectrolyte **29**. After illumination of a three-layer LBK film with UV light, i.e., *trans* to *cis* isomerization, some defects grew to form protrusions with a height of ~ 5 nm and a cross section of approximately 100 nm. The morphological changes were reversible upon irradiation with visible light, i.e., upon *cis* to *trans* isomerization.[63] The morphological changes were explained to be due to the increased cross-sectional area of the azobenzene moieties in the *cis*-form, which caused some stress in the film. The stress was released by giving a curvature to the film (see Figure 6.24).

Seki *et al.* have investigated the morphological changes in monolayers of partially azobenzene-substituted poly(vinylalcohol) **32** at the air/water interface and on hydrated mica by Brewster angle microscopy and by AFM.[110,111] They also found distinct protrusions when a relatively dense monolayer is irradiated with UV light. These protrusions have a height of ~ 10 nm and a width of ~ 200 to 300 nm. The protrusions disappear upon irradiation with visible light (or in the dark) due to the *cis* to *trans* isomerization.[110] If the coverage of the mica substrate is considerably lower, so that there is no stress built up during the photoisomerization, the increase of the domain size upon irradiation, which is responsible for the photomechanical effect of these polymers can be monitored. This increase in domain size can be followed directly at the air/water interface by Brewster angle microscopy.[111]

6.5.2 Photoreorientation

Photoreorientation of azobenzene chromophores by irradiation with polarized light is a very important photoinduced structural change. For azobenzene moieties, there is a widely accepted mechanism for the photoreorientation:[112–115] The azobenzene moieties that are parallel with their long axis (and therefore with their transition dipole) to the electric field vector

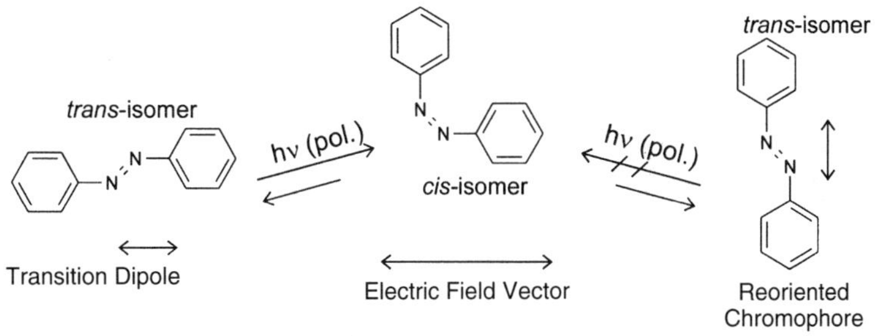

FIG. 6.25 Schematic representation of azobenzene photoreorientation by repeated photoisomerization upon irradiation with polarized light.

of the incoming light are isomerized. However, those molecules having an orientation perpendicular to the electric field vector are not isomerized (see Figure 6.25).

The thermal or photoinduced back reaction of the *cis*-azobenzene may result in an orientation different from the one that existed before isomerization (if there are no constraints posed by the matrix). The azobenzene moieties that are parallel to the incident light after this cycle can be isomerized again, but the moieties that are perpendicular cannot. Therefore, the population of chromophores oriented parallel to the incident light decreases with each isomerization cycle. As a result of this process, a new distribution for the azobenzene moieties is established, and the mean orientational direction is reoriented to be perpendicular to the polarized light. For an effective photoreorientation process, the number of isomerization cycles should be as high as possible. To this end, the photoisomerization should not be restricted in any way. Furthermore, the matrix should allow the orientational redistribution of the chromophores. Photoreorientation is a very promising effect for use in optical data storage systems, because the information is stable.[113,114,116,117] LBK films have been used extensively as model systems to study the photoreorientation process.

Employing UV-vis spectroscopy in combination with electrochemical methods, Wang *et al.* have shown that for an LBK film of azobenzene amphiphile, indeed only those chromophores that are parallel to the electric field vector of the incident light are isomerized by polarized irradiation. Furthermore, the authors have shown that the steric requirements of the *cis*-isomer favor a back reaction of the *cis*-isomer into a new orientation of the *trans*-isomer.[118]

The role of the supramolecular order of LBK films on the photoreorientation of azobenzene moieties upon polarized irradiation was investigated in detail. It was found that the photo(re)orientation proceeds readily in disordered, spin-coated films of azobenzene polymer **34** or azobenzene-containing polypeptide **38**, whereas the photoreorientation is hindered in highly ordered LBK films of these polymers.[71] In the case of polyacrylacrylat **34** and polypeptides **38** with shorter spacers (n = 2), the photoreorientation proceeds readily after the structure of the LBK film has been randomized by UV irradi-

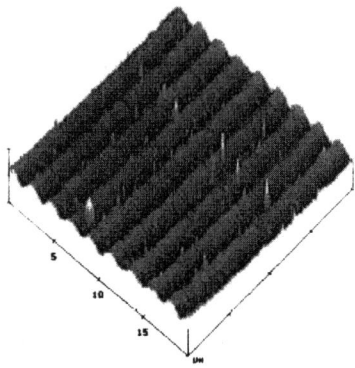

FIG. 6.26   Atomic-force microscopy 3-D topography image (20 μm × 20 μm × 200 nm) of the surface-relief grating on 100-layer mixed LBK film of polymer **44** and cadmium stearate deposited on glass (reproduced from reference 121 with permission. Copyright (1999) American Chemical Society).

ation[71] (see Figures 6.16 and 6.19). In the case of the polypeptide with the longer spacer **38** (n = 6), however, the coupling between the side chain and the still ordered main chain are too strong to efficiently reorient the azobenzene in the side chain.[71,119] Employing LBK films of copolypeptides **40**, Menzel *et al.* have shown that photoreorientation is hindered in well-ordered LBK films of the copolypeptide, even if the aggregation is suppressed by the introduction of the nonphotochromic biphenyl moiety. So the structure of the densely packed LBK films is decisive for the hindrance of the photoreorientation but not the aggregation of the chromophores.[93] Employing irradiation conditions for holographic experiments, the photoisomerization in azobenzene films can result in the formation of surface-relief gratings (see Chapter 14). Because the polarization of the writing beams has a distinct effect on the efficiency of the grating formation, it is supposed that photoreorientation of the chromophores plays a role in the grating formation.[120] Mendoça *et al.* have shown that surface-relief gratings also can be inscribed in LBK films of an azobenzene poly(methacrylate) **44** (see Figure 6.26)[121]

6.6 EXAMPLES OF LBK FILMS WITH A STRUCTURE TAILORED FOR THE DESIRED APPLICATION

6.6.1 Optical Data Storage

The majority of the applications that have been suggested for photoactive LBK films are in the area of optical data storage. The level of technical relevance of the proposed systems differs widely. The earliest reports suggested the use of the difference in the UV spectra due to the *trans* to *cis* isomerization for read-out of the inscribed information. More sophisticated read-out techniques, such as surface plasmon microscopy, have been developed to minimize the sensitivity problems of this approach. Another problem that arises from *trans* to *cis* photoisomerization in optical data storage is the transient nature of the stored information because of the unstable *cis*-isomer. It has been suggested that the combination of photoisomerization and electrochemical reduction of the *cis*-isomer could overcome this problem.[109] The most promising approach for optical data storage, however, is based on photo-induced birefringence as a consequence of photoorientation due to repeated photoisomerization of the chromophores. LBK films have been used as models for investigating the underlying mechanism of the photoreorientation, but only spin-coating films have the potential to be used in technical applications.

An example of tailoring the properties of photochromic moieties employing the LBK technique to obtain a material suitable for multifrequency optical data storage has been proposed by Hibino *et al.*[12] The authors investigated a number of spirobenzopyrans showing J-aggregation that differed based on their substitution and the matrix. By tailoring these parameters, they created a layered system in which the dye for each layer has a very narrow adsorption band with a position slightly different from that in all other layers. Up to ten different layers were deposited. The photoreaction shows a nonlinear dependency on the light intensity. Therefore, it is possible to write and nondestructively read at the same wavelength as long as the intensity of the read laser is low.

6.6.2 Sensors

Thin organic films have frequently been suggested for sensor applications.[122] The aim is to find an appropriate combination of functional units in a supramolecular assembly such that an external stimulus can be transferred

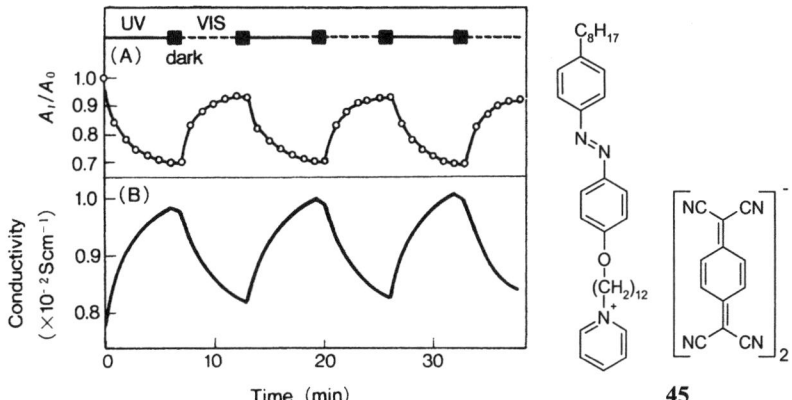

FIG. 6.27 Change in (A) absorbance at 356 nm and (B) conductivity of azobenzene amphiphile/TCNQ complex **45** LBK film (adapted from reference 123 with permission).

from the switching unit to a working unit that changes its function. Matsumoto *et al.* have proposed a sensor with a photochromic azobenzene dye as the switching unit. A photon received by the switching unit caused the azobenzene moiety to isomerize. The signal produced by this molecular deformation reached the working unit, a pyridinium $(TCNQ)_2$ complex, which changed the electrical conductivity of the film[121] (see Figure 6.27).[123]

6.7 SUMMARY

LBK films are highly ordered and densely packed supramolecular assemblies. The packing of the chromophores in LBK films influences their photo-reaction. Therefore, LBK films have to be tailored by the choice of the amphiphilic molecules and the transfer conditions to provide a matrix for the chromophores that has sufficient flexibility and free volume for the photo-reaction to occur. On the other hand, the LBK film's structure can be tailored in such a way that the photoreaction of the chromophore enhances the structure or the properties of the entire film. In this way, photoactive films can be designed to show a much more sophisticated photoresponse than that of a chromophore in an inert matrix.

REFERENCES

1. Spooner, S. P., and Whitten, D. G. Photoreactions in Monolayer Films and Langmuir-Blodgett-Assemblies. In: *Photochemistry in organized and constrained media*, V. Ramanmurthy, Ed. VCH: Weinheim (1991), p. 691.
2. Russel, J. C., Costa, S. B., Seiders, R. P. and Whitten, D. G. Photochemistry of Surfactant Stilbene in Organized Media: A Probe for Hydrophobic Sites in Micelles, Vesicles and other Assemblies. *J. Am. Chem. Soc.* **102**, 5678 (1980).
3. Quina, F. H., and Whitten, D. G. Photochemical Reactions in Organized Monolayer Assemblies. 4. Photodimerization, Photoisomerization, and Excimer Formation with Surfactant Olefins and Dienes in Monolayer Assemblies, Crystals, and Micelles. *J. Am. Chem. Soc.* **99**, 877 (1977).
4. Rennert, J. Photodimer Formation in the Condensed Phase of *trans*-Cinnamic Acid. *Photogr. Sci. Eng.* **15**, 60 (1971).
5. Stobbe, H. *Ber. Dtsch. Chem. Ges.* **52**, 666 (1919).
6. Mathauer, K., Schmidt, A., Knoll, W., and Wegner, G. Synthesis and Langmuir-Blodgett Multilayer-Forming Properties of Photo-Crosslinkable Polyglutamate Derivatives. *Macromolecules* **28**, 1214 (1995).
7. Seufert, M., Fakirov, C., and Wegner, G. Ultrathin Membranes of Molecular Reinforced Liquids on Porous Substrates. *Adv. Mater.* **7**, 52 (1995).
8. Wiegand, G., Jaworek, T., Wegner, G., and Sackmann, E. Heterogeneous Surfaces of Structured Hairy-Rod Polymer Film: Preparation and Methods of Functionalization. *Langmuir* **13**, 3563 (1997).
9. Kawamura, S., Tsutui, T., Saito, S., Murao, Y., and Kinae, K. Photochromism of Salicylidenenanilines Incorporated in a Langmuir-Blodgett Multilayer. *J. Am. Chem. Soc.* **110**, 509 (1988).
10. Brown, G. H. Ed. *Photochroism*, Wiley Interscience, New York (1971).
11. Polymeropoulos, E. E., and Möbius, D. Photochromism in Monolayers. *Ber. Bunsenges. Phys. Chem.* **83**, 1215 (1979).
12. Hibino, J., Hashida, T., and Suzuki, M. Multiple Memory Using Aggregated Photochromic Compounds. In *Photoreactive Materials for Ultrahigh Density Optical Memory*, M. Irie,

Ed. Elsevier: Amsterdam (1994), p. 25.

13. Möbius, D., Bucher, H., Kuhn, H., and Sondermann, J. Reversible Änderung der Fläche und des Grenzflächenpotentials monomolekularer Filme eines photochromen Systems. *Ber. Bunsenges. Phys. Chem.* **73**, 845 (1969).

14. Palto, S. P., Malthête, J., Germain, C., and Durand, G. On the Nature of Photoinduced Optical Anisotropy in Diacetylene Langmuir-Blodgett Films. *Mol. Cryst. Liq. Cryst.* **282**, 451 (1996).

15. Lackmann, H., Engelking, J., and Menzel, H. Photoreorientation of Non-Isomerizing Diphenyldiacetylene Chromophores in LB Films of a Polyelectrolyte Complex. *Mater. Sci. Eng.* C **8-9**, 127 (1999).

16. Lackmann, H., and Menzel, H. Change in Orientational Distribution of Non-Isomerizing Diphenyldiacetylene Chromophores in LB-Films. *Mol. Cryst. Liq. Cryst.* **345**, 131 (2000).

17. Brode, W. R., Gould, J. H., and Wyman, G. M. The Relation Between the Absorption Spectra and the Chemical Constitution of Dyes XXV. Phototropism and *cis-trans* Isomerism in Aromatic Azo Compounds. *J. Am. Chem. Soc.* **74**, 4641 (1952).

18. Liu, Z. F., Hashimoto, K., and Fujishima, A. A Novel Electrochemical Quantification Method for *trans/cis* Interconversion of Azo Compounds in a Solid Monolayer Film. *Chem. Lett.* 2177 (1990).

19. Fischer, E. J. The Calculation of Photostationary States in Systems A-B When Only A Is Known. *J. Phys. Chem.* **71**, 3704 (1967).

20. Zimmermann, G., Chow, L., and Paik, U. The Photchemical Isomerization of Azobenzene. *J. Am. Chem. Soc.* **80**, 3258 (1958).

21. Morigaki, K., Liu, Z. F., Hashimoto, K., and Fujishima, A. Electrochemical Counting of Photon Number Using the Assembled Monolayer-film of Azo Compound. *Sens. Actuators* B **13**, 226 (1993).

22. McRae, E. G., and Kasha, M. The Molecular Exciton Model. In: *Physical Processes in Radiation Biology.* L. Augenstein, R. Mason, and B. Rosenberg, Eds. Academic Press: New York (1964), p. 23.

23. Kasha, M., Rawls, H. R., and Ashraf El-Bayoumi, M. The Exciton Model in Molecular Spectroscopy. *Pure Appl. Chem.* **11**, 371 (1965).

24. Czikkely, V., Försterling, H. D., and Kuhn, H. Extended Dipole Model for Aggregates of Dye Molecules. *Chem. Phys. Lett.* **6**, 207 (1970).

25. Fukada K., and Nakahara, H. Electronic Spectra of Oriented Chromophore Systems by Incorporation of Azobenzene in the Polymer Main Chain and in Monolayer Forming Amphiphiles. *J. Colloid Interface Sci.* **98**, 555 (1984).

26. Shimomura, M., Ando, R., and Kunitake, T. Orientation and Spectral Characteristics of the Azobenzene Chromophore in the Ammonium Bilayer Assembly. *Ber. Bunsenges. Phys. Chem.* **87**, 1134 (1983).

27. Kunitake, T. Aqueous Bilayer Dispersions, Cast Multilayer Films and Langmuir-Blodgett Films of Azobenzene Containing Amphiphile. *Coll. Surf.* **19**, 225 (1986).

28. Penner, T. L., Schildkraut, J. S., Ringsdorf, H., and Schuster, A. Oriented Films from Polymeric Amphiphiles with Mesogenic Groups: Langmuir Blodgett Liquid Crystals? *Macromolecules* **24**, 1041 (1991).

29. Menzel, H., Weichart, B., Schmidt, A., Paul, S., Knoll, W., Stumpe, J., and Fischer, T. SAXS and UV-Vis Studies on the Structure and Structural Changes in LB Films of Polyglutamates with Azobenzene Moieties Tethered by Alkyl Spacers of Different Length. *Langmuir* **10**, 1926 (1994).

30. Furue, M., and Nozakura, S. Photoreduction of Bisviologen Compounds, Viologen-$(CH_2)_n$-Viologen, by 2-Propanol. *Bull. Chem. Soc. Jpn.* **55**, 513 (1982).

31. Shimomura, M., Utsugi, K., Horikoshi, J., Okuyama, K., Hatozaki, O., and Okuyama, N. Two-Dimensional Ordering of Viologen Polymers Fixed on Charged Surface of Bilayer Membranes: A Peculiar Odd-Even Effect on Redox Potential and Adsorption Spectrum. *Langmuir* **7**, 760 (1991).

32. Engelking, J., Ulbrich, D., and Menzel, H. Piezocchromic Effect and Orientational Order in Mono- and LB-Multilayers of Poly(p-phenylene sulfonate)-dioctadecyldimethylammonium Bromide Complexes. *Macromolecules* **33**, 9026, (2000).

33. Sato, T., Ozaki, Y., and Iriyama, K. Molecular Aggregation and Photoisomerization of

Langmuir-Blodgett Films of Azobenzene-Containing Long Chain Fatty Acids and Their Salts Studied by Ultraviolet-visible and Infrared Spectroscopy. *Langmuir* **10**, 2363 (1994).

34. Whitten, D. G. Photochemische Reaktionen oberflächenaktiver Moleküle in Systemen monomolekularer Schichten – Steuerung der Reaktivität durch die Umgebung. *Angew. Chem.* **91**, 472 (1979), *Angew. Chem. Int. Ed. Engl.* **18**, 440 (1979).

35. Loo, B. H., Liu, Z. F., and Fujishima, A. Scanning Tunneling Microscopic Images of an Azobenzene Derivative Differently Deposited on Highly Oriented Pyrolytic Graphite Surfaces. *Surf. Sci.* **227**, 1 (1990).

36. Kim, I., Rabolt, J. F., and Stroeve, P. *Colloids Surf. A* **171**, 167 (2000).

37. Lin, Z. F., Loo, B. H., Baba, R., and Fujishima, A. Excellent Reversible Photochromic Behaviour of 4-Octyl-4′(5-carboxylpentamethylene)-azobenzene in Organized Monolayer Assemblies. *Chem. Lett.* 1023 (1990).

38. Freimanis, J., Markawa, E., Matisowa, G., Gerca, L., Muzikante, I., Rutkis, M., and Siliush, E. Dynamic Monolayer Behavior of a Photo-responsive Azobenzene Surfactant. *Langmuir* **10**, 3311 (1994).

39. Menzel, H., and Rambke, B. The Influence of the Side Chain Architecture on the Thermal and Monolayer Forming Properties of "Hairy Rod"-like Polymers. *Macromol. Chem. Phys.* **197**, 2073 (1996).

40. Song, X., Geiger, C., Vaday, S., Perlstein, J., and Whitten, D. G., Supramolecular Aggregates of Photoreactive Aromatics. Structure, Photophysics and Photochemistry of Stilbene and Azobenzene Phospholipids. *J. Photochem. Photobiol. A: Chem.* **102**, 39 (1996).

41. Xing, L., and Mattice, W. L. Atomistic Simulations of Self-Assembled Monolayers That Contain Azobenzene. *Langmuir* **12**, 3024 (1996).

42. Liu, Z. F., Morigaki, K., Enomoto, T., Hashimoto, K., and Fujishima, A. Kinetic Studies on the Thermal *cis-trans* Isomerization of an Azo Compound in the Assembled Monolayer Film. *J. Phys. Chem.* **96**, 1875 (1992).

43. Kumar, G. S., and Neckers, D. C. Photochemistry of Azobenzene-Containing Polymers. *Chem. Rev.* **89**, 1915 (1989).

44. Xu, X. B., Majima, Y., and Iwamoto, M. Molecular Switching in Phospholipid-azobenzene Mixed Monolayers by Photoisomerization. *Thin Solid Films* **331**, 239 (1998).

45. Duda, G., and Wegner, G. A Helical Copolyglutamat as Solvent for Oleophilic Dyes in Langmuir-Blodgett Films. *Makromol. Chem., Rapid Commun.* **9**, 495 (1988).

46. Hickel, W., Duda, G., Wegner, G., and Knoll, W. Persistent Optical Storage in Ultrathin Polymeric Films and Read Out by Surface Plasmons. *Makromol. Chem., Rapid Commun.* **10**, 353 (1989).

47. Schoondorp, M. A., Schouten, A. J., Hulshof, J. B. E., Feringa, B. L., and Oostergetal, G. J. Langmuir-Blodgett Films of Amylose Acetal Dye Mixtures. 1. Monolayer Behaviour of Mixtures of Amylose Acetate and a Chiral p-Nitroazobis(benzene) Dye. *Langmuir* **8**, 1817 (1992).

48. Schoondorp, M. A., Schouten, A. J., Hulshof, J. B. E., and Feringa, B. L. Langmuir-Blodgett Films of Amylose Acetate/Dye Mixtures. 2. Langmuir-Blodgett Multilayers of Mixtures of Amylose Acetate and a Chiral p-Nitroazobis(benzene) Dye. *Langmuir* **8**, 1825 (1992).

49. Menzel, H. Langmuir-Blodgett Films of Photochromic Polyglutamates V. Mixtures of a Photochromic Polyglutamate and a Low Molecular Weight Azo Dye. *Macromolecules* **26**, 6226 (1993).

50. Ahuja, R. C., Maack, J., and Tachibana, H. Unconstrained *cis-trans* Isomerization of Azobenzene Moieties in Designed Mixed Monolayers at the Air/Water Interface. *J. Phys. Chem.* **99**, 9221 (1995).

51. Yabe, A., Kawabata, Y., Niino, H., Tanaka, M., Ouchi, A., Takahashi, H., Tamura, S., Tagaki, W., Nakahara, H., and Fukuda, K. *cis-trans*-Isomerization of the Azobenzenes Included as Guests in Langmuir-Blodgett Films of Amphiphilic β-Cyclodextrines. *Chem. Lett.* 1 (1988).

52. Markava, E., Gustina, D., Matisova, G., Kauda, I., Muzikante, I., Rutkis, M., and Gerca, L. Reversible *trans/cis* Photoisomerization in Langmuir-Blodgett Multilayers from Polyfunctional Azobenzenes. *Supramol. Sci.* **4**, 369 (1997).

53. Wang, Y. Q., Yu, H. Z., Mu, T., Luo, Y., Zhao, C. X., and Liu, Z. F. Photochromic and Electrochemical Behavior of a Crown-ether-derived Azobenzene Monolayer Assembly. *J. Electroanal. Chem.* **438**, 127 (1997).

54. Fujimaki, M., Matsuzawa, Y., Hayashi, Y., and Ichimura, K. Monolayers of Calix[4]resorcinarenes with Azobenzene Residues Exhibiting Efficient Photoisomerizability. *Chem. Lett.*, 165 (1998).

55. Fujimaki, M., Kawahara, S., Matsuzawa, Y., Kurita, E., Hayashi, Y., and Ichimura, K. Macrozyclic Amphiphiles. 3. Monolayers of O-octacarboxymethoxylated Calix[4]resorcinarenes with Azobenzene Residues Exhibiting Efficient Photoisomerizability. *Langmuir* **14**, 4495 (1998).

56. Seki, T., Fukuchi, T., and Ichimura, K. Humidity Sensitive Characteristic Stacking of Azobenzene in Monolayer of Urea Amphiphile on Hydrophilic Surface. *Langmuir* **16**, 3564 (2000).

57. Shimomura, M., and Kunitake, T. Preparation of Langmuir-Blodgett-Films of Azobenzene Amphiphiles as Polyion Complexes. *Thin Solid Films* **132**, 243 (1985).

58. Kimizuka, N., and Kunitake, T. Molecular Orientation and Domain Formation in Surface Monolayers of Azobenzene Containing Amphiphiles and Their Polyion Complexes. *Chem. Lett.* 827 (1988).

59. Nishiyama, K., and Fujihara, M. *cis-trans*-Reversible Photoisomerization of an Amphiphilic Azobenzene Derivative in its Pure LB Film Prepared as Polyion Complexes with Polyallylamin. *Chem. Lett.* 1257 (1988).

60. Tachibana, H., Yamaka, Y., Sakai, H., Abe, M., and Matsumoto, M. Structural and Morphological Changes and Polymerization Behaviors of Diacetylene Langmuir-Blodgett Films on Adding Water Soluble Polyallylamine in the Subphase. *Polymer* **42**, 1995 (2001).

61. Nishiyama, K., Kurihara, M. A., and Fujihira, M. Photochromism of Amphiphilic Azobenzene Derivative in its Langmuir-Blodgett Films Prepared as Polyion Complex with Ionic Polymers. *Thin Solid Films* **179**, 477 (1989).

62. Tachibana, H., Azumi, R., Tanaka, M., Matsumoto, M., Sako, S. I., Sakai, H., Abe, M., Kondo, Y., and Yoshino, N. Structures and Photoisomerization of the Polyion Complex Langmuir-Blodgett Films of an Amphiphile Bearing Two Azobenzene Units. *Thin Solid Films* **284-285**, 73 (1996).

63. Matsumoto, M., Miyazaki, D., Tanaka, M., Azumi, R., Manda, E., Kondo, Y., Yoshino, N., and Tachibana, H. Reversible Light Induced Morphological Change in Langmuir-Blodgett Films. *J. Am. Chem. Soc.* **120**, 1479 (1998).

64. Seki, T., and Ichimura, K. Formation and Langmuir-Blodgett Deposition of Monolayers of Poly(vinyl alcohol)s Bearing Azobenzene Side-Chains of Varied Spacer Length. *Thin Solid Films* **179**, 365 (1989).

65. Seki, T., and Ichimura, K. Formation and Langmuir-Blodgett Deposition of Monolayers of Polyvinylalcohol Having Azobenzene Side Chains. *Polym. Commun.* **30**, 108 (1989).

66. K. Ichimura, K. Photoalignment of Liquid-Crystal Systems. *Chem. Rev.* **100**, 1847 (2000).

67. Ichimura, K., Seki, T., Kawanishi, Y., Suzuki, Y., Sakuragi, M., and Tamaki, T. Photocontrol of Liquid Crystal Alignment by "Command Surfaces", in *Photoreactive Materials for Ultrahigh Density Optical Memory*. M. Irie, Ed., Elsevier: Amsterdam, (1994), p. 55.

68. Seki, T. Mono- and Multilayers of Photoreactive Polymers as Collective and Active Supramolecular Systems. *Supramol. Sci.* **3**, 25 (1996).

69. Dhanabalan, A., Dos Santos Jr., D. S., Mendonça, C. R., Misoguta, L., Balogh, D. T., Giacometti, J. A., Zilio, S. C., and Oliviera Jr., O. N. Optical Storage in Mixed Langmuir-Blodgett (LB). Films of Disperse Red-19 Isophorone Polyurethane and Cadmium Stearate. *Langmuir* **15**, 4560 (1999).

70. Geue, T., Stumpe, J., Möbius, G., Pietsch, U., Schuster, A., and Ringsdorf, H. Light Induced Modifications of Langmuir-Blodgett-multilayer Assemblies Containing Amphotropic Azocopolymers. *Mol. Cryst. Liq. Cryst.* **246**, 405 (1994).

71. Stumpe, J., Fischer, T., Ziegler, A., Geue, T., and Menzel, H. Photoorientation in LB Multilayers of Thermotropic Polymers. *Mol. Cryst. Liq. Cryst.* **299**, 245 (1997).

72. Sawodny, M., Schmidt, A., Urban, C., Stamm, M., Ringsdorf, H., and Knoll, W. Photoreactive Langmuir-Blodgett-Kuhn Multilayer Assemblies from Functional Liquid-

Crystalline Side Chain Polymers. I Homopolymers Containing Azobenzene Chromophores. *Polym. Adv. Technol.* **2**, 127 (1991).

73. Schmidt, A., Savodny, M., Knoll, W., Urban, C., Ringsdorf, H., Ahuja, R. C., and Möbius, D. Photoreactive Langmuir-Blodgett-Kuhn Multilayer Assemblies from Functionalized Liquid Crystalline Side Chain Polymers. II. Structural Characterization of Homo- and Copolymers Containing Azobenzene and Phenylbenzoate Moieties. *Acta Polym.* **45**, 217 (1994).

74. Sawodny, M., Schmidt, A., Stamm, M., Knoll, W., Urban, C., and Ringsdorf, H. Langmuir-Blodgett-Kuhn Multilayer Assemblies of Liquid-crystalline Azo-dye Side Chain Homo- and Copolymers. *Thin Solid Films* **210**, 500 (1992).

75. Sawodny, M., Schmidt, A., Urban, C., Ringsdorf, H., and Knoll, W. Photoreactions in Langmuir-Blodgett-Kuhn Assemblies of Liquid Crystalline Azo-dye Side Chain Polymers. *Prog. Colloid Polym. Sci.* **89**, 165 (1992).

76. Chen, X., Xue, Q. B., Yang, K. Z., Zhang, J. Z., and Zhang, Q. Z. Monolayer Behaviours for the Photochromic Liquid Crystalline Copolymers with Azobenzene and Cholesterol Side Groups. *Supramol. Sci.* **5**, 591 (1998).

77. Ueno, A., Takahashi, K., Anzai, J., and Osa, T. Photocontrol of Polypeptide Helix Sense by cis-trans Isomerism of Side Chain Azobenzene Moieties. *J. Am. Chem. Soc.* **103**, 6410 (1981).

78. Fissi, A., Pieroni, O., and Ciardelli, F. Photoreponsive Polymers: Azobenzene Containing Poly(L-Lysine). *Biopolymers* **26**, 1993 (1987).

79. Cooper, T. M., Natarajan, L. V., and Crane, R. L. Light-Sensitive Polypeptides. *Trends Polym. Sci.* **1**, 400 (1993).

80. Sisido, M. Molecular to Supramolecular Design of Synthetic Polypeptide. *Prog. Polym. Sci.* **17**, 699 (1992).

81. Malcolm, B. R., and O. Pieroni, O. The Photoresponse of an Azobenzene-Containing Poly(L-Lysine) in the Monolayer State. *Biopolymers* **29**, 1121 (1990).

82. Tedeschi, C., Fontana, M. P., Pieroni, O., Dei, L., Wilde, J., Pearson, C., Petty, M. C., and Tanner, B. K., Deposition and Characterisation of Langmuir-Blodgett Films of an Azobenzene-containing Poly-L-lysine. *Thin Solid Films* **335**, 197 (1998).

83. Orthmann, E., and Wegner, G. Herstellung ultradünner Schichten mit molekular kontrolliertem Aufbau aus polymeren Phthalocyaninen mit der Langmuir-Blodgett-Technik. *Angew. Chem.* **98**, 1114 (1986), *Angew. Chem. Int. Ed. Engl.* **25**, 1105 (1986).

84. Wegner, G. Ultrathin Films of Polymers. *Ber. Bunsenges. Phys. Chem.* **95**, 1326 (1991).

85. Menzel, H. Hairy Rod-like Polymers. In: *The Polymeric Materials Encyclopedia: Synthesis, Properties and Applications.* J. S. Salamone, Ed. CRC Press: Boca Raton (1996), p. 2917.

86. Menzel, H., and Hallensleben, M. L. Langmuir-Blodgett-Films of Photochromic Polyglutamates. *Polym. Bull.* **27**, 89 (1991).

87. Menzel, H., Hallensleben, M. L., Schmidt, A., Knoll, W., Fischer, T., and Stumpe, J. Langmuir-Blodgett-Films of Photochromic Polyglutamates IV. Spectroscopic and Structural Studies on LB-Films of Copolyglutamates Bearing Azobenzene Moieties and Long Alkyl Chains. *Macromolecules* **26**, 3644 (1993).

88. Menzel, H., Weichart, B., and Hallensleben, M. L. Langmuir-Blodgett-Films of Photochromic Polyglutamates II. Synthesis and Spreading Behaviour of Photochromic Polyglutamates with Alkylspacers and -tails of Different Length. *Polym. Bull.* **27**, 637 (1992).

89. Menzel, H., Weichart, B., and Hallensleben, M. L. LB-Films of Photochromic Polygluatamates III. Spectroscopic Studies on LB-Films of Photochromic Polyglutamates with Alkylspacers of Different Length. *Thin Solid Films* **223**, 181 (1993).

90. Büchel, M., Sekkat, Z., Paul, S., Weichart, B., Menzel, H., and Knoll, W. Langmuir-Blodget-Kuhn Multilayers of Polyglutamates with Azobenzene Moieties. Investigation of Photo-induced Changes in the Optical Properties and the Structure of the Film. *Langmuir* **11**, 4460 (1995).

91. Menzel, H., McBride, J. S., Weichart, B., and Rüther, M. Langmuir Blodgett Films of Photochromic Polyglutamates 8. Structure of the Monolayers at the Air/Water Interface. *Thin Solid Films* **284–285**, 640 (1996).

92. Menzel, H., Weichart, B., Büchel, M., and Knoll, W. Langmuir-Blodgett Films of

Photochromic Polyglutamates: Structure and Photochemically Induced Structural Changes. *Mol. Cryst. Liq. Cryst.* **246**, 397 (1994).

93. Menzel, H., Rüther, M., Fischer, T., and Stumpe, J. Discrimination of Structural Order and Chromophore Aggregation as Factors Effecting the Photo-reorientation of Azobenzene in Copolyglutamate LB Films. *Supramol. Sci.* **5**, 49 (1998).

94. Irie, M., Hirano, Y., Hashimoto, S., and Hayashi, K. Photoresponsive Polymers. Reversible Solution Viscosity Change of Polyamides Having Azobenzene Residues in the Main Chain. *Macromolecules* **14**, 262 (1981).

95. Irie, M., and Schnabel, W. Photoresponsive Polymers On the Dynamics of Conformational Changes of Polyamides with Backbone Azobenzene Groups. *Macromolecules* **14**, 1246 (1981).

96. Blair, H. S., Pogue, H. I., and Riordan, J. E. Photoresponsive Effects in Azopolymers. *Polymer* **21**, 1195 (1980).

97. Blair, H. S., and McArdle, C. B. Photoresponsive Polymers: 2. The Monolayer Behaviour of Photochromic Polymers Containing Aromatic Azobenzene Residues. *Polymer* **25**, 1347 (1984).

98. Reck, B., and Ringsdorf, H. Combined Liquid Crystalline Polymers: Mesogens in the Main Chain and as Side Groups. *Makromol. Chem., Rapid Commun.* **6**, 291 (1985).

99. Weener, J. W., and Meijer, E. W. Photoresponsive Dendritic Monolayers. *Adv. Mater.* **12**, 741 (2000).

100. Li, H., Zhang, L., Zhang, X., Shen, J., Ang, Y. Y., and Fei, H. A New Kind of Azo Polymeric LB Film for Reversible Optical Storage. *Polym. Bull.* **40**, 735 (1998).

101. Tachibana, H., Yoshino, N., and Matsumoto, M. Photoinduced Orientational Change in Langmuir-Blodgett Films of Azobenzene Complexed with Polyviologen. *Chem. Lett.,* 240 (2000).

102. Vollmer, M. S., Clark, T. D., Steinem, C., and Ghadiri, M. R. Photoschaltbare Wasserstoffbrücken-Verknüpfung in selbstorganisierten zylindrischen Peptidanordnungen. *Angew. Chem.* **111**, 1703 (1999), *Angew. Chem. Int. Ed. Engl.* **38**, 1598 (1999).

103. Steinem, C., Janshoff, A., Vollmer, M. S., and Ghadiri, M. R. Reversible Photoisomerization of Self-Organized Cylindrical Peptide Assemblies at Air-Water and Solid Interfaces. *Langmuir* **15**, 3956 (1999).

104. Möbius, G., Pietsch, U., Geue, T., Stumpe, J., Schuster, A., and Ringsdorf, H. Light Induced Modifications of Langmuir-Blodgett Multilayer Assemblies Containing Amphotropic Azopolymers. *Thin Solid Films* **247**, 235 (1994).

105. Schönhoff, M., Chi, L. F., Fuchs, H., and Lösche, M. Structural Rearrangements Upon Photoorientation of Amphiphilic Azobenzene Dyes Organized in Ultrathin Films on Solid Surfaces. *Langmuir* **11**, 163 (1995).

106. Uznanski, P., Kryzsewski, M., and Thulstrup, E. W. Polarized Absorption Spectroscopy of *trans*-Azobenzene and *trans*-Stilbene in Stretched Polyethylene Films. *Spectrochim. Acta Part A* **46**, 23 (1990).

107. Rothenhäusler, B., and Knoll, W. Surface-plasmon Microscopy. *Nature* **322**, 615 (1988).

108. Knoll, W. Polymer Thin Films and Interfaces Characterized with Evanescent Light. *Makromol. Chem.* **192**, 2827 (1991).

109. Liu, Z. F., Hashimoto, K., and Fujishima, A. Photoelectrochemical Information Storage Using An Azobenzene Derivative. *Nature* **347**, 658 (1990).

110. Seki, T., Tanaka, K., and Ichimura, K. Photomechanical Response in Monolayered Polymer Films on Mica at High Humidity. *Macromolecules* **30**, 6401 (1997).

111. Seki, T., Sekizawa, H., and Ichimura, K. Morphological Changes in Monolayer of a Photosensitive Polymer Observed by Brewster Angle Microscopy. *Polymer Commun.* **38**, 725 (1997).

112. Todorov, T., Nikolova, L., and Tomova, N. Polarization Holography 1: A New High Effiency Organic Material with Reversibel Photoinduced Birefringence. *Appl. Opt.* **23**, 4309 (1984).

113. Anderle, K., Birenheide, R., Eich, M., and Wendorff, J. H. Laser Induced Reorientation of the Optical Axis in Liquid Crystalline Side Chain Polymers. *Makromol. Chem., Rapid Commun.* **10**, 477 (1989).

114. Anderle, K., Birenheide, R., Werner, M. J. A., and Wendorff, J. H. Molecular Addressing?

Studies on Light-induced Reorientation in Liquid Crystalline Side Chain Polymers. *Liq. Cryst.* **9**, 691 (1991).

115. Barnik, M. I., Kozenkov, V. M., Shtykov, N. M., Palto, S. P., and Yudin, S. G. Photoinduced Optical Anisotropy in Langmuir-Blodgett-Films. *J. Mol. Electron.* **5**, 53 (1989).

116. Eickmanns, J., Bieringer, T., Kostromine, S., Berneth, H., and Thoma, R. Photoaddressable Polymers: A New Class of Materials for Optical Data Storage and Holographic Materials. *Jpn. J. Appl. Phys.* **38**, 1835 (1999).

117. Zilker, S. J., Bieringer, T., Haarer, D., Stein, R. S., van Egmond, J. W., and Kostromine, S. G. Holographic Data Storage in Amorphous Polymers. *Adv. Mater.* **10**, 855 (1998).

118. Wang, R., Iyoda, T., Hashimoto, K., and Fushiyama, A. Polarized Photoelectrochemical Reaction of an Azobenzene Derivative in Langmuir-Blodgett Films. *J. Photochem. Photobiol. A: Chem.* **92**, 111 (1995).

119. Stumpe, J., Fischer, T., and Menzel, H. Langmuir-Blodgett Films of Photochromic Polyglutamates 9. Relation Between Photochemical Modification and Thermotropic Properties. *Macromolecules* **29**, 2831 (1996).

120. Viswanathan, N. K., Kim, D. Y., Bian, S., Williams, J., Liu, W., Li, L., Samuelson, L., Kumar, J., and Tripathy, S. K. Humidity Sensitive Characteristic Stacking of Azobenzene in Monolayer of Urea Amphiphile on Hydrophilic Surface. *J. Mater. Chem.* **9**, 1941 (1999).

121. Mendoça, C. R., Dhanabalan, A., Balogh, D. T., Misoguti, L., dos Santos, D. S., Pereira-da-Silva, M. A., Giacometti, J. A., Zilio, S. C., and Oliveira, O. N. Optically Induced Birefringence and Surface Relief Gratings in Composite Langmuir-Blodgett (LB) Films of Poly4'-2-methacryloyloxy)ethylethylamino-2-chloro-4-nitroazobenzene (HPDR12) and Cadmium Stearate. *Macromolecules* **32**, 1493 (1999).

122. Roberts, G. G. Ed. *Langmuir-Blodgett-Films*. Plenum Press: New York (1990), p. 317.

123. Matsumoto, M. Information-Processing Using Photoisomeization in Langmuir-Blodgett Films. *FED-Janaru* **8** *(*Supplement 1), 30 (1997).

7

ELECTRONIC AND OPTICAL TRANSDUCTION OF PHOTOISOMERIZATION PROCESSES AT MOLECULAR- AND BIOMOLECULAR-FUNCTIONALIZED SURFACES

EUGENII KATZ
ANDREW N. SHIPWAY
ITAMAR WILLNER
Institute of Chemistry and The Farkas Center for Light-Induced Processes, The Hebrew University of Jerusalem, Jerusalem 91904, Israel

7.1 INTRODUCTION

Advances in supramolecular chemistry has led to ingenious molecular architectures exhibiting unique binding,[1] transport,[2,3,4] and catalytic[5,6] properties. Supramolecular architectures that reveal basic mechanical functions such as rotors[7,8,9,10,11] and translational movement[12] have been developed, and more complex assemblies also have been reported.[13,14,15] However, to realize the nanotechnologists' dreams of constructing complex devices, it is imperative that the supramolecular architectures exhibit external addressability and allow multidirectional communication with the macroscopic environment.

The ultimate goal of miniaturization is to use molecular assemblies as memory, processing, and mechanical devices. To reach these goals, it is essential to transform molecular structures between two or more states in response to external signals such as photonic, chemical, electrochemical, or magnetic stimuli, and to tailor readable output(s) such as electronic or optical signals that reflect the molecular state. The molecular units in a memory device, for instance, also must be individually addressable. To achieve this goal, the organization of molecular units on surfaces in two- and three-dimensional structures is of particular importance. Researchers have assembled functional monolayers and multilayers on electrodes[16,17,18,19] and other transducers.[20,21] Another desirable function of signal-activated chemical assemblies is the property of chemical self-amplification, that is, the output signal (or readout signal) is amplified as compared to the input stimuli. Amplification may be achieved if one state of an isomerizable molecule exhibits catalytic properties. Some of the fundamental features of these nanodevices are schematically presented in Figure 7.1. Numerous molecular systems that can be switched between two states by photonic,[22,23] electrical,[24,25] pH,[26] and chemical[27,28] stimuli have been studied in solution, and their functions as "switches," "memories," and "logic gates" have been discussed. Ingenious interlocked molecular structures[29,30] where threaded molecular units are translated between distinct molecular sites by electrical,[31] pH,[32] and photonic signals,[33] have been reported. Recently, some of these molecular structures have been incorporated in device-like assemblies and their potential application in molecular-scale computation was noted.[34,35] Many of the approaches to organize nanoscale devices follow models inspired by nature, and indeed, nature has provided us with definitive proofs that such goals may be realized. For example, the vision process represents an optoelectronic system where optical signals are transduced as electrical stimuli that activate the neural response. Photomorphogenesis represents many biological processes in plants

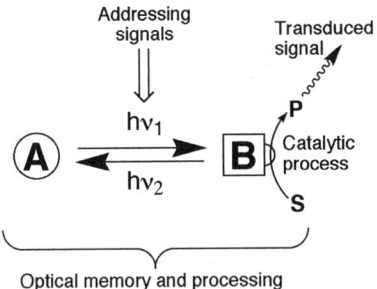

FIG. 7.1 Schematic representation of an amplified photoswitchable system, demonstrating some of the design principles of such devices.

where light-energy is transformed into movements of leaves or flowers. The light-triggered opening of ion-channels and the transformation of light into an ion flux represent biological devices that exhibit information storage and amplified read-out.

A broad class of materials, known as "photochromic" compounds, has been examined for decades as potential molecular units for information storage and optical memories.[36,37] Photochromic compounds undergo light-induced reversible isomerization between distinct states that differ in their properties. They possess some of the fundamental properties of optical "write-read-erase" functions, and recent research efforts suggest the possibility of using photochromic materials as 3D optical memories in the future.[38] In this chapter, we will address the development of optoelectronic systems based on photochromic materials. We will discuss the integration of photochromic materials with electronic transducers such as electrodes or piezoelectric crystals, and will emphasize the electronic read-out of photoswitchable systems. Complex and composite photoisomerizable molecular and biomolecular systems will be presented as "switches," "logic gates," "molecular machinery" and sensoric devices. It is our aim to present chemistry's contribution to the rapidly developing field of molecular and biomolecular optoelectronics and to highlight the future applications of these systems.

7.2 ELECTRONICALLY TRANSDUCED PHOTOCHEMICAL SWITCHING OF ORGANIC MONOLAYERS AND THIN FILMS

The immobilization of a photoisomerizable material that can be switched by light between redox-active and redox-inactive or conductive and insulating states offers an encouraging route toward integrated molecular memory devices. Figure 7.2 shows a photoisomer state "A" in which the molecular unit is redox-inactive and no electronic signal is transduced. Photoisomerization of the chemical component to state "B" generates a redox-active assembly, and the electron transfer between the electrode and the chemical modifier yields an amperometric (electrochemical) indicator of the state of the system.

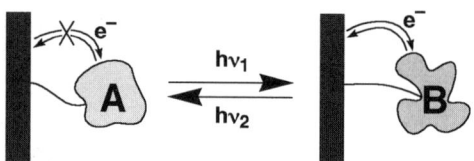

FIG. 7.2 Schematic representation of an optically switched redox monolayer.

The practical construction of such an optoelectronic system was shown with a phenoxynaphthacene quinone that is photoisomerizable between redox-active *trans*-quinone (**1a**) and redox-inactive *ana*-quinone (**1b**) states.[39,40,41,42] A carboxylic derivative of the phenoxynaphthacene quinone monolayer was assembled on a Au-electrode by the coupling of 6-[(4-carboxymethyl)phenoxy]-5,12-naphthacene quinone to a self-assembled cystamine monolayer [Figure 7.3(A)].[43,44,45] The cyclic voltammogram of the resulting monolayer shows an ill-defined redox-wave for **1a** [Figure 7.3. (B, curve a)] because a nondensely packed monolayer of the quinone is formed. The random orientation of the quinones relative to the electrode, as well as nonspecific adsorption of the quinone to the surface, yields a mixture of quinone units with different electrochemical features, leading to the broad voltammogram. Treatment of the **1a**-functionalized electrode with tetradecanethiol ($C_{14}H_{29}SH$) results in the plugging of pinhole defects in the monolayer, creating a densely packed mixed monolayer consisting of $C_{14}H_{29}SH$ and **1a**. Figure 7.3 (B), curve b, shows the effect of this thiol treatment. The quasi-reversible redox-wave of the electrode after treatment ($E°' = -0.62$ V vs SCE, curve b) is attributed to the two-electron redox process of **1a** in a rigid, aligned configuration. A coulometric assay of the charge associated with the reduction (or oxidation) of the **1a** component reveals a surface coverage of the quinone of 2×10^{-10} mole cm^{-2}. The electron transfer rate from the electrode to the quinone was estimated to be $k_{et} \approx 2.5$ s^{-1} by following the peak-to-peak separation of the redox-wave at different scan rates.[46] Figure 7.3 (B) shows cyclic voltammograms of the **1a/1b**-monolayer in the electrochemically active *trans*-quinone (**1a**)-state after irradiation, $\lambda > 430$ nm (curve b), and the electrochemically inactive *ana*-quinone (**1b**)-state produced upon irradiation, 305 nm $< \lambda <$ 320 nm (curve c). In the presence of the **1b**-monolayer, only the background current of the electrolyte is observed, implying that this photoisomer monolayer is redox-inactive within this potential range. By the cyclic photoisomerization of the monolayer between the **1a** and **1b** states, the transduced current is switched reversibly between "ON" and "OFF"-states [Figure 7.3 (B, inset)].

An important aspect of molecular optoelectronics is the amplification of the transduced response to the interface's state. Systems that perform this function could be used as electronic amplifiers for weak light signals or in the design of sensitive actinometers. One way to accomplish amplification is by coupling the electroactive component to an electron transfer cascade [Figure 7.4 (A)]. The mixed monolayer (consisting of $C_{14}H_{29}SH$ and the electrochemically inactive **1b**) provides an insulating layer, so that direct electron transfer

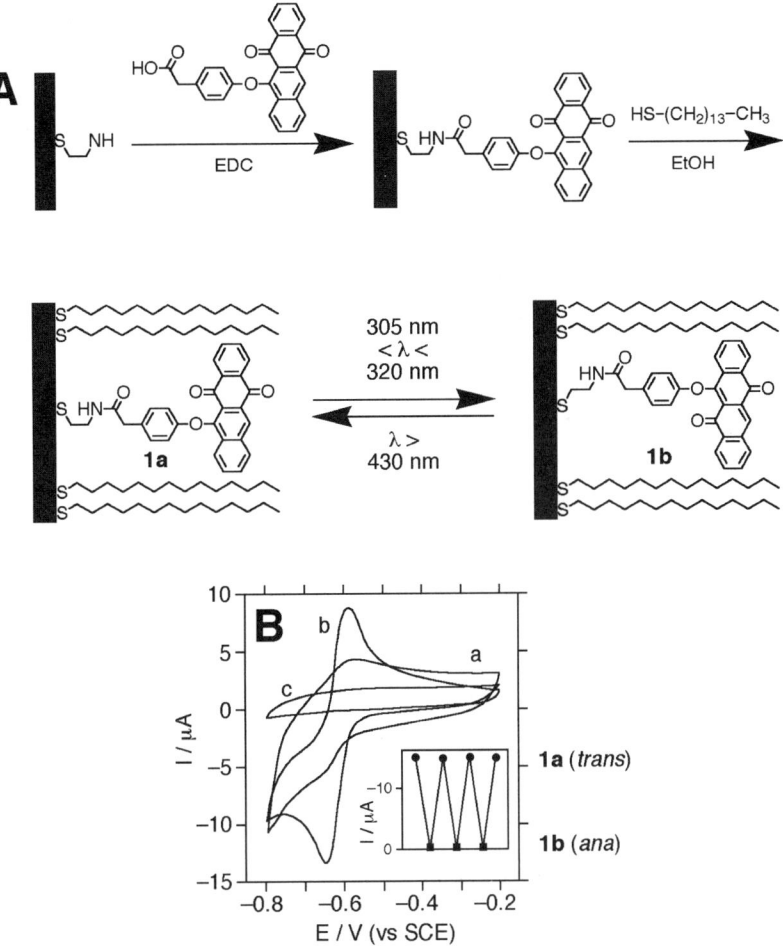

FIG. 7.3 (A) Assembly of a phenoxynaphthacene quinone/tetradecanethiol mixed monolayer on an Au-electrode and its photoisomerization. (B) Cyclic voltammograms of the *trans*-quinone monolayer (**1a**): (a) before rigidification with tetradecanethiol, (b) after rigidification with tetradecanethiol, and (c) cyclic voltammogram of the mixed monolayer after photoisomerization of the *trans*-quinone to the *ana*-quinone state. Cyclic voltammograms were recorded in 0.01 M phosphate buffer (pH 7.0) with a potential scan rate of 50 mV s^{-1}. Inset: Switching behavior of the cathodic peak current in the cyclic voltammogram of the mixed monolayer upon reversible photoisomerization. (Adapted from reference 43, Figure 1. Copyright 1996, American Chemical Society.)

from the electrode to a solution-state electron relay is inhibited. Electron contact between the electrode and solution can take place only using **1a** as a relay providing a gate for electron transfer through the insulating interface. In the electrochemically active *trans*-quinone state (**1a**), vectorial electron transfer from the redox-active units to the diffusional relay is possible, stimulating the electrocatalyzed reduction of the solubilized species. *N,N′*-Dibenzyl-4,4′-bipyridinium (BV^{2+}, **2**) may be used as a secondary diffusional electron relay.[43,44,45] It has a reduction potential of E°′ = –0.58 V vs SCE,[47] while the formal reduction potential of **1a** at pH=7.5 is E°′ = –0.65 V vs.

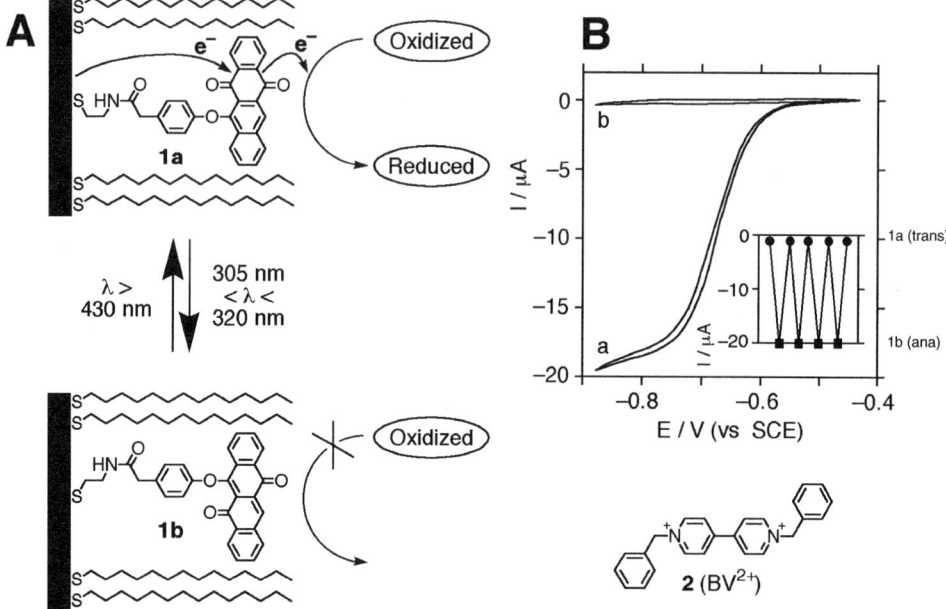

FIG. 7.4 (A) Use of diffusional electron relay to amplify the electrochemical signal of the photo-switchable phenoxynaphthacene quinone/tetradecanethiol mixed monolayer. (B) Cyclic voltammograms of the electrode in the presence of BV^{2+} (**2**) (1 mM): (a) in the *trans*-quinone state (**1a**) and (b) in the *ana*-quinone state (**1b**). Recorded at pH 7.5, scan rate 5 mV s^{-1}. Inset: Photoswitching behavior of the electrocatalytic current. (Adapted from reference 43, Figure 8, scheme 5. Copyright 1996, American Chemical Society.)

SCE.[43,44,45] The potential gradient allows the vectorial reduction of BV^{2+} by the electroactive **1a**-monolayer, and thus the activation of an electron transfer cascade can be achieved. The transduced current is enhanced ca. 10-fold in the presence of the electron relay. It should be noted that the cyclic voltammograms with and without the diffusional relay should be measured at the same slow potential scan-rate to allow the electron transfer cascade and demonstrate the current amplification. Figure 7.4 (B) shows the cyclic voltammograms of the photoisomerizable electrode in the presence of BV^{2+} and in the **1a**-state (curve a) and the **1b**-state (curve b). While the electro-catalytic current is present in the **1a**-state (achieved by electrode irradiation with visible light, $\lambda > 430$ nm), photoisomerization of the monolayer to the **1b**-state (305 nm $< \lambda < 320$ nm) results in a voltammogram showing only the background current (curve b), demonstrating that direct electron transfer to BV^{2+} is prohibited. The electrochemical response (electrocatalytic current) can be reversibly switched by cyclic photoisomerization of the quinone-functionalized electrode between redox-active (**1a**) and redox-inactive (**1b**) states [Figure 7.4 (B, inset)].

The reduction potential of the **1a**-monolayer is controlled by the pH of the electrolyte and is positively shifted as the pH decreases (e.g., $E^{\circ\prime} = -0.65$ V vs SCE at pH = 7.5 and $E^{\circ\prime} = -0.51$ V vs SCE at pH = 5.0),[43,44,45] whereas the reduction potential of **2** is pH-independent.[47] The pH-dependent

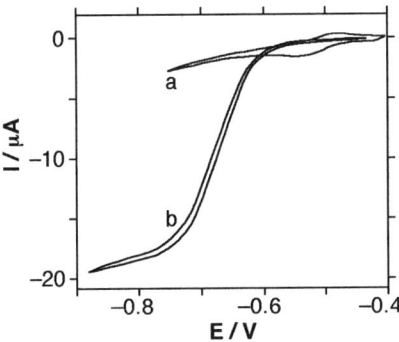

FIG. 7.5 Cyclic voltammograms of the *trans*-phenoxynaphthacene quinone (**1a**)/tetradecanethiol mixed monolayer in the presence of **2** (1 mM) at pH (a) 5.0, and (b) 9.2. Recorded in phosphate buffer (0.01M), with Na_2SO_4 (0.1 M), scan rate 5 mV s^{-1}.

shift of the redox potential of **1a** allows the use of pH as an additional controller of the interfacial electron transfer features of the **1a**-functionalized monolayer. At pH=5.0, the **1a**-monolayer is thermodynamically prohibited from stimulating electron transfer to BV^{2+} ($E^{o'}$ = −0.58 V vs SCE). Only a weak electrical response of the **1a**-monolayer is observed, without the activation of the electron transfer cascade [Figure 7.5 (curve a)]. At pH=9.2, the potential of **1a** is sufficiently negative to provide the efficient electrochemical reduction of **2** [Figure 7.5 (curve b)]. Thus, the phenoxynaphthacene quinone-functionalized monolayer electrode can be described as an "AND" gate with optical and pH inputs that act cooperatively in the activation of an electrochemical output.

Photoisomerizable properties of azobenzene-functionalized monolayers immobilized onto solid supports by covalent attachment,[48,49] by chemisorption of thiol groups,[50,51] and by the Langmuir-Blodgett (LB) method,[52,53] have been studied extensively using various spectral techniques and atomic force microscopy (AFM). These monolayers represent other examples of layered assemblies with electrochemical properties controlled by photoisomerization of the layer.[54] For example, 4-octyl-4'-(5-carboxy-pentamethylene-oxy)-azobenzene was deposited in the *trans*-state as a monolayer film onto a SnO_2 electrode using the Langmuir-Blodgett method.[54] The monolayer was reversibly photoisomerizable between *trans* (**3a**)- and *cis* (**3b**)-isomeric states [Figure 7.6 (A)], each of which displayed different electrochemical characteristics. Whereas the *cis*-isomer was readily reduced to the hydrazobenzene form (**3c**) [Figure 7.6 B, curve b)], the *trans*-isomer was electrochemically inactive in the potential range studied [Figure 7.6 B, curve a)]. The electrochemical oxidation of the hydrazobenzene returned the monolayer to the thermodynamically favored *trans*-isomer. Similar results were reported for a self-assembled monolayer of a thiol-functionalized azobenzene derivative.[55] The system has also been applied as an electrochemical actinometer because only one photogenerated isomer is electrochemically detectable.[56]

Another system providing photoswitchable redox-activated properties with amplification features via a secondary electrocatalytic vectorial electron

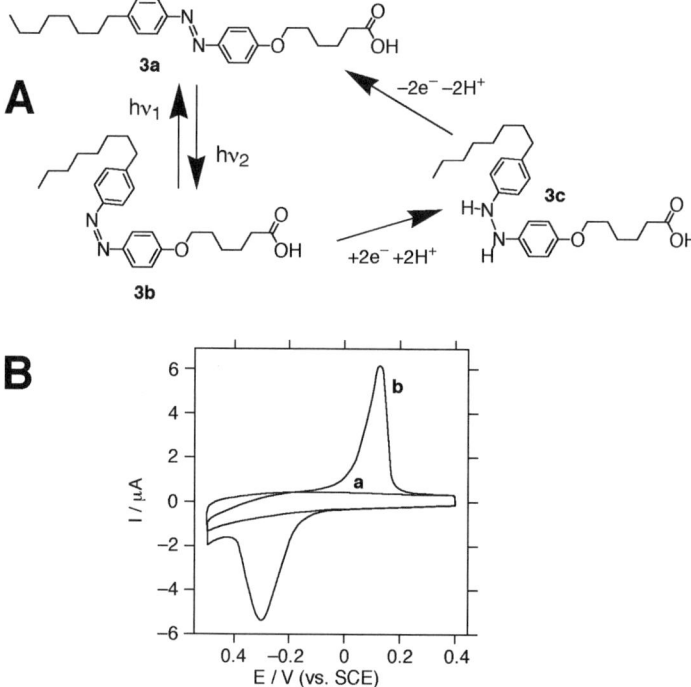

FIG. 7.6 (A) The isomerization of an azobenzene-based LB-monolayer between the electrochemically inactive *trans*-state (**a**) and the electrochemically lockable *cis*-state (**b**). (B) The cyclic voltammograms show traces for: (a) the *trans*- and (b) *cis*-isomers. Recorded at pH 7.0 at a potential scan rate of 20 mV s^{-1}.

transfer reaction has been exemplified by diarylethene molecules incorporated into long-chain thiol monolayer adsorbed on a Au-electrode due to hydrophobic interactions.[57] In the "closed" isomeric state, the monolayer demonstrates well-defined reversible cyclic voltammetry, whereas the "open" state is completely redox-inactive. The electrochemically active state provides electrocatalytic reduction of $[Fe(CN)_6]^{3-}$, thus enabling a vectorial electron cascade that amplifies the photonic input.

The photoisomerizable moiety and the signal transduction moiety in an active molecule can be distinct chemical groups.[58,59,60,61] For example, a Langmuir-Blodgett monolayer composed of molecules with two distinguishable parts—a photoisomerizable azobenzene unit and a 7,7,8,8-tetracyanoquinodimethane (TCNQ) salt (an organic conductor)—was assembled onto a solid support. Lateral conductivity of the monolayer was studied as a function of the photochemically controlled isomerization state of the azobenzene units [Figure 7.7 (A)]. The photochemically induced structural changes of the monolayer packing result in a variation of the conductivity, providing optical "write" and electrical "read" modes of signal transduction [Figure 7.7 (B)].

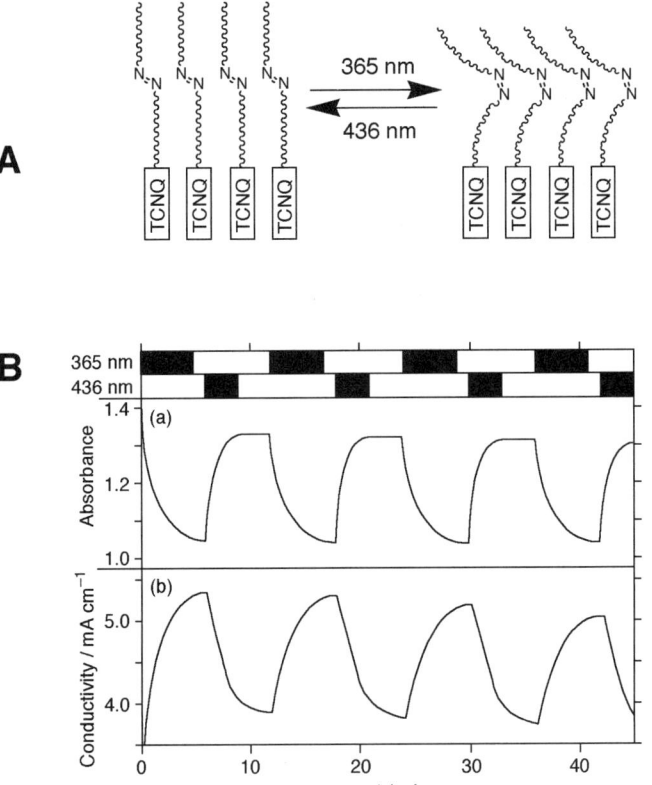

FIG. 7.7 (A) Schematic representation of a photoisomerizable azobenzene-TCNQ Langmuir-Blodgett monolayer. (B) Changes in (a) absorbance at 356 nm, and (b) conductivity of the monolayer upon photoisomerization.

7.3 ELECTRONICALLY TRANSDUCED PHOTOCHEMICAL SWITCHING OF ENZYME MONOLAYERS

The activation and deactivation of an enzymatic process by a photonic signal leads to the amplification of the optical stimulus by the biocatalyzed formation of a chemical product. Accordingly, photochemical switching of the biocatalytic functions of redox-proteins can lead to the activation and deactivation of biocatalytic electron transfer cascades that translate the photonic signal into an electrical output. Such systems represent "smart" biological interfaces where photonic signals are recorded by the photosensitive biomaterial and the encoded information is read back and amplified by the bioelectrocatalytic cascade.[62,63,64] The electronic retrieval of the recorded information requires the integration and coupling of the photoswitchable redox biomaterial with an electronic transducer element; it provides the basis for future optobioelectronic and sensoric devices.

Enzymes modified with photoisomerizable groups can be used as light-switchable biocatalysts.[65,66] Similarly, they can be used to photoregulate

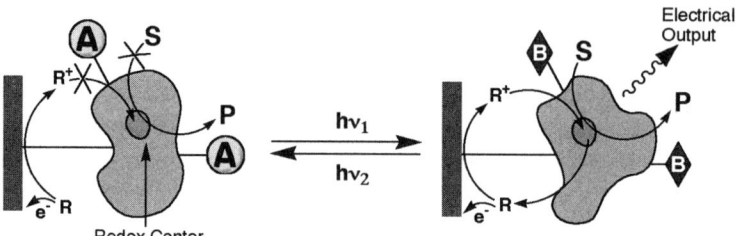

FIG. 7.8 Electronic transduction of photoswitchable bioelectrocatalytic functions of redox-enzymes by the tethering of photoisomerizable units to the protein. (R is a diffusional electron mediator that electrically contacts the redox site of the protein with the electrode support.)

biocatalytic electron transfer reactions at electrode interfaces (Figure 7.8). In configuration "A," the active-site environment of the enzyme is distorted, and the bioelectrocatalytic properties of the enzyme are blocked. Photoisomerization of the photoactive groups to state "B" restores the active-site structure, however, and the enzyme is activated. Electrical contact between the biocatalyst and the electrode and the activation of the bioelectro-catalytic process result in the transduction of a current to the macroscopic environment. By cyclic photoisomerization of the photoactive groups between the states "B" and "A," the amperometric transduction can be switched between "ON" and "OFF" states, respectively.

7.3.1 Redox-Enzymes Tethered with Photoisomerizable Groups

Glucose oxidase (GOx) has been employed as a redox-enzyme to tailor a photoisomerizable enzyme electrode for the photoswitchable oxidation of glucose.[67,68] Photoisomerizable nitrospiropyran (SP) units were tethered to lysine residues of GOx, and the resulting photoisomerizable protein was assembled on a Au-electrode as shown in Figure 7.9. A primary N-hydroxy-succinimide-functionalized thiol monolayer was assembled on the conductive support, and the photoisomerizable enzyme was covalently coupled to the monolayer to yield the integrated photoactive enzyme electrode. The enzyme monolayer was found to undergo reversible photoisomerization, and irradia-tion of the nitrospiropyran-tethered GOx (**4a**) with UV light (320 nm < λ < 380 nm) generated the protonated nitromerocyanine (MRH$^+$)-tethered GOx (**4b**). Further irradiation of the **4b**-monolayer with visible light (λ > 475 nm) restored the nitrospiropyran-tethered enzyme (**4a**).

The photoisomerizable enzyme monolayer electrode also revealed photo-switchable bioelectrocatalytic activity (Figure 7.10). In the presence of ferrocene carboxylic acid (**5**) as a diffusional electron transfer mediator, the nitrospiropyran-tethered GOx (**4a**) revealed a high bioelectrocatalytic activity, reflected by a high electrocatalytic anodic current. The protonated nitromerocyanine-GOx (**4b**) exhibited a two-fold lower activity, as reflected by the decreased bioelectrocatalytic current. By the reversible photoisomer-ization of the enzyme electrode between the **4a**- and **4b**-states, the current responses are cycled between high and low values (Figure 7.10, inset).

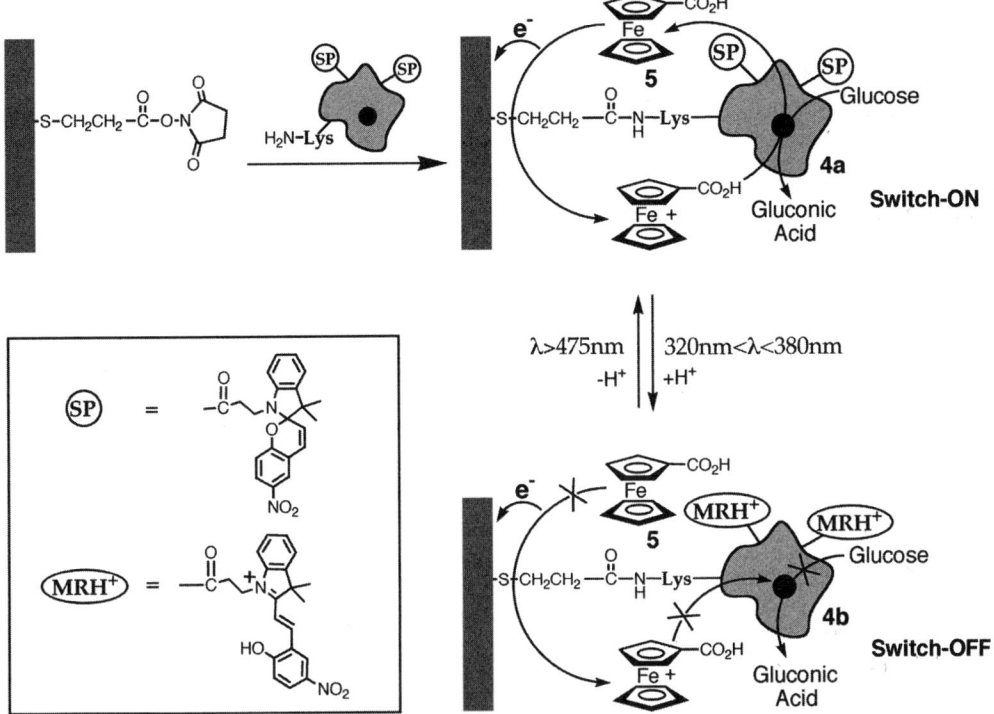

FIG. 7.9 Assembly of a photoisomerizable glucose oxidase monolayer electrode and the reversible photoswitchable activation/deactivation of the bioelectrocatalytic functions of the enzyme electrode.

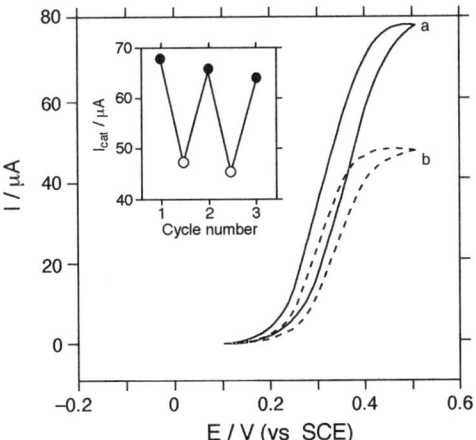

FIG. 7.10 Photostimulated bioelectrocatalyzed oxidation of glucose, 2.5×10^{-2} M, in the presence of ferrocene-carboxylic acid (**5**) 5×10^{-3} M, as diffusional electron mediator in the presence of (a) the nitrospiropyran-tethered GOx (**4a**) and (b) the protonated nitromerocyanine-tethered GOx (**4b**). Inset: Reversible photoswitchable amperometric transduction of the bioelectrocatalyzed oxidation of glucose by (**4a**) – (●) and (**4b**) – (○). (Adapted from reference 68, Figure 10. Copyright 1995, American Chemical Society.)

Although the tethering of photoisomerizable units to the protein leads to photoswitchable bioelectrocatalytic properties, the "OFF" state of the photoisomerizable GOx exhibits residual bioelectrocatalytic activity as the structural distortion of the protein in the **4b**-state is not optimized.

7.3.2 Enzymes Reconstituted onto Photoisomerizable FAD-Cofactors

To optimize the photoswitchable bioelectrocatalytic features of proteins, the site-specific functionalization or mutation of the active-site microenvironment is essential. This mutation has been accomplished by a semi-synthetic approach involving the reconstitution of the flavoenzyme glucose oxidase with a synthetic photoisomerizable flavin adenine dinucleotide (FAD)-cofactor (Figure 7.11).[69] The photoisomerizable carboxylic acid-functionalized nitrospiropyran **6** was covalently coupled to N^6-(2-aminoethyl)-FAD (**7**) to yield the synthetic photoisomerizable nitrospiropyran-FAD cofactor **8a** [Figure 7.11 (A)]. The native FAD cofactor was extracted from glucose oxidase, and the synthetic photoisomerizable-FAD cofactor (**8a**) was reconstituted into the apo-glucose oxidase (apo-GOx) to yield the photoisomerizable enzyme **10a** [Figure 7.11 (B)]. This reconstituted enzyme includes a photoisomerizable unit directly attached to the redox center of the protein, and hence the redox-enzyme is anticipated to reveal optimized photoswitchable bioelectrocatalytic properties. The resulting enzyme was assembled on a Au-electrode as described in Figure 7.11 (C) to yield an integrated optoelectronic assembly. The photoinduced bioelectrocatalytic oxidation of glucose was stimulated in the presence of ferrocene carboxylic acid (**5**) as a diffusional electron transfer mediator. The nitrospiropyran (SP) state of the reconstituted enzyme **10a** was inactive for the bioelectrocatalytic transformation, whereas photoisomerization of the enzyme-electrode to the protonated nitromerocyanine (MRH$^+$) state (**10b**) activated the enzyme for the bioelectrocatalyzed oxidation of glucose (Figure 7.12). By the cyclic photoisomerization of the enzyme monolayer interface between the nitrospiropyran (SP) and the protonated nitromerocyanine (MRH$^+$) states, the biocatalyzed oxidation of glucose was cycled between "OFF" and "ON" states, respectively (Figure 7.12, inset).

It also was found that the direction of the photobiocatalytic switch of the nitrospiropyran-FAD-reconstituted enzyme is controlled by the electrical properties of the electron transfer mediator.[65] With ferrocene dicarboxylic acid as a diffusional electron transfer mediator, the enzyme in the nitrospiropyran-FAD state (**10a**) was found to correspond to the "OFF" state biocatalyst, while the protonated nitromerocyanine state of the enzyme (**10b**) exhibits "ON" behavior. In the presence of the protonated 1-[1-(dimethylamino)ethyl]ferrocene, the direction of the photobioelectrocatalytic switch is reversed. The nitrospiropyran-enzyme state (**10a**) is activated toward the electrocatalyzed oxidation of glucose, while the protonated nitromerocyanine enzyme state (**10b**) is switched "OFF," and is inactive for the electrochemical oxidation of glucose. This control of the photoswitch direction of the photoisomerizable reconstituted enzyme was attributed to electrostatic interactions between the diffusional electron mediator and the photoisomerizable unit

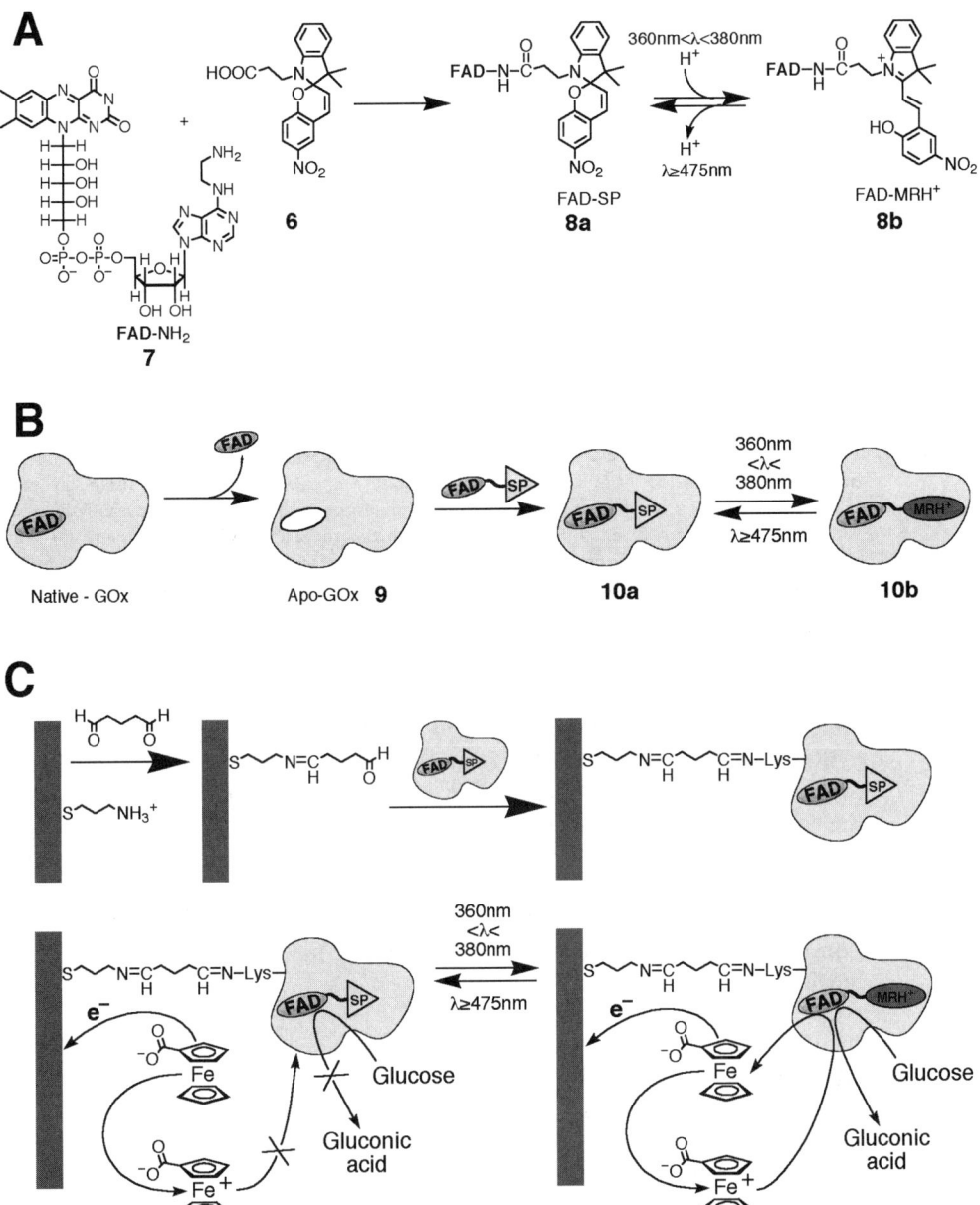

FIG. 7.11 (A) Synthesis of a synthetic photoisomerizable FAD-cofactor. (B) Reconstitution of apo-glucose oxidase with the synthetic nitrospiropyran-FAD photoisomerizable cofactor to yield a photo-isomerizable glucose oxidase. (C) Assembly of the nitrospiropyran-FAD-reconstituted GOx as a monolayer on the electrode and the reversible photoswitchable bioelectrocatalytic activation/deactivation of the enzyme electrode.

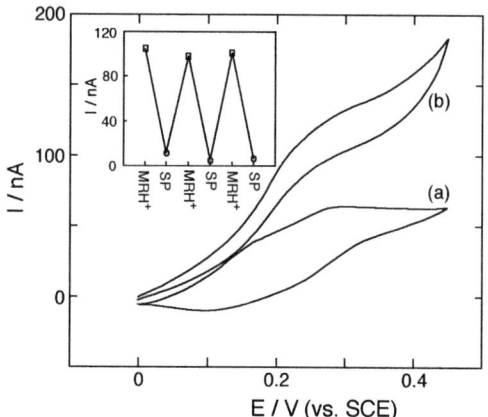

linked to the FAD. The protonated nitromerocyanine photoisomer state (MRH⁺) attracts the oxidized negatively charged electron mediator but repels the oxidized positively charged relay. As a result, the photoisomer state of the enzyme (**10b**) is switched "ON" in the presence of the negatively charged ferrocene derivative. With the positively charged electron transfer mediator, the opposite bioelectrocatalytic switch direction is observed. The **10b**-state of the enzyme repels the electron transfer mediator and its biocatalytic func-tions are blocked, but the **10a**-state allows the mediator to approach, and a bioelectrocatalytic current is observed.

7.4 TRANSDUCTION AT "COMMAND" INTERFACES

Another approach to the organization of integrated optoelectronic switches is schematically detailed in Figure 7.13; it involves the organization of a photoisomerizable "command" interface on a solid support.[70,71,72] A "com-

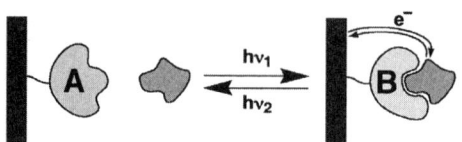

mand" interface controls the interfacial electrochemistry of a solution-state species by isomerization between two states. One of these states may block electron transfer (for instance, by electrostatically repelling the substrate), or alternatively, one state may promote electrical communication (for instance, by binding the substrate through weak interactions). In one photoisomer state, electron transfer to a redox-probe solubilized in the electrolyte solution is prohibited (e.g., by repulsive interactions), whereas in the complementary state of the monolayer, interfacial electron transfer is facilitated (e.g., through associative interactions). Various interactions, such as electrostatic interactions, host-guest, donor-acceptor, or bioaffinity interactions, can contribute to the selective contacting of the redox probe to one state of the photoisomerizable monolayer.

7.4.1 Electrochemical Processes of Organic Redox Molecules at "Command" Interfaces

Charged monolayers have been successfully employed as active interfaces for controlling electron transfer at electrode supports.[73,74] Negatively charged monolayers associated with electrodes have been shown to discriminate between the electrochemical reactions of a mixture of positively and negatively charged substrates.[75] Accordingly, a photoisomerizable nitrospiropyran monolayer that alters its electrical charge upon photoisomerization was constructed by assembling 1-(4-mercaptobutyl)-3,3-dimethyl-6'-nitrospiro[2'H-1-benzopyran-2',2-indoline]-(mercaptobutyl nitrospiropyran) on a Au-electrode [Figure 7.14 (A)].[76] While the nitrospiropyran monolayer (SP-state, **11a**) is neutral, photoisomerization of the monolayer (at pH=7.0) yields a protonated nitromerocyanine monolayer (MRH$^+$-state, **11b**). This cationic monolayer thus discriminates between charged redox-substrates. Positively charged redox probes are repelled by the functionalized electrode, while negatively charged species are attracted by the monolayer and display enhanced electron transfer. The photostimulated oxidation of the negatively charged substrate 3,4-dihydroxyphenyl acetic acid (**12**) and the positively charged substrate 3-hydroxytyramine (dopamine, **13**) at pH=7.0 was examined in the presence of the photoisomerizable monolayer electrode. Figure 7.14 (B, curve a) shows a cyclic voltammogram corresponding to the electrochemical oxidation of **12** by the **11a**-functionalized electrode. Photoisomerization of the monolayer to the **11b**-state results in the cyclic voltammogram shown in Figure 7.14 (B, curve b), and photoregeneration of the **11a**-monolayer restores the cyclic voltammogram shown in curve a. The enhanced response of the **11b**-electrode is attributed to the electrostatic attraction and concentration of **12** at the electrode. By reversible photoisomerization of the monolayer between the **11a**- and **11b**-states, the amperometric responses of the electrode are cycled between low and high values, respectively [Figure 7.14 (B, inset)]. With the positively charged electroactive substrate **13**, the direction of the transduced amperometric signals is reversed. Figure 7.14 (C) shows cyclic voltammograms of **13** in the presence of the photoisomerizable electrode. With the **11a**-functionalized electrode, a high amperometric response is observed (curve a), but photoisomerization of the monolayer to the protonated nitromerocyanine (**11b**)-state repels **13** from

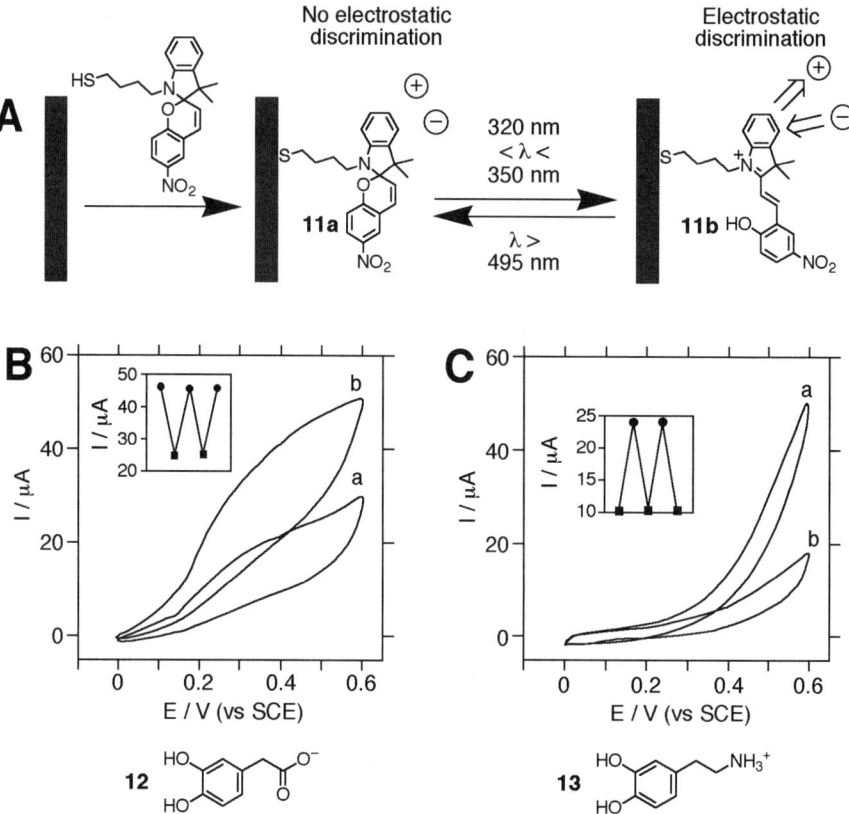

FIG. 7.14 (A) The assembly and photoswitchable states of a nitrospiropyran monolayer. (B) Cyclic voltammograms of the photoisomerizable electrode in the presence of dihydroxyphenylacetic acid (**12**, 0.5 mM): (a) in the "spiro" state (**11a**) and (b) in the "mero" state (**11b**). Inset: Photoswitching of the anodic current of the system at +470 mV vs. SCE. (C) Cyclic voltammograms of the photoisomerizable electrode in the presence of dopamine (**13**, 0.5 mM); (a) in the "spiro" state (**11a**) and (b) in the "mero" state (**11b**). Inset: Photoswitching of the anodic current of the system at +470 mV vs. SCE. Experiments were performed in 0.02 M phosphate buffer (pH 7.0) at a scan rate of 200 mV s^{-1}. (Adapted from reference 76, Figures 1 and 2. Copyright 1997, American Chemical Society.)

the interface. This repulsion retards the electrochemical oxidation of **13** (curve b). Back photoisomerization of the **11b**-monolayer to the **11a**-state regenerates the high amperometric response of the electrode (curve a). By reversible photoisomerization of the monolayer between the **11a**- and **11b**-states, the amperometric responses of the electrode are cycled between high and low values, respectively [Figure 7.14 (C, inset)].

The photoisomerization of a "command" interface resulting in different electrochemical kinetics of a soluble redox-probe also can be probed by faradic impedance spectroscopy.[77] A small electron transfer resistance (R_{et}) is found for the system when there is an attractive interaction between the charged redox-probe and the command interface, and a much larger one upon photoisomerization to the state when the repulsive interactions exist. This paradigm was demonstrated with a negatively charged redox-probe,

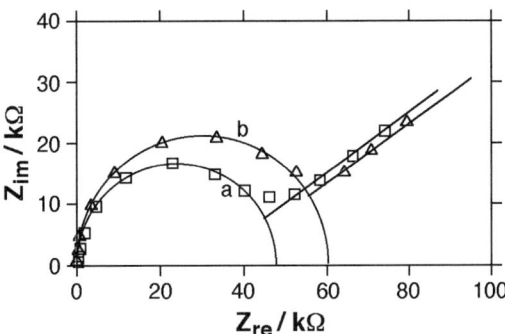

FIG. 7.15 Nyquist diagram for the faradic impedance measurements at the **11a/11b**-monolayer-modified Au electrode in the presence of 10 mM $[Fe(CN)_6]^{3-/4-}$: (a) **11b**-state, and (b) **11a**-state monolayer. Applied bias constant potential, 0.6 V, amplitude of the alternating voltage, 10 mV. Performed in 0.01 M phosphate buffer, pH 7.0. (Adapted from reference 77, Figure 4. Copyright 1998, American Chemical Society.)

$[Fe(CN)_6]^{3-/4-}$, electrochemically contacted at a photoisomerizable command interface (**11a/11b**). Figure 7.15 shows the impedance features (as Nyquist plots) of the nitrospiropyran (**11a**) and protonated nitromerocyanine (**11b**) electrodes in the presence of $[Fe(CN)_6]^{3-/4-}$ as a redox probe. The impedance spectra show a larger resistance to interfacial electron transfer when the monolayer is in the neutral dinitrospiropyran state (R_{et} = 60 kΩ) than when it is in the positively charged protonated merocyanine state (R_{et} = 48 kΩ) (Figure 7.15, curves b and a). The heterogeneous rate constants for electron transfer between the electrode and the redox probe were calculated to be 0.82×10^{-5} and 1.1×10^{-5} cm s^{-1} for the **11a** and **11b**-monolayer modified Au-electrodes, respectively.

The interactions between the photoisomerizable monolayer and charged substrates also may be controlled by the pH of the system.[76] At a pH below 8.6, the merocyanine is protonated (i.e., cationic),[78] so negatively charged redox-probes are attracted to the electrode and their electrochemical process is enhanced. At a pH above 8.6, however, the merocyanine is deprotonated (i.e., zwitterionic), so the electrochemistry of negatively charged species is not enhanced. The nitrospiropyran form of the electrode is neutral at every pH. The pH dependence of the charge on the merocyanine state results in a system where both the pH and the photoisomeric state of the monolayer can be used to control the interfacial electrochemistry of the redox probe. Thus, the system might be considered to perform a logical AND function— the enhanced oxidation of **12** is observed only when the monolayer is in the **11b**-state and the pH is below 8.6.

The light-induced control of the electroactivity of charged redox-active substrates allows the functionalized electrode to be used to control electrochemical transformations. A system composed of two oppositely charged diffusional redox probes—positively charged 2,5-bis[[2-(dimethylbutylammonio)ethyl]amino]-1,4-benzoquinone (**14**) and negatively charged pyrroloquinoline quinone (PQQ) (**15**)—was used to demonstrate this behavior

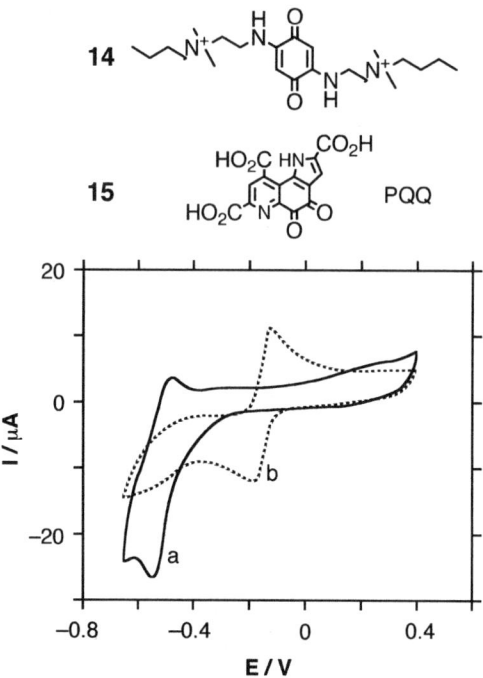

FIG. 7.16 Cyclic voltammograms at a photoisomerizable **11a/11b**-monolayer electrode in the presence of **14** (3×10^{-4} M) and **15** (3×10^{-4} M): (a) In the **11a**-state, and (b) in the **11b**-state. Background electrolyte, 0.02 M phosphate buffer, pH = 7.0; potential scan rate, 200 mV s^{-1}. (Adapted from reference 79, Figure 5. Copyright 1997, American Chemical Society.)

(Figure 7.16)[79] At the neutral **11a**-interface, the redox process of PQQ is irreversible; while at the positively charged **11b**-interface, it is reversible due to the electrostatic attraction of **15**. At the same time, repulsion of the positively charged **14** from the positively charged electrode surface yields the irreversible, inefficient reduction of **14**. With the neutral **11a**-monolayer electrode, however, **14** exhibits reversible redox features (Figure 7.16). By cyclic photoisomerization of the monolayer between the **11a** and **11b** states, reversible electrochemistry of either **15** or **14** is obtained.

In another example, a mixed monolayer composed of a photoisomerizable component and an electrochemical catalyst was applied to switch the electrocatalytic properties of a modified electrode between "ON" and "OFF" states. A Au-electrode surface functionalized with a nitrospiropyran monolayer and PQQ moieties incorporated into the monolayer was applied to control the electrocatalytic oxidation of 1,4-dihydri-β-nicotinamide adenine dinucleotide (NADH) by light.[80] The positively charged nitromerocyanine-state interface resulted in the repulsion of Ca^{2+} cations, which are promoters for the NADH oxidation by the PQQ,[81] thus resulting in the inhibition of the electrocatalytic process. In the nitrospiropyran state, the monolayer does not prevent the association of the PQQ catalyst and Ca^{2+} promoter, thus it provides efficient electrocatalytic oxidation of NADH. Similar outcomes have been achieved using a combination of the photo- and thermal effects resulting in the isomerization of a nitrospiropyran monolayer with the incorporated PQQ catalyst.[82] Other photoisomerizable materials such as an azobenzene alkanethiol derivative mixed with a ferrocene-redox component also have been used to control the electrocatalyzed electron transfer process between a command interface and a dissolved redox-probe.[83] A mixed monolayer composed of a *cis*-azobenzenealkanethiol (**16a**) and ferrocene units linked to a thiolated spacer (**17**) provided electron transfer to a diffusional redox-component ($[Fe(CN)_6]^{3-}$) [Figure 7.17 (A)]. This monolayer insulates the Au-electrode, preventing direct electron transfer to $[Fe(CN)_6]^{3-}$. Electron transfer to $[Fe(CN)_6]^{3-}$ mediated by the ferrocene units results in a diode-like response

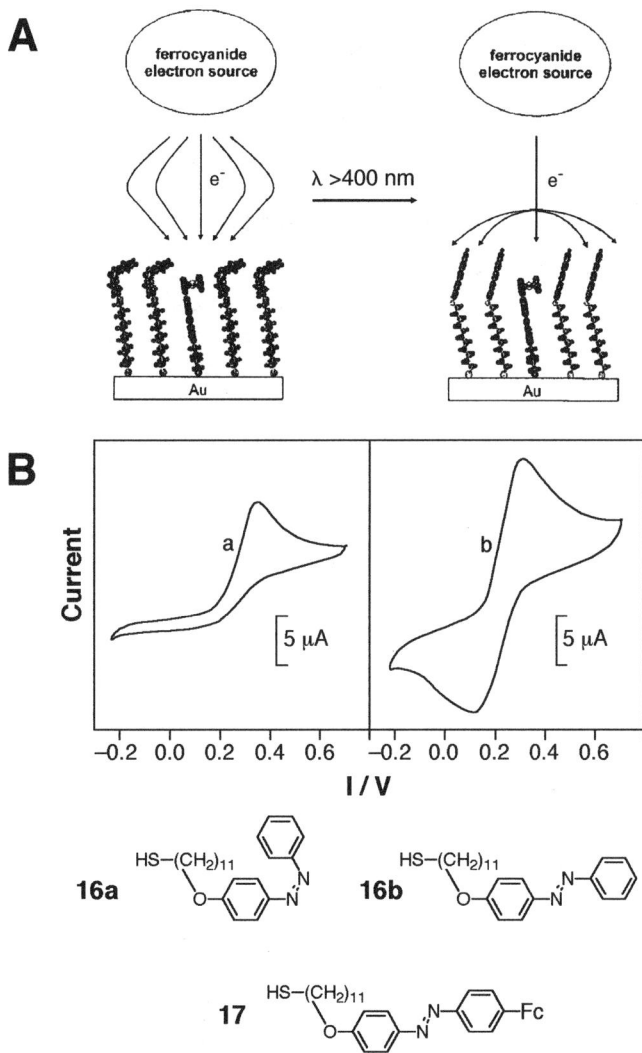

FIG. 7.17 (A) Photochemical switching of a monolayer from a close-packed insulating form to a pin-hole-containing form. (B) Electrochemical response for the two-component film (1:99 **17:16a**): (a) after the addition of $[Fe(CN)_6]^{4-}$ (0.5 mM), and (b) after subsequent irradiation at $\lambda > 400$ nm. (Adapted from reference 83, Figure 2. Copyright 1999, American Chemical Society.)

from the system. The cyclic voltammogram shows an irreversible electrochemical wave corresponding to the mediated electrochemical reduction of $[Fe(CN)_6]^{3-}$ [Figure 7.17 (B, curve a)]. Photoisomerization of the *cis*-azobenzene component (**16a**) to the *trans*-azobenzene component (**16b**) ($\lambda > 400$ nm) opens pinholes in the monolayer that allow the direct, reversible electrochemistry of $[Fe(CN)_6]^{3-}$ to take place [Figure 7.17 (B, curve b)].

Self-assembled monolayers of isomerizable compounds can give rise to surfaces with switchable permeability, thus resulting in different interfacial electrochemistry for a diffusional redox probe. A self-assembled monolayer of 4-cyano-4'-(10-thiodecoxy)stilbene on a Au-electrode demonstrated a higher blocking effect for the $[Fe(CN)_6]^{3-/4-}$ electrochemistry when it was in the *trans*-state (**18a**)[84] than when it was in the *cis*-state (**18b**). In the *cis*-state (**18b**), the monolayer packing was disturbed, giving rise to increased permeability (Figure 7.18). Reversible photoisomerization between two isomeric states allowed the electrochemical transduction of the structural changes of the monolayer resulting from the photoisomerization of the monolayer components.

7.4.2 Electrochemical Processes of Proteins and Enzymes at "Command" Interfaces

Redox-proteins usually lack direct electrical contact with electrodes because the redox site is embedded in the protein.[85] This situation insulates the electrical communication between the redox-site and the electrode by the spatial separation of the protein redox-site from the electrode support. Suitable func-

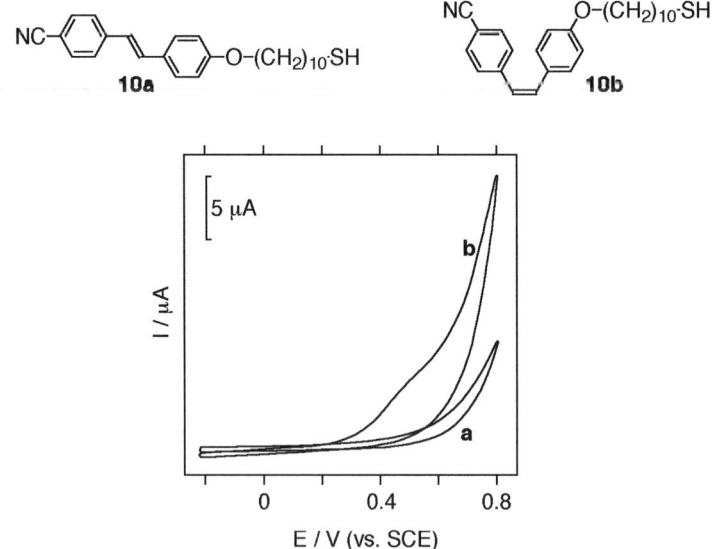

FIG. 7.18 Cyclic voltammograms of $[Fe(CN)_6]^{4-}$ (1.2 mM) at electrodes modified by **18a** (curve a) or **18b** (curve b) which exhibit different packing. Recorded in 0.1 M KCl at a potential scan rate of 100 mV s^{-1}. (Adapted from reference 84, Figure 2. Copyright 1995, American Chemical Society.)

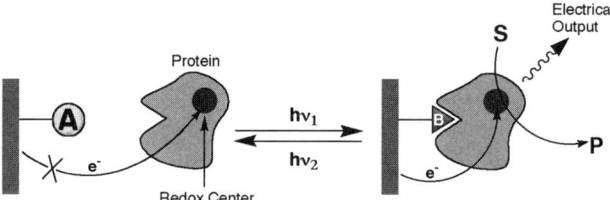

FIG. 7.19 Electronic transduction of photoswitchable bioelectrocatalytic functions of enzymes/ proteins by the application of a photoisomerizable command interface that controls the electrical contact between the redox enzyme/protein and the electrode.

tionalization of the electrode, however, is often able to stimulate electron transfer between the protein and the electrode. A photoswitchable electrode that stimulates this electron transfer in one state can be used as a "command" surface for the enzyme's biocatalytic process. Figure 7.19 shows a surface-bound photoisomerizable unit that binds an enzyme in the "B" state (but not in the "A" state), stimulating its biocatalytic process. For low molecular weight redox-proteins, e.g., the hemoprotein cytochrome c (Cyt c), the chemical functionalization of electrode surfaces with molecular promoter units has been found to facilitate electron transfer communication between the protein redox-center and the conductive supports.[86,87] The binding of the redox proteins to the promoter sites aligns the redox centers to the electrode surface. This alignment results in the shortening of the electron transfer distances and the electrical contacting of the redox sites. For example, pyridine units[86,87] or negatively charged promoter sites,[88] assembled on electrode surfaces, have been reported to align Cyt c on electrodes by affinity or electrostatic interactions, and to facilitate the electrical communication between the heme-site and the electrode. Cytochrome c is a positively charged hemoprotein at neutral pH. This finding suggests that by nanoengineering an electrode with a composite layer consisting of Cyt c binding sites and photoisomerizable units that are transformed from neutral to a positively charged state, the electrostatic photoswitchable binding and dissociation of Cyt c to and from the electrode could occur.[89] To photoregulate the electrical communication between Cyt c and the electrode, a mixed monolayer consisting of pyridine sites and photoisomerizable nitrospiropyran units (**19a**) was assembled on a Au-electrode [Figure 7.20 (A)]. This monolayer binds Cyt c to the surface, aligns the heme-center of the protein, and thus, stimulates electrical contact with the electrode [Figure 7.20 (B, curve a)]. Photoisomerization of the monolayer to the positively charged protonated nitromerocyanine state (**19b**) results in the electrostatic repulsion of Cyt c from the monolayer. As a result, the electrical communication between the hemoprotein and the electrode is blocked [Figure 7.20 (B, curve b)]. By cyclic photoisomerization of the monolayer between the nitrospiropyran (**19a**) and protonated nitromerocyanine (**19b**) states, the binding and dissociation of Cyt c to and from the monolayer occurs, and the electrical contact of the hemoprotein and the electrode is switched between "ON" and "OFF" states, respectively.

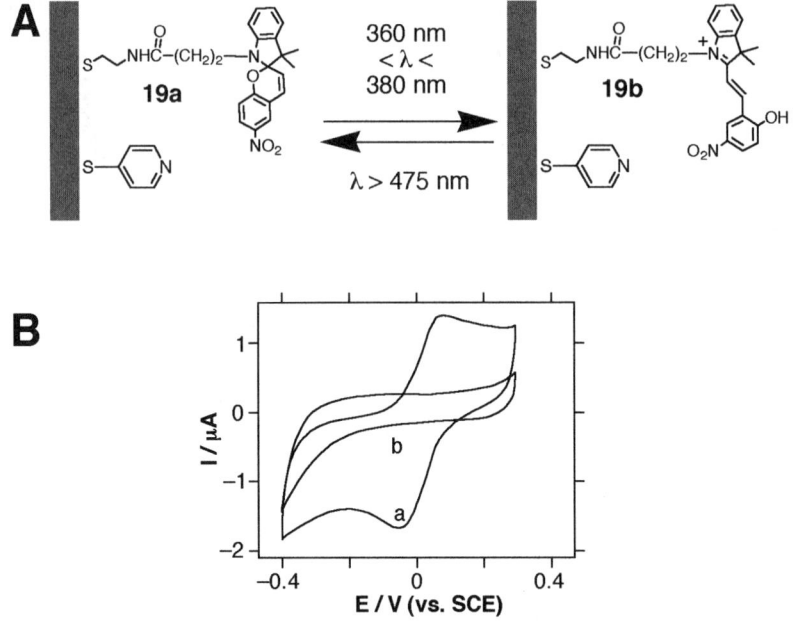

FIG. 7.20 (A) An interface for the light-controlled binding of Cyt c. The cytochrome electrochemistry is promoted by the pyridine component, but it is repelled from **11b**. (B) Cyclic voltammograms of cytochrome c, Cyt c, 1×10^{-4} M, in the presence of: (a) The pyridine-nitrospiropyran (**11a**)-mixed monolayer electrode. (b) The pyridine-protonated-nitromerocyanine (**11b**)-mixed monolayer electrode. Data recorded at a scan rate of 50 mV sec^{-1}. (Adapted from reference 89, Figure 1. Reproduced by permission of The Royal Society of Chemistry.)

Cytochrome c acts as an electron transfer mediator (cofactor) that activates many secondary biocatalyzed transformations by the formation of interprotein Cyt c-enzyme complexes.[90,91,92] Specifically, Cyt c transfers electrons to cytochrome oxidase (COx), which mediates the four-electron reduction of oxygen to water. The photoswitchable electrical activation of Cyt c enables the photostimulated triggering of the COx-catalyzed reduction of O_2 (Figure 7.21). In the presence of the pyridine-nitrospiropyran (**19a**) monolayer electrode, electron transfer to Cyt c activates the electron transfer cascade to COx, and the bioelectrocatalyzed reduction of O_2 to water proceeds.[92] The bioelectrocatalyzed reduction of O_2 is reflected by the transduced electrocatalytic cathodic current [Figure 7.21 (B, curve a)]. Photoisomerization of the monolayer to the protonated nitromerocyanine (**19b**) configuration results in the repulsion of Cyt c from the electrode interface. This repulsion blocks the interfacial electron transfer to Cyt c, and, thus, the COx-mediated reduction of O_2 also is inhibited [Figure 7.21 (B, curve b)]. Note that the electrocatalytic cathodic current transduced by the Cyt c-COx protein assembly is enhanced ca. 10-fold as compared to the amperometric current resulting from the Cyt. c alone. This amplification occurs because COx induces a bioelectrocatalytic process, and because the photonic activation of the Cyt c-COx system drives the reduction of O_2 with a high turnover. Thus, the amperometric response of the Cyt c-COx-layered

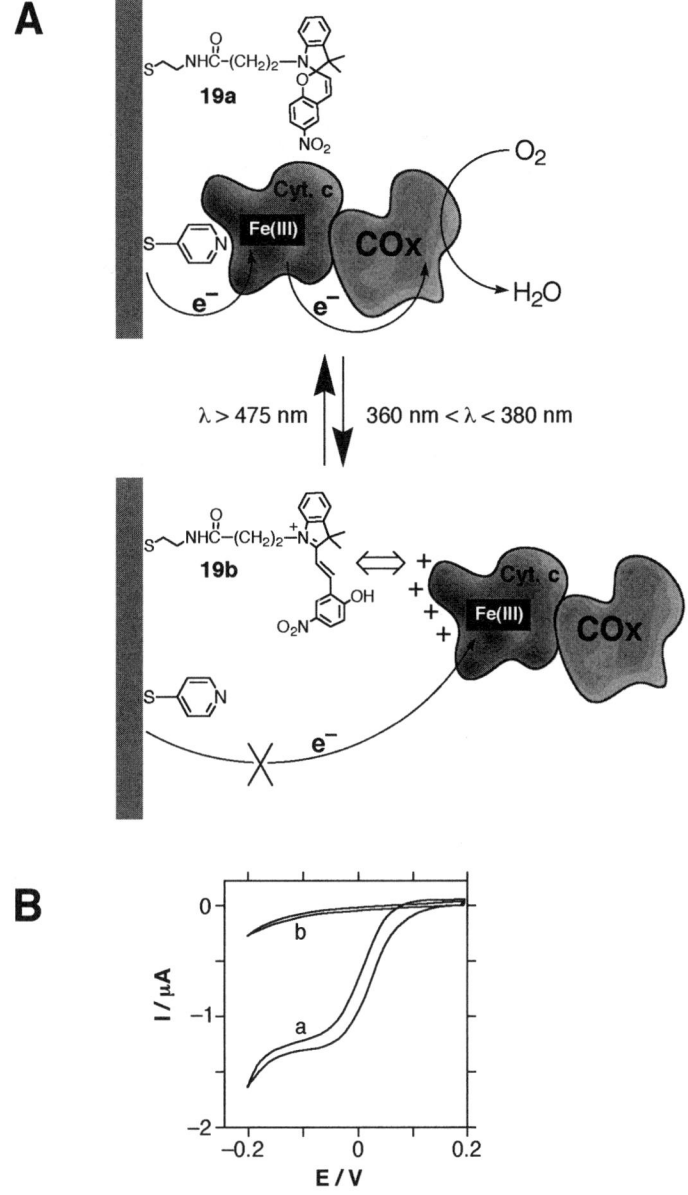

FIG. 7.21 (A) Reversible photoswitchable activation/deactivation of the electrical contact between cytochrome c and the electrode and the secondary activation/deactivation of the COx-biocatalyzed reduction of oxygen using a photoisomerizable (**11a/11b**)-pyridine mixed thiolated monolayer as a command interface. (B) Cyclic voltammograms of the Cyt c/COx system corresponding to the photo-stimulated bioelectrocatalyzed reduction of O_2 in the presence of the photoisomerizable monolayer electrode: (a) Bioelectrocatalyzed reduction of O_2 by Cyt c/COx in the presence of pyridine-nitro-spiropyran (**11a**) mixed monolayer electrode. (b) Cyclic voltammogram of the Cyt c/COx system under O_2 in the presence of the pyridine-protonated-merocyanine (**11b**) mixed monolayer electrode. Data recorded at scan rate 2 mV sec^{-1}. (Adapted from reference 89, Figure 2. Reproduced by permission of The Royal Society of Chemistry.)

electrode represents the amplified amperometric readout of the photonic information recorded in the monolayer. By the cyclic photoisomerization of the monolayer between the nitrospiropyran (**19a**) and protonated nitromerocyanine (**19b**) states, the amperometric responses from the system are reversibly cycled between "ON" and "OFF" states, respectively. This photocontrol over the Cyt c-COx bioelectrocatalytic cascade represents a system analogous to the vision process. Photoisomerization of rhodopsin in the eye stimulates the binding of protein G to the light-active membrane, which activates a biocatalytic cascade that generates c-GMP and triggers the neural response.

The electrostatic control of the electrical contact between redox-proteins and electrodes by means of "command" interfaces was further demonstrated by the photochemical switching of the bioelectrocatalytic properties of glucose oxidase (Figure 7.22).[93] Ferrocene units were tethered to the protein backbone of glucose oxidase to yield an "electrically wired" enzyme that is activated for the bioelectrocatalyzed oxidation of glucose. The enzyme is negatively charged at neutral pH values ($pI_{GOx} = 4.2$[94]) and, hence, could be

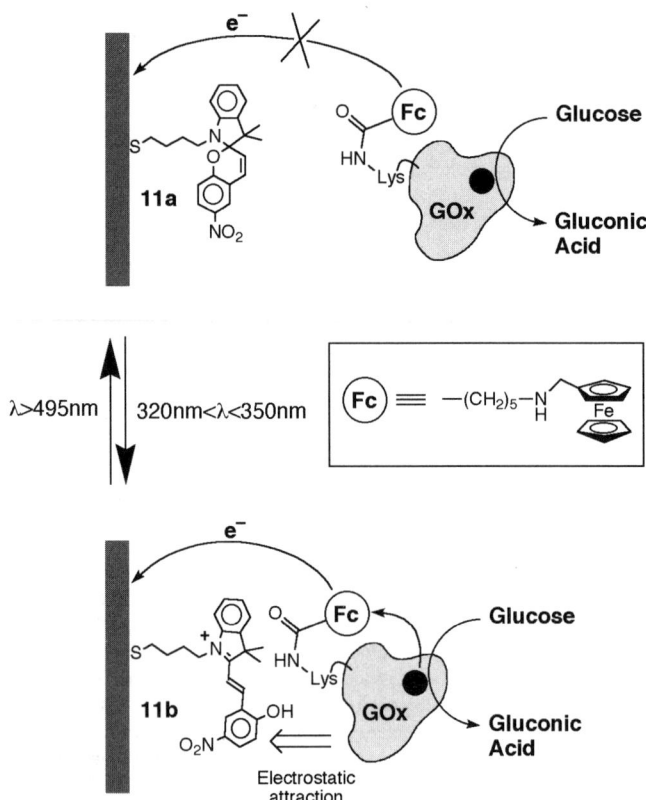

FIG. 7.22 Photochemical control of the electrical contact between a ferrocene-tethered glucose oxidase and the electrode using a thiolated nitrospiropyran as a command interface.

electrostatically attracted by positively charged surfaces. Accordingly, a thiolated nitrospiropyran (**11a**) monolayer was assembled on a Au-electrode. In this state, inefficient electrical interactions between the ferrocene-modified GOx and the modified electrode exist, and only moderate bioelectrocatalyzed oxidation of glucose occurs. Photoisomerization of the monolayer to the protonated nitromerocyanine state (**11b**) results in the electrostatic attraction of the biocatalyst to the electrode support. The concentration of the enzyme at the electrode surface leads to effective electrical communication between the biocatalyst and the electrode, and yields the enhanced bioelectrocatalyzed oxidation of glucose. The transduced current represents an amplified signal resulting from the photonic activation of the monolayer. By the cyclic photoisomerization of the monolayer between the protonated nitromerocyanine state (**11a**) and the nitrospiropyran state (**11b**), the enzyme can be switched between surface-associated and dissociated configurations, respectively, leading to reversible "ON"-"OFF" control over the bioelectrocatalytic activity of the enzyme.

7.4.3 Surface-Reconstituted Enzymes at Photoisomerizable Interfaces

Another example of a "command" interface involves the photostimulated electrostatic control of the electrical contacting of the redox-enzyme and the electrode in the presence of a diffusional electron mediator (Figure 7.23).[95] A mixed monolayer, consisting of photoisomerizable thiolated nitrospiropyran units (**11a**) and a synthetic FAD-cofactor, was assembled onto a Au-electrode. Apo-glucose oxidase (apo-GOx) was reconstituted onto the surface FAD-sites to yield an aligned enzyme electrode. The surface-reconstituted GOx (2×10^{-12} mole cm^{-2}) by itself lacks electrical communication with the electrode. In the presence of the positively charged diffusional electron mediator **20**, the bioelectrocatalytic functions of the enzyme electrode are activated and controlled by the photoisomerizable component co-immobilized in the monolayer assembly (Figure 7.23). With the monolayer in the neutral nitrospiropyran state (**11a**), the positively charged electron mediator **20** is oxidized at the electrode. This process allows **20** to act as an electron transfer mediator and to activate the bioelectrocatalyzed oxidation of glucose, as reflected by an electrocatalytic anodic current (Figure 7.24, curve a). Photoisomerization of the monolayer to the protonated nitromerocyanine state (**11b**) results in the electrostatic repulsion of **20** from the electrode, and the mediated electrical communication between the enzyme and the electrode is blocked. As a result, the bioelectrocatalytic functions of the GOx layer are switched "OFF" (Figure 7.24, curve b). By the cyclic photoisomerization of the monolayer between the nitrospiropyran and protonated nitromerocyanine states, the modified bioelectrocatalytic interface is reversibly switched "ON" and "OFF", respectively (Figure 7.24, inset).

7.4.4 Mechanical Photoisomerizable Monolayers

Self-assembled monolayers can adopt highly ordered, close-packed structures, reminiscent of crystal structures. As these structures have a very high depend-

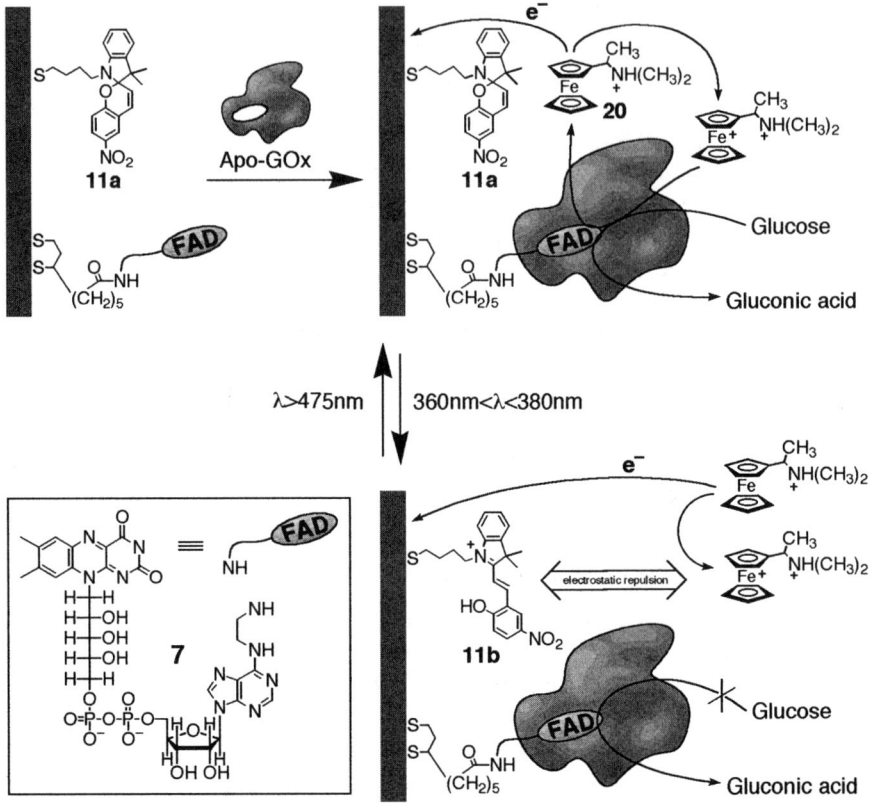

FIG. 7.23 Surface-reconstitution of apo-glucose oxidase on a mixed monolayer associated with an electrode consisting of an FAD-cofactor and photoisomerizable nitrospiropyran units, and reversible photoswitching of the bioelectrocatalytic functions of the enzyme electrode in the presence of a positively charged diffusional relay.

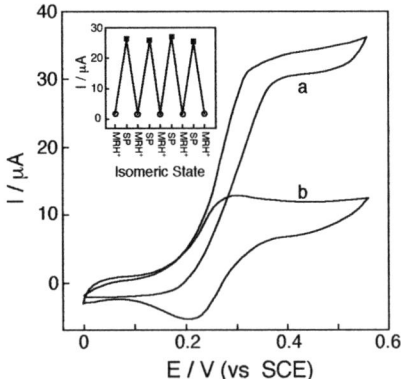

FIG. 7.24 Photoswitchable bioelectrocatalyzed oxidation of glucose, 8×10^{-2} M, by a composite monolayer consisting of GOx reconstituted onto FAD-units and nitrospiropyran photoisomerizable units in the presence of (**20**) as diffusional electron mediator: (a) In the presence of the nitrospiropyran state (**11a**). (b) In the presence of the protonated nitromerocyanine state (**11b**). Inset: Cyclic amperometric transduction of photonic signals recorded by the photoisomerizable monolayer electrode through the bioelectrocatalyzed oxidation of glucose.

ence on the structure of the individual molecular units, a conformational or configurational change of the self-assembled subunits upon photoisomerization can result in a substantial change in the monolayer structure and properties. Specifically, the hydrophobicity of the interface may change, or the close-packed structure may be disturbed, leading to pinhole defects. These changes can originate from conformational changes of the molecules resulting in the exposure of different sites of the molecules to the outer interface and from different packing modes of the photoisomers in the monolayer.

A self-assembled monolayer of 4-cyano-4′-(10-thiodecoxy)stilbene on a Au-electrode was shown to have different hydrophobicities. In the *trans-* (**18a**) and *cis-* (**18b**) isomeric states (Figure 7.25). When the monolayer is in the *cis*-state (**18b**), the polar cyano-groups are hidden in the monolayer and the hydrophobic spacers are exposed to the solution, thus providing the hydrophobic properties of the interface. After irradiation of the monolayer ($\lambda > 350$ nm), the *cis*-stilbene units are photoisomerized to the *trans*-state (**18a**), and the polar cyano-groups are mainly exposed to the solution. In this case, the interface becomes more hydrophilic, and the monolayer displays markedly different wetting properties. This difference allows the photo-patterning of the surface.[96]

Another mechanism providing structural changes of monolayers is the lateral cross-linking of molecular units in the monolayer, forming dimers [Figure 7.25 (B)].[97,98] This cross-linking decreases the barrier for the interfacial electrochemistry of a diffusional redox-probe. The effect possibly originates from the formation of pinholes between the dimers. The reversible photochemical switch between the dimeric and monomeric states of the monolayer allows the electrochemical transduction of the structural changes in the monolayer.

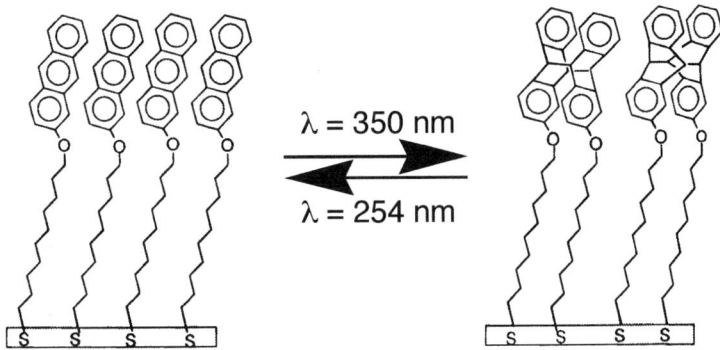

FIG. 7.25 Photochemical switching of a monolayer by dimerization.

7.5 RECOGNITION PHENOMENA AT SURFACES USING PHOTOISOMERIZABLE GUEST OR HOST COMPONENTS

Another method to control the interfacial electrochemistry between a redox-label and an electrode is by the formation and dissociation of host-guest complexes where one of the species is photoisomerizable. The advantage of the controllable formation of a strong complex lies in its detection. While "command" surfaces, relying on weak interactions, must necessarily use electrochemical methods, controllable binding may be measured by a large number of methods, including microgravimetric measurements, surface plasmon resonance (SPR), ion-sensitive field-effect transistors (ISFETs), ellipsometry, and electrochemical techniques.

7.5.1 Affinity Interactions at Interfaces Using Soluble Photoisomerizable Guest Components

Figure 7.26 shows the reversible association of a photoisomerizable guest to a chemically modified surface. In configuration "A" of the molecular component, no affinity interactions with the modified surface exist, and the system is in a mute state. Photoisomerization of the substrate to state "B" activates the affinity binding of the molecular component to the surface. The binding interaction may then be transduced by electronic, optical, or spectroscopic means.

This approach for tailoring molecular optoelectronic assemblies has been demonstrated using several photoisomerizable substrates.[99,100,101,102,103] One example utilizes *trans*-N-methyl-N'-[1-phenylazobenzyl]-4,4'-bipyridinium (**22a**), which yields the *cis*-isomer (**22b**) after irradiation at $\lambda = 355$ nm and is restored by illumination of **22b** at $\lambda > 430$ nm. The two photoisomers differ substantially in their binding constants with β-cyclodextrin (β-CD)— $K_a = 1700$ M^{-1} for **22a** and $K_a = 180$ M^{-1} for **22b**.[99,100] Accordingly, amino-β-cyclodextrin (**23**) was synthesized and assembled on a Au-electrode [Figure 7.27 (A)].[99,100] The association of **22a** to the β-CD receptor monolayer is reflected by its high amperometric response when associated with the electrode [Figure 7.27 (B, curve a)]. Photoisomerization of the substrate to **22b**

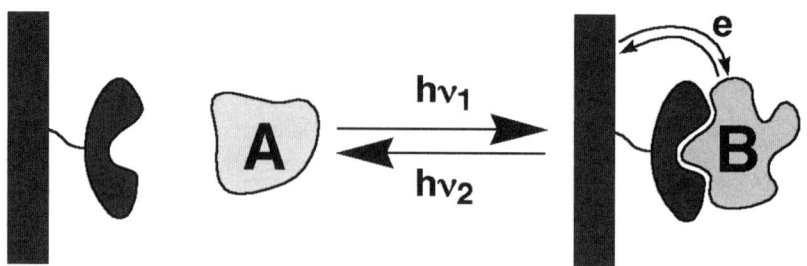

FIG. 7.26 The interaction of a solution-state species with a receptor-functionalized interface may be controlled by the isomerization of the free component.

results in a substantially lower amperometric response (curve b), mainly attributed to residual **22a** existing in a photostationary equilibrium. By cyclic photoisomerization of the substrate between **22a** and **22b**, high- and low-current signals are measured at the β-CD-functionalized electrode, respectively [Figure 7.27 (B, inset)]. The association of **22a** to the receptor monolayer is supported by the fact that the cathodic (and anodic) peak currents observed in the cyclic voltammogram vary directly with the scan rate ($I_p \propto v$), implying that the redox-active species is confined to the electrode surface.

The formation of donor-acceptor complexes between bipyridinium salts (electron acceptors) and xanthene dyes (electron donors) (e.g., eosin, Rose Bengal) has been studied extensively.[104] Crystal structures of these complexes have been identified, and the structural features of the donor-acceptor complexes in solutions have been characterized using NMR spectroscopy. The xanthene dye/bipyridinium donor-acceptor complexes are stabilized by

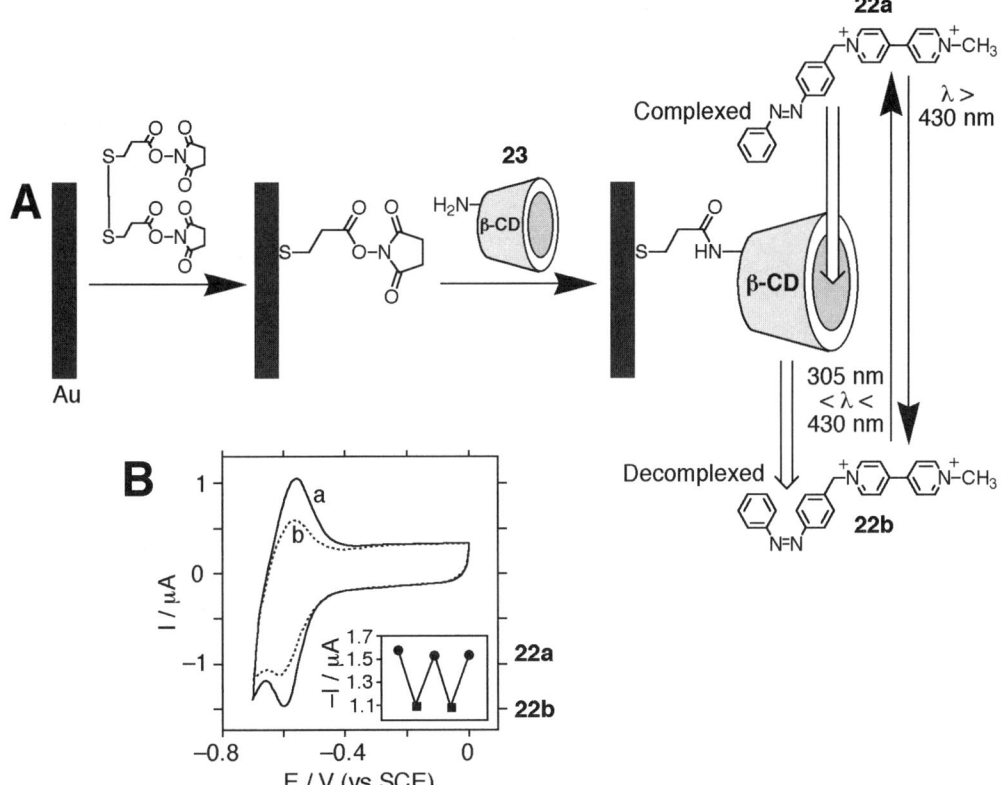

FIG. 7.27 (A) The assembly of a β-cyclodextrin monolayer on a Au-electrode. (B) Cyclic voltammo-grams of the β-CD-functionalized Au-electrode in the presence of: (a) **22a** and (b) **22b** at 1 μM. Inset: Photoswitching behavior of the cathodic peak current. Experiments were performed in 0.01 M phosphate buffer (pH 10.8) at a scan rate of 100 mV s⁻¹. (Adapted from reference 99, Figure 1. Reproduced by permission of The Royal Society of Chemistry.)

charge-transfer interactions, π-π overlap, and attractive electrostatic interactions between the electron donor and electron acceptor units.[105] It also has been demonstrated that the complexation features of photoisomerizable bipyridinium and *bis*-pyridinium electron acceptors to the xanthene dye are controlled by the photoisomer state of the electron acceptor.[101,102,103,106] The formation and the dissociation of the supramolecular donor-acceptor complex between the xanthene dye and the bipyridinium unit can therefore be triggered by the light-induced transformation of the latter component to photoisomers exhibiting high or low affinities for the electron donor, respectively.

The photoswitchable complexation and decomplexation properties of π donor-acceptor complexes between xanthene dyes and photoisomerizable bipyridinium salts has been used to generate an optoelectronic interface (Figure 7.28).[101,102,103] Eosin isothiocyanate (**24**) was covalently linked to an electrode surface via a thiourea bond [Figure 7.28 (A)]. The electron acceptor 3,3'-*bis*(N-methylpyridinium)azobenzene (**25**) was used as the photoisomerizable component [Figure 7.28 (B)]. The association constants of the π donor-acceptor complexes generated between eosin and **25a** or **25b** in solution are $K_a = 8.3 \times 10^3 \text{ M}^{-1}$ and $K_a = 3.4 \times 10^3 \text{ M}^{-1}$, respectively. The analysis of complexation on the functionalized surface was accomplished by quartz crystal microbalance (QCM) measurements. The frequency change (Δf) of a piezoelectric quartz crystal on which a mass change Δm occurs is given by the Sauerbrey equation (Equation 7.1),[107] where f_0 is the base frequency of the crystal, ρ_q is the quartz density, μ_q is the shear modulus of the crystal, and A is the surface area.

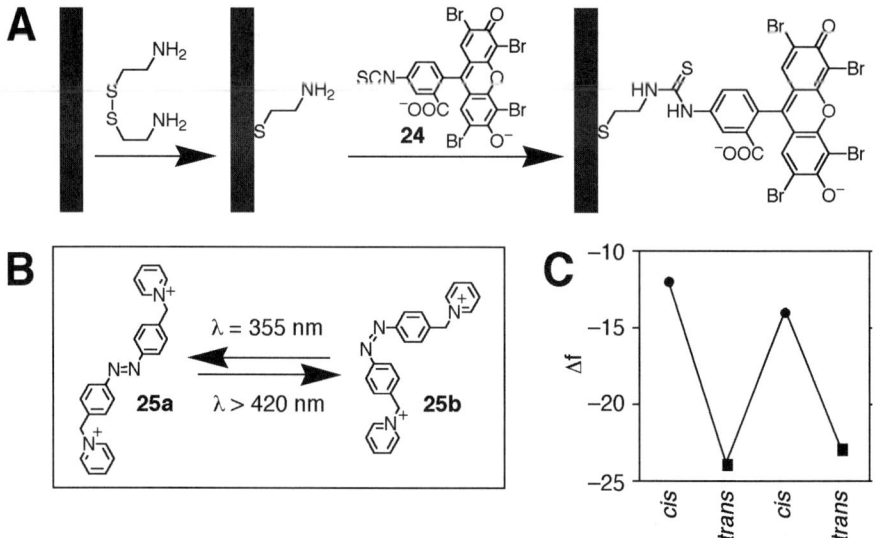

FIG. 7.28   (A) The assembly of an eosin monolayer on a QCM electrode. (B) Structures of the two states of the photoisomerizable acceptor **25**. (C) Frequency changes of an eosin-modified quartz crystal upon photoisomerization of **25** (at 74 μM) between the *trans-* and *cis*-states.

$$\Delta f = -2f_0{}^2 \frac{\Delta m}{A(\mu_q \rho_q)^{0.5}}$$

Photoisomerization of the electron acceptor to the **25b**-state yields an acceptor of lower affinity for the eosin-functionalized interface. Thus, dissociation of the π donor-acceptor complex occurs, causing a decrease in the mass on the crystal (i.e., an increase of the crystal's resonant frequency). While the association of **25a** is accompanied by a frequency decrease of $\Delta f = -25$ Hz, the binding of **25b** to the modified surface results in a frequency change of only $\Delta f = -13$ Hz, indicating a lower affinity for the π-donor interface. By the cyclic photoisomerization of the electron acceptor between the states **25a** and **25b**, their binding to the modified transducers can be switched between low- and high-affinity interactions, respectively. This effect is transduced by frequency changes of the piezoelectric crystal [Figure 7.28 (C)].

7.5.2 Affinity Interactions at Interfaces Using Immobilized Photoisomerizable Host Components

The immobilization of photoisomerizable host molecules onto an electrode surface can be used for the construction of novel ion-selective electrodes. A photoisomerizable calix[4]arene derivative was incorporated into a polymeric membrane on an electrode surface and the two different isomeric states of the host molecule provided responses selective to Li+ or Na+ ions depending on the state.[108] Another photoisomerizable host molecule has been constructed from spiropyran and crown-ether subunits.[109] Different binding affinities for Li+ were found depending on the isomeric state of the photoisomerizable component (**26**). The researchers suggested that this effect was caused by the coordination of the Li+ ion to the O− of the zwitterionic merocyanine-form (Figure 7.29). The photocontrolled host molecules were immobilized in a polymeric film onto an electrode surface, resulting in a photochemically switchable ion-selective electrode.

7.5.3 Reversible Bioaffinity Interactions at Photoisomerizable Interfaces

The electronic transduction of the formation of antigen-antibody complexes has been used as the basis for immunosensor devices.[110,111] Several trans-

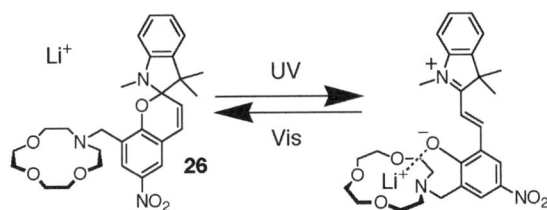

FIG. 7.29 Photocontrolled binding of Li+ by the "crowned spiropyran" **26** incorporated into a membrane at an electrode surface.

duction means, including electrochemical (potentiometric,[112,113,114,115] amperometric,[116,117,118,119,120,121,122] and impedometric[123,124] signals), microgravimetric,[125–131] and SPR,[132,133,134,135] have been used to follow the formation of antigen-antibody complexes on surfaces. Figure 7.30 shows the schematic amperometric, impedometric, microgravimetric, and SPR transduction of the formation of antigen-antibody complexes on solid supports. The formation of an antigen-antibody complex on an electrode insulates the conductive support and introduces a barrier for electron transfer at an electrode interface, thus blocking the amperometric response redox-labels solubilized in the electrolyte solution [Figure 7.30 (A)]. The use of a large redox probe such as a redox-labeled enzyme ensures that the binding of the antibody to the electrode blocks its bioelectrocatalytic function effectively. Thus, the formation of the antigen-antibody complex on the electrode retards the transduced bioelectrocatalytic current. Impedance spectroscopy, and specifically faradic impedance spectroscopy,[136,137] is a useful method to probe the electron transfer resistance and capacitance at the electrode surface. Upon the application of an alternating voltage close to the reduction potential of the redox label, the complex impedance (with real, $Z_{re}(\omega)$, and imaginary, $Z_{im}(\omega)$, compo-

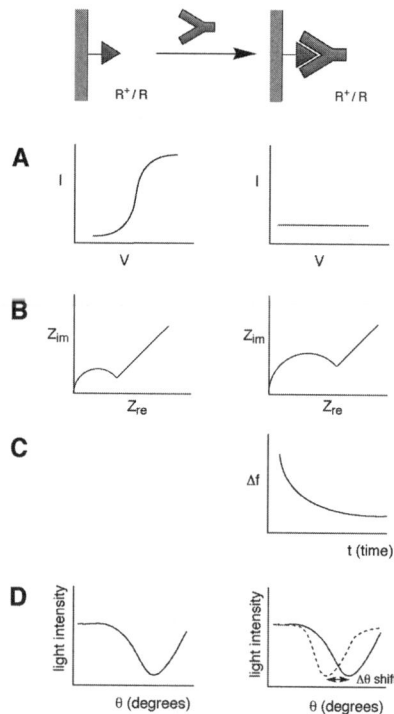

FIG. 7.30 Electronic and optical transduction of the formation of antigen-antibody affinity complexes on transducers: (A) Amperometric transduction at an electrode (R^+/R is a redox label in the electrolyte solution). (B) Transduction by faradic impedance spectroscopy. (C) Microgravimetric quartz crystal microbalance (QCM) transduction in the presence of a piezoelectric quartz crystal. (D) Surface plasmon resonance transduction.

nents) is measured as a function of the applied frequency. The interfacial electron transfer resistance at the electrode support, R_{et}, is derived from the respective Nyquist plot (Z_{im} vs Z_{re}) where the semicircle diameter of the impedance spectrum corresponds to the electron transfer resistance at the electrode surface. Accordingly, the formation of the hydrophobic antigen-antibody complex on the electrode surface insulates the electrode, as reflected by an increase in the interfacial electron transfer resistance and an enlarged semicircle diameter in the respective impedance spectrum [Figure 7.30 (B)].

Changes in the mass associated with a piezoelectric crystal result in changes in its resonant frequency (Equation 7.1). An increase in the quartz crystal mass as a result of the formation of the antigen-antibody complex is accompanied by a decrease in the resonance frequency of the crystal [Figure 7.30 (C)]. Various immunosensor devices based on the frequency changes of quartz crystals have been reported.[125–131] Surface plasmon resonance spectroscopy[138] is a further means to follow the formation of antigen-antibody complexes. Thin metal layers (e.g., Au or Ag) give rise to a surface plasmon. The interaction of the metal layer with polarized light causes excitation of the plasmon, resulting in the absorption of the light energy at a specific angle of incidence. The angle of minimum reflectivity is controlled by the dielectric constant and the thickness of the dielectric layer associated with the metal support. The formation of antigen-antibody complexes at metal surfaces alters the dielectric constant of the interface and increases the thickness of the dielectric layer, resulting in a shift in the minimum reflectivity angle of the plasmon resonance absorbance [Figure 7.30 (D)].[132–135]

Antigen-antibody affinity interactions usually exhibit very high binding constants ($K_a \approx 10^7 - 10^{10}$ M^{-1}). This result turns most immunosensor devices into single-cycle bioelectronic systems, where the regeneration of the sensor after interaction with the analyte is impossible. The concept of photoswitchable binding between a photoisomerizable substrate and a receptor has been used to address this problem (Figure 7.31).[139,140] A photoisomerizable antigen is assembled on the sensing interface as a monolayer. In one photoisomer configuration (state "A"), the monolayer exhibits affinity for the antibody.

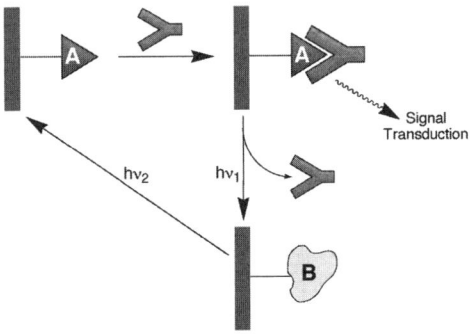

FIG. 7.31 Assembly of a reversible immunosensor using a photoisomerizable antigen-functionalized transducer.

The formation of the antigen-antibody complex on the surface is transduced to the environment, thus enabling the sensing of the antibody. After the sensing cycle, the monolayer is photoisomerized to state "B," which lacks antigen affinity properties for the antibody. This enables the antibody to wash off of the monolayer interface. The resulting monolayer is finally re-isomerized from state "B" back to state "A," regenerating the original antigen interface. Thus, a two-step photochemical isomerization of the mono-layer, with an intermediate rinsing process, enables the recycling of the sensing interface, and the functionalized transducer acts as a reusable immunosensor.[139,140]

The reversible cyclic sensing of dinitrophenyl-antibody (DNP-Ab) was accomplished by this method, by the application of a photoisomerizable dinitrospiropyran monolayer on a Au-support. A thiolated dinitrospiropyran photoisomerizable monolayer (**27a**) was assembled on Au-electrodes, Au-coated quartz crystals, or Au-coated glass slides (Figure 7.32). The dinitro-spiropyran monolayer acts as an antigen for the DNP-Ab, while the protonated dinitromerocyanine monolayer state (**27b**) lacks the antigen affinity for the DNP-Ab (Figure 7.32). The association of DNP-Ab to the dinitrospiropyran antigen monolayer can be transduced electrochemically using amperometry[141] or faradic impedance spectroscopy,[77] by QCM,[141] or by SPR.[142]

The association of the DNP-Ab to the antigen-monolayer-functionalized electrode leads to the electrical insulation of the electrode support and to the introduction of an electron barrier at the electrode surface. Thus, in the presence of an "electrically wired" enzyme (e.g., ferrocene-tethered glucose oxidase), the bioelectrocatalytic current is inhibited upon the formation of

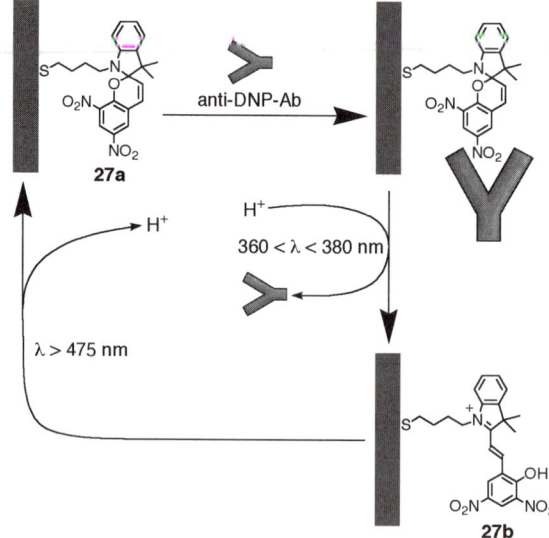

FIG. 7.32 Preparation of a photoisomerizable dinitrospiropyran monolayer on a transducer and the reversible sensing of DNP-Ab.

the DNP-Ab/dinitrospiropyran complex [Figure 7.33 (A, curve b)].[141] Similarly, the association of the DNP-Ab to the antigen-monolayer increases the interfacial electron transfer resistance, R_{et}.[77] In the presence of $[Fe(CN)_6]^{3-}/[Fe(CN)_6]^{4-}$ as redox-probe, the electron transfer resistance increases from 60 ± 2 kΩ to 80 ± 2 kΩ upon the formation of the DNP-Ab/(**27a**)-complex [Figure 7.33 (B, curves b and c, respectively)]. Photo-isomerization of the DNP-Ab/dinitrospiropyran monolayer interface (360 nm < λ < 380 nm) to the protonated dinitromerocyanine, followed by the rinsing off of the DNP-Ab, regenerates the original amperometric response of the monolayer-functionalized-electrode [Figure 7.33 (A, curve c)] and a low electron transfer resistance (R_{et}= 47 ± 2 kΩ) in the faradic imped-ance spectrum [Figure 7.33 (B, curve a)], indicating that the antibody was removed from the electrode support. The electron transfer resistance at the protonated dinitromerocyanine-functionalized electrode is lower than at the dinitrospiropyran-modified electrode using $[Fe(CN)_6]^{3-}/[Fe(CN)_6]^{4-}$ as redox-probe because the positively charged protonated dinitromerocyanine monolayer interface electrostatically attracts the redox label. Further photo-chemical isomerization of the protonated dinitromerocyanine monolayer to the dinitrospiropyran interface (λ > 475 nm) regenerates the original sensing interface. The reuse of the interface is demonstrated in Figure 7.33 (A and B, insets).

Microgravimetric transduction provides a further means to probe the binding interactions of the DNP-Ab with the photoisomerizable interface [Figure 7.34 (A)].[141] The binding of DNP-Ab to the dinitrospiropyran results in a mass increase on the crystal and a decrease in the crystal frequency of 120 Hz [Figure 7.34 (A, curve a)]. From the extent of frequency decrease, the surface coverage of the DNP-Ab on the dinitrospiropyran antigen layer was calculated to be ca. 3.8×10^{-12} mole cm^{-2} (Eq. 7.1). The frequency of the protonated dinitromerocyanine monolayer is only slightly affected upon interaction with the DNP-Ab (Δf = -40 Hz) ([Figure 7.34 (A, curve b)]. By the cyclic photoisomerization of the monolayer between the protonated dini-tromerocyanine and dinitrospiropyran states, the DNP-Ab can be washed off from the transducer, and the sensing interface regenerated to yield a reusable immunosensor device [Figure 7.34 (B)].

SPR provides an optical transduction means to follow the reversible binding and dissociation of the DNP-Ab to and from the photoisomerizable monolayer, respectively. Upon the binding of DNP-Ab to a dinitrospiropyran layer, the minimum reflectivity angle was shifted by 0.54°. Photoisomeriza-tion of the interface to the protonated dinitromerocyanine state, followed by washing off of the DNP-Ab, restores the minimum reflectivity angle of the bare monolayer. Theoretical fitting of the experimental data according to the Fresnel equation indicates that the refractive index and thickness of the antigen-antibody complex layer are 1.41 and 70 Å, respectively. From this layer thickness, it was suggested that ca. 60% of a random densely packed monolayer is formed upon the association of the DNP-Ab to the dinitro-spiropyran antigen monolayer. Figure 7.35 shows the light-induced reversible binding of DNP-Ab to the dinitrospiropyran antigen monolayer as reflected by the SPR-elucidated thickness of the sensing interface.[142]

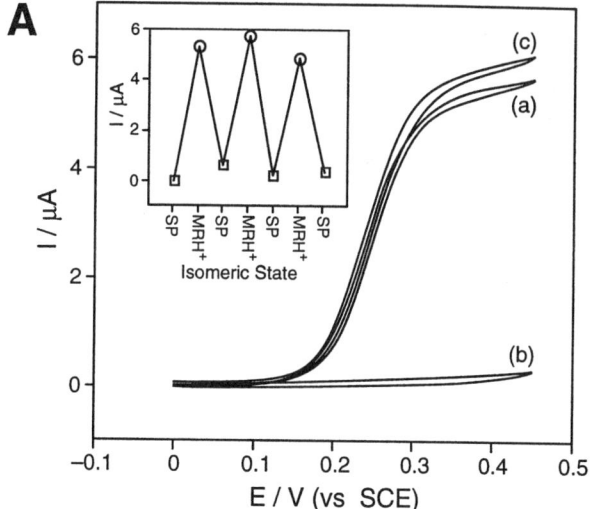

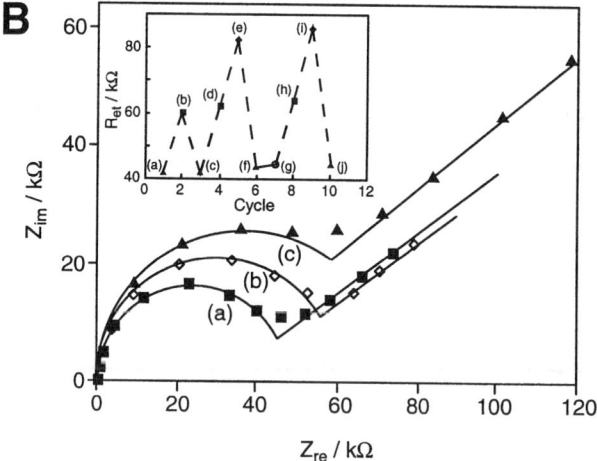

FIG. 7.33 (A) Cyclic voltammograms of: (a) The dinitrospiropyran (**27a**) monolayer electrode. (b) After addition of DNP-Ab to the dinitrospiropyran (**27a**) monolayer electrode. (c) After photoisomerization of the dinitrospiropyran/DNP-Ab to the protonated dinitromerocyanine (**27b**) monolayer and the washing off of the antibody. All data were recorded in the presence of ferrocene-tethered GOx as a redox biocatalyst and glucose, 5×10^{-2} M, scan rate 5 mV sec^{-1}. Inset: Cyclic amperometric sensing of the DNP-Ab by the dinitrospiropyran photoisomerizable monolayer electrode. (B) Faradic impedance spectra (Nyquist plots) of: (a) The protonated dinitromerocyanine (**27b**) monolayer electrode. (b) The dinitrospiropyran (**27a**) monolayer electrode. (c) The dinitrospiropyran (**27a**) monolayer electrode upon addition of the DNP-Ab. Impedance spectra were recorded in the presence of $[Fe(CN)_6]^{3-/4-}$, 1×10^{-2} M, as a redox label. Inset: Interfacial electron transfer resistances, R_{et}, at the functionalized electrodes upon the cyclic photoisomerization of the monolayer and the reversible sensing of the DNP-Ab: (a) and (c)—Monolayer in the protonated dinitromerocyanine-state (**27b**). (b) (d) and (h)—Monolayer in the dinitrospiropyran-state (**27a**). (e) and (i)—After binding of DNP-Ab to the dinitrospiropyran (**27a**) monolayer electrode. (f) and (j)—After photoisomerization of the dinitrospiropyran (**27a**)/DNP-Ab monolayer electrode to the protonated dinitromerocyanine (**27b**) and the washing off of the DNP-Ab. (g)—Addition of the DNP-Ab to the protonated dinitromerocyanine monolayer electrode (**27b**).

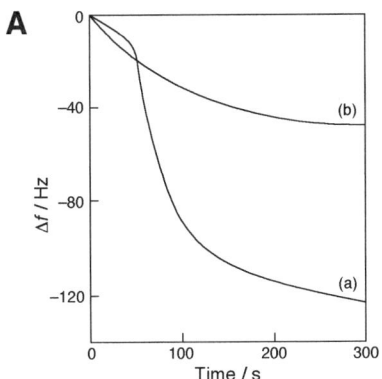

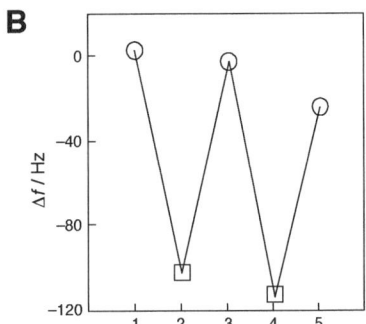

FIG. 7.34 (A) Time-dependent frequency changes of: (a) The dinitrospiropyran (**27a**) monolayer associated with a Au-quartz crystal upon addition of DNP-Ab. (b) The protonated dinitromerocyanine (**27b**) monolayer on a Au-quartz crystal upon addition of the DNP-Ab. (B) Cyclic microgravimetric sensing of the DNP-Ab by the photoisomerizable dinitrospiropyran monolayer Au-quartz crystal. (□) Frequency of the crystal after addition of the DNP-Ab to the dinitrospiropyran (**27a**) functionalized crystal. (○) Frequency of the crystal after photoisomerization of the (**27a**)/DNP-Ab monolayer to the protonated dinitromerocyanine (**27b**) washing off of the DNP-Ab, and back isomerization to **27a**-monolayer state.

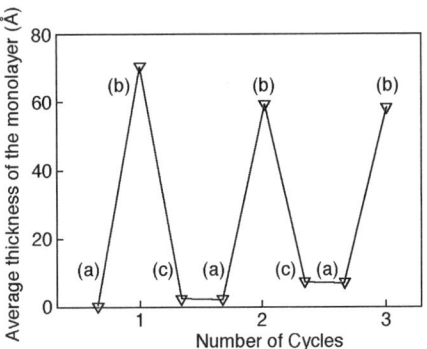

FIG. 7.35 Average thickness of the monolayer (extracted from the respective SPR-spectra): (a) The dinitrospiropyran (**27a**) monolayer. (b) The **27a**-monolayer after binding of the DNP-Ab. (c) After photoisomerization of the **27a**/DNP-Ab monolayer to the **27b**-state, and washing off of the DNP-Ab.

7.5.4 Complex Photochemical Biomolecular Switches

The ability to electrochemically transduce the photoswitchable electrocatalytic functions of photoisomerizable redox-proteins associated with electrodes, and to regulate the binding between a substrate and its receptor by light (e.g., the binding of a photoisomerizable antigen with the respective antibody), enables one to design complex photochemical bioswitches by the integration of several light-controlled biomaterials. A biphasic photochemical bioswitch exhibiting "ON," "Partial ON," and "OFF" states was constructed by assembling a dinitrospiropyran-functionalized glucose oxidase (GOx) on a Au-electrode (Figure 7.36).[143] The dinitrospiropyran-modified GOx monolayer electrode exhibits bioelectrocatalytic properties, and in the presence of ferrocene carboxylic acid (5) as a diffusional electron mediator, the biocatalytic interface electrocatalyzes the oxidation of glucose. This process is

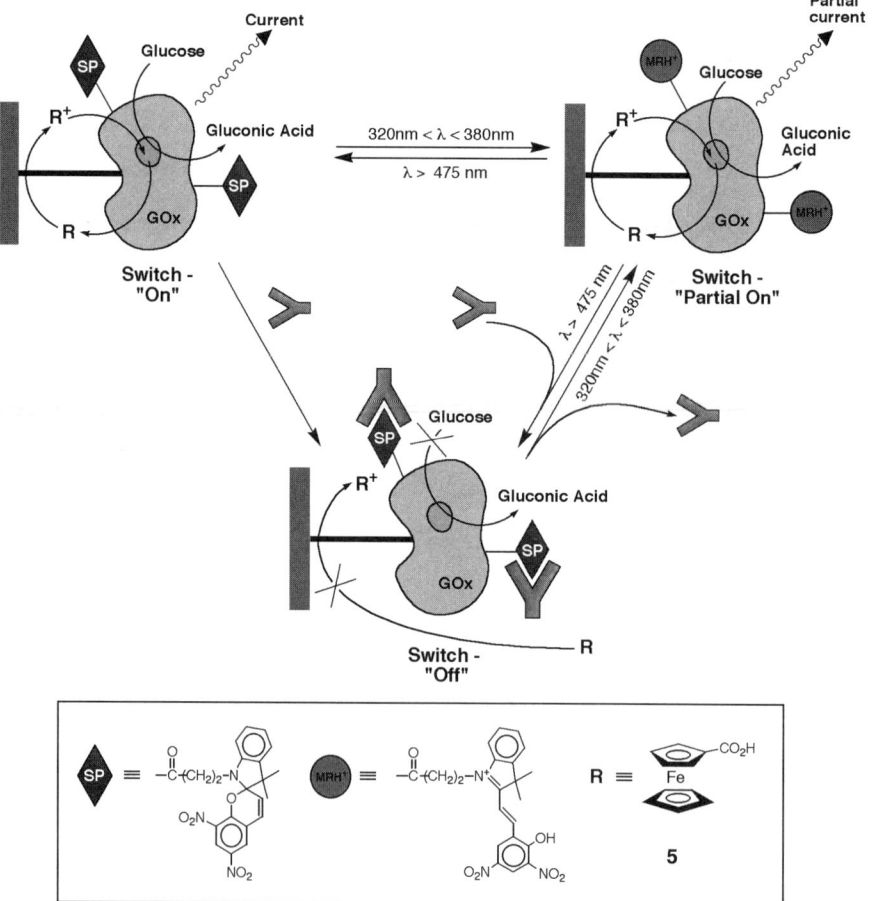

FIG. 7.36 Biphasic optobioelectronic switch by coupling of DNP-Ab with a dinitrospiropyran-functionalized GOx monolayer electrode.

reflected by a high electrocatalytic current [Figure 7.37 (A, curve a)]. By the photoisomerization of the enzyme monolayer between the dinitrospiropyran state and the protonated dinitromerocyanine state, the system can be cycled between "ON" and "Partial ON" states [Figure 7.37 (A, curves a and b, respectively)]. Because the dinitrospiropyran units tethered to the redox-enzyme act as photoisomerizable antigen sites for DNP-Ab, the light-controlled association and dissociation of the antibody to and from the interface may control the bioelectrocatalytic functions of the enzyme-layered electrode. Interaction of DNP-Ab with the "ON" state of the electrode results in the association of the antibody to the antigen sites, which blocks the electrical contacting of the enzyme and the electrode by the diffusional electron mediator (Figure 7.36). This prohibits the bioelectrocatalyzed oxidation of glucose by the enzyme-layered electrode, and the amperometric response of the system is fully switched "OFF" [Figure 7.37 (A, curve c)]. Photoisomerization of the GOx/DNP-Ab array to the protonated dinitromerocyanine state results in the dissociation of the antibody from the interface and "Partial ON" bioelectrocatalytic behavior. If the DNP-Ab is then washed from the system and the protonated dinitromerocyanine GOx-electrode is photoisomerized back to the dinitrospiropyran state, the "ON" behavior is

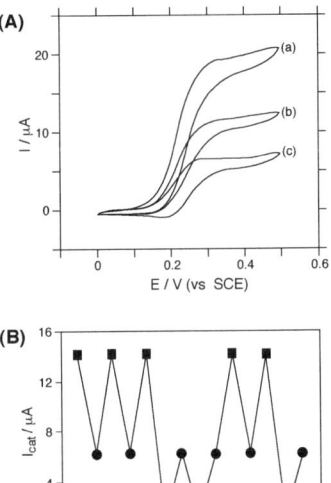

FIG. 7.37 (A) Cyclic voltammograms of: (a) The dinitrospiropyran-tethered GOx monolayer electrode. (b) The protonated dinitromerocyanine-tethered GOx monolayer electrode. (c) The dinitrospiropyran-tethered GOx monolayer electrode in the presence of the DNP-Ab. Data were recorded in the presence of ferrocene-carboxylic acid, 4×10^{-4} M, as diffusional electron mediator and glucose, 5×10^{-2} M, scan rate 2 mV sec^{-1}. (B) Biphasic switchable amperometric transduction of photonic signals recorded by the photoisomerizable-GOx/DNP-Ab system: (■) The electrode in the dinitrospiropyran-GOx state. (●) The electrode in the protonated dinitromerocyanine-state. (▲) The electrode in the presence of the dinitrospiropyran-GOx monolayer with associated DNP-Ab. (Adapted from reference 143, Figures 1 and 3. Reproduced by permission of The Royal Society of Chemistry.)

regenerated. Thus, the bioelectrocatalytic functions of the electrode may be cycled between three states: "ON," "Partial ON," and "OFF" [Figure 7.37 (B)].

7.6 INTERLOCKED COMPOUNDS AS MECHANICAL COMPONENTS AT PHOTOISOMERIZABLE INTERFACES

A thin film containing mechanically interlocked compounds and photo-isomerizable units can provide a reversible translocation of the molecular components between two distinct states. Different distances between the redox-components and the conductive support achieved in the two states of the assembly can provide the different kinetics of the interfacial electron transfer, thus an electrochemical means for probing the position of the translocated component.[144] An optoelectronic surface-bound switchable rotaxane is schematically described in Figure 7.38 (A).[145] The assembly consists of a ferrocene-functionalized cyclodextrin (**28**) molecule threaded on a "string" containing a photoisomerizable azobenzene unit and a long alkyl chain. The cyclodextrin is sterically confined to the "string" by an anthracene "stopper." When the azobenzene is in the *trans* configuration, it is complexed by **28**, but photoisomerization to the *cis*-state leaves complexation sterically impossible, so the **28** moves to the alkyl component. Back photoisomerization restores the original state. Spatial separation between the ferrocene redox-label and the electrode surface is anticipated to retard the interfacial electron transfer rate.[144] Figure 7.38 (B, curve a) shows the chronoamperometric response of the surface-bound **28**-rotaxane in the *trans*-state. A fast current decay ($k_{et}^{(1)} = 65$ s^{-1}) is observed, implying that the redox unit is close to the electrode surface. Photoisomerization of the monolayer to the *cis*-state ($320 < \lambda < 380$ nm) results in the chronoamperometric transient shown in Figure 7.38 (B, curve b). A substantially lower electron transfer rate-constant is observed ($k_{et}^{(2)}=15$ s^{-1}), indicating that the redox-active component is located farther from the electrode surface. Further photoisomerization of the *cis*-azobenzene unit back to the *trans*-azobenzene configuration ($\lambda > 420$ nm) restores the chronoamperometric response characteristic for the receptor positioned on the *trans*-azobenzene unit. By cyclic photoisomerization of the monolayer between the *trans*- and *cis*-states, the threaded receptor can be reversibly moved between the *trans*-azobenzene site and the alkyl chain [Figure 7.38 (C)].

7.7 CONCLUSIONS

Substantial progress has been accomplished in the fabrication of molecular and biomolecular optoelectronic devices. Light-activated molecular and bio-molecular systems have been integrated with electronic transducers, and the optical switching of the systems has been electronically transduced to the macroscopic environment. In particular, the photonic switching of an electron-transfer cascade has allowed the amplified electronic transduction of an input

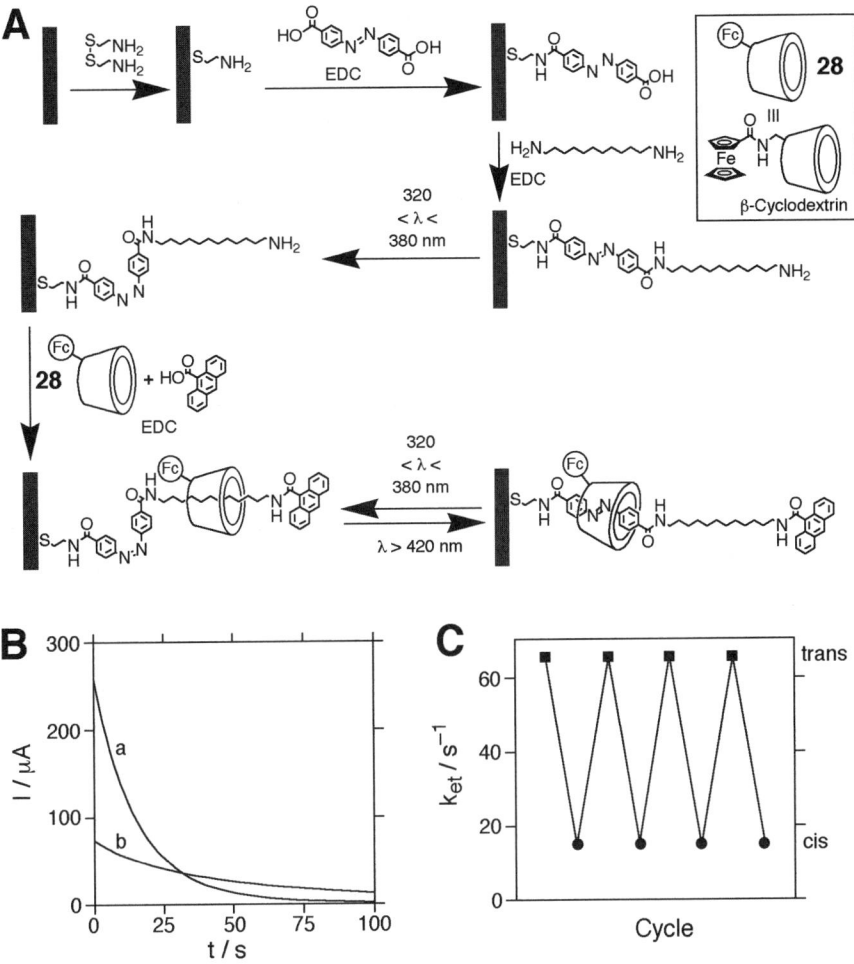

FIG. 7.38 (A) The assembly and photoisomerization of a switchable rotaxane on a Au-electrode. (B) The chronoamperometric response of the monolayer in (a) the *trans*-state, and (b) the *cis*-state. (C) The electron transfer rate constant between the electrode and the ferrocene component (**28**) over four photoisomerization cycles. Electrochemistry was performed in 0.1 M phosphate buffer (pH 7.3).

signal. The advances in extremely sensitive surface characterization techniques such as Fourier transform infrared spectroscopy (FTIR), scanning probe microscopies (AFM, STM), surface plasmon resonance (SPR), x-ray photoelectron spectroscopy (XPS) and electrochemistry (cyclic voltammetry, impedance spectroscopy, chronoamperometry) enable the identification of the structure and composition of nanostructured systems. Integrated molecular and biomolecular optoelectronic systems have already shown workable functionality as optical memories, processing devices, and sensors. Using the simple concepts of chemical synthesis, self-assembly, and electronic contacting, systems mimicking complex biological photoactive processes have been designed.

Despite these advances, however, the field still faces weighty problems. The future development of the subject requires the nanoengineering of dense, individually addressable molecular units. Nanotechnology provides means to construct nanosized wires,[146,147] dot-junctions[148,149] and circuits of high complexity.[150] Scanning probe microscopy provides "nanometric pens" where chemical solutions act as "inks" for nanoscale patterning of surfaces.[151,152,153] Similarly, scanning nearfield optical microscopy (SNOM) provides nanometric light sources for addressing molecular domains, as well as a microscopic scanning tool for imaging the fluorescence of nano-domains.[154,155] The further development of the battery of molecular functionalities and nanometer-sized tools will eventually expand our skills in molecular engineering and the construction of programmed functional arrays of higher complexity.

The use of molecular species for super-dense information storage is an intriguing prospect.[156] Multilayers and crystals of functional molecules may hold orders of magnitude more information than current systems and could be quickly addressed and read-out by multiple or focused light beams and CCD arrays. The use of additional building blocks such as dendrimers,[157,158] nanoparticles,[159,160] and biomaterials such as DNA[161] for structural and functional purposes may be envisaged. The subject of molecular and biomolecular optoelectronic systems will be an interdisciplinary scientific challenge for chemists, physicists, biologists, and materials scientists, and although future optoelectronic systems may differ substantially from the infant assemblies described here, this groundwork will make the future possible.

ACKNOWLEDGMENTS

The support of the Israel Science Foundation is gratefully acknowledged. A.N.S. thanks the Valazzi-Pikovsky fellowship trust for financial support.

REFERENCES

1. Gómez-López, M., and Stoddart, J. F. Molecular and supramolecular nanomachines. In *Handbook of Nanostructured Materials Vol 5: Organics, Polymers and Biological Materials* (Nalwa, H. S., Ed.) Academic Press, San Diego **2000**.
2. Shinkai, S., Miname, I., Kusano, Y., and Manabe, O. Photoresponsive crown ethers. 5. Light-driven ion transport by crown ethers with a photoresponsive anionic cap. *J. Am. Chem. Soc.* **1982**, *104*, 1967–1972.
3. Shinkai, S., Ishira, M., Ueda, K., and Manabe, O. Photoresponsive crown ethers. Part 14. Photoregulated crown-metal complexation by competitive intramolecular tail(ammonium)-biting. *J. Chem. Soc. Perkin Trans. 2* **1985**, 511–518.
4. Kimura, K., Mizutani, R., Yokoyama, M., Arakawa, R., Matsubayashi, G., Okamoto, M., and Doe, H. All-or-none type photochemical switching of cation binding with malachite green carrying a bis(monoazacrown ether) moiety. *J. Am. Chem. Soc.* **1997**, *119*, 2062–2063.
5. Breslow, R. Artificial enzymes. *Science* **1982**, *218*, 532.
6. Hosseini, M. W., Lehn, J.-M., and Mertes, M. P. Efficient molecular catalysis of ATP-hydrolysis by protonated macrocyclic polyamines. *Helv. Chim. Acta* **1983**, *66*, 2454.
7. Guenzi, A., Johnson, C.A., Cozzi, F., and Mizlow, K. Dynamic gearing and residual stereoisomerism in labeled bis(9-triptycyl)methane and related molecules. Synthesis and

stereochemistry of bis(2,3-dimethyl-9-triptycyl)methane, bis(2,3-dimethyl-9-triptycyl)carbinol, and bis(1,4-dimethyl-9-triptycyl)methane. *J. Am. Chem. Soc.* **1983**, *105*, 1438–1448.

8. Kelly, T. R., Bowyer, M. C., Bhaskar, K. V., Bebbington, D., Garcia, A., Lang, F., Kim, M. H., and Jette, M. P. A molecular brake. *J. Am. Chem. Soc.* **1994**, *116*, 3657–3658.

9. Schoevaars, A. M., Kruizinga, W., Zijlstra, R. W. J., Veldman, N., Spek. A. L., and Feringa, B. L. Toward a switchable molecular rotor. Unexpected dynamic behavior of functionalized overcrowded alkenes. *J. Org. Chem.* **1997**, *62*, 4943–4948.

10. Koumura, N., Zijlstra, R. W. J., van Delden, R. A., Harada, N., and Feringa, B. L. Light-driven monodirectional molecular rotor. *Nature* **1999**, *401*, 152–155.

11. Kelly, T. R., Tellitu, I., and Sestelo, J. P. In search of molecular ratchets. *Angew. Chem. Int. Ed. Engl.* **1997**, *36*, 1866–1868.

12. Philp, D., and Stoddart, J. F. Self-assembly in natural and unnatural systems. *Angew. Chem. Int. Ed. Engl* . **1996**, *35*, 1155–1196.

13. Menzer, S., White, A. J. P., Williams, D. J., Belohradsky, M., Hamers, C., Raymo, F. M., Shipway, A. N., and Stoddart, J. F. Molecular meccano. 25. Self-assembly of functionalized [2]catenanes bearing a reactive functional group on either one or both macrocyclic components – From monomeric [2]catenanes to polycatenanes *Macromolecules* **1998**, *31*, 295–307.

14. Ashton, P. R., Baldoni, V., Balzani, V., Claessens, C. G., Credi, A., Hoffman, H. A. D., Raymo, F. M., Stoddart, J. F., Venturi, M., White, A. J. P., and Williams, D. J. Molecular meccano. 57. Template-directed syntheses, spectroscopic properties, and electrochemical behavior of [n]catenanes. *Eur. J. Org. Chem.* **2000**, *7*, 1121–1130.

15. Amabilino, D. B., Asakawa, M., Ashton, P. R., Ballardini, R., Balzani, V., Belohradsky, M., Credi, A., Higuchi, M., Raymo, F. M., Shimizu, T., Stoddart, J. F., Venturi, M., and Yase, K. Aggregation of self-assembling branched [n]rotaxanes. *New J. Chem.* **1998**, *22*, 959–972.

16. Murray, R. W. Chemically modified electrodes. *Acc. Chem. Res.* **1980**, *13*, 135-141.

17. Abruña, H. D. Coordination chemistry in two dimensions: chemically modified electrodes. *Coord. Chem. Rev.* **1988**, *86*, 135–189.

18. Finklea, H. D. Electrochemistry of organized monolayers of thiols and related molecules on electrodes. In *Electroanalytical Chemistry* (Bard, A. J., Rubinstein, I., Eds.), Marcel Dekker, New York **1996**, Vol. 19, pp. 109–335.

19. Mandler, D., and Turyan, I. Applications of self-assembled monolayers in electroanalytical chemistry *Electroanalysis* **1996**, *8*, 207–213.

20. Knoll, W. Self-assembled microstructures at interfaces. *Current Opinion Colloid Interface Sci.* **1996**, *1*, 137–143.

21. Ulman, A. *An Introduction to Ultrathin Organic Films: From Langmuir-Blodgett to Self-Assembly*, Academic Press, San Diego **1991**.

22. Gilat, S. L., Kawai, S. H., and Lehn, J.-M. Light-triggered molecular devices – Photochemical switching of optical and electrochemical properties inmolecular wire type. *Chem. Eur. J.* **1995**, *1*, 275–284.

23. Kawai, S. H., Gilat, S. L., Posinet, R., and Lehn, J.-M. A dual-mode molecular switching device – Bisphenolic diarylethenes with integrated photochromic and electrochromic properties. *Chem. Eur. J.* **1995**, *1*, 285–293.

24. Goulle, V., Harriman, A., and Lehn, J.-M. An electro-photoswitch – redox switching of the luminescence of a bipyridine metal-complex. *J. Chem. Soc. Chem. Commun.* **1993**, 1034–1036.

25. Zelikovich, L., Libman, J., and Shanzer, A. Molecular redox switching based on chemical triggering of iron translocation in triple-stranded helical complexes. *Nature* **1995**, *374*, 790–792.

26. Bissel, R. A., De Silva, A. P., Gunaratne, H. Q. M., Lynch, P. L. M., Maguire, G. E. M., and Sandanayake, K. R. A. S. Molecular fluorescent signaling with fluor spacer receptor system – Approaches to sensing and switching devices via supramolecular photophysics. *Chem. Soc. Rev.* **1992** *21*, 187–195.

27. Ashton, P. R., Blower, M., Philp, D., Spencer, N., Stoddart, J. F., Tolley, M. S., Ballardini, R., Ciano, M., Balzani, V., Gandolfi, M. T., Prodi, L., and McLean, C. H. The control of translational isomerism in catenated structures. *New J. Chem.* **1993**, *17*, 689–695.

28. Leigh, D. A., Moody, K., Smart, J. P., Watson, K. J., and Slawin, A. M. Z. Catenane chameleons: Environment-sensitive translational isomerism in amphiphilic benzylic amide [2]catenanes. *Angew. Chem. Int. Ed. Engl.* **1996**, *35*, 306–310.

29. Livoreil, A., Dietrich-Buchecker, C. O., and Sauvage, J.-P. Electrochemically triggered swinging of a [2]-catenate. *J. Am. Chem. Soc.* **1994**, *116*, 9399–9400.

30. Amabilino, D. B., and Stoddart, J. F. Interlocked and intertwined structures and super-structures. *Chem. Rev.* **1995**, *95*, 2725–2828.

31. Bissell, R. A., Córdova, E., Kaifer, A. E., and Stoddart, J. F. A chemically switchable molecular shuttle. *Nature* **1994**, *369*, 133–137.

32. Martínez-Díaz, M. V., Spencer, N., and Stoddart, J. F. The self-assembly of a switchable [2]rotaxane. *Angew. Chem. Int. Ed. Engl.* **1997**, *36*, 1904–1907.

33. Murakami, H., Kawabuchi, A., Kotoo, K., Kunitake, M., and Nakashima, N. A light-driven molecular shuttle based on a rotaxane. *J. Am. Chem. Soc.* **1997**, *119*, 7605–7606.

34. Shipway, A. N., Katz, E., and Willner, I. Molecular memory and processing devices in solution and on surfaces. in: *Structure and Bonding*, Sauvage, J.-P. (Ed.), **2001**, Vol. 99, pp. 237–281.

35. Balzani, V., Gómex-López, M., and Stoddart, J. F. Molecular machines. *Acc. Chem. Res.* **1998**, *31*, 405–414.

36. *Organic Photochromic and Thermochromic Compounds* (Crano, J. C., Guglielmetti, R. Eds), Plenum Press, New York, 1998.

37. *Photochromism* (Dürr, H., Bouas-Laurent, H., Eds), Elsevier, Amsterdam, 1990.

38. Irie, M., Kobatake S., and Horichi M. Reversible surface morphology changes of a pho-tochromic diarylethene single crystal by photoirradiation. *Science* **2001**, *291*, 1769–1772.

39. Zelichenok, A., Buchholz, F., Fischer. E., Ratner, J., Krongauz, V., Anneser, H., and Brauchle, C. Photochemistry and multiple holographical recording in polymers with photochromic phenoxy-naphthacene-quinone side-groups. *J. Photochem. Photobiol. A* **1993** *76*, 135–141.

40. Fang, Z., Wang, S. Z., Yang, Z. F., Chen, B., Li, F. T., Wang, J. Q., Xu, S. X., Jiang, Z. J., and Fang, T. R. Synthesis and photochromism in solution of phenoxynaphthacenequinone derivatives. *J. Photochem. Photobiol. A* **1995**, *88*, 23–30.

41. Tajima, M., Keat, L. E., Matsunaga, K., Yamashita, T., Tokoro, M., and Inoue, H. Photochromism of 1-phenoxyanthraquinones. *J. Photochem. Photobiol. A* **1993**, *74*, 211–219.

42. Gritsan, N. P., and Klimenko, L. S. Photochromism of quinonoid compounds – Properties of photoinduced ana-quinones. *J. Photochem. Photobiol. A* **1993**, *70*, 103–117.

43. Doron, A., Portnoy, M., Lion-Dagan, M., Katz, E., and Willner, I. Amperometric transduc-tion and amplification of optical signals recorded by a phenoxynaphthacenequinone monolayer electrode: Photochemical and pH-gated electron transfer *J. Am. Chem. Soc.* **1996**, *118*, 8937–8944.

44. Doron, A., Katz, E., Portnoy, M., and Willner, I. An electroactive photoisomerizable monolayer-electrode: A command surface for the amperometric transduction of recorded optical signals. *Angew. Chem. Int. Ed. Engl.*, **1996**, *35*, 1535–1537.

45. Willner, I., Doron, A., and Katz, E. Gated molecular and biomolecular optoelectronic sys-tems via photoisomerizable monolayer electrodes. *J. Phys. Org. Chem.* **1998**, *11*, 546–560.

46. Laviron, E., General expression of the linear potential sweep voltammogram in the case of diffusionless electrochemical systems. *J. Electroanal. Chem.* **1979**, *101*, 19–28.

47. Bild, C. L., and Kuhn, A. T. Electrochemistry of the viologens. *Chem. Soc. Rev.* **1981**, *10*, 49–82.

48. Sekkat, Z., Wood, J., Geerts, Y., and Knoll, W. A smart ultrathin photochromic layer. *Langmuir* **1995**, *11*, 2856–2859.

49. Siewierski, L. M., Brittain, W. J., Petrash, S., and Foster, M. D. Photoresponsive mono-layers containing in-chain azobenzene. *Langmuir* **1996**, *12*, 5838–5844.

50. Caldwell, W. B., Campbell, D. J., Chen, K. M., Herr, B. R., Mirkin, C. A., Malik, A., Durbin, M. K., Dutta, P., and Huang, K. G. A highly ordered self-assembled monolayer film of an azobenzenealkanethiol on Au(111) – Electrochemical properties and structural characterization by synchrotron in-plane X-ray-diffraction, atomic-force microscopy, and surface-enhanced Raman-spectroscopy. *J. Am. Chem. Soc.* **1995** *117*, 6071–6082;

51. Wang, R., Iyoda, T., Tryk, D. A., Hashimoto, K., and Fujishima, A. Electrochemical modulation of molecular conversion in an azobenzene-terminated self-assembled monolayer film: An in situ UV-visible and infrared study. *Langmuir* **1997**, *13*, 4644–4651.

52. Vélez, M., Mukhopadhyay, S., Muzikante, I., Matisova, G., and Vieira, S. Atomic force microscopy studies of photoisomerization of an azobenzene derivative on Langmuir-Blodgett monolayers. *Langmuir* **1997**, *13*, 870–872.

53. Sato, T., Ozaki, Y., and Iriyama, K. Molecular aggregation and photoisomerization of Langmuir-Blodgett-films of azobenzene-containing long-chain fatty-acids and their salts studied by ultraviolet-visible and infrared spectroscopies. *Langmuir* **1994**, *10*, 2363–2369.

54. Liu, Z.-F., Hashimoto, K., and Fujishima, A. Photoelectrochemical information-storage using an azobenzene derivative. *Nature* **1990**, *347*, 658–660.

55. Yu, H.-Z., Wang, Y.-Q., Cheng, J.-Z., Zhao, J.-W., Cai, S.-M., Inokuchi, H., Fujishima, A., and Liu, Z.-F. Electrochemical behavior of azobenzene self-assembled monolayers on gold. *Langmuir* **1996**, *12*, 2843–2848.

56. Liu, Z.-F., Morigaki, K., Hashimoto, K., and Fujishima, A. Electrochemical actinometry using the assembled monolayer film of an azo compound. *Anal. Chem.* **1992**, *64*, 134–137.

57. Nakashima, N., Nakanishi, T., Nakatani, A., Deguchi, Y., Murakami, H., Sagara, T., and Irie, M. Photoswitching of a vectorial electron transfer reaction at a diarylethene modified electrode. *Chem. Lett.* **1997**, 591–592.

58. Tachibana, H., Nakamura, T., Matsumoto, M., Komizu, H., Manda, E., Niino, H., Yabe, A., and Kawabata, Y. Photochemical switching in conductive Langmuir-Blodgett films. *J. Am. Chem. Soc.* **1989**, *111*, 3080–3081.

59. Tachibana, H., Azumi, R., Nakamura, T., Matsumoto, M., and Kawabata, Y. New types of photochemical switching phenomena in Langmuir-Blodgett-films. *Chem. Lett.* **1992**, 173–176.

60. Tachibana, H., Nishio, Y., Nakamura, T., Matsumoto, M., Manda, E., Niino, H., Yabe, A., and Kawabata, Y. Control of photochemical switching phenomena by chemical modification. *Thin Solid Films* **1992**, *210/211*, 293–295.

61. Tachibana, H., Manda, E., Azumi, R., Nakamura, T., Matsumoto, M., and Kawabata, Y. Multiple photochemical switching device based on Langmuir-Blodgett-films. *Appl. Phys. Lett.* **1992**, *61*, 2420–2421.

62. Willner, I., and Willner, B. Chemistry of photobiological switches in bioorganic photochemistry. In *Biological Applications of Photochemical Switches, Bioorganic Photochemistry*, vol. 2 (H. Morrison, Ed.) Wiley, New York, **1993**, pp. 1–110.

63. Willner, I. Photoswitchable biomaterials – en route to optobioelectronic systems. *Acc. Chem. Res.* **1997**, *30*, 347–356.

64. Willner, I., Katz, E., Willner, B., Blonder, R., Heleg-Shabtai, V., and Bückmann, A. F. Assembly of functionalized monolayers of redox proteins on electrode surfaces: novel bioelectronic and optobioelectronic systems. *Biosens. Bioelectron.* **1997**, *12*, 337–356.

65. Blonder, R., Katz, E., Willner, I., Wray, V., and Bückmann A. F. Application of a nitrospiropyran-FAD reconstituted glucose oxidase and charged electron mediators as optobioelectronic assemblies for the amperometric transduction of recorded optical signals: control of the "ON"-"OFF" direction of the photoswitch. *J. Am. Chem. Soc.* **1997**, *119*, 11747–11757.

66. Lion-Dagan, M., and Willner, I. Nitrospiropyran-modified chymotrypsin – A photostimulated biocatalyst in an organic solvent: effects of bioimprinting. *J. Photochem. Photobiol. A* **1997**, *108*, 247–252.

67. Lion-Dagan, M., Katz, E., and Willner, I. Amperometric transduction of optical signals recorded by organized monolayers of photoisomerizable biomaterials on Au-electrodes. *J. Am. Chem. Soc.* **1994**, *116*, 7913–7914.

68. Willner, I., Lion-Dagan, M., Marx-Tibbon, S., and Katz, E. Bioelectrocatalyzed amperometric transduction of recorded optical signals using monolayer-modified Au-electrodes. *J. Am. Chem. Soc.* **1995**, *117*, 6581–6592.

69. Willner, I., Blonder, R., Katz, E., Stocker, A., and Bückmann, A. F. Reconstitution of apoglucose oxidase with a nitrospiropyran-modified FAD cofactor yields a photoswitchable biocatalyst for amperometric transduction of recorded optical signals. *J. Am. Chem. Soc.*

1996, *118*, 5310–5311.

70. Katz, E., Willner, B., and Willner, I. Light-controlled electron transfer reactions at photo-isomerizable monolayer electrodes by means of electrostatic interactions: Active interfaces for the amperometric transduction of recorded optical signals. *Biosens. Bioelectron.* **1997**, *12*, 703–719.

71. Willner, I., and Willner, B. Layered molecular optoelectronic assemblies. *J. Mater. Chem.* **1998**, *8*, 2543–2556.

72. Willner, I., and Willner, B. Photoswitchable biomaterials as grounds for optobioelectronic devices. *Bioelectrochem. Bioenerg.* **1997**, *42*, 43–57.

73. Lane, R. T., and Hubbard, A. T. Electrochemistry of chemisorbed molecules. II. The influence of charged chemisorbed molecules on the electrode reactions of platinum complexes. *J. Phys. Chem.* **1973**, *77*, 1411–1421.

74. Takehara, K., and Ide, Y. Electrochemical properties of a gold electrode modified with a mixed monolayer. *Bioelectrochem. Bioenerg.* **1992**, *27*, 207–219.

75. Malem, F., and Mandler, D. Self-assembled monolayers in electroanalytical chemistry – application of ω-mercapto carboxylic-acid monolayers for the electrochemical detection of dopamine in the presence of a high-concentration of ascorbic acid. *Anal. Chem.* **1993**, *65*, 37–41.

76. Doron, A., Katz, E., Tao, G., and Willner, I. Photochemically-, chemically- and pH-controlled electrochemistry at functionalized spiropyran monolayer electrodes. *Langmuir* **1997**, *13*, 1783–1790.

77. Patolsky, F., Filanovsky, B., Katz, E., and Willner, I. Photoswitchable antigen-antibody interactions studied by impedance spectroscopy. *J. Phys. Chem. B* **1998**, *102*, 10359–10367.

78. Kato, S., Aizawa, M., and Suzuki, S. Photo-responsive membranes. I. Light-induced potential changes across membranes incorporating a photochromic compound. *J. Membr. Sci.* **1976**, *1*, 289–300.

79. Shipway A. N., and Willner, I. Electronically transduced molecular mechanical and information functions on surfaces. *Acc. Chem. Res.* **2001**, *34*, 421–432.

80. Katz, E., Lion-Dagan, M., and Willner, I. Control of electrochemical processes by photo-isomerizable spiropyran monolayers immobilized onto Au-electrodes: Amperometric transduction of optical signals. *J. Electroanal. Chem.* **1995**, *382*, 25–31.

81. Katz, E., Lötzbeyer, T., Schlereth, D. D., Schuhmann, W., and Schmidt, H.-L. Electrocatalytic oxidation of reduced nicotinamide coenzymes at gold and platinum electrode surfaces modified with a monolayer of pyrroloquinoline quinone. Effect of Ca²⁺ cations. *J. Electroanal. Chem.* **1994**, *373*, 189–200.

82. Katz, E., and Willner, I. Thermal and photochemical control of an electrochemical process at an isomerizable spiropyran monolayer-modified Au-electrode. *Electroanalysis* **1995**, *7*, 417–419.

83. Walter, D. G., Campbell, D. J., and Mirkin, C. A. Photon-gated electron transfer in two-component self-assembled monolayers. *J. Phys. Chem. B* **1999**, *103*, 402–405.

84. Wolf, M. O., and Fox, M. A. Photochemistry and surface properties of self-assembled monolayers of *cis*-4-cyano-4′-(10-thiodecoxy)stilbene and *trans*-4-cyano-4′-(10-thiodecoxy)stilbene on polycrystalline gold. *J. Am. Chem. Soc.* **1995**, *117*, 1845–1846.

85. Heller, A. Electrical wiring of redox enzymes. *Acc. Chem. Res.* **1990**, *23*, 128–134.

86. Allen, P. M., Hill, H. A. O., and Walton, N. J. Surface modifiers for the promotion of direct electrochemistry of cytochrome c. *J. Electroanal. Chem.* **1984**, *178*, 69–86

87. Armstrong, F. A., Hill, H. A. O., and Walton, N. J. Direct electrochemistry of redox proteins. *Acc. Chem. Res.* **1988**, *21*, 407–413.

88. Tarlov, M. J., and Bowden, E. F. Electron-transfer reaction of cytochrome c adsorbed on carboxylic-acid terminated alkanethiol monolayer electrodes. *J. Am. Chem. Soc.* **1991**, *113*, 1847–1849.

89. Lion-Dagan, M., Katz, E., and Willner, I. A bifunctional monolayer electrode consisting of 4-pyridyl sulfide and photoisomerizable spiropyran: Photoswitchable electrical communication between the electrode and cytochrome c. *J. Chem. Soc., Chem. Commun.* **1994**, 2741–2742.

90. Jin, W., Wollenberger, U., Bier, F. F., Makower, A., and Scheller, F. W. Electron transfer

between cytochrome c and copper enzymes. *Bioelectrochem. Bioenerg.* **1996**, *39*, 221–225.

91. Cass, A. E. G., Davis, G., Hill, H. A. O., and Nancarrow, D. J. The reaction of flavo-cytochrome b₂ with cytochrome c and ferricinium carboxylate. Comparitive kinetics by cyclic voltammetry and chronoamperometry. *Biochim. Biophys. Acta* **1985**, *828*, 51–57.

92. Powis, D. A., and Wattus, G. D. Electrochemical reduction of dioxygen using a terminal oxidase. *FEBS Lett.* **1981**, *126*, 282–284.

93. Willner, I., Doron, A., Katz, E., Levi, S., and Frank, A. J. Reversible associative and disso-ciative interactions of glucose oxidase with nitrospiropyran monolayers assembled onto Au-electrodes: Amperometric transduction of recorded optical signals. *Langmuir* **1996**, *12*, 946–954.

94. Wilson, R., and Turner, A. P. F. Glucose oxidase – an ideal enzyme. *Biosens. Bioelectron.* **1992**, *7*, 165–185.

95. Blonder, R., Willner, I., and Bückmann, A. F. Reconstitution of apo-glucose oxidase on nitrospiropyran and FAD mixed monolayers on gold electrodes: Photostimulation of bioelectrocatalytic features of the biocatalyst. *J. Am. Chem. Soc.* **1998**, *120*, 9335–9341.

96. Wolf, M. O., and Fox, M. A. Photoisomerization and photodimerization in self-assembled monolayers of *cis*- and *trans*-4-cyano-4′-(10-mercaptodecoxy)stilbene on gold. *Langmuir* **1996**, *12*, 955–962.

97. Li, W., Lynch, V., Thompson, H., and Fox, M. A. Self-assembled monolayers of 7-(10-thiodecoxy)coumarin on gold: Synthesis, characterization, and photodimerization. *J. Am. Chem. Soc.* **1997**, *119*, 7211–7217.

98. Fox, M. A., and Wooten, M. D. Characterization, adsorption, and photochemistry of self-assembled monolayers of 10-thiodecyl 2-anthryl ether on gold. *Langmuir* **1997**, *13*, 7099–7105.

99. Lahav, M., Ranjit, K. T., Katz, E., and Willner, I. A β-amino-cyclodextrin monolayer-modified Au electrode: A command surface for the amperometric and microgravimetric transduction of optical signals recorded by a photoisomerizable bipyridinium-azobenzene diad. *Chem. Commun.* **1997**, 259–260.

100. Lahav, M., Ranjit, K. T., Katz, E., and Willner, I. *Isr. J. Chem.* **1997**, *37*, 185–195.

101. Marx-Tibbon, S., Ben-Dov, I., and Willner, I. Electrochemical and quartz-crystal-micro-balance transduction of light-controlled supramolecular interactions for monolayer-functionalized electrodes. *J. Am. Chem. Soc.* **1996**, *118*, 4717–4718.

102. Ranjit, K. T., Marx-Tibbon, S., Ben-Dov, I., Willner, B., and Willner, I. Photostimulated interactions of bipyridinium-azobenzene with a β-aminocyclodextrin monolayer-function-alized electrode: An optoelectronic assembly for the amperometric transduction of recorded optical signals. *Isr. J. Chem.* **1996**, *36*, 407–419.

103. Ranjit, K. T., Marx-Tibbon, S., Ben-Dov, I., and Willner, I. Formation of supramolecular donor-acceptor complexes between bis(pyridiniomethyl)azobenzenes and eosin in solutions and at solid interfaces: Transduction into optical and microgravimetric signals. *Angew. Chem. Int. Ed. Engl.* **1997**, *36*, 147–150.

104. Willner, I., Eichen, Y., Rabinovitz, M., Hoffman, R., and Cohen, S. Structure and thermo-dynamic and kinetic properties of eosin biptridinium complexes. *J. Am. Chem. Soc.* **1992**, *114*, 637–644.

105. Willner, I., Eichen, Y., Doron, A., and Marx, S. Effects of electrostatic and π-π-interactions on the stabilities of xantene dye-4,4′-bipyridinium complexes – Structural design of a geared supramolecular machine. *Isr. J. Chem.* **1992**, *32*, 53–59.

106. Willner, I., Marx, S., and Eichen, Y. Photoswitchable association of an azobenzene bipyri-dinium diad to eosin – Photostimulated ON-OFF guest binding. *Angew. Chem. Int. Ed. Engl.* **1992**, *31*, 1243–1244.

107. Buttry, D. A., and Ward, M. D. Measurement of interfacial processes at electrode surfaces with the electrochemical quartz crystal microbalance. *Chem. Rev.* **1992**, *92*, 1355–1379.

108. Deng, G., Sakaki, T., Kawahara, Y., and Shinkai, S. Light-switched ionophoric calix[4]arenes. *Tetrahedron Lett.* **1992**, *33*, 2163–2166.

109. Kimura, K., Yamashita, T., and Yokoyama, M. Photochemical switching of ionic-conduc-tivity in composite films containing a crowned spiribenzopyran. *J. Phys. Chem.* **1992**, *96*, 5614–5617.

110. Skladal, P. Advances in electrochemical immunosensors. *Electroanalysis* **1997**, *9*, 737–745.

111. Ghindilis, A. L., Atanasov, P., Wilkins, M., and Wilkins, E. Immunosensors: Electrochemical sensing and other engineering approaches. *Biosens. Bioelectron.* **1998**, *13*, 113–131.

112. Ghindilis, A. L., Skorobogat'ko, O. V., Gavrilova, V. P., and Yaropolov, A. I. A new approach to the construction of potentiometric immunosensors. *Biosens. Bioelectron.* **1992**, *7*, 301–304.

113. Pfeifer, U., and Baumann, W. Direct potentiometric immunoelectrodes. 2. An immunoelectrode sensitive to 3-indole acetic acid methylester. *Fresenius J. Anal. Chem.* **1993**, *345*, 504–511.

114. Engel, L., and Baumann, W. Direct potentiometric immunoelectrodes. 3. A graphite based atrazine immunoelectrode. *Fresenius J. Anal. Chem.* **1993**, *346*, 745–751.

115. Engel, L., and Baumann, W. Direct potentiometric immunoelectrodes. 4. An immunoelectrode for the trace-level determination of atrazine by separated incubation and potential measurement steps. *Fresenius J. Anal. Chem.* **1993**, *349*, 447–450.

116. Ho, W. O., Athey, D., and McNeil, C. J. Amperometric detection of alkaline phosphatase activity at a horseradish peroxidase enzyme electrode based on activated carbon – Potential application to electrochemical immunoassay. *Biosens. Bioelectron.* **1995**, *10*, 683–691.

117. Tiefenauer, L. X., Kossek, S., Padeste, C., and Thiébaud, P. Towards amperometric immunosensor devices. *Biosens. Bioelectron.* **1997**, *12*, 213–223.

118. Ivnitski, D., and Rishpon, J. A one-step, separation-free amperometric enzyme immunosensor. *Biosens. Bioelectron.* **1996**, *11*, 409–417.

119. Wittstock, G., Emons, H., and Heineman, W. R. Electron transfer through an immunoglobulin layer via an immobilized redox mediator. *Electroanalysis* **1996**, *8*, 143–146.

120. Zhang, S., Jiao, K., and Chen, H. Investigation of voltammetric enzyme – linked immunoassay based on a new system of OAP-H_2O_2-HRP. *Electroanalysis* **1999**, *11*, 511–516.

121. Blonder, R., Katz, E., Cohen, Y., Itzhak, N., Riklin, A., and Willner, I. Application of redox enzymes for probing the antigen – antibody association at monolayer interfaces: Development of amperometric immunosensor electrodes. *Anal. Chem.* **1996**, *68*, 3151–3157.

122. Katz, E., and Willner, I. Amperometric amplification of antigen – antibody association at monolayer interfaces: Design of immunosensor electrodes. *J. Electroanal. Chem.* **1996**, *418*, 67–72.

123. De Silva, M. S., Zhang, Y., Hesketh, P. J., Maclay, G. J., Gendel, S. M., and Stetter, J. R. Impedance based sensing of the specific binding reaction between staphylococcus enterotoxin-B and its antibody on an ultra-thin platinum film. *Biosens. Bioelectron.* **1995**, *10*, 675–682.

124. Maupas, H., Soldatkin, A. P., Martelet, C., Jaffrezic-Renault, N., and Mandrand, B. Direct immunosensing using differential electrochemical measurements of impedimetric variations. *J. Electroanal. Chem.* **1997**, *421*, 165–171.

125. Ben-Dov, I., Willner, I., and Zisman, E. Piezoelectric immunosensors for urine specimens of *Chlamydia trachomatis* employing quartz crystal microbalance microgravimetric analyses. *Anal. Chem.* **1997**, *69*, 3506–3512.

126. Suleiman, A. A., and Guilbault, G. G. Piezoelectric (PZ) immunosensors and their applications. *Anal. Lett.* **1991**, *24*, 1283–1292.

127. Suleiman, A. A., and Guilbault, G. G. Recent developments in piezoelectric immunosensors. *Analyst* **1994**, *119*, 2279–2282.

128. König, B., and Grätzel, M. *Anal. Chim. Acta* **1993**, *276*, 323–333.

129. König, B., and Grätzel, M. Long-term stability and improved reusability of a piezoelectric immunosensor for human erythrocytes. *Anal. Chim. Acta* **1993**, *280*, 37–41.

130. König, B., and Grätzel, M. Detection of human T-lymphocytes with a piezoelectic immunosensor. *Anal. Chim. Acta* **1993**, *281*, 13–18.

131. König, B., and Grätzel, M. A piezoelectric immunosensor for hepatitis viruses. *Anal. Chim. Acta* **1995**, *309*, 19–25.

132. Liedberg, B., Nylander, C., and Lundström, I. Biosensing with surface-plasmon resonance – How it all started. *Biosens. Bioelectron.*, **1995**, *10*, R1–R9.

133. Sasaki, S., Nagata, R., Hock, B., and Karube, I. Novel surface plasmon resonance sensor chip functionalized with organic silica compounds for antibody attachment. *Anal. Chim. Acta*, **1998**, *368*, 71–76.

134. Gruen, L. C., McKimm-Breschkin, J. L., Caldwell, J. B., and Nice, E. C. Affinity ranking of influenza neuraminidase mutants with monoclonal-antibodies using an optical biosensor: Comparison with ELISA and slot blot assays. *J. Immunol. Methods*, **1994**, *168*, 91–100.

135. Zeder-Lutz, G., Altschuh, D., Geysen, H. M., Trifilieff, E., Sommermeyer, G., Van and Regenmortel, M. H. V. Monoclonal antipeptide antibodies – Affinity and kinetic rate constants measured for the peptide and the cognate protein using a biosensor technology. *Molec. Immunol.* **1993**, *30*, 145–155.

136. Bard, A. J., and Faulkner, L. R. *Electrochemical Methods: Fundamentals and Applications*, Wiley, New York, **1980**.

137. Stoynov, Z. B., Grafov, B. M., Savova-Staynov, B. S., and Elkin, V. V. *Electrochemical Impedance*, Nauka Publisher, Moscow, **1991**.

138. Raether, H. *Surface Plasmon on Smooth and Rough Surfaces and on Gratings*, Springer Tracts in Modern Physics, Vol. 11, Springer-Verlag, Berlin, **1988**.

139. Willner, I., Blonder, R., and Dagan, A. Application of photoisomerizable antigenic monolayer electrodes as reversible amperometric immunosensors. *J. Am. Chem. Soc.* **1994**, *116*, 9365–9366.

140. Willner, I., and Willner, B. Electronic transduction of photostimulated binding interactions at photoisomerizable monolayer electrodes: Novel approaches for optobioelectronic systems and reversible immunosensor devices. *Biotechnology Progress* **1999**, *15*, 991–1002.

141. Blonder, R., Levi, S., Tao, G., Ben-Dov, I., and Willner, I. Development of amperometric and microgravimetric immunosensors and reversible immunosensors using antigen and photoisomerizable antigen monolayer electrodes. *J. Am. Chem. Soc.* **1997**, *119*, 10467–10478.

142. Kaganer, E., Pogreb, R., Davidov, D., and Willner, I. Surface plasmon resonance characterization of photoswitchable antigen-antibody interactions. *Langmuir* **1999**, *15*, 3920–3923.

143. Willner, I., Lion-Dagan, M., and Katz, E. Photostimulation of dinitrospiropyran-modified glucose oxidase in the presence of DNP-antibody – A biphase-switch for the amperometric transduction of recorded optical signals. *Chem. Commun.* **1996**, 623–624.

144. Katz, E., and Willner, I. Kinetic separation of amperometric responses of composite redox-active monolayers assembled onto Au-electrodes: Implication to the monolayer structure and composition. *Langmuir* **1997**, *13*, 3364–3373.

145. Willner, I., Pardo-Yissar, V., Katz, E., and Ranjit, K. T. A photoactivated "Molecular Train" for optoelectronic applications: light-stimulated translocation of a β-cyclodextrin receptor within a stoppered azobenzene-alkyl chain supramolecular monolayer assembly on a Au-electrode. *J. Electroanal. Chem.* **2001**, *497*, 172–177.

146. Braun, E., Eichen, Y., Sivan, U., and Ben-Yoseph, G. DNA-templated assembly and electrode attachment of a conducting silver wire. *Nature* **1998**, *391*, 775–778.

147. Richter, J., Seidel, R., Kirsch, R., Mertig, M., Pompe W., Plaschke, J., and Schackrt, H. K. Nanoscale palladium metallization of DNA. *Adv. Mater.* **2000**, *12*, 507–510.

148. Cullum, B. M., Mobley, J., Bogard, J. S., Moskovitch, M., Phillips, G. W., and Vo-Dihn, T. Three-dimensional optical random access memory materials for use as radiation dosimeters. *Anal. Chem.* **2000**, *72*, 5612–5617.

149. Dvornikov, A. S., Malkin, J., and Rentzepis, P. M. Spectroscopy and kinetics of photochromic materials for 3D optical memory devices. *J. Phys. Chem.* **1994**, *98*, 6746–6752.

150. Richter, J., Mertig, M., Pompe, W., Monch, I., and Schackert, H. K. Construction of highly conductive nanowires on a DNA template. *Appl. Phys. Lett.* **2001**, *78*, 536–538.

151. Piner, R. D., Zhu, J., Xu, F., Hong, S., and Mirkin, C. A. "Dip-pen" nanolithography. *Science* **1999**, *283*, 661–663.

152. Hong, S., Zhu, J., and Mirkin, C. A. Multiple ink nanolithography: Toward a multiple-pen nano-plotter. *Science* **1999**, *286*, 523–525.

153. Piner, R. D., and Mirkin, C. A. Effect of water on lateral force microscopy in air. *Langmuir* **1997**, *13*, 6864–6868.

154. Barbara, P. F., Adams, D. M., and O'Connor, D. B. Characterization of organic thin film materials with near-field scanning optical microscopy (NSOM). *Ann. Rev. Mater. Sci.* **1999**, *29*, 433.

155. Dunn, R. C. Near-field scanning optical microscopy. *Chem. Rev.* **1999**, *99*, 2891–2928.

156. Kawata, S., and Kawata, Y. Three-dimensional optical data storage using photochromic materials. *Chem. Rev.* **2000**, *100*, 1777–1788.

157. Matthews, O. A., Shipway, A. N., and Stoddart, J. F. Dendrimers – Branching out from curiosities into new technologies. *Prog. Polym. Sci.* **1998**, *23*, 1–56.

158. Newkome, G. R., Moorefield, C. N., and Vögtle, F. *Dendritic Molecules*, VCH, Weinheim, Germany, **1996**.

159. Shipway, A. N., Katz, E., and Willner, I. Nanoparticle arrays on surfaces for electronic, optical and sensoric applications. *ChemPhysChem* **2000**, *1*, 18–52.

160. Shipway, A. N., Lahav, M., and Willner, I. Nanostructured gold colloid electrodes. *Adv. Mater.* **2000**, *12*, 993–998.

161. Seeman, N. C. DNA components for molecular architecture. *Acc. Chem. Res.* **1997**, *30*, 357–363.

PHOTOCHEMISTRY AND ORGANIC NONLINEAR OPTICS

8

PHOTOISOMERIZATION EFFECTS IN ORGANIC NONLINEAR OPTICS: PHOTO-ASSISTED POLING AND DEPOLING AND POLARIZABILITY SWITCHING

ZOUHEIR SEKKAT
Department of Applied Physics, Osaka University, Suita, Osaka 565-0871, Japan
School of Science and Engineering, Al Akhawayn University in Ifrane,
53000 Ifrane, Morocco

ABSTRACT

The connection between photochemistry and organic nonlinear optics emerged in the beginning of the past decade when the correlation between Pockels electro-optic and light second harmonic activity and photoisomerization of nonlinear optical (NLO) azo dyes in polymers was established. This chapter discusses three effects of photoisomerization in NLO azo-polymers, including photo-assisted poling, which creates polar orientation with the synergetic application of a dc field far below the polymer glass transition temperature (Tg); sub-Tg photo-induced depoling of polymers; and nonlinear polarizability switching of NLO azo dye by photoisomerization.

8.1 INTRODUCTION

Nonlinear optics describe the interaction between light waves in nonabsorbing media. The field of organic nonlinear optics emerged when Davydov and coworkers established the correlation between enhanced nonlinear activity and charge transfer character in conjugated molecules.[1] Considerable exploration of molecular structures was subsequently undertaken, and measurements of polarity, conjugation, and charge transfer character were performed.[2] The nonlinear optical (NLO) response of a macroscopic material depends on both the molecular optical response (polarizability) and the molecular orientation (see Appendix page 284). If the chromophore and/or its local environment is photoresponsive, photoisomerization can control the molecular orientation and/or its polarizability, thereby controlling the optical nonlinearity of the macroscopic material.

In this chapter, I will discuss the nonlinear optical response, e.g., second- and third-order nonlinearities, of polymers containing NLO azobenzene derivatives when these polymers are manipulated by photoisomerization. In fact, the poling of films of polymers by photoisomerization has emerged in the last decade, including the so-called photo-assisted and all-optical poling techniques, whereby photoisomerization creates molecular polar orientation far below the polymer glass transition temperature (Tg).[3–7] Other interesting photoisomerization effects in NLO polymers also have been reported namely, photo-induced depoling,[8] which is depoling of poled polymers by photoisomerization, and nonlinear polarizability switching by reversible photoisomerization.[9] The photoisomerization of NLO azobenzene derivatives is discussed in the first chapters of this book, and all-optical poling is discussed in a separate chapter by Nunzi and coworkers. In this chapter, I will discuss photo-assisted poling, photo-induced depoling, and nonlinear polarizability switching.

8.2 PHOTO-ASSISTED POLING

In the last few years, there has been great interest in orienting polar chromophores in polymer hosts for various applications. For example, nonpolar orientation of the dipolar chromophores leads to birefringence with potential application in birefringent devices (see Chapter 3, page 63). A noncentrosymmetric polar order of nonlinear chromophores results in second-order nonlinearities with application in electro-optical (EO) modulators and harmonic generators. Thermal poling is often the most effective method of inducing polar order in a polymer. In this process, polar chromophores are oriented in a static electric field, at temperatures near Tg, and are frozen in this orientation by cooling to room temperature with the field applied.[10,11] Photo-assisted poling (PAP) occurs at temperatures far below Tg when the nonlinear chromophores undergo reversible *trans↔cis* photoisomerization in the presence of a static electric field (e.g., donor-acceptor substituted azobenzenes). The photoisomerization can enhance the mobility of the chromophores in the glassy state sufficiently to allow poling to occur below Tg.

The phenomenological theory of PAP assumes that at the operating temperature, the mobility of the *trans*-isomer is strongly reduced and that the mobility of an intermediate state, say the *cis*-isomer, is enhanced by photoisomerization to allow poling to occur when a dc field, E_0, is applied. Polar orientation of the *trans*-isomer occurs via the *cis→trans* thermal and via photo-induced back isomerization when this isomer retains even partial memory of the *cis*-isomer's orientation with the dc field applied. Subsequently, an efficient polar orientation may be built after successive cycles of *trans↔cis* isomerization. The general equations of PAP are those reported in Chapter 3 (page 71) with an additional polar interaction energy $U = -\mu E_0 \cos\theta$, where μ is the permanent dipole moment of the isomer, and θ is the polar angle between μ and E_0. This energy produces coupling between odd and even order parameters, thereby producing the polar orientation that occurs in PAP. A detailed theoretical study of the steady-state and transient properties of PAP has been reported.[12] This study shows good agreement between theory and experiment, and discusses the effect of all physical parameters involved in PAP of photoisomers, including the isomers' anisometry, pump intensity polarization, strength of the poling field, and memory of molecular orientation during isomerization.

PAP has been demonstrated both by attenuated total reflection (ATR) EO modulation and by second harmonic generation (SHG). Figure 8.1 shows the first PAP experiments reported for DR1 in films of PMMA.[3,4] This figure shows the evolution of the EO Pockels coefficient r_{33} of a film of the DR1-PMMA side-chain polymer, which is shown in Figure 8.2, during a typical

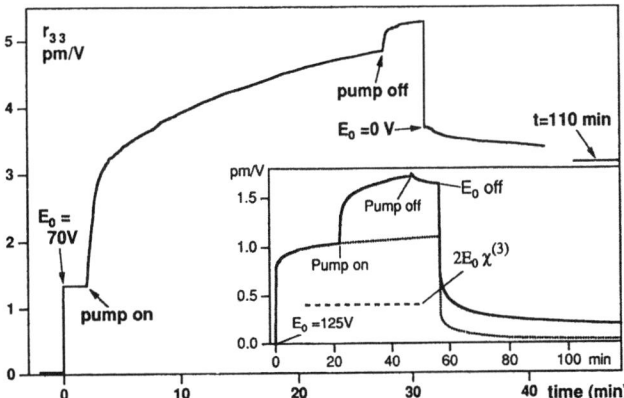

FIG. 8.1 Experimental curve from reference 4 showing the photo-assisted poling cycle at room temperature of a 1.25 μm thick film of the DR1-PMMA copolymer (Tg = 125°C) from ATR EO modulation at 830 nm. The irradiating pump being turned on and off is indicated by arrows. The sudden jumps up and down of the signal when the dc poling field, E_0, is switched on and off are due to the third-order nonlinearity $\chi^{(3)}$. When the circularly polarized laser pump beam (~ 30 mW/cm² from an Ar⁺ laser on the film; wavelength 488 nm near the maximum of absorption of the polymer film) is turned on, the growth of r_{33} is indicative of chromophore polar orientation, e.g., of photo-induced $\chi^{(2)}$. The inset shows a similar experiment[3] performed with DR1-doped PMMA. Adapted from reference 4, and redrawn by permission of Gordon & Breach http://www.tandf.co.uk.

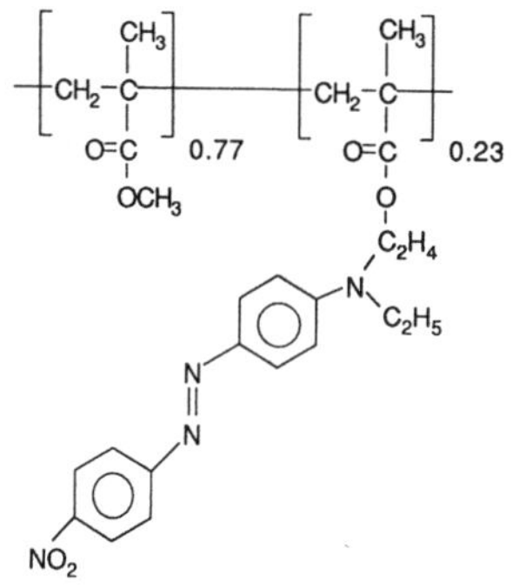

FIG. 8.2 Chemical structures of the DR1-PMMA copolymer (Tg = 125°C).

PAP cycle at room temperature; the inset shows the same experiment for a DR1/PMMA guest host system. Details of ATR EO modulation experiments can be found elsewhere.[13] Briefly, the EO coefficients of polymer films are determined by applying an ac voltage ($V_1\cos\Omega t$) between two metallic layers that sandwich the polymer film, and by recording the modulation of the reflectivity at Ω, for the Pockels effect, and at 2Ω, for the Kerr effect. A computer program evaluates the amplitudes of modulation of the film's thickness, h, and the refractive indices, n_i ($i = x$, y, z which represent the principal axes of the film), from which the Pockels, e.g., r_{ij}, and Kerr, e.g., s_{ij}, EO coefficients are obtained. These are given by

$$2\chi^{(2)}_{iiz} = - r_{iz}\, n_i^4 = 2n_i\Delta n_i(\Omega)\frac{h}{V_1} \tag{8.1}$$

and

$$3\chi^{(3)}_{iizz} = - s_{iz}n_i^4 = 4n_i\Delta n_i(2\Omega)\frac{h^2}{V_1^2} \tag{8.2}$$

where $\Delta n_i(\Omega$ and $2\Omega)$ is the variation of the n_i refractive index at frequencies Ω and 2Ω due to the applied ac field. The variation of the poling, e.g., $\chi^{(2)}$, can be followed by continuous recording of the ATR modulation at Ω near the largest incidence transverse magnetic, TM, e.g., p-polarized, guided mode which gives Δn_z. When the measuring ac (E_1) and the poling dc (E_0) electric fields are applied simultaneously to the polymer film, the Δn_z (z is equivalent to 3) is given by:

$$2n_z\Delta n_z(\Omega) = (2\chi^{(2)}_{333} + 6\chi^{(3)}_{3333}E_0)E_1\cos\Omega t$$
$$+ \frac{3}{2}\,\chi^{(3)}_{3333}E_1^2\cos 2\Omega t + non\ modulated\ terms \tag{8.3}$$

At time $t = 0$, a dc field is applied; it produces an electric field–induced Pockels effect (EFIPE), which is solely due to a third-order effect ($\chi^{(3)} E_0$) in the case of the copolymer because the molecules are not oriented by the dc field alone at room temperature, but which also contains a part due to the rotation, in a polar manner, of free chromophores in the guest-host system (induced $\chi^{(2)}$). The value of $\chi^{(3)}$ is measured from the modulations of ATR modes at 2Ω. When the circularly polarized pump beam is applied, propagating perpendicularly to the plane of the sample, the photoinduced polar orientation of the chromophores progressively builds up the Pockels coefficient. After the successive removal of the pump beam and the dc field, the $\chi^{(3)}$ part falls immediately, and $\chi^{(2)}$ relaxes rapidly in the guest-host system, although it is highly stable in the side-chain copolymer. Photoisomerization provides the NLO azo, chromophores with enough mobility to undergo poling well below the polymer Tg. PAP has been shown to depend on the polymer molecular structure (*vide infra*).[14]

PAP experiments in guest-host DR1-PMMA[15,16] and in a low Tg, e.g., 42°C, azo-polyester[17] have shown that the stability of the order induced by PAP in those polymers is not as good as that induced by thermal poling. Other polymers may exhibit comparable stabilities of the polar order after thermal poling or PAP depending on the polymers' molecular structure. The stability of the induced order is an issue in poled polymers.[18] In guest-host systems, the chromophores have enough mobility for spontaneous reorientation, and in functionalized polymers, such as side-chain and donor-embedded polymers, the long-term stability of the induced order improves, in principle, with increasing difference between the use temperature, T, and the Tg of the polymer. For this reason, attention in poled nonlinear materials has shifted to high-Tg polymers such as polyimides,[19] polyquinolines,[20] and polyarylene ethers.[21] In practice, it is often observed that as the Tg of a polymer increases, particularly above 200°C, the polymer becomes increasingly difficult to thermally pole. One reason for this difficulty is that the conductivity of many polymers increases rapidly around Tg, preventing the maintenance of high poling fields. The higher the Tg of the polymer, the larger the increase in conductivity often is. In addition, it is challenging to maintain low optical losses when poling at high temperatures. These difficulties could, in principle, be alleviated by the development of low-temperature poling techniques such as PAP.

PAP has been reported in a number of different molecular systems, e.g., Langmuir-Blodgett-Kuhn layers,[22] liquid crystalline polymers,[23] and nonlinear optical polymers.[24-27] In particular, Delaire and coworkers have studied PAP of photochromes, such as spiropyrans and diarylethenes, which are not push-pull NLO azobenzene derivatives.[28] In some cases, PAP can induce polar order at room temperature in NLO azo-polymers with an efficiency comparable to that of thermal poling at Tg. Indeed, in a comparative assessment of thermal and photo-assisted and all-optical in-plane poling of polymer-based EO modulators, Zyss and coworkers found that the poling efficiencies of photo-assisted and thermal poling are quite close and exceed that of all-optical poling of a DR1-PMMA copolymer of 30% molar dye concentration.[29] They have reported a Pockels EO coefficient $r_{33} \sim 2.7$ *and* 2.1 *pm/V* at 1.32 µm for thermal poling, performed at 120°C near the Tg of the polymer, versus PAP at room temperature, and inferred that significant improvement of PAP efficiency is possible by adjusting the pump laser spot and the electrodes region. In addition, Steier and coworkers utilized PAP in the maskless fabrication of polymer EO devices for integrated optics by simultaneous direct laser writing, with spot sizes adjustable from 1 to 50 µm, and electric poling of

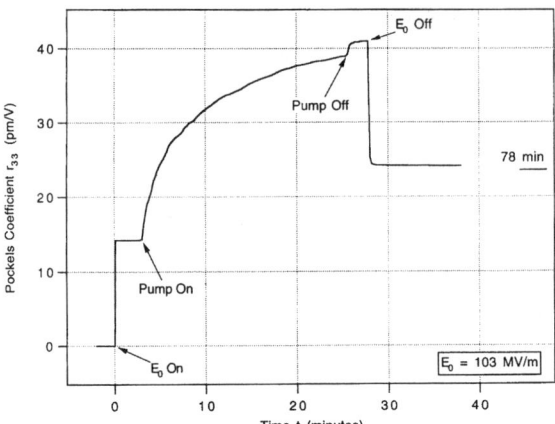

FIG. 8.3 Same as for Figure 8.1 (page 273) for PAP at room temperature of a 0.9 μm thick film of PI-3b (Tg = 210°C) from ATR EO modulations at 633 nm. The growth of r_{33}, which is indicative of photo-induced polar orientation, and $\chi^{(2)}$ are seen well below the polymer Tg when the pump beam (s-polarized, ~ 30 mW/cm^2 from a frequency-doubled laser diode on the film, wavelength 532 nm which allows adequate penetration in the sample) is turned on. The values of r_{33}/E_0 are close for both the DR1-PMMA copolymer and PI-3b considering a factor of 2 to 2.5 for the resonance enhancement of r_{33} of PI-3b. Adapted from reference 25, and redrawn by permission of OSA.

channel waveguides.[30] In particular, they have utilized the PAP technique to fabricate active EO channel waveguides in a disperse red 19 (DR19) containing thermoset polymer with an EO coefficient r_{33} ~ 17 *pm/V* at 1.3 μm, which is largely exceeding the 5 *pm/V*, at the same wavelength and poling voltage, e.g., 100 V/μm, that can be achieved in this material by conventional electrode thermal poling.

Besides the technological use of sub-Tg PAP of high temperature nonlinear optical polymers, the effects of Tg, the local free volume, and the polymer molecular structure on PAP are of fundamental importance. High Tg polymers may be poled by PAP. Indeed, Figure 8.3 shows that efficient PAP occurs 190°C below the Tg of a DR1-polyimide polymer (PI-3b in Figure 8.4; Tg = 210°C);[25] and independent studies on other polyimides containing azo dyes also showed that high Tg polyimides may be poled by PAP.[26] The poling shown in Figure 8.3 has been achieved by PAP with a linear pump polarization, showing that a circular polarization is not necessary for poling by PAP; a feature that has theoretical foundation (*vide infra*).[12]

To study the polymer structural effects on PAP, including the size of the chromophore and the mode of connection of the chromophore to the main chain, SHG-PAP studies have been performed both on a true side-chain polyimide system in which the chromophore is attached to the polymer backbone via flexible tether (PI-3a in Figure 8.4; Tg = 228°C), and on donor-embedded polyimide systems (PI-1 and PI-2 in Figure 8.4; Tg = 350 and 252°C, respectively), in which the azo chromophore is incorporated into the polymer backbone through the donor substituent without any flexible connector or tether.[14] Details of SHG-PAP can be found in reference 14. In both the side-chain and the donor-embedded polyimides, polar order can be generated by thermal poling

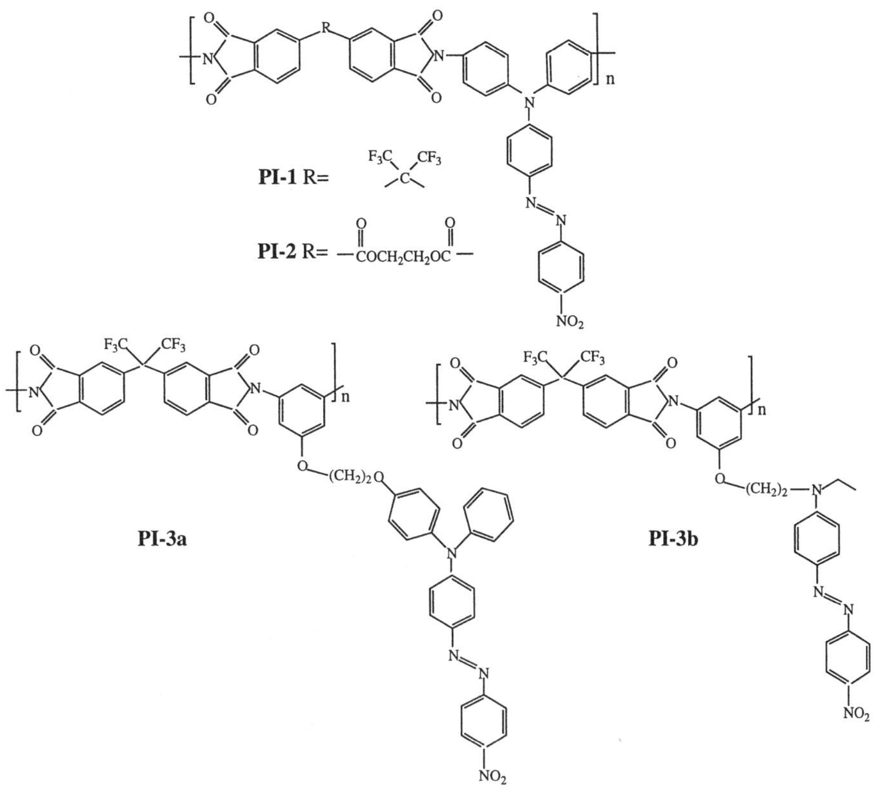

FIG. 8.4 Chemical structures of the azo-polyimide polymers. PI-1, Tg = 350°C, and PI-2, Tg = 252°C, are donor-embedded systems. PI-3a, Tg = 228°C, and PI-3b, Tg = 210°C, are side-chain polymers. Along the main chain of PI-2, there is some flexibility, which lowers the Tg in the vicinity of PI-3.

near Tg, and while some PAP can occur even at room temperature in the side-chain systems, PAP does not induce any polar order in the donor-embedded systems even at 150°C. PI-1 and PI-2 behave similarly in the PAP experiments. It is noteworthy that in spite of the relatively small difference in Tg between PI-2 and PI-3a (252 versus 228°C), PAP of PI-2 even at 150°C does not produce any polar order. This example clearly shows to what extent polymer molecular structure influences sub-Tg polymer molecular movement, a feature that is in contrast to the predictions of the Williams-Landel-Ferry (WLF) theory, which models sub-Tg behavior primarily by the difference Tg–T.[31] The polymer molecular structure also strongly influences sub-Tg molecular movement in polymers.

In PI-3a, PAP produces a relatively small polar order as opposed to the efficient poling achieved by PAP in true side-chain NLO polymers, such as PMMA or polyimides with Tgs in the 120–265°C range, which are flexibly tethered by DR1-type chromophores. The molecular size of the diarylene azo chromophore of PI-3a is substantially larger than that of the DR1-type molecules in the polymers studied previously (see PI-3b in Figure 8.4), a feature that requires more free volume for chromophore movement thereby decreasing mobility. The size of the azo chromophore adds an additional complexity to the polymer structural effects on PAP. For the DR1-copolyimide studied in reference 26, the efficiency of the polar order obtained by PAP is comparable to that obtained for PI-3b; it is apparently due to the large free volume available to the chromophore, because the structure of this NLO DR1-copolyimide implies a chromophore for every other repeating unit. The fact that PAP does not induce any polar order in donor-embedded polyimides PI-1 and PI-2, even at elevated temperatures, is somewhat surprising considering how easily both light-induced nonpolar orientation and photo-induced depoling occur in these polymers (*vide infra*). This finding suggests that the processes of light-induced nonpolar orientation and photo-assisted poling are seemingly different, each requiring some correlated motion of main chain and side chain but to a different extent.

8.3 PHOTO-INDUCED DEPOLING

Together with PAP, photo-induced depoling (PID) is another interesting phenomenon at the interface of photochemistry and organic nonlinear optics. Indeed, PID of poled polymers occurs when NLO chromophores, which are oriented in a polar manner, undergo photoisomerization without applied dc field. The chromophores lose their initial polar orientation after photo-isomerization and reorientation in azimuthal directions around the initial polar axis, thereby erasing $\chi^{(2)}$. PID has been observed both by photo-induced destruction of $\chi^{(2)}$ EO Pockels and by second harmonic generation. The first published PID experiments have been reported for DR1 in PMMA,[8] and the theory of PID is discussed in detail in reference 25.

Figure 8.5 shows the effect of photoisomerization on the SHG signal of a film of a blend of DR1-PMMA copolymer (see Figure 8.2) and PMMA, with a 10% w/w of DR1-PMMA in PMMA, which was initially poled by corona at 120°C. Corona poling is discussed in detail by Knoesen in Chapter 9.

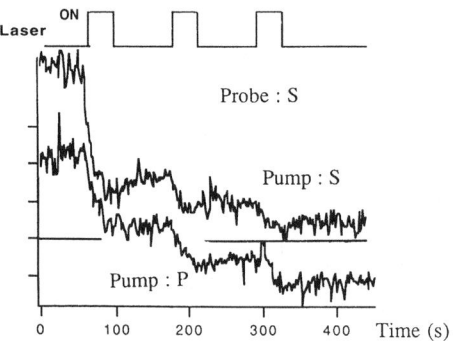

FIG. 8.5 Photo-induced depoling of the DR1-PMMA copolymer blended with PMMA with 10% w/w DR1-PMMA in PMMA, with p- and s-polarized blue light irradiation (~ 120 mW/cm² from a 488 nm Ar⁺ laser). The second harmonic probe was s-polarized. The irradiating pump being turned on and off is indicated. Adapted from reference 8, and redrawn by permission of ACS.

Photoisomerization induces an irreversible power-dependent erasure of the SHG signal with faster rates for higher irradiation intensities, regardless of the light polarization. When the irradiation light is switched off, the drop of the SHG signal stops, but it resumes when this beam is switched on again. The SHG signal is completely erased after a few minutes of irradiation time. Semi-empirical modified neglect of differential overlap calculations of the second-order polarizability β of the chromophore by a finite field method gave and $\beta^0_{trans} = 44.6 \times 10^{-30}$ *esu* and $\beta^0_{cis} = 8.4 \times 10^{-30}$ *esu*. Even though the second-order polarizability β of the chromophore decreases when *trans*-DR1 isomerizes to *cis*-DR1, the irreversibility of the SHG signal can be explained only by the photo-induced disorientation of the chromophores. The optical density of the film sample, which was decreased by poling due to the polar orientation of the chromophores perpendicular to the plane of polarization of the probe beam, was restored after PID, showing that the chromophores have been reoriented back into the film's plane. It is noteworthy that PID erases the polar orientation of the chromophores and creates a centrosymmetric anisotropy that depends on the pump polarization.[25]

Similar PID results have been reported in very high Tg (up to 350°C) azo-polyimides.[14] It has been shown previously that irradiation of the donor-embedded polyimide derivatives with polarized light alone at room temperature induces a quasi-permanent nonpolar orientation, which can be thermally erased only by heating the polymer above Tg.[32] While, the lifetimes of the polar order generated by thermal poling of the donor-embedded polyimides were found to be on the order of tens of years to centuries at room temperature,[33] photoisomerization can efficiently depole these polymer films in a matter of minutes at room temperature. Indeed, Figure 8.6 shows the effect of p-polarized irradiation on the SH signal of PI-2, which had been previously thermally

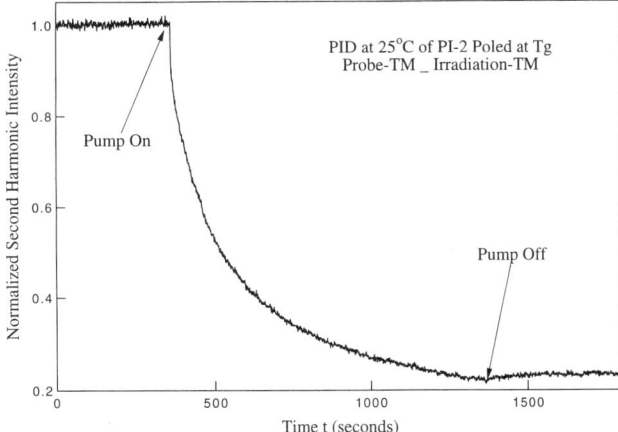

FIG. 8.6 Photo-induced depoling of the PI-2 polymer film (thickness, 140 nm; and O.D., 0.3 at 543.5 nm), with TM-polarized green light irradiation (~ 8 mW/cm² from a 543.5 nm He-Ne laser). The second harmonic probe was TM-polarized. The same depoling occurs for TE-polarized pump irradiation. The irradiating pump being turned on and off is indicated the arrows. Similar PID results were observed in PI-3. Adapted from reference 14, and redrawn by permission of John Wiley & Sons, Inc.

corona poled without irradiation at 250°C for 10 min. Transverse electric, TE, e.g., s-polarized, irradiation produces the same depoling behavior (not shown). Similar results have been obtained for PID of PI-3. These depoling experiments were performed one month after the initial poling of the sample, and it was shown that it is possible to selectively depole, by PID, and repole the same region of the film. This light-driven relaxation of poled functional polymers is polarization-independent, is not due to heating, and is very effective in erasing extremely stable induced order, a feature that makes it possible to use PID in the fabrication of active EO devices. Indeed, Dalton and coworkers utilized PID to fabricate integrated waveguide polarization splitter devices on a submicron scale in a Y-branch poled EO polymer consisting of azo dye attached to PMMA.[34] In addition to PAP and PID, nonpolar photo-orientation of initially isotropic films also has been used to produce micrometer-size passive channel waveguides in DR19 containing thermoset polymers.[35]

8.4 POLARIZABILITY SWITCHING BY PHOTOISOMERIZATION

In the studies described next, I will discuss a third aspect of the connection between photochemistry and organic nonlinear optics: the switching of nonlinear polarizability of NLO azo dye in polymers by photoisomerization shape change. Recent studies performed by absorption saturation and the z-scan technique (or degenerate four waves mixing), indicated that photoisomerization of azo chromophores in polymer hosts can contribute to the observed third-order nonlinearities of such polymer films through the nonlinear refractive index.[36,37] It has been shown that photoisomerization of NLO azo chromophores actually manipulates the third-order nonlinearity, e.g., the $\chi^{(3)}$, of azo-polyimide polymers.[9] This observation has been made directly through resonant electric field induced second harmonic (EFISH), which results from electronic non-linearity contributions dominated by the azo chromophore at second harmonic wavelength within the UV-vis spectral region. If polar order is present, the EFISH signal of an NLO azo-polymer has contributions from $\chi^{(2)}$ and $\chi^{(3)}$, and it is equivalent to the EO EFIPE effect discussed in the previous section. For centrosymmetric systems, only $\chi^{(3)}$ effects are present. In the donor-embedded systems PI-1 and PI-2, $\chi^{(2)}$ does not contribute to the observed EFISH signal at room temperature, because application, with or without photoisomerization, of a strong corona field to either polymer even at 150°C does not result in any polar order, a feature that permits the study in isolation of the effect of photoisomerization on $\chi^{(3)}$.

The effect of the irradiation intensity on the EFISH signal observed in PI-1 at room temperature is shown in Figure 8.7, and similar results were obtained for PI-2. The photo-induced change of the molecular geometry of the NLO azo chromophore in going from the *trans* to the *cis* form results in a drastic change in the $\chi^{(3)}$ of the PI-1 and PI-2 films. The drop of the nonlinearity at the moment when the sample is irradiated by the pump light is not due to photo-induced change in electric field. There was no change in either the corona voltage or the current during irradiation. The observed decrease in

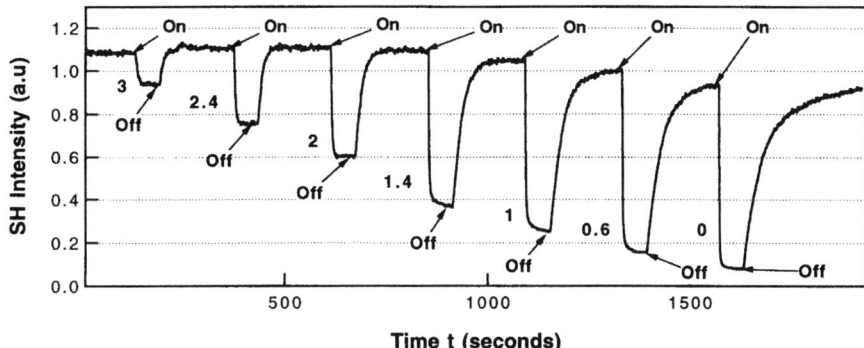

FIG. 8.7 Effect of the irradiation light intensity on the EFISH signal of the PI-1 polymer (film thickness, 720 nm; and O.D., 0.79 at 543.5 nm), with TM-polarized green light irradiation (~ 8 mW/cm² from a 543.5 nm He-Ne laser). The second harmonic probe was TM-polarized. Arrows show when the irradiating pump was turned on and off, and the numbers on the figure refer to the value of the optical density used to attenuate the intensity of the irradiating light. Adapted from reference 9, and redrawn by permission of ACS.

EFISH upon pump irradiation is too large to be due to a slight change in voltage or current, which would be undetectable. The EFISH signal can be nearly erased under high pump irradiation intensity, an effect that would require the voltage to drop to zero in presence of photoconductivity. The fast response rules out charge injection and charge migration. Such processes are much slower than the observed fast decrease in nonlinearity,[38] and photo-isomerization occurs within the pico-second time scale.[39] In addition, while all the NLO azo-polymers studied so far show a fast photo-induced decrease in nonlinearity by EFISH, these polymers do not show the photo-induced decrease in nonlinearity when they are studied by EO measurements through EFIPE. There is no evidence of such a decrease in PAP experiments (see Figures 8.1 and 8.3 for the DR1-functionalized PMMA and polyimide copolymers, respectively).

Figure 8.7 demonstrates that when the irradiation light is turned off, the EFISH signal increases due to the recovery in the $\chi^{(3)}$ caused by the *cis→trans* thermal back reaction. It also can be seen from Figure 8.7 that the EFISH signal is not completely recovered because of the existence of slow components of the *cis→trans* spontaneous recovery in PI-1 and PI-2.[40] This reversible erasure of the $\chi^{(3)}$ can be conducted many times leading to all-optical modulation of the SH light (see Figure 8.8 for PI-2). The slower component of the *cis→trans* recovery is not shown in the experiments described in Figure 8.8 because the initial EFISH level in this figure does not correspond to a fully relaxed sample.

The contribution of photo-induced nonpolar orientation through even order parameters (see Equation 8A.9 in the appendix, page 286) to the effects observed in the EFISH decrease is negligible. The same decrease is observed for a TE- or TM-polarized probe regardless of the pump polarization (not shown). The photochemically induced molecular shape change of the NLO dye blows out the strong optical field driven anharmonic movement of the

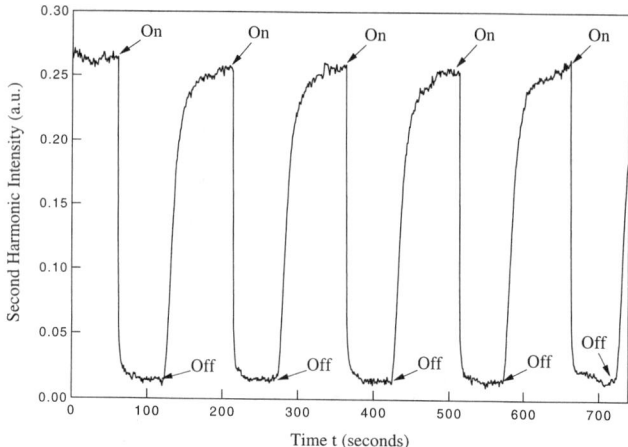

FIG. 8.8 Nonlinear polarizability switching by reversible photoisomerization of the NLO azo chromophore in PI-2 observed by EFISH, a feature that leads to all-optical light modulation of the SH signal of the PI-2 polymer. The irradiation and sample conditions are those explained in the caption to Figure 8.6, and the moments of turning on and off the irradiation are indicated. Adapted from reference 9, and redrawn by permission of ACS.

electronic cloud of the NLO dye, i.e., blows out γ. The effect is reversed upon back-isomerization of the dye to the trans form. This change in nonlinearity must be rationalized by the conformation of the *cis*-isomer of the NLO dye, which is more globular and less conjugated (twisted) than the *trans* form. It has been shown experimentally that twisted organic compounds exhibit smaller γ,[41] and theoretical calculations indicate that both the ground-state dipole moment and the second-order molecular polarizability of DR1, a chromophore structurally related to the chromophores in PI-1 and PI-2, are appreciably greater for the *trans* form (*vide infra*). Furthermore, in the case of conjugated organic compounds, γ can be expressed as the product of the linear polarizability, α, and a nonlinear term, f, corresponding to an anharmonicity factor.[42] So, the decrease upon photoisomerization of the linear absorption of the azo chromophore at the SH wavelength, ca. 526.5 nm, which is due to a smaller α for the *cis* form, demonstrates that γ also should decrease upon *trans*→*cis* isomerization as much as the anharmonicity factor f decreases with the twisted *cis* form of the chromophore.

The photochemical reduction of γ is further rationalized by a theoretical model that neglects photo-orientation effects to EFISH and assumes that γ of the chromophore decreases upon *trans* to *cis* molecular shape change. The model predicts an EFISH intensity at the photostationary state, which varies hyperbolically relative to the irradiating light intensity.[9] Indeed, hyperbolic functions were adjusted to the experimental data showing the variation of the SH intensity at the steady state of the irradiation versus the irradiation intensity for both PI-1 and PI-2 (not shown). This finding demonstrates that the azo chromophores in PI-1 and PI-2 behave consistently with the model, and validates the concept of the reversible rapid photochemical erase of γ of these isomerizable NLO dyes.

8.5 CONCLUSION

The coupling of photochemistry and organic nonlinear optics is quite rich in phenomena pertaining to the manipulation of molecular orientation and nonlinear polarizability by photoisomerization, whereby photo-induced polymer dynamics at sub-Tg temperatures play a central role. Even though polar and nonpolar photo-orientation are already used in the fabrication of light-patterned active and passive EO devices, a large amount of research is yet to come, always combining photochemistry and polymer dynamics and focusing on nonphotoisomerizable NLO chromophores in photo-responsive media. In such NLO photoresponsive systems, the freedom for molecular movement and orientation would be optically provided through the photoresponse of the local environment of the chromophore.

ACKNOWLEDGMENTS

The work on PAP and PID in DR1-PMMA was initiated with M. Dumont, J. A. Delaire, R. Loucif-Saïbi, and K. Nakatani at the University of Paris-Sud at Orsay; and the work on azo-polyimides was done with W. Knoll and J. Wood at the Max-Planck Institute for Polymer Research in Mainz, and with A. Knoesen at the University of California at Davis where polarizability switching was initiated, and with R. D. Miller, W. Volksen, and V. Y. Lee at the IBM-Almaden Research Center at San Jose. It is my pleasure to thank them all for their excellent collaboration. Research support came from the Office of Naval Research during my stay at the University of California at Davis, and from the Max-Planck-Society during my stay at the Max-Planck-Institute for Polymer Research in Mainz. Financial support also came from the National Science Foundation through the Center on Polymer Interfaces and Macromolecular Assemblies, and the Japan Society for the Promotion of Science under the Research for the Future Program.

APPENDIX 8A
FROM MOLECULAR TO MACROSCOPIC NONLINEAR OPTICAL PROPERTIES

The interaction of a nonlinear optical material with electromagnetic fields is described by the properties of the material and the powers of the applied fields. The electric polarization by unit volume at the optical frequency ω can be written formally in the dipole approximation as a series expansion in powers of the electric fields:[2]

$$P_i(\omega) = P_i^0 + \chi_{ij}^{(1)}(-\omega, \omega)\, E_j^\omega + \chi_{ijk}^{(2)}(-\omega, \omega_1, \omega_2) E_j^\omega E_k^{\omega_2}$$
$$+ \chi_{ijkl}^{(3)}(-\omega, \omega_1, \omega_2, \omega_3) E_j^{\omega_1} E_k^{\omega_2} E_l^{\omega_3} + \ldots \qquad (8A.1)$$

Summation over repeated indices is implied and $\chi^{(n)}$ is the nth-order susceptibility tensor that describes the interaction between the electric fields and the material. The first two terms on the right-hand side of Equation 8.A1 give the spontaneous polarization and linear optics effects. The last two terms lead to various phenomena in nonlinear optics. They include SHG and EO Pockels and Kerr effects. The EO susceptibilities are obtained by combining optical and static fields; therefore, the susceptibilities that describe the EO Pockels and Kerr effects are $\chi_{ijk}^{(2)}(-\omega, \omega, 0)$ and $\chi_{ijkl}^{(3)}(-\omega, \omega, 0, 0)$, respectively. In a centrosymmetric bulk material, $\chi^{(2)}$ will be zero, thus it will not produce EO Pockels or generate a second harmonic light.

The description of the interaction between light and molecules can be expressed in analogy to the bulky polarization

$$p_I = \mu_I + \alpha_{IJ}F_J + \beta_{IJK}F_J F_K + \gamma_{IJKL}F_J F_K F_L + \ldots \qquad (8A.2)$$

where F_J, F_K, and F_L are the local fields acting on the molecule; μ_I the ground state dipole moment; α_{IJ} the linear polarizability, and β_{IJK} and γ_{IJKL} are the second- and third-order polarizabilities, respectively. The indices, I, J, K, and L define the coordinates system of the molecule. In molecular materials such as guest-host, and many side-chain functionalized polymers, the interaction between the optical moieties are weak, and the bulk polarizability can be related to the microscopic polarizability through a sum over the molecular units.

Considering that the statistical molecular orientation is described by an orientational distribution function, $G(\Omega)$ (where Ω is the set of Euler angles, derived from the laboratory axes, which define the molecular unit oriental direction, and where it is different from the frequency of the ac field of section 2), the nonlinear-optical properties of polymer films can be related to the orientational order of the nonlinear-optical molecules. For cigar-shaped polar molecules with a ground state permanent dipole moment μ lying along the z-axis, and bulk azimuthal symmetry, the distribution function will depend only on the polar angle θ, and it can be expressed in the standard basis of Legendre polynomials:[43]

$$G(\theta) = \sum_{n=0}^{\infty} \frac{2n+1}{2} A_n P_n (\cos \theta) \tag{8A.3}$$

where $P_n (\cos \theta)$ is the Legendre polynomial of the nth order and where A_n is the corresponding order parameter given by:

$$A_n = \langle P_n(\cos \theta) \rangle = \int_{-1}^{1} d(\cos \theta) G(\theta) P_n(\cos \theta) \tag{8A.4}$$

An axially symmetric molecule is characterized by its linear polarizability in the principal axes α_{xx}^{ω} and $\alpha_{yy}^{\omega} = \alpha_{\perp}^{\omega}$ and $\alpha_{zz}^{\omega} = \alpha_{//}^{\omega}$. It is a good approximation to assume that its second- and third-order polarizability tensors each have only one component β_{zzz} and γ_{zzz} respectively, which is parallel to the z-principal axis of the molecule. For linear and nonlinear optical processes, the macroscopic polarization is defined as the dipole moment per unit volume, and it is obtained by the linear sum of the molecular polarizabilities averaged over the statistical orientational distribution function $G(\Omega)$. This is done by projecting the optical fields on the molecular axis; the obtained dipole is projected on the laboratory axes and orientational averaging is performed. The components of the linear and nonlinear macroscopic polarizabilies are then given by:

$$\begin{aligned}
P_i^{(1)} (\omega) &= \chi_{ij}^{(1)} (-\omega, \omega) E_j^{\omega} \\
&= \int d\Omega N(\Omega) \{ \alpha_{zz}^{\omega} \cos(\hat{i}, \hat{z}) \cos(\hat{j}, \hat{z}) + \alpha_{xx}^{\omega} \cos(\hat{i}, \hat{x}) \cos(\hat{j}, \hat{x}) + \alpha_{yy}^{\omega} \\
& \quad \cos(\hat{i}, \hat{y}) \cos(\hat{j}, \hat{y}) \} E_j^{\omega} \\
P_i^{(2)}(\omega) &= \chi_{ijk}^{(2)} (-\omega, \omega_1, \omega_2) E_j^{\omega_1} E_k^{\omega_2} \\
&= \int d\Omega N(\Omega) \beta_{zzz} \cos(\hat{i}, \hat{z}) \cos(\hat{j}, \hat{z}) \cos(\hat{k}, \hat{z}) E_j^{\omega_1} E_k^{\omega_2} \\
P_i^{(3)}(\omega) &= \chi_{ijkl}^{(3)} (-\omega, \omega_1, \omega_2, \omega_3) E_j^{\omega_1} E_k^{\omega_2} E_l^{\omega_3} \\
&= \int d\Omega N(\Omega) \gamma_{zzz} \cos(\hat{i}, \hat{z}) \cos(\hat{j}, \hat{z}) \cos(\hat{k}, \hat{z}) \cos(\hat{l}, \hat{z}) E_j^{\omega_1} E_k^{\omega_2} E_l^{\omega_3}
\end{aligned} \tag{8A.5}$$

The indices i, j, k, and l define the coordinates system of the macroscopic material with

$$N(\theta) = \frac{N}{2\pi} G(\theta) \tag{8A.6}$$

where N is the density of the molecules. The nonvanishing components of the tensors $\chi^{(1)}$, $\chi^{(2)}$, and $\chi^{(3)}$ are given by

$$\begin{cases} \chi_{33}^{(1)} = N\,\overline{\alpha}^{\omega}\,(1 + 2\delta^{\omega}\,A_2) \\ \chi_{11}^{(1)} = \chi_{22}^{(1)} = N\,\overline{\alpha}^{\omega}\,(1 - \delta^{\omega}\,A_2) \end{cases} \tag{8A.7}$$

$$\begin{cases} \chi_{333}^{(2)} = N\beta_{zzz}^{*}\left(\dfrac{3}{5}\,A_1 + \dfrac{2}{5}A_3\right) \\[2mm] \chi_{113}^{(2)} = N\beta_{zzz}^{*}\left(\dfrac{1}{5}A_1 - \dfrac{1}{5}\,A_3\right) \end{cases} \tag{8A.8}$$

$$\begin{cases} \chi_{3333}^{(3)} = N\gamma_{zzzz}^{*}\left(\dfrac{1}{5} + \dfrac{4}{7}A_2 + \dfrac{8}{35}A_4\right) \\[2mm] \chi_{2233}^{(3)} = \chi_{1133}^{(3)} = N\gamma_{zzzz}^{*}\left(\dfrac{1}{15} + \dfrac{1}{21}A_2 - \dfrac{4}{35}A_4\right) \\[2mm] \chi_{1111}^{(3)} = \chi_{1122}^{(3)} = 3N\gamma_{zzzz}^{*}\left(\dfrac{1}{15} - \dfrac{2}{21}A_2 + \dfrac{1}{35}A_4\right) \end{cases} \tag{8A.9}$$

where $\overline{\alpha}^{\omega} = (\alpha_{\parallel}^{\omega*} + 2\alpha_{1}^{\omega*})/3$ is the isotropic linear polarizabiity, and $\delta^{\omega} = (\alpha_{\parallel}^{\omega*} - \alpha_{1}^{\omega*})/(\alpha_{\parallel}^{\omega*} + 2\alpha_{1}^{\omega*})$ is the molecular anisotropy. Local field effect factors have been included in the molecular polarizabilities, and they are indicated by $*$.

REFERENCES

1. Davydov, B. D., Derkacheva, L. D., Dumina, V. V., Zhabostinskii, M. E., Zolin, V. F., Koreneva, L. G., and Sanokhina, M. A. *Opt. Spectrosc.* (USSR) **1971**, *30*, 274.
2. Zyss, J., and Chemla, D. S. in *Nonlinear Optical Properties of Organic Molecules and Crystals*, Zyss, J., Chemla, D. S., Eds. (Academic, New York, **1987**), 1, p3.
3. Sekkat, Z., and Dumont, M. *Appl. Phys.* B. **1992**, *54*, 486.
4. Sekkat, Z., and Dumont, M. *Mol. Cryst. Liq. Cyrs. Sci. Technol.—Sec B: Nonlinear Optics*, **1992**, *2*, 359.
5. Sekkat, Z., and Dumont, M. *Synth. Metals* **1993**, *54*, 373
6. Charra, F., Kajzar, F., Nunzi, J. M., Raimond P., and Idiart, E. *Opt. Lett.* **1993**, 12.
7. Fiorini, C., Charra, F., Nunzi, J. M., and Raimond, P. J. *Opt. Soc. Am.* **1997**, *B14*, 1984.
8. Loucif-Saibi, R., Nakatani, K., Delaire, J. A., Dumont, M., and Sekkat, Z. *Chem. Mater.* **1993**, *5*, 229.
9. Sekkat, Z., Knoesen, A., Lee, V. Y., and Miller, R. D. J. *Phys. Chem.* B **1997**, *101*, 4733.
10. Singer, K. D., Sohn, J. E., and Lalama, S. J. *Appl. Phys. Lett.* **1986**, *49*, 248.
11. Mortazavi, M. A., Knoesen, A., Kowel, S. T., Higgins, B., and Dienes, A. J. *Opt. Soc. Am.* B. **1989**, *6*, 733.
12. Sekkat, Z., and Knoll, W. J. *Opt. Soc. Am.* **1995**, *B12*, 1855.
13. Dumont, M., Morichère, D., Sekkat, Z., and Levy, Y. *SPIE* **1991**, *1559*, 127.
14. Sekkat, Z., Knoesen, A., Lee, V. Y., and Miller, R. D. J. *Polym. Science. B. Polym. Phys.* **1998**, *36*, 1669.
15. Blanchard, P. M., and Mitchell, G. R. *Appl. Phys.Lett.* **1993**, *63*, 2038.
16. Bauer, S., Bauer-Gogonea, S., Ylmaz, S., Wirges, W., and Gerhard-Multhaupt, R. in *Polymers for Second Order Nonlinear Optics*, Lindsay and Singer, Eds. **1995**, ACS Symposium Series 601, Ch. 22, pp. 304–316.
17. Sekkat, Z., Kang, C.-S., Aust, E. F., Wegner, G., and Knoll, W. *Chem. Mater.* **1995**, *7*, 142.

18. Kaatz, Ph., Pretre, Ph., Meier, U., Stalder, U., Bosshard, C., Gunter, P., Zysset, B., Stahelin, M., Ahlheim, M., and Lehr, F. *Macromolecules* **1996**, *29*, 1666.
19. Miller, R. D., Burland, D. M., Jurich, M. C., Lee, V. Y., Moylan, C. R., Thackara, J., Twieg, R. J., Verbiest, T., and Volksen, W. *Macromolecules* **1995** *28*, 4970.
20. Cai, Y. M., and Jen, A. K. Y. *Appl. Phys. Lett.* **1995**, *67*, 299.
21. Fu, C. Y. S., Lackritz, H., Priddy, D. B. Jr, and McGrath, J. E. *Chem. Mater.* **1996**, *8*, 514.
22. Palto, S. P., Blinov, L. M., Yudin, S. G., Grewer, G., Schönhoff, M., and Lösche, M. *Chem. Phys. Lett.* **1993**, *202*, 308.
23. Anneser, H., Feiner, F., Petri, A., Bräuchel, C., Leigeber, H., Weitzel, H. P., Kreuzer, F. H., Haak, O., and Boldt, P. *Adv. Mater.* **1993**, *5*, 556.
24. Hill, R. A., Dreher, S., Knoesen, A., and Yankelevich, D. *Appl. Phys. Lett.* **1995**, *66*, 2156.
25. Sekkat, Z., Wood, J., Aust, E. F., Knoll, W.,Volksen, W., and Miller, R. D. *J. Opt. Soc. Am. B.* **1996**, *13*,1713.
26. Chauvin, J., Nakatani, K., and Delaire, J. A. *Proc. SPIE* **1997**, *2998*, 205.
27. Grossmann, S., Weyrauch, T., and Haase, W. *J. Opt. Soc. Am.* **1998**, *B15*, 414.
28. Atassi, Y., Delaire, J. A, and Nakatani, K. *J. Phys. Chem. PC* **1995**, *99*, 16320.
29. Donval, A., Toussaere, E, Brasselet, S., and Zyss, *J. Optical Materials* **1999**, *12*, 215.
30. Steier, W. H, Chen, A., Lee, S.-S., Garner, S., Zhang, H., Chuyanov, V., Dalton, L. R., Wand, F., Ren, A. S., Zhang, C., Todorova, G., Harper, A., Fettermann, H. R., Chen, D., Udupa, A., Bhattacharya, D., and Tsap, B. *Chemical Physics* **1999**, *245*, 487.
31. Williams, M. L., Landel, R. F., and Ferry, J. D. *J. Am. Chem. Soc.* **1955**, *77*, 3701.
32. Sekkat, Z., Wood, J., Knoll, W., Volksen, W., Miller, R. D., and Knoesen, A. *J. Opt. Soc. Am.* **1997**, *B14*, 829.
33. Verbiest, T., Burland, D. M., Jurich, M. C., Lee, V. Y., Miller, R. D., and Volksen, W. *Science* **1995** *268*, 1604.
34. Lee, S.-S., Garner, S., Chen, A., Chuyanov, V., Steier, W. H, Guao, L., Dalton, L. R., and Shin, S.-Y. *Appl. Phys. Lett.* **1998**, *73*, 3052.
35. Lee, S.-S., Garner, S., Steier, and Shin, S.-Y. *Appl. Opt.* **1999**, *38*, 530.
36. Egami, C., Suzuki, Y., Sugihara, O., Okamoto, N., Fujirama, H., Nakagawa, K., and Fujiwara, H. *Appl. Phys. B.* **1997**, *64*, 471.
37. Dong, F., Koudoumas, E., Courtis, S., Shen, Y., Qiu, L., and Fu, X. *J. Appl. Phys.* **1997**, *81*, 7073.
38. Moerner, W. E., Grunnet-Jepsen, A., and Thompson, C. L. *Annual Review of Materials Science* **1997**, *27*, 585.
39. Lednev, I. K., Ye, T. Q., Hester, R. E., and Moore, J. *J. Phys. Chem.* **1996**, *100*, 13338.
40. Sekkat, Z., Prêtre, Ph., Knoesen, A., Volksen, W., Lee, V. Y., Miller, R. D., Wood, J., and Knoll, W. *J. Opt. Soc. Am. B.* **1998**, *15*, 401.
41. Heflin, J. R., Cai, Y. M., and Garito, A. F. *J. Opt. Soc. Am. B.* **1991**, *8*, 2132.
42. Brédas, J. L., Adant, C., Tackx, P., and Persoons, A. *Chem. Rev.* **1994**, *94*, 243.
43. Kuzyk, M. G., Sohn, J. E., and Dirk, C. W. *J. Opt. Soc. Am.* **1990**, *7*, 842.

9

PHOTOISOMERIZATION IN POLYMER FILMS IN THE PRESENCE OF ELECTROSTATIC AND OPTICAL FIELDS

ANDRÉ KNOESEN
Department of Electrical and Computer Engineering, University of California Davis, Davis, CA 95616-5294, USA

9.1 INTRODUCTION

Photoisomerization modifies the polarizability of a molecule leading to changes in the linear optical properties (i.e., indices of refraction and absorption), and also to changes in the second-order and third-order nonlinear optical properties provided the isomer exhibits such characteristics. In this chapter, I describe a nonlinear optical experimental technique to probe the change in nonlinear optical properties during photoisomerization in polymer thin films. Electrostatic and optical electric fields manipulate the geometrical shape and orientation of isomers in the polymer film, and second harmonic generation is used to detect the changes in the second- and third-order nonlinear properties.

9.2 PHOTOISOMERIZATION AND NONLINEAR POLARIZABILITY

An electric field applied to a molecule with a permanent dipole causes a shift in charge center. If field-induced displacement is x and the total charge is q, then the induced dipole moment is:

$$\mu = qx$$

If the electric field is small, the relationship between the induced dipole and the electric field is to a good approximation linear and given by $\mu = \alpha E$. For simplicity, the tensorial nature of the induced dipole is being neglected. Averaging over several induced dipoles gives a macroscopic polarization P and the electric flux density is

$$D = \varepsilon_0 E + P \approx \varepsilon_0 (1 + \chi) E \approx \varepsilon_0 \varepsilon_1 E$$

where χ is the dielectric susceptibility. The relative permittivity ε_1 depends on the frequency of the electric field, and the relationship between D and E can be generalized to lossy dielectrics so that

$$D(\omega) \approx \varepsilon_0 \varepsilon_1(\omega) E(\omega)$$

where the linear permittivity term $\varepsilon_1(\omega)$ is a complex quantity. In the optical regime, the index of refraction is $\Re e \sqrt{\varepsilon_1(\omega)}$, and the absorption coefficient is $\Im m \sqrt{\varepsilon_1(\omega)}$. If the geometrical shape of the molecule changes, the induced dipole is altered, which results in a change index of refraction and absorption. Such a change can be induced by photoisomerization. If the electric field is large, the relationship between the induced dipole and the electric field becomes nonlinear and is described to a good approximation by a series expansion

$$\mu \approx \alpha E + \beta E^2 + \gamma E^3$$

where α is the first-order microscopic polarizability, β is the second-order microscopic polarizability, and γ is the third-order microscopic polarizability. The second-order microscopic polarizability is zero only if the molecule has an inversion symmetry. Averaging over N molecules, the macroscopic polarizability is:

$$P = N \mu \approx \varepsilon_0 \chi^{(1)} E + \varepsilon_0 \chi^{(2)} E^2 + \varepsilon_0 \chi^{(3)} E^3$$

The macroscopic second-order nonlinearity $\chi^{(2)}$ is zero if the molecules are randomly oriented within the polymer. From the preceding relationships, it follows that during photoisomerization changes occur in the nonlinear optical properties $\chi^{(2)}$ and $\chi^{(3)}$, as well as the linear term $\chi^{(1)}$.

In this chapter, the interest is in an electric field that consists of a strong static electric field E_0 and an optical beam $E_1 \cos(\omega t)$. Substituting $E = E_0 + E_1 \cos(\omega t)$) into the expression for polarizibility produces:

$$P = \varepsilon_0 \left(\chi^{(1)} E_0 + \frac{1}{2}\chi^{(2)} E_1^2 + \chi^{(2)} E_0^2 + \frac{3}{2}\chi^{(3)} E_0 E_1^2 + \chi^{(3)} E_0^3 \right)$$

$$+ \varepsilon_0 E_1 \left(\chi^{(1)} + 2\chi^{(2)} E_0 + \frac{3}{4}\chi^{(3)} E_1^2 + 3\chi^{(3)} E_0^2 \right) \cos(\omega t)$$

$$+ \varepsilon_0 E_1^2 \left(\frac{1}{2}\chi^{(2)} + \frac{3}{2}\chi^{(3)} E_0 \right) \cos(2\omega t) + \frac{1}{4}\varepsilon_0 E_1^3 \chi^{(3)} \cos(3\omega t)$$

$$= P^0 + P^\omega + P^{2\omega} + P^{3\omega}$$

The nonlinear terms cause interactions between the electrostatic term, E_0 and the optical field that creates nonlinear electric field contributions at frequencies 0, ω, 2ω and 3ω. The objective will be to monitor changes in the polarization. Experimentally, it is easier to probe nonlinear changes in $P^{2\omega}$ and $P^{3\omega}$ than in P^0 and P^ω, because the latter have background contributions from $\chi^{(1)}$. Typically, the optical field is a short pulse with wavelength in the 1 micron region, the second harmonic conveniently falls in the green where highly sensitive detectors exist, and the third harmonic, being in the UV, is attenuated by a variety of mechanisms. For this reason, second harmonic generation is of specific interest to probe changes caused by isomerization. The term that generates second harmonic in the presence of electrostatic and optical field is:

$$P^{2\omega} = \varepsilon_0 E_1{}^2 \left(\frac{1}{2}\chi^{(2)} + \frac{3}{2}\chi^{(3)} E_0 \right) \cos(2\omega t) = \varepsilon_0 E_1{}^2 \chi^{(2)}_{eff} \cos(2\omega t)$$

The second harmonic contribution from $\chi^{(3)}$ requires an electrostatic electric field E_0 and is known as electric field induced second harmonic (EFISH). The intensity of the generated second harmonic, assuming that the film is very thin and lossless, is

$$I_{2\omega} \approx A d^2 \chi^{(2)^2}_{eff} I_\omega{}^2$$

where A is a proportionality constant and d is the film thickness. The film is considered thin if $|n_2 - n_1|d << \lambda_{2\omega}$ where n_2 and n_1 are the indices of refraction at the second harmonic wavelength $\lambda_{2\omega}$ and fundamental wavelength λ_ω, respectively.

The linear optical properties of the isomers of 4-[ethyl(2-hydroxy-ethyl) amino]-4-nitrobenzene (disperse red 1 or DR1) have attracted the attention of researchers for many years, and during the last decade, the nonlinear optical properties of DR1 placed in a polymer environment also have been of interest.[1] The *trans-* and *cis-*isomers of DR1 are illustrated in Figure 9.1. The photoisomerization reaction begins by elevating molecules to excited electronic states, followed by nonradiative decay back to the ground state in either the *cis* or *trans* forms, as illustrated in Figure 9.2. The ratio of *cis/trans* states is dependent on the quantum yield of the appropriate photoisomerization reaction (e.g., ϕ_{tc} and ϕ_{ct} for the direct *trans*$\Rightarrow$*cis* and reverse *cis*$\Rightarrow$*trans* photoisomerization reactions, respectively). As the *trans*-isomer is more stable than the *cis*-isomer, molecules in the *cis* form convert to the *trans* form by

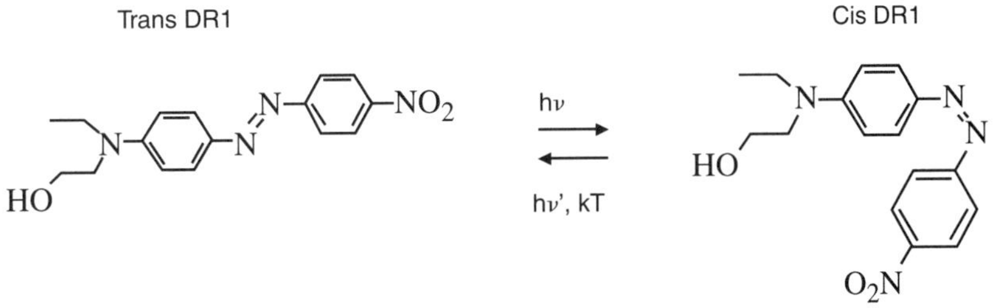

FIGURE 9.1   The *trans* and *cis* state of 4-[ethyl(2-hydroxy-ethyl) amino]-4-nitrobenzene (disperse red 1 or DR1).

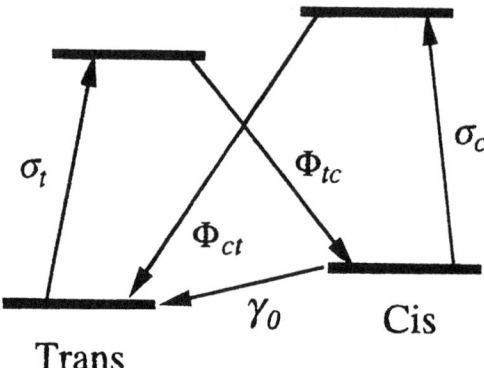

FIGURE 9.2 Simplified model of the molecular states. Only two excited states have been represented, but each of them may represent a set of actual levels: We only assume that the lifetime of all these levels is very short. σ_t and σ_c are the cross sections for absorption of one photon by a molecule in the *trans* or the *cis* state, respectively. γ_0 is the thermal relaxation rate. ϕ_{ct} and ϕ_{tc} are the quantum yields of photoisomerization and represent the probability per absorbed photon of the photochemical conversion.

one of two mechanisms: (1) a spontaneous thermal back reaction, or (2) a reverse *cis⇒trans* photoisomerization cycle. The *trans⇒cis* photoisomerization of azobenzenes has been shown to occur within a picosecond time scale in a liquid,[2] and faster than a nanosecond time scale in polymers.[3] The spontaneous thermal back reaction in a polymer is on the order of minutes to hours.

DR1 molecule has optical nonlinear properties because the donor and acceptor groups are linked by a delocalized π-electron system creating an anharmonic oscillator that gives rise to a large second-order nonlinear microscopic polarization β and a third-order nonlinearity γ. The change in the molecular shape of the DR1 when it photoisomerizes from the *trans*- to the *cis*-isomer significantly modifies the charge transfer that takes place between the donor and acceptor groups, and this change reduces the optical nonlinearities. The contracted DR1 *cis*-isomer has a β_{cis} that is approximately five times smaller than the β_{trans} of the *trans*-isomer.[4] It was experimentally observed that the contracted DR1 *cis*-isomer has a γ_{cis} that is neglible compared to the γ_{trans} of the extended *trans*-isomer.[5]

The dominant macroscopic second-order polarization element is

$$\chi^{(2)} = N \beta < \cos^3 \theta >$$

where $< \cos^3 \theta >$ is an order parameter that can be expressed in terms of a Langevin function of order three.[6] The number density of the embedded nonlinear molecules, their hyperpolarizability, and their degree of noncentrosymmetrical orientational order determine the macroscopic second-order $\chi^{(2)}$ element.

The dominant macroscopic third-order polarization element in DR1 is of the form

$$\chi^{(3)} = N \gamma \left(\frac{1}{5} + \frac{4}{7} < P_2 > + \frac{8}{35} < P_4 > \right)$$

where $<P_2>$ and $<P_4>$ are order parameters that can be expressed in terms of nth order Legendre polynomial.[7] All the order parameters are zero if the molecules are randomly oriented. An EFISH signal will be produced even if the molecules are not ordered. The macroscopic third-order nonlinear effect $\chi^{(3)}$ properties will change if the DR1 molecules are ordered, but more importantly, a $\chi^{(2)}$ can exist only if a noncentrosymmetric order exists. A rigorous treatment of microscopic and macroscopic optical nonlinearities in macromolecular systems, is given in Reference 7.

9.2 ALIGNMENT OF ISOMERS IN POLYMERS WITH ELECTRIC FIELDS

We are focusing on the alignment of molecules by electric fields that are in the electrostatic or optical regime. Electric fields can align molecules in a polymer host provided the molecules have enough mobility. Raising the temperature can increase the mobility of molecules. This mobility will be referred to as thermal-induced mobility. The process of photoisomerization itself can increase the mobility of the isomers in a polymer host. This process will be called photoinduced mobility. Compared to the extended shape of the *trans* state, the contracted shape of *cis* state gives the molecule a larger mobility in a polymer host.

Linear polarized light can orient molecules that can be photoisomerized through a process that will be called photoinduced alignment. This process relies on repeated excitation of the cycle *trans*⇒*cis*⇒*trans*. The probability of excitation depends on $\cos^2\Theta_p$ where Θ_p is the angle between the polarization of the pump and the molecule. Molecules aligned perpendicular to the pump polarization are not excited, and those parallel to the pump polarization are most efficiently excited. Photoinduced mobility makes it possible for the *cis*-molecules to reorient themselves. After the molecule relaxes back to the *trans* state, and if the molecule is more aligned with the pump polarization, it can be more efficiently excited again to the *cis* state. Through this competition process eventually, after a large number of isomerization cycles, the majority of molecules become oriented perpendicular to the electric field of the pump light. The order that is induced is centrosymmetric in the absence of any other alignment forces.

Electrostatic fields can orient a molecule that has an electrostatic dipole moment. During the application of the electric field, the mobility of the molecules is increased by either increasing the temperature, which is known as temperature-assisted poling, or photoisomerization, which is known as photo-assisted poling. Raising the temperature within the vicinity of the glass transition temperature, but not so high that the conductivity in polymer drastically increases, results in the optimum poling. In photoinduced poling, the alignment of the molecules occurs in the *cis* state.

The polymer film is typically a micron thick, and the electric field must ideally approach the dielectric breakdown of the thin film, which is on the order of 200 V/μm. At least one of the poling electrodes must be optically transparent such that the molecules in the polymer film can be probed with an optical beam. Ideally during poling, the polymer must be isolated from additional effects such as charges trapping at polymer metal interfaces. The simplest poling geometry is to apply a voltage to two conducting layers deposited on either side of the polymer film. In practice, it is difficult to achieve large electric field with this approach since breakdown occurs in localized regions in the thin film. With an electrode with large lateral conductivity, charge will rush to the breakdown region and the electric field is reduced throughout the film. In temperature assisted poling a trade-off is typically made between highest temperature where poling can be performed but yet the polymer has a low enough conductivity to maintain a large electrostatic field. This balance is difficult in contact poling because of the high lateral conductivity of the poling electrodes. Furthermore, it is also difficult to deposit conducting electrodes (e.g., a gold, aluminum, or indium tin oxide) directly onto a thin polymer film without causing some damage to the polymer interface. The charges trapped at the polymer metal interface also can influence the orientation of the molecular dipoles.[8] For such reasons, it is often difficult to get the poling field higher than 100 V/μm in contact poling.

Corona onset poling is a poling technique that achieves an electric field that approaches the dielectric breakdown strength of the polymer film, meets the requirement of a transparent electrode, and eliminates charge trapping at interfaces between polymers and metals.[9] Spraying electrostatic charges onto a polymer film using a corona discharge creates a transparent electrode. A layer of charge is created on the polymer surface that has a low lateral conductivity, and processing conditions can be controlled such that a large electric field can be maintained in a thin polymer film over a large area without being plagued by electric field reduction caused by localized breakdown effects.

Corona is a self-sustained, partial discharge that is highly localized within insulating surfaces between two conductors. Partial discharge is a complex phenomenon[10] that can exhibit chaotic behavior. A variety of different partial discharges have been identified (e.g., corona, constricted glows, electron avalanches, localized Townsend discharges, streamers, and dielectric-barrier discharges). This variety is in large part due to the various of geometries and material conditions under which a partial discharge can occur. The geometry of interest here is a wire-to-plane electrode gap where the plane electrode is covered by dielectric surfaces shown in Figure 9.3. The polymer is deposited onto a glass interface. The type of glass is chosen so that it has a conductivity higher than the polymer. For *in situ* optical studies, an indium-tin-oxide (ITO) layer deposited onto the glass is used, but note that the ITO is not adjacent to the polymer; it is deposited on the opposite glass interface. The region of poling corresponds to the length of the wire and has a width that is approximately twice the height of the corona wire above the polymer film. The electric fields that can be achieved in this configuration have been measured to be about 200 V/μm, which is close to the dielectric breakdown of thin polymer films.[11]

The region in which the partial discharge occurs is restricted to a small region between the two conductors to which the voltage is applied. The discharge zone is that region where the electric field strength is large enough to allow the release of electrons by various collision processes. In the wire-to-

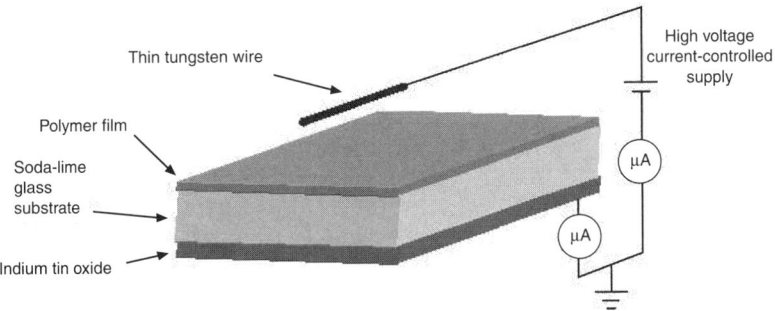

FIGURE 9.3 Corona poling experimental configuration.

plane geometry, the discharge zone is confined to the immediate vicinity of the wire. While the electric field in the region outside the discharge zone is too small to create a discharge, the existence of charge in this region influences the activity within the discharge zone. The corona discharge is considered to be "cold," because the mean energy of the electrons in the discharge zone is considerably less than the mean energy of the molecules in the gas. The discharge does not heat the surroundings, and a lack of equilibrium exists between the charged and neutral species. Corona discharge refers the special case when the insulating surfaces are far removed from the discharge zone. The corona is classified as positive and negative depending on the polarity of the wire or point. The behaviors of positive and negative corona are different. This is a result of different sources that initiate and sustain the discharge.[10] However, most important is that once the discharge is initiated, charges are deposited onto insulating surfaces and very quickly a very large electric field is created within the polymer film.

The electrical contact to both polymer interfaces must have a low lateral conductivity. The polymer-air interface of corona discharge meets this requirement, but the other polymer interface must do so as well. For this reason a dielectric substrate with a resistivity much less than the polymer in contact with the polymer film is preferred over a highly conducting layer. A ground electrode is attached to the opposite side of the dielectric substrate. For a fixed applied voltage V between the air-polymer interface and the substrate-ground interface, the voltage drop across the polymer film is

$$V_1 = \frac{1}{1+\delta} V$$

where $\delta = d_2\rho_2/d_1\rho_1$ and where d_i and ρ_i are the thickness and resistivity, respectively, of the layer i. The polymer thickness d_1 is on the order of microns and the substrate thickness d_2 is on the order of millimeters. The resistivity for PMMA is in the $1 \times 10^{17} - 1 \times 10^{18}$ $\Omega \cdot$ cm in the 25–100°C range.[12] The applied voltage V is dropped over the polymer film to a very good approximation if the substrate has $\rho_2 < 1 \times 10^{13}$ $\Omega \cdot$ cm. A good substrate choice is soda-lime glass microscope slides,[9] which are $1 \times 10^9 - 1 \times 10^{12}$ $\Omega \cdot$ cm in the 25–100°C range.[12] The resistivity of fused quartz is too high and should not be used.

Some groups have used a corona poling geometry that places an electrode with large conductivity in contact with the polymer.[13–16] However, no advantage is gained in terms of poling field strength, instead several problems are encountered such as poling uniformity, surface damage effects, etc. The poling uniformity could be improved in this configuration by a control grid, but it is best to avoid this poling geometry if at all possible. Without a grid good poling uniformity can be obtained by the configuration in Figure 9.3 provided that a lossy dielectric substrate is used.

The optimal thermal poling of nonlinear polymers is found when the applied voltage remains close to onset of the corona discharge. For this reason, the corona poling technique for electro-optic polymers was termed corona poling at elevated temperatures (COPET).[9] At larger voltages, surface damage occurs, without giving any additional advantage in terms of enhanced poling fields. While in general the corona onset condition is different for positive and negative corona, for small radii conductors the onset for condition is about the same for both polarities. The corona onset condition for a wire is given by an empirical approximate relationship[17]

$$E_c = 3 \times 10^6 \, m \left(\delta + 3 \times 10^{-2} \sqrt{\frac{\delta}{a}} \right) \, V/m$$

where m is a wire-roughness parameter ($m = 1$ for a smooth wire, typical values are 0.5 to 0.9), $\delta = (T/T_0)(P/P_0)$ is the relative air density, a is the radius of the wire, P is the pressure of air (in atmospheric units), T is the air temperate (in Kelvin), $T_0 = 298 \, K$, and $P_0 = 1$ atmospheric pressure. The wire-to-plane corona discharge can be modeled accurately,[18] however, the corona-onset voltage can be estimated by assuming a coaxial geometry shown in Figure 9.4, for which simple close form expressions exist.[19] The field in the coaxial geometry approximates the wire-to-plane geometry if $a \ll b$. For the coaxial geometry, the corona onset voltage is $V_{onset} = E_c \, a \, \ln(\frac{a}{b})$.

Typical values for $b \approx 1 \, cm$ and $a \approx 100 \, \mu m$, for which V_{onset} is in the 4–6 kV range for poling temperatures 25–150°C. This estimate corresponds to what is observed in practice for the wire-to-plane poling geometry.

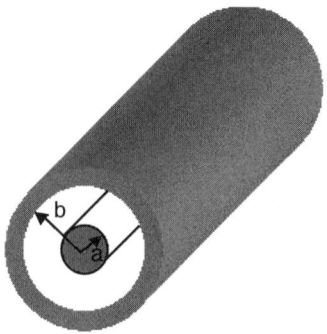

FIGURE 9.4 In the coaxial configuration, the corona onset voltage is $V_{onset} = E_c \, a \, \ln(a/b)$. The corona conditions approach the wire-to-plane geometry if $a \ll b$.

A typical thermal assisted poling procedure is described next. The films are spin coated onto a soda-lime glass substrate to a thickness of less than a micron. On the opposite side of the substrate surface that is in the contact with the polymer, the glass substrate has an indium tin oxide coating. A high-voltage current-controlled power supply is connected to a thin tungsten wire (40 μm diameter) held at 1 cm above the polymer film. For *in situ* optical measurements, the samples are placed in a custom-designed holder that controls the temperature and provides electrical isolation. In those cases where poling is performed at temperatures that exceed 200°C, nitrogen is flowed through the cell during poling. This helps to stabilized the poling current. A thermocouple is placed on the edge surface of the polymer film. The temperature distribution that exists on the sample is characterized, and differences between the thermocouple reading measured on the edge of the sample and where the second harmonic measurements are taken are calibrated out. The current is measured between the power supply and the ground, and also between the conducting plane in contact with the glass substrate and ground. The current is monitored in two locations to ensure that the current from the corona wire is conducted through the polymer film and substrate to ground, and is not finding a low resistance path elsewhere. The thermocouple is electrically isolated from the discharge (e.g., by wrapping it in Teflon tape) to avoid current flow from the corona discharge to the temperature controller through the thermocouple. A typical poling procedure in the vicinity of the glass transition temperature consists of raising the temperature of the polymer sample to the desired level. Next, a 5–6 kV high-voltage is applied to produce a current in the 1–4 μA range. The optimum poling current is determined by optimizing the second harmonic signal. During poling it is typical to find current variations of 10%. It is also typical to observed a 20% drop of poling current after a few minutes. If the poling is performed at an elevated temperature, after 30 minutes of poling at a fixed temperature, the heater is turned off, and the temperature of the sample is allowed to gradually drop to room temperature during a time interval of 1 to 2 hours. Once the sample reaches room temperature, the high-voltage source is turned off.

9.4 SECOND HARMONIC IN SITU INVESTIGATION OF PHOTOISOMERIZATION

Second harmonic is a convenient *in situ* probe of the dynamics of photo-isomers in a polymer environment if the photoisomers have a second-order optical nonlinearity. A corona discharge is used to apply a large electrostatic field to align the molecules. When a change in the orientation of the photoisomers is induced through photoisomerization, second harmonic generation is used to follow the changes in the state and orientation of the photoisomers.

The photoisomerization is induced by irradiation with green light (532 nm) from either a continuous laser or pulsed laser (frequency doubled Nd:YAG Q-switched laser operated in single pulse mode). The second harmonic is generated by a pulsed optical source such as produced by Nd:YLF (1053 nm) or Nd:YAG (1064 nm) Q-switched nanosecond laser. The probe beam is p-polarized and incident at the Brewster angle of incidence such that the reflected light is minimized. The film is thin enough so that interference effects between the fundamental and generated second harmonic waves have a small effect on the second harmonic generated in the thin film. The probe beam is focused to a spot smaller than the pump beam. A video camera is used to monitor the alignment of the beams and to ensure that the polymer film is not damaged during poling or exposure to the laser beams. Typically, the second harmonic signal is so large in the thin films that low fundamental pulse energy is used, completely eliminating any possibility of damage to the films by the laser intensity. The second harmonic signal is spatially separated from the fundamental beam, detected by a photomultiplier, and collected by a boxcar averager; the data is periodically transferred to a computer.

The dynamics and orientation of an isomer in the presence of electric fields depend on the molecular structure of the unit building blocks of the polymer and how the isomer interacts with the polymer. The isomers either can be simply mixed into the polymer to form a guest-host system, can dangle from a polymer chain to form a side-chain polymer, or can be embedded into the polymer to form a main-chain polymer. Examples will be given of various degrees of steric hindrance of the isomer in the polymer environment. As can be expected, embedding the dipolar isomer into polymer chain makes it the most difficult for the isomer to rotate within the polymer in the presence of an electric field. The chemical structures of two such polymers are shown in Figure 9.5.[20] They are polyimides that contain DR1 chromophores embedded into the polymer main chain. PI-1 has a rigid backbone (Tg = 350°C). PI-2 has some flexibility in the backbone (Tg = 252°C). The polymer films are deposited as previously described. A large electrostatic field is applied to the film at room temperature using a corona discharge. The second harmonic signal generated by PI-1 at room temperature is shown in Figure 9.6. The signal is generated by EFISH, because the isomers do not rotate within the polymer. Initially, all the DR1 molecules

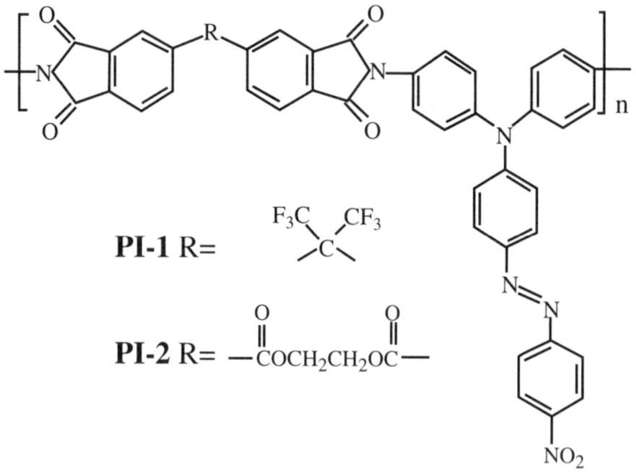

FIGURE 9.5   Chemical structures of mainchain azo-polyimides.

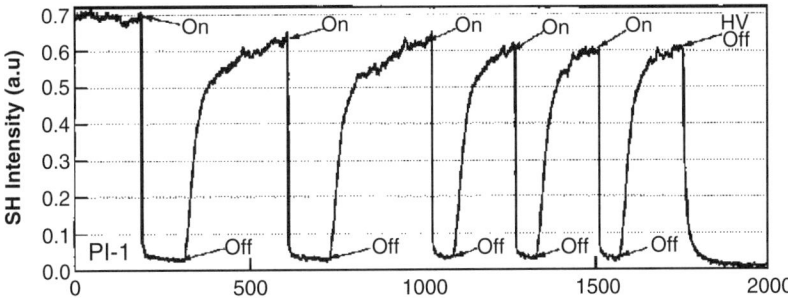

FIGURE 9.6 All-optical light modulation of the electric field enhanced second harmonic signal produced by PI-1. The arrows indicate when the green irradiating pump light is on or off. HV Off indicates when the electric field was turned off.

are in the *trans* state, but then photoisomerization of the azobenzene units was periodically induced by switching on and off a 543.5 nm green laser pump beam (2 mm diameter spot, 8 mW/cm²). When the green pump beam is turned on, the *trans⇒cis* transitions take place, and the population of *cis*-molecules increases and reaches a stationary state. The second harmonic signal decreases drastically because $\gamma_{cis} << \gamma_{trans}$. It was found that the polarization of the pump beam has a negligible influence. When the green pump beam turns off, the *cis*-isomers convert back to *trans*-isomers and the second harmonic intensity recovers by a spontaneous thermal back reaction. The signal never recovers to a level higher than when the experiment was started. Also, when the corona voltage is turned off, the electric field is removed, and the second harmonic intensity returns to zero. This was the case in both PI-1 and PI-2. These two observations indicate that during the photoisomerization process, the DR1 molecules did not align along the electric field and that the rearrangement of the azo chromophore in these polymers requires some correlated motion of a substantial region of the polymer backbone. Because the DR1 molecules were not aligned, $\chi^{(2)}$ plays a negligible role in the generation of the second harmonic. Further evidence that $\chi^{(2)}$ does not play a role is that applying a strong corona field to either polymer, even at 150°C, does not result in any polar order. Compared to PI-1, the spontaneous thermal back reaction was faster in PI-2 because of the more flexible backbone. Biexponential fits to the observed γ_{trans} recovery during the *cis⇒trans* thermal back-isomerization in several recovery cycles over a 500 s period yielded mean values of τ_{fast} and τ_{slow}. For PI-1 $\tau_{fast} = 28s$ and $289s < \tau_{slow} < 390s$. For PI-2, $\tau_{fast} = 14s$ and $67s < \tau_{slow} < 167s$. As can be expected, the optically induced reduction of $\chi^{(3)}$ depends on the irradiating light intensity. Figure 9.7 shows the $\chi^{(3)}$ signal of PI-1 under various irradiation intensities; the numbers on the curves refer to the optical density used to attenuate the irradiation intensity. This figure shows that the $\chi^{(3)}$ strongly depends on the irradiation light intensity through an intensity-dependent photostationary state composed of a mixture of *cis*- and *trans*-isomers of the NLO chromophore. The chromophore exhibits an appreciable molecular hyperpolarizability γ at the

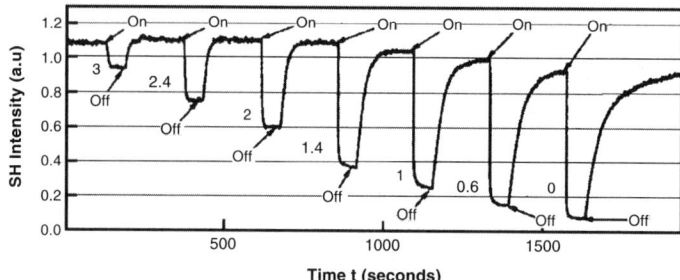

FIGURE 9.7 The intensity of irradiating light effects the EFISH signal generated by PI-I. The arrows indicate when the green irradiating pump light is on or off, and the numbers on the figures refer to the optical density used to attenuate the intensity of the irradiating light. The probe is p-polarized and the pump is s-polarized.

FIGURE 9.8 Chemical structures of side chain azo polyimides.

second harmonic wavelength only when the azo dye is in the *trans* form. The $\chi^{(3)}$ signal can be nearly extinguished by high irradiation intensity that results in a high *cis*-isomer content.

The nonlinear optical properties are more complex if the isomer can rotate within the polymer matrix. Figure 9.8 shows two polymides PI-3a (Tg = 228°C) and PI-3b (Tg = 210°C) that have the azo chromophore attached to the polymer backbone via a flexible tether. Compared to the donor-embedded photoisomers in PI-1 and PI-2, the flexible side-chain system permits more movement of the photoisomer. If one performs the same experiment as for the donor-embedded photoisomers, when the green pump is on, the second harmonic intensity gradually increases (see Figure 9.9). When the pump is turned off, the second harmonic intensity increases to a level higher than it was before the green pump was turned on. This increase in signal indicates that in the *cis* state the molecule orients along the electric field, and this order leads to a $\chi^{(2)}$ contribution to $\chi^{(2)}_{eff}$. In contrast to the donor-embedded systems,

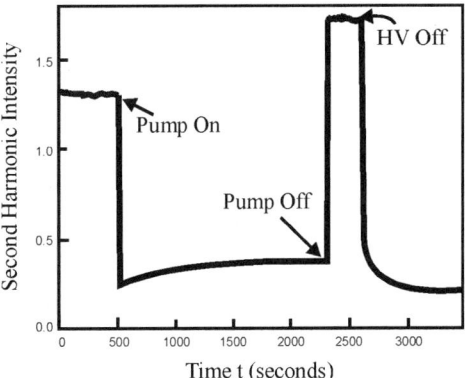

FIGURE 9.9   All optical light modulation of the electric field enhanced second harmonic signal produced by PI-3a. The pump on and pump off refer to green irradiating pump light being on or off. HV off indicates when the electric field was turned off. In contrast to the behavior of PI-1 shown in Figure 9.6, when the pump is turned off the second harmonic intensity increases to a level higher than it was before the green pump was turned on.

when the voltage is removed from the flexible side-chain system, the second harmonic intensity does not decay to zero. A nonzero component remains, because the photoisomers remained aligned within the polymer even with the electrostatic field removed, and this alignment produces a $\chi^{(2)}$ that is independent of an electrostatic field.

The *trans⇒cis* transition in azobenzenes takes place within picoseconds in a liquid.[2] In a polymer, the movement of the isomer is restricted and the question arises of how fast is this transition. In a liquid *trans⇒cis* transition, rate measurements are performed by flowing molecules dissolved in liquid through a sample cell and measuring the *trans⇒cis* by a pump-probe technique using a high repetition short pulse laser. The molecules in the *cis* state are removed from the sample volume, and the slow *cis⇒trans* thermal recovery takes place outside the sample volume. In this experimental arrangement, new *trans*-state molecules are continuously being moved into the sample volume. While a similar experiment is difficult to perform in a polymer, the following experiment obtains a bound on the *trans⇒cis* transition rate in a polymer.[3] The experiment was performed on the flexible side-chain polymer system shown in Figure 9.10, which consists of DR1 molecules tethered to poly(methyl methacrylate) (PMMA). A corona discharge is used to apply a large electrostatic field, at room temperature, to a film deposited onto a soda lime substrate. The photoisomerization was performed with a single 10 ns green pulse generated by a doubled Q-switched Nd:YAG laser. The results of the experiment are shown in Figure 9.11. The upper plot shows the detected second harmonic signal, the middle plot is a representation of the write and erase pulses that were applied to the film, and the lower plot is a representation of the applied corona voltage. The incident energy density was approximately 130 mJ/cm². Region 1 corresponds to a second harmonic signal before the green pulse hits the film. The signal is due to some alignment of the DR1 *trans*-isomers in the polymer film giving rise to a $\chi^{(2)}$-effect and an EFISH

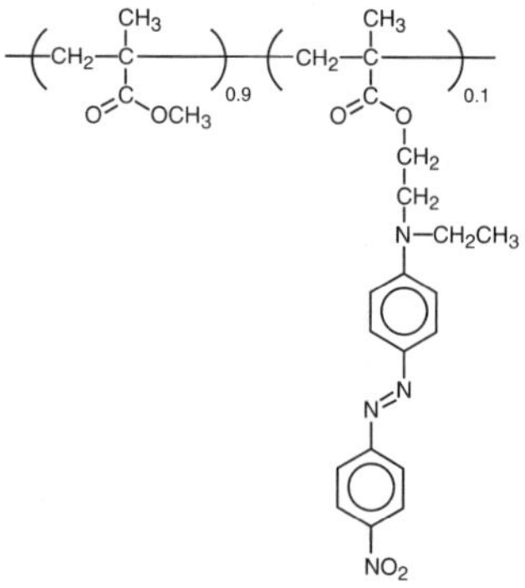

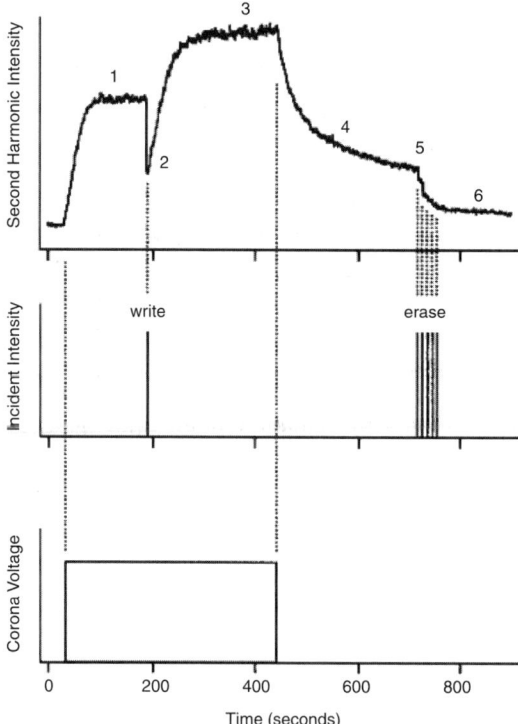

FIGURE 9.10   Chemical structure of side-chain azo-poly(methyl methacrylate) (DR1- PMMA).

FIGURE 9.11   The photoisomerization of DR1-PMMA by a single 10 ns green pulse in the presence of an electrostatic field. The upper plot shows the detected second harmonic signal, the middle
plot is a representation of the write and erase pulses that were applied to the film, and the lower plot is
a representation of the applied corona voltage. On the upper plot, several features are clearly notice-
able: (1) is the electric field induced second harmonic (EFISH) value reached after applying an electric
field to the unpoled film, (2) is the lower SH signal due to the *trans-cis* isomerization, (3) is the combined
photopoled and EFISH signal, (4) is the decay of the SH signal due to the removal of the corona field
which results in the photopoled signal only (i.e., no EFISH), (5) is the drop in SH signal due to the erase
pulses, and (6) is the SH signal following the erase.

contribution. In region 2, a single 10 ns pulse hits the film, and the second harmonic signal immediately decreases because of the photoinduced *trans*⇒*cis* transition and $\beta_{cis} < \beta_{trans}$ and $\gamma_{cis} \approx 0$. In region 3, the increased mobility of the *cis*-isomer enables the molecule to change its orientation and to improve its alignment along the electrostatic field. While the alignment of the molecule improves, the *cis*-isomer thermally relaxes to the stable *trans*-isomer. Local heating of the film also will increase the mobility of the isomer, but the film was thin enough that this effect could be ruled out. In regions 4 and 5, the corona discharge is turned off. In region 4, the corona discharge is stopped, and the electrostatic charge deposited onto the polymer decays, leaving a signal of $\chi^{(2)}$ due to the oriented *trans*-isomers. In region 5, five 10 ns erasure pulses hit the film in the absence of the applied electric field, resulting in region 6, which corresponds to the second harmonic intensity after the optical erase cycle. The use of multiple erase pulses in region 5 indicates that more thermal energy was required for the erase cycle than the write cycle. The local heating increases the entropy to such a level that sufficient randomization of the orientation of the nonlinear molecules occurs. This experiment shows that the *trans*⇒*cis* transition in a polymer occurs in less than a few nanoseconds; the pump pulse is on for only 20 nanoseconds. This experiment does suggest that photoinduced mobility in combination with the reorientation of the *cis*-state molecules to the electrostatic field relaxation to the *trans* state plays a more important role than photoinduced alignment along the polarized pump beam. However, some photoinduced alignment as a result of successive cycles of *trans*⇒*cis* and followed by fast reverse *cis*⇒*trans* photoisomerization cannot be completely ruled out.

9.5 CONCLUSION

The polymer environment plays an important role in the mobility of isomers. In this chapter, it was shown that second harmonic generation can provide detailed information of photoisomerization in a polymer film in the presence of a large electrostatic field. Photoisomerization can control the optical non-linearities of materials that have applications in optical switching and optical storage applications.

ACKNOWLEDGMENT

This work was supported by the MRSEC Program of the National Science Foundation under Award DMR-9808677 for CPIMA II.

REFERENCES

1. Singer, K. D., Sohn, J.E., and Lalama, S.J. Second harmonic generation in poled polymer films. *Applied Physics Letters*, vol. 49, (no. 5), 4 Aug. **1986**, p. 248–250.
2. Lednev, I. K., Tian-Qing Ye, Hester, R. E., and Moore, J. N. Femtosecond time-resolved UV-visible absorption spectroscopy of trans-azobenzene in solution. *Journal of Physical Chemistry*, vol. 100, (no. 32), ACS, 8 Aug. **1996**, p.13338–13341.

3. Hill, R. A., Dreher, S., Knoesen, A., and Yankelevich, D. R. Reversible optical storage utilizing pulsed, photoinduced, electric-field-assisted reorientation of azobenzenes. *Applied Physics Letters*, vol. 66, (no. 17), 24 April **1995**, p. 2156–2158.

4. Loucif-Saibi, R., Nakatani, K., Delaire, J. A., Dumont, M., and Sekkat, Z. Photoisomerization and second harmonic generation in disperse red one-doped and -functionalized poly(methyl methacrylate) films. *Chem. Mater.* **1993**, *5*, 229.

5. Sekkat Zouheir, Knoesen André, Lee Victor Y., and Miller Robert D. Observation of reversible photochemical blow out of the third-order molecular hyperpolarizability of push-pull azo dye in high glass transition temperature. *J. Phys. Chem. B* **1997**, p. 4733–4739.

6. Singer, K. D., Lalama, S. L., Sohn, J. E., and Small, R. D. Electro-optic Organic Materials, Chapter II-8, in *Nonlinear Optical Properties of Organic Molecules and Crystals*, Volume 1, Edited by D. S. Chemla and J. Zyss, Academic Press, Orlando, 1987.

7. Kuzyk, M. G. Relationship between the Molecular and the Bulk Response, Chapter 3 in *Characterization Techniques and Tabulations for Organic Nonlinear Optical Materials*, Edited by M. G. Kuzyk, C. W. Dirk, Marcel Dekker. New York, 1998.

8. Blum, R., Sprave, M., Sablotny, J., and Eich, M. High-electric-field poling of nonlinear optical polymers. *Journal of the Optical Society of America B (Optical Physics)*, vol. 15, (no. 1), Opt. Soc. America, Jan. **1998**, p. 318–328.

9. Mortazavi, M. A., Knoesen, A., Kowel, S. T., Higgins, B. G., and Dienes, A. Second-harmonic generation and absorption studies of polymer-dye films oriented by corona-onset poling at elevated temperatures. *Journal of the Optical Society of America B (Optical Physics)*, vol. 6, (no. 4), April **1989**, pp. 733–741.

10. Van Brunt, R. J. Physics and chemistry of partial discharge and corona: recent advances and future challenges,. *IEEE Transactions of Dielectrics and Electrical Insulation*, vol. 1, no. 5, Oct. **1994**, p. 761–784.

11. Knoesen, A., Molau, N. E., Yankelevich, D. R., Mortazavi, M. A., and Dienes, A. Corona-poled nonlinear polymeric films: *In situ* electric field measurement, characterization and ultrashort-pulse applications, *International Journal of Nonlinear Optical Physics*, vol. 1, no. 1, **1992**, p. 73–102.

12. Suzuki, A., Matsuoka, Y., and Ikushima, A. J. Substrate effect on poling process of nonlinear optical polymer films. *Japanese Journal of Applied Physics Part 2-Letters*, vol. 30, no. 8B, 15 Aug. **1991**, L1493–L1495.

13. Dao, P. T., Williams, D. J., McKenna, W. P., and Goppert-Beraducci, K. Constant current corona charging as a technique for poling organic nonlinear optical thin films and the effect of ambient gas. *Journal of Applied Physics*, vol. 73, (no. 5), 1 March **1993**, p. 2043–2050.

14. Stahelin, M., Walsh, C. A., Burland, D. M., Miller, R. D., Twieg, R. J., and Volksen, W. Orientational decay in poled 2nd-order nonlinear-optical guest-host polymers: Temperature dependence and effects of poling geometry. *J. Appl. Phys.* vol. 73 (no. 12), 15 June **1993**, p. 8471–8479.

15. Shi, W., Fang, C., Pan, Q., Gu, Q., Xu, D., Hu, H., Wei, H., and Yu, J. Poling optimization and optical loss measurement of polyetherketone polymer films. *Solid State Communications*, vol. 116, **2000**, p. 67–71.

16. Ribeiro, P. A., Balogh, D. T., and Giocomettim, J. A. Corona poling and electroactivity in side-chain methacrylate copolymer. *IEEE Transactions on Dielectrics and Electrical Insulation*, vol. 7, No. 4, Aug. **2000**, 572–576.

17. Peek, F. W. *Dielectric Phenomena in High-Voltage Engineering*, New York: McGraw-Hill, 1929.

18. Wintle, H. J. Unipolar wire-to-plane corona: accuracy of simple approximations. *Journal of Electrostatics*, vol. 28, (no. 2), July **1992**, p. 149–159.

19. Shen L. C., and Kong, J. A. *Applied Electromagnetism*, PWS Publishing Company, Boston, Third Edition, 1995.

20. Sekkat, Z., Pretre, P., Knoesen, A., Volksen, W., Lee, Y. Y., Miller, R. D., Wood, J., and Knoll, W. Correlation between polymer architecture and sub-glass-transition-temperature light-induced molecular movement in azo-polyimide polymers: influence on linear and second- and third-order nonlinear optical processes. *Journal of the Optical Society of America B (Optical Physics)*, vol. 15, (no. 1), Opt. Soc. America, Jan. **1998**, p. 401–413.

10

PHOTOASSISTED POLING AND PHOTOSWITCHING OF NLO PROPERTIES OF SPIROPYRANS AND OTHER PHOTOCHROMIC MOLECULES IN POLYMERS AND CRYSTALS

JACQUES A. DELAIRE
ELENA ISHOW
KEITARO NAKATANI
PPSM, CNRS UMR 8531, Ecole Normale Superieure de Cachan, 61, avenue du Président Wilson, 94235 Cachan Cedex, France

ABSTRACT

Spiropyrans, spirooxazins, fulgides, and dithienylethenes give rise to thermally metastable or stable photoisomers when imbedded in polymer films. Furthermore, the NLO properties drastically change when going from the colorless to the colored isomer of these photochromes. In this chapter, we will present the change in NLO properties associated with photochromism of the previously mentionned photochromes, the photoassisted poling of spiropyrans and fulgides in polymethylmethacrylate (PMMA) films and the photoswitching of the second-order NLO properties of PMMA films doped with different photochromes (spiropyrans, fulgides, and dithienylethenes) and also of anil crystals. Potential applications of these phenomena in optical signal processing and optical data storage will be described.

10.1 INTRODUCTION

It is well known that organic molecules exhibit the largest nonlinear optical (NLO) responses.[1–3] Among the different classes of organic solid materials, single crystals and polymers both offer advantages and disadvantages. Single crystals have the best NLO coefficients, but are not easily obtainable;[4] polymers can be easily functionalized with active molecules and processed in different shapes (thin or thick films) compatible with integrated optics requirements, but they sometimes suffer from a lack of stability.[5–10] A main advantage of organic materials is the possibility of easily combining different properties. For example, azo compounds combine a photochromic property with a (second-order) NLO property. It has been shown in Chapter 3 how many advantages can be drawn from the coupling of both properties. The aim of this chapter is to extend the study of this coupling in the case of photochromic materials other than azo compounds.

Photochromes in polymers and crystalline matrices have been studied mainly for their photoinduced change in absorbance, which can lead to wide applications in solar protection or optical recording.[11] During the last decade, it has been shown that other linear and nonlinear optical properties can be associated with photochromism in polymers and crystals. For example, photoinduced birefringence and dichroism have been widely studied in azopolymers[12,13] and also in some other polymers doped with spiropyrans.[14–16] More recently, photoassisted poling appeared to be an attractive method to induce the noncentrosymmetry required for second-order NLO in solid materials. Also based on photochromism, photoswitching of NLO properties allows one to modulate the second-order NLO signal by a UV-visible light signal. Photochromism is defined as a photoinduced reversible transformation between two forms of a chemical species having different absorption spectra.[17] The most common photochromes are colorless (or pale yellow) in their most stable form and turn to blue, orange, purple, or red upon UV or near-visible irradiation. A back thermal reaction sometimes reverts the colored to the colorless form. This reaction appears to be a drawback for some applications like erasable optical recording media. In the case of

Disperse Red 1 (DR1) used for photoassisted poling in polymers, the back thermal reaction advantageously allows one to end up with materials containing the most stable *trans*-isomers. In any case, the back reaction can be induced by light and plays an important role in both orientation and disorientation processes. Photoassisted poling does not necessarily require a stable photoisomer, as shown for azo compounds, which are very efficient. On the contrary, photoswitching of second-order NLO properties makes sense only when the photoisomer is thermally stable.

As concerns photochromes in a solid matrix, a question that immediately arises is to what extent the nature of the matrix impedes the photochromic reaction. This problem has been studied in detail[18,19] but it is beyond the scope of this review. There is a general rule that states photochromic reactions are sluggish in polymer matrices compared to fluid solutions. This statement is true for some stilbene derivatives,[20] but it is not true for azo derivatives, especially for push-pull azobenzene derivatives like DR1, for which the *trans*→*cis* quantum yield equals 0.11 in PMMA at 20°C[21] compared to 0.24 in a liquid hydrocarbon mixture at −110°C[22]. Photochromism of spiropyrans shows an important matrix effect as the quantum yield for the conversion between the spiropyran and the photomerocyanin is equal to 0.8 in ethyl acetate and decreases to 0.102 in PMMA at room temperature. The same decrease is observed for the back photochemical reaction efficiency: 0.6 in ethyl acetate,[23] compared with 0.02 in PMMA at room temperature.[24] Conversely, the matrix effect is much less for furylfulgides: the quantum yields are almost the same in solutions as in polymer matrices.[25] Although most of photochromic molecules exhibit photochromism in polymers and sol-gels, few of them exhibit this property in the crystalline state, due to topochemical reasons. However, some anils[26] and dithienylethenes[27] are known to be photochromic in the crystalline state.

This paper reviews experimental work done on the NLO properties of photochromes other than azo derivatives, which have been described in the preceding chapters of this book. In the first part, the molecular second-order NLO polarizabilities will be given: the NLO properties of the materials will indeed depend on these values. In the second part, photoassisted poling of different photochromes in polymer matrices will be described. The photoswitching of second-order NLO properties of poled polymers or crystals will then be described in the third section. Both second and third parts will end with some applications of these optical phenomena. We will conclude with the prospectives of these materials in NLO.

10.2 MOLECULAR SECOND-ORDER NONLINEAR OPTICAL POLARIZABILITIES OF PHOTOCHROMIC MOLECULES

10.2.1 Definitions

Comprehensive treatments of the physics of NLO originating from interaction of atoms and molecules with light can be found elsewhere.[2,3] Here, we will simply define the quantities that we will use further. Electric field (E_a), such as an applied dc field or a propagating electromagnetic wave, always induces displacements of the electronic cloud of molecules. In an applied oscillatory field, the electrons will oscillate at the applied frequency. At low field strengths, the magnitude of such an induced polarization will be proportional to the applied field. At high field strengths, the polarization is no longer linear. To account for this nonlinearity, it is common to develop the dipole moment μ of the molecule as a power series expansion in the local electric field E_L:

$$\mu = \mu_0 + \alpha\ E_L + \beta\ E_L\ E_L + \gamma\ E_L\ E_L\ E_L + \dots \tag{10.1}$$

In this equation, μ_0 is the permanent dipole moment of the molecule, α is the linear polarizability, β is the first hyperpolarizability, and γ is the second hyperpolarizability. α, β, and γ are tensors of rank 2, 3, and 4 respectively. Symmetry requires that all terms of even order in the electric field of the Equation 10.1 vanish when the molecule possesses an inversion center. This means that only noncentrosymmetric molecules will have second-order NLO properties. In a dielectric medium consisting of polarizable molecules, the local electric field at a given molecule differs from the externally applied field due to the sum of the dipole fields of the other molecules. Different models have been developed to express the local field as a function of the externally applied field E_a,[2,28] but they will not be presented here. In disordered media, it is useful to define a local field factor f such as $E_L = f \cdot E_a$. In the limits of spherical cavities and optical frequency, this local field factor f can be given by the Lorentz-Lorentz local field correction: $f = \dfrac{(n^2 + 2)}{3}$, where n^2 is the square of the refractive index.

Molecules possessing the largest β coefficients contain donor and acceptor substituents linked through a participating π-backbone. Paranitroaniline is a classic model for these push-pull nonsymmetric molecules. During the past twenty years, molecular engineering of this kind of structure has focused on increasing the β values: this engineering has been done by using more potent donating or accepting moieties or increasing the conjugation length between the substituents.[29,30,85] Photochromic compounds, which are generally π-conjugated systems, have potential NLO properties. As the colored form is more π-conjugated, we may expect a higher β value. The ability to switch between two different NLO responses while irradiating a material having both NLO and photochromic properties is attractive for applications in optical signal processing.

10.2.2 Methods of Determining Molecular Hyperpolarizabilities: Experiments and Calculations

The first experimental method for determining β measurements of neutral dipolar molecules is the electric field induced second harmonic generation (EFISH).[31] This method consists of measuring the light intensity at a frequency twice the fundamental frequency ω of the infrared nanosecond laser generated by a solution submitted to a static electric field $E°$. In the case where the static electric field is applied along the z-axis in the laboratory framework, and the polarization of the laser is also along the same axis, the macroscopic polarization $P_{2\omega}$ induced in the solution by the electric field of the incident laser wave E^ω is given by

$$P_Z^{2\omega} = \Gamma\, E^0\, E_Z^\omega E_Z^\omega \tag{10.2}$$

where Γ is the macroscopic third-order susceptibility (only one component Γ_{ZZZZ}) which can be related to the first molecular and second molecular hyperpolarizabilities β and γ

$$\Gamma = N f_{2\omega}\, f_\omega^2\, f_0 \left(\gamma + \frac{\mu\beta_z}{5kT} \right) \tag{10.3}$$

where $f_{2\omega}$, f_ω and f_0 are the local field factors at frequencies 2ω, ω, and zero respectively; μ is the ground state dipole moment; and β_z is the vectorial component of β along the ground state dipole moment, supposed to be oriented along z-axis in the molecular framework ($\beta_z = \beta_{zxx} + \beta_{zyy} + \beta_{zzz}$). Equation 10.3 shows two contributions to Γ in the parentheses: the first one is coming from second-order hyperpolarizability γ, the second one, $\dfrac{\mu\beta_z}{5kT}$, is coming from the orientation of dipoles along the electrostatic field. For usual π-conjugated chromophores, γ is negligible compared with $\dfrac{\mu\beta_z}{5kT}$. In the case of a solution containing one solute A, both solvent and solute being NLO active, the Γ coefficient varies linearly with the mass fraction x_A of solute A

$$\Gamma = \Gamma_0 + x_A \Gamma_A' \tag{10.4}$$

where Γ_0 is the macroscopic third-order susceptibility of the solvent and Γ_A' is the macroscopic third-order susceptibility of solute A, from which β can be determined. In the case of photochromes, as it is generally difficult to isolate colorless and colored forms (noted A and B respectively), one can perform EFISH experiments on the mixture of both forms A and B at the photo-stationary state. In this case Γ is written:

$$\Gamma = \Gamma_0 + x_A \Gamma_A' + x_B \Gamma_B' \tag{10.5}$$

From the third term Γ_B', and provided that x_A, x_B, Γ_0 and Γ_A' are known, the first hyperpolarizability β_B of the unstable colored form can be determined.

A second experimental method, Hyper-Rayleigh Scattering (HRS),[32,33] allows β measurements for nondipolar and/or ionic molecules, which cannot be measured with the standard EFISH. The intensity of the scattered light $I_{2\omega}$ is proportional to the square of the incident intensity and can be written as

$$I_{2\omega} = g \left(N_{\text{solvent}} \left\langle \beta_{IJK\,\text{solvent}}^2 \right\rangle + N_{\text{solute}} \left\langle \beta_{IJK\,\text{solute}}^2 \right\rangle \right) I_\omega^2 \tag{10.6}$$

where N_{solvent} and N_{solute} are the densities of solvent and solute molecules respectively, $\langle \beta_{IJK}^2 \rangle$ is the mean value of the square of some components of the hyperpolarizability tensor in the laboratory framework, and g is a constant value depending on experimental conditions. The averaged $\langle \beta^2 \rangle$ value of the rank-6 tensor $\beta \otimes \beta$ may be expressed in general as linear combinations of polynomial expressions depending on β_{ijk}, the tensor components in the molecular framework. These expressions can be found in the literature; we will simply point out that with this method it is possible to get different independent tensor components of β, which is not possible with EFISH.

Finally, this method can also be applied to a mixture of two species A and B (for example the two isomers of a photochrome); in such a case, Equation 10.6 transforms into:

$$I_{2\omega} = g \left(N_{solvent} \left\langle \beta^2_{IJK solvent} \right\rangle + N_A \left\langle \beta^2_{IJK_A} \right\rangle + N_B \left\langle \beta^2_{IJK_B} \right\rangle \right) I^2_\omega \tag{10.7}$$

During the past decade, theoretical calculations of hyperpolarizabilities[29,34] have been performed to help synthetic chemists design optimum NLO structures. Although extremely accurate calculations are still out of reach, it is now possible to predict the influence of structural changes on the NLO coefficients. In the case of photochromes, theoretical calculations may be useful for predicting β values of thermally unstable colored forms. The theoretical methods generally employed to calculate molecular hyperpolarizabilities are of two types; those in which the electric field is explicitly included in the Hamiltonian, frequently labeled as *Finite Field (FF)*; and those which use standard time dependent perturbation theory, labeled *Sum Over State (SOS)* method.

The basis of the FF method is the following: According to Equation 10.1, it is clear that the second partial derivative of the dipole moment with respect to the field, evaluated at zero field, gives the first hyperpolarizability:

$$\beta_{ijk} = \frac{1}{2} \frac{\partial^2 \mu_i}{\partial E_j \partial E_k} \bigg|_{E=0} \tag{10.8}$$

An alternative formulation examines the molecular energy expansion, rather than the previous dipole expansion, with respect to the field:

$$\beta_{ijk} = \frac{1}{2} \frac{\partial^2 \mu_i}{\partial E_j \partial E_k} = \frac{1}{2} \frac{\partial^2}{\partial E_j \partial E_k} \left(-\frac{\partial W}{\partial E_i} \right) = -\frac{1}{2} \frac{\partial^3 W}{\partial E_i \partial E_j \partial E_k} \bigg|_{E=0} \tag{10.9}$$

This method is valid only in the static field limit (zero frequency), which is a weakness. However, recent advances of a derived procedure (Coupled Perturbed Hartree-Fock) permit the frequency dependence of hyperpolarizabilities to be computed.[29] The FF method mainly uses MNDO (modified neglect of diatomic differential overlap) semi-empirical algorithm and the associated parametrizations of AM-1 and PM-3, which are readily available in the popular MOPAC software package.[35]

The SOS perturbation theory expression (Equation 10.10) for the hyperpolarizability shows that one needs dipole matrix elements between ground and excited states, together with excitation energies and excited state dipole moments to compute β.[36–38]

$$\beta_{ijk}(-2\omega; \omega, \omega) = \frac{-e^3 \pi^2}{h^2} \sum_{n'} \sum_{n} \left\{ [\langle ijk \rangle + \langle ikj \rangle] \left(\frac{1}{(\omega_{n'} + 2\omega)(\omega_n + \omega)} + \right. \right.$$

$$\left. \frac{1}{(\omega_{n'} - 2\omega)(\omega_n - \omega)} \right) + [\langle jki \rangle + \langle kji \rangle] \left(\frac{1}{(\omega_{n'} + \omega)(\omega_n + 2\omega)} + \right.$$

$$\left. \frac{1}{(\omega_{n'} - \omega)(\omega_n - 2\omega)} \right) + [\langle kij \rangle + \langle jik \rangle] \left(\frac{1}{(\omega_{n'} + \omega)(\omega_n + \omega)} + \right.$$

$$\left. \left. \frac{1}{(\omega_{n'} - \omega)(\omega_n - \omega)} \right) \right\} \tag{10.10}$$

with: $\langle ijk \rangle = \langle 0|r_i|n' \rangle \langle n'|r_j|n \rangle \langle n|r_k|0 \rangle$

The term $-e^3 \langle ijk \rangle$ represents the product of three transition dipole moments along coordinates i, j, and k, between the ground state $|0\rangle$ and the excited state $|n'\rangle$, the excited state $|n'\rangle$ and another excited state $|n\rangle$, and finally between the excited state $|n\rangle$ and the ground state. This method has been applied to paranitroaniline by mixing of 60 excited states.[39] By varying the number of excited states, it has been shown that the value of β rapidly converges below approximately 50 excited states.[40] The most widespread Hamiltonian used for this kind of computation is the CNDO/S (Completely Neglected Differential Overlap/ Spectroscopy) or the CNDOVSB,[40] which is a CNDO/S Hamiltonian adjusted to reproduce optical and dipolar data for six molecules with high β values. It is worthwhile to note that the laser frequency ω is an input parameter in the SOS formulation; therefore, first hyperpolarizabilities are computed for some given frequency, and the dispersive character of β can be easily computed. For linear molecules with a large charge transfer in the x direction, Equation 10.10 simplifies into the *two-level approximation*:[41]

$$\beta_{xxx}(-2\omega; \omega, \omega) = \frac{6\pi^2}{h^2} \mu_{01}^2 \, \Delta\mu \left(\frac{\omega_1^2}{(\omega_1^2 - 4\omega^2)(\omega_1^2 - \omega^2)} \right) \tag{10.11}$$

Here, $\hbar\omega$ is the energy of the laser photon, $\hbar\omega_1$ and μ_{01} are the energy difference and the transition dipole moment of the transition from the ground state to the first excited state respectively, and $\Delta\mu$ is the difference in dipole moments between the ground state and first excited state. This equation is often used to determine hyperpolarizability at zero frequency β_0 from experimental values of β, according to

$$\beta_{xxx} (-2\omega; \omega, \omega) = \beta_0 \, F(\omega, \omega_1) \tag{10.12}$$

with $\beta_0 = \dfrac{6\pi^2}{h^2\omega_1^2} \mu_{01}^2 \, \Delta\mu$ and the dispersion factor $F(\omega, \omega_1)$ given by:

$$F(\omega, \omega_1) = \frac{\omega_1^4}{(\omega_1^2 - 4\omega^2)(\omega_1^2 - \omega^2)} \tag{10.13}$$

The values of β_0 determined this way are directly comparable with the values obtained by the FF method. Furthermore, because of the dispersion factor, only comparison between values of β_0 of different molecules is meaningful.

10.2.3 Influence of Photochromism on Molecular Hyperpolarizabilities

Molecular hyperpolarizabilities of the two isomers of different photochromes are presented in Tables 10.1 and 10.2. As Disperse Red 1 (DR1) is probably the most studied photochrome due to its very interesting NLO properties in polymers, its nonlinear coefficients have also been tabulated.[42] The spiropyran/photomerocyanine group has also been investigated in detail in polymers and in solutions as well, and the NLO properties have been described already.[43,24] More recently, Atassi et al. have shown that it was possible to observe NLO response with a furylfulgide system,[44,47] and Lehn et al.[46] have shown that some diarylethene compounds can photochemically switch between a low and a high level of NLO response. The first two systems are

TABLE 10.1 Dipole moments[a] and first hyperpolarizabilities[b] of photochromes

Photochrome (most stable or colorless form)	Photochrome (colored form)	Method [b]	Ref.
Trans Disperse Red One $\mu = 8.6$ D $\beta = 44.6 \times 10^{-30}$ esu	Cis Disperse Red One $\mu = 6.3$ D $\beta = 8.4 \times 10^{-30}$ esu	AM1/FF	42
Spiropyran $\mu = 7.5$ D $\beta = 1.9 \times 10^{-30}$ esu	Photomerocyanine $\mu = 13.6$ D $\beta = -40 \times 10^{-30}$ esu	AM1/FF	24
Furylfulgide Aberchrome 540® $\mu = 7.2$ D $\beta\mu = 6.6 \times 10^{-48}$ esu	Dihydrobenzofurane derivative $\mu = 6.6$ D $\beta\mu = 91 \times 10^{-48}$ esu	EFISH at 1907 nm	47
N-salicylidene-4-bromoaniline (yellow, stable) $\mu = 2.9$ D $\beta = 2.3 \times 10^{-30}$ esu	(red, unstable) $\mu = 2.7$ D $\beta = 1.3 \times 10^{-30}$ esu	AM1/FF	48

(a) All dipole moments given in this table have been calculated with the AM1/FF method. (b) The method of determination of β is given in the third column. The values obtained by the AM1/FF method are β_0 values. All values are the modulus of the vectorial part of β projected along the direction of the ground state dipole moment.

thermally reversible (on a second timescale for *cis*-DR1 in PMMA[42] and on a day timescale for photomerocyanine in PMMA[24]). On the contrary, the colored forms of the last two systems are thermally stable, but they can be reversed to the colorless ones by visible light irradiation.

TABLE 10.2   **Dithienylethene derivatives studied for their NLO properties. Conversion ratio under steady-state illumination of OF in dioxan, and products of hyperpolarisabilities times dipole moments of both OF and CF measured at 1.907 μm in dioxan (from reference 45)**

Dithienylethene derivatives	[CF]/[OF] %	μβ (OF) 10^{-48} esu	μβ (CF) 10^{-48} esu	Ref.*
1,2 bis(5-p-anisyl-2,4-dimethyl 3-thienyl) perfluorocyclopentene OF1 CF1	44	31	137	49
1,2-dicyano-1,2-bis(2,4,5-trimethyl-3-thienyl)ethene OF2 CF2	60	31	78	50
1,2-bis(2-methylbenzo[b]3-thienyl)perfluorocyclopentene OF3 CF3	21	13	55	51
2,3-bis(2,4,5-trimethyl-3-thienyl)maleic anhydrid OF4 CF4	27	22	12	50

*The reference given here describes the synthesis of the photochrome.

The products μβ of furylfulgide and diarylethene in their open and closed forms have been determined experimentally by the EFISH technique, in a mixture of both forms.[47] The main result of these data is the large increase of β in going from a colorless to a colored form of a photochrome: this increase is true for spiropyran/photomerocyanine, but also for fulgides and diarylethenes. *Trans*-DR1 has larger β and μ than those of *cis*-DR1; the same holds for photomerocyanine compared with spiropyran, but furylfulgides (and probably diarylethenes) show no significant change in μ upon irradiation. Table 10.1 also includes the β values of both isomers of an anil compound, N-salicylidene-4-bromoaniline, which will be detailed further for its NLO switching properties in the crystalline state.[48]

Recently, another photochromic system based on nitrobenzyl pyridine (NBP) derivatives has been proposed for its potential in switching the second

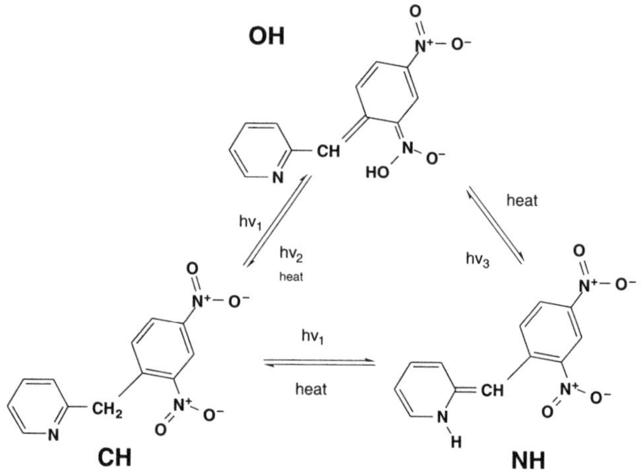

FIG. 10.1 Photochemical and thermal interconversion processes of "CH," "OH," and "NH" tautomers of dinitrobenzyl pyridine (after ref. 52).

harmonic generation efficiency.[52] The photochromic reaction, given later, implies the change from a colorless "CH" form to a deep blue "NH" form. The mechanism of the process (Figure 10.1) involves a proton transfer in two steps through an intermediate "OH" form. The lifetime of the colored "NH" form varies between the ms range in solution to several days in polymers.

To stimulate both colorless CH form and colored NH forms, Lehn and coworkers have synthesized NBP derivatives quaternized by a benzyl group, which can be easily deprotonated to give a stable conjugated form, which is an ideal model of the NH form (Figure 10.2).

The hyperpolarizability coefficients of these model compounds, determined by the HRS method, are given in Table 10.3.

From Table 10.3, it is obvious that the β_0 values of the blue-colored neutral forms 2 and 4 are much higher than those of the colorless "CH" forms 1 and 3. Although the error on the measured hyperpolarizabilities of compounds 1 and 3 is large, the order of magnitude of β is comparable to that of nitrobenzene, in agreement with the similarity of their π-systems. Conversely, compounds 2 and 4 have extended π-conjugation, which is accompanied by higher β values. It was concluded that NBP derivatives have potential for optical modulation of the hyperpolarizability and thus for modulating the second harmonic generation efficiency of NLO devices.[52]

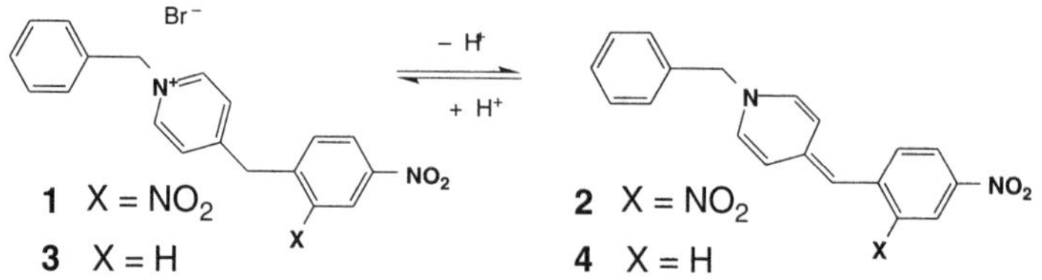

FIG. 10.2 Model nitrobenzyl pyridine derivatives (from ref. 52).

TABLE 10.3   Wavelength of maximum absorption λ_{max} (in nm), first hyperpolarizabilities β (in 10^{-30} esu) and static hyperpolarizabilities β_0 (in 10^{-30} esu) calculated using the two-level model (adapted from ref. 52).

Compound	λ_{max}	β	β_0
1	254	2 ± 1	1
2	582	450 ± 5	62
3	262	5 ± 2	3
4	548	1170 ± 100	52

10.3 PHOTOASSISTED POLING OF PHOTOCHROMES OTHER THAN AZO DERIVATIVES IN POLYMERS

The first example combining photochromism and NLO is probably the thread-like structure of photomerocyanine reported by Meredith, Krongauz, and Williams.[53] Formation of globules based on crystalline particles of 40 nm size occurs under dc electric field during irradiation of spiropyran units deposited on a glass plate (Figure 10.3). Aggregation of highly dipolar photomerocyanine under dc electric field is responsible for a noncentrosymmetric ordering, causing the medium to exhibit NLO effects. Related photomerocyanine compounds in liquid crystalline phase were further studied,[54–56] and $\chi^{(2)}$ values as high as 2×10^{-9} esu (ca. 1 pm. V^{-1}) were observed. Blends including push-pull molecules such as 4-dimethylamino-4'-nitrostilbene showed enhanced NLO properties.

10.3.1 Principle

The most widely employed method (Figure 10.4) to obtain a poled structure in a polymer is to heat a thin film up to its glass transition temperature (T_g),

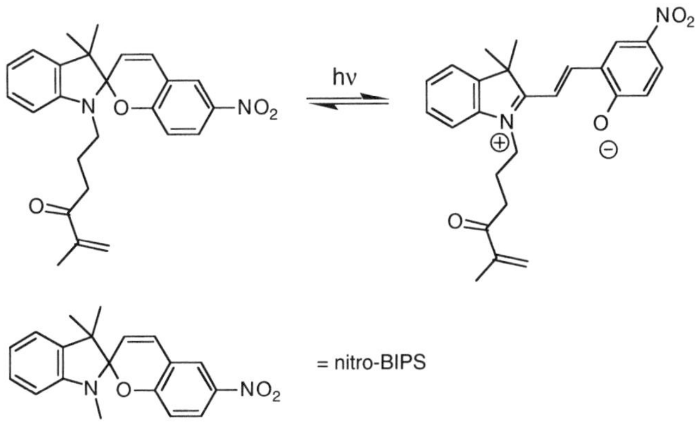

FIG. 10.3   Light-induced activation of second-order NLO properties based on aggregation of photomerocyanine (right). This product is obtained from photochromic nitro-BIPS type spiropyran (left) (from ref. 53).

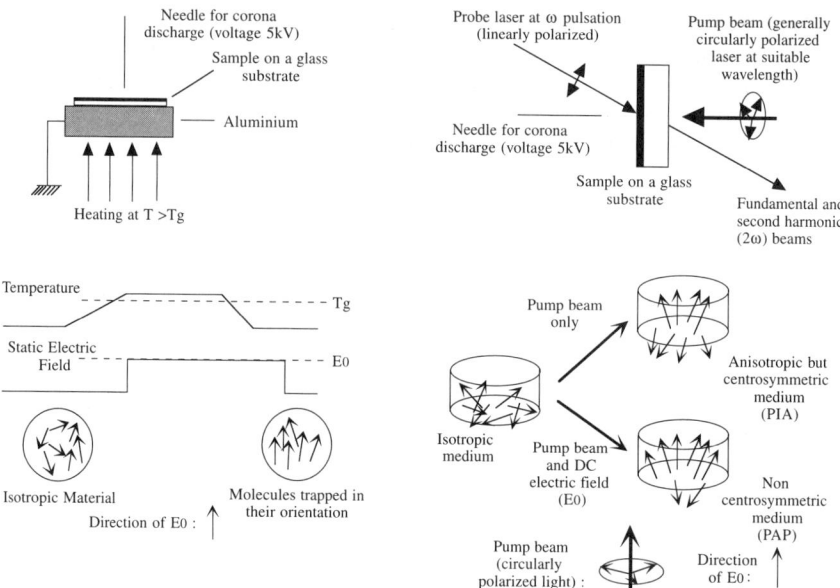

FIG. 10.4 Poling methods for amorphous polymers. Upper left: setup for thermally assisted poling (TAP), which allows heating up to glass transition temperature (T_g) and application of a poling electric field (E_0) by corona discharge. Down left: time sequence for TAP, showing the application of E_0 when the sample is cooled from T_g down to room temperature. Upper right: setup for photoassisted poling (PAP) by the application of E_0 and a pump beam at a suitable wavelength for photochromic reaction. Probe beam (e.g., Nd[3+]:YAG nanosecond pulse laser, 1064 nm) allows second harmonic generation (SHG) measurements. Down right: schematic view of PAP mechanism compared to PIA. In both cases, molecular dipoles line up perpendicular to the pump beam's polarization. E_0 breaks the symmetry between "upward" and "downward" orientations.

to apply a high dc voltage, and finally, to cool the sample down to room temperature with the voltage still on.[37] A static electric field (E^0) of a few MV. cm^{-1} is applied across the sample, either by putting the sample between two electrodes or by applying corona discharge. This process is called **thermally assisted poling** (TAP).

When the material is photochromic, a process based on **photoinduced anisotropy** (PIA) can be employed. PIA is an alignment process of dipolar photochromic molecules under the illumination of a polarized light beam. However, compared to PIA, the application of a dc electric field (poling field) in addition to the pumping beam is necessary, because the symmetry of the latter does not allow the creation of a noncentrosymmetric order. The experimental setup commonly used is shown in Figure 10.4. This configuration allows the application of optical pumping and subsequent and/or simultaneous dc electric field. In appropriate materials, the so-called **photoassisted poling** (PAP) process allows one to create a noncentrosymmetric order even below the T_g. Compared to TAP, heat is replaced by optical pumping, which enables the molecules to get into motion.

The fundamental mechanism of PAP has been described mainly by Dumont and Sekkat.[58–64] It is based on the three processes responsible for PIA (angular hole burning, angular redistribution, and rotational diffusion). However, the random and isotropic feature of angular redistribution in PIA is broken by the dc electric field: The torque of this field on the dipole moment of the molecules makes them rotate along the field direction. Angular hole burning is thus enhanced by angular redistribution. In addition, the rotational diffusion, depicted by a Smoluchowski diffusion operator, is influenced by the presence of a dc electric field.

10.3.2 Measurement Techniques

Second harmonic generation (SHG),[65] attenuated total reflection (ATR)[66] and Stark (electroabsorption) spectroscopy[67,68] were employed to find PAP. The response of the interaction between an electric field (E_a) and a material can be described by Equation (10.14).

$$P = P_0 + \chi^{(1)} E_a + \chi^{(2)} E_a^2 + \chi^{(3)} E_a^3 + \cdots \qquad (10.14)$$

where P is the polarization, P_0 the permanent polarization, $\chi^{(1)}$ the linear susceptibility, and $\chi^{(2)}$ and $\chi^{(3)}$ the quadratic and cubic (nonlinear) susceptibilities.

When E_a is associated with a ω frequency [$E_a = E_{a0} \cos(\omega t)$] electromagnetic wave, one gets a 2ω response (SHG) provided that $\chi^{(2)}$ is not equal to zero (Equation 10.15).[69] Figure 10.4 shows a crossed-beam set-up which allows probing of SHG during PAP.

$$\chi^{(2)} E_a^2 = \frac{1}{2} \chi^{(2)} E_{a0}^2 \left[1 + \cos(2\omega t) \right] \qquad (10.15)$$

In ATR, applying a combination of poling dc (E_0) and measuring ac [$E_1 \cos(\Omega t)$] electric fields modifies the angular positions of the Fabry-Perot dips . The shift is related to a small change of the refractive index ($\Delta n = 10^{-5}$). $\Delta n_i(\Omega)$ and $\Delta n_i(2\Omega)$ (with i = x, y, z), the Ω and 2Ω modulated signals, are measured and related to the nonlinear susceptibilities, and to the Pockels (r_{iz}) and Kerr (s_{iz}) coefficients (Equations 10.16 and 10.17).

$$2\chi^{(2)}_{iiz} = -r_{iz} n_i^4 = 2n_i \Delta n_i(\Omega) E_1^{-1} \qquad (10.16)$$

$$3\chi^{(3)}_{iizz} = -s_{iz} n_i^4 = 4n_i \Delta n_i(2\Omega) E_1^{-2} \qquad (10.17)$$

Along the poling direction (z), the refractive index change is given by Equation 10.18.[62]

$$2n_z \Delta n_z(\Omega) = \left(2\chi^{(2)}_{333} + 6\chi^{(3)}_{3333} E_0 \right) E_1 \cos(\Omega t) + \frac{3}{2} \chi^{(3)}_{3333} E_0^2 \cos(2\Omega t) + \ldots \quad (10.18)$$

As written in Equation 10.19, Stark spectroscopy is based on the absorption spectrum change (ΔA) induced by an electric field (of frequency Ω).[68]

$$\Delta A(\Omega) = E(\Omega) \frac{\Delta\mu.S_1}{hc} \, v \, \frac{\partial(A/v)}{\partial v} \qquad (10.20)$$

A is the absorbance, v the wavenumber, and $\Delta\mu$ the dipole moment change between the ground and excited states.

From measurement of $\Delta\mu$, A, and ΔA, the polar order parameter S_1 is determined. S_1 is connected to the Legendre polynomes and to $\chi^{(2)}$ (Equations 10.20 and 10.21).

$$S_1 = \frac{\langle\cos\theta\rangle - \langle\cos^3\theta\rangle}{1 - \langle\cos^2\theta\rangle} \tag{10.20}$$

$$\chi^{(2)}_{zxx} \propto \langle\cos\theta\rangle - \langle\cos^3\theta\rangle \tag{10.21}$$

10.3.3 Examples

Spiropyran/photomerocyanine (nitro-BIPS, Figure 10.3) in PMMA[24,65,70–72] was investigated with the experimental setup of Figure 10.4. Compared to DR1, the "unstable" photoisomer (photomerocyanine) is "more stable," and the gap between the dipolar and polarizability properties between the isomers is larger (see Section 10.2). This finding is of particular interest, because separate contributions of the photoassisted process and the spontaneous thermal diffusion process could be found and compared.[71] By analogy to what is observed with mechanical stress, the angular mobility was proved to be dependent on the time sequence of the application of optical pumping and of a dc electric field. The shorter the delay between both, the higher the efficiency.[70,71] In addition, a simultaneous application of both perturbations yields a more efficient orientation than a consecutive one (Figure 10.5),

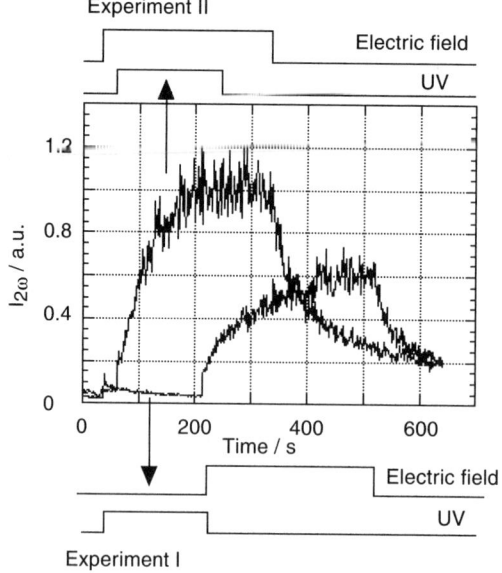

FIG. 10.5 PAP on a spiropyran (nitro-BIPS, Figure 10.3) doped PMMA film (25% w/w) probed by SHG ($I_{2\omega}$) at 1064 nm, pumping at 355 nm 10 mW. cm⁻². Experiment II (simultaneous poling and pumping) yields a higher SHG signal than Experiment I (poling after pumping) (Reprinted from ref. 24). Copyright 2002, American Chemical Society.

which reveals a synergistic effect of dc field and optical pumping.[24] According to these results, the volume created around the photochromic species during photoreaction allows easier rotation, and thus poling of the chromophores. The precise mechanism of this effect is not known as we cannot determine whether the photomerocyanine molecules rotate during the ring opening or after. In the case of Aberchrome 540® (Table 10.1) in PMMA (doped),[73] the photoinduced effect is reported to be less significant compared to thermal diffusion (see Figure 10.6). This fact is attributed to the more globular structure of the colored isomer of Aberchrome 540®, which allows this molecule to rotate in the matrix even without photoexcitation.

Photoassisted poling of photomerocyanine in PMMA has also been studied in ATR experiments. Dumont et al.[70,71] observed the same synergistic effect and concluded that the strictly speaking photoassisted poling does exist during the spiropyran→photomerocyanine photoisomerization process. As a matter of fact, this kind of conclusion is impossible in the case of DR1: Because of the short lifetime of the *cis* state, one observes only an overall photoassisted poling, including thermally assisted poling in the *cis* state. Poling of DR1 is observed for the stable *trans* state and results from many isomerization cycles.

During the photoassisted poling of spiropyran/photomerocyanine, it has also been observed that the longer the delay between the end of the pumping UV light and the application of the electric field, the less efficient the poling is.[70,71] This effect has been explained by a progressive decrease of the angular mobility of molecules, after the pump switch-off. This has been verified by the study of the relaxation of the orientation after the field switch-off: the

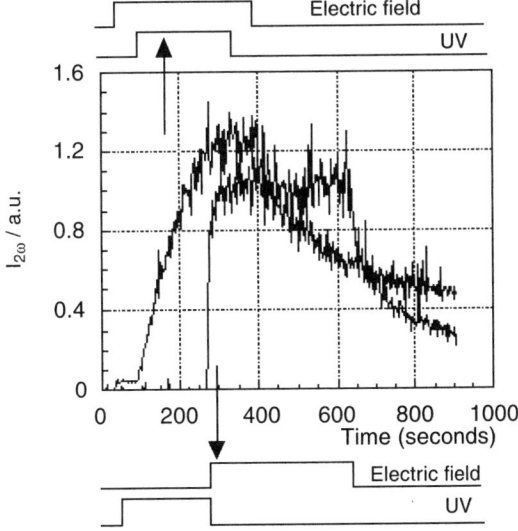

FIG. 10.6 PAP of Aberchrome 540® doped in PMMA film (10% w/w) probed by SHG ($I_{2\omega}$) at 1064 nm, pumping at 355 nm 10 mW. cm⁻². Experiment II (simultaneous poling and pumping) yields almost the same SHG signal as Experiment I (poling after pumping) (Reprinted from ref. 73 by permission of Taylor & Francis Ltd. http://www.tandf.co.uk/journals).

longer the delay, the slower the relaxation. A model proposed by S. Bauer-Gogonea et al.[74] for the interpretation of experiments with Disperse Red 1 in a styrene-maleic anhydride copolymer describes the decay of the experimental relaxation curves with only one set of parameters.[75]

$$f(t) = \exp\left[-\left(\int_0^t \frac{dx}{\tau(x,d)}\right)^\beta\right] \qquad (10.22)$$

where the time constant τ is a function of time (the integrand x) and of the delay d between pump and field switch-off. β expresses the inhomogeneity of the polymer as is usual for describing different kinds of relaxation with stretched exponentials.

10.3.4 Applications

As TAP needs to heat the polymer films at a temperature near T_g, there is a risk of destroying the organic chromophores during TAP, especially in the case of high T_g polymers. For this reason, PAP, which can be efficiently performed at sub-T_g temperatures, is of particular interest.

Another application of PAP is the patterning of quasi-phase matched configurations. In dispersive media like polymers, phase mismatch occurs, due to the difference of refractive index between ω and 2ω. As a consequence, SHG intensity is a sinusoidal function of the propagation length. Half period is the coherence length. A way to avoid the signal decrease after one coherence length is to create an alternating poling pattern every half period. PAP may be used for this purpose, because small areas of a material can be accurately selected. In this field, the combination of photo and thermal processes provides a technique suitable for building particular dipole orientation ("orientation grating") aimed for waveguide applications.[76] Such a device was built on a styrene-co-anhydride–based polymer with DR1 side chains by simultaneous applying a laser beam and alternating the electric field while translating (Figure 10.7). Using such a technique, full width at half maximum (FWHM), which characterizes the spatial resolution, was 7 μm for a beam diameter of 1.5 μm. Further potentialities of this poling technique are reviewed by Gerhard-Multhaupt and Bauer-Gogonea.[77]

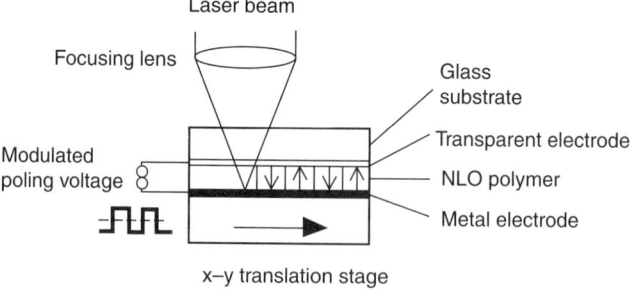

FIG. 10.7 Photo-thermal poling set-up. Poling voltage modulation is synchronized with the translation. Local illumination (He-Ne laser, 2 mW, 543 nm) provides heat to reach T_g (Reprinted from ref. 76). Copyright 2002, American Institute of Physics.

10.4 PHOTOSWITCHING OF NLO PROPERTIES IN ORGANIZED SYSTEMS AND MATERIALS

10.4.1 Introduction

The merit of photochromes is that conversion between both photoisomers is chemically reversible, hence enabling photoswitching of various properties going from one molecular species to the other.[78] As pointed out in Section 10.2, the two molecular species in a photochromic couple can exhibit different molecular NLO properties. Attempts to extend this property to an organized system (monolayer) and to a material (polymer, crystal) are reviewed in this section.

10.4.2 Photoswitching in Polymers

In all examples mentioned herein, polymers were poled thin films.

There is a great enhancement of the hyperpolarizability when spiropyran is switched to photomerocyanine (see Section 10.2). According to the β values, one could expect an increase of SHG intensity up to three orders of magnitude if the conversion was complete. A previously poled spiropyran (nitro-BIPS, Figure 10.3) doped PMMA was photoswitched by UV irradiation, and this resulted in a 10-fold increase of the SHG signal (Figure 10.8).[24] A subsequent visible irradiation within the absorption band of photomerocyanine induced the reverse reaction and dropped the SHG signal down to almost zero. Despite clear evidence of photoswitching, this system suffers from drawbacks related to the following: i) the high β species is not thermally stable; ii) the recycling ability of spiropyran is rather low.

Another problem is that switching of the SHG signal is possible only for a limited number of cycles: the higher SHG value reached after UV irradiation decreases with the performed number of cycles. This decrease is interpreted as an effect of disorientation. Indeed, in the absence of a poling field, the chromophores have no reason to recover exactly the same orientation after one isomerization cycle. In the worst cases, the poled order is completely lost after one switching attempt. Usually, this type of disorientation occurs faster in side-chain polymers than in doped ones.[42,79]

The same kind of experiment (Figure 10.8) was also performed on a furylfulgide compound (Aberchrome 540®, Table 10.1).[73] The lowering of the higher plateau cannot be avoided, but it is somehow slowed down. Compared to spiropyran, the ring opening and closure reactions need less free volume, so the matrix is less disturbed by molecular movements. Another advantage of furylfulgide is the thermal stability of the high β form.

In the search for reversible photoswitchable systems, dithienylethenes seem promising, as the change in geometry during the photochromic reaction is small and can prevent disorientation. A PMMA film doped with OF1 (10% w/w) was poled at 120°C under a corona discharge (6 kV) and submitted to a sequence of alternate irradiations in the visible (514 nm) and in the UV (325 nm) (see Figure 10.9).[45]

The experiment described in Figure 10.9 shows a photomodulation of the SHG signal measured at 532 nm, with an increase during ring closure and a

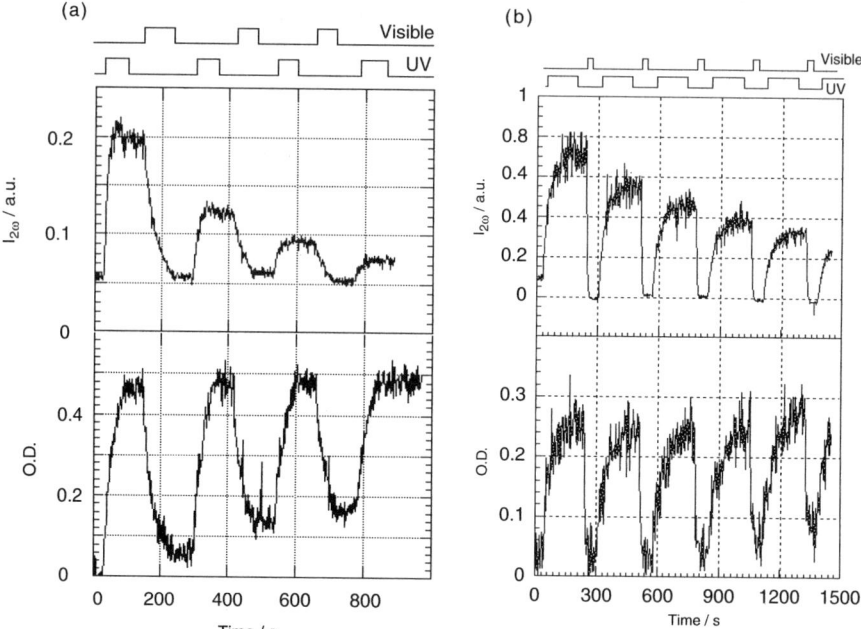

FIG. 10.8 Photoswitching of SHG signal (upper curve) in photochrome doped PMMA thin films (previously poled). UV irradiation is performed at 355 nm and visible irradiation at 514 nm. Optical density (lower curve) is measured within the absorption band of the colored isomer (532 nm) to follow the photo-reaction. No electric field is applied during these experiments. (a) Spiropyran (nitro-BIPS, Figure 10.3) 25% w/w and (b) Furylfulgide (Aberchrome 540®) 10% w/w. (Reprinted from ref. 73 by permission of Taylor & Francis Ltd. http://www.tandf.co.uk/journals).

decrease during ring opening, in agreement with our prediction. Indeed, the hyperpolarizability of the closed form is higher than that of the open form (see Table 10.2). Unfortunately, as in other polymer films, there is an irreversible decay of the overall SHG signal, which is due to the loss of polar order in the film.

10.4.3 Photoswitching of NLO Properties of Noncentrosymmetric Photochromic Crystals

A series of N-salicylidene aniline (SA) derivatives (Figure 10.10) have been investigated for their photochromic and NLO properties. For many years, scientists have known of the photochromism of some of these molecules in the crystalline state.[26] Among them, those that crystallize in a noncentrosymmetric space group were selected, and the NLO properties of the most stable isomer (usually "OH") were studied by SHG, along with the effect of photochromism on this property.[48]

The compound 2-Me-4-NO$_2$ is found in two different forms, according to the preparation method. When the melt was cooled down rapidly, a thermochromic red compound (R) was obtained; whereas for slow cooling rates, a photochromic yellow crystal (Y) was obtained. To the best of our knowledge, only one paper mentions this compound, and reports it as a thermochrome.[80] Both forms differ from their mp values (134.4°C for Y and

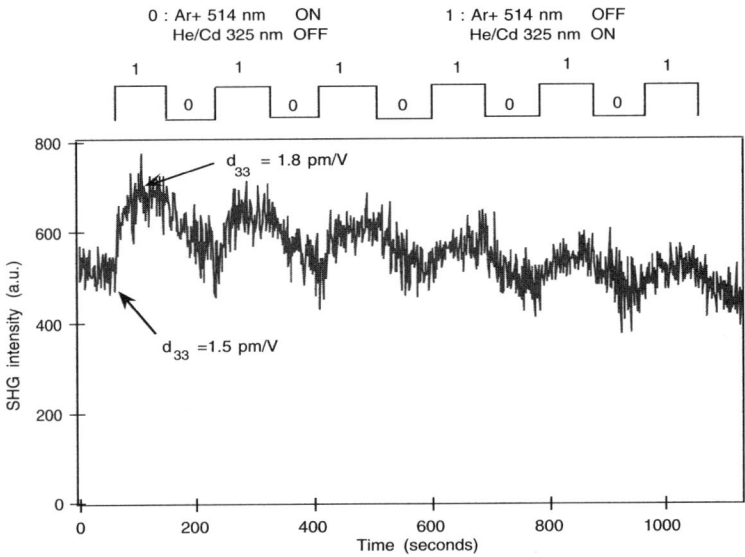

FIG. 10.9 Photomodulation of the SHG signal (fundamental at 1.064 μm) for a PMMA film doped with OF1 (10% w/w) under succesive irradiations at 325 nm and 514 nm. (Reprinted from ref. 45 by permission of Taylor & Francis Ltd. http://www.tandf.co.uk/journals).

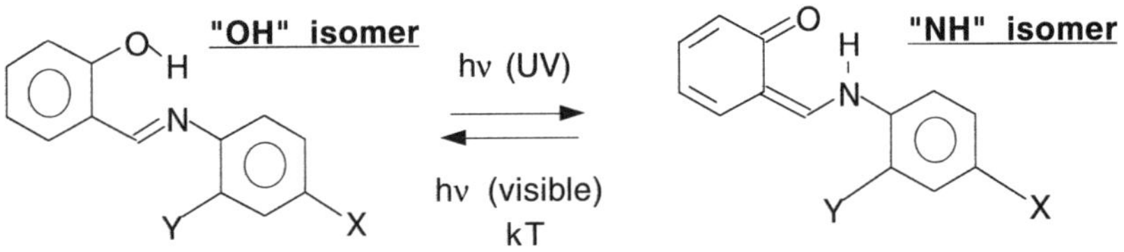

FIG. 10.10 N-salicylidene aniline derivatives.
N-salicylidene 2-chloro-aniline (2-Cl, X=H, Y=Cl), N-salicylidene 2-bromo-aniline (2-Br, X=H, Y=Br), N-salicylidene 4-bromo-aniline (4-Br, X=Br, Y=H), and N-salicylidene 2-methyl-4-nitro-aniline (2-Me-4-NO$_2$, X=NO$_2$, Y=CH$_3$).[48]

132.1°C for R, by DSC) and by their powder X-ray patterns. It was possible to switch from R to Y and vice versa via the melt, and scratching Y irreversibly led to R. Besides the characteristic red color, R showed an additional IR band at 1630 cm^{-1} compared to Y, and we may conclude that R is at least partially in the "NH" form, whereas Y is in "OH" form.

All compounds showed SHG activity as polycrystalline powder (Table 10.4).[81] Though the values of SHG are not very high, they are significant enough to confirm the NC structure for the "OH" isomer of 2-Cl, 2-Br and 4-Br.[26,82,83] In fact, the halogen substitution is not expected to yield a very important charge transfer. To understand the properties of 2-Me-4-NO$_2$ species, which has a significant electron attracting substituent, it will be necessary to investigate the crystalline structure.

SHG intensity change at 1064 nm could be observed for the three halogenated compounds, with a relative variation of SHG around 0.6 for the two bromine compounds (Table 10.5, Figures 10.11 and 10.12).[81]

TABLE 10.4   Powder SHG efficiencies at 1064 nm and 1907 nm fundamental beams (unit : urea reference powder)[81]

	2-Cl	2-Br	4-Br	2-Me-4-NO$_2$ (Y)	2-Me-4-NO$_2$ (R)
1064 nm	1.2	1.5	0.7	1.1	0.7
1907 nm	1.2	1.3	2.2	0.2	0.6

TABLE 10.5   SHG intensity change after UV irradiation. Relative variation of SHG between "high" and "low" states[81]

	2-Cl	2-Br	4-Br
1064 nm	0.15	0.55	0.64
1907 nm	not observed	not observed	0.10

However, the observed difference might arise at least partially from the reabsorption of the second harmonic beam (532 nm), and not from the difference of β values itself. At 1907 nm (SHG at 954 nm), absorption does not interfere, and switching could have been performed for 4-Br (Figure 10.12). According to theoretical calculations (β = 2.3×10^{-30} esu for the "OH" isomer and 1.3×10^{-30} esu for the "NH" isomer),[48] there should be some significant difference of properties between both isomers, but the low difference observed might arise from a low conversion ratio.

10.4.4 Potential Applications

Obtaining an efficient NLO switching material is not easy, because reversibility of molecular change provided by a reversible chemical reaction does not always imply the reversibility of macroscopic physical properties. Further studies should focus on the photochromic molecule itself as well as on the type of material used. On a molecular scale, a photochrome exhibiting large differences between the two properties of the two forms and efficient conversion should be examined. Concerning the material, there is a subtle trade-off between the rigidity so that disorientation can be avoided, and flexibility so that the photochromic reaction can efficiently take place.

Regarding potential applications, in terms of writing and reading stored information on photochromic materials, the nonresonant character of NLO enables reading outside the absorption band. Thus, erasure during reading can be avoided. Another possible application deals with the quasi-phase matching structure that can be obtained by switching. To the best of our knowledge, quasi-phase matching based on alternation of two molecular species, one obtained by isomerization of the other, has never been realized.

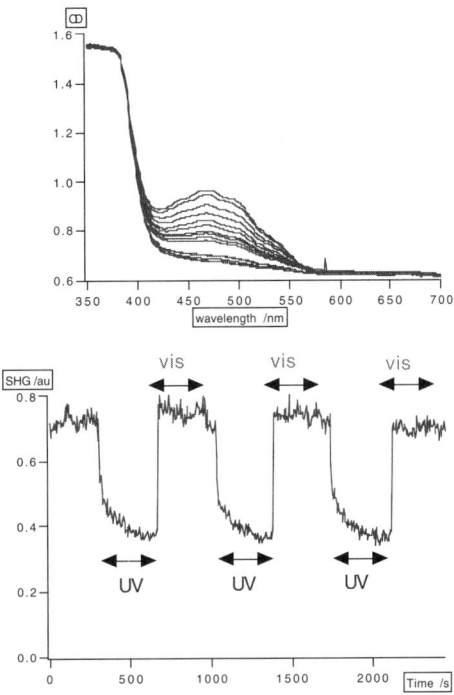

FIG. 10.11 Absorption change during color fading for 2-Br (during 3 h), and SHG change at 1064 nm (UV 355 nm 5 mW, vis 514 nm 19 mW) (Modified from ref. 45 by permission of Taylor & Francis Ltd. http://www.tandf.co.uk/journals).

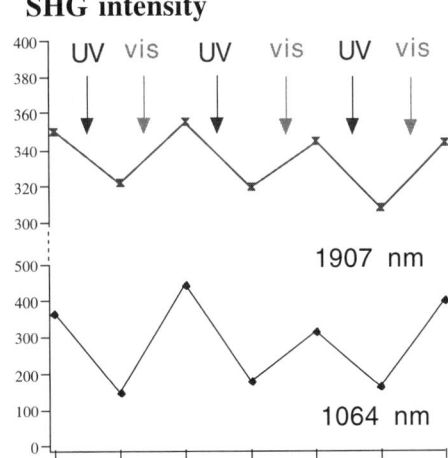

FIG. 10.12 SHG switching of 4-Br at 1064 nm and 1907 nm (UV 365 nm, vis 490 nm) (after ref. 81).

10.5 CONCLUSION

Although DR1 and other push-pull azo compounds are the best chromophores for inducing photoassisted poling in polymers, this orientational effect
induced by polarized light at room temperature has also been demonstrated
with spiropyran/photomerocyanine, and also with a furylfulgide in polymers.
As photomerocyanine is a stable photoisomer in PMMA (at least one day), it
is possible to study the mechanism of photoassisted poling much more easily
than with DR1. For example, the synergy between photoisomerization reaction
and orientation by the electric field has been proven to exist. Furthermore,
the perturbation of the polymer matrix through the photochemical reaction
relaxes with time, on a time scale of hours at room temperature in PMMA.
Therefore, the spiropyran/photomerocyanine system appears very suitable for
studying orientation and relaxation dynamics in polymers, which is still an
outstanding problem. As concerns applications, all the poled samples relaxed
their orientation very quickly, and a lot of efforts still need to be done to
achieve orientational stability with these photochromic systems.

Photoswitching of the second-order NLO response of polymer films has
been demonstrated with at least three photochromes: spiropyrane, furylfulgide, and dithienylethene. However, as previously demonstrated for
PMMA/DR1 poled polymer samples, irradiation with (natural or polarized)
light unavoidably leads to disorientation of chromophores in the absence of
any restoring electric field. As a result, in spite of the efforts to have the most
stable photochromes, the photoswitching experiments in polymers are irreversible, i.e., dipoles lose "memory" of their initial orientation after a few
irradiation cycles during which they transform from the initial colorless (and
almost NLO inactive) to the colored (and NLO active) form. To circumvent
this drawback, we have shown that the use of photochromic and NLO active
crystals of anils may lead to reversible photoswitching of the NLO response.

The coupling between different molecular properties in the same material
may lead to new functionalities like all optical switching or data storage.
Recently, a new photochromic molecule-based magnet has been elaborated
which contains a cationic spiropyran located in between the magnetic layers
of bimetallic ammonium and phosphonium salts.[84] It has been shown that
the reversible photoreaction of the spiropyran cation does not modulate the
critical temperature of the long-range ferromagnetic order, but modifies its
hysteresis loop in a significant way. Unfortunately, this solid compound had
no second-order NLO susceptibility, probably due to a centrosymmetrical
structure. Anyway, molecular magnets switched by light are hoped for in the
near future. In this search for new functionnalities, the development of new
photochromes, their incorporation in different solid materials (crystals,
organic glasses, sol-gels, polymers, and Langmuir-Blodgett layers), and the
study of their linear and nonlinear optical properties are strongly needed.

REFERENCES

1. Burland, D. M., ed., special issue on Optical Nonlinearities in Chemistry, *Chem. Rev. 94*,
1–278 (1994), and references therein.

2. Chemla D. S., and Zyss, J., eds., *Nonlinear Optical Properties of Organic Molecules and Crystals*, Academic Press, Orlando, FL, 1987.

3. Williams, D. J. *Nonlinear Optical Properties of Organic and Polymeric Materials*, ACS Symposium Series 233, American Chemical Society, Washington, DC, 1985.

4. Zyss, J., ed., *Molecular Nonlinear Optics, Materials, Physics and Devices*, Academic Press, San Diego, 1994.

5. Singer, K. D., Sohn, J. E., and Lalama, S. J. *Appl. Phys. Lett.* *49*, 248 (1986).

6. Singer, K. D., Kuzyk, M. G., and Sohn, J. E. *J. Opt. Soc. Am. B 4*, 968 (1987).

7. Mortazavi, M. A., Knoesen, A., Kowel, S. T., Higgins, B. G., and Dienes, A. *J. Opt. Soc. Am. B 6*, 733 (1989).

8. Kuzyk, M. G., Singer, K. D., Zahn, H. E., and King, L. A. *J. Opt. Soc. Am. B 6*, 742 (1989).

9. Hampsch, H. L., Torkelson, J. M., Bethke, S. J., and Grubb, S. G. *J. Appl. Phys.* *67*, 1037 (1990).

10. Hayden, L. M., Saufer, G. F., Ore, F. R., Pasillas, P. L., Hoover, J. M., Lindsay, G. A., and Henry, R. A. *J. Appl. Phys.* *68*, 456 (1990).

11. McArdle, C. B. ed., *Applied Photochromic Polymer Systems*, Blackie, Glasgow, 1992.

12. Xie, S., Natansohn, A., and Rochon, P. *Chem. Mater.* *5*, 403, (1993).

13. Sekkat, Z., and Knoll, W. *Advances in Photochemistry* 22, 117 (1997).

14. Hosotte, S., and Dumont, M. *SPIE Proc.* 2852, 53 (1996).

15. El Osman, A., and Dumont, M. *Polym. Prepr.* 39, 1036 (1998).

16. Dumont, M., and El Osman, A. *Chem. Phys.* *245*, 437 (1999).

17. Crano, J. C., and Guglielmetti, R. J. eds., *Organic Photochromic and Thermochromic Compounds*, Vols. 1 and 2, Plenum Press, New York, 1999.

18. Smets, G. *Advances in Polymer Science 50*, 17 (1983).

19. Kongrauz V. A. in *Photochromism, Molecules and Systems*, H. Dürr and H. Bouas-Laurent eds., Elsevier, Amsterdam, 1990, p. 793.

20. Delaire, J. A., Delouis, J. F., Nakatani, K., Atassi, Y., and Chauvin, J. *Photonics Science News 5*, 130 (2000).

21. Loucif-Saïbi, R., Dhenaut, C., Nakatani, K., Delaire, J. A., Sekkat, Z., and Dumont, M. *Mol. Cryst. Liq. Cryst.* *235*, 251 (1993).

22. Gabor, G., and Fischer, E. *J. Phys. Chem.* *75*, 581 (1971).

23. Arsenov, V. D., Mal'tsev, S. D., Marevtsev, V. S., Cherkashin, M. I., Freidzon, Y. S., Shibayev, V. P., and Plate, N. A. *Vysokomol. Soyed. A 24*, 2298 (1982).

24. Atassi, Y., Delaire, J. A., and Nakatani, K. *J. Phys. Chem.* *99*, 16320 (1995).

25. Whittal, J., in ref.11, p. 80.

26. Hadjoudis, E. in Photochromism, Molecules and Systems, H. Dürr and H. Bouas-Laurent eds., Elsevier, Amsterdam, 1990, p. 685.

27. Yamada, T., Kobatake, S., Muto, K., and Irie, M. *J. Am. Chem. Soc.* *122*, 1589 (2000).

28. Burland, D. M., Miller, R. D., and Walsh, C. A. in Ref. 1, p. 31.

29. Kanis, D. R., Ratner, M. A., and Marks T. J. *Chem Rev.* *94*, 195 (1994).

30. Nalwa, H. S., Watanabe, T., and Miyata, S., in *Nonlinear Optics of Organic Molecules and Polymers*; Nalwa, H. S., and Miyata, S., eds., CRC Press: Boca Raton, 1997, p. 89.

31. Levine, B. F., and Bethea C. G. *J. Chem. Phys.* *60*, 3856 (1974).

32. Clays, K., and Persoons, A. *Phys. Rev. Lett.* 66, 2980, 1991; *Rev. Sci. Instrum.* 63, 3285 (1992).

33. Zyss, J., and Ledoux, I. *Chem Rev.* *94*, 77 (1994).

34. Pugh, D., and Morley, J. O. in *Nonlinear Optical Properties of Organic Molecules and Crystals, Vol. 1*; Zyss, J., and Chemla, D. S., eds. Academic Press, Orlando, 1987, Chap. II-2, p. 193.

35. Kurtz, H. A., Stewart, J. J. P., and Dieter, K. M. *J. Comput. Chem.* 11, 82 (1990).

36. Armstrong, J. A., Bloembergen, N., Ducuing, J., and Pershan, P. S. *Phys. Rev.* *127*, 1918 (1962).

37. Ward, J. F. *Rev. Mod. Phys.* 37, 1 (1965).

38. Orr, B. J., and Ward, J. F. *Mol. Phys.* 20, 513 (1971).

39. Morell, J. A., and Albrecht, A. C. *Chem. Phys. Lett.* 64, 46 (1979).

40. Pugh, D., and Morley, J. O. *Nonlinear Opt. Prop. Org. Mol. Cryst.* 1, 193 (1987).

41. Oudar, J. L. *J. Chem. Phys.* 67, 446 (1977).

42. Loucif-Saïbi, R., Nakatani, K., Delaire, J. A., Dumont, M. and Sekkat, Z. *Chem. Mater.* 5,

229 (1993).

43. Kongrauz, V. in *Applied Photochromic Polymer Systems*, McArdle, C. B., ed., Blackie, Glasgow, 1992; p. 21.

44. Nakatani, K., Atassi, Y., and Delaire, J. A. *Nonlinear Optics 15*, 351 (1996).

45. Delaire, J. A., Fanton-Maltey, I., Chauvin, J., Nakatani, K., and Irie, M. *Mol. Cryst. and Liq. Cryst. 345*, 233 (2000).

46. Gilat, S. L., Kawai, S. H., and Lehn, J.-M. *Chem. Eur. J. 1*, 275 (1995).

47. Atassi, Y., Chauvin, J., Delaire, J. A., Delouis, J. F., Fanton-Maltey, I., and Nakatani, K. *Pure Appl. Chem. 70*, 2157 (1998).

48. Nakatani, K., and Delaire, J. A. *Chem. Mater. 9*, 2682 (1997).

49. Irie, M., Sakemura, K., Okinaka, M., and Uchida, K. *J. Org. Chem. 60*, 8305 (1995).

50. Irie, M., and Mohri, M. *J. Org. Chem. 53*, 803 (1988).

51. Hanazawa, H., Sumiya, R., Horikawa, Y., and Irie, M. *J. Chem. Soc. Chem. Com.* 206 (1992).

52. Houbrechts, S., Clays, K., Persoons, A., Prikamenou, Z., and Lehn, J.-M. *Chem. Phys. Lett. 258*, 485 (1996).

53. Meredith, G. R., Krongauz, V., and Williams, D. J. *Chem. Phys. Lett. 87*, 289 (1982).

54. Yitzchaik, S., Berkovic, G., and Krongauz, V. *Chem. Mater. 2*, 162 (1990).

55. Yitzchaik, S., Berkovic, G., and Krongauz, V. *Macromolecules 23*, 3539 (1990).

56. Yitzchaik, S., Berkovic, G., and Krongauz, V. *Adv. Mater. 2*, 33 (1990).

57. Eich, M., Looser, H., Yoon, D., Twieg, R., Bjorklund, G., and Baumert, J. *J. Opt. Soc. Am. B 6*, 1590 (1989).

58. Dumont, M. *Mol. Cryst. Liq. Cryst. 282*, 437 (1996).

59. El Osman, A., and Dumont, M. *SPIE Proc. 3417*, 36 (1998).

60. Sekkat, Z., and Dumont, M. *Synth. Met. 54*, 373 (1993).

61. Sekkat, Z., Prêtre, P., Knœsen, A., Volksen, W., Lee, V. Y., Miller, R. D., Wood, J., and Knoll, W. *J. Opt. Soc. Am. B 15*, 401 (1998).

62. Sekkat, Z., and Dumont, M. *Appl. Phys. B*, 54, 486 (1992).

63. Sekkat, Z., and Dumont, M. *Mol. Cryst. Liq. Cryst. B 2*, 359 (1992).

64. Dumont, M., Hosotte, S., Froc, G., and Sekkat, Z. *SPIE Proc. 2042*, 2 (1993).

65. Delaire, J. A., Atassi, Y., Loucif-Saïbi, R., and Nakatani, K. *Nonlinear Optics 9*, 317 (1995).

66. Dumont, M., Sekkat, Z., Loucif-Saibi, R., Nakatani, K., and Delaire, J. A. *Mol. Cryst. Liq. Cryst. Sci. Technol. Sect. B: Nonlinear Optics 5*, 395 (1993).

67. Palto, S. P., Blinov, L. M., Yudin, S. G., Grewer, G., Schönhoff, M., and Lösche, M. *Chem. Phys. Lett. 202*, 308 (1993).

68. Blinov, L. M., Barnik, M. I., Weyrauch, T., Palto, S. P., Tevesov, A. A., and Haase, W. *Chem. Phys. Lett. 231*, 246 (1994).

69. In the symmetry of poled materials poled along an axis (z), there are only two different quadratic nonlinear coefficients, namely $\chi^{(2)}_{zzz}$ and $\chi^{(2)}_{zxx}$. In the case of SHG, half values of these coefficients are also noted respectively d_{33} and d_{31}.

70. Dumont, M., Froc, G., and Hosotte, S. *Nonlinear Optics*, 9, 327 (1995).

71. Hosotte, S., and Dumont, M. *Synth. Met. 81*, 125 (1996).

72. Nakatani, K., Atassi, Y., Delaire, J. A., and Guglielmetti, R. *Nonlinear Optics 8*, 33 (1994).

73. Nakatani, K., Atassi, Y., and Delaire, J. A. *Nonlinear Optics 15*, 351 (1996).

74. Bauer-Gogonea, S., Bauer, S., Wirges, W., and Gerhard-Multhaupt, R. *J. Appl. Phys. 76*, 2627 (1994).

75. Bauer-Gogonea, S., Bauer, S., Wirges, W., Gerhard-Multhaupt, R., and Wintle, H. J. in *"Organic Thin Films for Photonic Applications,"* OSA-ACS meeting, Portland, Oregon, OSA Technical Digest Series 21, 133 (1995).

76. Yilmaz, S., Bauer, S., and Gerhard-Multhaupt, R. *Appl. Phys. Lett. 64*, 2770 (1994).

77. Bauer-Gogonea, S., and Gerhard-Multhaupt, R. in *Electrets*; Laplacian Press: Morgan Hill, 1999, 3rd edition, vol 2, Chapter 14.

78. Coe, B. J. *Chem. Eur. J. 5*, 2464 (1999).

79. Aoki, H., Ishikawa, K., Takezoe, H., and Fukuda, A. *Jpn. J. Appl. Phys. A 35*, 168 (1996).

80. Gallagher, P. *Bull. Soc. Chim. Fr.*, 683 (1921).

81. Poineau, F., Nakatani, K., and Delaire, J. A. *Mol. Cryst. Liq. Cryst. 344*, 89–94 (2000).

82. Bregman, J., Leiserowitz, L., and Osaki, K. *J. Chem. Soc.*, 2086 (1964).

83. Lindeman, S. V., Shklover, V. E., Struchkov, Yu. T., Kravcheny, S. G., and Potapov, V. M. *Cryst. Struct. Commun.* *11*, 49 (1982).
84. Bénard, S., Rivière, E., Pei Yu, Nakatani, K., and Delouis, J. F. *Chem. Mater.* *13*, 159 (2001).
85. Alain, V., Rédoglia S., Blanchard-Desce, M., Lebus, S., Lukaszuk, K., Wortmann, R., Gubler, U., Bosshard, C., and Günter, P. *Chem. Phys.* *245*, 51 (1999).

11

ALL OPTICAL POLING IN POLYMERS AND APPLICATIONS

ALEKSANDRA APOSTOLUK[*]
CÉLINE FIORINI-DEBUISSCHERT[†]
JEAN-MICHEL NUNZI[*]

[*]*Université d'Angers, Laboratoire POMA, UMR-CNRS 6136, ERT Cellules Solaires PhotoVolotaïques Plastiques, 2, Boulevard Lavoisier 49045 Angers Cedex, France*
[†]*CEA Saclay DRT-LIST-DECS-SE2M Laboratoire Composants Organiques, 91191 Gif sur Yvette Cedex, France*

ABSTRACT

The self-induced generation of a second harmonic light was first observed by Margulis and Österberg[1] in an optical fiber illuminated with intense light at 1.06 µm, and later illuminated at the same time with the beam at the fundamental frequency (1.06 µm) and at 0.532 µm (second harmonic, SH), revealing a way of inducing a second-order nonlinear susceptibility $\chi^{(2)}$ by a purely optical method in a centrosymmetric material. However, in the case of the optical fiber, the induced nonlinearities remain relatively small. The induction of non-centrosymmetry in azo-dye materials has been an object of extensive studies[2–5] because of their various potential applications in electro-optic devices and integrated optics, as they possess large second-order nonlinear properties. Side-chain polymer matrices containing organic moieties with large second-order polarizabilities β combine the possibility of inducing significant nonlinearities and the ease of processing of polymers. Structuring materials with nonlinear optical properties in a way that the nonlinearity is modulated spatially in them on a wavelength scale permits us to assure phase matching between light waves and opens up the possibility of making wave guides with frequency doubling properties. It may also lead to the development of new technologies for the production of blue coherent light sources. Organic thin films having large $\chi^{(2)}$ nonlinearities are now extensively studied to optimize all optical poling and phase-matching conditions (optimization of the interaction length) to obtain cheap, easy-to-process and fabricate materials that will be used in tuneable optical devices providing the light at any desired wavelength.

Spontaneous orientation is not a natural tendency for most molecules, and so the main difficulty lies in the realization of noncentrosymmetric structures. After a brief review of the different standard poling techniques, we present a detailed description of the optical poling technique. This technique is based on a purely optical process enabling us to take full advantage of the rich processing capabilities of optical tools. This technique offers broad possibilities for phase-matching conditions, and it enlarges the achievable poled geometries, leading to the possibility of a full control of the induced symmetry of the macroscopic second-order susceptibility $\chi^{(2)}$.

11.1 STANDARD POLING TECHNIQUES

11.1.1 Electric Field Poling (Corona Poling)

The orientation of dipolar chromophore molecules under a static electric field is now a widely used technique for preparing of noncentrosymmetric polymer materials. Two main static field poling techniques have been developed: contact electrode poling and corona poling. The main problem with the first technique is the dielectric breakdown which requires a special design of the electrodes to avoid this effect. The latter, which is the most common, consists of applying a high voltage between a needle and a conducting bottom electrode. Ionization of the surrounding ambient air leads to deposition of electric charges at the surface of the polymer film, resulting in a high internal electric field. When the polymer film is heated around its glass transition temperature, molecules are free to rotate: Through dipolar coupling between their permanent dipole and the static field, a molecular reorientation in the direction of the dc field Figure 11.1 results. The induced polarization is frozen when the sample is cooled, still maintaining the electric field.

Electric field poling method is widely used for *in situ* studies of the dynamics of the dipole moment orientation in amorphous polymers and for poling of new materials, which may potentially possess a great number of defects. This method gives a good efficiency of orientation, but its application is limited only to dipolar molecules.[6] Molecules of the octupolar (threefold) symmetry cannot be poled, as they do not possess a permanent dipole moment.

When the polymer film is heated up to its glass transition temperature, the material conductivity facilitates the breakdown risk, and the sample surface may be seriously damaged.[7] Moreover, the conductivity of a polymer film containing ionic chromophores diminishes the amplitude of the applied electric field.

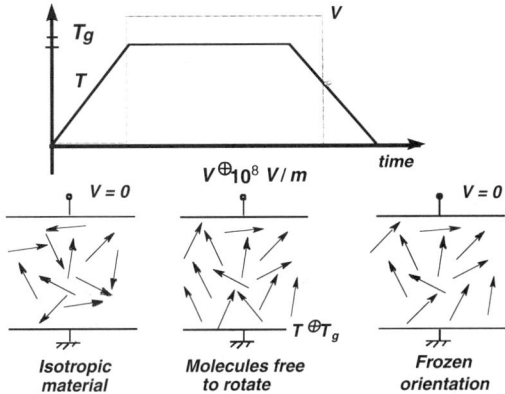

11.1.2 Photo-Assisted Electric Field Poling

11.1.2.1 Photoinduced Anisotropy

Photoinduced anisotropy in organic polymers doped by azo-dye molecules has been widely studied and used for holographic recording.[8,9] Microscopically, the mechanism implies a selective excitation of the molecules whose axis is parallel to the polarization direction of the incident beam. The orientational redistribution following each excitation-relaxation cycle leads to an alignment of the molecules in the plane perpendicular to the incident light polarization, which minimizes the interaction energy of the molecule with the irradiating field. Amplitude of the exciting field changes sign with a frequency of the order of 10^{15} Hz, and its temporal average ($<E_\omega(t)>_t$) is zero: It cannot break the centrosymmetry of the material. It privileges the direction of the orientation, but not the sense. The process is schematically shown in Figure 11.2.

11.1.2.2 Photo-Assisted Poling

In photo-assisted poling, a scheme similar to corona poling is used (Figure 11.3). The organic thin film is resonantly irradiated by a circularly polarized beam at a wavelength in the dye molecule absorption band, thus enabling the

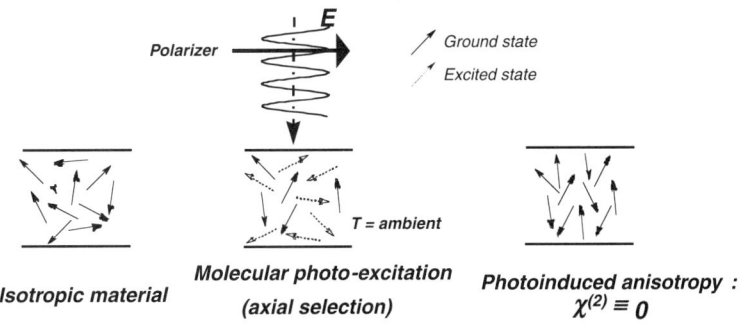

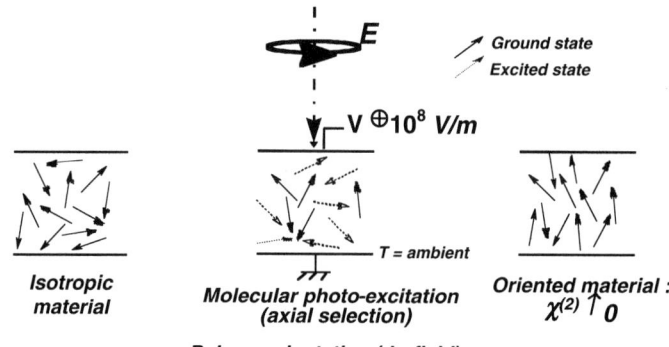

FIG. 11.3   Photo-assisted poling scheme.[10]

reorientation of the in-plane chromophores at ambient temperature. Chromophores whose transition dipole moment is not perpendicular to the polarization direction of the incident beam are excited across the polymer thickness. The dipole moment ordering occurs only over the penetration zone of the pump beam. Application of a dc field perpendicular to the light polarization plane forces a polar reorientation of the chromophores along the direction of the static electric field. This results in the recording of a quasi-permanent second-order susceptibility $\chi^{(2)}$. This method requires effective photo-induced movements and is performed at ambient temperature, much below T_g.

11.2 ALL OPTICAL POLING

The main differences between various poling methods are presented in Table 11.1 The only purely optical method is all optical poling (AOP)

A key issue in the field of nonlinear optical polymers for second-order processes is to achieve a noncentrosymmetric order. One challenge is to realize noncentrosymmetric order by optical means to take full advantage of its vast processing capabilities. Indeed, the all optical poling of polymers offers an interesting alternative to the fabrication of noncentrosymmetric structures. Of particular interest is the possibility of controlling the spatial and tensorial properties of polymers using only optical methods.[10] From a practical point of view, polymers present all the possibilities offered by molecular engineering

TABLE 11.1   Characteristics of poling methods

Orientation Method	Corona Poling	Photo-Assisted Poling	All Optical Poling
Excitation mechanism	Thermal activation		Optical activation
Orientation mechanism (breaking of the centrosymmetry)		Electric field	Optical field

for the tailoring of their properties in a broad sense. Using spin-coated films of dyed polymers and copolymers, the possibility of achieving an efficient and quasi-permanent all optical poling of the molecules has been demonstrated.[11,4] Optimizing the preparation conditions, the same orientation efficiency as using the more standard corona poling method can be achieved.[12] In this paper, we explain the influence on the all optical poling efficiency of so-called seeding parameters such as the relative phase and relative intensities between the writing beams at ω and 2ω frequencies. The mechanisms responsible for the permanent orientation of the molecules are identified, and a simplified model is proposed which, based on three significant parameters, accounts for the essential physics relevant to the all optical poling process and permits optimization of its efficiency.

The interference between a fundamental wave, E_ω, and its SH, $E_{2\omega}$, leads to the presence of a polar field, $E(t) = E_\omega(t) + E_{2\omega}(t)$, inside the optical medium. The temporal average of the field cube $<E^3(t)>_t$, is nonzero. This results in the selective axial excitation of molecules that are not oriented perpendicular to the polarization direction of the irradiating beams at the fundamental and the second harmonic frequencies and the induction of the quasi-permanent polar net orientation of dye molecules, thus breaking the centrosymmetry of the material and recording of the $\chi^{(2)}$ susceptibility with a spatial period satisfying the phase matching condition for SHG (Second Harmonic Generation). At the microscopic level Figure 11.4, the all optical poling mechanism involves a selective polar excitation of molecules, depending on the polarity of the seeding beam combination, $E(t) = E_\omega(t) + E_{2\omega}(t)$. The orientation redistribution following each excitation-relaxation cycle leads to a quasi-permanent orientation of dye molecules.[4] Then the orientation diffusion tends to restore the initial isotropic equilibrium.

The all optical poling experiment consists of two periods: the writing period (so-called seeding) and the readout period of the induced second-order susceptibility $\chi^{(2)}$. During the seeding period, the mutually coherent fundamental ω-beam and second harmonic 2ω-beam simultaneously irradiate the sample, and due to similar one- and two-photon absorptions, the $\chi^{(2)}$ grating is printed

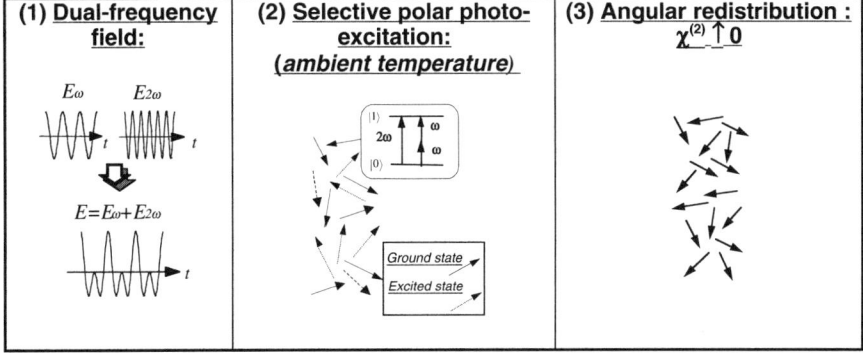

FIG. 11.4 Schematic description of the physical origin of photoinduced orientation.[10]

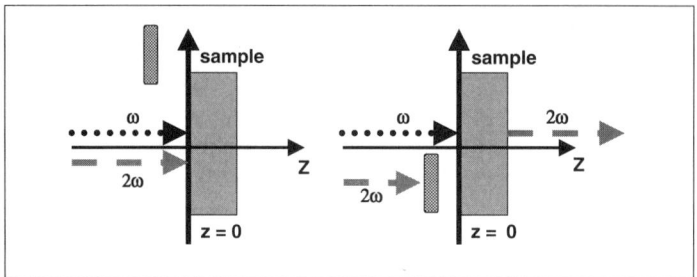

FIG. 11.5   Registration and readout of the photoinduced susceptibility $\chi^{(2)}$.

inside the medium. In the reading phase, the second harmonic seeding beam is cut off, and only the fundamental beam illuminates the sample. The second harmonic generated signal is observed at the sample output. The whole process is schematically shown in Figure 11.5.

The dye molecules whose transition dipole moment is not perpendicular to the polarization direction of the incident fundamental (ω) and second harmonic (2ω) beams will be excited and will undergo subsequent *trans-cis-trans* cycles.[5] The rate of the excitation process is proportional to $\cos^2\theta$ (absorption of one photon) and to $\cos^4\theta$ (absorption of two photons), where θ stands for the angle between the transition dipole moment of a molecule and the beams' polarization direction. These molecules, which will undergo many isomerization cycles, finally will be aligned perpendicularly to the polarization direction of the irradiating light. They will not be excited any more. In this way, the amount of molecules oriented perpendicularly increases, and in consequence, the number of molecules aligned parallelly diminishes. The alignment of molecules perpendicular to the light wave polarization is centrosymmetric (see Figure 11.6). Conversely, the amount of molecules aligned parallelly to the beams' polarization direction is noncentrosymmetric, so there is a difference in the number of molecules oriented upward and downward, and it contributes to the printing of the second-order nonlinear susceptibility $\chi^{(2)}$ in the

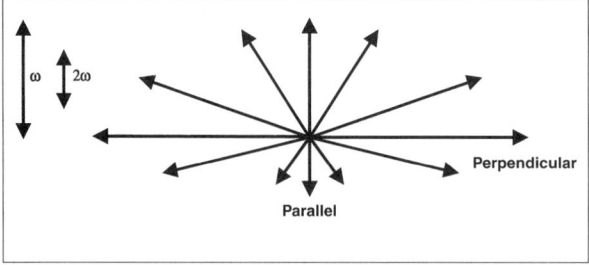

FIG. 11.6   Induced molecular orientation in the material after the all optical poling process. The difference in the amount of molecules oriented upward or downward is the largest in the direction parallel to the polarization direction of the seeding beams. In the direction perpendicular to the polarization of the seeding beams the number of molecules oriented in positive or negative sense remains the same, i.e., it is centrosymmetric.[13]

material only when two linearly parallelly polarized beams at fundamental and second harmonic frequencies illuminate the sample. Yu et al.[13] show that the presence of a third beam at second harmonic frequency polarized perpendicularly to the writing beams during the seeding phase of the all optical poling enhances the efficiency of the process. This third beam re-excites the molecules that have undergone successive isomerization cycles and are already oriented perpendicularly to the polarization direction of the writing beams. This beam reorients these molecules, and they are be realigned parallel to the writing beams, and in this way, the efficiency of the all optical poling increases.

11.2.1 Quasi-Permanent All Optical Encoding of Noncentrosymmetry

11.2.1.1 Photoinduced Transient Second Harmonic Generation

11.2.1.1.1. Theory of the Six-Wave Mixing Process: Phenomenological Description of the Material Response

The photoinduced second harmonic generation (SHG) phenomenon is the result of encoding a $\chi^{(2)}$-grating in the optical medium, through the superposition of a light wave and its second harmonic. At a certain point M inside the material, the recorded $\chi^{(2)}$ is proportional to the cube of the field time average $<E(M,t)^3>$, where $E(M,t)$ stands for the time-dependent combination at point M of two plane waves at ω and 2ω frequencies: $E(M,t) = \Re[E_\omega(M)expi\,(\omega t)\,+\,E_{2\omega}(M)expi\,(2\omega t)]$, where $E_\omega(M)$ and $E_{2\omega}(M)$ are the complex amplitudes of the seeding fields.[14,15] So we obtain:

$$\chi^{(2)} \propto \langle E(M,t)^3\rangle = E_\omega^2\,(M)\,E_{2\omega}^*\,(M) + cc \qquad (11.1)$$

To avoid the possible "painting over" of the encoded interference pattern, it is essential for the writing waves to be mutually coherent. This condition is satisfied, for example, in a KDP crystal, where the wave at 2ω is produced through frequency doubling of the fundamental wave.

Readout of the $\chi^{(2)}$ with a third beam $E'_\omega(M)$ at fundamental frequency produces a nondisappearing second harmonic polarization, with the complex amplitude $P_{2\omega}(M) = \frac{1}{2}\,\varepsilon_0\,\chi^{(2)}\,E_\omega'^2$. This formula may also be written as:

$$P_{2\omega}(M) \propto \frac{1}{2}\,\varepsilon_0\,E_\omega'^2\cdot(E_\omega(M)^2\,E_{2\omega}^*(M) + E_\omega^*(M)^2\,E_{2\omega}(M)) \qquad (11.2)$$

The entire seeding and readout scheme can be viewed as a six-wave mixing (SWM) process. Such a description can be generalized to a parametric conversion process, where writing waves are the sum of three fields of ω_1, ω_2 and ω_3 frequencies, satisfying $\omega_3 = \omega_1 + \omega_2$.[14,16]

11.2.1.1.2 Seeding Type Process: Quasi-Permanent Recording of $\chi^{(2)}$

As can be inferred from Equation 11.1, for a transparent medium at frequency ω and taking into account the losses by absorption with coefficient α at frequency 2ω, the light-induced second-order susceptibility $\chi_{ind}^{(2)}$ can be written as

$$\chi_{ind}^{(2)} \propto \; < E^3(M, t) >_t = \{E_\omega^2 \, E_{2\omega}^* \exp(-i\Delta k.M) + E_\omega^{*2} \, E_{2\omega} \exp(i\Delta k.M)\} \exp\left(-\frac{\alpha}{2} z\right)$$

$$(11.3)$$

where $\Delta k = 2k_\omega - k_{2\omega}$ is the wavevector mismatch, and E_ω and $E_{2\omega}$ are the complex amplitudes of the waves at ω and 2ω frequencies, respectively. The waves are given by: $E_i(M) = E_i \cos(\omega_i t - k_i z)$, where z is the propagation direction coordinate with $z = 0$ taken on the front face of the sample. Equation 11.3 assumes that the forward and backward induced susceptibilities [i.e., $\chi^{(5)}(2\omega;\omega,\omega,\omega,\omega,-2\omega) \equiv \chi^{(5)}(2\omega;\omega,\omega,-\omega,-\omega + 2\omega)$] are identical.

If $\Delta\Phi$ denotes the phase difference between writing fields at ω and 2ω frequencies, for linear and parallel seeding beams polarizations, Equation 11.3 can be written as

$$\chi_{ind}^{(2)}(z) = \chi_{eff}^{(2)} \cos(\Delta\Phi + \Delta k.z) \exp\left(-\frac{\alpha}{2} z\right) \qquad (11.4)$$

with $\chi_{eff}^{(2)} \propto |E_\omega^2 \, E_{2\omega}^*|$

Probing the induced susceptibility $\chi_{ind}^{(2)}$ with another beam at fundamental frequency ω' and complex amplitude $E_{\omega'}'$ and wavevector $k_{\omega'}'$, a nonvanishing second-order polarization is generated inside the material: $P_{2\omega'}(M) = \frac{1}{2} \varepsilon_0 \chi_{ind}^{(2)} E_{\omega'}'^2$

Evolution along the propagation direction of the amplitude E^{SHG} of the field at $2\omega'$ frequency, generated with k^{SHG} wavevector, is described by the classical wave equation[15]

$$\frac{dE^{SHG}}{dz} = -\frac{\alpha'}{2} E^{SHG} + i \frac{\omega \chi_{ind}^{(2)}(z) E_{\omega'}'^2}{2nc} \exp{-i(2k_{\omega'}' - k^{SHG})z} \qquad (11.5)$$

where n and α' are, respectively, the refractive index and the absorption coefficient at frequency $2\omega'$.

For the same writing and reading fundamental beams, wave Equation 11.5 will be simplified. For the SH signal generated along the propagation direction of the 2ω seeding beam, the wavevector mismatch is given by: $\Delta k = 2k_\omega - k_{2\omega} = 2k_{\omega'}' - k^{SHG}$ and Equation 11.5 yields[2]

$$I_{2\omega}^{SHG}(z=l) = ||E^{SHG}(z=l)||^2 = \frac{\omega^2 \, d_{eff}^2 \, l^2}{4 \, n^2 \, c^2 \, 10} I_\omega^2 \left(1 + \mathrm{sinc}^2\left(2\pi \frac{l}{l_c}\right) + 2\,\mathrm{sinc}\left(2\pi \frac{l}{l_c}\right)\cos\right.$$

$$\left.\left(2\Delta\Phi + 2\pi \frac{l}{l_c}\right)\right) \qquad (11.6)$$

where $d_{eff} = \frac{\chi_{eff}^{(2)}}{2}$, l is the sample thickness, l_c (defined as $\Delta k \cdot l_c = 2\pi$) is the sample coherence length, and OD is its optical density at frequency 2ω; the $sinc$ function is defined as: $\mathrm{sinc}\, x = \frac{\sin x}{x}$.

The sample coherence length l_c is calculated from the refractive index dispersion Δn: $l_c = \frac{\lambda_{2\omega}}{\Delta n}$; Δn can be estimated from the absorption spectrum using the Kramers-Kronig relations. For the DR1-MMA 35/65 copolymer at 1064 nm fundamental wavelength, $\Delta n \approx 0.3$, which yields $l_c \approx 1.7 \, \mu m$.[2]

For thick samples ($l \gg l_c$), the SH signal $I_{2\omega}^{SHG}$ originates only from the phase-matched part coming from the second term in Equation 11.3. This term corresponds to the nonmodulated part in Equation 11.6

$$I_{2\omega}^{SHG} = \frac{\omega^2 d_{eff}^2}{4n^2 \, c^2 \, 10^{DO}} \, I_\omega^2 \, l^2 \tag{11.7}$$

which describes the SH generation in a phase-matched material.

11.2.1.1.3 Seeding Type Preparation Setup

This section describes the experimental study of quasi-permanent all optical poling. The seeding type preparation setup corresponds to a copropagating configuration Figure 11.5. The light source is the Nd^{3+}:YAG laser. An external-cavity KDP crystal (type 2) allows for partial frequency doubling of the fundamental beam. The resulting beam is a coherent superposition of the generated SH beam and of the residual fundamental beam. A polarizer is used to ensure parallel polarization of the two laser beams. The relative phase between the writing beams at ω and 2ω frequencies can be adjusted by tilting a BK7 plate of known thickness and refractive index dispersion.

The experiment consists of a seeding-type process with alternate writing and probing periods. During writing periods, the two writing beams at ω and 2ω frequencies are simultaneously incident onto the sample. For measurement, the writing period is interrupted at regular intervals by insertion of a green-blocking RG 630 Schott filter leaving only the ω beam incident onto the sample. The photoinduced second-order susceptibility $\chi^{(2)}$ is probed using SH generation inside the sample. A photomultiplier (PM) tube is used to measure the SH signal. The shutter in front of the PM is opened in synchronization with the insertion of the green blocking filter. A set of calibrated filters is also used to ensure a correct scaling of the SH signal.

11.2.1.2 Preparation of the Samples

Samples used in our experiments are spin-coated films of the azo-dye molecule Disperse Red 1 [4-(N-(2-hydroxyethyl)-N-ethyl)-amino-4′-nitroazobenzene, DR1] in a poly(methylmetacrylate) (PMMA) matrix. We used a grafted polymer system, which enhanced the temporal stability of the induced polar order after poling. Moreover, a copolymer permits a high concentration of chromophore molecules inside the polymer matrix (up to 50% in weight), in contrast to guest-host (or doped) systems, where the concentration of the nonlinear molecules is limited to 10% in weight due to phase segregation effects. Disperse Red 1 is considered to be a two-level molecule undergoing a photoinduced isomerization from the *trans* to the *cis* state (Figure 11.7).

The studied copolymer (DR1-MMA 35/65) is presented in Figure 11.8. It is obtained by free radical polymerization of a 65/35 molar mixture of methylmetacrylate (MMA) and N-ethyl-N-(metacryloxyethyl)-4′-amino-4-nitroazobenzene (DR1 derivative). Its glass transition temperature, measured by Differential Scanning Calorimetry (DSC), is found to be 130°C. Polymer films are prepared by spin coating the solution of DR1-MMA in 1,1,2-trichloroethane (50 g/l) onto clean glass substrates.

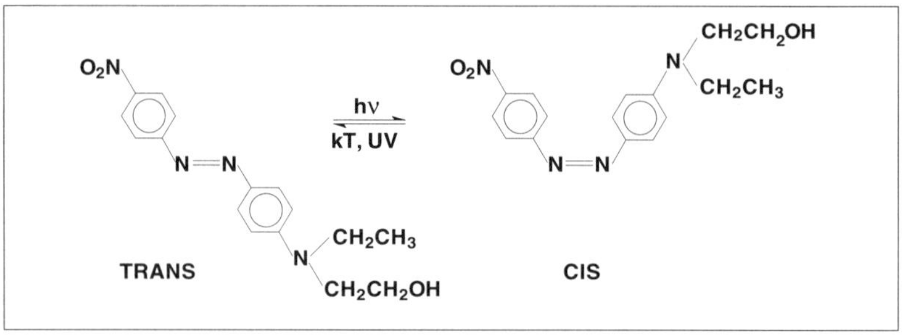

FIG. 11.7 Photoisomerization reaction of the Disperse Red 1 molecule.

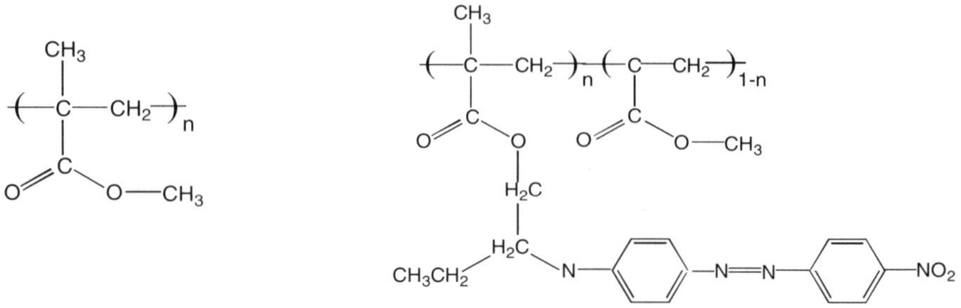

FIG. 11.8 Chemical structure of the PMMA and DR1-MMA, n = 0.35.

11.2.2 Relevant Parameters for an Efficient All Optical Poling

A typical dynamic growth of the photoinduced SH signal in the copolymer film is illustrated in Figure 11.9. No particular optimization of the seeding conditions was performed in this case. An initial increase of the second-harmonic generated signal is observed. After about one hundred minutes of the seeding step, the signal reaches its saturation. As will be explained in the following sections, this value strongly depends on both the relative phase difference and the relative intensities of the seeding beams at ω and 2ω frequencies.

After encoding of the $\chi^{(2)}$, we observe two regimes of relaxation: an initial, very rapid decay, occurring just after stopping of the seeding process, and the second, much slower, multiexponential decay, comparable to the decay observed in polymers poled by the corona poling method. Orientation losses vary from 10% to 20% after 15 hours in the dark at ambient temperature.

11.2.2.1 Influence of the Relative Phase of the Writing Beams

After several tests, it is possible to estimate approximately the optimal relative intensities needed to obtain large photoinduced nonlinearities within relatively short preparation periods. The dependence of the generated SH signal is a function of the phase difference $\Delta\Phi$ between the writing beams at ω and 2ω frequencies. The relative phase difference $\Delta\Phi$ can be varied by tilting a BK7 plate of known thickness and refractive index dispersion.[17]

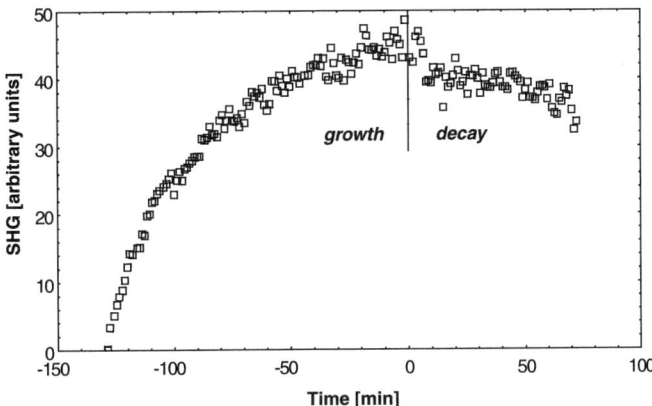

FIG. 11.9 Real-time growth and decay of the SHG signal (in arbitrary units).[4] Negative times correspond to the seeding step. At time zero, the seeding is stopped. Positive times are associated with the study of the temporal stability of the induced $\chi^{(2)}$ susceptibility. The sample is a spin-coated thin film of the DR1-MMA 35/65 copolymer, 0.5 μm-thick, OD = 1.5 at 532 nm. The peak intensity of the ω beam was a few GW/cm², the SH seeding beam was much weaker: between 10 and 100 MW/cm².

Within the limit of small angles of incidence, the phase difference between the writing beams is derived from the incidence angle θ of the waves onto the BK7 plate $\Delta\Phi = \dfrac{\pi(\Delta n)_{BK7}t\,\theta^2}{n^2\lambda_{2\omega}}$, where t is the BK7 plate thickness, and $(\Delta n)_{BK7}$ is its refractive index dispersion $(\Delta n)_{BK7} = n_{2\omega} - n_\omega = 1.2 \times 10^{-2}$.[17]

The results obtained for a 0.1 μm thick film of the DR1-MMA 35/65 are shown in Figure 11.10.[2] The signal saturation is reached after 20 minutes of the seeding step. The sinusoidal variation of the SH intensity generated at saturation as a function of $\Delta\Phi$, with a period of 180°, is obtained. The theoretical fit, shown as a continuous curve in Figure 11.10, is a result of a numerical simulation based on Equation 11.6 using the parameters of the experiment. As can be seen, the experimental results are in a good agreement with the theoretical model. The sample coherence length derived from the fitting curve is equal to 1.75 μm, which confirms the value estimated on the basis of Kramers-Kronig relations.

The measurement of the SH signal intensity does not give access to the absolute orientation of the molecules. However, it is possible to demonstrate that the sign of the susceptibility $\chi^{(2)}$ changes from one oscillation to the other Figure 11.11. After reaching the saturation of the SH signal (seeding performed with an optimal phase difference, $\Delta\Phi = \Delta\Phi_{max}$), a steep phase jump of 180° after saturation of the growth ($\Delta\Phi = \Delta\Phi_{max} + \pi$) results in a very sharp decline of the SH signal. There follows a reconstruction of the SH signal to a value comparable to the one previously obtained (before the phase jump). As shown in the insert of Figure 11.11, no variation of the induced birefringence is observed. After the first poling step ($\Delta\Phi = \Delta\Phi_{max}$), the molecules are oriented in one direction and sense. The sharp decrease of the signal after the phase jump ($\Delta\Phi = \Delta\Phi_{max} + \pi$) corresponds to a disorientation of the molecules. It originates from the reorientation in opposite sense to the one previously

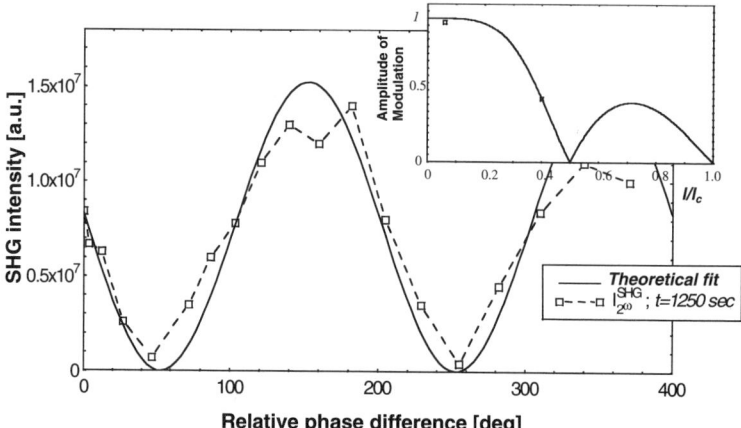

FIG. 11.10 Intensity of the generated SH signal (in arbitrary units) after 20 minutes of seeding time, in function of the relative phase difference $\Delta\Phi$ between the writing beams at frequencies ω and 2ω.[4] The reference (i.e. $\Delta\Phi = 0$) is arbitrarily taken at normal incidence of the writing beams onto the BK7 plate. The solid line represents a theoretical fit to Equation 11.6. The curve shown in the insert shows the modulation amplitude C in function of the sample thickness; $C = \dfrac{I_{max}^{SHG} - I_{min}^{SHG}}{I_{max}^{SHG} + I_{max}^{SHG}}$ where I_{max}^{SHG} and I_{min}^{SHG} are, respectively, the extreme values, in function of $\Delta\Phi$, of the SH intensity generated at saturation.

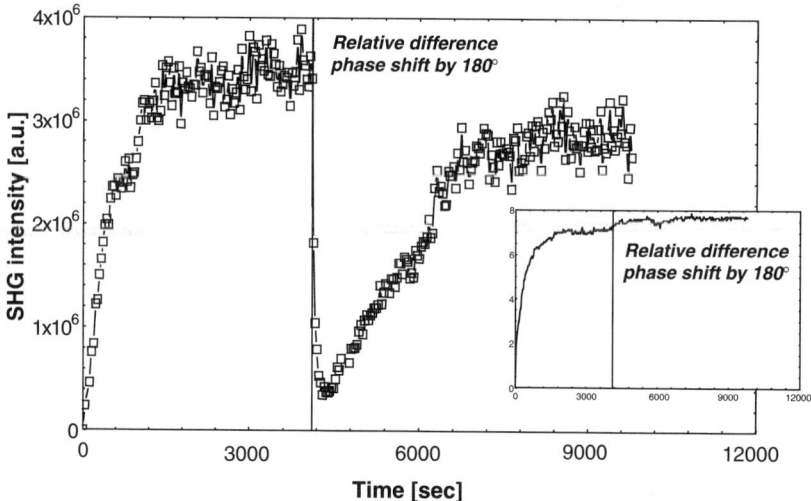

FIG. 11.11 Influence of the relative phase difference $\Delta\Phi$ on the sign of the photoinduced order polarity.[4] The generated SH signal intensity is given in arbitrary units. Like for Figure 10, the sample was a 0.1 μm thick film of the DR1-MMA 35/65 copolymer. The experiment was performed for the optimal relative phase difference ($\Delta\Phi_{OPT}$). As it was measured preliminarily, in the case of two different poled areas, both the parameters $\Delta\Phi = \Delta\Phi_{OPT}$ and $\Delta\Phi = \Delta\Phi_{OPT} + \pi$ lead to the same generated SH signal intensity. The insert shows variations of the photoinduced birefringence during the seeding and reconstruction periods.

induced; therefore, the SH signal generated after saturation reaches a level comparable to the one observed before the phase jump.

For thicker DR1-MMA 35/65 copolymer samples, a sinusoidal dependence of the SH signal in function of the relative phase difference $\Delta\Phi$ is also observed. But the contrast between the extreme values of the SH signal intensities is lower than for the 0.1 µm thick film. Modulation amplitude C is given by: $C = \dfrac{I_{max}^{SHG} - I_{min}^{SHG}}{I_{max}^{SHG} + I_{min}^{SHG}}$, where I_{max}^{SHG} and I_{min}^{SHG} stand for the extreme values (in function of $\Delta\Phi$) of the generated SH intensity at saturation. This equation permits us to get an insight into the phase-matching process.

For thick samples ($l \gg l_c$), the amplitude of modulation C falls down to zero. For such samples, the generated SH wave results only from the phase-matched contribution (see Equation 11.7). Still, for polymers containing dye molecules, such as the DR1-MMA 35/65, the study of samples thicker than 0.5 µm is difficult due to the sample absorption at 2ω frequency. The SH generated signal is partly reabsorbed in the sample. Owing to the absorption at 2ω, the intensity of the 2ω seeding beam varies when propagating through the sample, so the seeding conditions are not homogeneous, which results in an inhomogeneous polarization induced in the material. Using PMMA rods with a low content of grafted dye molecules, the phase-matching process can be studied in thick samples ($l \gg l_c$). These rods are fabricated in cylindrical tubes via free radical polymerization of a low concentration MMA-DR1 monomers in MMA. The α,α'-azoisobutyronitrile (AIBN) is used as a free radical polymerization initiator. The tubes are cut into pieces of the desired length and then polished.[4]

The results of experiments on two samples of 5 and 7 mm thickness presented in Figure 11.12 show that the SH signal does not depend any more on the relative phase difference between seeding beams. It can be inferred from Equation 11.6

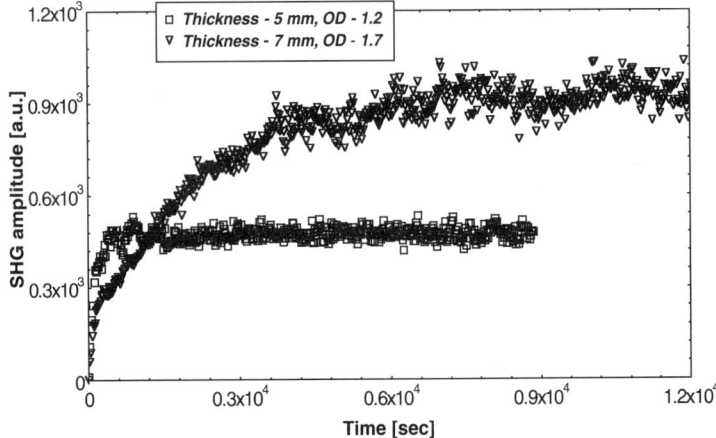

FIG. 11.12 Amplitude of the SH generated signal in two polymer rods of low concentration of DR1.[4] The dye concentration (approximately $2.8*10^{-3}\%$ in monomer) is the same in both samples (l = 5 mm and l = 7 mm).

that for $l >> l_c$ the SH signal results only from the phase-matched contribution. A linear variation of the amplitude of the generated SH signal as a function of the sample thickness should result. The experimental ratio of the SH intensities generated in both samples is $\sqrt{\left(\dfrac{I_{7mm}^{SHG}}{I_{5mm}^{SHG}}\right)_{exp}} = 1.8$, where $I_{5\,mm}^{SHG}$ and $I_{7\,mm}^{SHG}$ stand for the SH intensity generated at saturation, respectively, in the 5 mm and 7 mm thick polymer rods. This ratio is in a good agreement with the theoretical ratio estimated from the sample thicknesses.

11.2.2.2 Influence of the Seeding-Beam Intensities

As we face a cubic interference phenomenon, the initial growth of the photoinduced $\chi^{(2)}$ is proportional to $|E_\omega^2 \, E_{2\omega}|$ (Figure 11.13). As presented in Figure 11.14, the SH signal generated at saturation strongly depends on the relative intensities of the seeding beams at ω and 2ω frequencies. For each given intensity of the beam at fundamental frequency, there is a corresponding optimal value of the SH seeding beam intensity needed to reach a maximum signal after saturation. Above this intensity, a very rapid initial growth of the induced $\chi^{(2)}$ is observed, but the generated SH signal reaches a saturation very quickly at a far lower level.

Using samples with different thicknesses, a very strong and quite narrow dependence of the SH signal generated at saturation in function of the seeding ratio $\gamma = \left| \dfrac{E_{2\omega}}{E_\omega^2} \right|$ is observed (Figure 11.15). The optimal seeding ratio γ_{max}, measured on the front side of the sample, depends strongly on the sample thickness. From Figure 11.15 we obtain:

$$\begin{cases} \gamma_{max}^{l=0.3\mu m} \approx 2.3 \times 10^{-10} \; mV^{-1} \\ \gamma_{max}^{l=0.7\,\mu m} \approx 8.9 \times 10^{-10} \; mV^{-1} \end{cases}$$

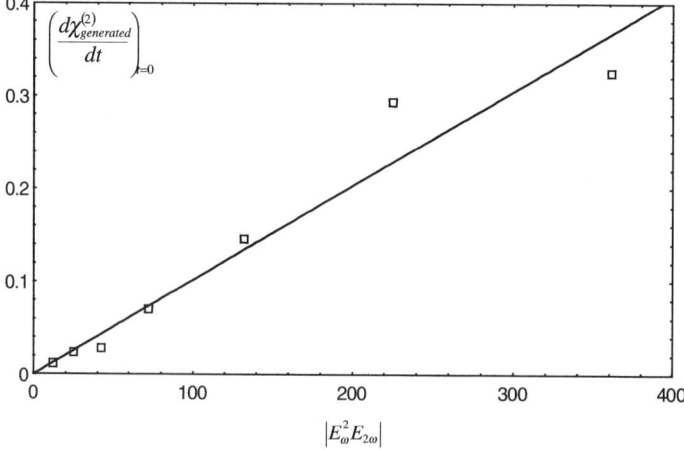

FIG. 11.13 Dependence of the initial growth of the photoinduced $\chi^{(2)}$ susceptibility in function of the writing beams intensities (in arbitrary units).[4] The sample is a film of the DR1-MMA 35/65 copolymer, 0,1 μm-thick.

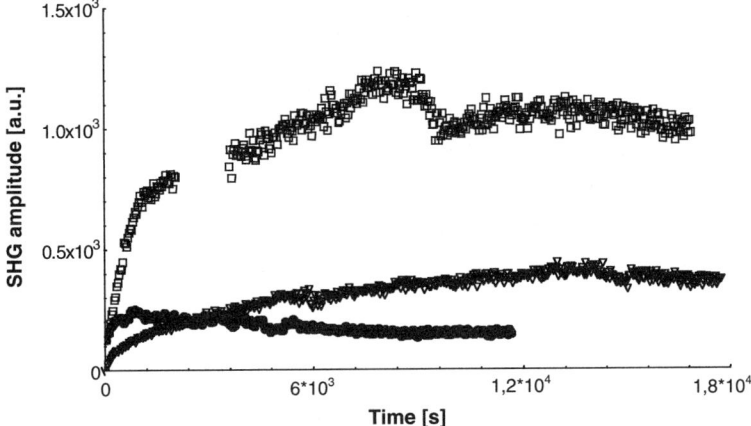

FIG. 11.14 Influence of the relative intensities of the seeding beams at ω and 2ω frequencies on the efficiency (growth and decay) of the poling process.[4] The sample was a 0.3 μm thick film of the DR1-MMA 35/65 copolymer. The fundamental beam intensity was constant and the second harmonic beam intensity was varied by slightly changing the KDP adjustment. Each measurement was performed with an optimized relative phase difference between the writing beams. Seeding beam intensities at 2ω frequency were measured by the photomultiplier tube and are given in arbitrary units: squares $-$ $I_{2\omega,SEED}$ = 1,5*10^{10}, triangles $-$ $I_{2\omega,SEED}$ = 2*10^9, circles $-$ $I_{2\omega\ SEED}$ = 3*10^{11}.

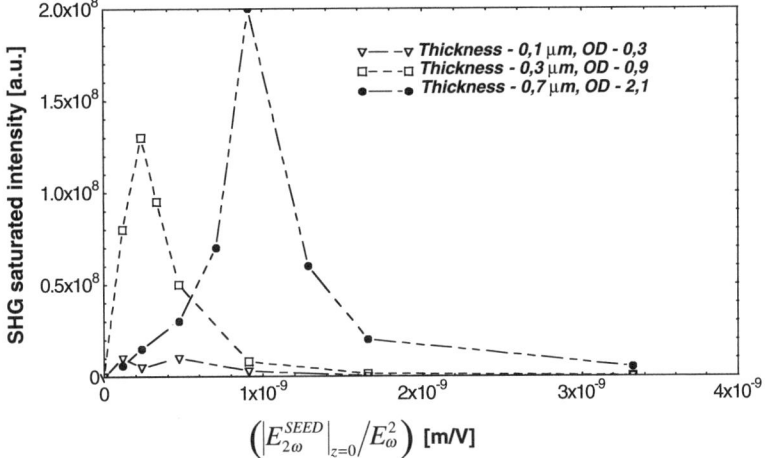

FIG. 11.15 Intensity of the SH generated signal at saturation in function of the relative intensities of the writing beams.[4] Samples are spin-coated films of the DR1-MMA 35/65 copolymer with various thicknesses. Each experimental point is measured for an optimized value of the relative phase difference $\Delta\Phi$ between the seeding beams.

The difference in these values results from the fact that the DR1-MMA 35/65 copolymer strongly absorbs light at the 2ω frequency. It means that the generated SH signal is attenuated as the SH beam propagates inside the thin film. As a consequence, the signal measured by the photomultiplier comes from the contribution of the second-order susceptibility $\chi^{(2)}$ recorded on the back side of the sample. As was already mentioned, the SH seeding beam is also absorbed, which results in inhomogeneous seeding conditions along the propagation direction (z axis): $\gamma(z) = \gamma_0 \exp(-\frac{\alpha}{2} z)$, with γ_0 being the seeding ratio γ measured on the front side of the sample. Taking into account the optical densities of the samples in Figure 11.15, the optimal ratio γ_{max} for each sample (measured on the front side of the sample) corresponds to an identical $\gamma' = \gamma \sqrt{10^{-OD}}$ ratio measured on the back side of the sample: $\gamma'^{l=0.3\mu m}_{max} \approx \gamma'^{l-0.7\,\mu m}_{max} \approx 0.8 \times 10^{-10} \ mV^{-1}$.

The influence of the seeding intensities can be easily explained if the physical origin of the induced $\chi^{(2)}$ is considered. In a two-level molecule, like DR1, the probability of excitation P_{01}, which is responsible for the orientational hole-burning in the initially isotropic distribution of molecules, is the sum of three terms

$$P_{01} \propto \frac{1}{4} (\mu_{01} \cdot E_{2\omega})(\mu_{01} \cdot E_{2\omega})^* + \frac{(\mu_{01} \cdot E_\omega)(\Delta\mu \cdot E_\omega)(\mu_{01} \cdot E_\omega)^*(\Delta\mu \cdot E_\omega)^*}{16(\hbar\omega)^2} +$$

$$\frac{(\mu_{01} \cdot E_{2\omega})^*(\mu_{01} \cdot E_\omega)(\Delta\mu \cdot E_\omega) + (\mu_{01} \cdot E_{2\omega})(\mu_{01} \cdot E_\omega)^*(\Delta\mu \cdot E_\omega)^*}{8(\hbar\omega)} \tag{11.8}$$

with μ_{01} the transition dipole moment and $\Delta\mu$ the difference between the dipole moments in the excited and in the ground state ($\Delta\mu = \mu_1 - \mu_0$).

The first two terms, related, respectively, to one- and two-photon absorption, are axial ones: They are quadratic with the amplitude of the optical fields. They only induce a birefringence and a dichroism inside the sample. The last term, depicting an interference between one- and two-photon absorption, is polar. It is responsible for the centrosymmetry breaking into the material. To optimize the poling process, it is thus necessary to maximize the relative weight of this polar term to achieve the largest possible contrast of the interference fringes. It means that the relative phase as well as the relative intensities of the seeding beams must be optimized. On a molecular level, it yields the following condition.

$$P_{01}^{1-photon} = P_{01}^{2-photon} \tag{11.9a}$$

where $P_{01}^{1-photon}$ and $P_{01}^{2-photon}$ represent the excitation probabilities by one-photon (first term in Equation 11.8) and two-photon absorption (second term in Equation 11.8), respectively. It can be demonstrated that for two-level molecules this condition depends only on $\Delta\mu$, the difference between the dipole moments in the fundamental and excited states. Locally, Equation 11.9a simplifies to

$$\left| \frac{E_{2\omega}}{E_\omega^2} \right| = \frac{\Delta\mu}{2\hbar\omega} \ \sqrt{\langle \cos^2\theta \rangle} = \frac{\Delta\mu}{2\sqrt{3}\hbar\omega} \tag{11.9b}$$

where $\langle \rangle$ denotes the statistical average.

In the case of the DR1, the difference between dipole moments in the fundamental and excited state was measured by two photon absorption experiments:[18] $\Delta\mu = 15.5$ D ($52 \cdot 10^{-30}$ C·m). This yields an optimal ratio of approximately $0.8 \times 10^{-10} mV^{-1}$, which is consistent with the experimental values γ' determined on the back side of the samples. As mentioned previously, due to the absorption of the 2ω beam, the generated SH signal measured by the PM tube comes mainly from the contribution of the second-order susceptibility $\chi^{(2)}$ encoded on the back side of the sample. The value of the optimal ratio can also be estimated by a complementary measurement of photo-induced birefringence: The optimized ratio γ_{max} leading to an optimized poling of the sample corresponds to seeding ω and 2ω beam intensities inducing separately comparable birefringence inside the sample.

11.2.3 Characterization of the Photoinduced $\chi^{(2)}$ Susceptibility

11.2.3.1 Magnitude of the Photoinduced $\chi^{(2)}$ Susceptibility

When the seeding conditions are optimized, the generated SH signal is clearly visible on a white screen. In the case of a 0.3 μm thick film of the DR1-MMA 35/65, it is found that after saturation, the intensity ratio between the 2ω generated generated beam $I_{2\omega}^{SHG}$ and the 2ω seeding beam $I_{2\omega}$ is almost: $\dfrac{I_{2\omega}^{SHG}}{I_{2\omega}} \approx 10^{-3}$. Furthermore, the generated SH signal intensity is comparable to the one obtained from the same sample poled by the corona poling method in optimized conditions, close to the glass transition temperature.[19]

To determine the magnitude of the photoinduced $\chi^{(2)}$, a quartz plate with known refractive index ($n_{2\omega} \approx 1.5$) and susceptibility d_{11} ($d_{11} = 0.5$ pm/V)[20] is used as a reference. The magnitude of the photoinduced susceptibility d_{eff} in the DR1-MMA 35/65 is estimated from the ratio between the SH intensity generated in the reference quartz plate and in the polymer sample, for equal ω-reading beam intensities:[15]

$$d_{eff} = d_{11}^{Quartz} \sqrt{\frac{I_{SHG}^{Copol}}{I_{SHG}^{Quartz}}} \frac{2n_{DR1}\sqrt{10^{OD}}}{n_Q \Delta k_Q l} \tag{11.10}$$

where $\Delta k_Q = \dfrac{2\pi}{\lambda_{2\omega}} \Delta n_Q = 0.15 \times 10^6 \text{m}^{-1}$ with $\Delta n_Q = 1.3 \times 10^{-2}$ [20] being the refractive index dispersion of the quartz. l and OD are, respectively, the thickness and the optical density of the copolymer film. I_{SHG}^{Quartz} and I_{SHG}^{Copol} stand, respectively, for the intensities of the SH signal generated in the quartz plate and in the polymer sample, and n_Q and n_{DR1} are the refractive indices at 2ω frequency of the quartz plate and the copolymer.

For the 0.3 μm thick DR1-MMA 35/65 copolymer film, poled in optimized seeding conditions, a value of $d_{eff} = \dfrac{\chi^{(2)}_{eff}}{2} \approx 76 \; pm/V$ can be obtained (this is a resonance value). This value is confirmed by resolution of the wave equation (Equation 11.6) using the seeding beams' influences.

Each element of the macroscopic second-order susceptibility $\chi^{(2)}$ tensor may be expressed in terms of the molecular second-order polarizability β.[21] For uniaxial (dipolar) molecules, like DR1, we have $\chi^{(2)}_{xxx} = \dfrac{N}{\varepsilon_0} f(2\omega) f(\omega)^2 \langle \beta_{iii} \rangle_{xxx}$, where N is the chromophore concentration, f are the local field factors, and <> denotes the statistical average. The local field factors can be obtained from the local Onsager relations $f(2\omega) \approx f(\omega) = \dfrac{n^2+2}{3}$. We thus have $\chi^{(2)}_{xxx} = \dfrac{N}{\varepsilon_0} \left(\dfrac{n^2+2}{3}\right)^3 \beta_{iii} \langle \cos^3\theta \rangle$.

For the DR1-MMA 35/65, $N = 1.4*10^{21}$ molecules/cm^3 and $\beta \approx 0.8*10^{-48}$ Cm^{3}V^{-2}.[22] It implies a mean orientation ratio $<\cos^3\theta>$ equal to 0.3, which is comparable to what can be achieved using the dc electric field poling technique such as corona poling.

It should be noted that an identical efficiency of orientation can be achieved using a low-power microsecond laser after focusing the writing beams at ω and 2ω frequencies.[23] It shows that for organic materials the all optical poling process is sensitive only to the energy dissipated by the writing beams, independent of the regime of pulse duration.

11.2.3.2 Spatial profile of $\chi^{(2)}$

The spatial profile of the photoinduced susceptibility $\chi^{(2)}$ is interesting to study. The seeding of copolymer samples was performed with collimated writing beams. The diameter of the prepared area was approximately 1 millimeter. Readout of the induced $\chi^{(2)}$ susceptibility was carried out using a much smaller beam diameter to scan the seeded area. In this aim, the probing beam at fundamental frequency was strongly attenuated before focusing on the sample: The ω-probing beam diameter was about 100 μm. The sample was placed in the focal plane of an afocal system and held on a translating stage, permitting micrometric displacements in both x and y directions, perpendicular to the propagation direction along z axis. The analysis of the SH signal intensity generated along the seeded area is given in Figure 11.16.

As can be deduced from Figure 11.16, the observed spatial profile is uniform along the prepared area, reflecting the Gaussian profile of the seeding beams. It points out that for such organic materials all optical poling is a local effect, in contrast to the semi-conductor doped glasses[24] or optical glass fibers.[25]

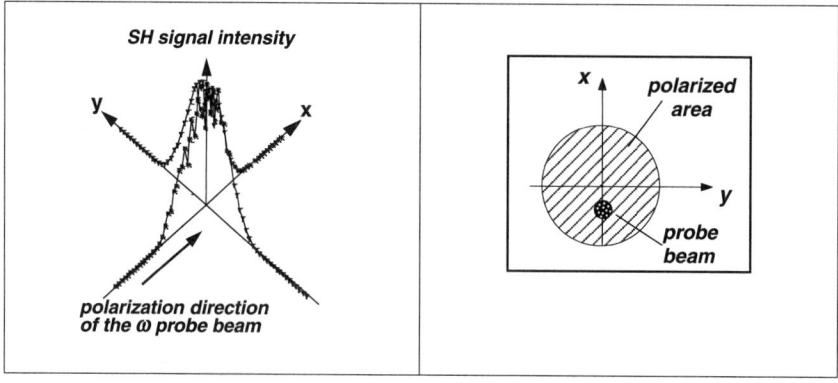

FIG. 11.16 Spatial profile (in the plane of the film) of the photoinduced second-order susceptibility $\chi^{(2)}$.[4] The sample was a 0.3 μm thick spin-coated film of the DR1-MMA 35/65 copolymer. The polarization direction of the probe beam is identical to the polarization direction of the writing beams. The probe beam diameter was about 100 μm. This profile yields a full width at half height of about 1 mm, which is consistent with the diameters of the writing beams.

11.2.3.3 Stability of the Photoinduced Order

11.2.3.3.1 All Optical Poling Dynamics: Model

The photoinduced polar order results from the redistribution of molecules following their repeated selective polar excitation. Polar excitation involves simultaneous one- and two-photon absorption processes on highly coherent electronic excited states. The excitation probability can be written as the sum of three terms: two axial terms related to one- and two-photon absorption, respectively, and the polar term corresponding to the interference between one- and two-photon absorption. Practically, in addition to molecular orientational diffusion, polar ordering also results from the competition between polar and axial excitations.

For uniaxial molecules, the polar order depends only on the polar Euler angle θ. We consider only linear polarizations of the writing beams. Variation of the orientational distribution of molecules $N(\Omega_0)$ initially oriented in the solid angle $\Omega_0(\theta_0, \varphi_0)$ after summing up all the contributions to the light-induced polar and axial excitations is given by:

$$\frac{dN(\Omega_0)}{dt} = -\xi(A \cdot \cos^2 \theta_0 + B \cdot \cos^4 \theta_0 + C \cdot \cos^3 \theta_0) \cdot N(\Omega_0)$$

$$+\xi \iint (A \cdot \cos^2 \theta_1 + B \cdot \cos^4 \theta_1 + C \cdot \cos^3 \theta_1) \cdot N(\Omega_1) \cdot P(\Omega_1 \rightarrow \Omega_0) \cdot d\Omega_1 + D \cdot \nabla^2 N(\Omega_0)$$

$$(11.11)$$

The first term describes the orientational hole-burning process, the second term corresponds to the reorientation after nonradiative relaxation, and the third term depicts the orientational diffusion. ξ stands for the quantum efficiency of the molecular reorientation. (It contains such parameters as the resonant excitation, nonradiative relaxation, transient photochromism, laser repetition rate, matrix rigidity, etc.) $P(\Omega_1 \rightarrow \Omega_0)$ is the probability (normalized) for an excited molecule to leave its initial orientation along the direction $\Omega_1(\theta_1, \varphi_1)$ and turn into the direction $\Omega_0(\theta_0, \varphi_0)$. As there is no direct access possible to the reorientation process, we assume an isotropic angular redistribution: Each molecule can rotate by any arbitrary angle with the same probability. As a consequence, the reorientation probability is taken as a constant: $P(\Omega_1 \rightarrow \Omega_0) = 1/4\pi$. D is the orientational diffusion constant of the material. A, B, and C are the excitation rates due to one-photon absorption at frequency 2ω, the two-photon absorption at ω, and the interference term, respectively. The latter term lies at the origin of the photoinduced susceptibility $\chi^{(2)}$. The phase dependence in $\cos(\Delta\Phi + kz)$ due to propagation along the z direction is contained in C. $\Delta\Phi$ is the phase difference between writing beams measured on the front side of the sample ($z = 0$). Excitation rates A, B, and C are related to the field amplitudes at fundamental $E_\omega(z)$ and SH $E_{2\omega}(z)$ frequencies. In the case of a two energy level molecule with the transition dipole moment μ_{01} and the permanent dipole moment difference $\Delta\mu$, we have

$$\begin{cases} A \propto \mu_{01}^2 \|E_{2\omega}^2\| = B\gamma^2 \\ \\ B \propto \frac{\mu_{01}^2 \Delta\mu^2}{(2\hbar\omega)^2} \|E_\omega^4\| \\ \\ C \propto \frac{\mu_{01}^2 \Delta\mu}{(\hbar\omega)} \|E_\omega^2 E_{2\omega}^*\| \cos(\Delta\Phi + \Delta k.z) = 2 B \gamma \cos(\Delta\Phi + \Delta k.z) \end{cases} \quad (11.12)$$

where γ is a purely nonlinear optical parameter describing the relative weight between one- and two-photon absorption contributions. A ratio $\gamma = 1$ corresponds to equal one- and two-photon excitation probabilities. Owing to the absorption of light at 2ω, γ varies with the propagation inside the sample

$$\gamma = \gamma_0 \cdot \exp\left(-\frac{\alpha}{2} \cdot z\right) \propto \left\| \frac{E_{2\omega}}{E_\omega} \right\| \tag{11.13}$$

where α is the absorption coefficient of the sample. Under slow diffusion conditions, optimum intensity ratio is achieved for equal one- and two-photon absorptions: $\gamma = 1$. If both the relative phase and relative intensity ratio of the seeding beams are optimized ($\gamma = 1$), the maximal photoinduced susceptibility $\chi^{(2)}$ can be achieved.

11.2.3.3.2 Dark Relaxation Effect

The dynamics of the growth and decay of the induced polar order that was given in Figure 11.9 shows that after, when the seeding process is stopped, the decay is initially very fast and occurs within a few seconds after interruption of the preparation process, and after it becomes much slower, with a multiexponential nature. Figure 11.17 presents the evolution of the light-induced $\chi^{(2)}$ and of the light-induced birefringence Δn obtained after stopping the seeding process. Fit of the slow-decay curve (with a monoexponential law) reveals a ratio close to three between the lifetime τ of the light-induced susceptibility $\chi^{(2)}$ and the lifetime $\tau_{\Delta n}$ of the induced birefringence, $\tau \approx 3\tau_{\Delta n}$.

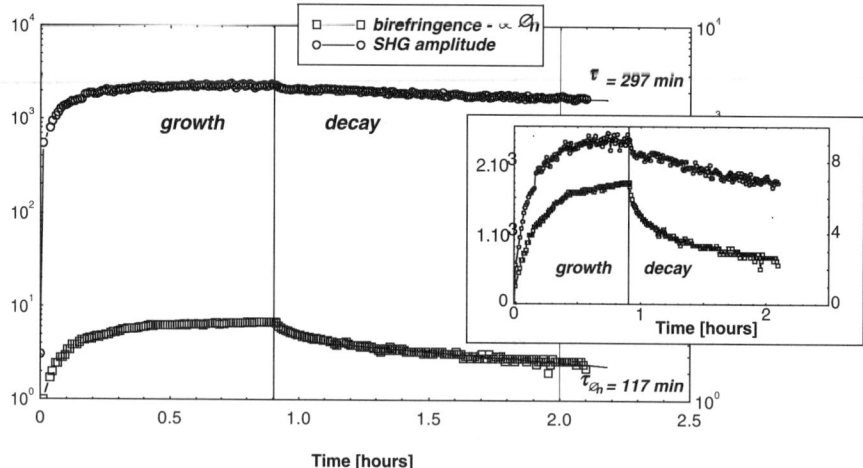

FIG. 11.17 Temporal growth and decay of the second-order susceptibility $\chi^{(2)}$ and birefringence Δn in a semi-logarithmic scale.[4] (In the insert a linear scale is used). The sample is a 0.1 μm thick film of the DR1-MMA 35/65 copolymer, "seeded" with optimized preparation conditions. After stopping the seeding process, SH generation is probed periodically by the reading beam at ω frequency. The laser is switched off between every measurement, and the break period between each measurement lasts 30 seconds.

It shows that at the origin of the observed relaxation there must be an orientational diffusion of the molecules inside the matrix. Indeed, although it is usually applied only to free rotators, the diffusion equation[26] $\frac{dN(\Omega)}{dt} = D\nabla^2 N(\Omega)$, where D is the diffusion constant of the material, correctly describes the essential relaxation features of the quasi-permanent photoinduced polar distribution $N(\Omega)$.

For linear parallel polarizations of the writing beams at ω and 2ω frequencies and uniaxial (rod-like) molecules, the distribution $N(\Omega)$ depends only on the polar angle θ. So the eigensolutions of the diffusion equation are the Legendre polynomials, with eigenvalues $-j(j+1)$. Therefore, the different order parameters A_j have different times of relaxation, τ_j: $\tau_j = \frac{1}{D_j} = \frac{1}{j(j+1)D}$. Considering only the molecular diffusion process, the light-induced susceptibility $\chi^{(2)}$ and the birefringence Δn are written as:[21]

$$\chi^{(2)}_{xxx} \propto \frac{3}{5} A_1(t) + \frac{2}{5} A_3(t) \propto \frac{3}{5} A_1 \exp(-2Dt) + \frac{2}{5} A_3 \exp(-12Dt) \quad (11.14a)$$

$$\Delta n = n_x - n_y \propto \chi^{(1)}_{xx} - \chi^{(1)}_{yy} \propto = \int_\Omega N(\Omega)(\cos^2\theta - \sin^2\theta \sin^2 \varphi)d\Omega \propto A_2(t) \propto A_2 \exp(-6Dt) \quad (11.14b)$$

The experimental relationship: $\tau \approx 3 \, \tau_{\Delta n}$ shows that at the experimental time scale (a few minutes after stopping the seeding process), the A_3 contribution coming from the nondipolar part of the writing field $<E^3>_\tau$ can be neglected. The experimental ratio is a little less than three between $\tau \approx 1/2D$ and $\tau_{\Delta n} \approx 1/6D$, due to the multiexponential decay behavior. Moreover, the free-rotator diffusion equation gives us only a restricted insight into the molecular relaxation process, because the rotational movements are inhibited in grafted systems.

The photoinduced susceptibility shown in Equation 11.14a is the sum of two terms: one with $\exp(-2Dt)$ (relaxation of the first-order parameter A_1) decay and the second with $\exp(-12Dt)$ (relaxation of the third-order parameter A_3) decay. Hence, the first very rapid decay may contain the fast $\exp(-12Dt)$ contribution. However, as can be seen from Figure 11.14, the relative magnitude of this initial very fast decay does not depend on the optimization of the intensity ratio between the writing beams. So, this first rapid decay may not be due to the decay of the third-order parameter A_3. In addition, because the hyperpolarizability β of DR1 is different in the *cis* and in the *trans* state,[27] the first very rapid decay also contains a contribution connected with the lifetime of the metastable *cis* form, which is due to molecules coming back to the *trans* form without any net orientation.[27,28] A better model would have to account for a distribution of diffusion constants for molecules embedded with various free volumes, which may explain the multiexponential behavior of the decay.[29]

11.2.3.3.3 Photostimulated Relaxation Effect

When either of the two writing beams illuminates the material after the seeding process is stopped (Figures 11.18 and 11.19), the relaxation dynamics of the induced susceptibility $\chi^{(2)}$ is much faster than for the dark relaxation dynamics case. Then again, as shown in Figure 11.19, no relaxation of the

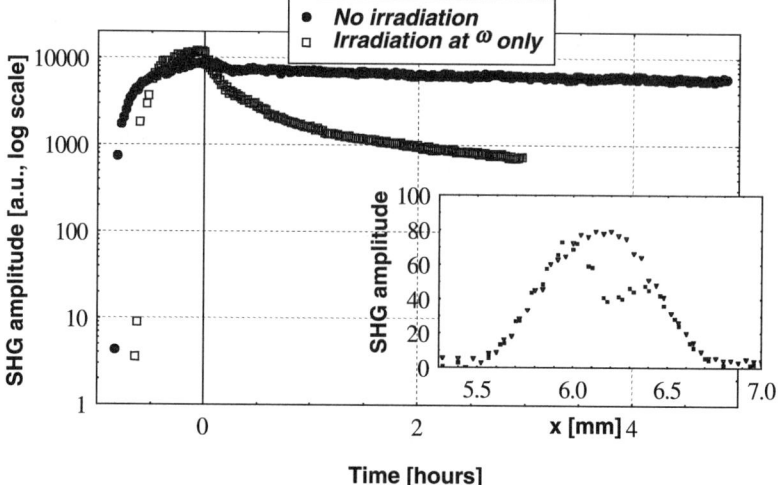

FIG. 11.18 Erasure of the photoinduced polar orientation with monochromatic irradiation at ω frequency.[4] Negative times correspond to the seeding process. Circles and squares correspond, respectively, to the dark and to the photostimulated decays. The insert shows the spatial profile of the photoinduced $\chi^{(2)}$ susceptibility, obtained by SH generation inside the sample with a beam at fundamental frequency strongly attenuated before focusing. The spatial profile was measured before (triangles) and after irradiation (squares) of the central prepared area with an intense and focused beam at ω frequency (about 10^3 more intense than the beam used for the $\chi^{(2)}$ reading).

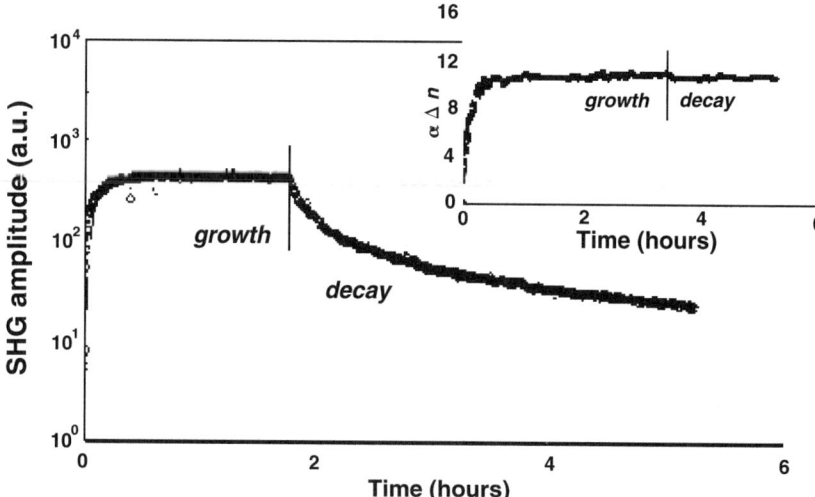

FIG. 11.19 Erasure of the photoinduced polar orientation with a monochromatic light at 2ω frequency.[4] The sample was a 0.1 μm thick film of the DR1-MMA 35/65 copolymer. After seeding, the sample was continuously irradiated with a 2ω beam (fluence used was a few MW/cm², which corresponds to the 2ω seeding beam intensity). The curve in the insert presents the evolution of the photoinduced birefringence in the same conditions.

induced birefringence for a 2ω single-beam irradiation is observed. Thus, irradiation of a previously seeded area with a monochromatic beam at ω or 2ω frequency erases only the induced polar orientation.[28,30] This erasure process can be easily understood using the excitation probability of the molecules (Equation 11.8). Indeed, the first and second terms of this equation show that under monochromatic irradiation, the molecules are again excited and reoriented. It is only an axial excitation, so the previously induced polar order is destroyed. Persistence of the birefringence during monochromatic illumination shows that the birefringence related to the seeding process (growth period) is basically due to the axial terms of the excitation probability.

Erasure mechanisms occurring under an intense monochromatic irradiation are the main source limiting the practical application of such systems. Yet, experiments performed on sol-gel systems and cross-linkable polymers[31] provided an attractive route to improve the photostability of the photoinduced polar order. In all optical poling of ionic polymers, the electrostatic interactions reveal an interesting way of stabilizing the induced orientational order.[32] Another solution could be to reach the conditions for self-preparation of the optically induced orientation. But to exploit the real harmonic conversion potentials of such a phenomenon, it is essential to determine the phase difference between the SH generated signal and the 2ω seeding beam. Any phase difference may lead to an overpainting of the interference pattern, finally leading to a limited phase matched length in the sample and automatically lower SHG efficiency.

11.2.4 Optimization of the Poling Efficiency

11.2.4.1 Fringe Contrast

As was mentioned before, due to the holographic character of this multifrequency field process, for thin samples, the relative phase $\Delta\Phi$ between the ω and 2ω writing fields strongly influences the efficiency of the light-induced orientation phenomenon. As shown in Figure 11.10, for a certain seeding time, the SHG signal reveals a sinusoidal variation with a period of $180°$ in function of the relative phase difference between the seeding beams. It is a proof that the induced $\chi^{(2)}$ is modulated in space exactly with the period necessary for phase-matched frequency doubling.[2]

As expected from a cubic interferences process, the initial growth of the induced $\chi^{(2)}$ is proportional to $|E_{2\omega}^{*} E_{2\omega}|$ (Figure 11.13). As shown in Figure 11.15, the value of the induced $\chi^{(2)}$ reached at saturation strongly depends on the relative intensities of the seeding waves at ω and 2ω frequencies. This fact can be easily understood by considering the physical origin of the recording of $\chi^{(2)}$. The excitation probability at the origin of the orientational hole burning in the isotropic distribution of molecules is the sum of three terms: one related to one-photon absorption, the second one related to two-photon absorption, and the third one, an interference term (Equation 11.11). Only the last term is polar and permits centrosymmetry breaking inside the sample. Its relative weight determines the contrast of the $\chi^{(2)}$-grating, so in other words, the efficiency of the SHG process. Locally, optimization of the contrast of interference fringes requires equal excitation probabilities through one- and two-photon absorptions. This condition depends only on $\Delta\mu$—the difference between the dipolar moments of the molecule in its fundamental and excited state: $|E_{2\omega}/E_\omega^2|$ = $\Delta\mu/2\hbar\omega$. For DR1, it gives an optimal ratio equal to $|E_{2\omega}/E_\omega^2| \approx 0.9 \times 10^{-10}$, which is consistent with the experimental value deduced from Figure 11.15.[33]

11.2.4.2 Phase Matching

In the general case of second harmonic generation, it can be shown that the conversion efficiency $I_{2\omega}/I_\omega$ is strongly dependent on the phase synchronization factor $sinc^2(\Delta k \; l/2)$ where $\Delta k = k_{2\omega} - 2k_\omega$ and l are, respectively, the phase mismatch vector between the beams at fundamental and SH frequencies and the thickness of the sample. Different techniques have been developed to optimize this term.

11.2.4.2.1 Indices

Phase matching is most often achieved using birefringence: $n_{2\omega} = n_\omega$. Organic materials are well suited in this respect, due to their large anisotropy and strong dispersion of the refractive indices.[34] Phase matching can be achieved by angle tuning or by temperature tuning.

11.2.4.2.2 Quasi-Phase Matching

One of the most interesting prospects with all optical poling is that it enables a spatially modulated self-organization of the molecules over the propagation length. Indeed, as illustrated in Figure 11.20, the polarity of the optical field E resulting from the coherent superposition of the two fields at fundamental and second harmonic frequencies depends strongly on the relative phase between these two fields. Because of the material dispersion (different refractive indices for fundamental and second harmonic beam), it results in a polarity oscillating between positive and negative extremes. For normal incidence on the sample (copropagating configuration) of both the writing fields at fundamental and second harmonic frequencies, we get exactly the period for phase-matched frequency doubling.

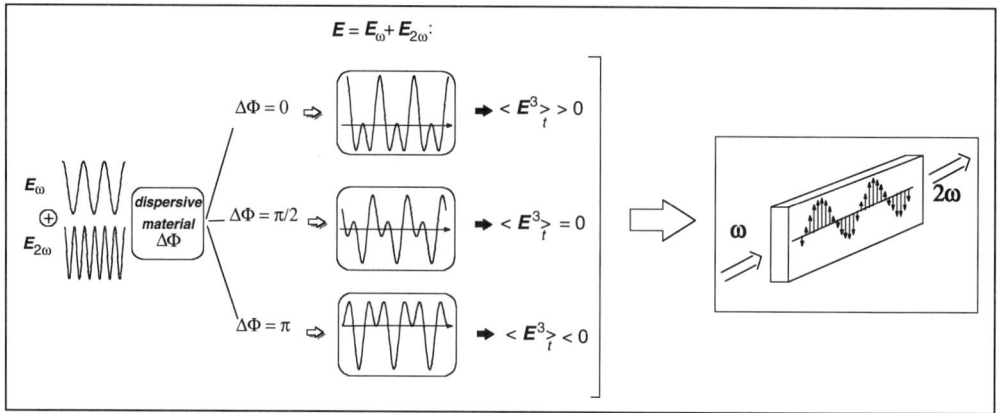

FIG. 11.20 Periodic self-organization for perfectly phase-matched second harmonic generation.[10]

The modulation periodicity Λ of the molecular order can be easily changed by varying the writing fields wave-vector mismatch, so that the all optical orientation processes can be used more generally for the design and the realization of devices for parametric amplification and frequency conversion: $\omega_P = \omega_S + \omega_I$. Two configurations are possible, as schematically presented in Figure 11.21. The waveguide configuration is particularly interesting, because it permits us to maintain high-power densities over large propagation lengths, whereas the frequency conversion efficiency depends quadratically on both the pump power density and the propagation length.

Efficient SHG requires phase matching, so that the harmonic fields generated in different areas along the material interfere constructively at the output. One of the ways of obtaining phase matching is quasi-phase matching (QPM). The main idea of quasi-phase matching is to modulate periodically the $\chi^{(2)}$ where the pertinent grating vector compensates for the phase mismatch between the propagating beams. Though the basis of this effect has been known since 1962,[35] the technical difficulties of producing quasi-phase matched structures in inorganic as well as organic materials have been overcome only recently.[36,37] It is an important technique for nonlinear frequency conversion, because it allows for efficient operation at any specified wavelength and requires only creation of an appropriate modulation period of the second-order nonlinearities for a specific wavelength. (Inaccurate modulation period results in significant mismatch and therefore smaller frequency conversion efficiency). In a QPM device, the phase mismatch between the fundamental wave ($k = k_\omega$) and its second harmonic ($k = k_{2\omega}$), $\Delta k = k_{2\omega} - 2k_\omega$, can be eliminated by modulating the second-order nonlinearities (precisely speaking the $\chi^{(2)}$ susceptibility) at a period that is the coherence length. The coherence length is

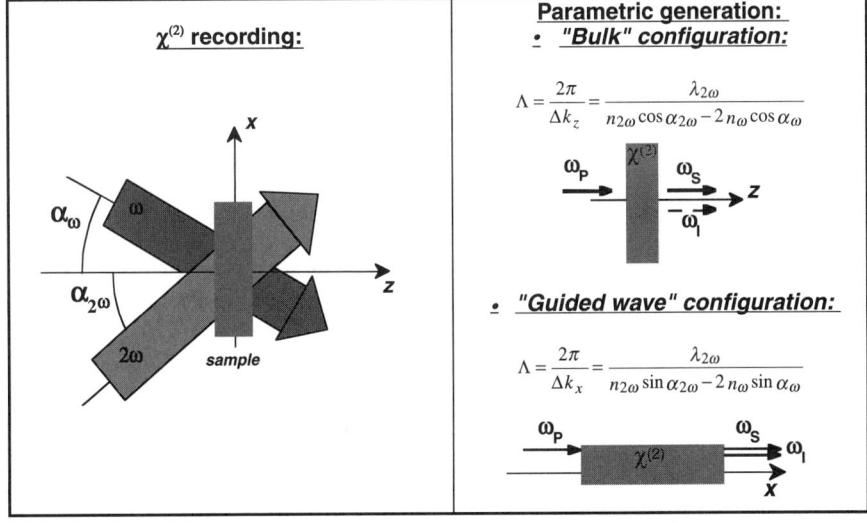

FIG. 11.21 Tailoring of the modulation period of the photoinduced order.[10]

generally a few microns in the case of organics. The propagating fundamental beam drives a nonlinear polarization wave inside the medium, which lies at the origin of the second harmonic lightwave. As the medium is a dispersive one, a phase shift between the two waves occurs. This phase shift is a function of the length over which the waves interact with each other. When the fundamental beam propagates through the first half of l_c, the SHG intensity is built up and reaches its maximal value. After the propagation over the second half of the coherence length, the SHG power decreases to zero as a result of back energy flow. Inversion of the sign of the second-order nonlinear coefficient after the first half-l_c region causes the phase of the propagating polarization wave to be shifted by 180°, which results in constructive interference over the next half-l_c region. Placing many domains with alternated sign of the second-order nonlinear optical susceptibility $\chi^{(2)}$ gives a buildup of the second harmonic generated signal intensity. The energy flow between the first and the second harmonic wave occurs as their relative phase drifts due to the difference between their phase velocities. Experimentally, QPM structures can be achieved by periodic electrode poling, which implies quite a complicated electrode patterning. In comparison with standard corona poling, photo-assisted poling offers interesting perspectives for this purpose because only the irradiated area is poled, which permits the use of a single pair of electrodes.[12,38-40] Another optional approach consists of heating the polymer locally by means of a focused laser beams, while the electric field is also applied to a single pair of electrodes.[41]

11.2.4.2.3 Non-colinear Optical Poling

Recently a new technique for avoiding the influence of the relative phase difference on the SHG intensity has been demonstrated. The samples used were thin polymer films of the DR1-MMA 35/65 copolymer (as described in section 11.2.2). Their thickness was determined with a Dektak profilermeter to be about 0.6 μm. The refractive index difference for the fundamental wave at 1.06 μm and the second harmonic wave at 532 nm is equal to $\Delta n = n_\omega - n_{2\omega} = 0.3$ and the coherence length $l_c = \lambda_{2\omega}/(n_\omega - n_{2\omega}) = 1.7$ μm. The absorption spectrum of the DR1-MMA 35/65 is presented in Figure 11.22.

The experimental setup consisted of a Q-switched mode locked Nd:YAG nanosecond pulsed laser delivering a fundamental (1.06 μm) beam at the repetition rate of 10 Hz. An external-cavity KDP crystal (type 2) permitted partial frequency doubling of the fundamental beam. The resulting beam was a superposition of the generated SH and the residual fundamental beam. To obtain the maximal second harmonic generation efficiency, the beam intensity ratio ($I_\omega/I_{2\omega}$) was optimized and energies per pulse were equal to 20.2 mJ and 12.9 μJ, respectively at 1.06 μm and 532 nm. The fundamental and second harmonic beams were slightly focused into 2 mm diameter spot at the sample location. A Glan polarizer was used to ensure that the beams incident on the sample were vertically polarized. The experiment consisted of a seeding process with alternate writing and probing periods. During the seeding process, the two coherent copropagating beams at frequencies ω and 2ω simultaneously irradiated the sample. The light-induced noncentrosymmetry was investigated by generation of the second harmonic inside the sample. To stop the 2ω beam

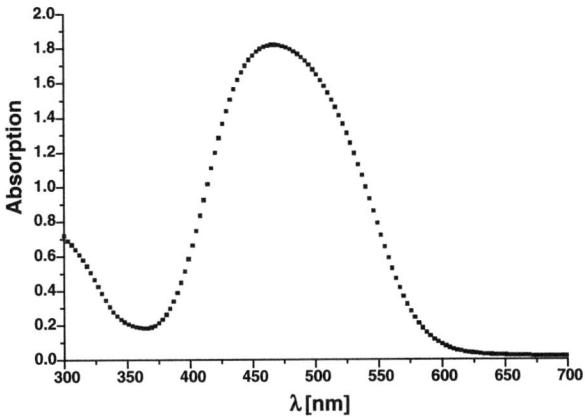

FIG. 11.22 Absorption spectrum of the DR1-MMA 35/65 spin-coated polymer thin film. The sample thickness was 0.6 μm and the optical density at 532 nm was 1.17.

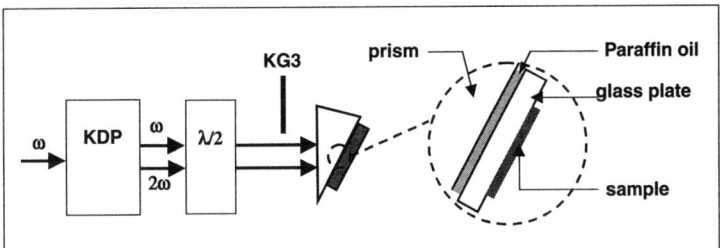

FIG. 11.23 Experimental setup for the preparation of periodically poled thin films.

during the $\chi^{(2)}$ readout phase, a RG 630 Schott red filter was used, inserted in front of the sample, leaving only the fundamental beam. The SH signal was detected by a photomultiplier tube (PM). The generated signal was observed by an oscilloscope. A shutter set in front of the PM was kept closed during the seeding period, so that the PM would be saved from potential damage. The freestanding sample was put between two lenses set in an afocal system, so as to be able to vary the intensity incident onto the sample by translating the sample holder. To investigate the relative phase influence on the SH generated signal in different seeding conditions, the sample was placed on an isosceles shaped glass prism with acute angle equal to 22.1°. The original configuration is schematically presented in Figure 11.23.

The seeding phase was performed for about 30 minutes to reach saturation of the SH signal. After that, the 2ω beam was cut off by the red filter, and the readout process was carried out, with only the ω beam irradiating the sample. In case of the sample placed on a glass prism, the irradiating copropagating ω and 2ω beams were perpendicular to the front side of the prism. No variation of the signal as a function of the phase change could be observed, in contrast to the case of the freestanding sample. The scheme of the beams' arrangement at the prism output is shown in Figure 11.24.

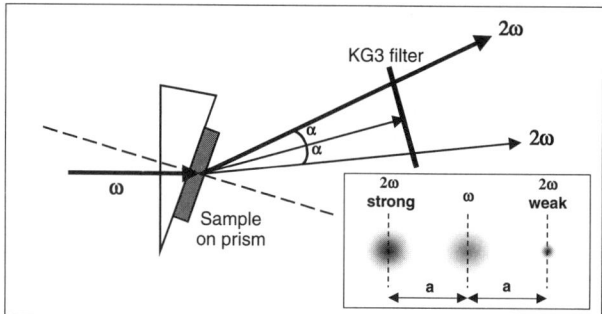

FIG. 11.24 Beams arrangement at the prism output. The two observed green signals are placed symmetrically aside the infrared signal.

During the readout period of the photoinduced $\chi^{(2)}$ (when only the beam at the fundamental frequency irradiates the sample), we observed two green signals at the prism output, the strong one, refracted in the direction of the 2ω writing beam, and the second one, diffracted and weaker. These two beams are placed symmetrically on both sides of the fundamental beam at the output of the prism. (The angles between the axis appointed by the fundamental beam and both observed signals are equal.) This configuration is shown in the insert in Figure 11.24. The stronger signal observed during the readout period is phase matched transversally and longitudinally, the second one, weaker—is only transversally phase matched. The presence of the weaker signal permits us to perform the real-time monitoring of the all optical poling. In Figure 11.25, the dynamics of the all optical poling process is presented. The SHG signal is monitored in two ways—in seeding-readout sequences, when during reading of the induced $\chi^{(2)}$ the green beam is cut off and in the real-time regime.

After 30 minutes of seeding, the DR1-MMA 35/65 sample was taken off from the prism, and we tried to measure the diffracted light from a He-Ne red laser (632.8 nm, Spectra Physics) with normal illumination onto the sample. No diffraction was observed, which can be explained by the fact that there is no refractive index grating as there is no difference in the refractive index between the domains where the azo molecules are oriented in the "positive" or "negative" sense. In the case of the diffraction of the Nd:YAG laser observed also after 30 minutes of seeding were performed on a sample placed on a prism, the sample was taken from the prism and irradiated with the beam at the fundamental frequency (1.06 µm). We observed two weak diffracted beams at 532 nm wavelength, placed symmetrically around the central fundamental beam.

As all optical poling and second harmonic generation in glasses was performed by some research groups,[25,42] we made the seeding of the prism 2 hours to check if there were any contributions to the observed SH signal originating from the prism. No SH signal was observed, so we conclude that the two observed 2ω beams in the experiment with the sample on a prism are generated by the polymer film only.

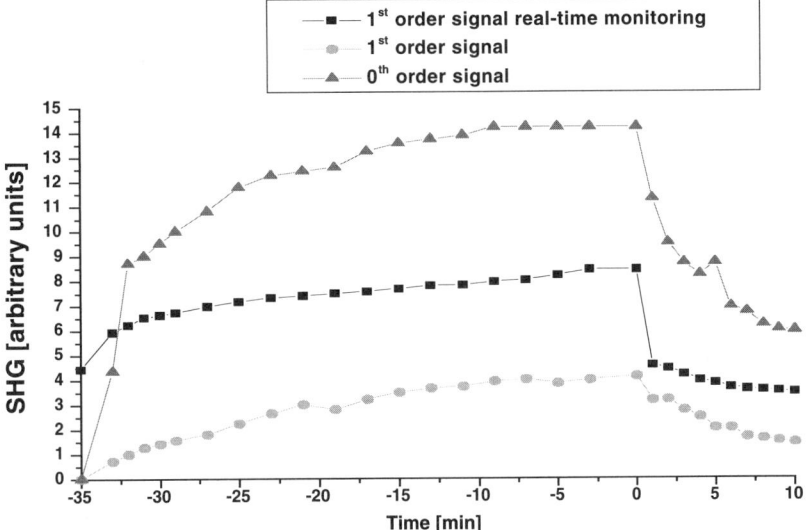

FIG. 11.25 Dynamics of growth and decay of the different SH signals at the prism output. Negative times correspond to the seeding phase. At time zero, 2ω seeding beam is stopped. Positive times correspond to the decay of the induced $\chi^{(2)}$ grating. Circles and triangles represent the growth and decay of the weak and strong SHG signal, respectively, and the measurement was made in a classical way, i.e., with cutting the green seeding beam during readout. Squares represent the real-time monitoring of the seeding process: i.e., the nonperturbative measurement of the intensity of the weak signal.[48]

In the direction parallel to the plane of the sample we obtain a modulated, quasi-phase matched structure with domain inversion. The periodicity of this arrangement depends on the wavelengths of the writing beams and on the acute angle of the prism. So changing the angle of the prism and the seeding beam wavelengths, wave-guided structures with frequency conversion for any wavelength may be obtained. This structure is schematically presented in Figure 11.26.

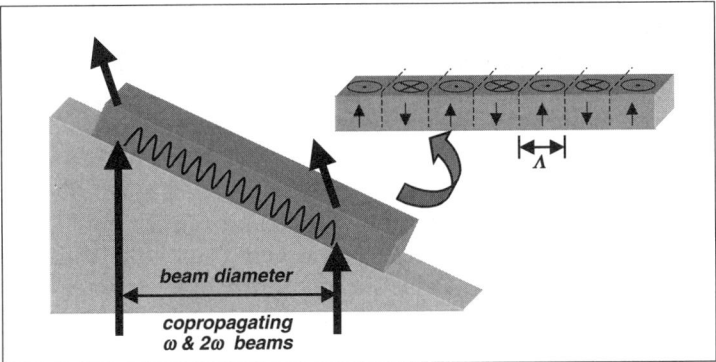

FIG. 11.26 Schematic presentation of the polarized structure of the polymer thin film. The curve shows the amplitude of the induced polarization in the film. Λ stands for the period of the induced grating. The refractive index of the zones polarized either upward or downward is the same. Scheme is not to the scale.[48]

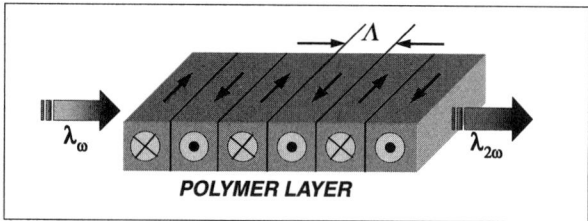

FIG. 11.27 Potential application of the induced quasi-phase matched structure as a polymer waveguide for frequency conversion.[48]

11.2.4.3 Applications

Nonlinear optical polymer films with second harmonic generation and waveguide properties are promising materials because of the ease of fabrication of thin films and their potentially large NLO susceptibilities $\chi^{(2)}$. A polymer film with a periodic reversal of the sign of the $\chi^{(2)}$ domains provides strong beam confinement over long propagation distances and gives the possibility of efficient SHG (Figure 11.27). It is a step toward a cheap, easily engineered optical device that could be designed to provide broadband light at any desired wavelength especially in the infrared transparency region, as quasi-phase matched second harmonic generation allows for any arbitrary frequency to be doubled. In addition to the wide number of potential applications of organic materials, the all optical poling technique provides vast possibilities for designing polar microstructures for electronics and optoelectronics.

Charra et al.[43] first suggested that the all optical poling technique was appropriate for optical data storage. Si et al.[44] showed that it was possible to store an image in a thin film of a thermally cross-linked azo-polyurethane containing Disperse Red 19. They achieved a resolution better than 60 μm and the NLO coefficient d_{33} induced by all optical poling was equal to 12.7 pm/V. The image storage turned out to be reversible and to have a high spatial resolution. The thermal cross-linking of the azo-polyurethane improved the optical storage stability. A second harmonic beam with larger power could erase the image.

Another experiment of optical data storage in an all optically poled film was also carried out in polyimide films with large glass transition temperature by Si et al.[45] They performed thermally assisted optical poling, which enhanced the light-induced d_{33} coefficient (from 2.6 pm/V obtained through all optical poling at room temperature to 6.9 pm/V through all optical poling at the optimal elevated temperature) and improved the stability of the induced polar order. To erase the stored image, the polymer film had to be heated above its T_g. The proposed technique of optical data storage does not require any processing before or after storage.

11.2.5 Perspectives

In comparison to other poling methods, all optical poling has several advantages: It does not require the use of any electrodes, and the phase matching condition for SHG is automatically fulfilled. Additionally, the coupling of two coherent beams at fundamental and second harmonic frequencies can result in a self-organized multipolar orientation with a periodic structure. Such a photoinduced multipolar organization process is based on two essential properties of the time-averaged tensorial product $<E^3>_t$. First, $<E^3>_t$ has a nonzero cubic time average, and second, it can be decomposed in irreducible components of dipolar and multipolar orders. Thus, coupling of the $<E^3>$ field tensor with a noncentrosymmetric molecule is permitted even in the absence of a permanent dipole moment (e.g., as it is in the case of molecules with octupolar symmetry.[3]) Interestingly, the photoinduced SH signal dependence with the beams polarizations can give a direct insight into the molecular geometry. It is possible to control the macroscopic $\chi^{(2)}$ symmetry from a dipolar to an octupolar configuration only by the ellipsometric adjustment of the writing beams.[46,47]

ACKNOWLEDGMENTS

We are indebted to Dr Paul Raimond for the synthesis of the compounds used in this study. We thank the Direction of International Relations of the Center of Nuclear Studies CEA in Saclay for enabling the visit of Aleksandra Apostoluk in the Center.

REFERENCES

1. Österberg, U., and Margulis W. (1986). Dye laser pumped by Nd:YAG laser pulses frequency doubled in a glass optical fiber. *Opt. Lett.* **11**, 516–518.
2. Fiorini, C., Charra, F., Nunzi, J.-M., and Raimond, P. (1995). Photoinduced non centrosymmetry in azo-dye polymers. *Nonlinear Optics* **9**, 339–347.
3. Fiorini, C., Charra, F., Nunzi, J.-M., Samuel, I. D. W., and Zyss, J. (1995). Light-induced second-harmonic generation in an octupolar dye. *Opt. Lett.* **20**, 2469–2471.
4. Fiorini, C., Charra, F., Nunzi, J.-M., and Raimond, P. (1997). Quasi-permament all-optical encoding of noncentrosymmetry in azo-dye polymers. *J. Opt. Soc. Am. B* **14**, 1984–2003.
5. Fiorini, C., Charra, F., Etilé, A.-C., Raimond, P., and Nunzi, J.-M. (2000). Photoinduced non-centrosymmetry in azo-polymers using dual frequency interferences. In *Advanced functional molecules and polymers* (H. S. Nalwa, ed.), pp. 167–216, Gordon and Breach Science Publishers, Amsterdam.
6. Etilé, A.-C. (1999). Étude des potentialités de divers mécanismes d'orientation tout optique en milieu polymère. *PhD thesis*, University of Paris-Sud XI, Paris Orsay.
7. Sui, Y., Yu, X., Yin, J., Zhong, X., Li, Q., Chen, Y., Zhu, Z., and Wang, Z. (2001). All-optical poling of a side-chain poly(urethane-imide) film and surface morphology studies. *Opt. Comm.* **191**, 439–444.
8. Todorov, T., Nikolova, L., and Tomova, N. (1984). Polarization holography: a new high-efficiency organic material with reversible photoinduced birefringence. *Appl. Opt.* **23**, 4309–4312.
9. Rochon, P., Gosselin, J., Natansohn, A., and Xie, S. (1992). Optically induced and erased birefringence and dichroism in azo-aromatic polymers. *Appl. Phys. Lett.* **60**, 1, 4–5.
10. Fiorini, C., Nunzi, J.-M., Charra, F., and Raimond, P. (1997). All-optical patterning of 3D microstructures in azo polymers: Towards a full control of the molecular order. In: *Photosensitive Optical Materials and Devices*, Mark P. Andrews, ed. Proceedings SPIE, vol 2998, 286–294.
11. Charra, F., Kajzar, F., Nunzi, J.-M., Raimond, P., and Idiart, E. (1993). Light-induced second-harmonic generation in azo-dye polymers. *Opt. Lett.* **18**, 941–943.

12. Sekkat, Z., and Dumont, M. (1992). Photoassisted poling of azo dye doped polymeric films at room temperature. *Appl. Phys. B* **54**, 486–489.

13. Yu, X., Zhong, X., Li, Q., Luo, S. and Chen, Y., Sui, Y., and Yin, J. (2001). Method of improving optical poling efficiency in polymer films. *Opt. Lett.* **26**, 220–222.

14. Baranova, N. B., and Zeldovich, B. Ya. (1987). Extension of holography to multifrequency fields. *JETP Lett.* **45**, 562–565.

15. Yeh, P. (1993). *Introduction to Photorefractive Nonlinear Optics*, Chap. 12, pp. 377, Wiley, New York.

16. Fiorini, C., Charra, F., Nunzi, J.-M., and Raimond, P. (1994). Six-wave mixing probe of light-induced second harmonic generation: example of dye solutions. *J. Opt. Soc. Am. B* **11**, 2347–2358.

17. Pfaender, H. G. (1996) *Schott guide to glass*. 2nd ed. Chapman and Hall, London.

18. Delysse, S., Raimond, P., and Nunzi, J. M. (1997). Two–photon absorption in non-centrosymmetric dyes. *Chem. Phys.* **219**, 341–351.

19. Broussoux, D., Chastaing, E., Esselin, S., Le Barny, P., Robin, P., Bourbin, Y., Pocholle, J. P., and Raffy, J. (1989). Organic materials for nonlinear optics. In: *Revue Technique Thomson-CSF*. Gauthier-Villars, Paris, pp. 151–190.

20. Philip, H. R. (1985). Silicon dioxide type α. In *Handbook of optical constants of solids*, Academic Press, p. 179.

21. Singer, K. D., Kuzyk, M. G., and Sohn, J. E. (1987). Second-order nonlinear-optical processes in orientationally ordered materials: relationship between molecular and macroscopic properties. *J. Opt. Soc. Am. B* **4**, 968–976.

22. Zyss, J., and Chemla, D. S. (1987). Quadratic nonlinear optics and optimisation of the second-order nonlinear optical response of molecular crystals. In *Nonlinear Optical Properties of Organic Molecules and Crystals*. (Chemla D. S., Zyss J., eds.), Academic Press, Boston.

23. Chalupczak, W., Fiorini, C., Charra, F., Nunzi, J.-M., and Raimond, P. (1996). Efficient all-optical poling of an azo-dye copolymer using a low-power laser. *Opt. Comm.* **126**, 103–107.

24. Driscoll, T. J. and Lawandy, N. M. (1994). Optically encoded second-harmonic generation in bulk silica-based glasses. *J. Opt. Soc. Am. B* **11**, 355–371.

25. Mizrahi, V., Hibidino, Y., and Stegeman, G. (1990). Polarization study of photoinduced second harmonic generation in glass optical fibers. *Opt. Comm.* **78**, 283–288.

26. Eichler, H. J., Günter, P., and Pohl, D. W. (1986). *Laser Induced Dynamic Gratings*, Springer Series in Optical Science, Springer Werlag, Berlin, Heidelberg.

27. Loucif-Saibi, R., Nakatani, K., Delaire, J. A., Dumont, M., and Sekkat, Z. (1993). Photoisomerization and second harmonic generation in disperse red one-doped and -functionalized poly(methyl methacrylate) films. *Chem. Mater.* **5**, 229–236.

28. Rau, H. (1989), "Photoisomerization of azobenzenes," In *Photochemistry and photophysics* (J. F. Rabek, ed.), CRC Press Inc., Boca Raton, vol. 2, chap. 4.

29. Victor, J. G., and Torkelson, J. M. (1987). On measuring the distribution of local free volume in glassy polymers by photochromic and fluorescence techniques. *Macromolecules* **20**, 2241–2250.

30. Aoki, H., Ishikawa, K., Takezoe, H., and Fukuda, A. (1996). Photoinduced destruction of polar structure in dye-pendant polymers studied by second-harmonic generation. *Jpn. J. Appl. Phys.* **35**, 168–174.

31. Xu, G., Si, J., Liu, X., Yang, Q., Ye, P., Li, Z., and Shen, Y. (1998). Permanent optical poling in polyurethane via thermal crosslinking. *Opt. Comm.* **153**, 95–98.

32. Bock, H., Advincula, R. C., Aust, E. F., Käshammer, J., Meyer, W. H., Fiorini, C., Nunzi, J.-M., and Knoll, W. (1997). Proceedings, *Nonlinear Optical Materials and Applications*, Cetraro, Italy.

33. Fiorini, C. (1995). Propriétés optiques non-linéaires du second ordre induites par voie optique dans les milieux moléculaires. *PhD thesis*, University of Paris-Sud XI, Paris Orsay.

34. Bosshard, Ch. *et al.* (1995), *Organic Nonlinear Optical Materials*, Gordon and Breach Publishers, 1995, and references therein.

35. Armstrong, J. A., Bloembergen, N., Ducuing, J., and and Pershan, P. S. (1962). Interactions between light waves in a nonlinear dielectric. *Phys. Rev.* **127**, 1918–1939.

36. Sato, M., Yaguchi, H., Shoji, I., Onabe, K., Ito, R., Shiraki, Y., Nakagawa, S., and Yamada, N. (2000). Second-harmonic generation from GaP/AlP multilayers on GaP (111) substrates based on quasi-phase matching for the fundamental standing wave. *Jpn. J. Appl. Phys.* **39**, 334–336.

37. Jäger, M., Stegeman, G. I., Yilmaz, S., Wirges, W., Brinker, W., Bauer-Gogonea, S., Bauer, S., Ahlheim, M., Stähelin, M., Zysset, B., Lehr, F., Diemeer, M., and Flipse, M. C. (1998). Poling and characterization of polymer waveguides for modal dispersion phase-matched second harmonic generation. *J. Opt. Soc. Am. B* **15**, 781–788.

38. Sekkat, Z. and Dumont, M. (1992). Poling of polymer films by photoizomerisation of azo-dye chromophores. *Nonlinear Optics* **2**, 359–362.

39. Loucif-Saibi, R., Nakatani, K., Delaire, J. A., Dumont, M., and Sekkat, Z. (1993). Influence of photoisomerization of azobenzene derivatives in polymeric thin films on second harmonic generation: towards applications in molecular electronics. *Mol. Cryst. Liq. Cryst.* **235**, 533–540.

40. Dumont, M., Froc, G., and Hosotte, S. (1995). Alignment and orientation of chromophores by optical pumping. *Nonlinear Optics* **9**, 327–338.

41. Yilmaz, S., Bauer, S., and Gerhard-Multhaupt, R. (1994). Photothermal poling of nonlinear optical polymer films. *Appl. Phys. Lett.* **64**, 2770–2772.

42. Antonyuk, B. P. (2000). All optical poling of glasses. *Opt. Comm.* **181**, 191–195.

43. Charra, F., Idiart, E., Kajzar, F., Nunzi, J.-M., Pfeffer, N., and Raimond, P. (1993). Two–photon induced isomerization and memory effect in functionalized polymers. SPIE **2042**, 333–346.

44. Si, J., Kitaoka, K., Mitsuyu, T., and Hirao, K. (1999). Optical image storage based on all-optical poling in polymer films. *Jpn. J. Appl. Phys.* **38**, 390–392.

45. Si, J., Mitsuyu, T., Ye, P., Shen, Y., and Hirao, K. (1998). Optical poling and its application in optical storage of a polyimide film with high glass transition temperature. *Appl. Phys. Lett.* **72**, 762–764.

46. Etilé, A.-C., Fiorini, C., Charra, F., and Nunzi, J.-M. (1997). Phase-coherent control of the molecular polar order in polymers using dual-frequency interferences between circularly polarized beams. *Phys. Rev. A* **56**, 3888–3896.

47. Brasselet, S., and Zyss, J. (1998). Multipolar molecules and multipolar fields: probing and controlling the tensorial nature of nonlinear molecular media. *J. Opt. Soc. Am. B* **15**, 257–288.

48. Apostoluk, A., Chapron, D., Sahraoui, B., Gadret, G., Fiorini, C., Raimond, P., and Nunzi, J.-M. (2002). Novel real-time monitoring technique of the all-optical poling process. In: *Nonlinear materials: Growth, Characterization, Devices, and Applications.* (D. D. Lowenthal, and Y. Y. Kalisky, Eds.) Proceedings SPIE, vol. 4628, p. 39–45.

12

PHOTOINDUCED THIRD-ORDER NONLINEAR OPTICAL PHENOMENA IN AZO-DYE POLYMERS

VICTOR M. CHURIKOV
CHIA-CHEN HSU
Department of Physics, National Chung Cheng University, Ming Hsiung, Chiayi 621, Taiwan, R.O.C.

ABSTRACT

In this chapter we discuss photoinduced third-order nonlinear optical phenomena in azo-polymer thin films. Optical control of third-order nonlinearity in azo-dye polymers via one-photon or two-photon excitation is demonstrated. This effect is attributed to the reversible change of third-order susceptibility $\chi^{(3)}$ due to *trans* to *cis* isomerization of azo molecules. Evidence for the photoinduced change of $\chi^{(3)}$ was obtained in experiments on third harmonic generation (THG), electric field induced second harmonic generation (EFISH), and degenerate four wave mixing (DFWM). The photoinduced third-order nonlinear phenomena observed in these experiments can be qualitatively interpreted by a microscopic model, which includes mechanisms for angular hole burning and molecular angular redistribution of azo molecules.

Photoreactive Organic Thin Films
Copyright 2002, Elsevier Science (USA). All rights of reproduction in any form reserved.

12.1 INTRODUCTION

Azo-benzene molecules are widely recognized as attractive candidates for many nonlinear optical applications. A highly deformable distribution of the π-electron gives rise to very large molecular optical nonlinearities.[1] Photo-isomerization of azo molecules allows linear and nonlinear macroscopic susceptibilities to be easily modified,[2-4] giving an opportunity to optically control the nonlinear susceptibilities. In this chapter, we will discuss third-order nonlinear optical effects related to photoisomerization of azo-dye polymer optical materials.

Macroscopic polarization P of a dielectric medium can be expanded to a power series of the applied field strength

$$P = \chi^{(1)} : E + \chi^{(2)} : EE + \chi^{(3)} : EEE + \dots \qquad (12.1)$$

where the $\chi^{(n)}$ coefficients are n+1-rank tensors. The first term describes linear effects; the second and third ones are related to quadratic and cubic nonlinear phenomena, respectively. A similar equation can be written for microscopic molecular polarization p

$$p = \alpha : E + \beta : EE + \gamma : EEE + \dots \qquad (12.2)$$

where the coefficients α, β, and γ are linear molecular polarizability and first and second hyperpolarizabilities, respectively.

For symmetry reasons, the first macroscopic nonlinear coefficient $\chi^{(2)}$ is zero in unordered polymer materials. On the other hand, azo-dye polymers can exhibit very large $\chi^{(3)}$ values, which is interesting for applications in optical limiting and optical switching devices. We will consider the relationship between microscopic and macroscopic third-order susceptibilities. The most general equation for this relationship can be written as[1]

$$\chi^{(3)}_{IJKL} (\omega, \omega_1, \omega_2, \omega_3) = N f_I (\omega) f_J (\omega_1) f_K (\omega_2) f_L (\omega_3) c_{IJKL} \qquad (12.3)$$

where

$$c_{IJKL} = (1/N_g) \sum_{i,j,k,l} (\sum_{s=1}^{N_g} \cos \theta^{(s)}_{Ii} \cos \theta^{(s)}_{Jj} \cos \theta^{(s)}_{Kk} \cos \theta^{(s)}_{Ll}) \gamma_{ijkl} \qquad (12.4)$$

N is the number density of molecule, and N_g is the number of equivalent sites (in a crystal). They are related by $N = N_g/V$, where V is the unit cell volume. The angle $\theta^{(s)}_{Ii}$ expresses the relation between the crystallographic (or laboratory) axis I and molecular axis i_s for a particular constituent of the unit cell. The local field factors $f_I (\omega)$, etc., are related to the appropriate laboratory axis and frequency. For disordered materials, $N_g=1$.

Under the influence of an optical pump, the molecular angular distribution described by Equation 12.4 can be considerably modified. In turn, this results in modification of the $\chi^{(3)}_{IJKL}$ tensor components. Further, we discuss the influence of a polarized pump beam on third-order nonlinear phenomena such as third harmonic generation (THG) [(described by $\chi^{(3)}_{IJKL} (-3\omega,\omega,\omega,\omega)$ coefficient], electric field induced second harmonic generation (EFISH) [$\chi^{(3)}_{IJKL} (-2\omega, \omega, \omega, 0)$][3] and degenerate four-wave mixing (DFWM) $\chi^{(3)}_{IJKL} [-\omega,\omega, -\omega,\omega]$.[4]

12.2 THIRD HARMONIC GENERATION

12.2.1 Theoretical Model

In this section, basic equations necessary for understanding photoinduced modification of $\chi^{(3)}_{IJKL}$ tensor are given. The contributions of angular hole burning (AHB) and angular redistribution (AR) mechanisms are described.

The typical problem in studying photoinduced processes is to find the molecular angular distribution produced by the optical pump and to calculate the modified macroscopic linear or nonlinear susceptibilities. The most general theoretical model of photoinduced molecular ordering in azo-polymers is presented in References 5 and 6. Formal dynamics equations[5,6] include photoexcitation, molecular reorientation, and orientational diffusion for both *trans* and *cis* populations. The approach developed in References 2, 5, and 6 is, in general, applicable for studying third-order nonlinear phenomena. However, aside from the problems associated with many unknown parameters in the equations, the interpretation of THG experiments can be complicated by secondary effects such as THG reabsorption, self-focusing or defocusing, and phase-mismatch of interacting ω and 3ω waves. Therefore, the main goal of this section is to understand the experimental results qualitatively in order to pave the way for further more detailed investigations.

In the absence of a pumping optical field, the azo-benzene molecules are mostly in the more stable *trans*-conformation (Figure 12.1). In the *trans*-form, the molecules are highly anisotropic and can be considered as rod-like molecules with only one major polarizability component parallel to the molecular axis. This assumption has been proved to be very successful in the study of nonlinear optical properties of this type of molecules.[7] For simplicity, it is supposed that the pump is linearly polarized. If the intensity is not very high, the probability $\mathrm{Pr}(\theta_p)$ of a rod-like molecule to be excited by a linearly polarized light in unit time is proportional to $\cos^2\theta_p$, where θ_p is the angle between the pump beam polarization and the molecular axis (Figure 12.2). This selective excitation leads to the AHB in the distribution of *trans*-molecules, because after excitation, the molecules tend to relax to the *cis*-state with smaller polarizabilities. The reverse *cis-trans* photoisomerization also exists, in competition with the direct *trans-cis* one, but it can be ignored if the pump intensity is not very high and the pump frequency is on the red side of the *trans*-absorption spectrum.[2,5] Moreover, considering the reverse *cis-trans* photoisomerization complicates the calculations without improving the understanding of the physics of the phenomenon. So, a simple three-level scheme for *trans-cis* isomerization (Figure 12.3) is applied in the model. Although the thermal *cis-trans* relaxation is not exponential in polymers, for the sake of simplicity, some effective relaxation rate η_{CT}, is introduced. Furthermore, we assume the third-order susceptibility of the azo-polymer is formed by only *trans*-molecules (the contribution of the *cis*-molecules and the polymer matrix is ignored) and the molecular second hyperpolarizability of a *trans*-molecule has only one nonzero component $\gamma = \gamma_{zzzz}$, where the subscript z refers to the direction of the molecular axis. The general expression of Equation 12.3 for $\chi^{(3)}$ responsible for THG by linearly polarized light then can be written as

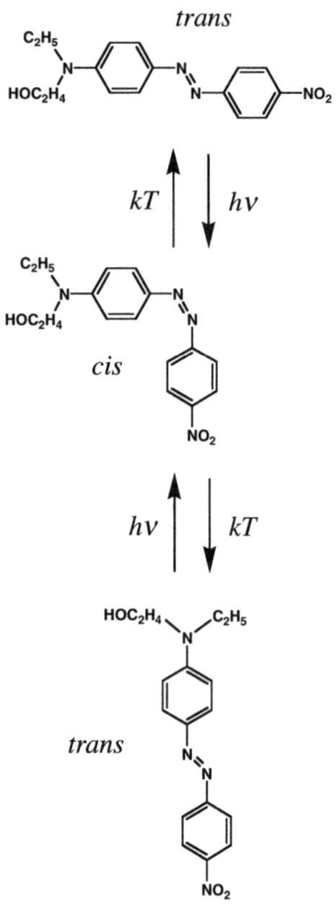

trans

C₂H₅

kT ↑ ↓ *hv*

cis

hv ↑ ↓ *kT*

trans

FIG. 12.1 Structure of DR1 molecule and picture of *trans-cis* photoisomerization and *cis-trans* thermal relaxation.

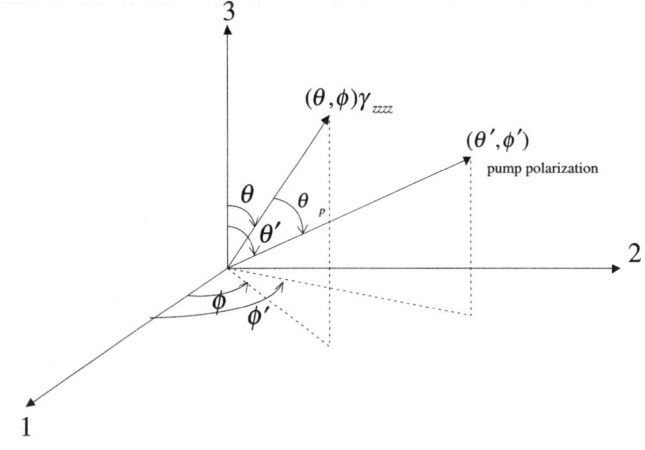

FIG. 12.2 Diagram of molecular orientation relative to the pump and probe electric fields in the case of arbitrary orientation of the pump and probe electric field vectors.

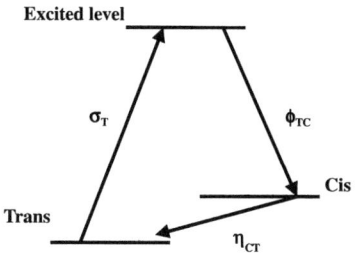

FIG. 12.3 Simplified three-level scheme for *trans-cis* photoisomerization of a DR1 molecule. Here σ_T is the absorption cross section of molecules, whose dipole moments are parallel to the pump field, ϕ_{TC} is the quantum yield for *trans* to *cis* transition, and η_{CT} is the thermal relaxation rate from *cis* to *trans*.

$$\chi^{(3)}_{3333}(-3\omega, \omega, \omega, \omega) = f^4 \gamma \int N_T(\Omega)\cos^4\theta\, d\Omega \qquad (12.5)$$

where f^4 is a collection of local field factors and $f^4 = f_\omega^3 f_{3\omega}$ in the case of THG. The subscript 3 represents the laboratory axis 3. $N_T(\Omega)$ is the number density of *trans* molecules oriented in the direction Ω, which also can be denoted as (θ,ϕ) (Figure 12.2). In the three-level model, the dynamics of the *trans*-population is described by equation

$$dN_T(\Omega)/dt = -N_T(\Omega)\phi_{TC}\Pr(\theta_p) + N_C(\Omega)\eta_{CT} \qquad (12.6)$$

where $N_C(\Omega)$ is the number density of *cis*-molecules, which relax to *trans*-molecules oriented in the direction Ω; ϕ_{TC} is the quantum yield for the *trans* to *cis* transition; η_{CT} is the thermal relaxation rate from *cis* to *trans*; and the probability of excitation $\Pr(\theta_p) = \sigma_T I_p \cos^2\theta_p$, where σ_T is the absorption cross section of *trans*-molecules parallel to the pump field, and I_p is the average pump intensity. In Equation 12.6, the population of the excited level is ignored because of very short lifetime. Assuming that there is no molecular rotation during the *trans-cis-trans* cycle, one has $N_T(\Omega)+N_C(\Omega)=N/(4\pi)$, where N is the number density of dye molecules in original isotropic distribution, and the stationary angular molecular distribution is

$$N_T(\Omega) = (N/4\pi)(1 + a^2\cos^2\theta_p)^{-1} \qquad (12.7)$$

with $a^2 = \sigma_T\phi_{TC}I_p/\eta_{CT}$. For simplicity, the molecular angular distribution is supposed to be axially symmetrical with respect to the pump polarization axis. Under the axial symmetry condition, the molecular angular distribution $N_T(\Omega)=N_T(\theta_p)$ can be expressed in terms of the order parameters:

$$N_T(\theta_p) = (N/4\pi)\sum_l (2l + 1) A_l P_l(\cos\theta_p) \qquad (12.8)$$

Here, A_l are the order parameters, and $P_l(\cos\theta_p)$ are the Legendre polynomials. Using orthogonal polynomials properties and the function in Equation 12.7 one can obtain the stationary order parameters in the AHB approximation:

$$A_l = (1/2)\int_0^\pi \frac{P_l(\cos\theta_p)\sin\theta_p\, d\theta_p}{1 + a^2\cos^2\theta_p} \qquad (12.9)$$

In Equation 12.8 only terms with A_0, A_2, and A_4 are needed to express the $\chi^{(3)}$ in a system with center of symmetry. (A_1 and A_3 are zero for the symmetry reason.)[1] Plain calculations give the following expressions for the order parameters required:

$$A_0 = \frac{\arctan a}{a}$$

$$A_2 = \frac{3}{2a^2}\left[1 - \frac{\arctan a}{a} - \frac{a \arctan a}{3}\right]$$

$$A_4 = \frac{5}{8a^2}\left[\frac{a \arctan a}{5} \times \left[\frac{35}{a^4} + \frac{30}{a^2} + 3\right] - \frac{7}{a^2} - \frac{11}{3}\right] \quad (12.10)$$

Applying the addition theorem,[8] the Legendre polynomials $P_l(\cos\theta_p)$ can be expressed in terms of products of the spherical harmonics as below

$$P_l(\cos\theta_p) = 4\pi/(2l+1) \sum_{m=-l}^{m=l} Y_{lm}^*(\theta, \phi) Y_{lm}(\theta', \phi') \quad (12.11)$$

where $Y_{lm}(\theta, \phi)$ and $Y_{lm}(\theta', \phi')$ are the spherical harmonics, and (θ', ϕ') specifies the direction of the pump polarization (Figure 12.2). Consequently, Equation 12.8 can be written as

$$N_T(\theta_p) = N_T(\theta, \phi) = N \sum_{l=0} \sum_{m=-l}^{m=l} A_l Y_{lm}(\theta, \phi) Y_{lm}^*(\theta', \phi') \quad (12.12)$$

Combining Equation 12.2 with Equations 12.5, one can obtain:

$$\chi_{3333}^{(3)} = Nf^4\gamma \int_0^{2\pi}\int_0^\pi \sum_l \sum_{m=-l}^{m=l} A_l Y_{lm}(\theta, \phi) Y_{lm}^*(\theta', \phi')\cos^4\theta \sin\theta\, d\theta\, d\phi \quad (12.13)$$

After calculations, we obtain

$$\chi_{3333}^{(3)} = Nf^4\gamma\left[\frac{A_0}{5} + \frac{4}{7}A_2 P_2(\cos(\theta')) + \frac{8}{35}A_4 P_4(\cos(\theta'))\right] \quad (12.14)$$

The $\chi^{(3)}$ value, as shown in Equation 12.14, depends on the pump beam polarization direction and intensity. If we consider the special cases for the pump to be parallel ($\theta' = 0°$) or to be perpendicular ($\theta' = 90°$) polarized with the 3-axis, the $\chi_{3333}^{(3)}$ tensor component can be written as:

$$\chi_{3333}^{(3)}(\theta' = 0°) = Nf^4\gamma\left[\frac{A_0}{5} + \frac{4}{7}A_2 + \frac{8}{35}A_4\right] \quad (12.15a)$$

$$\chi_{3333}^{(3)}(\theta' = 90°) = Nf^4\gamma\left[\frac{A_0}{5} - \frac{2}{7}A_2 + \frac{3}{35}A_4\right] \quad (12.15b)$$

Equations 12.15a and 12.15b clearly demonstrates the anisotropy of the stationary $\chi^{(3)}$ caused by the AHB effect. In the case of the $\chi_{3333}^{(3)}$, which is the only $\chi^{(3)}$ tensor component involved in the THG experiment, the probe and signal beams are all polarized along the 3-axis. Therefore, in Equations 12.15a and 12.15b, the pump polarization is either parallel ($\theta' = 0°$) or perpendicular ($\theta' = 90°$) to the probe polarization. Defining $\chi_{3333}^{(3)}(\theta' = 0°) = \chi_\parallel^{(3)}$ and $\chi_{3333}^{(3)}(\theta' = 90°) = \chi_\perp^{(3)}$ and combining Equations 12.15 and 12.10, we obtain

$$\chi_{\parallel}^{(3)} = \frac{f^4 \gamma N}{3a^2} \left[1 - \frac{3}{a^2} + \frac{3 \arctan a}{a^3} \right] \tag{12.16a}$$

$$\chi_{\perp}^{(3)} = \frac{3f^4 \gamma N}{8} \left[\arctan a \times \left[\frac{1}{a} + \frac{2}{a^3} + \frac{1}{a^5} \right] - \frac{5}{3a^2} - \frac{1}{a^4} \right] \tag{12.16b}$$

Equations 12.16a and 12.16b are the rigorous expressions of $\chi^{(3)}$ components contributed from the *trans* population in the AHB approximation. They remain valid for any pump intensity in the absence of reverse *cis-trans* photoisomerization. In practice, however, low-intensity pumping is the most interesting and important case, because the largest anisotropy in molecular distribution can be obtained far from excitation saturation, and at the low pump intensity, the photodegradation of chromophores[5] is considerably reduced.

In the low-intensity approximation, arctan a is expanded in power series and only terms linear on a^2 are kept in Equation 12.17. The $\chi^{(3)}$ components can be then written as

$$\chi_{\parallel}^{(3)} = \chi_{iso}^{(3)} + \Delta\chi_{\parallel}^{(3)} = \chi_{iso}^{(3)} - (1/7) f^4 a^2 \gamma N \tag{12.17a}$$

$$\chi_{\perp}^{(3)} = \chi_{iso}^{(3)} + \Delta\chi_{\perp}^{(3)} = \chi_{iso}^{(3)} - (1/35) f^4 a^2 \gamma N \tag{12.17b}$$

where $\chi_{iso}^{(3)} = (1/5) f^4 \gamma N$ is the original isotropic cubic susceptibility. Therefore, one can see that $\Delta\chi_{\parallel}^{(3)}/\Delta\chi_{\perp}^{(3)} = 5$. This result shows that the photoinduced anisotropy of $\chi^{(3)}$ due to the AHB mechanism is appreciably larger than that of the linear polarizability in low-intensity approximation ($\Delta\chi_{\parallel}^{(1)}/\Delta\chi_{\perp}^{(1)} = 3$).[2]

The foregoing can be applied to the case of two-photon (TP) excitation. However, there are difficulties with obtaining the general expressions for the order parameters in elementary functions in this case. For this reason, for the moment, we do not go beyond the low-intensity approximation in consideration of the TP-induced $\chi^{(3)}$ anisotropy. The TP absorption depends on the peak intensity of the pumping beam, and so the development of the distribution function $N_T(\Omega)$ gives

$$N_T(\Omega) = (N/4\pi)(1 - b^2 \cos^4 \theta_p) \tag{12.18}$$

with $b^2 = \sigma_{T2}\phi_{TC}I_p^2/\eta_{CT}T_p^2R^2$), where σ_{T2} is the TP absorption cross section, I_p is the average pump intensity, T_p is the laser pulse width, and R is the laser pulse repetition rate. The proper calculations for the TP excitation give $\Delta\chi_{\parallel}^{(3)}/\Delta\chi_{\perp}^{(3)} = 35/3$.

Another mechanism responsible for the optically induced anisotropy is angular redistribution (AR) of molecules. This mechanism has been widely developed to explain photoinduced birefringence and dichroism.[9] In most experimental cases, there is evidence of some rotation of molecules during the photoisomerization cycle (see Reference 2, for example). This rotation results in AR, because the molecules remain longer in states with lower excitation probability, and so more molecules are accumulated perpendicular to the pump polarization. The AR process is initiated by the AHB, and these two processes should be studied simultaneously in the framework of general theory.[5] However, it is difficult even for the strict theoretical model to fit the THG experiments completely because of concurrent change of the THG reabsorption, nonuniformity of pumping along the sample thickness, and possible photoinduced phase-mismatch change. So, for the moment, we give a simplified consideration of the AR mechanism, sufficient just for qualitative understanding of the experimental results.

The AR could be clearly observed at low intensities, because in this case, the *cis* population is negligible, so the stationary AHB and reabsorption change effects are considerably reduced. For simplicity, the lifetime of *cis*-state is ignored, and the *trans-cis-trans* transition is considered to be a nearly instantaneous process. Reverse *cis-trans* photoisomerization is also ignored here. In this approximation, the dynamics of *trans* angular distribution is described by the following equation:

$$dN_T(\Omega)/dt = -N_T(\Omega)\phi_{TC}\mathrm{Pr}(\theta_p) + \phi_{TC} \int\int N_T(\Omega_1)\,\mathrm{Pr}(\theta_{p1})R(\Omega_1 \to \Omega)d\Omega_1 + DV^2N_T(\Omega)$$

$$(12.19)$$

All parameters in Equations 12.19 have the same meaning as in Equation 12.6, $R(\Omega_1 \to \Omega)$ is the probability of the molecule to be reoriented from the direction Ω_1 to Ω during the photoisomerization cycle, and D is the orientational diffusion constant. To solve Equation 12.19, we use the formal approach developed in Reference 10. Although the redistribution probability function has a peak centered on the original orientation,[5] we assume for simplicity that after excitation any molecule may rotate from its original direction Ω_1 to any other direction Ω with the same probability. The orientation probability is then $R(\Omega_1 \to \Omega) = 1/4\pi$. This simplification affects the AR dynamics and order parameters value, but it does not change the final result qualitatively. Combining Equation 12.8 with Equation 12.19, one can obtain a set of coupled differential equations.[10] Keeping only A_0, A_2, and A_4 in Equation 12.8 and putting $dA_i/dt = 0$ in the stationary state, we have

$$A_0 = 1 \text{ (the \textit{trans} population does not change)}$$

$$A_2 = \frac{2}{15} \times \frac{(20\xi + \frac{39}{77})}{\frac{16}{245} - (6\xi + \frac{11}{21})(20\xi + \frac{39}{77})}$$

$$A_4 = -\frac{8}{315} \times \frac{1}{\frac{16}{245} - (6\xi + \frac{11}{21})(20\xi + \frac{39}{77})}$$

$$(12.20)$$

with $\xi = D/\sigma_T\phi_{TC}I_p$ parameter, which describes the ratio between molecular orientational diffusion and optically induced reorientation rates. Finally, calculations of $\chi^{(3)}$ due to the AR give:

$$\chi^{(3)}_{\parallel} = f^4\gamma N\left[\frac{1}{5} + \frac{4}{7}A_2 + \frac{8}{35}A_4\right] = \chi^{(3)}_{iso} + \Delta\chi^{(3)}_{\parallel} \qquad (12.21a)$$

$$\chi^{(3)}_{\perp} = f^4\gamma N\left[\frac{1}{5} - \frac{2}{7}A_2 + \frac{3}{35}A_4\right] = \chi^{(3)}_{iso} + \Delta\chi^{(3)}_{\perp} \qquad (12.21b)$$

Analysis of Equation 12.20 shows that A_4 is much smaller than A_2 at any value of the ξ parameter and can be omitted in most practical cases. Because the A_2 is negative, the AR mechanism gives a positive change of $\chi^{(3)}$ for the probe beam polarized perpendicular to the pump beam: $\Delta\chi_{\perp}^{(3)}/\Delta\chi_{\parallel}^{(3)} \approx -(1/2)$. Although this ratio is not entirely accurate because of ignoring A_4, it still gives good qualitative understanding of the AR influence on the properties of $\chi^{(3)}$ susceptibility. More accurate values of the order parameters and, therefore, of the $\chi^{(3)}$ components can be obtained by taking into account the *cis*-population and choosing the more appropriate form of the $R(\Omega_1 \to \Omega)$ function.

One can see that the AR of molecules can reduce the decrease of $\chi_{\perp}^{(3)}$ due to AHB and, at low-intensity pump, even result in the increase of $\chi_{\perp}^{(3)}$. However, in reality the situation is more complicated. If the dye molecule absorbs considerably at the frequency of THG, the AR also results in the decrease of absorbance for THG copolarized with the pump and the increase of absorbance for the THG perpendicularly polarized with the pump. Therefore, the increased absorption can reduce the expected growth of THG generated by the probe, which is cross-polarized with the pump.

12.2.2 Experiment

The 10% guest-host polymer thin films were made by dissolving DR1 together with PMMA in chloroform. The chemical structure and absorption spectrum of DR1 are shown in Figure 12.4. The experiment involved illumination of the sample by a ω or 2ω beam and probing with a low-intensity ω beam, which generated a THG signal. The THG signal temporal behavior was monitored during the pumping process. To explore the symmetry properties of third-order susceptibility, the experiments were done with parallel and orthogonal polarizations of the pumping and probing beams. Because the illumination of the DR1-PMMA thin film by near-resonant light can change

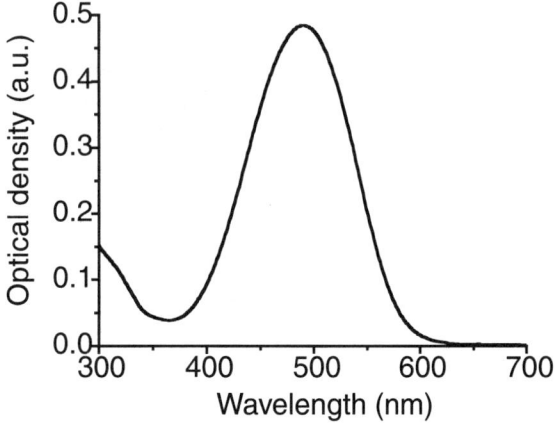

FIG. 12.4 The absorption spectrum of the DR1-PMMA thin film.

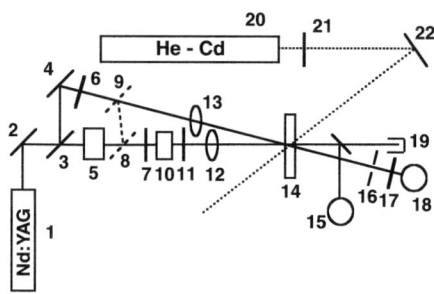

FIG. 12.5 Setup for Nd:YAG and He-Cd laser experiments: 1, Nd:YAG laser; 2, 4, 1064 nm mirrors; 3, glass plate; 5, BBO crystal; 6, 7, half-wave ($\lambda/2$) plates; 8, 9, 355 nm mirror pair; 10, polarizer; 11, color filter; 12, 13, lenses; 14, thin film sample; 15, fast photodiode; 16, iris pinhole; 17, 21, neutral density filters; 18, PMT; 19, pump damper; 20, He-Cd laser; and 22, mirror.

its absorption spectrum,[11] the evolution of transmittance of the sample at 355 nm was studied separately to retrieve its contribution to the full change of the signal at 355 nm.

The experimental setup is shown schematically in Figure 12.5.[12–13] A Q-switched Nd:YAG laser with 7 ns pulse width and 10 Hz repetition rate was used as a light source. The fundamental beam was partially frequency-doubled and tripled in a β-barium borate (BBO) crystal. A small amount of the fundamental beam was branched off by a glass plate to be used as a probe for THG and was focused into the sample by a lens. The pumping beam was focused onto the sample by another lens. The angle between pumping and probing beams was about 15°. The power of the pumping beam was changed by a half-wave ($\lambda/2$) plate put before the polarizer. The polarization of the pumping beam was horizontal in all measurements. The color filter extracted the light of desired wavelength for pumping the sample. The plane of polarization of the probing beam was rotated by a 1064 nm $\lambda/2$ plate. The peak intensity of the probing beam was about 1 GW/cm^2, the average intensities of the pump were varied in the range of 0–10^3 W/cm^2 for 1064 nm and 0–0.3 W/cm^2 for 532 nm. The probe beam was weak enough to ignore its own pumping effect. The efficiency of THG by probing beam before pumping was ~10^{-9}. The THG signal was detected by a photomultiplier tube (PMT). The PMT response was always kept in linear range by using calibrated neutral density filters. The power of the pump was monitored by a fast photodiode. The signal from the PMT was integrated by a boxcar integrator and averaged over 20–100 shots by computer software.

To clarify whether the 532 nm pulse pumping is essential to produce the $\chi^{(3)}$ anisotropy, the samples were also pumped by a continuous wave (cw) 442 nm He-Cd laser. The use of an incoherent pump also excludes polar alignment of molecules[10] and, therefore, 355 nm sum-frequency generation that can occur with coherent ω and 2ω beams.

For transmittance measurements, a 355 nm mirror pair was set as shown in Figure 12.5 in dotted lines. In this case, the PMT detected only an external THG signal (generated by the BBO crystal) transmitted through the pumped area of the sample.

12.2.3 Results

Before starting the THG experiments, the behavior of the external 355 nm light passing through the sample pumped by 532 nm or 442 nm laser was studied. Some small decrease of transmission is observed for both pump wavelengths and both polarizations of the probe (Figure 12.6). This change

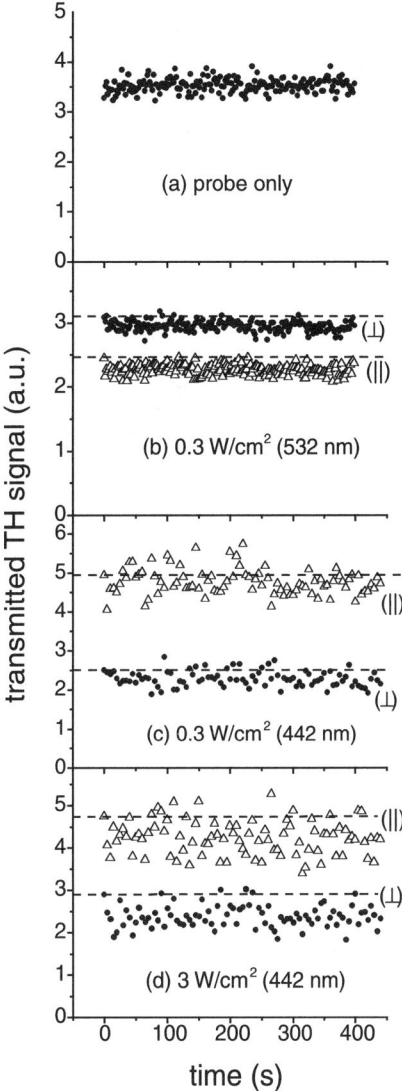

FIG. 12.6 355 nm light transmitted through the sample without pumping (a) and through the sample pumped by the SH of a Nd:YAG laser (b) or by a cw He-Cd laser (c,d). The pump and probe are copolarized (‖) and perpendicularly polarized (⊥). The average pumping intensities are indicated in each diagram. The curves for ‖- and ⊥-cases are given in arbitrary units as they were obtained and not normalized to the same initial level. In this and other figures, the first point (t=0) and dashed line represent the initial THG level obtained by averaging of 10 points before pumping.

of transmission is apparently due to the rise of the *cis* population and *trans* angular redistribution as observed in previous works.[2,6,14] However, the relatively small absorption change and laser pulse fluctuations make it difficult to draw any definite conclusions about the signal behavior on the whole. No visible change of transmission was detected at the intensities ≤ 0.1 W/cm^2.

Evolution of the THG intensity in the sample pumped by 532 nm laser light and probed by both parallel and perpendicular polarized fundamental beam is shown in Figure 12.7. As mentioned in the previous paragraph, the change of transmission at the THG wavelength in the pumped sample is relatively small, so the observed variations of the THG are mainly due to the change of third-order susceptibility $\chi^{(3)}$. The THG dynamics observed can be interpreted as a competition between the AHB and AR mechanism. The AHB is stationary after a few seconds of pumping, but the AR process takes a few minutes.[15] So, the fast drop of THG generated by a copolarized probe [proportional to $(\chi_{\parallel}^{(3)})^2$] or cross-polarized probe $((\chi_{\perp}^{(3)})^2)$] in the first few seconds

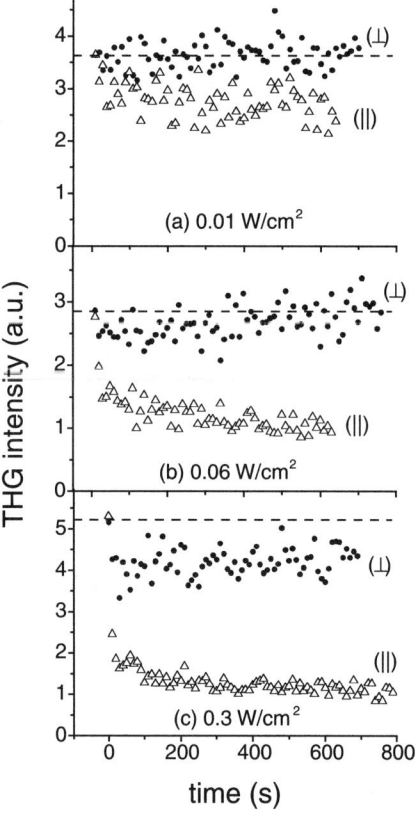

FIG. 12.7 THG intensity as a function of time in the sample pumped by a SH of Nd:YAG laser. The sample was probed by the fundamental copolarized (∥) and perpendicularly polarized (⊥) with the pumping beam. The curves for ∥- and ⊥-cases are normalized to the same initial level and represent a relative change of the THG level.

can be attributed to the AHB, whereas the slow rest of the curves is due to the AR and photodegradation processes. Analysis of Equations 12.16b and 12.21b shows, that the "AR," $\chi_{\perp}^{(3)}$, is changed very slightly at the pump intensities higher than 10 mW/cm^2, but the "AHB," $\chi_{\perp}^{(3)}$, decreases almost linearly with pump intensity up to 100 mW/cm^2. For this estimation, we have used ϕ_{TC}=0.11^2, η_{CT}~0.5s^{-1} [11], σ_T~3×10^{-17} cm^2 at 532 nm, obtained from our thickness and transmission measurements, and D~10^{-3} s^{-1}, obtained from the lifetime of photoinduced $\chi^{(2)}$ in the identical guest-host DR1-PMMA thin film.[10,16] At low pump intensities, the AHB contribution is relatively small, because only a tenuous drop of $\chi_{\parallel}^{(3)}$ is observed in the first few seconds of pumping (Figure 12.7a) and, further, the slow $\chi_{\parallel}^{(3)}$ decrease and $\chi_{\perp}^{(3)}$ growth take place. At medium and high intensities, the AHB becomes progressively more appreciable, which results in a fast decrease of $\chi_{\perp}^{(3)}$ in the beginning of pumping with a subsequent slow increase. The behavior of the THG in Figure 12.7 is similar to the dynamics of the absorption caused by the *trans* population.[2,6,14] This evidence shows that mainly *trans* molecules are responsible for the $\chi^{(3)}$ effects in DR1-PMMA.

The curves of the temporal behavior of the THG in the sample pumped by fundamental frequency beam (1064 nm) of different intensities are shown in Figure 12.8. At average intensities higher than 100W/cm^2, the decrease of the THG by copolarized probe is observed. This decrease is similar to that seen in Figure 12.7 and can be attributed to TP photoisomerization, because the one-photon absorption of DR1-PMMA is very small at 1064 nm wavelength. In Figure 12.8a, only slow decrease of $\chi_{\parallel}^{(3)}$ is clearly seen, indicating the prevalence of the AR contribution at low pump intensities. In Figure 12.8c, the fast decrease of THG [with the rate of about (5 s)$^{-1}$] in the first few seconds is mainly due to the AHB mechanism. In fact, both AHB and AR mechanisms take place at any pumping intensity, but the relative importance of each mechanism is different at high and low intensities.

The saturated change of $\chi_{\parallel}^{(3)}$ versus the pump intensity is plotted in Figure 12.9. Because the initial THG signal (before pumping) slightly varied from point to point of the sample, the normalized $\chi_{\parallel}^{(3)}$ change ($\Delta\chi_{\parallel}^{(3)}/\chi_{\parallel 0}^{(3)}$) is used. The initial THG intensity I_0 is proportional to $(\chi_{\parallel 0}^{(3)})^2$ and the reduced intensity $I=I_0-\Delta I$ is proportional to $(\chi_{\parallel}^{(3)})^2 = (\chi_{\parallel 0}^{(3)}) - \Delta\chi_{\parallel}^{(3)})^2$ with the same coefficient. Therefore, one can obtain $\Delta\chi_{\parallel}^{(3)}/\chi_{\parallel 0}^{(3)} = 1 - \sqrt{I/I_0}$. The experimental results are compared with the theoretical dependence obtained in the AHB approximation (Eq. 12.16a). The AHB model gives a good idea about the $\chi_{\parallel}^{(3)}$ change at medium pump intensities, but there are some deviations from the theoretical curve at low and high intensities. At low intensities (<10 mW/cm^2), the $\Delta\chi_{\parallel}^{(3)}$ is about two to three times larger than predicted by the AHB, probably because of the molecular AR. At the high pump intensities (>0.1 W/cm^2), the likely reason for the discrepancy between the theory and experiment is reverse *cis-trans* photoisomerization, which is ignored in our model. At high pump intensities, the reverse photoisomerization restrains the decrease of the *trans* population,[5] and consequently, the decrease of the $\chi^{(3)}$. Another possible limiting factor for the total $\Delta\chi^{(3)}$ is nonzero $\chi^{(3)}$ of the *cis* population, whose contribution increases with the *trans* population reduction.

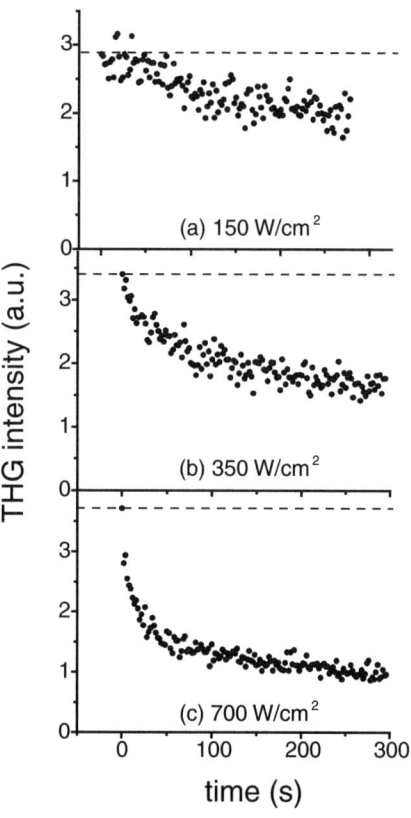

FIG. 12.8 THG in the DR1-PMMA thin film pumped by the fundamental beam as a function of time. The average intensity of the pumping beam is shown in each diagram.

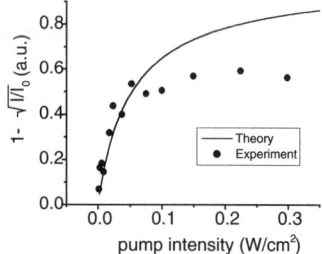

FIG. 12.9 Normalized value $\Delta\chi_{\parallel}^{(3)}/\chi_{\parallel 0}^{(3)} = 1 - \sqrt{I/I_0}$ versus the pumping intensity in the cases of excitation by 532 nm pump. The solid curve is the AHB theoretical dependence of $\Delta\chi_{\parallel}^{(3)}/\chi_{\parallel 0}^{(3)}$ obtained from Equation 12.17a with the parameters $\phi_{TC}=0.11$, $\eta_{CT}\sim0.5$ s^{-1}, $\sigma_T\sim3\times10^{-17}$ cm^2 for DR1-PMMA.

Figure 12.10 shows the dynamics of the THG in the sample pumped by a 442 nm He-Cd laser. We would like to point out some differences between THG dynamics in Figure 12.7 and Figure 12.10. First, although the pump intensities in Figure 12.7c and Figure 12.10a are the same, the growth of THG probed with the perpendicular polarized fundamental beam is much more prominent, and no significant decrease of the signal is observed in the

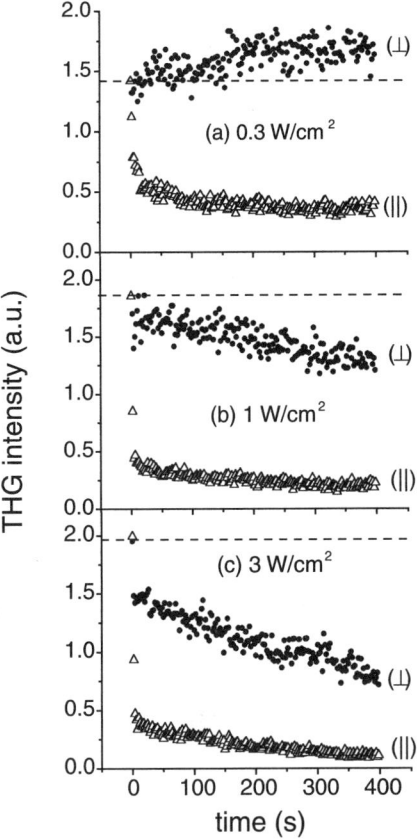

FIG. 12.10 THG intensity as a function of time in the sample pumped by a cw He-Cd laser. The sample was probed by the fundamental copolarized (∥) and perpendicularly polarized (⊥) with the pumping beam.

case of 442 nm pumping. Second, the THG decrease for the copolarized pump and probe is a little smaller in the 442 nm case. These differences are probably due to some increase of reverse *cis-trans* photoisomerization because of a higher *cis* absorption cross section at the 442 nm wavelength. The higher rate of reverse photoisomerization could result in reduction of the *cis* population and, consequently, smaller AHB effect. Another reason for more efficient AR at 442 nm could be the continuous generation regime of the He-Cd laser. Note, there is an increase of THG using a cross-polarized pump-probe arrangement at low pump intensities, whereas the external 355 nm signal drops with time with the same pump and probe configuration (compare Figure 12.6c and Figure 12.10a). This result can be explained in part by the fact that the *trans* molecules reoriented perpendicularly to the pump polarization bring additional absorption for the 355 signal cross-polarized with the pump and, at the same time, increase $\chi_{\perp}^{(3)}$ according to Eq. 12.21. The increase of absorption may also be contributed from the *cis* population, but the final level of the $\chi_{\perp}^{(3)}$ is higher than its initial level (Figure 12.10a) which shows the absorption change for cross-polarized pump and probe (Figure 12.6c) is mainly due to the AR.

At higher pump intensities, a fast decrease of both $\chi^{(3)}$ components is observed in the first few seconds, indicating the transient AHB process due to *trans-cis* photoisomerization (Figure 12.10b, c). The subsequent slow decrease of the THG in Figure 12.10b, c is caused by the photodegradation of the sample. It is worth mentioning that no visible THG intensity change was observed in the sample pumped by the 1350 nm laser beam of ~10 GW/cm² peak intensity (700 W/cm² of average intensity), in contrast to the case of the 1064 nm pump (see Figure 12.7). The SH of the 1350 nm laser is far from the absorption band of DR1, so the TP absorption is too small to produce considerable change of $\chi^{(3)}$.

It is important that no THG intensity change was observed in the pure PMMA sample. Moreover, the THG efficiency of the pure PMMA sample was about 300 times smaller than that of the 10% DR1-PMMA sample with the same thickness. This proves the rationality of our ignoring the $\chi^{(3)}$ of the polymer matrix in our theoretical model.

The photoinduced THG change is reversible due to the *cis* population dark relaxation and molecular orientational diffusion. After a short pumping period (about 10 seconds), the THG signal fully recovers in half a minute, regardless of pump wavelength and intensity. This recovery is because no considerable molecular reorientation or photodegradation occurs during the 10 seconds of pumping, and the THG intensity change is due only to the AHB, which disappears very fast after the pump is switched off. After a few minutes of pumping, considerable AR takes place and some photodegradation is possible, especially at high pump intensities. Figure 12.11 gives an idea about the evolution of different polarization components during pumping and relaxation stages for different pump wavelengths and intensities. There is obvious isotropic angular dependence of the THG signal before pumping (first column from the left). The THG signal angular distribution becomes highly anisotropic after a few minutes of pumping with the 532 nm or 442 nm beam. One can see the difference in the relative level of $\chi_\perp^{(3)}$ after 8 minutes of pumping. In Figure 12.11a, the $\chi_\perp^{(3)}$ drops a little, but in Figure 12.11b, the $\chi_\perp^{(3)}$ is at least the same or even higher than before pumping. In Figure 12.11c (highest pump intensity), the $\chi_\perp^{(3)}$ decrease is most considerable (about two times in THG intensity). These observations are consistent with Figure 12.7c and Figure 12.10a, c. After 1 min. of dark relaxation (third column), the *cis* population relaxes back to *trans*, and the THG anisotropy due to the AR of *trans* molecules is observed. It is clearly seen that $\chi_\perp^{(3)}$ rises above its level before pumping in Figure 12.11a, b, but in Figure 12.11c, it remains much lower than the initial level.

The further transformation of the THG polarization dependence (column 4) is mainly due to the orientational diffusion. After 15 more min. of dark relaxation, the angular dependence tends to be more isotropic, but the $\chi^{(3)}$ anisotropy still exists. Spectroscopic study shows that some photobleaching the sample occurs in all three cases presented in Figure 12.11, but at 3 W/cm², the degradation is very obvious and contributes to $\Delta\chi^{(3)}$ most significantly. At pump intensities of 0.1 W/cm² or lower, no visible bleaching is observed during a typical pumping period (~10 min.).

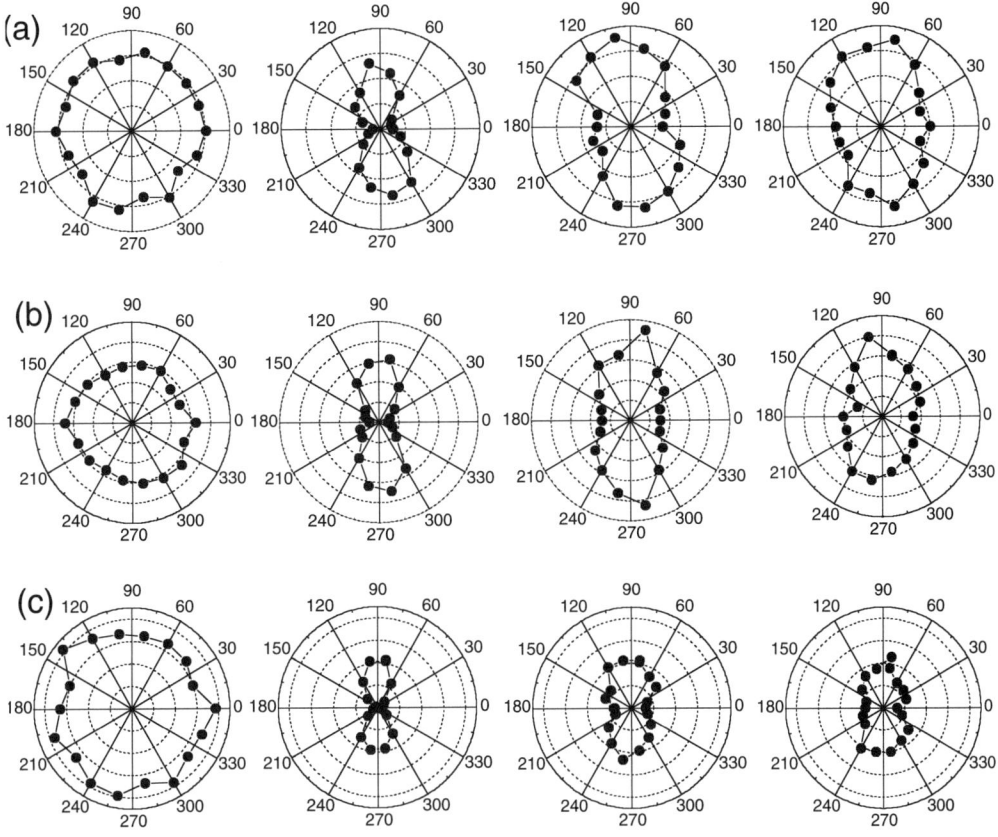

FIG. 12.11 THG intensity as a function of polarization angle φ of the probing beam before pumping (first column from the left), after 8 min. of pumping with the 532 nm and 442 nm light of 0.3 W/cm² average intensity (a,b) and 442 nm light of 3 W/cm² intensity (c) (second column), after 1 min of dark relaxation (third column), and after following 15 min. of dark relaxation (fourth column). The zero angle corresponds to the plane of polarization of a pump.

12.3 ELECTRIC FIELD INDUCED SECOND HARMONIC GENERATION

12.3.1 Theory

Another effect sensitive to optically induced $\chi^{(3)}$ change is the electric field induced second harmonic generation (EFISH) experiment. In the presence of external electric static field, an isotropic medium becomes noncentrosymmetrical and the second harmonic generation becomes possible:

$$P_i^{2\omega} = \chi_{ijkl}^{(3)}\,(-2\omega,\,\omega,\,\omega,\,0)\,E_j^\omega\,E_k^\omega\,E_l^0 \tag{12.22}$$

This expression is for polarization at 2ω frequency, generated by optical wave with ω frequency in the presence of the static electric field E_l^0.

Sekkat et. al. have employed EFISH technique to study photoinduced third-order nonlinearity of some high glass transition temperature (T_g) polyimide polymers containing azo chromophores.[3,17] The initial centrosymmetry of the polymer film is preserved at room temperature due to its high T_g even when a dc field is applied to the polymer films, either with or without simultaneous irradiation.[3,17] Therefore, the second harmonic output in their experiment is totally due to the contribution from the second hyperpolarizability γ not from the first hyperpolarizability β, and the third-order nonlinearity is given by

$$\chi_{EFISH}^{(3)} = \sum_{i=1}^{2} N_i f^4 < \gamma_i >_{jj33} \tag{12.23}$$

where $i = 1$ and 2 denotes the polyimide backbone and NLO azo chromophore species, respectively; N_i values are number densities expressible as inverse specific volume; and γ_i is the electronic second hyperpolarizability. The subscript j denotes the laboratory axes (1, 2, 3), and the static field is directed along the 3-axis and the local field factor $f^4 = f_\omega^2 f_{2\omega} f_0$. Because the nonlinear contributions from the polyimide backbone and cis form molecules are much smaller than that of $trans$ form azo molecules, the $\chi_{EFISH}^{(3)}$ contributions from the polyimide backbone and cis form molecules are ignored, and the $\chi_{EFISH}^{(3)}$ can be expressed as below:

$$\chi_{3333}^{(3)} (-2\omega, \omega, \omega, 0) = f^4 \gamma \int N_T(\Omega)\cos^4 \theta d\Omega \tag{12.24a}$$

$$\chi_{3113}^{(3)} (-2\omega, \omega, \omega, 0) = f^4 \gamma \int N_T(\Omega)\cos^2 \theta \sin^2 \theta \cos^2 \phi d\Omega \tag{12.24b}$$

Following the calculations from Equation 12.5 to Equation 12.12 described in the section 12.2.1, one can obtain:

$$\chi_{3333}^{(3)} (-2\omega, \omega, \omega, 0) = f^4 \gamma \int_0^{2\pi} \int_0^\pi \sum_l \sum_{m=-l}^{m=l} A_l Y_{lm}(\theta', \phi') Y_{lm}^*(\theta, \phi)\cos^4 \theta$$
$$\sin \theta d\theta d\phi \tag{12.25a}$$

$$\chi_{3113}^{(3)} (-2\omega, \omega, \omega, 0) = N f^4 \gamma \int_0^{2\pi} \int_0^\pi \sum_l \sum_{m=-l}^{m=l} A_l Y_{lm}(\theta', \phi') Y_{lm}^*(\theta, \phi)\cos^2 \theta$$
$$\sin^3 \theta d\theta \cos^2 \phi d\phi \tag{12.25b}$$

As with Equation 12.14, we can find the stationary $\chi_{EFISH}^{(3)}$ caused by the AHB mechanism:

$$\chi_{3333}^{(3)} (-2\omega, \omega, \omega, 0) = N f^4 \gamma \left[\frac{A_0}{5} + \frac{4}{7} A_2 P_2 (\cos(\theta')) + \frac{8}{35} A_4 P_4 (\cos(\theta')) \right] \tag{12.26a}$$

$$\chi_{3113}^{(3)} (-2\omega, \omega, \omega, 0) = N f^4 \gamma \left[\frac{A_0}{15} + \frac{1}{21} A_2 P_2 (\cos(\theta')) - \frac{4}{35} A_4 P_4 (\cos(\theta')) \right] \tag{12.26b}$$

Equations 12.26a and 12.26b show that SH output from the EFISH experiment can be optically modulated by changing the intensity and the polarization direction of the pump beam.

Sekkat et. al. have observed photoinduced AR in these azo-polyimide polymers even at temperatures far below the polymer T_g.[18,19] One can use the same kind of approach described in the previous section (from Eq. 12.19 to Eq. 12.21) to obtain the photoinduced $\chi^{(3)}_{EFISH}$ due to the AR mechanism. The expressions of $\chi^{(3)}_{3333}$ and $\chi^{(3)}_{3113}$ are the same as Equation 12.26, except for using the order parameters in Equation 12.20. The photoinduced SH output due to the AR mechanism in the EFISH experiment should be anisotropy, just like in the THG experiment.

In this section, we are going to discuss the results of optical control of EFISH signal in azo-polyimide polymers obtained by Sekkat et al.[3,17] and compare these results with our results from the THG described in the previous section.

12.3.2 Experiment

The chemical structure of the polyimide polymers (named PI-1 and PI-2) studied by Sekkat et al.[3,20] is shown in Figure 12.12. They prepared the polymer samples by spin-casting onto glass substrates. PI-1 was cast from a cyclohexanone solution and PI-2 from 1,1,2,2- tetrachloroethane. The T_g values of PI-1 and PI-2 were determined to be 350°C and 252°C, respectively, by scanning calorimetry method. The thicknesses of the PI-1 and PI-2 films were, respectively, approximately 0.72 μm and 0.14 μm, and their respective optical densities were approximately 0.79 and 0.3 at 543.5 nm. Details of the preparation and characterization of the samples can be found in References 3 and 20. In their EFISH experiment, a typical corona poling technique was used to pole the samples, with a dc electric field about 2–3 MV/cm across a 1–2 μm thick polymer film.[21] They used the SHG output from the EFISH experiment to *in situ* monitor the photochemical change in the third-order susceptibility of the PI-1 and PI-2 polymers.

In the EFISH experiment, a green light pump beam (wavelength 543.5 nm) from a helium-neon laser was used to initiate the photoisomerization of the photochromic units. This beam was linearly polarized either in *s* or *p* polarization tuning by a λ/2 plate, and the intensity for both polarizations was about 8 mW/cm². The pump beam was incident on the sample at

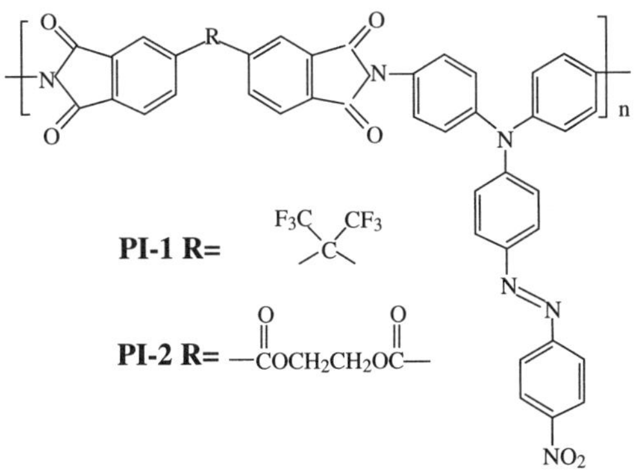

PI-1 R=

PI-2 R=

FIG. 12.12 Chemical structures of the azo polyimide polymers. After Ref. 3.

approximately 24 degrees to the normal, with a 2 mm diameter spot size. A Q-switched Nd:YLF laser (Quantronix Model 527) at 1053 nm with a 1 kHz repetition rate was used as the EFISH probe. The probe was incident at 47 degrees near the Brewster angle to obtain the maximum SH signal. The probe was focused on a smaller spot than that of the pump beam. A video camera was used to monitor the alignment of the beams and to ensure that the polymer film was not damaged during exposure. The probe was linearly polarized either in s or p direction. The polarization was controlled by using a $\lambda/2$ plate and a polarizer leading to equal power for both s and p polarizations, and an analyzer was used for the SH light. The SH signal was detected by a photomultiplier, collected by a boxcar averager, and recorded by using a computer.

12.3.3 Results and Discussion

Figure 12.13 shows the pumping effect on the SH signal observed by Sekkat et al. in PI-1 at room temperature.[3] In this figure, the top graphs and bottom graphs refer, respectively, to p- and s-polarized probes for both p (left)- and s (right)-pumping. They also obtained similar results (not shown) for PI-2. As mentioned previously, the SH signal is contributed only from γ of azo molecules, but not from β of azo molecules due to the high T_g value of the polyimides. The decay of SH signal right after turning on the pump beam is caused by the AHB effect of azo dye, which transforms its confirmation from

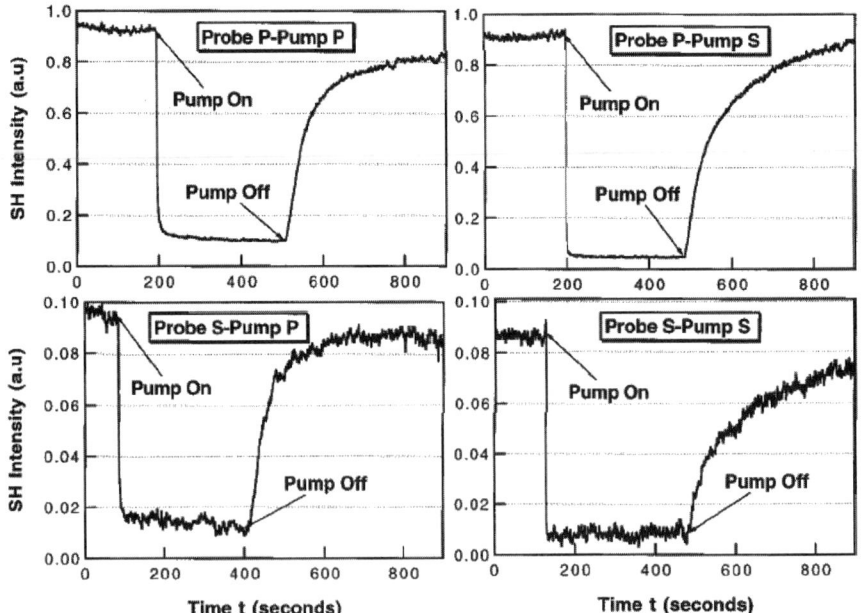

FIG. 12.13 Reversible rapid erase of the EFISH signal of the PI-1 polymer for four combinations of the irradiation and the probe polarizations, e.g., p (top)- or s (bottom)-probes and p (left)- or s (right)-irradiation. After Ref. 3.

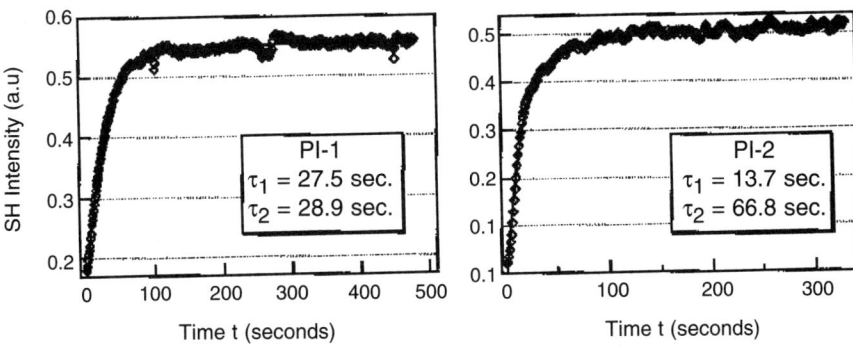

FIG. 12.14 Example of biexponential fits (solid lines) to EFISH recovery (markers) during the *cis-trans* thermal back-isomerization of the PI-1 (left) and PI-2 (right) polymers. The insets show the time constants of these biexponentials. After Ref. 3.

the *trans* form to the *cis* form within the picosecond time scale.[22] The AHB effect results in a drastic change of the $\chi^{(3)}$ values of the PI-1 and PI-2 films. The EFISH intensity ratio between the *p*- and *s*-polarized probe beams is about 9. This finding agrees with the theoretical prediction of Equation 12.26, which gives a value of 3 for the relative ratio of $\chi^{(3)}$ probed by *p*-(corresponding to $\chi^{(3)}_{3333}$) and *s*- polarized (corresponding to $\chi^{(3)}_{3113}$) beams.

Figure 12.13 also shows the recovery of the EFISH signal when the pump beam is turned off. This recovery is due to the regaining of $\chi^{(3)}$ value caused by the *cis*→*trans* thermal back reaction, which increases the population of *trans* molecules. Biexponential fits to the $\chi^{(3)}$ recovery curves yielded fast recovery time about 28 and 14 s, for PI-1 and PI-2, respectively. The longer recovery time for PI-1 ranged from 289 to 390 s, while that for PI-2 varied from 66.8 to 167 s. It agrees well with the previous results obtained in UV-vis absorption measurements for the same samples.[19] The slow thermal back-isomerization probably results from the strong coupling between the azo chromophore and the polymer backbone.[19] Figure 12.14 illustrates the biexponential fits of the recovery of the $\chi^{(3)}$ signal for both PI-1 (left) and PI-2 (right), and the time constants determined from the fits are given in the insets. This kind of photoinduced reversible change of molecular confirmation can be repeated many times, resulting in all-optical modulation of the SH output [see Figure 12.15 for both PI-1 (top) and PI-2 (bottom)].

The effect of pumping intensity on the EFISH signal is shown in Figure 12.16. This figure plots the SH signal of PI-2 (top) and PI-1 (bottom) under various pumping intensities (*p*-probe and *s*-pump for both the upper and the lower figure), where the numbers on the curves refer to the optical density used to attenuate the pumping intensity. Similar to the case of THG, the EFISH signal decreases as the pumping intensity increases, which is due to the decrease of the *trans* population at higher pumping intensity.

From Figure 12.13, one finds that both *p*- and *s*- polarized pump beams produce the same effect on the SH signal decay for both *p*- and *s*- polarizations of the probe beam. This finding means the photoinduced change of EFISH signal does not depend on the pump's polarization. This result is

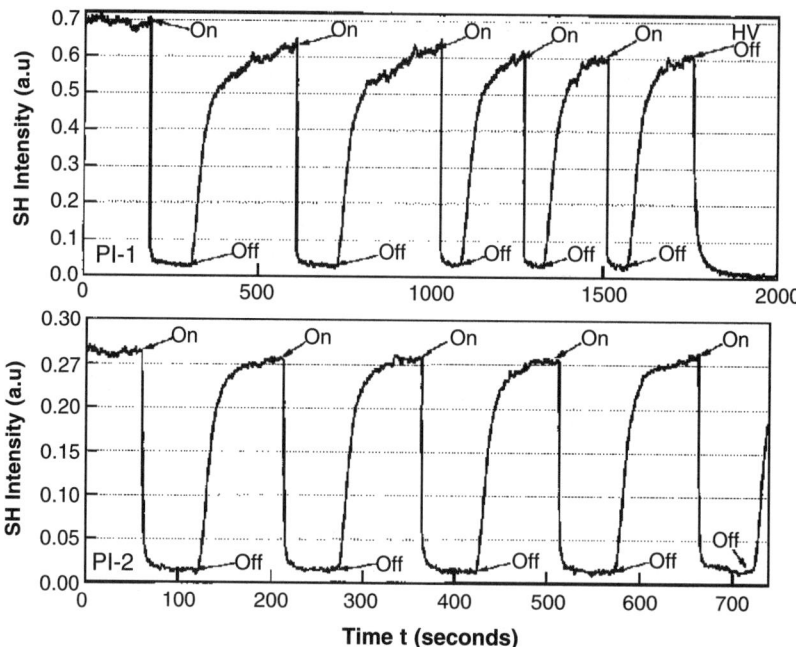

FIG. 12.15 All-optical light modulation of the SH signal of the PI-1 (top) and PI-2 (bottom) polymers. The moments of turning the irradiating light on and off are indicated by arrows. HV Off in the top graph refers to the moment where the corona voltage was turned off. After Ref. 3.

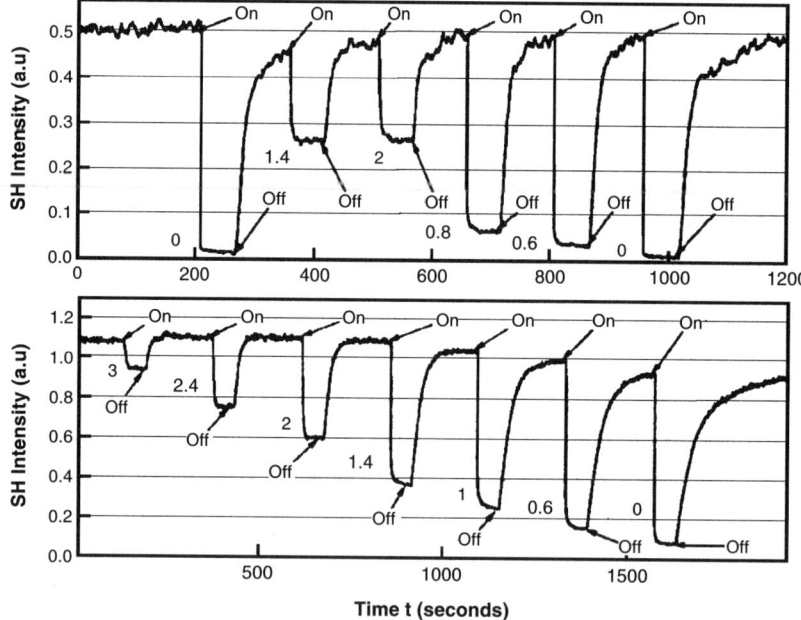

FIG. 12.16 Effect of the irradiating light intensity on the EFISH signal of the PI-2 (top) and PI-1 (bottom) polymers. The moments of turning the irradiating light on and off are indicated by arrows, and the numbers on the figures refer to the optical density used to attenuate the intensity of the irradiating light. After Ref. 3.

different from what we found in the THG experiment, in which an anisotropic distribution of photoinduced $\chi^{(3)}$ was obtained (see section 12.2.3). Saturation effect is not the reason for obtaining this isotropic change of EFISH signal, because similar results were also found for lower pumping intensity.[3] To ascertain the reason, we need to compare the $\chi^{(3)}$ expressions in the EFISH and THG experiments. In the EFISH experiment, because the pump beam was arranged to incident on the sample approximately 24 degrees, the angle between the 3-axis and the p-pump beam's polarization direction is 66 degrees ($\theta' = 66°$). In the case of s-pumping beam, the angle between the 3-axis and the polarization direction of s-pump beam is 90 degrees ($\theta' = 90°$). Substituting them into Equation 12.26, one can find:

$$\chi^{(3)}_{3333,\, p\text{-}pump}(-2\omega,\omega,\omega,0) = Nf^4\gamma\left[\frac{A_0}{5} - \frac{1}{7}A_2 - \frac{1}{35}A_4\right] \quad (12.27a)$$

$$\chi^{(3)}_{3333,\, s\text{-}pump}(-2\omega,\omega,\omega,0) = Nf^4\gamma\left[\frac{A_0}{5} - \frac{2}{7}A_2 + \frac{3}{35}A_4\right] \quad (12.27b)$$

$$\chi^{(3)}_{3113,\, p\text{-}pump}(-2\omega,\omega,\omega,0) = Nf^4\gamma\left[\frac{A_0}{15} - \frac{1}{82}A_2 + \frac{1}{70}A_4\right] \quad (12.27c)$$

$$\chi^{(3)}_{3113,\, s\text{-}pump}(-2\omega,\omega,\omega,0) = Nf^4\gamma\left[\frac{A_0}{15} - \frac{1}{42}A_2 - \frac{3}{70}A_4\right] \quad (12.27d)$$

In this equation, the subscripts *p-pump* and *s-pump* represent the polarization direction of the pump beam, and the values of $\chi^{(3)}_{3333}$ and $\chi^{(3)}_{3113}$ are related to the SH signal of p- and s-probes, respectively. To compare with $\chi^{(3)}$ in the THG experiment (Eq. 12.15), we plot $\chi^{(3)}$ values of these two experiments as a function of a in Figure 12.17. One can clearly see that for both p- and s-probe configurations, the differences between $\chi^{(3)}_{p\text{-}pump}$ and $\chi^{(3)}_{s\text{-}pump}$ of the EFISH experiment are much smaller than that of the THG experiment, especially for the s-probe configuration. Consequently, the photoinduced $\chi^{(3)}$ distribution was observed to be almost isotropic in this particular EFISH experiment.

Although photoinduced change of $\chi^{(3)}$ can be detected with both EFISH and THG techniques, there are still some differences existing in these two methods.[1] The photoinduced $\chi^{(3)}$ anisotropy can be easily observed in the THG experiment, but it is more difficult to do in the EFISH case due to the peculiar experimental geometry. 2) The EFISH method can be used only in high T_g azo-polymer thin films, otherwise the modulation on first hyperpolarizability (β) of azo molecules will also contribute to the EFISH signal. On the other hand, in the THG method, only the THG signal corresponds to the γ value of azo molecules, and the photoinduced change of β effect does not contribute to the THG signal. Therefore, the THG method can be employed in both low and high T_g polymer thin films. 3) No dc electric field is needed in the THG technique. It is thus simpler and easier than the EFISH technique. 4) In the EFISH method, one can measure the photoinduced change of $\chi^{(3)}_{3333}$ and $\chi^{(3)}_{3113}$ values, but in the THG method only the change of $\chi^{(3)}_{3333}$ component is detectable. 5) Photoinduced $\chi^{(1)}$ effect such as photoinduced change of absorption will influence the detected signal from both

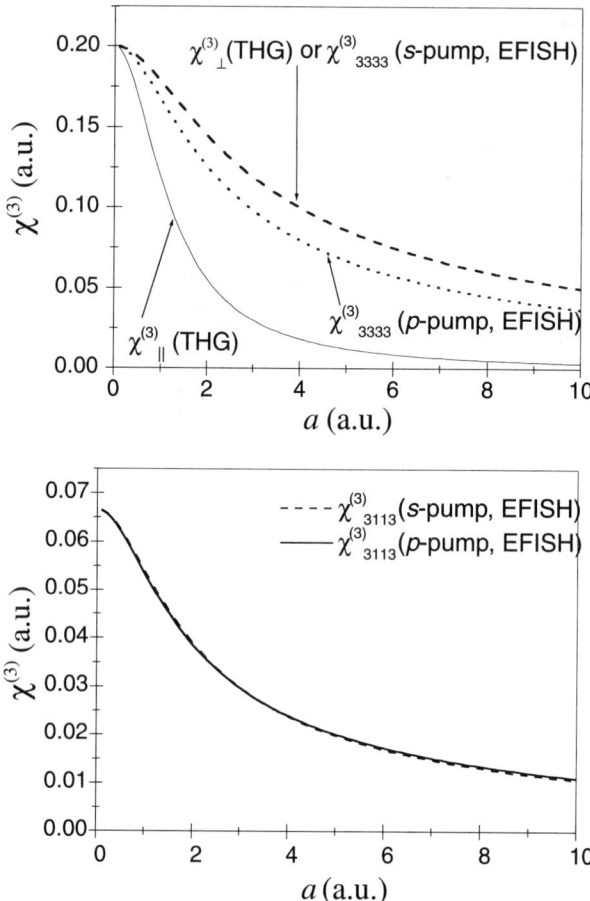

FIG. 12.17 Theoretical dependences of $\chi^{(3)}$ components on $a = \sqrt{\sigma_T \phi_{TC} I_p / \eta_{CT}}$ parameter for different polarization configurations in THG and EFISH experiments.

techniques. One should carefully choose the SH and TH wavelengths and also the sample thickness to reduce the influence of the photoinduced change of absorption.

12.4 DEGENERATE FOUR-WAVE MIXING

12.4.1 Introduction

Degenerate four-wave mixing (DFWM) in azo-dye polymers is interesting and may be useful for many applications, because it allows achieving phase conjugation of the signal wave using low-intensity writing and reading beams with the same frequency. Because DFWM is a third-order nonlinear process, the efficiency of conjugation is essentially $\chi^{(3)}$ dependent and can be strongly affected by the optical pump. Dynamics of the DWFM signal in an azo-benzene polymer pumped by strong pulses of a frequency doubled Nd:YAG laser were studied by Wang et al.[23] They observed a decrease of the conjugated signal during the SH pulse passage with subsequent recovery to the original level. This decrease was attributed to the shift of *trans* and *cis* populations caused by strong near-resonant optical excitation. In their experiment, the DFWM and external pump wavelengths were substantially different (633 and 532 nm, respectively). However, the writing and reading beams involved in the DFWM process can also affect the conjugation efficiency, if their wavelength is near the absorption peak of the sample.

The stationary $\chi^{(3)}$ value depends not only on the pump intensity, but also on the cis population relaxation rate, which is strongly temperature dependent.[24,25] This dependency provides an opportunity for controlling the temperature of third-order susceptibility $\chi^{(3)}$ and, therefore, provides the efficiency of the phase conjugation. In this section, we describe recent experiments on vector phase conjugation (VPC) as an example of the temperature-tuned DFWM process. Original results were presented by Awangku Yusof et al. in Reference 4. VPC is a time-reversal of both the spatial and polarization information of a wavefront.[26–31] A VPC may be used for correcting not only phase distortions of a light beam, but also polarization distortions produced by anisotropic distorters. Mohajerani and Mitchell[32] showed that in a methyl red (MR) PMMA system, the reflectivities for two orthogonally polarized components of a probe beam show contrasting temperature dependence as a sample is heated from room temperature to beyond the T_g (105°C) of the polymer. One can get the same reflectivities for these two polarized components at a particular temperature (T_{vpc}), and obtain the VPC in this system by controlling the temperature of the sample.

12.4.2 Experiment

The materials used in Reference 4 consisted of thin layers of MR in PMMA. These were fabricated by dip-coating cleaned glass microscope slides in a solution of MR and PMMA dissolved in dichloromethane to produce films with average thickness of about 4 μm.[32] The material absorbs strongly at 488 nm. The origin of the nonlinear optical properties of such azo-dye impregnated polymer films has been discussed in previous sections. In particular, models have been proposed describing the physical origin of photoinduced anisotropies suitable for the generation of intensity and polarization gratings in the materials, which can be investigated by DFWM.[6,32,33]

The arrangement employed for the VPC experiment is described in Reference 4. A cw argon-ion laser at 488 nm was used in a standard DFWM geometry. The *s*-polarized output beam was first split by a beam-splitter to provide the pump and the probe beams. The transmitted beam from the beam-splitter was then divided into the two *s*-polarized pump beams each with a power of approximately 0.35 mW. The reflected beam from the beam-splitter was used as the probe beam, whose intensity was about 7% of the total intensity in both pump beams. The forward pump beam and the probe, which constituted 'writing' beams, were overlapped at the sample. Their optical path length difference was much smaller than the laser coherence length, so that they were coherent at the sample. The backward pump beam was optically delayed so that it was incoherent with the writing beams and acted as the 'read' beam. With this arrangement, a single transmission grating was formed with the writing beams in the sample, and this grating would diffract the counter-propagating pump beam to generate the phase conjugation signal. A $\lambda/2$ plate was used to rotate the probe's polarization either parallel or perpendicular to the pump polarization direction. The conjugate beam, which retraced exactly the optical path of the probe beam, was picked off by a polarization preserving beam-splitter, analyzed by a linear polarizer, and then detected by a photodetector. The signal was detected using a phase-sensitive detection system, which involved mechanically chopping the read beam and using a lock-in amplifier to measure the chopped conjugation signal. All the beams were overlapped at the film sample with a beam spot size of about 1.5 mm radius, giving a total irradiance of approximately 11 mW/cm^2 on the film. The angle between the probe beam and the forward pump beam was approximately 9°, producing a grating with a period of approximately 5 μm.

Two different optical configurations were employed (Fig. 12.18) to measure the reflectivities for both polarization components of the probe beam. In the first configuration as shown in Figure 12.18a, all of the incident beams were s-polarized, and an intensity grating was formed. In the region where the writing beams interfered, the periodic spatial variation of intensity produced a corresponding spatially periodic variation of refractive index. This

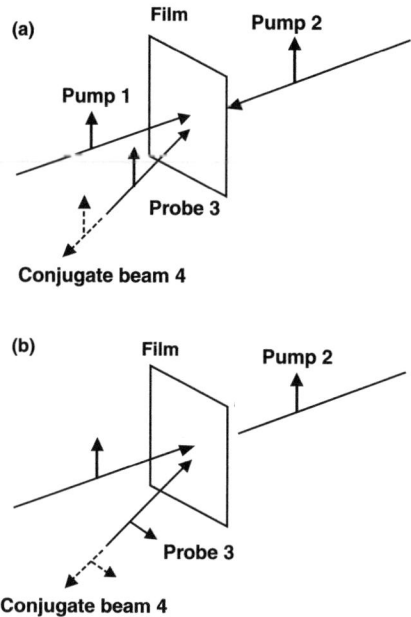

FIG. 12.18 (a) Configuration 1: geometry for the intensity grating formation with s-polarized pump and probe beams; (b) Configuration 2: geometry for the polarization grating formation with the pump beams s-polarized and p-polarized probe beam. After Ref. 4.

grating diffracted the backward pump beam 2 to form the conjugate beam (beam 4). In this configuration, the conjugate beam had the same polarization state as the probe beam, i.e., s-polarized.[33] The ratio of the intensity of the conjugate beam to that of the probe beam in this arrangement, I_4/I_3, was recorded as the reflectivity R_{int}.

For the second configuration shown in Figure 12.18b, the probe polarization was p-polarized, which was perpendicular to the pump polarization. The periodic spatial variation of the polarization state of the resultant electric field vector in the region of overlap of the writing beams produced a polarization grating in the medium.[9] Diffraction of the pump beam 2 from this grating resulted in a p-polarized DFWM signal, which had the same polarization state as the probe beam.[33] The intensity ratio of the conjugate beam to the probe beam in this configuration, I_4/I_3, was recorded as the reflectivity R_{pol}. The sample was placed in a temperature control oven, and the temperature of the sample was varied from 30°C to 65°C, and the reflectivities R_{int} and R_{pol} were measured.

12.4.3 Experimental Results and Discussion

Figure 12.19 plots the reflectivity of the conjugation beam as a function of temperature. At lower temperature (T = 30°C), the p-polarized reflectivity, R_{pol} is larger than the s-polarized reflectivity, R_{int}. However, at higher temperatures such as 65°C, R_{pol} is smaller than R_{int}. Figure 12.19 shows that the two reflectivities are equal at T $\cong$ 50°C. At this temperature, the film should work as a VPC. To verify this property, Awangku Yusof et. al. measured the fidelity ratio F_{VPC}, of the sample at different temperatures.[4] F_{vpc} is a parameter determining how well the material exhibits as a VPC. F_{vpc} will be unity for

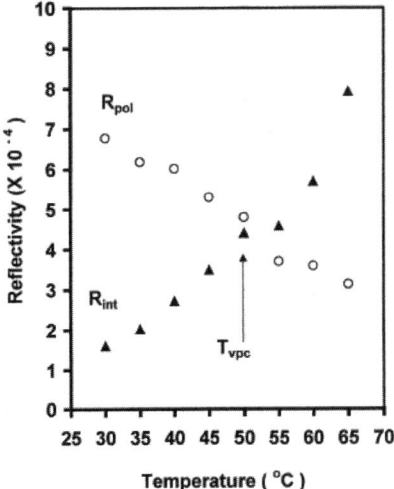

FIG. 12.19 The reflectivity of the intensity grating, R_{int}, and of the polarization grating, R_{pol}, as a function of sample temperature. At about T=50°C, the reflectivities are equal, and this is the VPC temperature T_{vpc}. After Ref. 4.

a perfect VPC, but zero for a silvered mirror. They found the F_{vpc} close to unity at a sample temperature of approximately 50°C. This finding indicates that the material is indeed working as a VPC at this temperature.

Now we are going to give a physical insight into an origin of the mechanisms responsible for experimental results mentioned above. The physics of the temperature tuning of the VPC can be qualitatively understood in the framework of the model presented in the Section 12.1. The amplitude of the DFWM signal can be written as:

$$P_I^{conj} = \chi_{IJKL}^{(3)} (-\omega, \omega, -\omega, \omega) E_J^{pump} E_K^{pump} E_L^{probe} \qquad (12.28)$$

The $\chi_{3333}^{(3)}$ component is then responsible for intensity grating efficiency, and $\chi_{3113}^{(3)}$ the component is responsible for the polarization grating efficiency. Therefore, the conjugation efficiencies for two different DFWM configurations are

$$R_{int} = I^{conj}/I^{probe} = C_1 (\chi_{3333}^{(3)})^2 I_1^{pump} I_2^{pump} \qquad (12.29a)$$

$$R_{pol} = I^{conj}/I^{probe} = C_2 (\chi_{1331}^{(3)})^2 I_1^{pump} I_2^{pump} \qquad (12.29b)$$

where I^{conj}, I^{probe}, I_1^{pump}, I_2^{pump} are intensities of conjugated, probe, forward, and back pump beams, respectively; C_1 and C_2 are coefficients, which depend on experimental geometry, refractive index, and absorption of the sample.

After the DFWM process starts, the $\chi^{(3)}$ of the dye-polymer sample is contributed by only the original *trans*-population, and $R_{int} \gg R_{pol}$, because in an isotropic medium $\chi_{3333}^{(3)} = 3$. After a few seconds, $\chi_{3333}^{(3)}$ drops several times and so does R_{int}. This drop is due to a decrease of the number of *trans* molecules with dipole moment near the 3-axis because of the AHB and AR effects. The change of $\chi_{1331}^{(3)}$ due to the AHB is much smaller than that of $\chi_{3333}^{(3)}$ and even can increase due to the molecular reorientation and the rise of the index perpendicular to the pump polarization. When the temperature of the sample increases, the *cis* thermal relaxation rate η_{CT} increases[24,25], so the *cis* population, and therefore, the AHB are reduced. Indeed, the parameter $a = \sqrt{\sigma_T \phi_{TC} I_{pT}/\eta_{CT}}$ (see Section 12.1) drops when $\eta_{CT} = 1/\tau_C$ increases. This results in increasing of stationary $\chi_{3333}^{(3)}$ with the temperature. In Figure 12.20, the temporal dependence of R_{int} at two different temperatures is shown. At $T=35°C$, the sharp peak is observed in the first few seconds and after that, the signal decays to a much lower stationary value. At $T=65°C$, the R_{int} grows steadily to its stationary value without any decay.

On the other hand, heating of the sample results in a higher rate of thermal disorientation of molecules and, hence, distorts the polarization grating and decreases R_{pol}. At some temperature T_{vpc}, the stationary values of R_{int} and R_{pol} are equal to each other. In this case, pure VPC is realized.

12.5 PROSPECTIVE AND CONCLUSIONS

As mentioned in the beginning, the azo-polymers take a special place among prospective materials for nonlinear optical applications. This chapter attempted to summarize research activities aimed at all-optical controlling third-

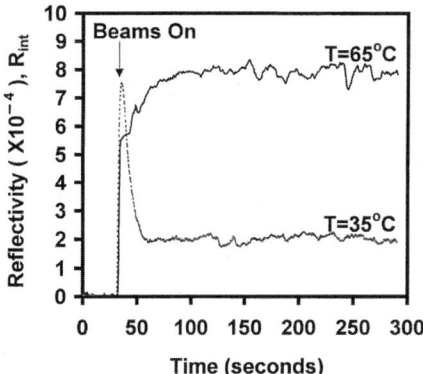

FIG. 12.20 Intensity grating reflectivity R_{int} as a function of time, at sample temperatures of 35°C and 65°C. Note that at T =35°C, the reflectivity shows a sharp peak when the beams are switched on at time 30 s, as shown by arrow. After Ref. 4.

order nonlinear effects in such materials. The main conclusion, which can be drawn from the works presented here, is that there is a wide-range of variation of third-order susceptibility $\chi^{(3)}$ in azo-dye polymers via excitation by near-resonant light. Evidence for the photoinduced change of $\chi^{(3)}$ was obtained in experiments on DFWM, EFISH, and THG. We have shown that the results of these experiments can be qualitatively interpreted in the framework of a stationary model of photoinduced $\chi^{(3)}$ anisotropy and discussed peculiarities of each experimental technique, in which the photoinduced change of $\chi^{(3)}$ can be observed. Our model is an extension of earlier proposed theory of optical manipulation of azo molecules[2,6] to third-order phenomena. The model considers the angular hole burning and molecular angular redistribution as a result of photoisomerization of dye molecules.

The theoretical model presented in Section 12.2 is in good qualitative agreement with the experiment, but for more accurate interpretation of the experimental results, further theoretical development is needed. For instance, one should consider the THG in a medium with nonuniform $\chi^{(3)}$, because in an absorbing medium, intensity of the pump is changed along the probe beam propagation direction. Furthermore, the AHB and AR mechanism should be studied together in the framework of general dynamic equations[5,6] including the contribution of the *cis* population and reverse *cis-trans* photo-isomerization. Provided an improved theoretical model, the techniques discussed could be powerful means for studying molecular dynamics in polymers with high third-order susceptibility.

Along with the employment of third-order effects for molecular dynamics studies, another promising direction is using them for light modulation and information storage. The main problem for the use of azo-polymers in light-processing devices is the slow response of $\chi^{(3)}$ on optical excitation, because the AHB and AR processes require in the range from milliseconds to minutes, depending on particular material and pump intensity. Development of new materials with a faster response would help to solve this problem. On the other hand, materials with irreversible isomerization are necessary for permanent information storage.

The fact that isomerization and relaxation processes are sensitive to the temperature of a specimen allows a combined optical-temperature control of azo-polymer third-order susceptibility. The temperature-tuned VPC[4] is just one example of this. Combination of optical excitation and temperature variations may result in more effective governing linear and nonlinear optical properties of azo-dye polymer materials.

REFERENCES

1. Prasad, P. N., and Williams, D. J. (1991). *Introduction to Nonlinear Optical Effects in Molecules and Polymers*. John Wiley & Sons, New York.
2. Dumont, M., Hosotte, S., Froc, G., and Sekkat, Z. (1994). Orientational manipulation of chromophores through photoisomerization. *In Photopolymers and Applications in Holography, Optical Data Storage, Optical Sensors, and Interconnects*. (R. A. Lessard, ed.), Proc. SPIE **2042**, 2–13.
3. Sekkat, Z., Knoesen, A., Lee, V. Y., and Miller, R. (1997). Observation of reversible photochemical "blow out" of the third-order molecular hyperpolarizability of push-pull azo dye in high glass transition temperature polyimides. *J. Phys. Chem. B* **101**, 4733–4739.
4. Awangku Yusof, A. A. R., O'Leary, S. V., and Mitchell, G. R. (1999). Temperature-tuned vector phase conjugation in azobenzene dye impregnated polymer films. *Opt. Commun.* **169**, 333–340.
5. Dumont, M., and Osman, A. E. (1999). On spontaneous and photoinduced orientational mobility of dye molecules in polymers. *Chem. Phys.* **245**, 437–462.
6. Sekkat, Z., and Dumont, M. (1993). Photoinduced orientation of azo dyes in polymeric films: Characterization of molecular angular mobility. *Synth. Met.* **54**, 373–381.
7. Page, R. H., Jurich, M. C., Reck, B., Sen, A., Twieg, R. J., Swalen, J. D., Bjorklund, G. C., and Willson, C. G. (1990). Electrochromic and optical waveguide studies of corona-poled electro-optic polymeric thin films. *J. Opt. Soc. Am. B* **7**, 1239–1250.
8. Arfken, G. B and Weber, H. B. (1995). *Mathematical methods for physicists*. Academic Press, San Diego.
9. Nikolova, L., and Todorov, T. (1984). Diffraction efficiency and selectivity of polarization holographic recording. *Optica Acta* **31**, 579–588.
10. Fiorini, C., Charra, F., Nunzi, J. M., and Raimond, P. (1997). Quasi-permanent all-optical encoding of noncentrosymmetry in azo-dye polymers. *J. Opt. Soc. Am. B* **14**, 1984–2003.
11. Sekkat, Z., Morichere, D., Dumont, M., Loucif-Saibi, R., and Delaire, J. A. (1992). Photoisomerization of azobenzene derivates in polymeric thin films. *J. Appl. Phys.* **71**, 1543–1545.
12. Churikov, V. M., and Hsu, C. C. (2000). Optical control of third harmonic generation in azo-doped polymethylmethacrylate thin films. *Appl. Phys. Lett.* **77**, 2095–2097.
13. Churikov, V. M., and Hsu, C. C. (2000). Optically induced anisotropy of third-order susceptibility in azo-dye polymers. *J. Opt. Soc. Am. B* **18**, 1722–1731.
14. Dumont, M., Froc, G., and Hosotte, S. (1995). Alignment and orientation of chromophores by optical pumping. *Nonlin. Opt.* **9**, 327–338.
15. Osman, A. E., and Dumont, M. (1998) Dynamical and spectroscopic study of photoinduced orientation of dye molecules in polymers. In *Photopolymer Device Physics, Chemistry and Applications*. (R. A. Lessard Ed.), Proc. SPIE **3417**, 36–46.
16. Churikov, V. M., Hung, M. F., Hsu, C. C., Shiau, C. W., and Luh, T. Y. (2000). Encoding of macroscopic second order nonlinearity via all-optical polar alignment in substituted norbornene polymer thin films. *Chem. Phys. Lett.* **332**, 19–25.
17. Sekkat, Z., Pretre, P., Knoesen, A., Volksen, W., Lee, V. Y., Miller, R. D., Wood, J., and Knoll, W. (1998). Correlation between polymer architecture and sub-glass-transition-temperature light-induced molecular movement in azo-polyimide polymers: influence on linear and second- and third-order nonlinear optical processes. *J. Opt. Soc. B* **15**, 401–413.
18. Sekkat, Z., Wood, J., Aust, E. F., Knoll, W., Volksen, W., and Miller, R. D. (1996). Light-

induced orientation in a high glass transition temperature polyimide with polar azo dyes in the side chain. *J. Opt. Soc.* B **13**, 1713–1724.

19. Sekkat, Z., Wood, J., Knoll, W., Volksen, W., Miller, R. D., and Knoesen, A. (1997). Light-induced orientation in azo-polyimide polymers 325°C below the glass transition temperature. *J. Opt. Soc.* B **14**, 829–833.

20. Miller, R. D., Burland, D. M., Jurich, M. C., Lee, V. Y, Moylan, C. R., Thackara, J., Tweig, R. J., Verbiest, T., and Volksen, W. (1995). High temperature NLO chromophores and polymers. *Macromolecules* **28**, 4970.

21. Mortazavi, M. A,. Knoesen, A, Kowel, S. T., Higgins, B., and Dienes, A. J. (1989). Second-harmonic generation and absorption studies of polymer-dye films oriented by corona-onset poling at evaluated temperatures. *Opt. Soc. Am. B* **6**, 733–741.

22. Lednev, I. K., Ye, T.-Q., Hester, R. E., and Moore, J. (1996). Femtosecond time resolved UV-visible absorption spectroscopy of trans-azobenzene in solution. *J. Phys. Chem.* **100**, 13338–13341.

23. Wang, Y., Zhao, J., Si, J., Ye, P., Fu, X., Qiu, L., and Shen, Y. (1995). Dynamic studies of degenerate four-wave-mixing in an azobenzene-doped polymer film with an optical pump. *J. Chem. Phys.* **103**, 5357–5361.

24. Loucif-Saibi, R., Nakatani, K., Delaire, J. A., Dumont, M., and Sekkat, Z. (1993). Photo-isomerization and second harmonic generation in disperse red one-doped and -functionalized poly(methyl methacrylate) films. *Chem. Mater.* **5**, 229–236.

25. Tompkin, W. R., Malcuit, M.S., Boyd, R.W. (1990). Enhancement of the nonlinear optical properties of fluorescein doped boric-acid glass through cooling. *Appl. Opt.* **29**, 3921–3926.

26. Basov, N. G., Efimkov, V. F., Zubarev, I. G., Kotov, A. V., Mikhailov, S. I., and Smirnov, M. G. (1978). Wave front reversal accompanying VRMB of depolarised pumping. *JETP Lett.* **28**, 197.

27. McMichael, I. (1988). Externally pumped polarization-preserving phase conjugator. *J. Opt. Soc. Am. B* **5**, 863–865.

28. Tompkin, W. R., Malcuit, M. S., Boyd, R. W., and Sipe, J. E. (1989). Polarization properties of phase conjugation by degenerate four-wave mixing in a medium of rigidly held dye molecules. *J. Opt. Soc. Am. B* **6**, 757–760.

29. Damzen, M. J., Camacho-Lopez, S., and Green, R. P. M. (1996). Wave-mixing and vector phase conjugation by polarization-dependent saturable absorption in Cr_4^+:YAG. *Phys. Rev. Lett.* **76**, 2894–2897.

30. Martin, G., Lam, L. K., and Hellwarth R. W. (1980). Generation of a time-reversed replica of a nonuniformly polarized image-bearing optical beam. *Opt. Lett.* **5**, 185–187.

31. Brignon, A., Sillard, P., and Huignard, J.-P. (1996). Vector phase conjugation in Cr_4^+:YAG by four-wave mixing with linearly-polarized pump beams. *Appl. Phys. B* **63**, 537–540.

32. Mohajerani, E. and Mitchell, G. R. (1993). Temperature optimisation of optical phase conjugation in dye doped polymer films. *Opt. Commun.* **97**, 388–396.

33. Tomov, I. V., VanWonterghem, B., Dvornikov, A. S., Dutton, T. E., and Rentzepis, P. M. (1991). Degenerate four-wave mixing in azo-dye-doped polymer films. *J. Opt. Soc. Am. B* **8**, 1477–1482.

IV

OPTICAL MANIPULATION AND MEMORY

13

PHOTOINDUCED MOTIONS IN AZOBENZENE-BASED POLYMERS

ALMERIA NATANSOHN*
PAUL ROCHON†
Department of Chemistry, Queen's University, Kingston, ON K7L 3N6, Canada
†*Department of Physics, Royal Military College, Kingston, ON K7K 5L0, Canada*

ABSTRACT

In this chapter, we review our work on photoinduced motions in azobenzene-based polymers. We observed in 1991 the photoinduced orientation of the azobenzene groups in amorphous high-Tg homopolymers containing azobenzene groups in each structural unit, and we measured the orientation as dichroism and birefringence. Over the next few years, we designed novel polymers containing a variety of azobenzenes bound in a variety of ways, and we elucidated some of their optical properties as a function of their structure. The roles of the azobenzene dipole, the chromophore concentration, and the spacer length, and the type of bonding to the main chain have been investigated. In addition, other types of bonding, including bonding to the main chain, were shown to produce similar optical effects. One of the most significant findings was that non-azo groups can be moved in concert with the azo groups well below Tg, and this cooperative motion is driven by a polar interaction. In 1994, we reported formation of surface-relief gratings in our azo polymers and suggested a mechanism of formation based on pressure gradients produced by different isomerization activity on different parts of the polymer film. Both the photoinduced birefringence and the surface grating formation were shown to have a variety of possible applications in photonic devices. Photorefractivity was demonstrated on some azo polymers containing carbazole or other photoconductive moieties. More recently, we reported a photoinduced helical supramolecular structure obtained on some smectic azo-based polymers. The handedness of the photoinduced helix depends on the light polarization and can be switched back and forth simply by changing the polarization.

13.1 INTRODUCTION

In 1991, while working toward the development of a nonlinear optical (NLO) polymer research program, our then student Dr. Shuang Xie decided to try a preliminary synthesis of such a polymer. She bought Disperse Red 1 from Aldrich, reacted it with acryloyl chloride to produce a monomer, then polymerized it under free radical conditions and obtained a polymer, which we called pDR1A. This was to be a substitute for the real thing, because at the time everybody knew, or thought they knew, that, in order to be able to orient the chromophore independently of the backbone, the spacer length we had (two methylene groups) was not long enough. The glass transition temperature (Tg) was 91°C, which was rather high based on common knowledge at the time, but manageable, if we were to pole the polymer in an electric field above Tg. All things considered, this was to be just an intermediary (and probably unsuccessful) step toward designing novel NLO polymers.

Of course, while investigating the properties of pDR1A and its analogs, a world of surprises awaited us. We are only now getting more into their poling and NLO properties. What we concentrated on in the years since then, and what we are still investigating, are a variety of unexpected polymer motions derived from the well-known azobenzene photoisomerization. Because this volume is about photochromic polymers, it is not necessary to present any background or history of the phenomena. We will concentrate on telling our story and emphasizing what we believe are our contributions to this field. In some instances, however, we will have to establish the context, at least the context as we saw it, so some other contributions will be mentioned. We would like to emphasize, however, that the literature covered is by no means complete, because that is not the purpose of this chapter. An attempt to review the whole literature on photoinduced motions in azobenzene polymers has recently been submitted for publication to *Chemical Reviews*.[1]

13.2 PHOTOINDUCED MOTIONS

Under illumination, azobenzene undergoes photoisomerization.[2] If the azobenzene is bound to a polymer chain, the consequence of this repeated photoisomerization is a series of motions of the chromophores, and more. Even nonbound azobenzenes affect their environment if they are dissolved in a polymer matrix. These motions can be classified into roughly three levels. The first level is the chromophore motion at the molecular level. It is influenced by light polarization. With linearly polarized light, the photoisomerization is activated only when the chromophore axis has a component that is parallel to the light polarization, thereby excluding the direction that is perpendicular to the light polarization. In a continuous photoisomerization process, this direction will become enriched in chromophores, because those found along it are inert to light, photoisomerization, and subsequent motion. The concentration of chromophores aligned perpendicular to the light polarization steadily increases under illumination with polarized light, until a saturation level is attained. It is very important to emphasize that there are *two* directions perpendicular to the light polarization. One direction is in the plane of the film, and these chromophores can be monitored; the other is in the direction of the light propagation, which is perpendicular to the polymer film surface, that is, in the direction of the film thickness. The chromophores aligned preferentially in the homeotropic direction are usually invisible to the normal monitoring of the film. If the light is circularly polarized, there will be no preferred orientation within the film plane, but about the same amount of chromophores will probably align in the homeotropic direction. Until recently, we neglected to investigate these chromophores in our studies.

After a preferred orientation has been photoinduced in the polymer film and irradiation is terminated, this orientation can be conserved, or not, depending on the nature of the polymer film. We will discuss this orientation preservation when we discuss the polymers we synthesized.

The second level of motion is at the domain level, roughly the nanoscale level. It is present in all instances when constraints on the azo groups exist. This type of motion, like the next one, *requires* that the chromophore be bound to the polymer matrix. It also requires that the matrix has some degree of intrinsic order, which can be liquid crystalline or semicrystalline. When the chromophores align themselves into an ordered structure, photoisomerization and the first type of motion are hindered, since they tend to destroy the intrinsic order. Nevertheless, azobenzenes have a high quantum yield for photoisomerization, and the driving force for selection along the "blind" perpendicular orientation is very strong; thus what happens is reorientation of whole liquid crystalline or crystalline domains to a direction perpendicular to the light polarization. This is known as *cooperative motion* and is very common in ordered materials. The order parameter within the domain does not change, but an overall orientation of whole domains occurs, thus creating a very strong overall orientation—much stronger than in amorphous polymers. Because these motions occur at the level of liquid crystalline or crystalline domains within the material, the amount of moved material is greater than in the first case. The domain size varies with the thermal history of the material, but it usually is at the nanoscale level. This second kind of motion has very interesting consequences when circularly polarized light is irradiated on a polymer film containing organized domains. The orientation of the domains affects the polarization of the light, which, in turn, changes the orientation of the other domains along the film depth, thus creating a supramolecular chiral helical arrangement.[3,4]

Finally, the third type of motion is at an even larger scale; it can be called macroscopic motion. It also requires that the chromophore be bound to the polymer, and it involves massive motion of the polymer material. The driving force here is the pressure gradients created by interfering light and unequal isomerization patterns, or the electric field of the light, depending on which mechanism is more plausible. This motion can produce patterns on the film surface that are visible to the naked eye, their depth and spacing being at the micrometer scale. This was an extremely unexpected finding, and we were fortunate to be the first group to report it,[5] just before Tripathy's report appeared.[6]

To summarize, there are three types of photoinduced motions: at the molecular level, at the nanometer, or domain, level, and at the micrometer (macroscopic) level. All are the result of photoinduced isomerization of the azobenzene groups. One interesting direction of our research was to try to exploit these phenomena for their photonic applications. We have demonstrated, at least as proof of principle, a few possible photonic functions for the new materials we synthesized. Some of these are summarized in publications;[7,8] they will also be reviewed here.

13.2.1 Photoinduced Motions at the Molecular Level

The first pair of polymers[9,10] (Figure 13.1) studied in our laboratory were the polyacrylate (pDR1A) and the polymethacrylate (pDR1M) derived from Disperse Red 1. Their glass transition temperatures were 91° and 129°C, respectively. Disperse Red 1 and the polymers derived from it have a maximum absorbance at about 480 nm. The actual maximum wavelength depends on the conditions, such as in film or in solution, and on the possibility of association of the chromophores. This absorbance is associated with the $\pi - \pi^*$ transition of the *trans* azobenzene isomer. A thin film of one of the polymers can be subjected to laser irradiation (either 488 or 514 nm are close enough to the

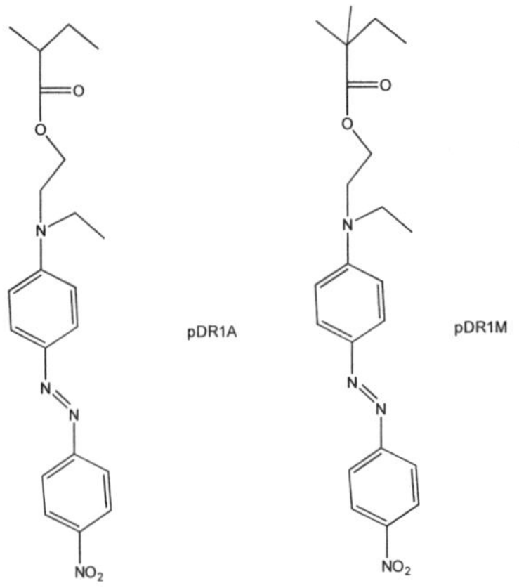

FIG. 13.1 pDR1A and pDR1M.

absorption maximum of the materials), which will induce photoisomerization. Because the absorbance of the *cis* azobenzene isomer, which is formed under irradiation, is in the same region of the spectrum,[11] both *trans-cis* and *cis-trans* photoisomerizations will be pumped while the film is illuminated. This allows efficient photoinduced motion, and—as explained in the previous section—the initially random orientation of the azobenzene chromophores will gradually change toward a preferred orientation perpendicular to the light polarization. We measured this as either dichroism or, preferably, birefringence.

A typical experiment consisted of three steps, and it was repeated a number of times on the same spot of the polymer film. The first step was to turn on the linearly polarized laser and measure the birefringence through cross-polarizers as a function of illumination time. The birefringence starts increasing relatively quickly and reaches a saturated level in a time that varies between less than a second and a few minutes. The factors affecting the time to reach saturation will be discussed later. The photostationary level of the birefringence also depends on many factors that will be discussed in this section. For pDR1M, for example, in optimum illumination conditions, the birefringence value is 0.1. After the photostationary state is reached, we turn the pumping laser off; this is the second step of the experiment. A relaxation process starts, and the photoproduced *cis* isomers thermally relax back to *trans*, which is the stable state. At the same time, motions accompanying this isomerization and heat dissipation allow some of the *trans* isomers aligned perpendicular to the light polarization to randomize their orientation. Thus, the birefringence decreases to some extent. Nevertheless, when working with amorphous, high-Tg polymers, some of the birefringence is conserved for a long time because, after this initial relaxation process, the photoinduced orientation is frozen in the material where motions cannot occur due to its glassy state. How long the birefringence can stay at this level is not yet clear, but our oldest sample, where we induced birefringence in 1991, is still showing about the same level in the years that have passed. This sample has been kept in a relatively dark drawer. How much of the birefringence is conserved again depends on a series of structural and experimental factors that will be discussed later. The third step of a typical experiment involves turning on again the same laser, but now its polarization is circular rather than linear. *Trans-cis-trans* photoisomerization starts again, although now there is no preferred direction within the plane of the film; thus the chromophores are allowed to return to their random orientations. Since the first process of photoinducing birefringence can be called *writing*, this process is obviously called *erasing*, because the photoinduced birefringence is eliminated by optical means. A typical curve (writing-relaxation-erasing) is illustrated in Figure 13.2.

At the time of our research, to our best knowledge, there were a few novel findings that we reported for the first time. Photoinduced birefringence by this Weigert effect had been reported and explained by a Bulgarian group in 1984[12] in a polymer material containing a donor-acceptor substituted azobenzene molecule dispersed in a physical mixture. In that case, the Tg of the composite material was low enough to allow complete randomization of the photoinduced orientation as soon as the laser was turned off. There was no long-term birefringence left in the polymer film. In 1987,[13] a group in West

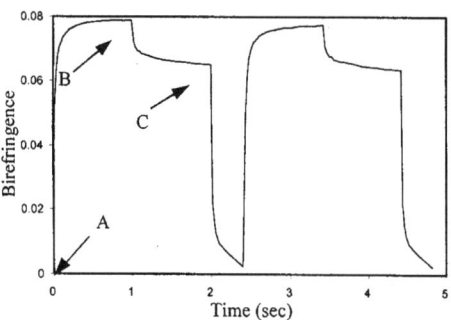

FIG. 13.2   A typical curve of photoinducing birefringence (*writing*, starting at A), relaxation (starting at B), and eliminating birefringence (*erasing*, starting at C). Reprinted with permission from reference 7. Copyright 1997 American Chemical Society.

Germany had started studying liquid crystalline polymers. Their film had to be prealigned (by a magnetic or electric field), and then the orientation was changed by the use of polarized light. Birefringence levels were very high, because of the intrinsic birefringence of the liquid crystalline state, and practically permanent, because of the thermodynamically stable ordered state. To eliminate this birefringence one had to heat the whole polymer film above the isotropization temperature; after cooling, the liquid crystalline phase reformed within broken, disordered domains.

By using amorphous, high-Tg polymers, we reported a combination of some desirable properties of the two formerly reported systems. There was a significant level of long-term birefringence (about 80% of the photostationary state) in a material with no liquid crystalline phase. Better yet, this birefringence could be optically (thus selectively) "erased." The same experiment could be repeated after "erasure" on the same micron-sized spot of the polymer film and the same results would be obtained: Birefringence could be photoinduced to the same level and with the same rate, it would relax in a similar way, and it could be erased again and again. In fact, we showed that fatigue, as bleaching, occurred after the spot had received a certain amount of total power. At very low irradiation intensities, more than 100,000 writing/erasing cycles could be performed on the same spot.[14]

The first question that had to be answered was, What were the optimal experimental conditions to photoinduce and photoeliminate birefringence. A typical experimental setup is shown in Figure 13.3. Since the polymer film is absorbing the pump light, we could calculate a maximum film thickness that would not completely absorb the pump beam.[15] This is approximately 4 divided by the initial film absorptivity; thus, in the case of pDR1A and pDR1M homopolymers, the useful thickness cannot exceed 300 nm. Obviously, polymers with lower absorptivities at the laser wavelength or with lower chromophore concentration can afford thicker useful films.

We then started synthesizing many novel azobenzene-containing polymers in order to investigate what were the most important structural factors affecting this photoinduced orientation. In order to be able to compare various polymers,

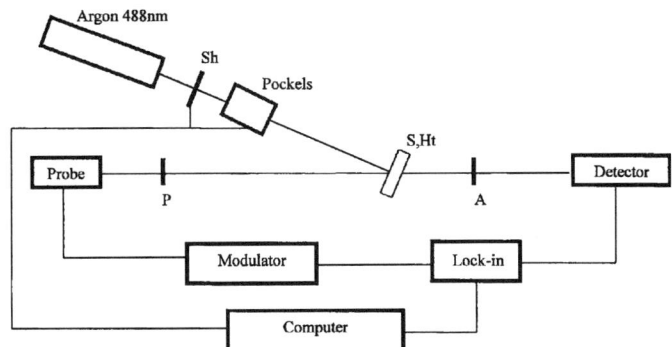

FIG. 13.3 Experimental setup for measuring photoinduced and photoeliminated birefringence. S, Ht = Sample on heating stage, Sh = Shutter, P = Polarizer, A = Analyzer.

the processes of photoinducing birefringence (A to B in Figure 13.2) and of relaxation (B to C in Figure 13.2) were modeled by two biexponential equations:

$$\Delta n = A[1-\exp(-k_a t)] + B[1- \exp(-k_b t)] \tag{13.1}$$

$$\Delta n = C\exp(-k_c t) + D\exp(-k_d t) + E \tag{13.2}$$

These equations describe the growth of the photoinduced orientation (13.1) and its relaxation (13.2) as a function of time. Most of the time it is useful to normalize ($\Delta n = 1$) the terms in order to find out the dynamic parameters, because various polymers produce various birefringence levels, which are also dependent on their structure. A brief review of the important structural factors follows.

13.2.1.1 The Bulkiness and Polarity of the Chromophore

While maintaining the same concept of a (meth)acrylate polymer with a side-group azobenzene chromophore, amorphous and with a high Tg, we wanted to investigate the role of the chromophore. Introducing a chlorine atom at the ortho position to the azo bond[16] (pDR13A, Figure 13.4), or replacing one of the benzene rings with naphthalene[17] (pNDR1M, Figure 13.4) slows down the process of photoinduced birefringence, probably because of the increased bulkiness of the chromophore. The photostationary level of birefringence does not appear to be affected by the bulkiness. The reason why pDR13A gives a significantly higher birefringence than pDR1A is most likely correlated to the difference in polarity and thus to the different absorbance at the laser wavelength. Increasing bulkiness at the amine bound to the azobenzene group, as in pNCARBA[18,19] in Figure 13.4, has similar consequences, plus an increase in birefringence stability after the pump laser has been turned off, for the polymers with shorter spacer lengths. PNCARBA and other similar polymers have additional useful properties that will be discussed later. Another increase in bulkiness, this time coupled with an increase of azo bond concentration and a significant decrease in polymer solubility, is afforded by p3RM[20] in Figure 13.4. This system is more complicated, and its behavior is still currently under investigation.

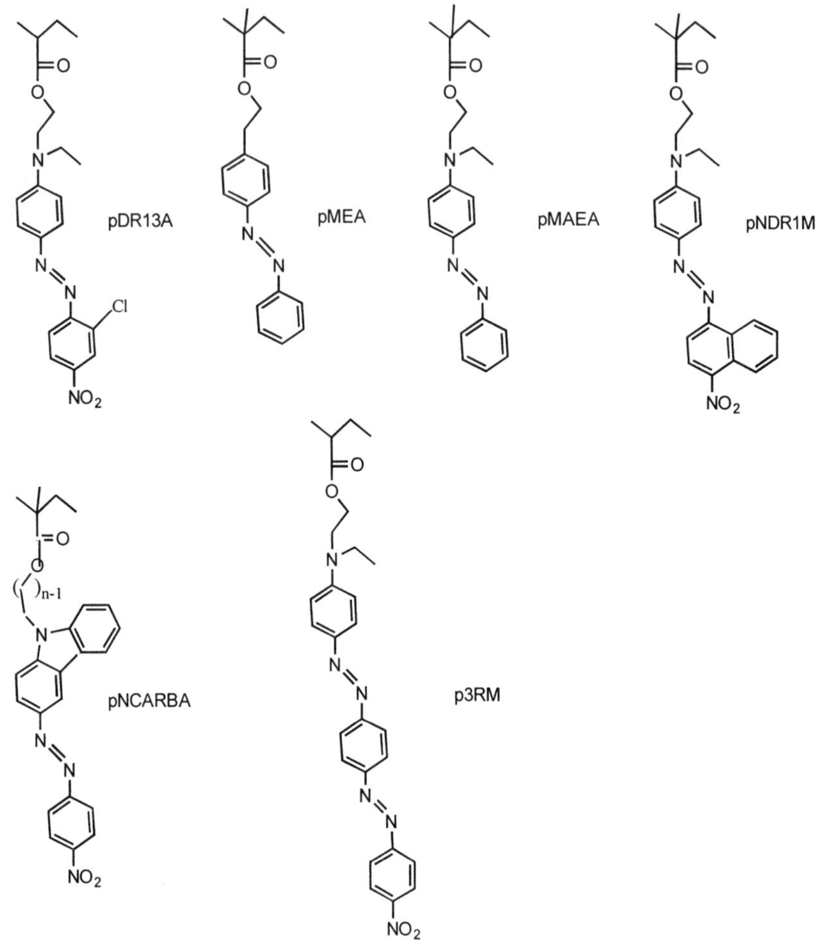

FIG. 13.4 pDR13A, pMEA, pMAEA, pNDR1M, pNCARBA, and p3RM.

The chromophore polarity has a drastic effect on the photoinduced birefringence. This is expected, because the polarity will affect the absorbance and thus the absorptivity at the laser wavelength. Moreover, the polarity affects the photoisomerization behavior itself[2]; thus the comparison among pDR1M (Figure 13.1), pMEA[21] and pMAEA[22] (Figure 13.4), although it does generate some numbers, is not quite fair. PMEA is of the azobenzene type, and the levels of photoinduced birefringence are about 10 times lower than pDR1M. On the other hand, much thicker films of pMEA can be used to photoinduce birefringence. PMAEA is at an intermediate level and is of the amino-azobenzene type, its behavior is closer to pDR1M, but its performance is a bit poorer, as expected.

13.2.1.2 The Type of Bonding of the Chromophore to the Main Chain

All polymers shown in Figures 13.1 and 13.4 have the chromophores bound to the main chain through one covalent bond. The type of bonding used to bond the chromophore to the main chain should definitely affect its mobility, thus the photoinduced birefringence of the polymer film. We studied different modes of binding the azo chromophore to the polymer chain, and the structures are shown in Figure 13.5. For example, one can use two bonds, as in the case of pANPP.[23] This significantly increases the Tg of the polymer (156°C for pANPP as compared to 91°C for the analogous pDR1A), increases the attainable level of photoinduced birefringence, and, as expected, slows down the birefringence growth process. Another possibility is to have the chromophore bound to the main chain, either through two short spacers (pMNAP,[24]) or directly, that is, part of the chromophore is a part of the main chain (pI.[25]) Then, things change significantly. The level of photoinduced birefringence cannot exceed 0.03–0.04, due to the restricted motion of the chromophore, and the rate at which birefringence can be photoinduced is

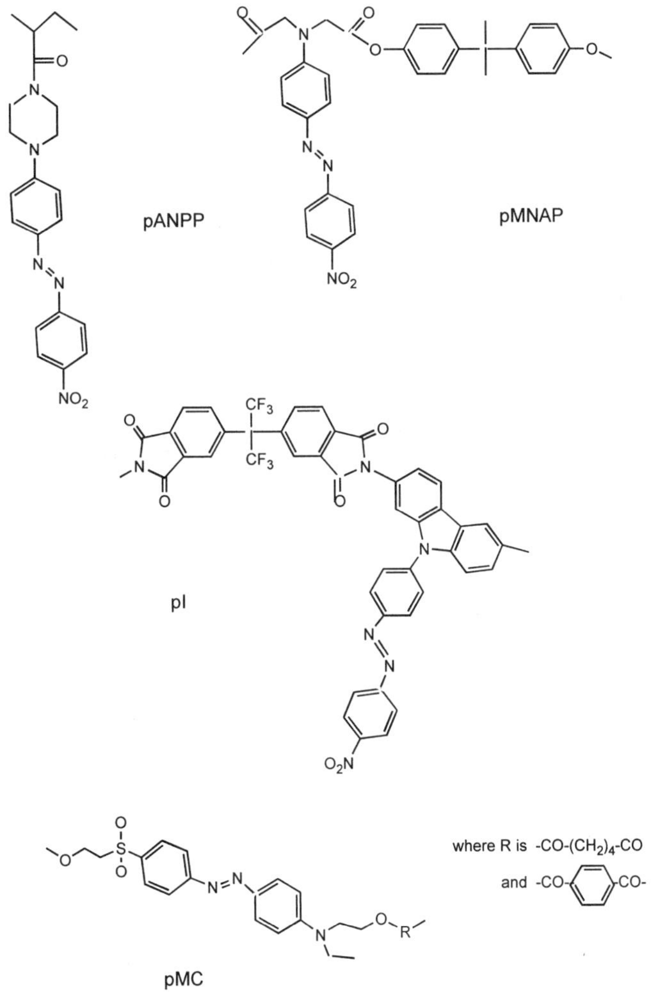

FIG. 13.5 pANPP, pMNAP, pI, and pMC.

very low. In the case of pMNAP, the polymer decomposes on heating, which prevents its use at higher temperatures (to increase the mobility). The polyimide pI is extremely thermally stable by design, but its drawback is that the imide bonds lower the electron-donating ability of the nitrogen end of the chromophore, thus shifting the maximum absorbance and lowering the photoinduced motion efficiency. The advantage of this polyimide is that the stability of the photoinduced birefringence is excellent.

To increase the stability of photoinduced birefringence, we designed new copolymers to contain photocrosslinkable groups, and we analyzed the effect of crosslinking these materials in various conditions on the long-term stability of the photoinduced birefringence.[26]

Based on the side-chain polymer results, we designed polymers containing the azo chromophore as part of the main chain to contain some spacers, because we expected that binding the chromophore at two ends would restrict its motion too much. Thus the series pMC[27] is actually a copolymer series, containing various ratios of adipic acid and terephthalic acid residues. To our surprise, these polymers turned out to provide very high azobenzene mobility: The best photoinduced birefringence levels were 0.07, and their stability was among the lowest of all polymers we studied. One part of the explanation could be the relatively low molecular weight of the polymers, but other factors—for example, the relatively high flexibility of the spacers—are surely significant. Other azo polymers, polyurethanes containing two types of azo groups in the main chain,[28] shown in Figure 13.5, behave in a more "normal" way, both for photoinduced birefringence and for surface relief gratings inscription.

13.2.1.3 The Azo Group Concentration

It is normal to expect that the higher the chromophore concentration in the polymer, the higher the birefringence that can be obtained from that material. This is indeed the case for a few series of copolymers and blends. DR1M was copolymerized with methyl methacrylate (MMA)[29] and with styrene.[30] PDR1M and pDR1A were mixed with pMMA to produce compatible blends.[31] DR1A was also copolymerized with MMA.[29] In all these cases, the level of photoinduced birefringence is proportional to the azo content of the material. Some unexpected results were obtained for p(DR1M-co-MMA) and p(DR1A-co-MMA).[32] For these copolymers, if one calculates the birefringence obtained per azo unit and plots it as a function of the azo content in the copolymer, the copolymers with very low azo content appear to produce much more birefringence than would be expected. The stability of the photoinduced birefringence in low-content azo samples also appears to be lower. This unexpected behavior opened a new area of research for us. If indeed these results were reliable, they pointed to the possibility of cooperative motion. Almost all azobenzenes we used have strong dipole moments. It is known that such dipoles tend to associate and form aggregates.[33] A detailed description of such aggregates is not available, but a clear indication that they exist is given by the change in absorbance wavelength as a function of azo content in the copolymers.[32] Such aggregates will behave differently from isolated chromophores when addressed by light. Their photoisomeriza-

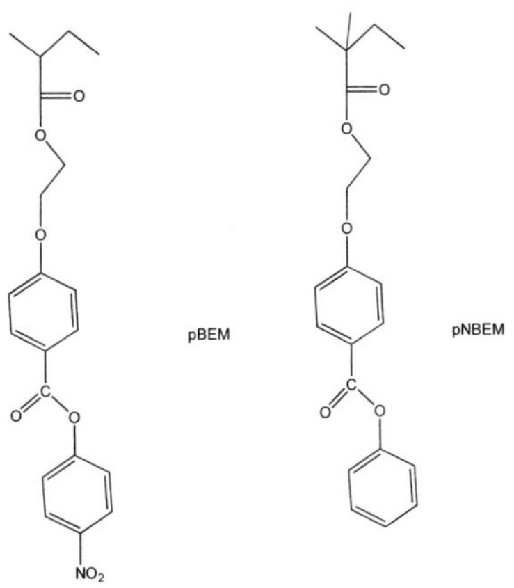

FIG. 13.6 pBEM and pNBEM.

tion will be hindered or amplified, and it is reasonable to assume that they will move in concert. This means that disrupting the initial random orientation of the chromophores will be more difficult if they are aggregated, that is, it would be easier to photoinduce birefringence in copolymers with less azo, where the probability that they are isolated is higher. This is indeed what was observed for copolymers with MMA, but—as a note of caution—it was not observed when the comonomer was styrene.[30] It is also important to note that aggregation, especially for samples with strong dipoles, like ours, is much more likely to occur between polymer molecules rather than within the same polymer. Intermolecular aggregation is available for both copolymers and blends, whereas intramolecular aggregation cannot take place in blends.

13.2.1.4 Cooperative Motion

In order to confirm the existence of this possible cooperative motion, we synthesized two series of copolymers. The first series contained DR1M and BEM[34] (Figure 13.6) structural units, and the copolymer compositions covered the whole possible range. A shift of absorption wavelength confirmed that dipole aggregation did indeed exist in this copolymer series, and the photoinduced birefringence measurements confirmed a significant contribution of the BEM chromophores to the photoinduced motion. This system is very interesting, in that pBEM or the BEM structural units have no absorbance around 500 nm; that is, they are completely inert to irradiation at either 488 or 514 nm. Obviously, even if it had any absorbance, the BEM unit, being an ester, does not isomerize. Nevertheless, the birefringence obtained in the copolymer films is much higher than would be expected for almost all compositions; when the copolymer contains 50% BEM units or more, the birefringence is practically constant at the value exhibited by pDR1M (Figure 13.7). The only source for this additional birefringence has to be the cooperative motion of the BEM units. Thus, light induces photoisomerization and orientation of the azo chromophores in a direction perpendicular to the light polarization; at the same time, the BEM units are coerced to move together with the azo

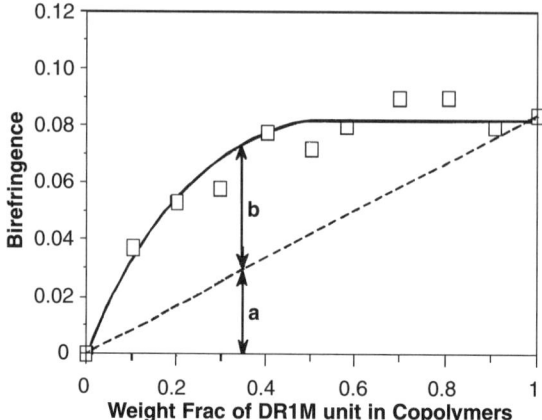

FIG. 13.7 Maximum photoinduced birefringence for p(DR1M-co-BEM) as a function of copolymer composition. Reprinted with permission from reference 34. Copyright 1996 American Chemical Society.

units. In fact, the orientation obtained from the BEM units can reach a maximum of 80% of the orientation obtained from the azo groups.

Photoinduced birefringence illustrates what is happening in bulk, but use of time-dependent polarization modulation infrared spectroscopy can offer a detailed insight at the molecular level.[35] Different infrared bands can be monitored in time, and the change in their orientation function allows one almost to "watch" the different groups move in real time. In order to do this, the process must be slowed down considerably, which is achieved by using a fraction of the pump intensity.

These studies confirm that the BEM groups move together with the DR1M groups, and the kinetics of the process depends on copolymer composition. While the photoorientation and relaxation of the azo groups followed the already established biexponential behavior, the BEM groups' photoinduced orientation could be described by a single exponential, and they did not relax. This suggests that the *cis-trans* isomerization that happens when the pump beam is turned off is the most important relaxation process in all polymer films.

To our knowledge, this was the first instance when cooperative motion was demonstrated in amorphous polymers below Tg, although its effects had been briefly mentioned before. Cooperative motion is well known in liquid crystalline and crystalline polymers, especially in the ordered phases, and it is caused by the thermodynamic tendency to maintain order. Below Tg, it was believed that there is not enough mobility to allow like groups to move together. This belief was similar to the belief that liquid crystallinity was necessary to move azobenzene groups under illumination. These are only two examples of how our research led to a revision in the understanding of polymer physical chemistry.

At this point, there was still one unanswered question: What is the principal driving force for this cooperative motion below Tg? Is it a steric factor, similar to the cooperative motion in crystalline and liquid crystalline polymers? Is it a dipolar interaction, as assumed in azobenzene aggregation? Is it a combination of both? To answer these questions, a second copolymer series was synthesized and analyzed in comparison with p(DR1M-co-BEM). The comonomer, NBEM,[36] Figure 13.6, was similar to BEM, but it lacked the nitro group; thus, its polarity was much smaller. The copolymer series analyzed was p(DR1M-co-NBEM), covering all compositional range. Both bulk (by photoinduced birefringence) and molecular level (by time dependent polarized infrared spectroscopy) behaviors were investigated. The results clearly showed that polarity is the determining factor in cooperative motion below Tg. NBEM also moves in concert with DR1M, but to a much lesser extent. This means that one can describe the photoinduced birefringence phenomenon as a disturbance of the electric field of the dipole arrangement in the film formed by spin coating. The groups must be distributed in such a way that their dipoles cancel each other to produce an electrically neutral film. When light starts to move some of the azo groups in a direction perpendicular to its polarization, the local director is reoriented and the remaining molecules are no longer at the minimum potentials but are in gradient electric fields. Those with strong dipole moments will tend to reorient toward the minimum of the local fields. Thus, the motion of the azo groups reorients the nonphotoactive molecules in the same direction. This behavior can occur on the molecular scale as well as on the larger domain scale, as explained in the next section.

An unexpected consequence of these findings was that some contradictions found in the literature (from 1991) could now be explained.[36] The so-called molecular addressing term coined by Anderle et al[37] was proven to be correct, that is, the azo groups *can* be selectively addressed by light, provided that the other groups present in the material are of significantly different polarity than the azo groups *and* the material is below its Tg.

13.2.2 Photoinduced Motions at the Domain Level

When the polymer already has crystalline or liquid crystalline domains, and therefore order already exists in the system, the behavior under illumination can be different. Since most azobenzene chromophores are mesogenic, they usually are part of the liquid crystalline phase or the crystalline phase. The earliest reports of photoinduced birefringence in liquid crystalline azo polymers[13,38] and in semicrystalline azo polymers[39] date from 1987. Whereas for the liquid crystalline polymers the emphasis was on inscribing holographic gratings, the photoinduced birefringence in the semicrystalline copolymers was as high as 0.21, which is significantly higher than the values obtained later on amorphous polymers.

The explanation for these high values of photoinduced birefringence is simple. Liquid crystalline and semicrystalline polymers consist of domains of highly ordered chromophores. These domains have no initial directional orientation, but polarized light can, through photoisomerization, change the orientation of whole domains to align them, more or less, along the direction perpendicular to its polarization. The order parameter does not really change upon illumination; what changes is the direction of alignment. Thus, the

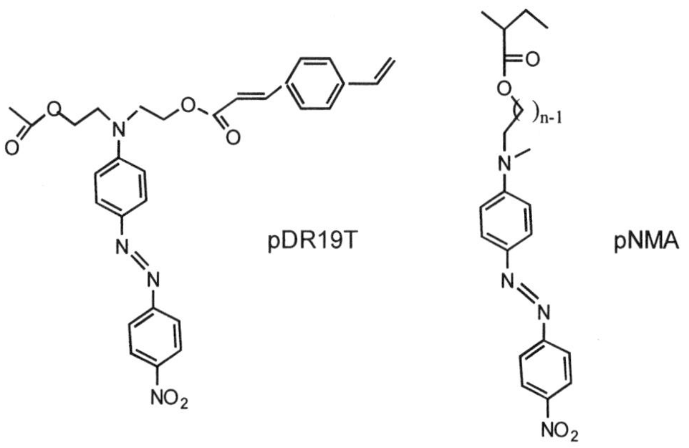

FIG. 13.8 pDR19T and pNMA.

birefringence levels in this case are intrinsic, not created by light. This is the case if the polymer film is already in its semicrystalline or liquid crystalline state. The most common case, however, is when the polymer film is amorphous as cast, usually below its Tg, and, upon solvent evaporation, there is not enough motion to allow crystallization or formation of the liquid crystalline phase. In this case, photoisomerization plays a similar role to heating the polymer above its Tg, that is, it provides the mobility required for the mesogens to form the ordered domains. One could argue that this is a novel type of phase transition, a photoinduced phase transition (or photoinduced crystallization).

We have quite a few pieces of evidence that this kind of phenomenon happens in semicrystalline and liquid crystalline polymers. Other reports in the literature suggest that this is what happens. It is not within the scope of this chapter to discuss all available arguments, pro and con. Instead, we will summarize only our results here. A semicrystalline polymer, pDR19T[40] (Figure 13.8) was subjected to linearly polarized light as an amorphous film. Figure 13.9 illustrates the differences between its behavior and the behavior of pDR1A. The level of photoinduced birefringence is much higher, 0.25 for pDR19T, but the time to achieve this level is much longer (almost 30 seconds, compared to less than 1 second for pDR1A). Both of these findings agree with the possibility that crystallization occurs in pDR19T while photoisomerization takes place. Crystalline domains are formed and oriented perpendicular to the light polarization. This process is much slower than the photoinduced orientation in a simple amorphous phase. After illumination, optical observation through cross-polarizers under microscope confirms the presence of an ordered phase in the region that was subjected to the laser light. Perhaps the most interesting fact is that when the light is turned off, the birefringence slightly *increases* (Figure 13.9). This is consistent with the tendency of the system to use the *cis-trans* and all other available thermal energy to form and grow the crystalline domains. In the amorphous polymers, the birefringence decreases. Another argument in favor of photoinduced crystallization is that the writing-erasing process is no longer reversible, as is in amorphous polymers.

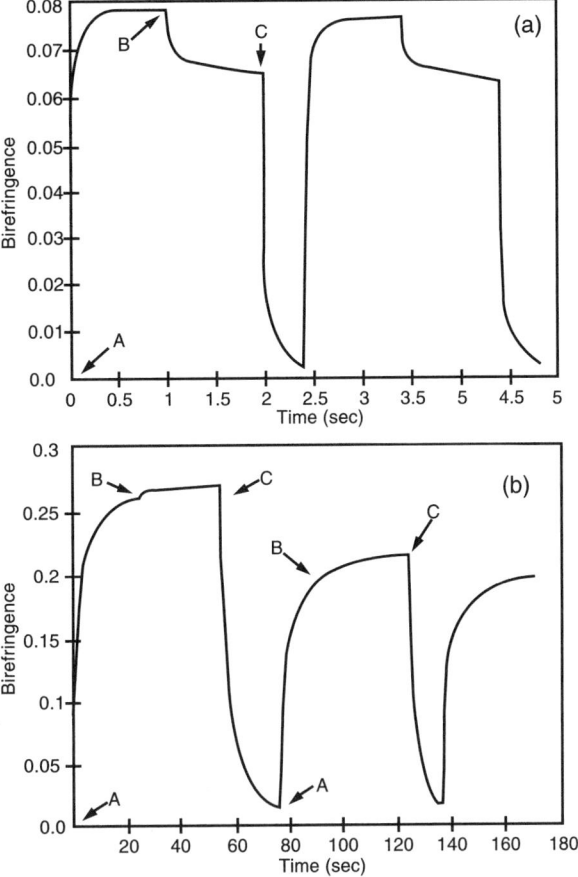

FIG. 13.9 Typical writing curves for (a) pDR1A (b) pDR19T. A = writing laser is turned on; B = writing laser is turned off; C = erasing laser is turned on. Reprinted with permission from reference 40. Copyright 1994 American Chemical Society.

Another example of formation of an ordered phase, this time a smectic liquid crystalline phase, under illumination followed by thermal treatment can be found in a series of liquid crystalline polymers pnMAN,[41] Figure 13.8. In this case, the intrinsic tendency to form ordered domains within the systems is lower. Therefore, when the polarized light is turned on after photoinducing birefringence, a usual, albeit smaller, relaxation occurs. Figure 13.10 illustrates this behavior. Birefringence is photoinduced at levels similar to amorphous polymers (0.12 in p4MAN), and it decays slightly upon termination of illumination. If the sample is heated after the light has been turned off, the birefringence increases; in this case, it can grow to 144% of its initial value. This increase is assigned to the formation and growth of smectic liquid crystalline domains along the photoinduced direction. If heating continues above the smectic-isotropic transition temperature, the birefringence is gradually lost (Figure 13.10a). If heating is terminated before the smectic phase become isotropic, the increased birefringence is conserved (Figure 13.10b).

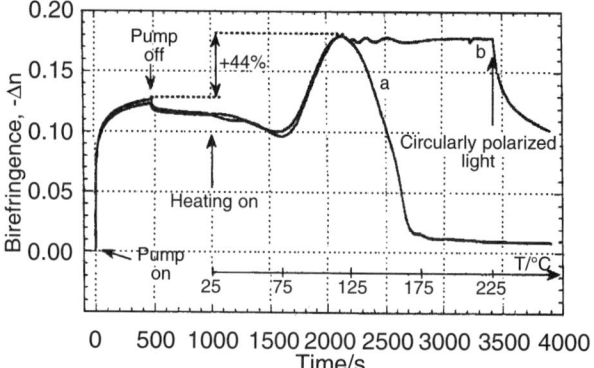

FIG. 13.10 Birefringence cycles on freshly prepared p4MAN films. For film a, the linearly polarized light is turned on at $t = 0$ for 500 s; after a 500 s relaxation period, the sample is *in situ* heated at a rate of 5°C/min to 175°C and then cooled to room temperature. For film b, the sample is cooled rapidly when the maximum birefringence is reached at approximately $t = 2200$ s as the temperature reached 120°C. Reprinted with permission from reference 41. Copyright 2000 American Chemical Society.

We would like to emphasize that, whereas photoinduced phase transitions to isotropic are fairly well known in the literature,[42] reports of photoinduced phase transitions *toward* the ordered phases are very unusual and not well explained, to our knowledge. We are currently investigating such phenomena.

The pnMAN series has intriguing thermochromic and photochromic properties due to chromophore aggregation.[41] Since all polymers in this series have the tendency to form smectic phases, the possibility of chromophore aggregation is greater than in all other polymers we have studied. Structurally, just replacing the usual ethyl group bound to the amine nitrogen with a methyl group allows a better parallel (or antiparallel) arrangement of the chromophores. It is reasonable to assume antiparallel aggregation, because of the strong dipole, and the consequence is a significant blue shift of the absorbance. Figure 13.11 shows an example of the change in the UV-Vis spectrum as a

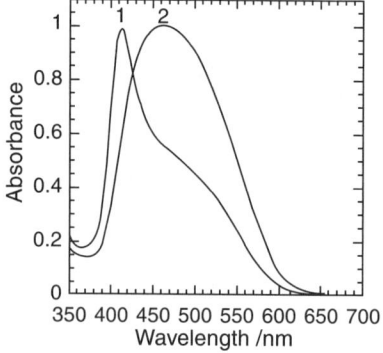

FIG. 13.11 UV-visible spectra of a thin film of p9MAN as a function of temperature. Spectrum 1 was recorded at room temperature, whereas spectrum 2 was recorded *in situ* at Tg + 20°C. Reprinted with permission from reference 41. Copyright 2000 American Chemical Society.

function of temperature for p9MAN. The polymer film, as cast, forms a highly aggregated phase, which is stable. This is due to the relatively low Tg of p9MAN (40°C), which allows the long spacers enough flexibility and mobility to provide aggregation during solvent evaporation. Heating up to 20°C above Tg increases motion to the point that the aggregates are broken. The maximum absorbance is so different for these two states that the polymer film color is initially orange and changes to red when the polymer is heated. The aggregated phase is thermodynamically favored, though, and upon cooling the orange state is restored. Things are slightly different for polymers with shorter spacers that do not have enough freedom to form the aggregated state by casting. The film as cast is red for all polymers with spacers shorter than nine methylene groups, and Figure 13.12 shows an example for p4MAN. Upon heating above Tg, the increased motion now allows aggregation, and the polymer color becomes orange. This, again, is the stable form of the film, and the orange polymer film can be maintained indefinitely in the absence of irradiation. When heating at higher temperatures, the smectic phase appears. It consists of broken domains of aggregated chromophores that have different orientations in the film plane. The absorbance is more symmetrical than in the cast film state, and the film color appears to be opaque red. The smectic phase is stable, and the color red is maintained for an indefinite length of time. Thus, the thermochromic behavior of this polymer series can be explained and controlled.

A more interesting phenomenon is the possibility to break the chromophore aggregation (or to photoinduce the smectic phase) using light. For the shorter spacers, the orange film is irradiated with linearly polarized light, preferably through a mask. The irradiated parts turn red, probably from either rotation of the chromophores in their aggregated state away from each other or, more likely, a phenomenon similar to that reported above: Light allows the smectic phase to appear, and the smectic phase is stable. One can thus "write" red on orange through a mask on a polymer film. The "writing" is stable if the sample is maintained in the dark.

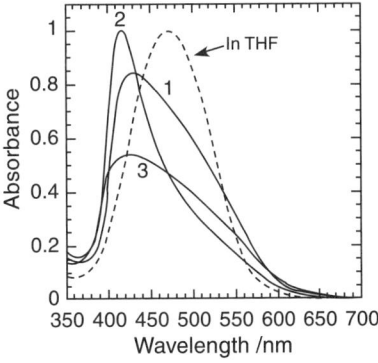

FIG. 13.12 UV-visible spectra of a thin film of p4MAN as a function of temperature. Spectrum 1 was recorded at room temperature, whereas spectra 2 and 3 were recorded *in situ* at Tg + 20°C and Tg + 60°C, respectively. Reprinted with permission from reference 41. Copyright 2000 American Chemical Society.

A newly discovered and exciting phenomenon that liquid crystalline polymers undergo takes place in the smectic phase and under irradiation with circularly polarized light. When a film of p4MAN, for example, is heated into its smectic phase, randomly distributed domains are formed on the surface and within the film depth. Addressing such a film with circularly polarized light should have as a main effect a preferential orientation of the azo chormophores perpendicular to the film surface (a homeotropic orientation), as discussed previously in this section. This does happen, to some extent. However, while the circularly polarized light goes through smectic domains, its polarization will be affected and slightly changed from circular to elliptic. Going along the light propagation direction, perpendicular to the film surface, the beam will encounter the next smectic domain, but now the light is not circularly but slightly elliptically polarized. This means that homeotropic orientation of the chromophores will still happen, but it will be accompanied by a reorientation of the smectic domain in a direction perpendicular to the long axis of the ellipse. At the same time, the light will be affected again in the same sense, that is, its ellipticity will increase. Coming now to the third domain within the depth of the film, there is still more orientation of the domain and more change of the light polarization, and so on. The end result is that the initially circularly polarized light will induce domain orientation of the same handedness as the light polarization. A supramolecular chirality will be produced in the illuminated areas, right circularly polarized light will produce a right-handed helical organization of the smectic domains within the film, and left circularly polarized light will produce a left-handed helical organization. Figure 13.13[4] illustrates the circular dichroism spectrum of two spots on a p4MAN film. Before irradiation, there was no chirality and no optical activity in the polymer film. This phenomenon has been previously reported and explained on other smectic liquid crystalline polymer films containing azobenzene groups.[3]

Another interesting aspect of this photoinduced supramolecular chirality is that the light is capable not only of winding up a helical structure of smectic domains, but it also is capable on unwinding the helix and rewinding it in the different direction. This phenomenon is illustrated in Figure 13.14.[4]

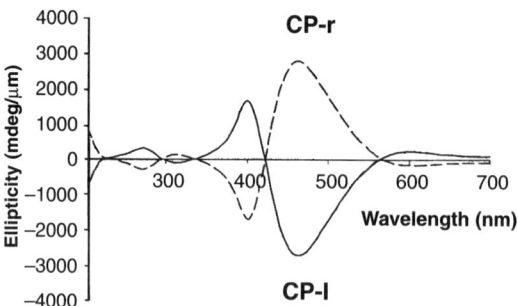

FIG. 13.13 CD spectra of thin films (140 nm) of p4MAN recorded after irradiation with circularly polarized light (514nm, 75 mW/cm²): induced with CP-l (—) and with CP-r (---), respectively. Reprinted with permission from reference 4. Copyright 2000 American Chemical Society.

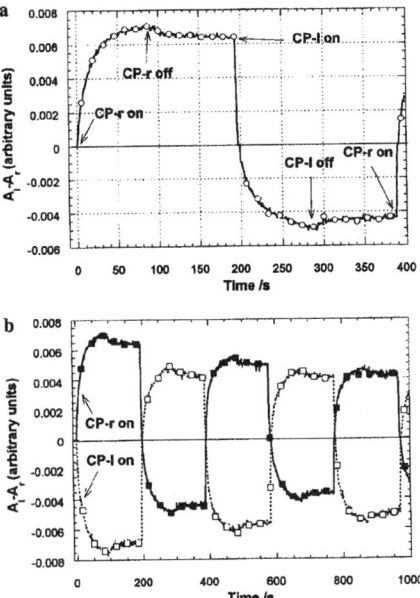

FIG. 13.14 (a) Cycle of chirality switching. At $t = 0$, the film is irradiated with CP-r during 100 s followed by a relaxation (pump off) during the same period. At $t = 200$ s, the pump beam is set to CP-l for 100 s. (b) Several cycles of chirality switching started initially with CP-r polarized light (full squares) or CP-l polarized light (empty squares). Reprinted with permission from reference 4. Copyright 2000 American Chemical Society.

It is apparent in Figure 13.14b that the absolute level of chirality decreases for each cycle, but heating the film above isotropization and then cooling to room temperature restores the initial level that can be attained. Thus, the apparent decrease in photoinduced chirality is due to the increased homeotropic alignment of the azobenzene chromophores, not to a bleaching effect, and it is thus reversible.

These findings are novel and potentially very useful. At this time they raise more questions than they answer, and our group is studying a series of copolymers to help clarify this phenomenon.

13.2.3 Photoinduced Motions at the Macroscopic Level

The third level at which the photoinduced isomerization of the azobenzene groups bound in polymers might affect the polymer material is very large, and it can be called macroscopic. In 1995 we[5] and immediately afterwards Tripathy's group[6] reported that interfering circularly polarized light creates surface-relief gratings of significant depth (up to microns) on azo polymer films. The phenomenon occurs below the Tg of the material, and heating the material to Tg restores the flat film surface.

We discovered this phenomenon by chance when attempting to compare the efficiency of the photoinduced birefringence process with the efficiency of the refractive index modulation under electric field in photorefractive polymers.

Photorefractive polymers were and are the subject of intense investigation because of their potential photonic functions.[43] The efficiency of storage in photorefractive polymers was always reported as diffraction efficiency; thus we decided to inscribe a volume holographic grating on our azo polymer film and measure its diffraction efficiency. The grating would be the result of oriented and disordered areas in the polymer film, which would themselves be the product of different polarizations of the incoming light. The setup for inscription consisted of an expanded beam split by a mirror, to give stability to the inscription. Since the beam intensity was lower, due to expansion, the time required for the experiment was longer. At the end of some 50 minutes of irradiation, we found—in addition to the expected volume phase grating— a grating on the surface of the polymer film that was visible to the naked eye. Atomic force microscopy revealed a quasi-sinusoidal deformation of the polymer film surface, reaching depths comparable to the initial film thickness (close to 1 µm). These gratings are not optically erasable, or at least we have never succeeded in optically erasing a previously inscribed grating. They can be erased thermally, by heating the film slightly above Tg and allowing the material to flow and reform the film with its initial thickness. One advantage of this stability is that one can inscribe, successively, many gratings on the same polymer film. Figure 13.15 illustrates a double grating viewed with an atomic force microscope. Very soon afterwards, we (and other research groups worldwide) discovered that practically all azo polymers are amenable to surface gratings inscription. In addition, the inscription conditions have been optimized. Currently—under similar conditions—only a few seconds of irradiation are required to obtain an acceptable grating.

A most important fundamental question is: What is the mechanism of formation of such gratings? Currently, there are four proposed mechanisms in the literature,[44–48] but we will describe only our proposed mechanism. A schematic representation is shown in Figure 13.16. The photoisomerization in a stiff matrix (the polymer below its Tg) is assumed to create pressure in the matrix because the *cis* isomer is about 4% bulkier than the *trans* isomer. Photoisomerization takes place continually for as long as the irradiating beam is on; therefore, the pressure is also exerted continually. The sinusoidal curve in the figure shows the light-intensity profile resulting from two interfering circularly polarized beams. Note that the intensity on the surface varies

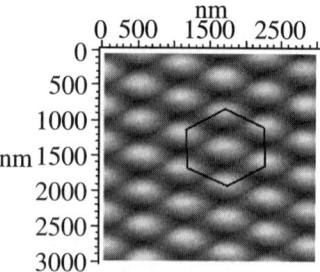

FIG. 13.15   Hexagonal surface relief pattern written on a thin film.

Proposed Mechanism

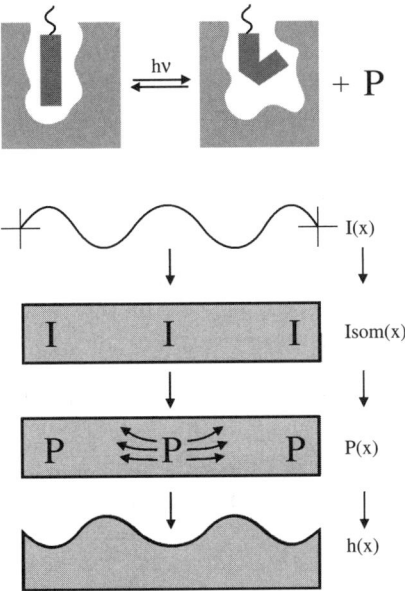

FIG. 13.16 Schematic representation of the mechanism of surface gratings formation, involving pressure gradients. P is the pressure, I represents the light intensity and isomerization, and x is the place on the polymer film.

because the beams have a finite angle between them. Below, the polymer film is represented as a rectangle. Where the light intensity is high, the isomerization is more intense (represented by an I); at low illumination intensities, there is less isomerization. Consequently, pressure gradients are created in the film, that is, there are regions on the polymer film of alternating higher pressure (P) and lower pressure (due to photoisomerization). If the polymer film is above its yield point, a flow of material will be created from the regions of higher pressure to the regions of the lower pressure. The yield point of the polymer in the region of absorption is lowered because the isomerization process plasticizes the polymer. As the relief grows, the flow rate is reduced because the surface-tension produces a counter pressure. Thus, when the process reaches a saturated state, the surface-relief profile shown at the bottom of the figure has peaks where the light intensity was the lowest and troughs where the light was most intense.

The key finding that has guided us toward this proposed model was the fact that copolymers of DR1M and MMA generated nice surface-relief gratings, whereas the blends of the respective homopolymers, although compatible, did not. The explanation was that the molecular weight of purchased pMMA used for blending was much higher than the molecular weights of our nitro-containing homo- and copolymers. As soon as we used low molecular pMMA, which was below the yield point, the blends started to behave in an identical manner to the copolymers.[44] A more careful analysis and modelling of the phenomenon, including an array of factors influencing the gratings-inscription process,[45] confirmed the validity of the proposed model.

Research in the area of photoinduced surface-relief gratings is extremely active. One review[49] covers the literature published until 1998, but the field has practically exploded. One recent paper[50] even reports the creation of a holographic grating with a single short laser pulse, although there are reports questioning the mechanism.[51]

One interesting and intriguing recent finding was discovered when analyzing the process of thermally erasing the surface-relief gratings. Concurrent with the disappearance of the surface-relief deformation, a volume-density grating is being produced under illumination.[52] The periodicity of the buried grating is essentially identical to that of the initial surface grating, and the high-density regions appear to have been created at the places where the peaks were present. The proposed (tentative) mechanism involves yet another hypothesis of formation of a liquid-crystalline-like phase, which will be the dense buried phase. The more intriguing aspect of this mechanism is that pDR1M, the material showing this behavior, has no known liquid crystalline phase in the absence of the particular conditions used in this experiment. The formation of the buried lateral grating is not generally applicable to all azo polymers. PMEA, for example, has less propensity to form a liquid crystalline phase and does not form this below-the-surface photoinduced structure. Obviously, our group is currently conducting more research in this area. Investigations into the early state of formation of surface-relief gratings are also informative with respect to the mechanism of grating formation. Pulses of various lengths were used to reveal the importance of the local lateral forces.[53]

13.3 POSSIBLE PHOTONIC DEVICES

This brief review of our findings in azo polymer behavior when subjected to laser illumination raises the obvious question: Are any of these phenomena useful? The answer is yes, and what follows is a brief review of the potential applications we saw for our novel materials. All of these are at the stage of proof-of-principle demonstration.

13.3.1 Reversible Optical Storage

Ever since we noted the reversibility of the azobenzene groups' photoinduced alignment, we were inclined to see this procedure as a reversible optical-storage process. One could write (photoinduce birefringence), store (in the dark), erase (with circularly polarized light), and rewrite on the same spot on the polymer film. The rates of writing and erasing, as well as the level of long-term storage, could be manipulated by using different polymer structures and illumination procedures. The long-term birefringence stability was demonstrated by our oldest sample (11 years), and the repeatability of the inscription was demonstrated on the same spot (about 100,000 cycles). Nevertheless, the obvious disadvantage of this process was the rather long time required to reach a saturation level. The shortest time we achieved was about 10 ms, which is too long for real-time storage. Even though reaching the saturation level is not required, the whole process is inherently slow. A realistic alternative for reversibly storing information on such films is to use holographic storage, by inscribing whole pages of information at the same time. This is the principle used by Bayer scientists in liquid crystalline polymers with very high Tg.[54] The photoinduced-birefringence process is still conducted in the amorphous phase, because it is faster than in the liquid crystalline phase. The order parameter, though, is dictated by the liquid crystalline phase, and the birefringence is very high, allowing the possibility of using a gray scale and multiple "writing" levels. The erasing becomes a bit problematic, but this is a technical problem that already has a few suggested solutions.[54]

13.3.2 Holographic Storage

The obvious use of the surface-relief gratings is in holographic storage, since the grating is the hologram of the mirror used to create it. Scientists in Denmark have built a demo that shows real-time holographic storage using azo polymers. The obvious advantage is the absence of any postinscription processing. The hologram is optically inscribed and is stable (in our case for seven years, as per our oldest sample). The fact that the holograms are stable and that more than one can be written (and read) on the same film surface confers a third dimension (and a huge memory capacity) for this process. An example of a technical question that still has to be answered is, What is the maximum number of images that can be independently recorded on the same piece of film?

13.3.3 Waveguides

Instead of one single point irradiated on the polymer film, one could inscribe a line, for example. The inscribed line would be birefringent while the rest of the polymer film would be isotropic. This line can act as a waveguide, by maintaining a beam of light within its confines.[55] We did not attempt to optimize this function with respect to losses, and so forth, but showed that such lines can be drawn and used as waveguides. The advantage of such a waveguide is that, even when it is part of a functioning circuit, it can be erased and redrawn in another part of the circuit and with different contacts.

13.3.4 Couplers

In order to couple light into the above-mentioned waveguide (or another, more classical, waveguide), one could optically inscribe a grating on the polymer film. The grating would diffract the incoming beam into the waveguide without the need of prisms or any other additional elements. This procedure could also be used on a different kind of photonic device, because one could deposit a small piece of azo polymer film by solvent evaporation at the site where coupling is necessary and then write a grating on it. The coupling was shown to be efficient.[56]

13.3.5 Wavelength and Angle Filters

The surface-relief gratings will diffract light along the waveguide when the incident light comes at a required angle. This angle depends on the geometry of the device and on the depth and periodicity of the grating. Thus, one can design a filter that will selectively eliminate (send into the waveguide) specific angles of the incident light.[57] Alternatively, one could select the appropriate wavelengths of light to be filtered.[8] The filters are very selective and efficient.

13.3.6 Polarization Separators

Holographic images can be recorded as surface-relief patterns, as mentioned earlier. But holographic images can also be recorded in the volume of the polymer film in the form of locally induced dichroism and birefringence. In the later case, the information that is recorded is not only the intensity variation in the image but also the local polarization state of the light in the image. One direct application that we have demonstrated was the fabrication of a film that detected the circular polarization content of a beam by diffracting the right circularly polarized light in one direction and the left circularly polarized light in the other direction. This was done by inscribing a polarization hologram produced when two contra-circularly polarized beams interfered on the film. In this case, there is only a small variation in light intensity, but the local polarization of the resultant light varies periodically as a function of position. This is recorded as a periodic variation of birefringence in the film and acts as a polarization-sensitive phase grating that selectively diffracts the probe beam, thereby separating the polarization states.[58]

13.3.7 Optoelectronic Switches

For a photonic circuit that is optically inscribed on a simple azo polymer film, apart from the waveguide and the coupling functions described above, one needs a switching function. The two-wave coupling phenomenon can be employed as a switch; thus, photorefractive properties would be desirable in order to obtain two-wave coupling in our polymers. For a polymer to show photorefractive properties, some conditions have to be met.[43] First, one needs electro-optic activity. Azobenzenes were some of the first chromophores to show such activity when poled in an electric field.[59] Second, one needs photoconduction; and third, one needs charge trapping. The last condition is ubiquitous in polymer films. Thus, in our case, the only requirement was the presence of a photoconductive moiety. We chose to employ carbazole,[18,19,25] and demonstrated that the polymers could be poled under an electric field to generate the photorefractive effect and then that two-wave mixing (thus switching) can be obtained. We did not work under optimum conditions, and we did not manage to obtain gains that offset the polymer absorbance in the region of interest. Some of the preliminary results were extremely encouraging.[60] Overall, we have demonstrated that a full photonic device can, in principle, be optically written and redesigned even while functioning on a simple azo polymer film (Figure 13.17).

13.3.8 Liquid Crystal Orientation

Another possible use for the surface-relief gratings is as alignment layer for liquid crystals. Since the depth and periodicity of the surface-relief gratings can be adjusted at will, one could use the troughs to provide the direction for alignment of various liquid crystals. The procedure is much cleaner than the current commercial procedure of rubbing a polyimide layer, and it is very efficient in providing the required orientation.[61] The alignment can be performed either before or after the cell has been filled with the liquid crystal. The main concern related to this procedure is the possible solubility of the azo polymer layer on the wall of the cell in the liquid crystal that has to be aligned. The obvious solution to this problem is a pretreatment that would crosslink and thus render the azo polymer layer insoluble prior to filling the liquid crystal. After the alignment, one could use a single linearly polarized argon laser beam to photoinduce a twisted alignment structure, which can, in turn, be erased by a circularly polarized beam. Thus the liquid crystal cell can be used for optically switched displays.[62] Another intriguing and promising use of the surface-relief gratings as support is in building possible photonic crystals by allowing polystyrene spheres to self-assemble on photoinscribed surfaces.[63]

13.3.9 Optical Switches

Finally, the possibility of pure optical switching has been demonstrated in principle by photoinducing supramolecular helical arrangements of smectic domains in some liquid crystal azo polymer films.[4] The photoinduced chirality is pronounced and stable (our oldest sample is still fairly young), but the process is again slow, because it requires reorientation of whole domains.

13.4 CONCLUSIONS

Research in azobenzene-based polymers is extremely active and challenging. It has generated in the past, and continues to generate, many unexpected

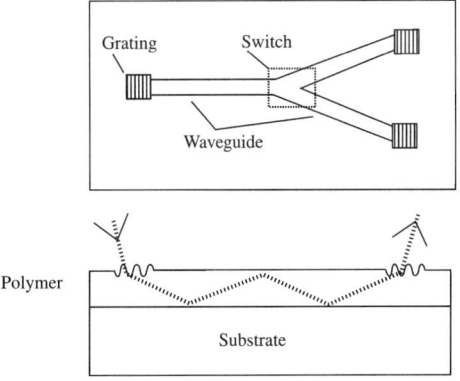

FIG. 13.17 Photonic device optically inscribed on a polymer film. Reprinted with permission from reference 7. Copyright 1997 American Chemical Society.

results and requires quite a bit of revision of our understanding of polymer-materials behavior under illumination.

ACKNOWLEDGMENTS

A. N., Canada Research Chair in Polymer Chemistry, thanks the CRC Program of the Government of Canada. We acknowledge funding for this research by NSERC Canada, Department of Defense, Canada, and the Office of Naval Research, United States. This research was performed by a series of very talented and industrious young scientists, working as students and postdoctoral associates. We are extremely grateful for their collaboration.

REFERENCES

1. Natansohn, A., Rochon, P. (submitted). Photoinduced motions in azo-containing polymers. *Chem. Rev.*
2. Rau, H., (1990). Photoisomerization of azobenzenes. *In* "Photochemistry and Photophysics" (J. K. Rabek, Ed.), Vol. 2, 119–141, CRC Press, Inc: Boca Raton, FL.
3. Ivanov, M., Naydenova, I., Todorov, T., Nikolova, L., Petrova, T., Tomova, N., Dragostinova, V. (2000). Light-induced optical activity in optically ordered amorphous side-chain azobenzene containing polymer. *J. Mod. Opt.* **47**, 861–867.
4. Iftime, G., Lagugné Labarthet, F., Natansohn, A., Rochon, P. (2000). Control of chirality of an azobenzene liquid crystalline polymer with circularly polarized light. *J. Am. Chem. Soc.* **122**, 12646–12650.
5. Rochon, P., Batalla, E., Natansohn, A. (1995). Optically induced surface gratings on azo polymer films. *Appl. Phys. Lett.* **66**, 136–138.
6. Kim, D. Y., Tripathy, S. K., Li, L., Kumar, J. (1995). Laser-induced holographic surface relief gratings on nonlinear optical polymer films. *Appl. Phys. Lett.* **66**, 1166–1168.
7. Natansohn, A., Rochon, P. (1997). Azobenzene-containing polymers: Digital and holographic storage. *In* "ACS Symposium Series: Photonic and Optoelectronic Polymers" (K. J. Wynne and S. A. Jenekhe, Eds.), Vol. 672, 236–249, American Chemical Society: Washington, DC.
8. Natansohn, A., Rochon, P. (1999). Photoinduced motions in azobenzene-based amorphous polymers. *Adv. Mater.* **11**, 1387–1391.
9. Rochon, P., Gosselin, J., Natansohn, A., Xie, S. (1992). Optically induced and erased birefringence and dichroism in azoaromatic polymers. *Appl. Phys. Lett.* **60**, 4–5.
10. Natansohn, A., Rochon, P., Gosselin, J., Xie, S. (1992). Azo polymers for reversible optical storage. 1. Poly[4′[[2-(acryloyloxy)ethyl]ethylamino]-4-nitroazobenzene]. *Macromolecules* **25**, 2268–2273.
11. Loucif-Saibi, R., Nakatani, K., Delaire, J. A., Dumont, M., Sekkat, Z. (1993). Photoisomerization and second harmonic generation in disperse red one—doped and—functionalized poly(methyl methacrylate) films. *Chem. Mater.* **5**, 229–236.
12. Todorov, T., Nikolova, L., Tomova, N. (1984). Polarization holography.1: A new high-efficiency organic material with reversible photoinduced birefringence. *Appl. Opt.* **23**, 4309–4312.
13. Eich, M., Wendorff, J., Reck, B., Ringsdorf, H. (1987). Reversible digital and holographic optical storage in polymeric liquid crystals. *Makromol. Chem., Rapid Commun.* **8**, 59–63.
14. Natansohn, A., Rochon, P. (1994) Reversible optical storage in azo polymers. *In* "Progress in Pacific Polymer Science" (K. P. Ghiggino, Ed.), Vol. 3, 295–305, Springer Verlag: Berlin Heidelberg.
15. Rochon, P., Bissonnette, D., Natansohn, A., Xie, S. (1993). Azo polymers for reversible optical storage 3. Effect of film thickness on net phase retardation and writing speed. *Appl. Opt.* **32**, 7277–7280.
16. Natansohn, A., Rochon, P., Xie, S. (1992). Azo polymers for reversible optical storage. 2. Poly[4′-[[2-(acryloyloxy)ethyl]ethylamino]-2-chloro-4-nitroazobenzene. *Macromolecules* **25**, 5531–5532.

17. Ho, M. S., Natansohn, A., Rochon, P. (1995). Azo polymers for reversible optical storage 7. The effect of the size of the photochromic groups. *Macromolecules* **28**, 6124–6127.

18. Ho, M. S., Barrett, C., Paterson, J., Esteghamatian, M., Natansohn, A., Rochon, P. (1996). Synthesis and optical properties of poly{4'-nitrophenyl-3-[N-[2-(methacryloyloxy)-ethyl]]carbazolyl diazene}. *Macromolecules* **29**, 4613–4618.

19. Barrett, C., Choudhury, B., Natansohn, A., Rochon, P. (1998). Azocarbazole polymethacry-lates as single-component electrooptic materials. *Macromolecules* **31**, 4845–4851.

20. Meng, X., Natansohn, A., Rochon, P. (1997). Azo polymers for reversible optical storage. 13. Photoorientation of rigid side groups containing two azo bonds. *Polymer* **38**, 2677–2682.

21. Natansohn, A., Rochon, P., Ho, M. S., Barrett, C. (1995). Azo polymers for reversible opti-cal storage 6. Poly[4-(2-methacryloyloxy)ethyl-azobenzene]. *Macromolecules* **28**, 4179–4183.

22. Ho, M. S., Natansohn, A., Barrett, C., Rochon, P. (1995). Azo polymers for reversible optical storage. 8. The effect of polarity of the azobenzene groups. *Can. J. Chem.* **73**, 1773–1778.

23. Meng, X., Natansohn, A., Rochon, P. (1996). Azo polymers for reversible optical storage. 12. Poly{1-acryloyl-4-[4-(4-nitrophenylazo)phenyl]piperazine}. *Supramol. Sci.* **3**, 207–213.

24. Meng, X., Natansohn, A., Rochon, P. (1996). Azo polymers for reversible optical storage. 11. Poly{4,4'-(1-methylethylidene)bisphenylene 3-[4-(4-nitrophenylazo)phenyl]3-aza-pen-tanedionate}. *J. Polym. Sci., Part B. Polym. Phys.* **34**, 1461–1466.

25. Chen, J. P., Lagugné Labarthet, F., Natansohn, A., Rochon, P. (1999). Highly stable optically induced birefringence and holographic surface gratings on a new azocarbazole-based poly-imide. *Macromolecules* **32**, 8572–8579.

26. Takase, H., Natansohn, A., Rochon, P. (2001). Effect of crosslinking on the photoinduced orientation of azo groups in thin polymer films. *J. Polym. Sci.: Part B: Polym. Phys.* **39**, 1686–1696.

27. Xu, Z.-S., Drnoyan, V., Natansohn, A., Rochon, P. (2000). Novel polyesters with amino-sul-fone azobenzene chromophores in the main chain. *J. Polym. Sci.: Part A: Polym. Chem.* **38**, 2245–2253.

28. Wu, Y., Natansohn, A., Rochon, P. (2001). Photoinduced birefringence and surface relief gratings in novel polyurethanes with azobenzene groups in the main chain. *Macromolecules* **34**, 7822–7828.

29. Xie, S., Natansohn, A., Rochon, P. (1994). Microstructure of copolymers containing disperse red 1 and methyl methacrylate. *Macromolecules* **27**, 1885–1890.

30. Iftime, G., Fisher, L., Natansohn, A., Rochon, P. (2000). Photoinduced birefringence in copolymers containing Disperse Red 1 and styrene. *Can. J. Chem.* **78**, 409–414.

31. Xie, S., Natansohn, A., Rochon, P. (1994). Compatibility studies of some azo polymer blends. *Macromolecules* **27**, 1489–1492.

32. Brown, D., Natansohn, A., Rochon, P. (1995). Azo polymers for reversible optical storage 5. Orientation and dipolar interactions of azobenzene side groups in copolymers and blends containing methyl methacrylate structural units. *Macromolecules* **28**, 6116–6123.

33. Menzel, H., Weichart, B., Schmidt, A., Paul, S., Knoll, W., Stumpe, J., Fischer, T. (1994). Small-angle X-ray scattering and ultraviolet-visible spectroscopy studies on the structure and structural changes in Langmuir-Blodgett films of polyglutamates with azobenzene moieties tethered by alkyl spacers of different length. *Langmuir* **10**, 1926–1933.

34. Meng, X., Natansohn, A., Barrett, C., Rochon, P. (1996). Azo polymers for reversible opti-cal storage. 10. Cooperative motion of polar side groups in amorphous polymers. *Macromolecules* **29**, 946–952.

35. Buffeteau, T., Natansohn, A., Rochon, P., Pezolet, M. (1996). Study of cooperative side group motions in amorphous polymers by time dependent infrared spectroscopy. *Macromolecules* **29**, 8783–8790.

36. Natansohn, A., Rochon, P., Meng, X., Barrett, C., Buffeteau, T., Bonenfant, S., Pézolet, M. (1998). Molecular addressing? Selective photoinduced cooperative motion of polar ester groups in copolymers containing azobenzene groups. *Macromolecules* **31**, 1155–1161.

37. Anderle, K., Birenheide, R., Werner, M. J. A., Wendorff, J. H. (1991). Molecular addressing? Studies on light-induced reorientation in liquid-crystalline side chain polymers. *Liq. Cryst.* **9**, 691–699.

38. Eich, M., Wendorff, J. (1987). Erasable holograms in polymeric liquid crystals. *Makromol. Chem. Rapid Commun.* **8**, 467–471.

39. Tredgold, R. H., Allen, R. A., Hodges, P., Khoshdel, E. (1987). Films formed from co-polymers containing azobenzene side groups. *J. Phys. D: Appl. Phys.* **20**, 1385–1388.

40. Natansohn, A., Rochon, P., Pézolet, M., Audet, P., Brown, D., To, S. (1994). Azo polymers for reversible optical storage 4. Cooperative motion of rigid groups in semicrystalline polymers. *Macromolecules* **27**, 2580–2585.

41. Lagugné Labarthet, F., Freiberg, S., Pellerin, C., Pézolet, M., Natansohn, A., Rochon, P. (2000). Spectroscopic and optical characterization of a series of azobenzene-containing side-chain liquid crystalline polymers. *Macromolecules* **33**, 6815–6823.

42. Ikeda, T., Horiuchi, S., Karanjit, D., Kurihara, S., Tazuke, S. (1990). Photochemically induced isothermal phase transition in polymer liquid crystals with mesogenc phenyl benzoate side chain.2. Photochemically induced isothermal phase transition behaviors. *Macromolecules* **23**, 42–48.

43. Moerner, W. E., Silence, S. M. (1994). Polymeric photorefractive materials. *Chem. Rev.* **94**, 127–155.

44. Barrett, C. J., Natansohn, A. L., Rochon, P. L. (1996). Mechanism of optically-inscribed high efficiency diffraction gratings in azo polymer films. *J. Phys. Chem.* **100**, 8836–8842.

45. Barrett, C. J., Rochon, P. L., Natansohn, A. L. (1998). Model of laser-driven mass transport in thin films of dye-functionalized polymers. *J. Chem. Phys.* **109**, 1505–1516.

46. Pedersen, T. G., Johansen, P. M. (1997). Mean-field theory of photoinduced molecular reorientation in azobenzene liquid crystalline side-chain polymers. *Phys. Rev. Lett.* **79**, 2470–2473.

47. Lefin, P., Fiorini, C., Nunzi, J. M. (1998). Anisotropy of the photo-induced translation diffusion of azobenzene dyes in polymer matrices. *Pure Appl. Opt.* **7**, 71–82.

48. Kumar, J., Li, L., Jiang, X. L., Kim, D.-Y., Lee, T. S., Tripathy, S. (1998). Gradient force: the mechanism for surface relief grating formation in azobenzene functionalized polymers. *Appl. Phys. Lett.* **72**, 2096-2098.

49. Viswanathan, N. K., Kim, D. Y., Bian, S., Williams, J., Liu, W., Li, L., Samuelson, L., Kumar, J., Tripathy, S. (1999). Surface relief structures on azo polymer films. *J. Mater. Chem.* **9**, 1941–1955.

50. Ramanujam, P. S., Pedersen, M., Hvilsted, S. (1999). Instant holography. *Appl. Phys. Lett.* **74**, 3227–3229.

51. Baldus, O., Leopold, A., Hagen, R., Bieringer, T., Zilker, S. J. (2001). Surface relief gratings generated by pulsed holography: a simple way to polymer nanostructures without isomerizing side-chains. *J. Chem. Phys.* **114**, 1344–1349.

52. Pietsch, U., Rochon, P., Natansohn, A. (2000). Formation of a buried lateral density grating in azobenzene polymer films. *Adv. Mater.* **12**, 1129–1132.

53. Henneberg, O., Geue, T., Saphiannikova, M., Pietsch, U., Chi, L. F., Rochon, P., Natansohn, A. L. (2001). Atomic force microscopy inspection of the early state of formation of polymer surface relief gratings. *Appl. Phys. Lett.* **79**, 2357–2359.

54. Hagen, R., Bieringer, T. (2001). Photoaddressable polymers for optical data storage. *Advanced Materials* **13**, 1805–1810.

55. Barrett, C., Natansohn, A., Rochon, P. (1997). Photoinscription of channel waveguides and grating couplers in azobenzene polymer thin films. *Proceedings SPIE* **3006**, 441–449.

56. Paterson, J., Natansohn, A., Rochon, P., Callender, C., Robitaille, L. (1996). Optically inscribed surface relief diffraction gratings on azobenzene-containing polymers for coupling light into slab waveguides. *Appl. Phys. Lett.* **69**, 3318–3320.

57. Rochon, P., Natansohn, A., Callender, C. L., Robitaille, L. (1997). Guided mode resonance filters using polymer films. *Appl. Phys. Lett.* **71**, 1008–1010.

58. Lagugné Labarthet, F., Rochon, P., Natansohn, A. (1999). Polarization analysis of diffracted orders from a birefringence grating recorded on azobenzene-containing polymer. *Appl. Phys. Lett.* **75**, 1377–1379.

59. Burland, D. M., Miller, R. D., Walsh, C. A. (1994). Second-order nonlinearity in poled-polymer systems. *Chem. Rev.* **94**, 31–75.

60. Iftime, G., Lagugné Labarthet, F., Natansohn, A., Rochon, P., Murti, K. (2002). Main chain–containing azo-tetraphenyldiaminobiphenyl photorefractive polymers. *Chem. Mater.* **14**, 168–174.

61. Li, X. T., Natansohn, A., Rochon, P. (1999). Photoinduced liquid crystal alignment based on

a surface grating in an assembled cell. *Appl. Phys. Lett.* 74, 3791–3793.

62. Li, X. T., Natansohn, A., Kobayashi, S., Rochon, P. (2000). An optically controlled liquid crystal device using azopolymer films. *IEEE J. Quantum Electronics* 36, 824–827.

63. Ye, Y. H., Badilescu, S., Truong, V. V., Rochon, P., Natansohn, A. (2001). Self-assembly of colloidal spheres on patterned substrates. *Appl. Phys. Lett.* 79, 872–874.

14

SURFACE-RELIEF GRATINGS ON AZOBENZENE-CONTAINING FILMS

OSVALDO N. OLIVEIRA, Jr.*   **JAYANT KUMAR**
LIAN LI   **SUKANT K. TRIPATHY**

Center for Advanced Materials, University of Massachusetts Lowell, 1 University Avenue, Lowell, MA 01854, USA
*Instituto de Física de São Carlos, USP, CP 369, 13560-970, São Carlos, SP, Brazil

ABSTRACT

An overview of the processes involved in direct photoinscription of surface-relief structures (SRGs) on azobenzene-containing polymer films is presented in this chapter. The phenomenon was first reported by two research groups independently in 1995. Since then, SRGs have been produced by various groups on amorphous side-chain and main-chain azopolymers, and on liquid-crystalline azopolymers, mostly in the form of spin-coated films. Sol/gel materials, Langmuir-Blodgett (LB) films, and films assembled by ionic layer-by-layer interactions have also been used. SRGs are inscribed by projecting an interference pattern of laser light onto the films, which causes mass transport over micrometer distances at temperatures well below the glass-transition temperatures of the polymers. Because the transport is associated with the molecular reorientation upon isomerization, the presence of an azochromophore is essential. When modest light intensities are used, mass transport is predominantly light driven, with negligible thermal effects. SRGs thus fabricated depend on the polarization of the writing beams and can be erased either optically or thermally. These gratings are π-shifted in relation to the interference pattern, which implies that the material moves away from the illuminated regions, and they have considerable depth only where there is variation of both light intensity and field gradient. These features can be explained by a model based on a field-gradient force. Applications of these SRGs include diffraction gratings for holography, waveguide couplers, alignment of liquid crystals and photonic devices.

14.1   INTRODUCTION

Ever since Todorov et al.[1] observed birefringence gratings (BGs) on azobenzene-containing polymers, the photoisomerization characteristics have been investigated extensively for a variety of azobenzene-containing polymeric materials with several possible applications in mind.[2] The basic exploited phenomenon is the ability of the azobenzene chromophore to be photoisomerized between its two forms, the lower energy *trans* and the more energetic *cis* forms. These photoisomerization properties make the azopolymers good candidates for optical storage devices,[3] optical switches,[4] holograms,[5] and so on. While investigating the storage characteristics of polymers with the azochromophore covalently attached to the main chain of an epoxy-based polymer, Tripathy, Kumar, and coworkers noticed an unusually large diffraction efficiency, which seemed unlikely to be caused only by the bulk phenomenon of BGs. It was then realized that an unexpected surface deformation occurred, even though the laser power used in writing the gratings was unlikely to lead to any thermal effects. It is worth emphasizing that in all these cases the experiments were conducted at room temperature, which was substantially below the glass-transition temperature (Tg) of the polymers. The formation of such surface-relief gratings (SRGs) was first observed in 1992, which prompted the group to pursue a systematic investigation to rule out experimental artifacts, and the first results were published in 1995.[6] This very effect was also observed independently by Rochon et al.,[7] with their discovery also being reported in 1995.[7]

In the seminal works just mentioned, SRGs were photoinscribed in a single-step procedure on spin-coated films of acrylate, styrene, or epoxy-based polymers containing azobenzene side groups. Because these polymers absorb in the visible region of the spectrum, with an absorption peak around 400–500 nm, excitation was usually performed with the 488 nm line of an Ar+ laser. Modest intensities were used, typically of the order of a few tens to some hundreds mW/cm^2. Inscription was carried out by exposing the polymer surface to an interference pattern at room temperature. A surface modulation depth of up to 600–800 nm could be achieved, depending on the writing conditions. The periodicity could be accurately controlled by varying the angle between the writing beams. For a fixed dosage, below the saturation value, the most efficient writing for epoxy-based polymers was observed for periods of about 1000 nm.[8,9] SRGs with spacings from 270 up to 460 nm could be easily inscribed. The SRGs were stable at temperatures below the Tgs of the polymers, but in many cases they could be photoerased even at room temperature by impinging a single beam of laser light with the appropriate polarization. They could be erased by heating the polymer sample close to or above its Tg, unless polymer crosslinking occurred. The inscription process could be entirely reversible, with several SRGs being written and erased on the same spot. In addition, multiple gratings could be recorded on the same spot, provided that the polarization of the writing beams was properly selected, which allowed a variety of complex patterns to be produced (see Section 14.6.4). The formation of SRGs was usually monitored by measuring the first order diffraction efficiency of a probe beam, such as a low power He-Ne laser beam, which is not significantly absorbed by the azobenzene-containing polymers and therefore does not affect the material properties. Because bulk gratings (birefringence gratings) also give origin to diffraction efficiencies, the presence of an SRG had to be confirmed by monitoring topographical changes in the polymer surface, which was performed with atomic force microscopy (AFM).[6–12]

Since these pioneering works, SRGs have been exploited with not only amorphous but also liquid-crystalline (LC) materials.[13–19] While the vast majority of the work has been done in spin-coated or cast polymer films, SRGs have also been inscribed on Langmuir-Blodgett (LB),[20] electrostatically assembled layer-by-layer (ELBL) films,[21–24] and gels.[25,26] The mechanisms for the production of some of these SRGs differ from the originally suggested all-light-induced mass transport. In some cases, it has already been established that the surface relief is mainly induced by thermal processes; other cases involved photodegradation. Recent papers[27–29] have reviewed the main findings and contributions related to the formation of SRGs on azopolymers, where the processes involved are discussed and the various models to explain the physical mechanisms are analyzed. Nevertheless, there is still some confusion as to the limitations of the models, and sometimes failure to acknowledge that very distinct processes might lead to the formation of SRGs on azopolymer films. In this chapter, we aim at clarifying a number of points in that connection, especially by establishing a clear-cut distinction between the cases where the SRGs are induced by light-driven mass transport only (or mainly) and those where thermal effects and photodegradation are predominant. Here, we shall refer to light-driven mass transport as the cases where ablation, photodegradation, and thermal effects are negligible. In addition, the models proposed in the literature[9,30-37] were generally conceived to explain different aspects of the SRGs, which might also cause misunderstandings that we hope to clarify here.

This chapter is organized as follows. The main processes in the formation of SRGs are discussed in Section 14.2, followed in section 14.3 by a description of the theoretical models proposed to explain the results from SRG recordings in Section 14.3. A number of parameters affect the SRG formation, and Section 14.4 is a summary and a brief review of results published in the literature. In fact, this section represents a short survey of results published in the literature. Section 14.5 discusses topics whose understanding is not complete at this time, and it points to a number of challenges facing scientists to explain fully the mass transport in the SRG formation of the various kinds. Mention will be made in Section 14.6 of possible applications of SRGs, which is a strong motivation for ongoing research.

14.2 PROCESSES OF SRG FORMATION

The discovery of SRG formation on azopolymers occurred in high Tg polymers through the exposure of the sample to an interference pattern from relatively low intensities of a writing laser.[6-12] As subsequent studies confirmed,[29,38-40] under low-intensity writing conditions the surface modulation basically arises from light-driven mass transport, with negligible thermal effects. That is to say, it is an entirely photonic process. It is based on the ability of azobenzene chromophores to undergo facile *trans-cis-trans* isomerizations, which are followed by molecular reorientation. Figure 14.1 shows the two configurations, *trans* and *cis*, which can be distinguished by their absorption characteristics. Figure 14.2 illustrates that molecular reorientation follows the isomerization associated with the thermal relaxation from the higher-energy *cis* to the lower-energy *trans* configuration. It indicates in particular that chromophores aligned perpendicularly to the laser-light polarization have much smaller transition dipole moments. This leads to birefringence and dichroism, which are exploited in optical storage.[3]

Figure 14.3 shows a typical experimental setup employed for inscribing SRGs. The 488 or 514 nm line of an Ar+ laser is split into two beams of equal intensity with appropriate polarizations, and it is made to impinge on the sample surface. The periodicity, Λ, of the resulting grating can be predicted

Photo-isomerization

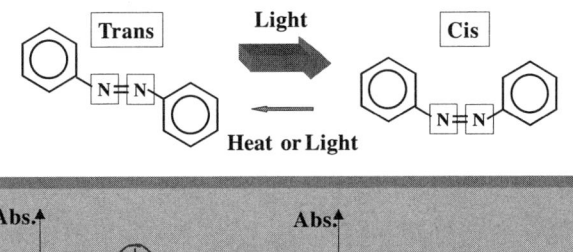

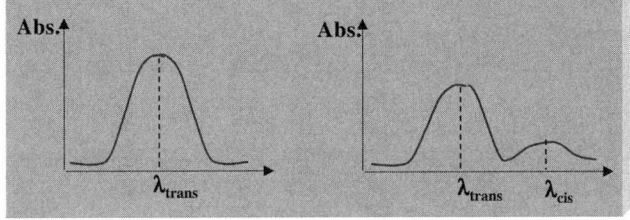

FIG. 14.1   The two configurations of the azobenzene group, namely *trans* and *cis*, are illustrated. They can be distinguished by monitoring the UV-vis absorption of the azobenzene-containing material. The lower figure schematically illustrates the differences in absorption characteristics. Interconversion between *cis* and *trans* can be achieved optically, with the chromophores being excited from the lower-energy *trans* configuration to the higher-energy *cis* configuration. The *cis-trans* isomerization can also be photoinduced, but normally a thermal decay is always present, because the chromophores tend to relax to the lower-energy *trans* configuration.

Orientation Mechanism

Photochemical *trans-cis* rate: $R = I \cos^2(\Phi)$

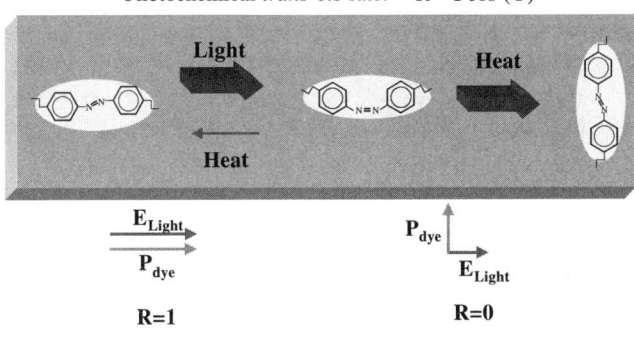

FIG. 14.2   Molecular reorientation resulted an important consequence of the *trans-cis-trans* isomerization cycles. When light with the appropriate wavelength is impinged, the probability of the chromophores undergoing *trans-cis* isomerization will be maximum (Rate R = 1) when their dipole moment is parallel to the laser-light polarization, and zero if it is perpendicular (R = 0). Upon thermal relaxation, the chromophores can adopt any orientation, including the perpendicular direction to the light polarization, in which case they will no longer be affected by the light electric field.

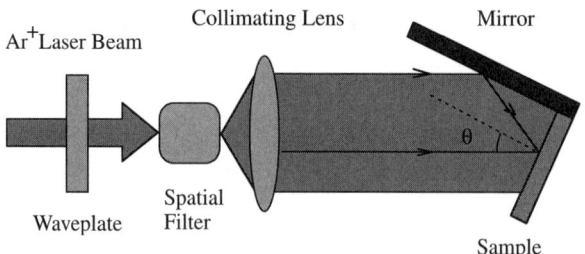

FIG. 14.3   Typical experimental setup for producing SRGs on an azobenzene-containing film. A laser beam, usually from an Ar+ (488 nm or 514 nm) laser, is polarized, spatially filtered, and collimated.

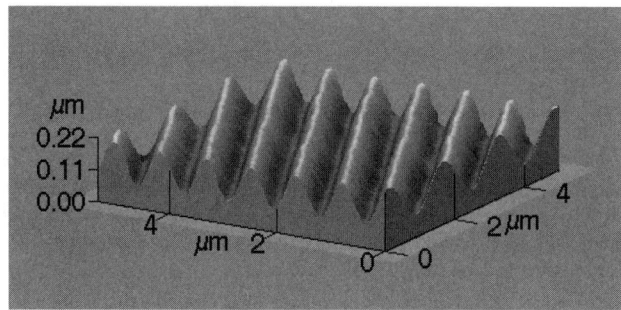

FIG. 14.4   A typical sinusoidal-shaped SRG inscribed on a side-chain azobenzene polymer.

using the relationship $\Lambda = 2\pi/k = \lambda/[2\sin(\theta)]$, and it can be varied by changing the angle between the writing beams, θ. λ is the wavelength of the laser beam.

A typical sinusoidal-shaped SRG inscribed on a side-chain polymer (PDO3) film is shown in Figure 14.4. Though this cannot be inferred from the figure, the peaks of the SRG correspond to the nonilluminated regions of the interference pattern, the reason for which is given in the theoretical model in Section 14.3. The depth of the SRG can be controlled with reasonable accuracy by varying the fluence. For example, Munakata et al.[41] showed that the SRG depth increased linearly with time at the beginning of the irradiation, before slowing down and reaching saturation at a certain depth, which depended on the film thickness. They also noted that the fabrication time could be shortened by varying the laser intensity. For a 1 μm thick film, the saturation depth, 300 nm, could be reached within 5 min with an intensity of 500 mW/cm^2 and within 30 min at 50 mW/cm^2.

Figure 14.5 shows several examples of amorphous azopolymers that have been employed in producing SRGs. Additionally, SRGs have also been produced from LC polymers and under a variety of experimental conditions, such as change of laser intensity[39,41,42] and film-fabrication techniques.[6,20,21,25] Attention should be drawn in particular to the wide range of Tgs of the polymers so far investigated, and to the laser power used for inscribing the gratings. The details of the results are discussed throughout the next section.

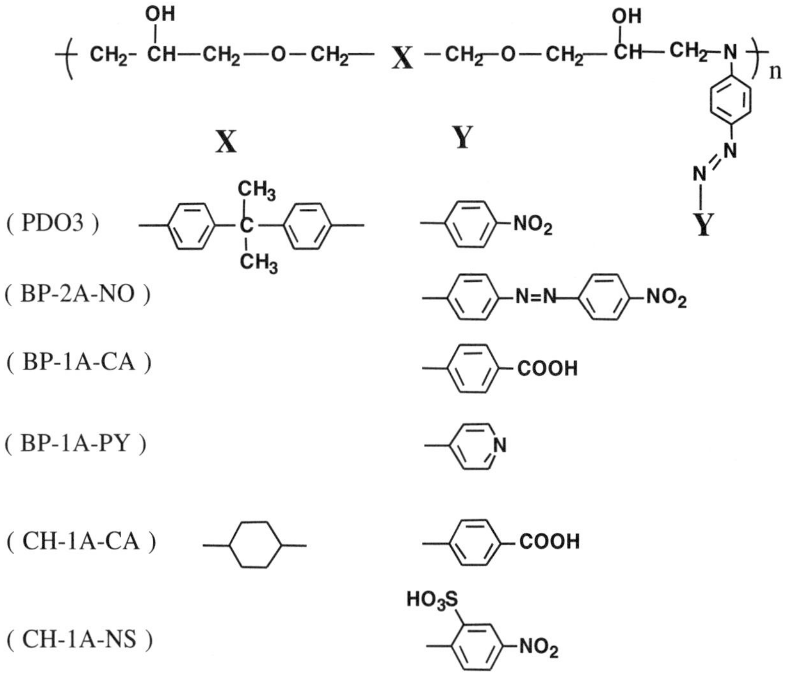

FIG. 14.5 A variety of azopolymers employed in spin-coated films on which SRGs were inscribed. Different substituents were used in the main-chain backbone (X) and also connected to the azobenzene chromophore (Y).

In some of these works, the mechanisms responsible for the SRGs might be completely different from the light-driven mass transport. Indeed, the following mechanisms have been suggested:

1. Light-driven mass transport. This leads to a π phase difference between the interference pattern and the grating, that is, the polymer molecules are moved away from the illuminated regions. This occurs for high Tg, amorphous azopolymers prepared via spin-coating and on gel films, at moderate writing intensities, and is highly dependent on the polarization of the writing beams. The SRGs thus formed can be optically or thermally erased.

2. Thermal ablation. The SRG is normally π-shifted in relation to the interference pattern,[43,44] and is the result of the application of a very intense pulse of laser light or a high-power continuous wave (cw) laser. The SRGs cannot be optically erased, but rather only thermally erased. In some experiments, some polarization dependence was observed,[45] which cannot be explained with present knowledge of the field. The dye does not need to be azobenzene for such SRGs (see Grzybowski et al.[46]).

3. Mass transport associated with photodegraded azochromophores, which occurs for high-power writing lasers. It has been observed with azopolymers as well as small dyes transferred onto a polymer film by the ELBL technique.[21-25] It is interesting that, for spin-coated high Tg polymer

films, this mechanism appears in conjunction with the light-driven mass transport. The predominance of one of the mechanisms can be investigated as a function of the laser power.[39] The dependence on the polarization of the writing lasers is not as strong as in the first item.

4. Mass transport that requires thermal treatment of the samples. The SRGs cannot be erased, even after heating to temperatures close to Tg in some cases.[18,19,47]

5. Mass transport in liquid-crystalline azopolymers. Here two important effects should be considered: thermal effects and the cooperative behavior of the mesogenic unities. The SRGs formed are in phase with the interference pattern, and they also show strong polarization dependence of the writing beams.[14,15,48]

14.2.1 Light-Driven Mass Transport

The most investigated systems, as far as SRGs are concerned, have been those in which gratings were inscribed on high Tg, amorphous azopolymers, under modest light intensities. The origin of such SRGs has been established as due to the light-driven mass transport, with negligible influence of thermal effects. As will be discussed in Section 14.3, in these cases theoretical models already provide qualitative explanations of the experimental results, such as polarization and film-thickness dependence of the SRGs. It is noted that significant SRG formation requires variations in both light intensity and resultant electric field on the azopolymer films. We shall start by mentioning some well-established experimental facts for SRGs formed on these systems, emphasizing the results that demonstrate that thermal effects and photo-degradation are not at the origin of the SRG fabrication.

SRGs arising from light-driven mass transport are only formed on azobenzene-containing materials, which means that the *trans-cis-trans* photo-isomerization is an essential prerequisite.[8] Moreover, even polymers containing other chromophores such as biphenyls that are amenable to photoisomerization do not form SRGs, probably because the free volume required by them (much larger than for azobenzenes) precludes significant isomerization in the solid films.[8] Evidence that thermal effects were not dominating also came from irradiation of a polymer film containing an absorbing, nonisomerizing dye (rhodamine 6G), which led to no SRG.[9] Another convincing proof of the optical origin of the light-driven mass transport was provided by measurements of the diffraction efficiency for the SRG as a function of recording power and energy.[40] As shown in Figure 14.6, for low energies the measured diffraction efficiency is independent of the laser power.

The azochromophore must be covalently attached to the polymer chain (either as a side chain or in the main chain). Attempts to produce SRGs with guest-host systems led to very small surface modulations (about 15 nm or less).[49,50] Though photoisomerization can be induced in these systems, efficient SRG formation is not possible, probably because the polymer chains are not induced to move unless covalently attached. However, SRG formation of a doped system in which the polymer also contained an azobenzene unit has also been studied[51] (see Section 14.4.4). The much less efficient inscription

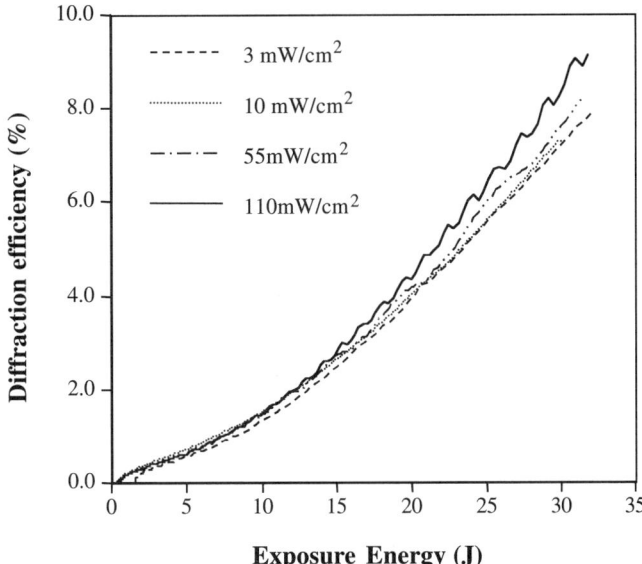

FIG. 14.6 Diffraction efficiency of the SRG as a function of recording power and energy, with the laser polarization set at 45° with respect to the s-polarization. From reference 40.

in a doped system was demonstrated in a direct comparison of two similar systems by Fiorini et al.,[50] whose results are shown in Figure 14.7. The high amount of dye (up to 30%) incorporated in the polymer is to be noted. The authors[50] claim that such a high percentage was possible only by using a functionalized dye with improved solubility.

As one should expect, the rheology of the polymer films plays an important role. SRGs can be inscribed on very high Tg (rigid) azopolymers, such as polyureas[52,53] and polydiacetylenes,[54] but the process is less efficient because of the rigidity of the chains. In addition, SRG formation is more difficult in high-molecular-weight polymers[30] and in azo-containing gels[25] for the same reason.

The SRG formation process is surface-initiated, at least for the light-driven mass transport. Convincing evidence of this has been obtained by protecting the surface of a spin-coated azopolymer film with some layers of transparent polymers adsorbed using the ELBL technique.[55] A top layer of transparent polymer as thin as 25 nm is able to completely prevent SRG formation on an azopolymer film that would otherwise be amenable to inscription of large-depth SRGs. Figure 14.8 shows the evolution of the diffraction efficiency of SRGs inscribed on an azopolymer film that had its surface coated with ELBL films of distinct thicknesses.[55] The data clearly indicate that the surface modulation decreases sharply with the number of deposited bilayers, as shown in Figure 14.9.

One important question lies in whether there is any difference in orientation of the chromophores between the peaks and valleys of SRGs at the film surface. A conclusive answer would probably bring important information on the mechanisms of mass transport. Let us first consider what should one

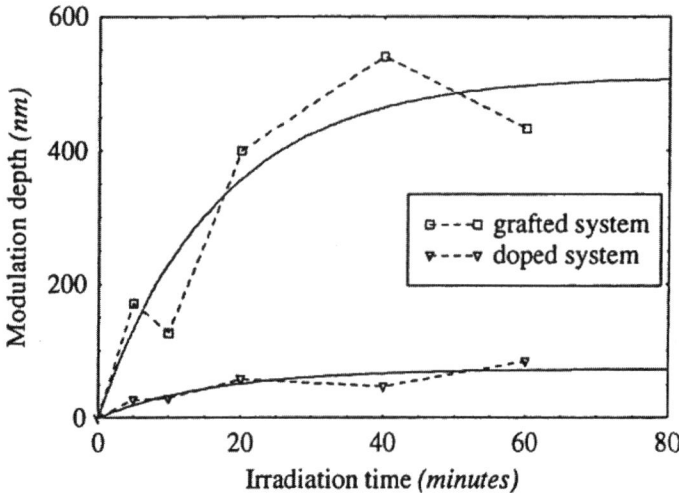

FIG. 14.7 Time evolution of the photoinduced modulation in SRGs recorded under the same experimental conditions in two identically concentrated dye-grafted (DRIG) and dye-doped (DRPR30) polymeric systems. DRIG is poly(methyl-methacrylate) (PMMA) grafted with the Disperse Red I [4-(N-(2-hydroxyethyl)-N-ethyl-)amino-4'-nitroazobenzene] (DRI); DRPR30 is a guest-host system with 30% in weight of the dye DRPR [4-(dibutylamino)-4'-nitroazobenzene] mixed with PMMA. The two polymer systems had similar Tg, ~110°C for DRIG and ~105°C for DRPR30, and the spin-coated films had almost the same thicknesses and absorption spectra. The SRG recording was performed using p-polarized Argon laser beams ($\lambda = 514$ nm), with an intensity of 100 mW/cm². The surface modulation was measured using AFM. From reference 50.

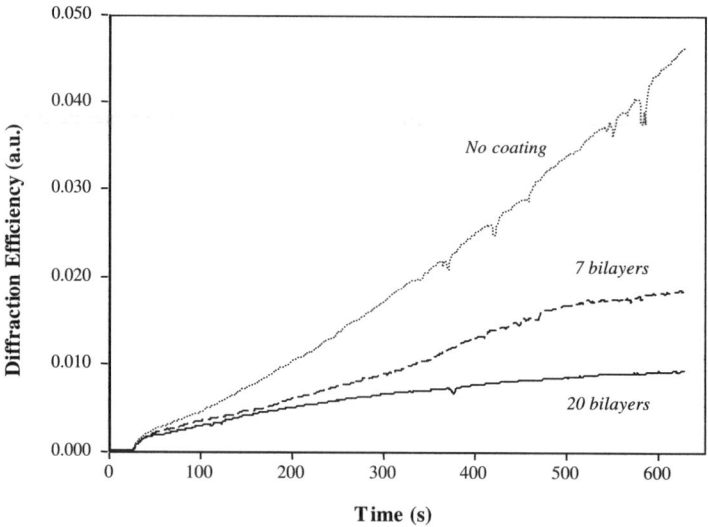

FIG. 14.8 Diffraction efficiency versus time while an SRG was being inscribed on a spin-coated side-chain azobenzene (CH-1A-CA) polymer film. The efficiency is seen to decrease when the film is coated with an ELBL film of poly(diallyl dimethylammonium chloride) (PDAC) alternated with sulfonated polystyrene (SPS), whose molecules do not undergo *trans-cis-trans* isomerization. For an ELBL with 20 bilayers, the efficiency dropped considerably, and practically no SRG could be inscribed. From reference 55.

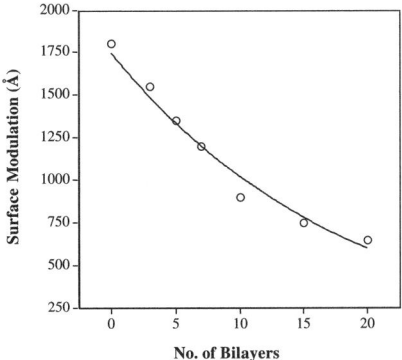

FIG. 14.9 Surface modulation for an SRG against the number of bilayers (PDAC/SPS) in an ELBL film deposited on the spin-coated CH-1A-CA azopolymer film. From reference 55.

expect from the writing of an SRG with a linearly polarized light. It is known that mass transport takes place in order to remove polymer molecules from the illuminated areas. The molecules that are driven outward from these regions are those with chromophores in the direction of the laser-light polarization. Therefore, in the valleys one should expect an excess of molecules with chromophores perpendicularly oriented in relation to the electric field. On the tops or peaks of the SRGs, the chromophores should be oriented preferably along the polarization direction. Because relaxation in the *trans-cis-trans* isomerization cycles is accompanied by molecular reorientation, however, probably the chromophore alignment would not be strong. According to this picture, chromophores should be more oriented (aligned perpendicularly to the polarization direction) in the valleys of the SRG. In the peaks, there should be less alignment. However, it is also known that thermal relaxation (optical as well as mechanical) occurs after the writing beams are switched off. With such relaxation, it is likely that all chromophore orientation should be lost. The chromophores should then be oriented at random, both at the peaks as in the valleys.

At least two groups have attempted to probe the state of chromophores using Raman microspectroscopy, but the results are conflicting. By investigating SRGs inscribed on spin-coated films of azopolymers, Labarthet et al.[56] showed that the polymer removed from the valleys has higher perpendicular orientation (to the laser-light polarization) than the material accumulated at the peaks. In intermediate regions, the final orientations were found alternately weaker and higher, which was attributed to competing forces and/or to more or less efficient cooperative (or dye-intermolecular) orientation effects during the SRG formation. An analysis using information entropy theory confirmed that the chromophore orientation should be different in the peaks and valleys.[57] Such results would agree with the rationale given above, when there is no mechanical relaxation after the recording of the SRGs.

In another study, Constantino et al.[58] observed by surface-enhanced resonant Raman scattering that the molecular architecture in an SRG inscribed on an LB film from an amorphous azobenzene-containing polymer was the same along and across the grooves, in contradiction to the findings of

Labarthet *et al.*[56,57] It could be argued that the systems investigated are different, but the main features in both cases are of SRGs formed via light-driven mass transport: The polymers are amorphous, and the SRGs were inscribed with modest light intensities. Further investigations will be required to resolve this discrepancy. Both studies nevertheless agree that a gradient exists of azobenzene-chromophores as the peaks and bottoms of the grooves were probed, which was only to be expected from a photoisomerization-driven mass transport.

14.2.2 Ablation, Photobleaching, and Photodegradation

Laser ablation has been used for quite some time in the production of submicron structures[41] or to pattern polymers for fabricating optical diffraction elements. Obviously, in this case SRGs can be inscribed on polymers that are not azofunctionalized. Egami *et al.*[59] justified their interest in laser ablation of azopolymers in the potential application in near-field optics. They stressed that using a highly resolving photosensitive material provides a means for the near-field distribution to be investigated *in vivo*. With laser ablation, the mechanism for SRG formation is completely distinct from the light-driven mass transport that has been the focus of this chapter. The first obvious difference between SRGs induced by phototransport and those originating from laser ablation is in the much higher power required for the latter. While modest light intensities, in some cases only a few mW/cm^2, suffice to record SRGs via light-driven mass transport, in experiments where thermal effects dominate the laser power is usually in excess of $1 \ W/cm^2$.

Leopold *et al.*[42] formed SRGs on spin-coated films of random copolymers (Tg approximately 140°C) consisting of a backbone of polymethylmethacrylate with different concentrations of chromophore and mesogen side chains. They used two s-polarized pulses from a frequency doubled Nd: YAG laser (at 532 nm with a pulse width of 6 ns) with an angle of 20° between the two laser beams, leading to a spacing of 1.5 μm on 0.3 to 1 μm thick films. SRGs formed by thermal ablation or thermal shock expansion are formed primarily when the energy employed is above a threshold value of the order of $500 \ mJ/(cm^2 \ \mu m)$. Under such conditions, the rise in local temperature in the sample can be considerable. Much below this value, the photoinduced birefringence decreases very rapidly after the laser pulses, and it decays to zero within 2s. That the effect is thermal was proven by attempting to record SRGs using a s:p or +45°:–45° to the plane of incidence for the writing pulsed lasers. Because there is a relatively small intensity modulation under these conditions, no surface gratings could be inscribed.

An azocopolymer was employed by Hattori *et al.*[49] to produce SRGs with pulsed lasers as well as with a cw laser. In the first experiment, s-polarized beams of a 532 nm line of the mode-locked Nd:YAG laser were made to impinge on the polymer surface. The average beam power was 0.4W, with a pulse duration of 100 ps and repetition rate of 82 MHz. In the second experiment, an SRG with 25 nm modulation amplitude was formed by exposing the film to an interference pattern of two s-polarized lights with $2.8 \ W/cm^2$ intensity, for 40 minutes. Infrared (IR) absorption spectroscopy indicated that, in both cases (pulsed or cw lasers), the grating formed is caused mainly by photobleaching, as the absorption strength from an azo bond of the exposed film decreased upon exposure. The possibility of simple thermal ablation was not checked. The modulation might have been small due to the polarization of the light, even though with photobleaching one might form SRGs with s-polarized light, but was still less efficient than p-polarized light.[21]

Urethane-urea copolymers, with Tg ~142°C, have been used in several instances for producing SRGs. Using spin-coated films of one such copolymer, Egami et al.[59] produced SRGs due to laser ablation, where the ablated depth was proportional to the incident laser energy at low energies (~2 J/cm²). The sinusoidal grating with a 40-nm amplitude was produced by impinging two coherent writing beams of an Ar+ laser (488 nm), with a total energy of less than 10 J/cm². The mechanism responsible for the SRG formation was evaporation of the copolymer, which was confirmed in experiments with quartz-crystal microbalance (QCM).[59] The temporal profiles of the QCM resonant frequencies demonstrate that mass loss occurs while the writing beams are switched on (see Figure 14.10). Egami et al.[43] recorded SRGs on 1.4 µm thick films with an Ar+ 488 nm laser using high energies (>> 2 J/cm²). The SRGs had amplitude of up to 80 nm, with the maximum in the surface profile of the SRG corresponding to the minimum in intensity (a π-shift); the recording was independent of the laser polarization. The SRGs were not photoerasable and could not be erased when the sample was heated for 10 min at 150°C (above Tg). The authors did mention that thermal treatment at higher temperatures could erase the SRG, but this was not studied systematically. In another study, Che et al.[44] obtained SRGs that were attributed to thermal ablation (laser power of 9.31 W/cm²), which were also π-shifted with respect to the interference pattern. These SRGs were neither

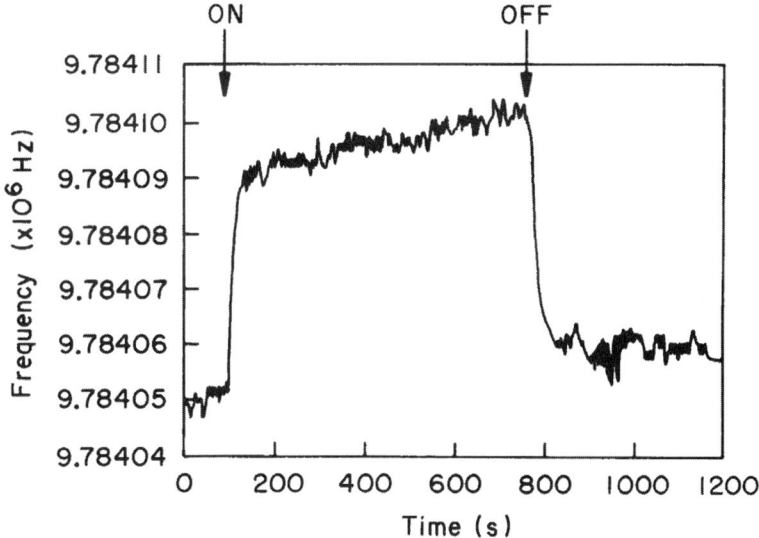

FIG. 14.10 Temporal profiles of QCM resonant frequencies when the laser beam is switched on and off. From reference 59.

photoerasable nor thermally erasable. According to these authors, the possibility of photobleaching and crosslinking should also be taken into account.

It has been demonstrated that photobleaching and photodegradation can play an important role on the SRG formation on ELBL films.[21,24] These results will be discussed separately when the influence of film architecture is analyzed in Section 14.4.9.

In order to produce surface-relief electro-optic gratings, Munakata *et al.*[41] compared two fabrication methods of SRG inscription. In the first, the SRG was produced with an interference pattern of cw laser, with relatively modest intensities. The gratings so recorded were photo- and thermally erasable, and efficient writing was polarization dependent. In the second method, a phase mask was employed to provide the periodic intensity modulation of a pulsed laser, the 3rd-order harmonic (at 355 nm) of a Nd:YAG laser. The SRG was produced with a single laser pulse, allowing a very short fabrication time (less than 1s). The threshold for ablation was 500 mJ/(cm^2·pulse), and the amplitude of the SRG increased with pulse energy. A depth of up to 300 nm could be achieved, leading to a smooth but not sinusoidal surface modulation.

It is obvious that thermal effects are of considerable importance for the fabrication of SRGs with high-intensity pulsed lasers, especially if LC polymers are employed (see Section 14.2.3). Indeed, the generated SRGs normally cannot be optically erased, and the estimated temperature increase is expected to affect the polymer matrix rheological properties on which the viscous flow of polymer material depends. However, results by Ramanujam *et al.*[45] indicate that the mechanisms for SRG formation can be much more complex. The authors were convinced that the 90 nm-depth SRG they inscribed on a cyanoazobenzene side-chain polyester film (Tg = 63°C), using a single 5 ns pulse, was not the result of ablation only. They provided the following evidence for such a belief:

1. The films can be reused after the gratings had been completely erased by keeping the sample at 80°C for one minute.
2. No SRG could be formed on films from main-chain polyesters containing azobenzene in the side chains, even though they had similar absorption coefficients.
3. SRGs could also be formed with orthogonally polarized beams where the total intensity across the film is constant.

No mention has been made of possible photodegradation, but this is a likely possibility. Furthermore, thermally induced phase transitions or even melting could play a role. In summary, this remains an open question, as the authors acknowledged in their paper.[45]

One can note some discrepancies in the literature. Some researchers have reported no dependence on polarization for ablation or photodegradation,[43] but for ELBL films there was some dependence, though not as strong as in SRGs originating from light-driven mass transport.[21] In addition, the phase of the SRG appears to be different in distinct papers. Egami *et al.*[43] and Che *et al.*[44] reported a π-shift. Bian *et al.*[39] reported a zero shift for SRGs recorded with high power lasers.

14.2.3 SRGs on LC Polymers

A special section is devoted to LC azopolymers, which have been investigated for more than a decade for the recording of birefringence gratings with high diffraction efficiency and storage of real 3-D objects, as illustrated by Wendorff and coworkers.[60,61] One of the LC disadvantages compared to amorphous materials, that is, the requirement that monodomains be inducted for the fabrication of optically homogeneous nonscattering films, has been overcome by methods in which the LC phase formation is suppressed.[19] Therefore, it is now possible to produce spin-coated films from LC materials where the films are amorphous or have relatively small LC domains that do not cause significant scattering. They were obvious candidates for production of SRGs, which has been exploited by a number of groups. There are fundamental differences to the behavior of amorphous polymers to be considered when analyzing SRG results for LC polymers. First, SRGs are normally inscribed after the polymer had been quenched to room temperature from the isotropic state. Phase transitions in the LC materials can be achieved photochemically while producing SRGs, as reported in the work by Yamamoto.[17] Second, one should expect it to be much more difficult to record an SRG on a LC material because the azobenzene chromophores will tend to align cooperatively. Indeed, experiments related to SRG formation on LC polymers are usually carried out with much higher powers of the writing lasers, which in consequence makes the appearance of thermal effects inevitable.

Bublitz et al.[62] have shown that LC and amorphous polymers indeed behave differently when a polarized beam is made to impinge onto films of these materials. They spread E1aP (amorphous) and P6a12 (LC), both polyester-based polymers, onto a water surface. Figure 14.11 shows the chemical structures of E1aP and P6a12. In their experiments, a drop of a chloroform solution of the polymer was made to spread over the water surface, in a manner similar to a Langmuir film experiment, with the chloroform evaporating a few seconds after spreading. Then, because the aqueous surface had turned hydrophobic, they added more chloroform solution in order to produce a thicker, circular film (~2 μm thick), as the film now did not spread. Upon shining linearly polarized light onto such films, their shape changed, becoming elongated in the direction of the electric field for E1aP but contracting for P6a12. No volume changes were noted. The difference in orientation was attributed to the difference in architectures of the main chain, as the chromophore was the same in both polymers. Bublitz et al.[62] claim that the side chains carrying the chromophores are aligned perpendicularly to the electric field of the impinging light for P6a12, with a lower degree of main-chain alignment in the same direction.[63] They reasoned that because of the rigidity of the main chain and shorter spacers in E1aP, a higher degree of main-chain alignment parallel to the incident polarization was achieved, following the perpendicular orientation of the azobenzenes. Therefore, in the latter case elongation along the field direction should be expected, as is observed experimentally.

Such differences might be at the origin of the distinct phase shifts between the SRG and the interference pattern observed for the gratings

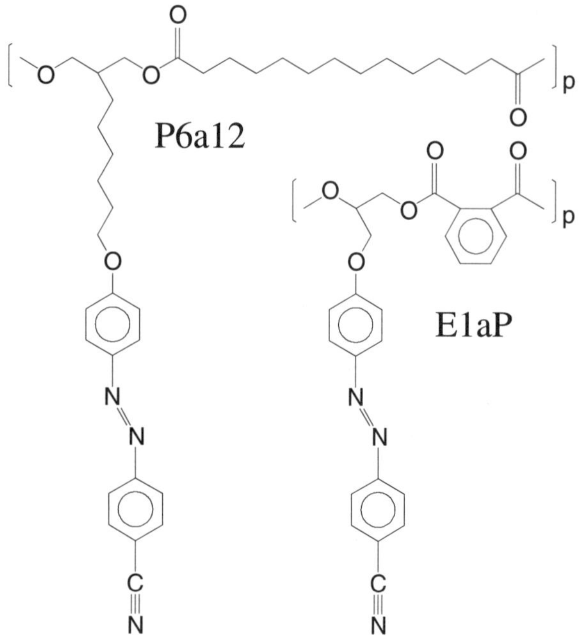

FIG. 14.11 Chemical structures of the amorphous E1aP polymer and LC P6a12. From reference 62.

inscribed on LC polymer films. Unlike the π-shift observed for the amorphous polymers, in LC polymers the maxima of the SRGs coincide with the maxima in the interference pattern. This was clearly shown by Holme *et al.*,[48] with a comparison of SRGs on amorphous and LC polymers. For the LC polyester structures, the molecules migrated into the irradiated regions, whereas the molecules tended to migrate away from illuminated areas in an amorphous oligomer. Another feature that distinguishes SRGs inscribed on LC polymers from those on amorphous polymers is the appearance of SRG with doubled frequency,[35] when certain recording conditions are used. These features could be explained using the theoretical model by Pedersen *et al.*,[33] based on attractive forces between dipoles aligned parallel to the grating lines.

In a series of papers, Yamamoto *et al.*[16,17,64,65] have shown the temperature at which the SRG is inscribed on LC polymer films to be extremely important; they suggest that the SRG formation is associated with phase transitions induced by photoisomerization. They used poly{6-[4-(4-ethoxyphenyl-azo)phenoxy]hexyl methacrylate} (PM6AB2), an LC polymer with Mw of 78,000 g/mol that exhibits the following phases: isotropic (I) above 150°C, nematic (N) between 68 and 150°C, and glassy (G) below 68°C. A higher diffraction efficiency could be reached when the gratings—BG as well as SRG—were inscribed in the nematic phase rather than in the glassy phase.[64] For instance, the exposure needed to achieve 20% of diffraction efficiency was about 13 J/cm^2 at 80°C and 300 J/cm^2 at room temperature, thus representing an increase by a factor of 23 in the LC phase.[17] Moreover, the surface modulation required to achieve 20% of diffraction efficiency was only 30 nm for the grating recorded at 80°C, but 70 nm for that recorded at room temperature (glassy state). These SRG modulations were measured at room temperature, and the authors did not specify whether larger SRG depths could be produced with writing in the N phase. Presumably, recording large-depth SRGs in the N phase would be more difficult as the mass transport is precluded by the tendency of the chromophores to align. It was also mentioned that, whereas the modulation was a fine-relief structure for the SRGs recorded at room temperature (larger depth), SRGs were somewhat distorted when inscribed at 80°C.[64]

The grating formation at 80°C was attributed[64] to alternation of I and N phases, where the I phase would be induced via photoisomerization in the illuminated regions. Such gratings would consist of a spatial modulation of the surface structure (SRG) and a periodic arrangement of N and I phases. At room temperature, the phase transition to I would be less favorable, and this explains the lower diffraction efficiency of the gratings. In another paper,[17] they further elaborated on the mechanism for grating formation. Three stages were assumed to explain the temporal evolution of the total diffraction efficiency (note that the gratings here include BG and SRG), as follows. In the first stage, small domains of isotropic material (I) would be created by *trans-cis* photoisomerization. These domains grow upon further irradiation. In the second stage, complete photoinduced phase transition would be reached in the bright fringes, thus leading to a maximum in the measured diffraction efficiency. In the third stage, further irradiation would cause the spreading of isotropic domains to nonilluminated areas, which would reduce the contrast between illuminated and nonilluminated regions and decrease the diffraction efficiency.

When the recording is carried out at 155°C, only BG is formed, with no surface modulation,[16] because one cannot get large-scale mass transport with such a high temperature. The appearance of a BG, in spite of the LC polymer being in an isotropic phase, was ascribed[16] to disorder-to-order changes induced by photoisomerization of azobenzenes. Yamamoto *et al.*[16] do acknowledge that there is no long-range order in the optically isotropic state, but the mesogens still tend to align parallel to each other in local regions (short-range order, cf. de Gennes and Prost.[66]) It was then argued that chromophores will align perpendicularly to the polarization direction owing to *trans-cis-trans* isomerization cycles.

SRGs have also been inscribed by Sánchez *et al.*[67] on side-chain azobenzene LC polyesters using biphotonic processes, in which a blue (Xe lamp) incoherent source is used in conjunction with a He-Ne laser. They illuminated a sample of P6a12 (see the structure in Figure 14.11) with blue light in order to promote *trans-cis* isomerization. Because the *cis* states of this polymer are long lived, about 2 h, the application of a pattern from s-polarized coherent red light (He-Ne laser) is able to selectively cause *cis-trans* isomerization with the *trans* molecules ending up preferentially oriented parallel to the polarization of the laser light. The anisotropy thus induced is accompanied by intensity and polarization gratings as well as SRGs. It should be stressed that if only the interference pattern of the He-Ne red light is projected on a sample (no previous blue light), nothing is observed because the chromophores do not absorb the red light in their *trans* state. The SRGs could be erased completely by turning the Xe lamp on again, with only one He-Ne beam on.

14.3 THEORETICAL MODELS

It should be emphasized from the start that the models proposed in the literature were generally conceived to explain distinct aspects of the formation of SRGs, and, moreover, they should not be expected to cope with all experimental observations. This is particularly true because the mechanisms involved in SRGs can be completely different, depending on the polymers employed or even on the experimental conditions adopted for writing, such as laser power. Before going into any specifics of the limitations and strengths of each model, we shall describe them briefly. For further information, the reader is referred to the original publications of such models, as indicated.

14.3.1 Free Volume Model

In the first model, proposed by Barrett *et al.*,[9] the driving force for the mass transport was assumed to arise from pressure gradients induced by photoisomerization of the chromophores due to the constructive interference of the two writing beams. Regions of high *trans-cis-trans* isomerization would be created next to regions of low isomerization. The pressure gradients appear because a free volume is required for the isomerization, which should be above the yield point of the material. The region with more isomerization possesses a higher pressure, thus inducing a viscoelastic flow of material from the high-pressure to the low-pressure areas, leading to sinusoidal SRGs. This flow was later investigated in detail using the Navier-Stokes equations for laminar flow of a viscous fluid.[30] The velocity components in the film were related to the pressure gradients, by definition of boundary layer conditions. A number of experimental results can be explained with these models. For instance, in the latter model the rate of grating inscription is predicted to increase with the intensity of the writing laser and to vary inversely with the molecular weight of the polymer below the limit of entanglement, consistent with experimental results. It also predicts the inscription rate to scale with the third power of the initial thickness of the film, which is obeyed for thin films. One important successful feature of these models was their ability to provide results that were consistent with known values of free volume required for the isomerization and bulk viscosity of the polymer matrix. In addition, they predict that the peaks in the SRG should correspond to the nonilluminated areas of the polymer film, that is, the SRGs should be π-shifted in relation to the interference pattern, which is indeed observed experimentally. They cannot account, however, for the difference in writing properties of s- and p-polarized lasers, which will be discussed later.

14.3.2 Field Gradient Force Model

This approach is based on the forces originating from the electric field gradient that is induced optically.[68] The main ingredients to be included take into account that experimental results showed no volume change in the film during the surface deformation, and this deformation is caused by lateral movement (in large scale) of polymer chains. The forces behind this movement are a combination of the change in the susceptibility (induced by light) and the field gradient. The time average of the gradient force density is given by:

$$f(r) = <[P(r,t)\cdot\nabla]E(r,t)> = <[\varepsilon_0\chi E(r,t)\cdot\nabla]E(r,t)> = \frac{1}{2}\varepsilon_0\chi'E(r)\cdot\nabla E(r) \quad (14.1)$$

where $P(r,t)$ is the polarization, $E(r,t)$ is the optical electric field, $< >$ represents the time average, ε_0 is the vacuum permittivity, and χ' is the optically induced change in the susceptibility of the film.[68] From Equation 14.1, one notes that the polymer chains are subjected to a force only in the direction where there is a component of the field gradient. This force is null when the polarization is perpendicular to the gradient. In addition, the dipolar interaction between the optical field and the azochromophores induces a change in susceptibility, which is due to the induced dichroism and birefringence. The induced change in susceptibility depends strongly on the direction of the total field (resulting from the superposition of fields in the interference pattern).

In experiments with a single laser beam, the direction of the resulting force, due to a combination of induced susceptibility and the field gradient, is parallel to the light polarization direction, whereas in the formation of SRGs it is parallel to the grating vector. For a single, Gaussian beam, the sample surface is taken on the x-y plane, with the laser polarization in direction x. Because the light penetrates in the polymer for 0.1–0.3 μm, which is much lower than the Rayleigh length ($\sim 25 \ \mu$m) for the Gaussian beam employed experimentally, and because the sample is placed on the focal plane, the force density exerted on the chromophores is:

$$f(x, y, z) = \frac{1}{4}\varepsilon_0\chi'\exp(-\alpha z)\frac{\partial I(x, y)}{\partial x} \ \mathbf{x}_0 \quad (14.2)$$

where $I(x,y)$ is the Gaussian distribution of intensities in the focal plane, and $\mathbf{x}_0$ is the unit vector in direction x.[39] The results from using a single beam (to be shown later), where the polymer chains moved away from the regions of higher intensity, suggest that the force is directed outward in relation to the center of the Gaussian beam, and along the polarization direction.

It is also known that SRGs arise from surface-initiated processes.[55] The free polymer surface can be treated as a thin, mobile layer, with a viscous drag between layers dominated by the force $f(x,y,z)$. The limit superficial velocity, v_s, due to this force is:

$$v_s(x,y,z) = \mu \ f(x,y,z) \quad (14.3)$$

where μ is a coefficient that depends on the viscoelasticity of the photoaltered polymer. At the molecular level, μ is related to the conformation of the chains, including the arrangement of the chromophores. It incorporates the effects from the viscous drag between the mobile layer and the sample bulk. For the sake of calculations to explain the experimental results, μ was taken as a constant, independent of the light intensity (for modest intensities) and of the time. In Equation 14.3, the transient region in which the velocity changes in the beginning of the deformation was neglected, because the time scale for the transient is much shorter than the writing time for an SRG. For an incompressible polymer, the rate with which the surface is deformed,

$v_z(x,y,0)$, can be obtained from the continuity equation, under the boundary condition that $v_z(x,y,d) = 0$, as:

$$v_z(x,y,0) = \frac{1}{4}h\mu\varepsilon_0\chi'\frac{\partial^2 I(x, y)}{\partial x^2} \tag{14.4}$$

For a Gaussian beam, linearly polarized with the field in the direction x, $f(x,y,z)$ has only an x component, as can be seen in Equation 14.2. Therefore, the surface deformation induced by the beam, in the approximation of small amplitudes, and also assuming that the intensity I is a function of x only in Equation 14.4, is given by:

$$S(x,y,t) = \int_0^t v_z(x,y,0)dt' = \frac{1}{4}h\mu\varepsilon_0\chi'\frac{\partial^2 I(x, y)}{\partial x^2} t \tag{14.5}$$

The surface deformation caused by a circularly polarized Gaussian beam, according to the above analysis, is:

$$S(r,t) = \frac{1}{4}h\mu\varepsilon_0\chi'\frac{\partial^2 I(r)}{\partial x^2} t \tag{14.6}$$

Equations 14.5 and 14.6 indicate that the profile of the surface deformation is proportional to the second derivative of the intensity with respect to the direction of the polarization, which is confirmed by the experimental results. The formation of SRGs can also be described using the model depicted in Equation 14.1–14.6 with an intensity $I(x) = sin(kx)$ in Equation 14.5, thus leading to:

$$S(x,t) = -\frac{1}{\Lambda^2}\pi^2 h\mu\varepsilon_0\chi'\sin(kx)t \tag{14.7}$$

where $\Lambda = 2\pi/k = \lambda/[2\sin(\theta/2)]$ is the period of the interference pattern, λ is the laser wavelength, and θ is the angle between the interfering beams. When there is an electric field pattern that varies in the space due to interference of two polarized beams, the induced change in susceptibility (χ') is treated as a tensor; this includes the material susceptibility before the writing process (χ_{xy}^0) and the spatial variation of the susceptibility change (χ_{xy}') owing to induced dichroism and birefringence.[38] The real (birefringence) and imaginary (dichroism) of the induced susceptibility can be measured experimentally by using writing lasers with different polarizations.

This theoretical treatment predicts a pattern for the SRG, given a pattern of superposition of the optical field, and it also provides a measure of the viscoelastic parameter μ of the azopolymer. Figure 14.12 illustrates the main features of the theoretical model, indicating in particular that the chromophores are moved from the illuminated to the nonilluminated regions. The simulations using Equations 14.6 and 14.7, along with the corresponding experimental results, are shown in Figures 14.13 and 14.14.

14.3.3 Mean-Field Theory Model

This model is based on a mean-field theory, according to which the mean-field potential tends to align chromophores along the director and also to

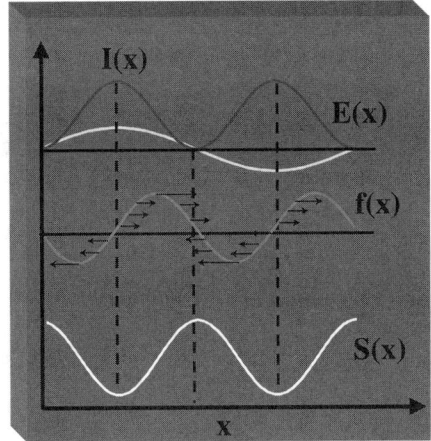

FIG. 14.12 Schematic representation of the field-gradient force acting on the chromophores, leading to a surface deformation ($S(x)$) that is π-shifted with respect to the light intensity pattern ($I(x)$) on the sample. The force $f(x)$ is zero when either the field $E(x)$ is zero or when there is no field gradient in the x-direction. Notice how the arrows, representing the direction of the force acting on the chromophores, change sign, leading to accumulation of chromophores in the nonilluminated regions.

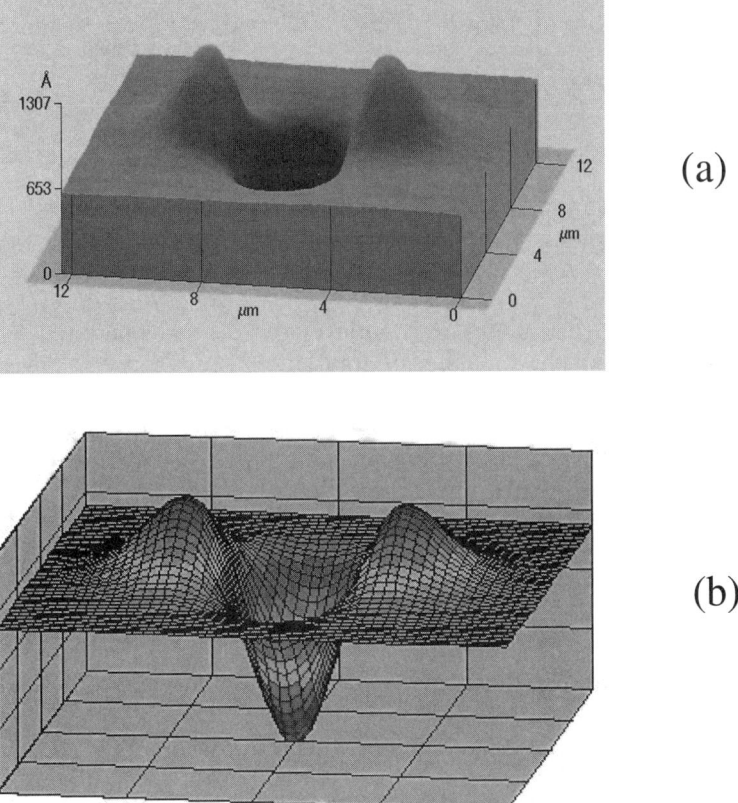

(a)

(b)

FIG. 14.13 (a) A 3D AFM image of a surface relief pattern inscribed with a linearly polarized, single Gaussian beam. (b) The theoretical simulation that reproduces the main features of the actual image.

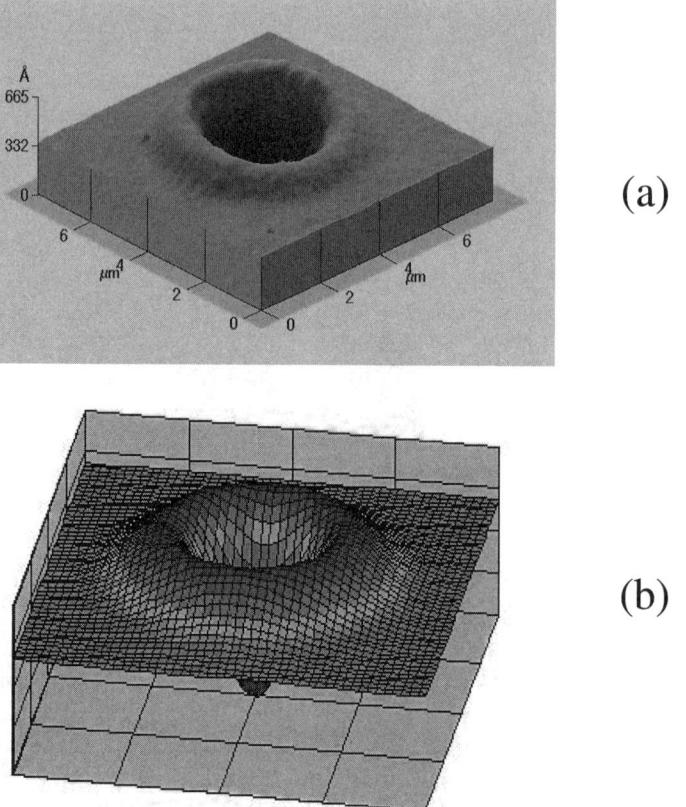

FIG. 14.14 (a) A 3D AFM image of a relief structure inscribed with a circularly polarized, single Gaussian beam. (b) The theoretical simulation that reproduces the main features of the actual image.

attract chromophores with a parallel orientation.[33] Because the model includes the breaking of translational invariance at the axis of the grating vector, the attractive interaction between chromophores will have distinct effects depending on whether the chromophores are aligned side by side or end to end. Mass transport is only possible with the chromophores side by side, since in this case the direction of the attractive force coincides with the direction of the broken translation invariance, that is, the grating vector. This is a key factor to explain the polarization dependence that is observed experimentally. With this model, in which the height profiles of the grating were calculated using a variational method, Pedersen *et al.*[33] were able to explain not only the polarization dependence but also the spatial periodicity of SRGs inscribed on films of LC polymers. Another important feature of the model is its prediction that the chromophores should be attracted to illuminated regions from an interference pattern on a sample. While this prediction is consistent with results obtained for LC polymers, it is contradicted by results on SRGs formed on amorphous polymers. Therefore, the model cannot be extended to the SRGs on the polymers for which the mass transport has been proven to be entirely, or at least predominantly, light driven.

Furthermore, one should also emphasize that this mean-field-based model does not take into account any possible thermal effects. This is an important limitation because SRGs on LC polymers are inscribed using high-laser powers, considerably higher than for the spin-coated, amorphous polymers, and therefore thermal effects might be expected to contribute. For the laser powers used in LC polymers, for example, it has been already shown[39] that thermal effects are not negligible for the amorphous polymers.

14.3.4 Model of Viscous Mass Flow

The fluid-mechanics model takes into account the depth dependence of the photoinduced driving force and the velocity distribution in the film surface.[36,37] The formulation for the SRG dynamics is derived analytically as a function of film thickness and interference wave number (corresponding to the pitch of interference fringe). The new feature in this model is that the thickness dependence for SRG formation can be explained even for large thicknesses. This model also explains the dependence of the SRG amplitude on the pitch of interference, which is observed experimentally. Note that in this model, the authors assumed a sinusoidal force was present, but they made no attempt to explain why this shape occurred. That is to say, the "initial" driving force was not discussed, but a full account of the SRG formation process might be obtained by assuming the field-gradient force model of reference 31.

Figure 14.15 shows the schematic diagram of the model by Fukuda and Sumaru,[37] where the change in the velocity of mass transport, as the substrate is approached, is depicted to decrease. The results of applying the model are shown in Figure 14.16, indicating the correct dependence of the surface modulation with the film thickness. Note that for thin films, the cubic dependence on the thickness is observed. Furthermore, the theoretical model correctly predicts the dependence of the surface modulation on the periodicity of the grating, Λ.

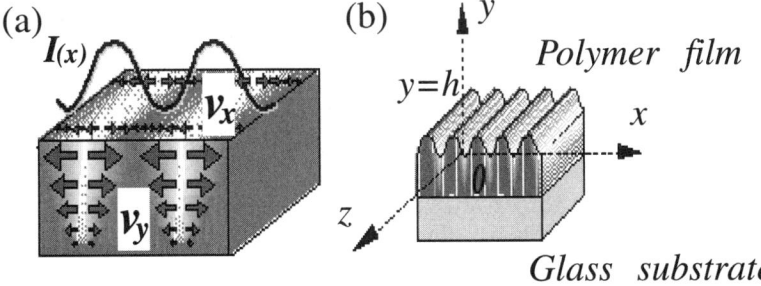

FIG. 14.15 (a) Schematic diagram of Fukuda and Sumaru's model[37] The velocity of the mass transport at the top layer (v_x) follows the directions predicted by the field-gradient model.[31] The velocity of mass transport for the layers underneath, v_y, decreases as they lie deeper in the sample bulk, and is zero at the substrate. (b) The coordinate system for the model.

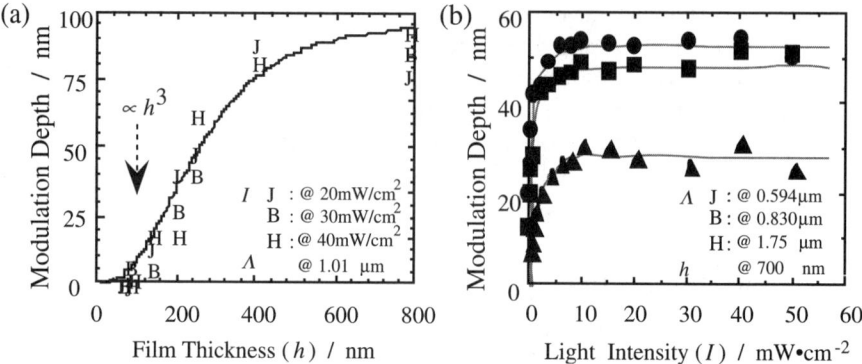

FIG. 14.16 Dependence of the SRG inscription rate as a function of (a) film thickness h, and (b) the light intensity I, for various spacings of the grating Λ. Irradiated total photon energy was fixed at 14 and 7.2 J/cm² for each experiment, respectively. From reference 37.

14.3.5 Diffusion Model

This model is based on solving the diffusion equation by assuming that the surface effects are governed by bulk diffusion of the molecules.[32] The authors studied one-directional migration effects along the axis perpendicular to the sample surface, taking into account concentration diffusion for two types of polarization excitation. Their basic assumptions were:

1. Rotation-translation coupling was neglected when the molecules were moving, that is, translation did not imply rotation and vice-versa.
2. A molecule that undergoes a *trans-cis-trans* cycle moves in a direction parallel to its orientation before the isomerization, similarly to the movement of a worm. The molecules are allowed to move randomly forward and backward.
3. The angular distribution of the molecules is assumed to be time independent, just as the time scale for orientation distribution (from 0.1 to a few seconds) is much smaller than that of diffusion (10 min or more).

The probability that an excited molecule will not collide during its migration decreases exponentially with the distance it moves divided by the diffusion length, which is typically of the order of the molecular size, and therefore much smaller than the grating period. The integral leading to diffusion flow contains the probability multiplied by the number of chromophores that cross a given area section per unit time. Such a number depends on the light intensity, the quantum efficiency of the translation process, the angle between the light polarization, and the molecular axis. Because of the latter dependence, the actual mass transport will depend on the polarization of the writing lasers; the theoretical results agree with the findings of Jiang *et al.*[10] in that p-polarized beams are more efficient in writing SRGs than s-polarized beams.

The mass-density flow is then input into the diffusion equation, from which the number density of chromophores is obtained and the surface modulation calculated. For the latter, the appearance of the SRG is taken as a consequence of the buildup of a concentration grating. Another feature of this model that is observed experimentally is that the azochromophore molecules are directed outwards from the illuminated regions, even though the authors did not mention this explicitly in their paper.[32]

14.3.6 Model Using Jones Matrix Formalism

It is important to emphasize that this model[34,69] explains the polarization dependence of the writing and reading beams, and it provides a full picture of what one should expect in terms of the diffraction efficiency for the various experimental conditions. Note, however, that the formation of the SRG, which contributes with a phase difference to be input in the Jones matrix formalism, was assumed a priori. The model is not, therefore, aimed at explaining the origin of the mass transport, unlike the case of the models in references 9, 30–33, 36, and 37.

14.3.7 Comments on the Models

The free volume model[9,30] has been able to explain several experimental features of SRGs on high Tg polymers, whose origin is in the light-driven mass transport. It cannot explain the polarization dependence because the mass transport based on the pressure difference would be insensitive to a p- or s-polarized writing beams, contrary to the experimental observations. It is therefore a model that provides insight into the dynamics of the mass transport but does not give the right clue for the driving force to set the azobenzene-containing polymer into motion. The model by Lefin et al.[32] explains the polarization dependence and provides a picture for the mass flow. Reservations have been raised about their proposed physical mechanisms involved in the wormlike movement,[38] particularly because of the macromolecular nature of the azopolymers.[48] However, the need for some degree of photoisomerization-induced plasticization is necessary in any model to explain the large-scale movement, and in this respect the wormlike movement could simply accompany or be a consequence of the pressure gradients. Nevertheless, as in the free-volume model,[9,30] the viscous flow is assumed to originate from bulk phenomena, while there is now evidence that the SRG formation is a surface-initiated process.[55]

The field-gradient force model,[31] on the other hand, does provide the driving force for the chromophores to move and drag the polymer chains with them, as it is based on a force associated with the field gradient that explains the polarization dependence entirely. In its original form,[31] it does not touch on the mechanisms for the dynamics of mass transport. Nor does it explain how the polymer is plasticized in order to allow for a substantial viscoelastic flow of the chains. In order to cover that, Viswanathan et al.[38] extended the original model of reference 31 and predicted quantitatively the polarization dependence of the SRG. They also showed that the shape of the experimentally observed SRG could be reproduced theoretically. In principle, the original model[31] could be combined with another one dealing with the viscous flow, such as the one by Barrett *et al.*,[9,30] Lefin *et al.*[32] or Sumaru and Fukuda.[36,37] Indeed, while presenting their model, Sumaru *et al.*[36] and Fukuda *et al.*[37] also demonstrated that their results were compatible with the field-gradient force model, and their results were consistent with the experimental data for a variety of polymer systems. A further feature brought by the model in references 36 and 37 was the thickness dependence.

It should be noted that these models were all conceived to explain the light-driven mass transport in the fabrication of SRGs. Therefore, they cannot be applied to LC polymers or to SRGs inscribed using high power lasers, because other effects, especially thermal ones, should be invoked. The model by Pedersen *et al.*,[33] based on a mean-field-theory, gives good results for the LC polymers. It predicts that the SRG should be in phase with the interference pattern, which may be valid for LC polymers owing to the tendency of their molecules to align cooperatively. However, this model should not be used to explain SRG formation for amorphous polymers because it contradicts the experimental observation of a π phase shift of the SRG with respect to the interference pattern.

In analyzing models, mention should also be made of those aimed at explaining experimental results for the diffraction efficiency in polymer films on which SRGs have been inscribed. This is the case of Naydenova *et al.*[35] and Labarthet *et al.*[34,69] However, it should be mentioned that in these models the mechanisms leading to the actual mass transport are not addressed. They are, therefore, of a different nature to those of references 9, 30–33, 36, and 37, discussed earlier.

14.4 FACTORS INFLUENCING THE RECORDING OF SRGS

14.4.1 Plasticization

One important tenet of all models proposed to explain the formation of SRGs on glassy polymers, albeit with important discrepancies among them, is that the polymer is somewhat "softened" (plasticized) upon photoisomerization. That is a necessary prerequisite for mass transport over macroscopic distances. A recent paper by Srikhirin *et al.*[70] brought experimental evidence of such a softening effect. Employing quartz crystal resonators, these authors measured the elastic compliance of thin polymer films doped with an azobenzene dye. They employed spin cast films of two polymers, poly(methylmethacrylate) (PMMA) – Tg = 103°C, and poly(ethylmethacrylate) (PEMA) – Tg = 71°C, doped with 4-[4'-[hexylphenyl]azo]phenol. The compliance decreases by a few percent upon irradiation with UV light at low intensity, but the material is softened when irradiated with high-intensity visible light. In the latter case, they attributed the softening to the rapid cycling of azobenzene molecules through the *trans*- and *cis*-states.

Evidence for the softening of the polymer matrix was also provided in the studies of Stracke *et al.*,[71] where the polar-order relaxation times obtained from pyroelectric coefficient measurements were longer than the relaxation times of the diffraction gratings. The difference was attributed to the excess free volume induced by the isomerization cycles in the optical storage processes. In reference 71, spin-coated films were obtained from an electrically poled styrene-maleic anhydride copolymer with side groups of disperse red chromophores.

14.4.2 Importance of the Dye and Degree of Azofunctionalization

In various works, the importance of photoisomerization has been probed. In the early works by Natansohn's group,[9] attempts were made to build SRGs on polymer films containing a nonisomerizing dye (rhodamine 6G), but no grating was produced. Likewise, Kim *et al.*[8] and Viswanathan *et al.*[28] reported that SRGs are not formed on functionalized polymers when the dye was not azobenzene. A direct comparison was obtained from epoxy-based polymers with the same backbone structure and differing only on the functionalization dye. While the azodyes led to considerable SRGs, the epoxy-based polymer containing side-chain biphenyls was not amenable to SRG formation, because the biphenyls cannot undergo isomerization. Also investigated were functionalized polymers where the chromophores (not azodyes) can undergo isomerization. This was the case of polymers with stilbene and imine chromophores, for which significant SRGs could not be recorded. This failure was attributed[28] to the lack of efficient *trans-cis-trans* isomerization cycling in these chromophores.[72] Consistent with these results, Darracq *et al.*[25] observed that SRGs could not be formed on gel films when the gel contained dyes that did not photoisomerize.

A systematic investigation has been made of the influence of the degree of functionalization on the formation of SRG for a series of maleimide high Tg polymers[73,74] and relatively lower Tg acrylate-based polymers.[74] This also allowed investigation into the effects gained from changing the structure of the main chain. The main conclusions drawn were the following:

1. Considerable modulation depth was obtained only for polymers with more than 30% of dye functionalization. At lower degrees of functionalization, photoisomerization processes are insufficient to plasticize the polymer matrix. Above 40–50%, the ability to form SRGs is independent of the degree of functionalization. One of the reasons for such saturation was suggested to be the effects of light absorption, because the penetration depth of the irradiated light should be a function of the azo functionalization, just as the coefficient of absorption is proportional to the degree of functionalization.[74] Sumaru and Fukuda's model[36] actually predicts the saturation of SRG inscription rate for high functionalizations, but it fails to explain the behavior at low degrees of functionalization. Presumably, the assumption of a viscosity independent of the degree of functionalization is not valid.[74]

2. Formation of SRGs depends on the intensity of the exposed light, and it becomes significant for laser powers of ~10mW/cm^2, below which only weak SRGs can be inscribed.

3. The efficiency in writing the SRG decreases with the temperature at which the writing is carried out. Interestingly, the temperature dependence was the same for all polymers that were amenable to the writing of SRG,

even though their Tg varied widely. For instance, the polymers investigated had Tgs varying by 60°C, but qualitatively showed similar behavior.

The fact that large-modulation SRGs are formed only above a certain threshold amount of functionalization was also shown by Andruzzi et al.,[75] for amorphous as well as LC polymers. They investigated copolymers derived from two photochromic monomers, 6-(4-oxy-4′-cyanoazobenzene)hex-1-yl methacrylate and 8-(4-oxy-4′-cyanoazobenzene)oct-1-yl methacrylate, and a nonphotochromic comonomer, (-)-menthyl methacrylate. Considerable amplitudes were obtained for all polymers with more than 30% of azofunctionalization. The maximum amplitudes were observed for copolymers possessing 75–80% of dye content, not the fully functionalized homopolymers.

Contrary to the expected trend that higher degrees of functionalization should lead to larger amplitudes of SRGs, Fiorini et al.[76] failed to observe a systematic increase in surface modulation when the concentration of DR1 in DR1-MMA copolymers was increased. They obtained a relative surface modulation—normalized by the film thickness—of 34, 40, and 28% for copolymers with chromophore contents of 15, 35, and 50%, respectively.

14.4.3 Polarization Dependence

In the early studies of SRGs on high-Tg polymers, it was already realized that the appearance of a large-depth SRG depended on the polarization of the writing lasers.[10] This was confirmed in several subsequent studies.[29,38,77–79] In fact, this polarization dependence was a key factor to establish the writing of SRGs on high Tg polymer films, under modest light intensities, as an all-photonic process. Large-depth SRGs cannot be produced with two s-polarized beams in spite of the high intensity modulation,[29] thus ruling out any major contribution from thermal effects or photodegradation. Even more interesting was the fact that the photoerasure characteristics of the SRGs were dependent on the polarization of the erasing beams as well as on that of the writing beams.[29,78] That is to say, the SRG keeps in its "memory" the polarization of the laser light with which it was inscribed.

There is ample evidence in the literature that—for the light-driven mass transport—writing is more effective when employing contra-circularly polarized, followed by p-polarized, light. Very weak SRGs are obtained if either s-polarized light or a combination of orthogonally polarized s:p light is used. Here s-polarized light means that the electric field vector is perpendicular to the grating vector, whereas the p-polarized would be parallel to it. Barrett et al.[9] argued that because the mass transport appears to be driven by isomerization pressure, the rate of isomerization of the chromophores is more important than their orientation. The authors state that "the interference of two circularly polarized beams creates alternating regions of high (elliptical) and low (linear) isomerization, while the interference of two linearly polarized beams creates alternating regions of low (linear) and no (completely destructive interference) isomerization. The effect of a pressure gradient between the regions is similar, but greatly diminished in the latter case (linear!)." SRGs have also been produced in azo-hybrid gels,[25] where a more efficient writing is achieved with p-polarized rather than with s- or s:p polarized beams.

Because the experimental conditions differed from experiment to experiment, with regard to the polymer and film thickness as well as with the writing conditions, a direct comparison is not always straightforward. Distinct azopolymers, for example, might display different absorption characteristics and viscoelastic properties, thus altering their sensitivity to the writing process. In this context, recent comprehensive studies of such polarization effects were carried out in references 28, 29, and 38, in which the samples and writing conditions were identical, with the exception of the polarization of the writing beams. We shall highlight some of the most important findings in the following discussion.

Several combinations of polarizations for the two interfering beams have been used, namely: s-:s-; s-:p-; p-:p-; +45:+45; +45:−45; RCP:RCP; LCP:LCP; and RCP:LCP, where ±45° indicates the angle between the polarization direction and the s direction. RCP and LCP stand for right and left circularly polarized light, respectively. Samples from an epoxy-based polymer containing a side-chain azobenzene, obtained under identical experimental conditions, were used. The writing procedures were also the same, with a laser power of 50 mW/cm^2 from an Ar$^+$ laser at 488 nm. SRGs with distinct surface modulation depths were obtained, depending on the polarization, as illustrated in Table 14.1. It is clear that the most efficient way to produce the gratings is the contra-circularly polarized light, consistent with other works.

The results were explained in detail with a model that had two important ingredients: (1) The calculation of the intensity-modulation amplitude and the spatial variation in magnitude and direction of the resultant electric field vector, as a function of the phase difference between the two beams. For that, a matrix approach was used, with the interference matrix being defined in terms of the electric field amplitude. (2) The formation of considerable SRG is only possible when there is a component of the electric field gradient in the direction of the grating vector. This follows the model proposed by Kumar *et al.*[31] that establishes prerequisites for the initializing driving force for the large-scale mass transport.

TABLE 14.1 Surface modulation of SRGs produced at 488 nm with an intensity of 50 mW/cm^2 under various polarization combinations on a 0.7 μm thick spin-coated film of an epoxy-based polymer (CH-1A-CA).

Polarization	SRG Depth
s:s	< 10 nm
s:p	< 20 nm
p:p	90 nm
+45:+45	< 10 nm
+45:-45	250 nm
RCP:RCP	< 10 nm
RCP:LCP	320 nm

There is practically no formation of SRG when there is no spatial variation of the optical field, as in the s:p and +45:+45 combinations. In addition,
there must be a component of the field gradient in the direction of the grating
vector, a requirement that is satisfied by the combinations RCP:LCP, p:p, and
+45:–45. In the cases where there is only spatial variation of the optical field,
but no component of the electric field in the grating vector, the SRG is very
weak but a BG, which is a bulk phenomenon, is formed, as in the s:s case.
These predictions were confirmed experimentally by monitoring the grating
formation with orthogonally polarized read beams.[29]

The polarization dependence is also shown to obey this model for SRGs
inscribed with a single writing beam where intensity modulation is achieved
using a beam with a Gaussian profile.[39,80] Essentially, the results indicate that
a more efficient SRG formation is seen with p-polarized light than with s-
polarized light, and again the circularly polarized light was the most efficient
in building the gratings. The polarization dependence on the erasure characteristics has also been investigated systematically in references 39 and 80.
This will be discussed in Section 14.4.10 where erasure is analyzed in detail.

Although the mechanisms for SRG formation in LC polymers are
somewhat different, polarization dependence has also been observed.[14,15,48]
Analogous to the amorphous polymers, the LC polymers also show strong
polarization dependence of the SRG recording process. Efficiency is higher
for circularly polarized light, followed by p:polarized light. Very small surface
deformation occurs if s: polarized beams are employed. Interestingly, even
in experiments where SRGs were inscribed with a high-power laser pulse, in
which the grating formation is certainly dependent on thermal effects and
photodegradation, polarization dependence has been observed.[45] These
results are not yet fully understood, particularly because many different
mechanisms might be operating at the same time. Cases where the polarization dependence is not very strong necessarily mean that processes other than
the light-driven mass transport are involved. For instance, Egami et al.[43]
investigated SRG formation on urethane-urea copolymers and noted that the
process was polarization independent; mass transport was attributed to photochemical reactions. In addition, for SRGs inscribed on layer-by-layer films,
the polarization dependence was not as strong as in the spin-coated films.
Therefore, additional mechanisms are involved in the mass transport, which
was confirmed through Fourier Transform Infrared (FTIR) and UV-vis
spectroscopy.[24]

14.4.4 Dependence on the Molecular Weight

Barrett et al.[30] showed that the inscription rate decreases drastically with the
molecular weight (Mw) on SRGs formed on acrylate polymers (pDRA,
pDR13A, and pMEA). This was attributed[9] to the increase in viscosity when
the molecular weight increases. Indeed, the bulk viscosity of polymers
increases with the first power of Mw up to the limit of entanglement, after
which it increases with at least the third power of Mw. In experiments with
blends of an azopolymer and high molecular PMMA, the authors confirmed
this expectation with large-scale mass transport to generate SRGs being pre-

cluded even when 5% of high Mw PMMA was mixed with the PDR1A films. Obviously these comparisons are not straightforward because the PMMA was not azofunctionalized and it has been well established that doped polymers are not amenable to the recording of large-depth SRGs. Considerable SRG depth, about 35 nm, was obtained for the doped system comprising 4-m-polyimide doped with an azobenzene dye, but in this case the polyimide itself already contained an azobenzene group.[51]

According to reference 28, in high-molecular-weight azo functionalized systems, the gratings formed are similar to the ones formed in low Mw (< 5000 g/mol) systems but are less efficient. Synthetic details for preparing these very high molecular weight (> 500,000 g/mol) azobenzene functionalized copolymers were reported in reference 55. The SRG formation behavior in the high molecular weight azo polymers was less efficient, possibly due to chain entanglement. In polymer systems where chain entanglement is not an important issue, one might expect to record efficient SRGs even for large molecular weights. This appears to be the case for very-high-molecular weight (ca. 10^6 Daltons) cellulose functionalized with an azobenzene dye, where considerable SRG formation was observed.[81]

With regard to the formation of SRGs with low-molecular-weight compounds, mention should be made of the SRGs recorded on layer-by-layer films, where Congo Red layers were alternated with a cationic polymer.[21,22,24] These results are discussed in Section 14.4.9. Another interesting study was presented by Fuhrmann and Tsutsui,[82] in which SRGs of 9 nm depth were inscribed on films of low-molecular-weight dye compounds. Unfortunately, the only data available are for s-polarized light, where one would not expect significant SRG. The possibility of producing large amplitude SRGs from these molecules remains to be seen.

14.4.5 SRG Inscription Requiring Thermal Treatment

There are cases in which thermal treatment is required, after the laser-writing process, for the SRGs to be produced. Obviously, in such cases the mass transport differs widely from that of light-driven transport for amorphous polymers discussed at length here. LC trisazomelamine (TAM) films were subjected to an interference pattern of s-polarized beams at room temperature, with a considerably high intensity (990 mW/cm²).[18,19] TAM is isotropic above 239°C, and it can be frozen into a glassy state (Tg = 32°C) if quenched from the smectic phase. Spin-coated films from chloroform solutions show no birefringence upon heating to 109°C, which demonstrates that the LC phase was suppressed up to this temperature. The writing process at room temperature led to almost no SRG, as one should expect since s-polarization was employed. However, heating the film after the writing procedure (in the absence of light) at a constant rate causes the diffraction efficiency to increase considerably owing to the appearance of an SRG. The depth of the SRG increased from 5 nm to 400 nm when the sample was heated to ca. 100°C. The authors[19] ascribed the SRG formation to the suppression of the thermotropic phase formation in the glassy state, combined with the tendency of mesogenic self-organization at high temperatures. It is not clear whether the reason for requiring thermal annealing is associated with the low-molecular nature of the LC, which is not a polymer. In addition, significant SRG could not be inscribed with the film in the LC state.

Postexposure baking was also required to form SRGs on films of LC polymethacrylate with styrylpyridine side chains.[83] The authors employed a photocrosslinkable polymer with Mw = 176,000 that exhibits smectic phase between 86 and 105°C. Films were spin coated, 100 nm thick. The pristine films did not absorb the laser light at 488 nm, but they could be made to absorb if protonated by HCl vapor. SRGs were then inscribed by exposing the films to two p-polarized interfering beams, with an intensity of 30 mW/cm². The gratings were inscribed with a pitch varying from 0.7 to 2 μm. An SRG with 15 nm modulation, corresponding to about 15% of the original film thickness, was formed after the thermal treatment that caused irreversible crosslinking. The SRGs thus formed were extremely stable and could not be erased even thermally.

14.4.6 Single-Beam Experiments

So far we have discussed cases where the intensity modulation for building SRGs was obtained by interference of two beams with appropriate polarization combination. Surface relief structures can also be produced with a single beam, provided that the appropriate polarization and intensity variation can be created. The main advantage of using a single beam with a well-defined intensity and polarization profile is in the analysis of theoretical models to explain the large-scale mass transport. This occurs because the phase shift of the SRG in relation to the interference pattern can be directly obtained in experiments with a single beam, and also because the intensity distribution and size of a Gaussian laser beam can be directly measured. The spatial relationship between the laser intensity profile and the induced surface deformation can then be unambiguously determined by means of atomic force microscopy.

Surface structures have been recorded by Bian et al.,[39] where two types of azobenzene functionalized polymers, PDO3 (Tg = 106°C) and HPAA-NO2 (Tg = 120°C) (Figure 14.17), were used. Spin-coated films were prepared for PDO3 and HPAA-NO2. A He-Ne laser beam with a well-defined Gaussian light intensity profile at 544 nm was focused by a spherical lens and impinged on the sample film that was placed at the focal plane of the lens. A one-dimensional (1D or cylindrical) Gaussian beam can be obtained by simply substituting the spherical lens with a cylindrical one.

For exposure at low intensities, large-surface modifications were produced only when a field gradient and a component of electric field were simultaneously present. The bottom of the relief corresponded to the maximum in light intensity, as predicted by Kumar et al.[31] This was true for a circular as well as a cylindrical Gaussian beam. For a circular Gaussian beam, a maximum deformation of 37 nm was obtained for a p-linearly polarized beam with 238 mW/cm² in the center of the beam and time of exposure of 50 min. For a circularly polarized beam, with intensity 230 mW/cm² in the center and time of exposure of 50 min, the maximum deformation was

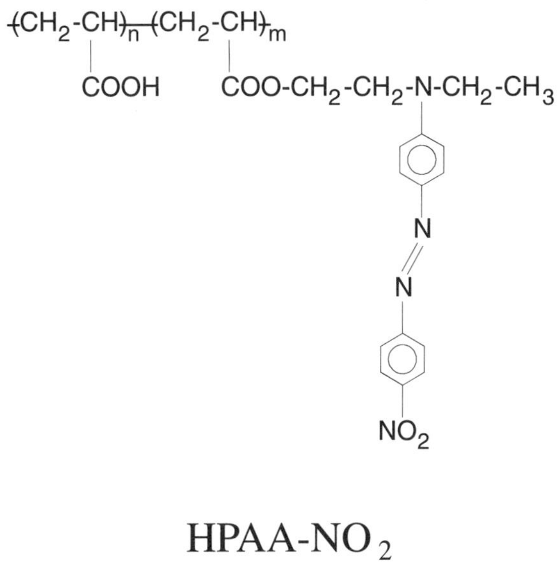

$$\{CH_2-CH\}_n\{CH_2-CH\}_m$$

COOH   COO-CH_2-CH_2-N-CH_2-CH_3

NO_2

HPAA-NO$_2$

FIG. 14.17 Structure of the HPAA-NO$_2$ polymer.

110 nm. Figures 14.13 and 14.14 show that the model in reference 31 predicts qualitatively the shape of the SRGs observed experimentally.

Similar single beam experimental results were reported by Bian *et al.*[84] in which large-amplitude surface reliefs were recorded on enzymatically synthesized poly(4-phenyl-azophenol) dye films. The strong dependence on polarization of the writing beam and the π-shift between the topographic grating and the interference pattern were also observed. Therefore, the results could be explained in the framework of the optical field-gradient force model.[31] Fiorini *et al.*[50] also observed that, in the formation of surface reliefs on amorphous azopolymer films with a single beam exposure, molecules migrate from the illuminated to the nonilluminated regions.

14.4.7 Dependence on the Beam Intensity

The nature of SRG recorded at lower and higher intensities can differ considerably. At higher intensities, photothermal and photobleaching effects may play a dominant role. In addition, some crosslinking in the polymer can occur due to thermal or photochemical effects, which usually affect the erasure behavior of the SRGs. With such high powers, thermal processes cannot be neglected, because a temperature increase of several degrees might be generated locally upon laser irradiation.[85] As emphasized by Andruzzi *et al.*,[75] such a temperature increase can plasticize the surface, facilitating SRG formation. At lower intensities, the mechanism for SRG formation has been established as due to the light-driven mass transport. For higher intensities, other effects must be taken into account. For instance, Bublitz *et al.*[62] suggested that frictional forces against the substrate should be considered, since the material directly in contact with the substrate cannot move. The free surface would expand or contract in the direction of the light polarization, depending on whether the polymer was amorphous or LC, respectively. It should be noted that if the Tg of the polymer is too low, one cannot record large modulation SRGs.[8]

Control of the beam energy allows one to obtain distinct features of SRGs. For instance, Bian et al.[39] showed that, for a PDO3 polymer film irradiated with a high-intensity Gaussian laser beam (hundreds of W/cm[2]) in the beam center, a central peak appears in the surface profile, in addition to the surface deformation induced by the optical-field gradient force. The polymer in the center of the exposed spot becomes bleached, indicating dye degradation and perhaps photocrosslinking. Photochemical and photothermal effects can result in the breaking of the N=N bond in the azochromophore and even other bonds in the polymer backbone and the subsequent rearrangements of the molecular structure of the polymer. Furthermore, the peaklike deformation in the center becomes predominant when the beam intensity increases.

In another experiment, Bian et al.[39] recorded in- and out-of-phase SRGs on the same sample, with an interference pattern of two cw Ar[+] laser beams with equal intensity at 488 nm. The two beams were focused by a lens to get higher intensity at the sample surface. The diameter of the two Gaussian beams was 0.6 mm. Because of the Gaussian intensity distribution of the laser beams, the SRG produced by the high-intensity mechanism will dominate in the central part of the exposed area, and the relief grating induced by the optical-field gradient forces will appear in the outer perimeter of the exposed region. The appearance of the two kinds of gratings was confirmed by the micrographs of the exposed samples, as shown in Figure 14.18.

There was a clear π phase shift between the two kinds of gratings. A modulation amplitude of 60 nm was obtained at the center of the in-phase grating. The growth of the in-phase grating did not depend strongly on the writing beam polarization, in contrast to the out-of-phase grating writing.

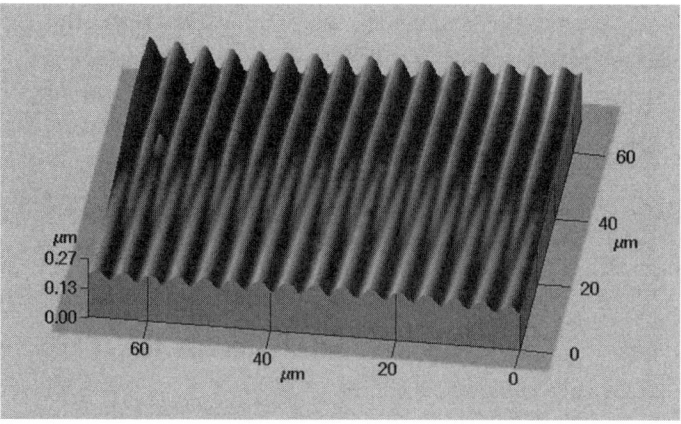

FIG. 14.18 In-phase and out-of-phase SRGs inscribed on the same polymer film due to different light intensities. From reference 39.

Below a certain threshold intensity, no in-phase grating could be formed even if the out-of-phase grating was completely saturated and the writing continued. This threshold intensity at 488 nm was about 24 W/cm^2 for PDO3 and 22 W/cm^2 for HPAA-NO2.[39] These systems also showed interesting optical erasure characteristics, which are discussed in Section 14.4.10.

14.4.8 Main Chain and Conjugated Azopolymers

Most experiments for recording SRGs have been carried out on side-chain azopolymers, but polymers with other architectures have also been used.[28] For example, polyureas containing mono- and bisazoaromatic groups in the main chain with relatively high Tgs (197 and 236°C) were amenable to SRG fabrication, even though the diffraction efficiency and modulation depth were considerably lower than for side-chain azopolymers.[52] In addition, the writing process in the main-chain polymers is considerably slower than with side-chain polymers. This is due to the polymer rigid backbone and the fact that azobenzene groups have restricted mobility as they are attached to the backbone. Similar results were obtained with high-molecular-weight azobenzene functionalized polydiacetylene (PDA).[54] This conjugated polymer did not display any Tg in Differential Scanning Calorimetry (DSC) measurements, as it possesses a rigid backbone with long persistence length. Again, the SRG formation was less efficient than in epoxy side-chain polymers. Large amplitude SRGs were also produced on an enzymatically synthesized poly(4-phenyl-azophenol) film, using both an interference pattern of laser beams and a Gaussian single beam.[84] The polarization dependence and the phase shift of the SRG relative to the interference pattern could be accounted for by the model in reference 31.

Ozaki et al.[86] showed that the addition of a conjugated polymer with a rigid main-chain structure to an acrylate azopolymer led to an increase in the SRG depth up to a limit of 3% of the conducting polymer (PPV derivative—poly(2-methoxy-5-dodecyloxy-p-phenylenevinylene). When the concentration of the PPV derivative in the blend was above 3%, the SRG modulation became less than with the pure azopolymer, and it was practically negligible for 10% of PPV derivative. The mechanisms leading to the improvement in SRG efficiency with the addition of conducting polymer were not discussed. They also observed that the SRG formation progresses even after terminating the laser light irradiation on the composite film.

A recent report by Xu et al.[87] indicated that more efficient writing of SRGs can be obtained with chromophores in the main chain, provided that azochromophores with adequate push-pull electron substituents are used. They produced SRGs on spin-coated films from polyesters containing aminosulfone azobenze in the main chain, employing circularly polarized Ar$^+$ laser at 488 nm and 52 mW/cm^2. The surface modulation increased with the rigidity of the main chain (and consequently with its Tg), with depths of up to 230 nm being obtained. This increase with the rigidity was attributed to differences in the viscoelastic properties of the polymers.[87] The SRGs on the polyesters exhibit larger amplitudes than those produced on main-chain polyureas,[52] which was ascribed to the presence of the push-pull electron substituents that would allow continuous photoactivation of the chromophores.[87] The latter authors argued that in the polyureas, because the azobenzene in the main chain did not possess push-pull groups, the photoactivation would be slowed, leading to shallower SRGs.[87]

14.4.9 Film Architecture—SRGs Recorded on Organized Films and Gels

The vast majority of the studies on SRGs have been carried out with spin-coated or cast films, but gratings have been recorded on LB films[20] and on ELBL films (also referred to as self-assembled films with alternated anionic/cationic layers).[21–24] The advantages of these techniques are in the precise control of film thickness and the possibility of altering the molecular architecture. For the ELBL technique, in particular, there is the added advantage of the simplicity of the experimental procedures. SRGs produced so far on ELBL films involve photodegradation[21–24] and have peculiarities that cannot be explained with the existing models.

Based on the requirements for building SRGs, one might wonder whether it should be at all possible to produce such gratings out of ELBL films, in which the layers containing the azobenzene chromophores are alternated with inert polymers. In principle, this could be considered as similar to the guest-host polymer systems where the azobenzene chromophores dope a polymeric matrix. As is well documented,[49,50] large-depth SRGs cannot be built on these films because the chromophores are unable to drag the polymer molecules during the isomerization cycles, as they are not covalently attached. Indeed, the first attempts to produce SRGs from layer-by-layer films were not successful.[88] However, we have recently been able to produce SRGs on ELBL films of two different systems: a polyazo compound[23] and an azodye, Congo Red,[21,22,24] whose layers were alternated with the polyelectrolyte PDAC. It was surmised that the ability to control the electrostatic interactions in the layer-by-layer films was a key feature for the successful inscription of SRGs. This means that the chromophores are now capable of dragging the molecules of the inert polymer.

In the qualitative analysis of the requirements for fabricating SRGs on ELBL films, we shall compare the cases of successful inscription to those where an inscription could not be made. The large modulations obtained on the ELBL films from azodyes show that even the inert polymer molecules can be dragged in the mass transport. This can happen not only in the azodye systems, as demonstrated for Congo Red,[21] but also with the azobenzene-containing polymer.[23] In addition, photodegradation can also contribute to the SRG formation. Upon photodegradation, products might be formed whose molecules occupy a volume that is significantly different from the azobenzene *trans*-isomer, as noted in photomodification studies of azobenzene-containing polyurethanes.[89] This would contribute to the mass transport. It should be mentioned that for SRGs involving photodegradation, the dependence on the polarization of the writing beams is not as strong as for the spin-coated films of amorphous azopolymers.[21] Such behavior should be expected as the field-gradient force is not the only operating mechanism for large-scale motion.

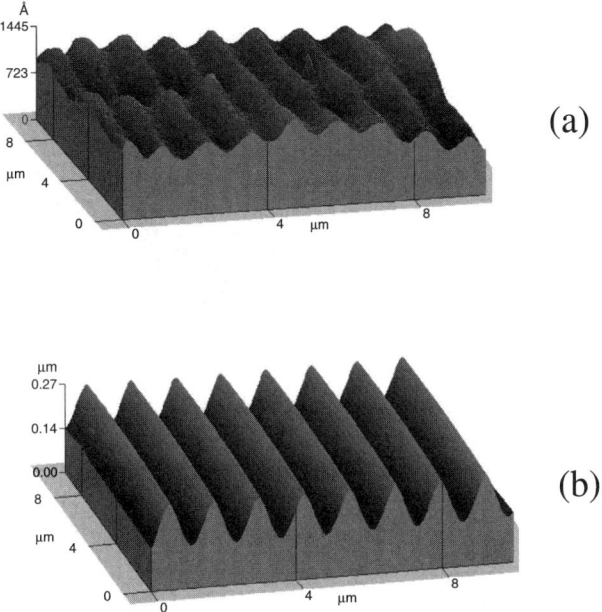

FIG. 14.19 AFM views of the SRG structures on an ELBL assembled PDAC/ PAA-AN film (a) and a spin-coated PAA-AN film (b).

SRGs are difficult to inscribe if the interactions in the ELBL films preclude photoisomerization or the movement of materials due to the formation of ionic complexes. For the polyazo system, we have found[23] that writing SRGs was successful only after optimization of experimental conditions, particularly the pH of the polymeric solutions. Large modulations were obtained for less-charged polymers, in which the electrostatic attraction between the molecules in the oppositely charged layers was decreased. Figure 14.19 shows that for the same azopolymer, it is easier to inscribe a SRG on the spin-coated film than on the ELBL film. The inevitable conclusion was that the chromophores are somewhat locked by these strong interactions, as already mentioned by Dante et al.[90] This is confirmed in optical storage experiments where the time to store information, dependent on the ability to photoisomerize the azochromophore, was much longer than in spin-coated or cast films.[23] By analyzing the results on the azodyes, we can conclude that the photoisomerization behavior of the ELBL films is also affected by the ability of the components to H-bond. For instance, under identical experimental conditions we could write SRGs on CR/PDAC ELBL films but not on ELBL films of CR and PAH (polyallylamine hydrochloride).[22] An AFM image in Figure 14.20 illustrates the SRG formed on an 80-bilayer ELBL film of CR/PDAC. The surface modulation was considerably increased if thicker films were employed, as demonstrated with the results of Figure 14.21.

We can attribute the failure in writing SRGs on CR/PAH films to the stronger tendency of PAH to H-bonding, which acts as a further source of hindrance to photoisomerization. The importance of H-bonding in ELBL

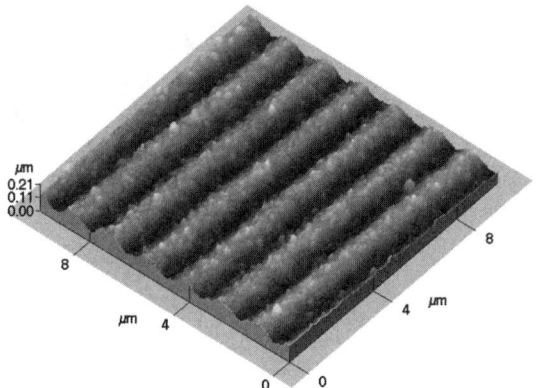

FIG. 14.20 A 3D AFM image of an SRG on an 80-bilayer CR/PDAC film (See color plate following p. 479).

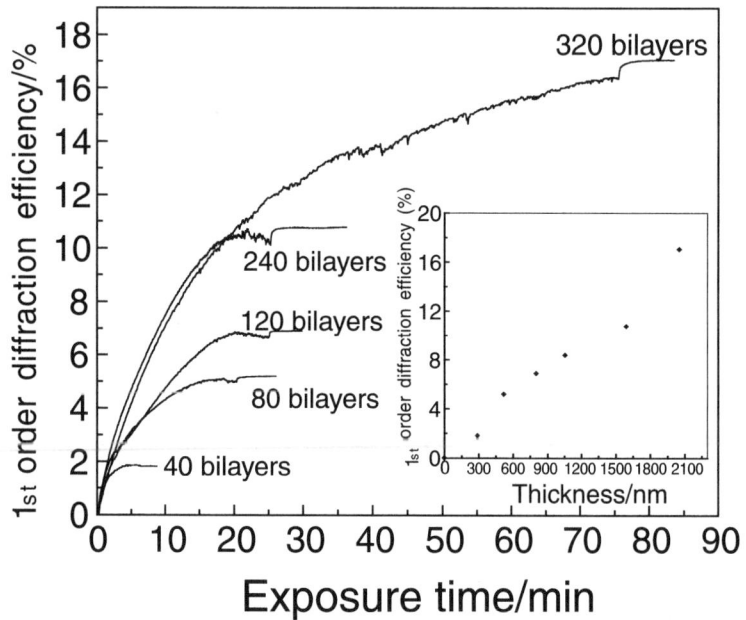

FIG. 14.21 Evolution of the diffraction efficiency during the writing of SRGs on CR/PDAC films with various thicknesses. In the inset, the diffraction efficiency is plotted against the film thickness.

films was first demonstrated by Stockton and Rubner,[91] who even deposited ELBL films entirely based on H-bonding. In subsequent work by Raposo and Oliveira,[92] it was also shown that the energies involved in the adsorption process of polyanilines are characteristic of H-bonding as well as ionic inter-actions. The H-bonding capability of these polymers might lead to non-self-limiting adsorption,[93] which would not occur if the interactions were purely electrostatic attraction and repulsion.

The ELBL films might allow specific interactions that are normally not observed in other films. For example, photodegradation of chromophores observed in CR/PDAC ELBL films does not occur to the same degree in cast films of the same components.[24] Another interesting effect was the dramatic change in optical storage characteristics of a layer-by-layer film of an azobenzene-containing polymer alternated with PAH that was dipped into pure water. Because the water disrupted the intermolecular H-bonds in the polymer system, photoisomerization was facilitated and the time to store information dropped by one order of magnitude.[94] In summary, the electrostatic interactions and the H-bonding in the ELBL films might impair large-scale mass transport by either hindering photoisomerization and/or decreasing the plasticizing effect when photoisomerization still occurs. Indeed, even when optical storage is made possible, we might still not observe SRG formation, as was the case of the CR/PAH layer-by-layer films. The presence of ionic bonds has been shown to be important in spin-cast films as well. Large-depth SRGs could not be produced on films from an aqueous solution containing an azofunctionalized polymer with sulfonate groups that are prone to strong ionic bonds (unpublished results).

SRGs have been recorded on LB films from an azopolymer, a polymethacrylate containing DR13 covalently attached, mixed with cadmium stearate, where the latter was employed to enhance monolayer transfer in the LB process.[20] SRGs with a modulation depth of 50–60 nm were inscribed using the interference of two p-polarized beams, at 532 nm from a cw Nd: YAG laser, with 180 mW/cm^2 for 3 min (see Figure 14.22). Mendonça *et al.*[20]

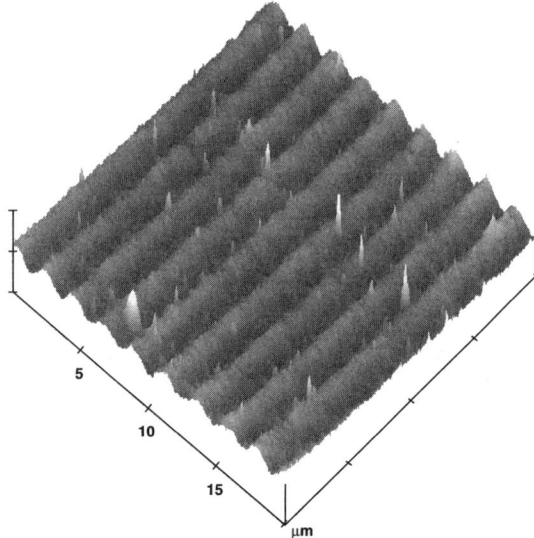

FIG. 14.22 A 3D AFM image of an SRG inscribed on an LB film of poly(4'-[[2-methacryloyloxy)-ethyl]ethylamino]-2-chloro-4-nitroazobenzene] (HPDR13). The writing conditions were: laser power = 180 mW/cm^2, p-polarized light. The LB film had 100 layers of HPDR13 mixed with cadmium stearate, where the latter was used only to improve the transferability of the monolayers from the water surface onto the substrate. The grating spacing was c. 2.6 μm, and the depth was 50–60 nm. From reference 20.

mentioned their surprise in finding that the nonactive cadmium stearate domains were apparently moved together with the azobenzene-containing macromolecules during the mass transport for forming the SRG. The fact that SRGs can be inscribed on polymeric films assembled by the ELBL process (where the polycationic layers do not absorb light) clearly demonstrates that the azobenzene-containing molecules can drag polymer chains with them. One might also wonder whether the mechanisms involved in the formation of SRGs on LB films are the same as in other types of film. Evidence suggests that the mechanisms are not altered, since no photodegradation was observed upon recording SRGs on the LB films.[20,58] The layered structure of the films appears to have no influence either, which is not the case of the ELBL films. In the latter, the electrostatic interactions affected the writing characteristics considerably (as discussed earlier).

SRGs have also been produced in azo-hybrid gels,[25,26] despite the usual high rigidity of the sol-gel matrix. Darracq et al.[25] could record SRGs immediately after the gel film was spin coated onto a glass substrate, since in this state the film does not form a sufficiently crosslinked network. This allows volume and surface deformations in the film to be photoinduced by photoisomerization of azobenzenes. If the film was heated for 24 h at 100°C, a condensed state was achieved, and no SRG could be inscribed. The hybrid films containing azo side groups were synthesized from silane-modified monomers, and the coating solutions were obtained from copolymerization of modified silane monomers with tetraethoxysilane. SRGs with modulations of up to 700 nm could be inscribed using the interference of two p-polarized beams (intensity of 6.5 mW and spot size 2 mm). The authors reported that the energy density was insufficient to cause photobleaching of the silica gels. The gratings were stable after switching off the writing laser, but could be erased if the sample was heated above 100°C. The gratings could not be rewritten after heating, due to the drastic increase in rigidity of the matrix following the sol-gel condensation. Darracq et al.[25] also observed that in the power range between 7 and 26 mW, the diffraction efficiency depended only on the energy (fluence). They attributed this result to the proportionality of the photoisomerization rate to the recording intensity, which also excludes thermal effects.

In a later work, Frey et al.[26] used the time-resolved grating translation technique to investigate the SRG and anisotropy gratings (index and absorption). Particular attention was paid to the phase shift between the SRG and the interference pattern. They showed that a π-shift exists, thus confirming previous results of SRGs obtained by light-driven mass transport. It was also shown that only weak SRGs are formed with s-polarization, in contrast to the more efficient p-polarization.

14.4.10 Post-writing Evolution and Erasure Characteristics

Even after the writing lasers had been turned off, in several cases considerable evolution of the gratings was observed. This occurred in samples thermally treated after writing SRGs, during erasure experiments,[78] or even when samples were simply left to rest.[13] When the writing beams are switched off, the BG that accompanies the SRG might relax rapidly, which would cause a fast decay in diffraction efficiency if the BG were the dominant effect. Such a decrease is easily explained by the angular relaxation process of the chromophores.[51] In some cases, however, the SRG continues to develop even after the writing beam has been switched off. It seems that the mass-transport process continues (even with no light), thus enhancing the surface modulation and consequently the diffraction efficiency. This effect has been observed for amorphous polymers under writing conditions where the mass transport is light driven.[51] The diffraction efficiency of gratings produced on an azocarbazole-based polyimide film increased significantly after the pump was turned off. It has been suggested that this increase could be due to the thermal and mechanical stabilization of the polymer mass transport.[51]

In SRGs inscribed on LC polymer films, time evolution has also been reported in reference 13. The depth of the SRG was seen to increase from 70 nm up to 300 nm, 16 h after the writing process was completed. During this period, the samples were left to rest under dark conditions. Ozaki et al[86] also showed that the SRG formation progresses even after terminating the laser light irradiation on a composite film of an azopolymer and a conducting polymer derived from poly(p-phenylene vinylene).

Interesting effects due to thermal treatments of a polymer sample with an inscribed SRG have been reported by Pietsch et al.[95] The SRG was initially recorded on a spin-coated film of functionalized polyacrylate (pDR1, Tg = 129°C). A 5 min inscription time at 488 nm and 1W/cm^2 led to a modulation depth of 250 nm. The SRG was then erased by heating the polymer in the range between 125 and 140°C, from which the activation energy of the erasing process could be estimated as 2.5 ± 0.3 eV. AFM measurements confirmed that the SRG had vanished. Then the sample was submitted to a second thermal treatment, which led to the formation of a chromophore density variation grating but no SRG. After several hours of annealing, the intensity of the first-order diffracted beam was the same as that obtained initially with the SRG. The scattering angles of the diffracted orders were the same as before, that is, the density grating had the same spacing as the initial SRG.

To explain their results, Pietsch et al.[95] argue that after the SRG had been erased, further heating would cause nucleation of an ordered phase in the place where the peaks were. A thermotropic self-organization would have occurred above Tg due to a photogenerated order (cf. Meier et al.[96]). Pietsch et al.[95] also assume that the nature and organization of the molecules in the troughs of the SRG would not be modified, and therefore a density grating would appear with the same period as the original SRG. They mention that no formation of LC phases has been reported for the polymers used, but a low-order LC phase could have been created because the azobenzene in pDR1M forms LC when its motion is decoupled from the main chain motion. It was suggested that this new phase appears because heating the polymer above Tg provides the required motion for alignment of the chromophores. An LC phase would indeed bring an amplification of the diffraction efficiency, as reported by Stracke et al.[19] In the latter work, annealing of an LC film above Tg caused the SRG depth to increase significantly, which was attributed to a thermally induced reorientation of meso-genic chromophores. Nevertheless, Pietsch et al.[95] stressed that they had no other evidence for the existence of an LC phase. In case such a phase is indeed formed, this would mean that organized phases (LC phases) could be photoinduced in a material with no thermotropic characteristics.[95]

The observations by Pietsch et al.[95] indicate that some kind of memory effect exists in the SRGs inscribed on the amorphous polymer, similar to what was already reported[78] on the study of erasure characteristics of SRGs (see below). Indeed, the increase in diffraction efficiency during the supposed erasure procedure was observed in SRGs that had been inscribed with contra-circularly polarized light, as was the case of the SRGs inscribed by Pietsch et al.[95] The main difference was that the SRG was enhanced during the photoerasure,[78] whereas no SRG was formed in the high-temperature anneal-ing by Pietsch et al.[95]

In bulk BGs, erasure is simply performed by randomization of the chromophore orientation. This can be achieved either by shining circularly polarized light onto the sample, or by heating the sample and thereby causing thermal relaxation of the chromophores. Erasing an SRG, on the other hand, involves distinct mechanisms since large-scale mass transport must occur during the erasure process, analogous to the writing process. SRGs inscribed via light-driven mass transport can usually be erased by impinging a single erasing beam with an intensity comparable to the writing beams on the SRG sample.[78] Erasing takes place because the chromophores experience a force from the resulting optical pattern. It is interesting that the erasure characteris-tics depend not only on the erasure beam but also on the writing beam. The latter property might have an important impact on the exploitation of memory effects because the gratings appear to be able to "memorize" the polarization states that created them, thus influencing the erasure behaviors.

SRGs inscribed at low-light intensities on side-chain polymers or on high-Tg main-chain azoaromatic polymer films can generally be photoerased.[53] The erasure characteristics of SRGs have been systematically investigated by Tripathy et al.,[78] Jiang et al.,[29] Bian et al.,[39] and Geue et al.[97] In the experi-ments done by Jiang et al.,[78] erasure was monitored via the decrease in diffraction efficiency with a low-power reading beam and by measuring the remaining surface deformation with AFM. For SRGs written with p-polarized beams, the decrease in diffraction efficiency depends on the polarization of the erasing beam. Erasure was faster for a beam linearly polarized parallel to the SRG grooves, in comparison to that of erasing beams that were circularly polarized or linearly polarized perpendicular to the grooves. The most inter-esting behavior, however, was observed for SRGs that had been inscribed with writing beams having polarization 45° with respect to the s-polarization. As Figure 14.23 shows, the erasure was again most efficient with the beam polarized in the direction of the grooves, but no erasure could be achieved if the erasing beam was linearly polarized perpendicularly to the grooves. The circularly polarized erasing beam gave an intermediate behavior. This latter behavior has important implications for real applications, because a grating written with the appropriate polarization (say the 45° mentioned above) might be used as a phase mask to fabricate other gratings. This is possible

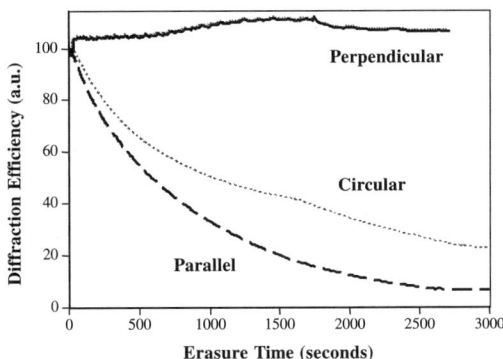

FIG. 14.23 Erasing processes of SRGs recorded with beams having polarization set at 45° to s-polarization, under three erasing polarizations: linear polarizations parallel to the grating grooves (blue line) and perpendicular to the grooves (red line) and circular polarization (green line).

because, provided the correct polarization combinations are selected, the other gratings can be inscribed with no fear of erasing the original grating.

SRGs recorded with circularly polarized beams showed an even more interesting erasure behavior.[78] The SRG can be erased by beams polarized parallel to the grooves and circularly polarized, as in the previous cases, but when "erasing" was tried with a beam polarized perpendicularly to the grating grooves an increase in diffraction efficiency and amplification of the surface modulation occurred (44% increase in the amplitude of the SRG, consistent with doubling of the diffraction efficiency).

SRGs cannot be photoerased in several instances, all of which are associated with a writing process that involves mechanisms other than—or in addition to—the purely light-driven mass transport. For example, even for amorphous, high-Tg polymers, SRGs might not be photoerasable if they are inscribed with a high-intensity exposure,[39] which involves photochemical bleaching and possibly crosslinking. In most cases SRGs can be thermally erased, regardless of the mechanisms for writing. There are a few exceptions, however, such as the films of LC polymethacrylate with styrylpyridine side chains, where the SRG formation requires post exposure baking, which leads to crosslinking.[83] Another exception was represented by the SRGs inscribed on an azocarbazole-based polyimide,[51] for which a remarkable thermal stability was demonstrated, with no erasure even after baking the polymer sample at 240°C for 1 h in air.

Geue et al.[97] investigated the thermal erasure of SRGs obtained via light-driven mass transport, at various temperatures, which allowed them to estimate the activation energy for erasure to be 2.6 eV. For erasing at T < Tg, the SRG was erased by flow of polymer material perpendicular to the initial surface, accompanied by the formation of an intrinsic density grating, as commented on in the work of reference 95. At T > Tg, the lateral density modulation was equalized by a lateral flow of material.

14.5 OPEN QUESTIONS AND CHALLENGES FOR THE NEAR FUTURE

As has been already emphasized, a good understanding of the origin of SRGs has been achieved for those cases where modest light intensities are impinged onto high-Tg amorphous azopolymers. Thanks to a predictable dependence on the writing-beams polarization, a theoretical model[31] has been able to reproduce qualitatively a number of features for this all-photonic, light-driven mass transport. Models have also provided a reasonable basis for the viscous flow[30,32,36,37] that leads to the SRG, even though the details for a quantitative analysis in terms of the parameters involved—e.g., molecular weight and Tg of the polymer—will still require considerable additional efforts. Among the several features that are explained by a combination of these models are: (1) the phase shift between the SRG profile and the interference pattern; (2) the thickness dependence of the SRG depth; and (3) the polarization dependence. However, some points still cannot be explained even for these light-driven mass transport SRGs. Perhaps the most important is the postwriting evolution, which can be due to mechanical relaxation that has not been treated in any of the models. The data available on this issue are insufficient for conclusive statements to be drawn, and this remains an open point. In this context, an interesting possibility is the formation of low-ordered or pseudo-LC phases on the amorphous polymers, as suggested by the results of Pietsch *et al.*[95]

Another line of research to be pursued is to treat quantitatively the effects of some experimental parameters on the polymer plasticization and consequent ability to generate SRGs. For that, a fully fledged comparison study must be done with polymers in which the rigidity (Tg), degree of functionalization, molecular weight, and polarity of main chain and azobenzene chromophores are varied systematically. Some efforts have already been done in this connection, as mentioned in several places in this chapter. They are insufficient for a quantitative analysis, however, especially because there is interdependence of various parameters—as demonstrated for storage characteristics of azocopolymers.[98] In addition, there are also some discrepancies with regard to some of these dependences.

The possibility of producing SRGs with very-high Mw compounds has been demonstrated only recently, contrary to expectations that mass transport would be completely impaired. There is evidence that high molecular weights can be used only if chain entanglement is not important in the given material, as is the case of cellulose,[81] but investigations with other materials are necessary to confirm this point. With regard to the use of oligomers or very-low-molecular-weight compounds, SRGs have been produced with dyes in ELBL films, which may require photodegradation.

As far as the SRGs recorded on LC materials are concerned, there are issues to be addressed before a complete picture is achieved. These gratings are produced using high light intensities, and thermal effects should not be neglected. It is true that the model by Pedersen *et al.*[33] can explain that the SRG profile is in phase with the interference pattern, unlike the case of high-Tg polymers. However, in their model no account was taken of the inevitable thermal effects. Moreover, recording of SRGs on LC materials probably involves phase transitions that are induced partially by thermal effects and partially by *trans-cis-trans* photoisomerization,[17] and this has not been treated theoretically.

In the experiments where SRGs were produced on high-Tg amorphous polymers using very-high intensities, the contributions of the thermal effects cannot be distinguished clearly from the photoinduced mass transport. In fact, for SRGs produced from thermal effects, photobleaching and/or photodegradation—regardless of the fabrication conditions and materials, for high-Tg or LC polymers, using spin-coated or ELBL films—there are issues that have not been understood, even qualitatively. To start with, there are discrepancies in the literature about the polarization dependence and on the phase shift between the SRG and the interference pattern. Some authors reported no polarization dependence while forming SRGs via laser ablation,[43] in contradiction to the results by Ramanujam *et al.*[45] Also, the SRGs recorded on amorphous polymers using high laser intensities, presumably involving photodegradation and thermal effects, were in phase with the interference pattern.[39] But in other works where ablation and photodegradation have been suggested as responsible for the SRG formation, a π-phase shift was reported.[43,44] In summary, the existing models cannot cope with the results for high-intensity exposures, and addressing this issue theoretically will certainly be a major undertaking owing to the participation of several mechanisms. For instance, in addition to ablation, SRGs might result from melting induced locally or from sudden thermal expansion due to the intense laser light.

In layer-by-layer films, molecular interactions make it difficult for SRGs to be inscribed, presumably because such interactions preclude photoisomerization. At this time there is no theoretical model that can predict how such interactions will affect the mass transport. To employ a phenomenological parameter to cater for such preclusion—for instance, using an effective, increased viscosity in the theoretical models dealing with viscous flow—is certainly insufficient because the SRG formation in ELBL films also includes photodegradation. Also lacking is a systematic investigation of the influence of film architecture on SRG processes. Because the SRGs on ELBL films might involve photodegradation, one cannot make direct comparisons with data for the light-driven mass transport in the spin-coated films. The SRGs so far reported on LB films, on the other hand, were not caused by photodegradation, which would appear to point to a light-driven mass transport. However, a full characterization in terms of dependence on film thickness, laser power, and polarization is not available yet. Data on these topics must be awaited before one can conclude that the layer-by-layer nature of the LB films does not affect the SRG formation.

14.6 POSSIBLE APPLICATIONS

The ability to sculpt surfaces of polymers well below Tg with light is certainly a novel and unexpected phenomenon with many potential technological applications. Moreover, as far as fundamental science is concerned, the study of SRGs allows one to probe surface and material-flow properties of polymers at a micron or submicron scale. The potential for technological applications was realized as soon as the first SRGs were observed. However, most efforts have been concentrated on the exploration of new materials and fabrication of SRGs under different experimental conditions. In particular, the aim has been to further the understanding of the complex processes involved. We hope we have given a flavor of such developments in the previous sections. A number of real applications have nevertheless been demonstrated that indicates the multitude of devices that can be fabricated by exploiting the capability of inscribing SRGs on azopolymer films. In the following sections we shall comment upon several of these initiatives.

14.6.1 Holography Applications

In holography, the amplitude and phase of light can be modulated, allowing recording and display of a 3D image. Dynamic holography, in particular, has large potential for real-time image processing.[65] An interesting development in this context has been reported by Ramanujam et al.;[45] they achieved an instant off-axis holographic process with SRGs produced by a pulsed laser. Exposures as short as 5 ns and an instant-display process were performed, not requiring any wet chemical processing, as the authors took advantage of the possibility of inscribing an SRG on an azofunctionalized polymer film. The holograms could be erased through a thermal treatment of the film, enabling the film to be reused. They also stressed that the very rapid inscription process opens the way for cheap mass replication of holograms if a micromolding technique is used. In fact, the so-called instant holography allows for the possibility of making a holographic movie of 2D objects— which might even be extended to 3D objects—since the pulsed laser used to inscribe the SRG is capable of a 20 Hz pulse rate.[45] Microscopic images have also been stored exploiting the topography of SRGs on LC polymers.[48]

Viswanathan et al.[78] described several efforts to apply SRGs on holography. Holographic image storage and retrieval on an azo-functionalized polymer film have been studied. A laser beam (at 488 nm) from an Ar+ laser was expanded and collimated by a lens system and then split by a beam splitter into two beams. An object mask was placed in the path of one of the beams. The beam carrying the object mask information interfered with another collimated beam at the recording medium plane. Exposing the interference pattern onto the azo polymer film formed a surface-relief hologram. An expanded He-Ne laser beam (at 633 nm) was used to reconstruct the holographic image.

Jiang et al.[78] successfully demonstrated that the SRG written on the azobenzene-functionalized polymer films can be used as phase masks, which have been extensively used to produce periodic light intensity modulation. A phase mask of 900 nm spacing was fabricated by interfering two 488nm Ar+ laser beams on an azo polymer film (PDO3). The polarization of the writing beams was set at an angle of 45° with respect to the s-polarization. The mask was then replicated onto another film of similar azopolymer by photoprinting using a single beam exposure at 514 nm with a polarization perpendicular to

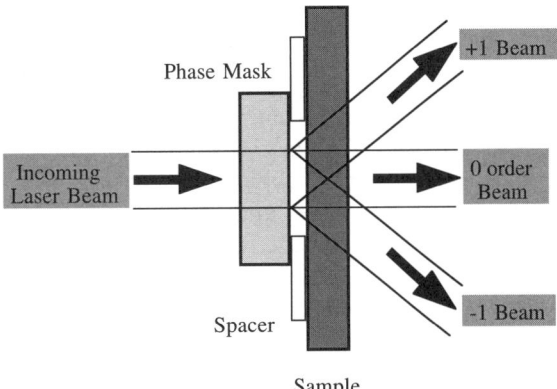

FIG. 14.24 Experimental setup employed for direct photoprinting from a phase mask.

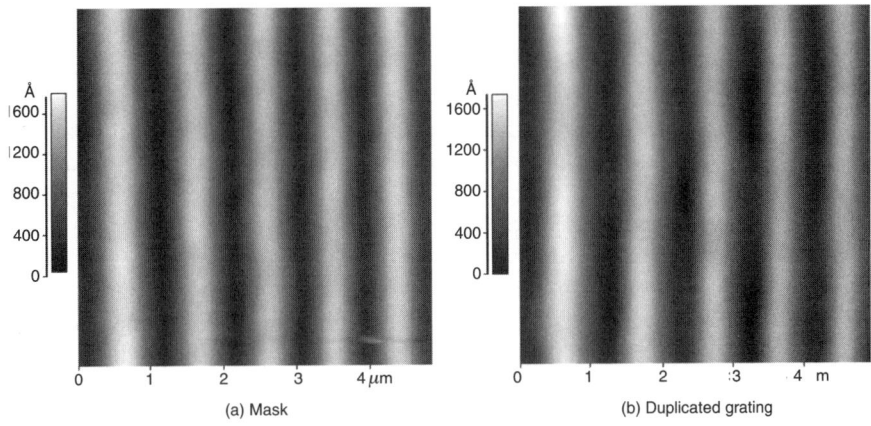

FIG. 14.25 AFM images of (a) a phase mask and (b) a replica grating.

the grooves of the mask. Figure 14.24 shows the experimental arrangement for photoprinting from a phase mask. The AFM images of the mask and the replica grating are shown in Figure 14.25. Under this polarization condition, no erasure of the mask during the photoprinting process was observed.[78] This is an example of how the polarization-dependence of the erasure characteristics of SRGs can be exploited.

14.6.2 Liquid Crystal Anchoring

SRGs inscribed on azopolymers have been used to control alignment of LC molecules. Li *et al.*[99] obtained a uniform LC alignment on the inscribed regions of a polymer film on which an SRG had been recorded. They suggested that useful applications in the in-plane switching LC displays could result from this alignment method. In yet another work by the same group, Li *et al.*[100] developed an LC device totally controlled using light, again using the capability of recording SRGs on an assembled cell filled with LC. The LC alignment is performed by irradiating an azobenzene polymeric film with a single linearly polarized argon laser beam. The alignment can be destroyed either by shining circularly polarized light on the assembly, or by heating the sample to a temperature in the vicinity of the Tg of the azopolymer. Such command surfaces for LC alignment have also been used by Kim *et al.*[101] and Parfenov *et al.*[102] The latter authors employed SRGs recorded on an azobenzene-containing polyurethane film; these SRGs were capable of orienting the LC director from the initial homeotropic alignment to a nematic alignment. This method has the potential to form command surfaces for LC displays and reversible and adaptable LC devices.

14.6.3 Waveguide Couplers

Another important use of diffraction grating in integrated optics is to filter and couple light into and out of optical waveguides.[11] SRGs inscribed on azopolymer films represent an alternative to the traditionally used gratings, which are formed using multistep lithographic/etching techniques,[103] or by direct methods such as UV laser ablation.[104] Paterson *et al.*[11] demonstrated that surface-relief-diffraction gratings recorded on an azofunctionalized, amorphous polymer could be used as light couplers into slab waveguides. Output coupling of both 633 and 830 nm light could be performed in the waveguides. Narrow-band waveguide filters have also been produced by inscribing SRGs on azopolymer films deposited on slab waveguides,[105] where bandwidths of less than 1 nm could be reached. Viswanathan *et al.*[28] showed that a coupling efficiency of ca. 20% could be obtained for an SRG with 250 nm depth recorded on an azopolymer film, using a diode laser at 830 nm as the light source. A Charge-Coupled Device (CCD) camera was used to capture the image when the laser light was coupled into the waveguide, as shown in the photograph shown in Figure 14.26. Exciting new possibilities have been suggested by Parfenov *et al.*,[102] by exploring different combinations of polymer waveguides and LC elements, especially because the gratings

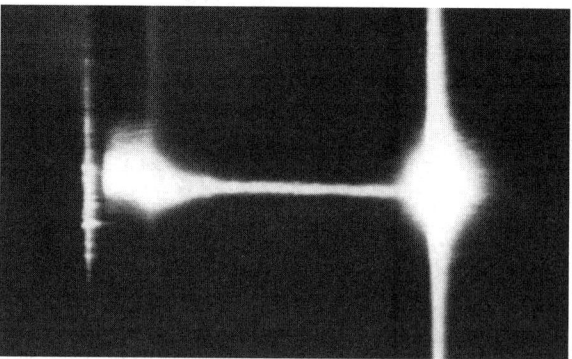

FIG. 14.26 Photograph of the near IR laser light being coupled into the azopolymer-based planar waveguide by an SRG on the left and propagated to the right.

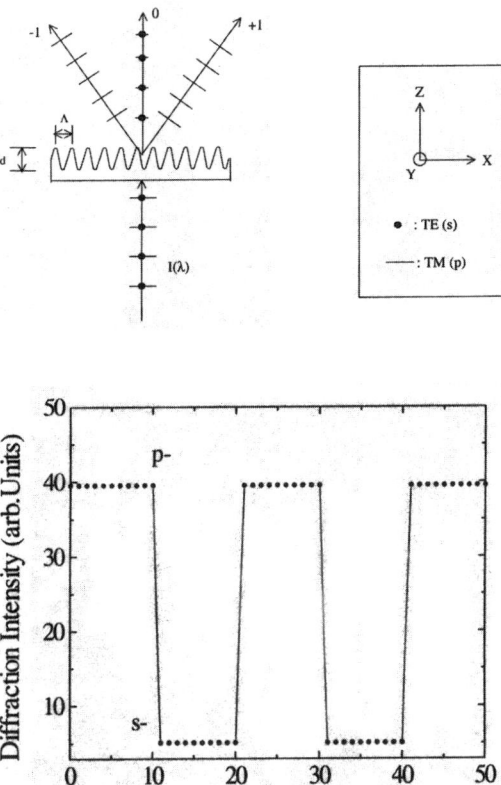

FIG. 14.27 The dependence of the diffraction efficiency on the probe beam polarization is observed for an unpolarized probe beam (at 633 nm). In the upper left part of the figure, a film is illustrated on which an SRG of amplitude d and period Λ has been inscribed. The unpolarized beam $I(\lambda)$ impinges on the film and the 1st order (+1 and −1) diffracted beams are detected. Polarization directions are given in the figure at the right, denoted by ● (TE) and — (TM), which are s- and p-polarizations, respectively. The figure in the lower part indicates that the diffraction intensity of p-polarized light is considerably larger than for s-polarized light. This suggests that the SRG can be used as a polarization discriminator.

can be rewritten optically, unlike the SRGs produced by other methods. SRGs can also be employed as a polarization discriminator, as demonstrated in Figure 14.27. This unique feature is due to the simultaneous and independent inscription of both anisotropic and relief gratings on the azopolymer films.

14.6.4 Intricate Surface Structures

The inscription of SRGs on polymer films also allows for fabricating diffractive optical elements that require intricate surface structures. For a review of unconventional methods to fabricate and pattern nanostructures, see Xia *et al.*[106] Viswanathan *et al.*[28] illustrated several such possibilities, such as the honeycomb pattern shown in Figure 14.28, the egg-crate-like structure of Figure 14.29, or the beat structure shown in Figure 14.30. These widely

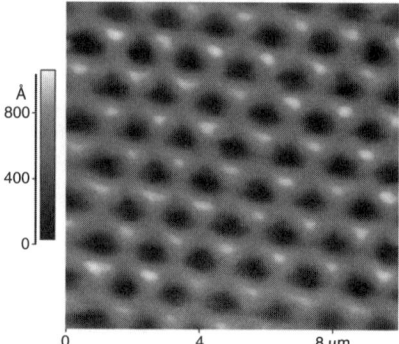

FIG. 14.28   Hexagon "honeycomb" structure produced on a azopolymer film by three laser beams with appropriate polarizations.

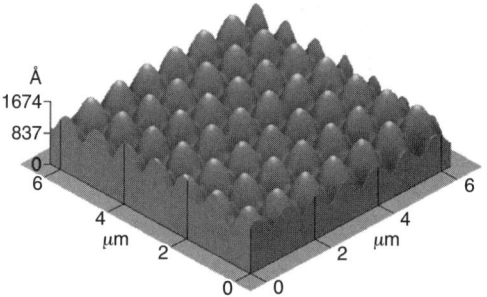

FIG. 14.29   Two sets of gratings recorded orthogonally to each other on the same spot of a PDO3 film.

distinct features were achieved by varying the writing conditions: The honeycomb pattern is generated by interfering three beams of appropriate polarization; the egg-crate structure is obtained by recording two gratings that are orthogonal to each other at the same place; and the beat structure is obtained by writing the SRG with two distinct wavelengths at a fixed writing angle, as the spacing depends upon the wavelength—in the case of Figure 14.30, these were 488 and 514 nm. Alternatively, two gratings with slightly different periods can be sequentially recorded for the appropriate amount of time to yield the beat structure. Superposition of gratings generating interesting structures has also been used in gel films,[25] where two perpendicular gratings were inscribed on the same spot by rotating the sample 90° between two successive recordings. All the properties of the double grating were essentially the same as the single gratings, including SRG stability under uniform illumination and erasure by heating the film above 100°C. It should also be mentioned that one might take advantage of the mechanical properties of polymers, which allow SRGs to be inscribed on flexible substrates, as demonstrated in the photograph shown in Figure 14.31.

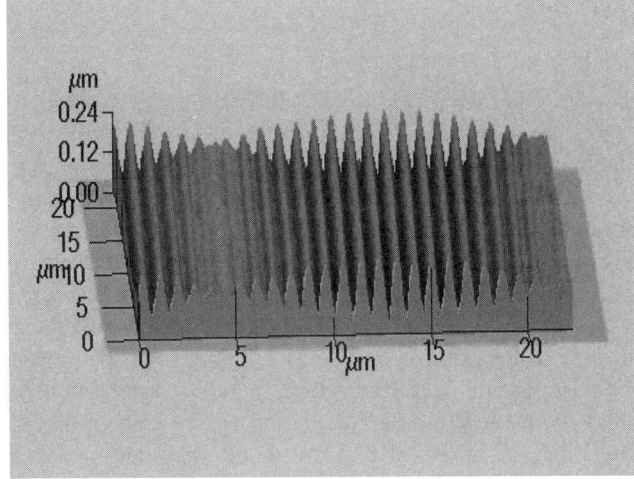

FIG. 14.30 A 3D AFM image of dual gratings (beat structure) sequentially written with the Ar+ laser beams at 488 and 514 nm at a fixed writing angle.

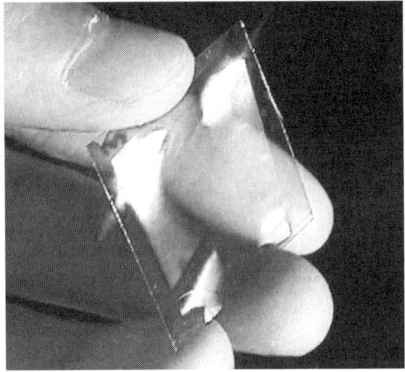

FIG. 14.31 Photograph of two SRGs inscribed on an azopolymer film using a flexible plastic substrate (See color plate).

14.6.5 Electro-Optic Devices and Second Harmonic Generation

Munakata *et al.*[41] showed that highly efficient electro-optic gratings can be obtained in azopolymer films, by poling the SRGs previously inscribed with a corona discharge. The SRGs were recorded using one of two methods, either via surface deformation using a cw laser, as in the conventional light-driven mass transport, or via laser ablation with a pulsed laser. In the latter case, the periodic modulation was achieved with a phase mask. Another application was second harmonic generation (SHG), in which an SRG-inscribed urethane-urea copolymer film was poled with corona discharge at 150°C, above the Tg of the copolymer.[44] An efficiency of 2.12% for SHG conversion at a guided fundamental power of 264 W was demonstrated. The film could be poled at a temperature greater than Tg, with no erasure of the SRG.[44]

14.7 FINAL REMARKS

In this overview, emphasis was placed on distinguishing the processes that lead to the formation of SRGs on azobenzene-containing materials. For spin-coated films from a variety of side-chain and main-chain azopolymers, SRGs produced under low-light intensities are mainly due to a light-driven mass transport. The main features of this transport can be accounted for by a theoretical model in which a field-gradient force acting on the chromophores is combined with a viscoelastic flow of polymer chains. The process is surface-initiated; when the velocity of mass transport is made to decrease as layers in the bulk are treated, the thickness dependence of the surface modulation can be explained. The SRGs thus formed are π-shifted with respect to the interference pattern, which means that the chromophores are moved away from the illuminated regions. In the case of SRGs fabricated using high intensities, thermal effects must be included, and the writing process is no longer dose dependent but intensity dependent. Processes other than the light-driven mass transport might also be involved in SRGs inscribed on LC polymers or on ELBL films. In the former, the SRGs have been reported to be in phase with the interference pattern, while in the latter photodegradation and photobleaching have been found to play important roles. We have also mentioned possible applications of SRGs, some of which have already been demonstrated by several groups. This makes the field extremely interesting for novel technological developments, especially in photonics and holography.

ACKNOWLEDGMENTS

This work was partially funded by the National Science Foundation (NSF) and Office of Naval Research (ONR) (USA). One of the authors (ONOJ) acknowledges a fellowship from Fundação de Amparo à Pesquisa do Estado de São Paulo (FAPESP) (Brazil). We are also grateful to the many colleagues from the Center for Advanced Materials at the University of Massachusetts, Lowell, who carried out extensive work on SRGs and whose names appear in the references.

REFERENCES

1. Todorov, T., Nikolova, L., and Tomova, N. Polarization holography 1: a new high-efficiency organic material with reversible photoinduced birefringence. *Appl. Opt.* 1984, 23, pp. 4309–4312.
2. Nuyken, O., Scherer, C., Baindl, A., Brenner, A. R., Dahn, U., Gärtner, R., Kaiser-Röhrich, S., Kollefrath, R., Matusche, P., and Voit, B. Azo-group-containing polymers for use in communication technologies. *Progress Polym. Sci.* 1997, 22, pp. 93–183.
3. Dhanabalan, A., Mendonça, C. R., Balogh, D. T., Misoguti, L., Constantino, C. J. L., Giacometti, J. A., Zilio, S. C., and Oliveira, Jr. O. N. Storage studies of Langmuir-Blodgett (LB) films of methacrylate copolymers derivatized with disperse red-13. *Macromolecules* 1999, 32, pp. 5277–5284.
4. Maack, J., Ahuja, R. C., Mobius, D., Tachibana, H., and Matsumoto, M. Molecular cis-trans switching in amphiphilic monolayers containing azobenzene moieties. *Thin Solid Films* 1994, 242, pp. 122–126.
5. Nikolova, L., Todorov, T., Ivanov, M., Andruzzi, F., Hvilsted, S., and Ramanujam, P. S.

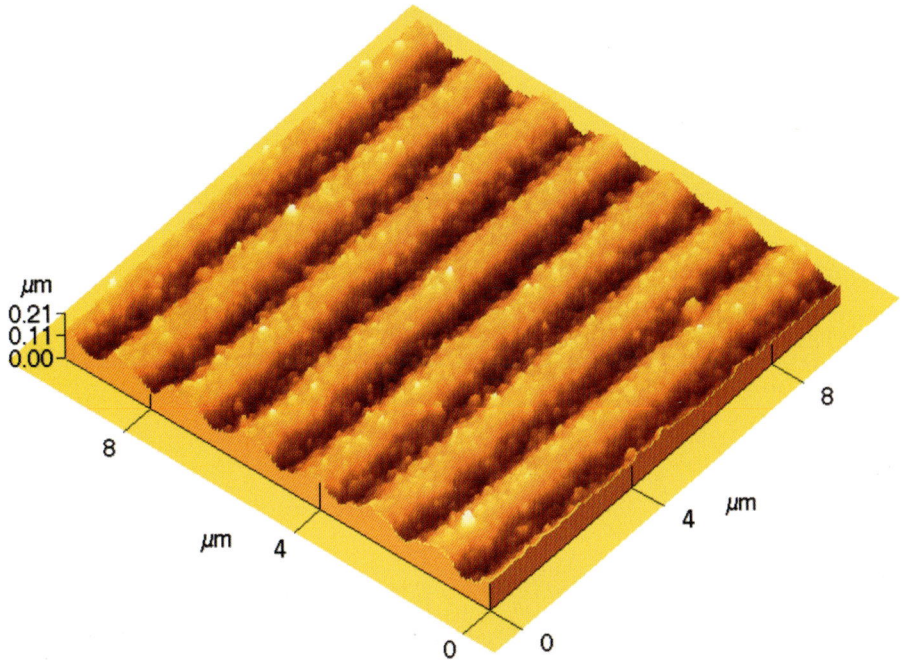

FIGURE 14.20 A 3D AFM Image of an SRG on an 80-bilayer PDAC/CR film.

FIGURE 14.31 Photograph of two SRGs inscribed on an azopolymer film using a flexible plastic substrate.

Polarization holographic gratings on side-chain azobenzene polyesters with linear and circular photoanisotropy. *Appl. Opt.* 1996, 35, pp. 3835–3840.

6. Kim, D. Y., Tripathy, S. K., Li, L., and Kumar, J. Laser-induced holographic surface-relief gratings on nonlinear optical polymer films., *Appl. Phys. Lett.* 1995, 66, pp. 1166–1168.

7. Rochon, P., Batalla, E., and Natansohn, A. Optically induced surface gratings on azoaromatic polymer films. *Appl. Phys. Lett.* 1995, 66, pp. 136–138.

8. Kim, D. Y., Li, L., Jiang, X. L., Shivshankar, V., Kumar, J., and Tripathy, S. K. Polarized laser induced holographic surface relief gratings on polymer films. *Macromolecules* 1995, 28, pp. 8835–8839.

9. Barrett, C. J., Natansohn, A. L., and Rochon, P. L. Mechanism of optically inscribed high-efficiency diffraction gratings in azo polymer films. *J. Phys. Chem.* 1996, 100, pp. 8836–8842.

10. Jiang, X. L., Kumar, J., Kim, D. Y., Shivshankar, V., and Tripathy, S. K. Polarization dependent recordings of surface relief gratings on azobenzene containing polymer films. *Appl. Phys. Lett.* 1996, 68, pp. 2618–2620.

11. Paterson, J., Natansohn, A., Rochon, P., Callender, C. L., and Robitaille, L. Optically inscribed surface relief diffraction gratings on azobenzene containing polymers for coupling light into slab waveguides. *Appl. Phys. Lett.* 1996, 69, pp. 3318–3320.

12. Ho, M. S., Barret, C., Paterson, J., Esteghamatian, M., Natansohn, A., and Rochon, P. Synthesis and optical properties of poly{(4-nitrophenyl)-[3-[N-[2-(methacryloyloxy)ethyl]-carbazolyl]]diazene}. *Macromolecules* 1996, 29, pp. 4613–4618.

13. Ramanujam, P. S., Holme, N. C. R., and Hvilsted, S. Atomic force and optical near-field microscopic investigations of polarization holographic gratings in a liquid crystalline azobenzene side-chain polyester. *Appl. Phys. Lett.* 1996, 68, pp. 1329–1331.

14. Holme, N. C. R., Nikolova, L., Ramanujam, P. S., and Hvilsted, S. An analysis of the anisotropic and topographic gratings in a side-chain liquid crystalline azobenzene polyester. *Appl. Phys. Lett.* 1997, 70, pp. 1518–1520.

15. Holme, N. C. R., Nikolova, L., Norris, T. B., Hvilsted, S., Pedersen, M., Berg, R. H., Rasmussen, P. H., and Ramanujam, P. S. Physical processes in azobenzene polymers on irradiation with polarized light. *Macromolecular Symposia* 1999, 137, pp. 83–103.

16. Yamamoto, T., Yoneyama, S., Tsutsumi, O., Kanazawa, A., Shiono, T., and Ikeda, T. Holographic gratings in the optically isotropic state of polymer azobenzene liquid-crystal films. *J. Appl. Phys.* 2000, 88, pp. 2215–2220.

17. Yamamoto, T., Hasegawa, M., Kanazawa, A., Shiono, T., and Ikeda, T. Holographic gratings and holograhic image storage via photochemical phase transitions of polymer azobenzene liquid-crystal films. *J. Materials Chem.* 2000, 10, pp. 337–342.

18. Stracke, A., Wendorff, J. H., Goldmann, D., and Janietz, D. Optical storage in a smectic mesophase: thermal amplification of light-induced chromophore orientations and surface relief gratings. *Liquid Crystals* 2000, 27, pp. 1049–1057.

19. Stracke, A., Wendorff, J. H., Goldmann, D., Janietz, D., and Stiller, B. Gain effects in optical storage: Thermal induction of a surface relief grating in a smectic liquid crystal. *Adv Mater*, 2000, 12, pp. 282–285.

20. Mendonca, C. R., Dhanabalan, A., Balogh, D. T., Misoguti, L., dos Santos, D. S., Pereira-da-Silva, M. A., Giacometti, J. A., Zilio, S. C., and Oliveira, O. N. Optically induced birefringence and surface relief gratings in composite Langmuir-Blodgett (LB) films of poly[4′-[[2-(methacryloyloxy)ethyl]ethylamino]-2-chloro-4-nitroazobenzene] HPDR13) and cadmium stearate. *Macromolecules* 1999, 32, pp. 1493–1499.

21. He, J. A., Bian, S. P., Li, L., Kumar, J., Tripathy, S. K., and Samuelson, L. A. Surface relief gratings from electrostatically layered azo dye films. *Appl. Phys. Lett.* 2000, 76, pp. 3233–3235.

22. Bian, S., He, J. A., Li, L., Kumar, J., and Tripathy, S. K. Large photoinduced birefringence in azo dye/polyion films assembled by electrostatic sequential adsorption. *Adv. Mat.* 2000, 12, pp. 1202–1205.

23. Lee, S. H., Balasubramanian, S., Kim, D. Y., Viswanathan, N. K., Bian, S., Kumar, J., and Tripathy, S. K. Azo polymer multilayer films by electrostatic self-assembly and layer-by-layer post azo functionalization. *Macromolecules* 2000, 33, pp. 6534–6540.

24. He, J. A., Bian, S., Li, L., Kumar, J., and Tripathy, S. K. Photochemical behavior and forma-

tion of surface relief grating on self-assembled polyion/dye composite film. *J. Phys. Chem. B* 2000, 104, pp. 10513–10521.

25. Darracq, B., Chaput, F., Lahlil, K., Levy, Y., and Boilot, J. P. Photoinscription of surface relief gratings on azo-hybrid gels. *Advanced Materials* 1998, 10, pp. 1133–1136.

26. Frey, L., Darracq,, B., Chaput, F., Lahlil, K., Jonathan, J. M., Roosen, G., Boilot, J. P., and Levy, Y. Surface and volume gratings investigated by the moving grating technique in sol-gel materials. *Opt Commun* 2000, 173, pp. 11–16.

27. Delaire, J. A., and Nakatani, K. Linear and nonlinear optical properties of photochromic molecules and materials. *Chem. Reviews* 2000, 100, pp. 1817–1845.

28. Viswanathan, N. K., Kim, D. Y., Bian, S. P., Williams, J., Liu, W., Li, L., Samuelson, L., Kumar, J., and Tripathy, S. K. Surface relief structures on azo polymer films. *J. Materials Chem.* 1999, 9, pp. 1941–1955.

29. Tripathy, S. K., Viswanathan, N., Balasubramanian, S., Bian, S., Li, L., and Kumar, J. Polarization dependent holographic write, read and erasure of surface relief gratings on azopolymer films, in Multiphoton and Light Driven Multielectron Processes in Organics: New Phenomena, Materials and Applications, pp. 421–436, eds F. Kajzar and M. V. Agranovich, 2000 Kluwer Academic Publishers, Netherlands.

30. Barrett, C. J., Rochon, P. L., and Natansohn, A. L. Model of laser-driven mass transport in thin films of dye-functionalized polymers. *Journal of Chemical Physics* 1998, 109, pp 1505–1516.

31. Kumar, J., Li, L., Jiang, X. L., Kim, D. Y., Lee, T. S., and Tripathy, S. Gradient force: The mechanism for surface relief grating formation in azobenzene functionalized polymers. *Appl. Phys. Lett.* 1998, 72, pp. 2096–2098.

32. Lefin, P., Fiorini, C., and Nunzi, J. M. Anisotropy of the photoinduced translation diffusion of azo-dyes. *Optical Materials* 1998, 9, pp. 323–328.

33. Pedersen, T. G., Johansen, P. M., Holme, N. C. R., and Ramanujam, P. S. Mean-field theory of photoinduced formulation of surface reliefs in side-chain azobenzene polymers. *Phys. Rev. Lett.* 1998, 80, 89–92.

34. Labarthet, F. L., Buffeteau, T., and Sourisseau, C. Analyses of the diffraction efficiencies, birefringence, and surface relief gratings on azobenzene-containing polymer films. *J. Phys. Chem. B* 1998, 102, pp. 2654–2662.

35. Naydenova, I., Nikolova, L., Todorov, T., Holme, N. C. R., Ramanujam, P. S., and Hvilsted, S. Diffraction from polarization holographic gratings with surface relief in side-chain azobenzene polyesters. *J. Opt. Soc. of America B—Optical Physics* 1998, 15, pp. 1257–1265.

36. Sumaru, K., Yamanaka, T., Fukuda, T., and Matsuda, H. Photoinduced surface relief gratings on azopolymer films: Analysis by a fluid mechanics model. *Appl. Phys. Lett.* 1999, 75, pp. 1878–1880.

37. Fukuda, T., Sumaru, K., Yamanaka, T., and Matsuda, H. Photo-induced formation of the surface relief grating on azobenzene polymers: Analysis based on the fluid mechanics. *Mol. Cryst. and Liq. Cryst.* 2000, 345, pp. 263–268.

38. Viswanathan, N. K., Balasubramanian, S., Li, L., Tripathy, S. K., and Kumar, J. A detailed investigation of the polarization-dependent surface-relief-grating formation process on azo polymer films. *Japanese J. Appl. Phys. Part 1—Regular Papers Short Notes & Review Papers* 1999, 38, pp. 5928–5937.

39. Bian, S. P., Williams, J. M., Kim, D. Y., Li, L. A., Balasubramanian, S., Kumar, J., and Tripathy, S. Photoinduced surface deformations on azobenzene polymer films. *J. Appl. Phys.* 1999, 86, pp. 4498–4508.

40. Tripathy, S. K., Kumar, J., Kim, D. Y., Li, L., and Xiang, X. L. Novel photoprocessing using photo-dynamic azobenzene polymers. *Naval Research Reviews*, 1997, 2, pp. 1–9.

41. Munakata, K., Harada, K., Yoshikawa, N., Itoh, M., Umegaki, S., and Yatagai, T. Direct fabrication methods of surface relief electro-optic gratings in azo-polymer films. *Optical Review* 1999, 6, pp. 518–521.

42. Leopold, A., Wolff, J., Baldus, O., Huber, M. R., Bieringer, T., and Zilker, S. J. Thermally induced surface relief gratings in azobenzene polymers. *J. Chem. Phys.* 2000, 113, pp. 833–837.

43. Egami, C., Kawata, Y., Aoshima, Y., Alasfar, S., Sugihara, O., Fujimura, H., and Okamoto,

N. Two-stage optical data storage in azo polymers. *Japanese J. Appl. Phys. Part 1—Regular Papers Short Notes & Review Papers* 2000, 39, pp. 1558–1561.

44. Che, Y. L., Sugihara, O., Egami, C., Fujimura, H., Kawata, Y., Okamoto, N., Tsuchimori, M., and Watanabe, O. Fabrication of surface relief grating with second-order nonlinearity using urethane-urea copolymer films. *Japanese J. Phys. Part 1—Regular Papers Short Notes & Review Papers* 1999, 38, pp. 6316–6320.

45. Ramanujam, P. S., Pedersen, M., and Hvilsted, S. Instant holography. *Appl. Phys. Lett.* 1999, 74, pp. 3227–3229.

46. Grzybowski, B. A., Haag, R., Bowden, N., and Whitesides, G. M. Generation of micrometer-sized patterns for microanalytical applications using a laser direct-write method and microcontact printing. *Anal. Chem.* 1998, 70, pp. 4645–4652.

47. Kawata, Y., Aoshima, Y., Egami, C., Ishikawa, M., Sugihara, O., Okamoto, N., Tsuchimori, M., and Watanabe, O. Light-induced surface modification of urethane-urea copolymer film used as write-once optical memory. *Jpn J Appl Phys 1*, 1999, 38, pp. 1829–1831.

48. Holme, N. C. R., Nikolova,, L., Hvilsted, S., Rasmussen, P. H., Berg, R. H., and Ramanujam, P. S. Optically induced surface relief phenomena in azobenzene polymers. *Appl. Phys. Lett.* 1999, 74, pp. 519–521.

49. Hattori, T., Shibata, T., Onodera, S., and Kaino, T. Fabrication of refractive index grating into azo-dye-containing polymer films by irreversible photoinduced bleaching. *J. Appl. Phys.* 2000, 87, pp. 3240–3244.

50. Fiorini, C., Prudhomme, N., de Veyrac, G., Maurin, I., Raimond,, P., and Nunzi, J. M. Molecular migration mechanism for laser induced surface relief grating formation. *Synthetic Met*, 2000, 115, pp. 121–125.

51. Chen, J. P., Labarthet, F. L., Natansohn, A., Rochon, P. Highly stable optically induced bire-fringence and holographic surface gratings on a new azocarbazole-based polyimide. *Macromolecules* 1999, 32, pp. 8572–8579.

52. Lee, T. S., Kim, D. Y., Jiang, X. L., Li, L. A., Kumar, J., and Tripathy, S. Synthesis and optical properties of polyureas with azoaromatic groups in the main chain. *Macromolecular Chem. and Physics* 1997, 198, pp. 2279–2289.

53. Lee, T. S., Kim,, D. Y., Jiang, X. L., Li, L., Kumar, J., and Tripathy, S. Photoinduced surface relief gratings in high-T-g main-chain azoaromatic polymer films. *J. of Polymer Sci. Part A—Polymer Chem.* 1998, 36, pp. 283–289.

54. Sukwattanasinitt, M., Lee, D. C., Kim, M., Wang, X. G., Li, L., Yang, K., Kumar, J., Tripathy, S. K., and Sandman, D. J. New processable, functionalizable polydiacetylenes. *Macromolecules* 1999, 32, pp. 7361–7369.

55. Viswanathan, N. K., Balasubramanian, S., Li, L., Kumar, J., and Tripathy, S. K. Surface-initiated mechanism for the formation of relief gratings on azo-polymer films. *J. Phys. Chem. B* 1998, 102, pp. 6064–6070.

56. Labarthet, F. L., Buffeteau, T., and Sourisseau, C. Molecular orientations in azopolymer holographic diffraction gratings as studied by Raman confocal microspectroscopy. *J. Phys. Chem. B* 1998, 102, pp. 5754–5765.

57. Labarthet, F. L., Buffeteau, T., and Sourisseau, C. Orientation distribution functions in uniaxial systems centered perpendicularly to a constraint direction. *Applied Spectroscopy* 2000, 54, pp. 699–705.

58. Constantino, C. J. L., Aroca, R. F., Mendonça, C. R., Mello, S. V., Balogh, D. T., Zílio, S. C., and Oliveira, Jr. O. N. Micro-Raman scattering imaging of Langmuir-Blodgett surface relief gratings. *Advanced Functional Materials* 2001, 1, pp. 1–5.

59. Egami, C., Kawata, Y., Aoshima,, Y., Takeyama, H., Iwata, F., Sugihara, O., Tsuchimori, M., Watanabe, O., Fujimura,, H., and Okamoto, N. Visible-laser ablation on a nanometer scale using urethane-urea copolymers. *Opt Commun*, 1998, 157, pp. 150–154.

60. Eich, M., and Wendorff, J. H. Erasable holograms in polymeric liquid-crystals. *Makromol. Chem. Rapid Commun.* 1987, 8, pp. 467–471.

61. Anderle, K., and Wendorff, J. H. Holographic recording using liquid-crystalline side-chain polymers. *Mol. Cryst. Liq. Cryst.* 1994, 243, pp. 51–75.

62. Bublitz, D., Helgert, M., Fleck, B., Wenke, L., Hvilsted, S., and Ramanujam, P. S. Photoinduced deformation of azobenzene polyester films. *Applied Physics B—Lasers and*

Optics 2000, 70, pp. 863–865.

63. Kulinna, C., Hvilsted, S., Hendann, C., Siesler, H. W., and Ramanujam, P. S. Selectively deuterated liquid crystalline cyanoazobenzene side-chain polyesters. 3. Investigations of laser-induced segmental mobility by fourier transform infrared spectroscopy. *Macromolecules*, 1998, 31, pp. 2141–2151.

64. Yamamoto, T., Hasegawa, M., Kanazawa, A., Shiono, T., and Ikeda, T. Phase-type gratings formed by photochemical phase transition of polymer azobenzene liquid crystals: Enhancement of diffraction efficiency by spatial modulation of molecular alignment. *J. Phys. Chem. B* 1999, 103, pp. 9873–9878.

65. Hasegawa, M., Yamamoto, T., Kanazawa, A., Shiono, T., and Ikeda, T. Photochemically induced dynamic grating by means of side chain polymer liquid crystals. *Chem. of Materials* 1999, 11, pp. 2764–2769.

66. de Gennes, P. G., and Prost, J. The *Physics* of Liquid Crystals, 2nd ed., New York, Oxford University Press, 1993, p. 41.

67. Sánchez, C., Alcalá, R., Hvilsted, S., and Ramanujam, P. S. Biphotonic holographic gratings in azobenzene polyesters: surface relief phenomena and polarization effects. *Appl. Phys. Lett.* 2000, 77, pp. 1440–1442.

68. Ashkin, A., Dziedzic, J. M., Bjorkholm, J. E., Chu, S. Observation of a single-beam gradient force optical trap for dielectric particles. *Opt. Lett.* 1986, 11, pp. 288–290.

69. Labarthet, F. L., Buffeteau, T., and Sourisseau, C. Azopolymer holographic diffraction gratings: Time dependent analyses of the diffraction efficiency, birefringence, and surface modulation induced by two linearly polarized interfering beams. *J. Phys. Chem. B* 1999, 103, pp. 6690–6699.

70. Srikhirin, T., Laschitsch, A., Neher, D., and Johannsmann, D. Light-induced softening of azobenzene dye-doped polymer films probed with quartz crystal resonators. *Appl. Phys. Lett.* 2000, 77, pp. 963–965.

71. Stracke, A., Bayer, A., Zimmermann, S., Wendorff, J. H., Wirges, W., Bauer-Gogonea, S., Bauer, S., and Gerhard-Multhaupt, R. Relaxation behaviour of electrically induced polar orientation and of optically induced non-polar orientation in an azo-chromophore side group polymer. *J. Phys. D-Appl. Phys.* 1999, 32, pp. 2996–3003.

72. Victor, J. G., and Tokelson, J. M. On measuring the distribution of local free-volume in glassy polymers by photochromic and fluorescence techniques. *Macromolecules*, 1987, 20, pp. 2241–2250.

73. Fukuda, T., Matsuda, H., Viswanathan, N. K., Tripathy, S. K., Kumar, J., Shiraga, T., Kato, M., and Nakanishi, H. Systematic study on photofabrication of surface relief grating on high-Tg azobenzene polymers. *Synethetic Met.* 1999, 102, pp. 1435–1436.

74. Fukuda, T., Matsuda, H., Shiraga, T., Kimura, T., Kato, M., Viswanathan, N. K., Kumar, J., and Tripathy, S. K. Photofabrication of surface relief grating on films of azobenzene polymer with different dye functionalization. *Macromolecules* 2000, 33, pp. 4220-4225.

75. Andruzzi, L., Altomare, A., Ciardelli, F., Solaro, R., Hvilsted, S., and Ramanujam, P. S. Holographic gratings in azobenzene side-chain polymethacrylates. *Macromolecules* 1999, 32, pp. 448–454.

76. Fiorini, C., Prudhomme, N., Etile, A. C., Lefin, P., Raimond, P., and Nunzi, J. M. All-optical manipulation of azo-dye molecules. *Macromolecular Symposia* 1999, 137, pp. 105–113.

77. Labarthet, F. L., Rochon, P., and Natansohn, A. Polarization analysis of diffracted orders from a birefringence grating recorded on azobenzene containing polymer. *Appl. Phys. Lett.* 1999, 75, pp. 1377–1379.

78. Jiang, X. L., Li, L., Kumar, J., Kim, D. Y., and Tripathy, S. K. Unusual polarization dependent optical erasure of surface relief gratings on azobenzene polymer films. *Appl. Phys. Lett.* 1998, 72, pp. 2502–2504.

79. Tripathy, S. K., Viswanathan, N. K., Balasubramanian, S., and Kumar, J. Holographic fabrication of polarization selective diffractive optical elements on azopolymer film. *Polym. Adv. Technol.* 2000, 11, pp. 1–5.

80. Bian, S., Li, L., Kumar, J., Kim, D. Y., Williams, J., Tripathy, S. K. Single laser beam-induced surface deformation on azobenzene polymer films. *Appl. Phys. Lett.* 1998, 73, pp. 1817–1819.

81. Constantino, C. J. L., Aroca, R. F., Yang, S., Zucolotto, V., Li, L., Liveira, Jr. O. N., Cholli, A. L., Kumar, J., Tripathy, S. K. Raman microscopy and mapping of surface-relief gratings recorded on azocellulose films. *J. of Macromolecular Sci.–Pure and Appl. Chem.* 2001, 38, pp. 1549–1557.

82. Fuhrmann, T., and Tsutsui, T. Synthesis and properties of a hole-conducting, photopatternable molecular glass. *Chem. of Materials* 1999, 11, pp. 2226–2232.

83. Yamaki, S., Nakagawa, M., Morino, S., and Ichimura, K. Surface relief gratings generated by a photocrosslinkable polymer with styrylpyridine side chains. *Appl. Phys. Lett.* 2000, 76, pp. 2520–2522.

84. Bian, S. P., Liu, W., Williams, J., Samuelson, L., Kumar, J., and Tripathy, S. Photoinduced surface relief grating on amorphous poly(4-phenylazophenol) films. *Chem. of Materials* 2000, 12, pp. 1585–1590.

85. Bartholomeusz, B. J. Thermal modeling studies of organic compact disk-writable media. *Appl. Optics* 1992, 31, pp. 909–918.

86. Ozaki, M., Nagata, T., Matsui, T., Yoshino, K., and Kajzar, F. Photoinduced surface relief grating on composite film of conducting polymer and polyacrylate containing azo-substituent. *Jpn J Appl Phys.* 2, 2000, 39, pp. L614–L616.

87. Xu, Z. S., Drnoyan, V., and Natansohn, A., and Rochon, P. Novel polyesters with aminosulfone azobenzene chromophores in the main chain. *J Polym Sci Pol Chem.* 2000, 38, pp. 2245–2253.

88. Wang, X. G., Balasubramanian, S., Kumar,, J., Tripathy, S. K., and Li, L. Azochromophore-functionalized polyelectrolytes. 1. Synthesis, characterization, and photoprocessing. *Chem. of Materials* 1998, 10, pp. 1546–1553.

89. Itoh, M., Harada, K., Matsuda, H., Ohnishi, S., Parfenov, A., Tamaoki, N., and Yatagai, T. Photomodification of polymer films: azobenzene-containing polyurethanes. *J Phys D Appl Phys.* 1998, 31, pp. 463–471.

90. Dante, S., Advincula, R., Frank, C. W., and Stroeve, P. Photoisomerization of polyionic layer-by-layer films containing azobenzene. *Langmuir* 1999, 15, pp. 193–201.

91. Stockton, W. B., and Rubner, M. F. Molecular-level processing of conjugated polymers. 4. Layer-by-layer manipulation of polyaniline via hydrogen-bonding interactions. *Macromolecules* 1997, 30, pp. 2717–2725.

92. Raposo, M., and Oliveira, Jr. O. N. Energies of adsorption of poly(o-methoxyaniline) layer-by-layer films. *Langmuir* 2000, 16, pp. 2839–2844.

93. Pontes, R. S., Raposo, M., Camilo, C. S., Dhanabalan, A., Ferreira, M., and Oliveira, Jr. O. N. Non-equilibrium adsorbed polymer layers via hydrogen bonding. *Phys. Status Solid A* 1999, 173, pp. 41–50.

94. Zucolotto, V., Mendonça, C. R., Dos Santos, D. S., Balogh, D. T., Zilio, S. C., and Oliveira, Jr. O. N. The influence of electrostatic and H-bonding interactions on the optical storage of layer-by-layer films of an azopolymer. *In press.*

95. Pietsch, U., Rochon, P., and Natansohn, A. Formation of a buried lateral density grating in azobenzene polymer films. *Advanced Materials* 2000, 12, pp. 1129–1132.

96. Meier, J. G., Ruhmann, R., and Stumpe, J. Planar and homeotropic alignment of LC polymers by the combination of photoorientation and self-organization. *Macromolecules* 2000, 33, pp. 843–850.

97. Geue, T., Schultz, M., Grenzer, J., Pietsch, U., Natansohn, A., and Rochon, P. X-ray investigations of the molecular mobility within polymer surface gratings. *J. Appl. Phys.* 2000, 87, pp. 7712–7719.

98. Iftime, G., Fisher, L., Natansohn, A., and Rochon, P. Photoinduced birefringence in copolymers containing Disperse Red 1 and styrene. *Can. J. Chem.* 2000, 78, pp. 409–414.

99. Li, X. T., Natansohn, A., and Rochon, P. Photoinduced liquid crystal alignment based on a surface relief grating in an assembled cell. *Appl. Phys. Lett.* 1999, 74, pp. 3791–3793.

100. Li, X. T., Natansohn, A., Kobayashi, S., and Rochon, P. An optically controlled liquid crystal device using azopolymer films. IEEE *Journal of Quantum Electronics* 2000, 36, pp. 824–827.

101. Kim, D. Y., Kumar, J., and Tripathy, S. K. Liquid crystal alignment using optically induced surface relief gratings on azo polymer films. *Polymer Preprints* 1998, 319, pp. 1107–1108.

102. Parfenov, A., Tamaoki, N., and Ohnishi, S. Photoinduced alignment of nematic liquid crys-

tal on the polymer surface microrelief. *J. Appl. Phys.* 2000, 87, pp. 2043–2045.

103. Johnson, L. F., Kammlott, G. W., and Ingersoll, A. Generation of periodic surface corrugations. *Appl. Optics*, 1978, 17, pp. 1165–1181.

104. Philips, H., Callahan, D., Sauerbrey, R., Szabo, G., and Bor, Z. *Appl. Phys. Lett* 1991, 58, pp. 276.

105. Stockermans, R. J., and Rochon, P. L. Narrow-band resonant grating waveguide filters constructed with azobenzene polymers. *Appl. Opt.* 1999, 38, pp. 3714–3719.

106. Xia, Y. N., Rogers, J. A., Paul, K. E., and Whitesides, G. M. Unconventional methods for fabricating and patterning nanostructures. *Chem Rev* 1999, 99, pp. 1823–1848.

15

DYNAMIC PHOTOCONTROLS OF MOLECULAR ORGANIZATION AND MOTION OF MATERIALS BY TWO-DIMENSIONALLY ARRANGED AZOBENZENE ASSEMBLIES

TAKAHIRO SEKI*
KUNIHIRO ICHIMURA[†]
Chemical Resources Laboratory, Tokyo Institute of Technology, 4259 Nagatsuta, Midori-ku, Yokohama 226-8503, Japan
[†]*Research Institute for Science and Technology, Science University of Tokyo, 2641 Yamazaki, Noda, Chiba 278-8510, Japan*

15.1 INTRODUCTION

The interplay between photochromic molecules and organized molecular or macromolecular assemblies is expected to provide wide opportunities for creating new classes of photonic materials. Regarding light-triggered molecular processes, one might think of the biological systems triggered by photo-receptor biomolecules such as rhodopsins and phytochrom. Information about light captured by a photochromic biological molecule is transferred and amplified to the physiological processes by surrounding proteins and lipid membranes in very complicated and sophisticated ways. Although artificial approaches do not seem to reach these levels, we can mimic, in part, the molecular amplification systems by means of designed molecular organizations. There is no doubt that efforts to create more and more smart and highly controllable amplification systems will be an ongoing subject in the future.

Needless to say, the most important feature of photochromic systems is their ability to give rise to dynamic processes that are important for their reversible structural changes. Light can induce motions of versatile molecular systems at any hierarchical-scale level from individual molecules to macroscopic materials. When a photoprocess takes place at a local site, it can be used as ultrahigh-density optical memory, which has been attracting current interest for use in future information technologies. On the other hand, if a triggering molecule induces secondary sequential events to larger scales, mesoscopic to macroscopic operations of molecules and polymers can be realized. This situation provides us with not only novel principles for photomemory exhibiting marked molecular amplifications, but also with dynamic molecular systems mimicking biorelated phototriggered events. This chapter deals with our recent progress studying phototriggered molecular systems designed on the basis of the latter aspects using photochromic molecular (mostly azobenzene [Az]) monolayers.

Monolayers (2D films) can be regarded as ultimate thin states of organic and polymeric materials and are quite fascinating for constructing molecular amplification systems for the following reasons. First, the simplest (thinnest) structure of a monolayer allows one to comprehend the details of molecular communications and motions, because the excessive complexities of three-dimensional materials are efficiently excluded. This can lead to many implications on the material design of wide varieties of related systems. Second, for surface-mediated transfer processes, which consist of molecule-to-molecule transfer of information based on molecular orientations inscribed on uppermost surface layers, the existence of monolayers on material surfaces is sufficient for performances of macroscale controls of materials. Note that the nature and behavior of buried photochromic moieties in thicker films play minor roles.

The most typical demonstration of this was presented in 1988 as the *command surface effect*,[1] in which a reversible switching of nematic liquid crystal (LC) alignment between homeotropic (perpendicular) and planar (parallel) molecular orientations was induced by E (*trans*)/Z (*cis*) photo-isomerization of an Az-containing monolayer. Since then, a great number of works have dealt with the surface-mediated photoalignment of LCs. Works on surface-assisted photoalignment of LC systems by surface photochromic layers have recently been summarized in a review article.[2] A large body of data has already been accumulated by Ichimura's group, indicating a tremendous applicability of this photoaligning principle to multiple classes of LCs. Not only nematic LCs, but also side-chain LC polymers,[3] lyotropic LCs,[4,5] chiral nematic (cholesteric) LCs,[6] and even highly viscous discotic LCs[7] can be controlled by the surface anisotropic isomerization process induced either by irradiation with linearly polarized light or by oblique irradiation with nonpolarized light. The present chapter, however, will not touch on such typical LC system, but instead will discuss the other dynamic molecular processes mediated by Az-monolayers. All the processes, of course, are essentially related to dynamic interplays between the organized molecular systems.

This chapter consists of three sections, following this introduction. The first section summarizes the photocontrol of liquid motion by self-assembled adsorbed Az monolayers (Section 15.2). The second section (Section 15.3) introduces controls of molecular organizations of polymer chains triggered by Langmuir-Blodgett-Kuhn (LBK) monolayers of an Az polymer. The final section (Section 15.4) describes the motional function of Az layers themselves, the photoinduced deformation of monolayers, and mass migration of Az polymer film systems. The events discussed in these three parts are illustrated in Figure 15.1.

15.2 PHOTOCONTROL OF LIQUID MOTION BY AZOBENZENE MONOLAYERS

It has been known for many years that surface-free energies of flat solid substrates are determined by atomic-level constitutions of their outermost surfaces[8,9]; as a result, alterations of chemical structures of outermost monomolecular layers by external stimuli—including pH changes, heat application, photoirradiation, and so on—lead to the modification of versatile interfacial phenomena. Photoirradiation seems to be the most convenient way to manipulate surface properties because of its ability to control them precisely in time and space. In fact, there have been a number of reports

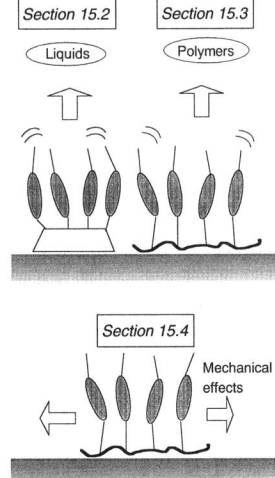

FIG. 15.1 Photocontrol of various soft materials by azobenzene-containing monolayers.

on photoinduced manipulation of interfacial phenomena, including wet-ability,[10,11,12,13,14,15,16] LC alignment,[2] dispersibility,[17] and biomaterial affinity[18,19] by the modification of substrate surfaces with photoreactive monolayers. When a photochemical reaction of a monolayer takes place reversibly, the surface-mediated photocontrol of interfacial phenomena just mentioned can be carried out reversibly to provide multiple functional materials applicable to new devices.

When a liquid droplet is placed on a solid surface, the droplet spreads symmetrically to hold an equilibrium contact angle of $\theta > 0°$. On the other hand, if a liquid droplet is put in contact with a substrate with a spatial inhomogeneity in surface tension, a force imbalance is generated that results in the movement of the droplet in order to compensate for the imbalanced force. The experimental observation of such a liquid motion has been reported by Chaudhury and Whitesides.[20] They demonstrated a water droplet running uphill by using a silica surface that was modified asymmetrically with a long-chain alkyl silylating reagent to generate a gradient of surface hydrophobicity. The movement stops immediately after running about 2 mm because the imbalance of surface tension is leveled off. This situation led us to the idea that we could generate spatially controlled changes in surface-free energies of a substrate by photoinduced structural alterations of the outermost surface of the substrate; we anticipated that the motion of a liquid could be guided by spatially controlled photoirradiation of the surface on which the liquid was placed.

15.2.1 Molecular Design and Preparation of Azobenzene Monolayers

Azobenzene is a promising photoreceptor moiety for the control of surface-free energy, because its E/Z photoisomerization exhibits a good photofatigue-resistance, accompanied by reversible changes in dipole moment.[21] An octacarboxymethoxylated calix[4]resorcinearene derivative substituted with four p-octylazobenzenes (CRA-CM) (Figure 15.2) was designed for the present purpose, taking the following points into account. The first is that CRA-CM molecules adsorb readily and efficiently on polar surfaces of substrates such as a silica plate from dilute solutions as a result of the formation of multisite hydrogen bonds to give densely packed monolayers.[22] The second unique feature of the self-assembled monolayer of CRA-CM is the fact that a sufficient free volume required for E-to-Z photoisomerization of the Az is ensured even though the cyclic skeleton of CRA-CM is closely packed in the adsorbed monolayer.[23,24] This is because the molecular packing is determined specifically by the base area of the rigid CRA-CM ring system, which is much larger than a sum of cross sections of the four Az moieties.

On the other hand, this procedure for the self-assembled monolayer formation has a drawback. A monolayer adsorbed on a silica plate surface is subjected to the detachment of CRA-CM molecules more or less in polar solvents such as water and alcohols,[22] since the adsorption stems from the hydrogen bond formation. This problem was overcome by the pretreatment of a silica plate surface with an aminoalkylation silylating reagent to introduce amino residues on the surface.[25,26] A photoresponsive monolayer of

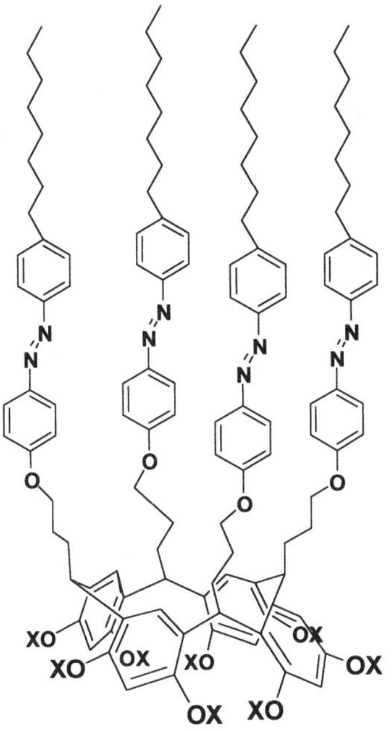

CRA-CM (X=CH$_2$COOH)

FIG. 15.2 Chemical structure of CRA-CM.

CRA-CM was prepared simply by immersing the aminated silica plate in a dilute solution of CRA-CM. The resulting monolayer exhibited high desorption resistance even toward polar solvents because of the interactions between the CO$_2$H groups of CRA-CM and the aminosilanized surface.[25,26] In this way, a densely packed CRA-CM monolayer exhibiting a conversion into Z-isomer of ca. 90% under 365 nm light irradiation was fabricated.

15.2.2 Spreading/Retraction Motion of Liquid Droplets

Because the outermost surface of a UV-exposed CRA-CM monolayer is likely terminated by Z-isomer of the Az with a higher dipole moment, an increase in surface-free energy is generated.[10,11,12] In fact, Zisman plots[27,28] revealed that a critical surface tension of a UV-exposed monolayer is 3.5 mJ m^{-2} higher than that of the monolayer before the illumination. Photoirradiation of the Z-rich surface with blue (436 nm) light recovered E-isomer, so that reversible changes in surface tension are achieved by alternating irradiation with UV and blue light. This situation is visualized by reversible spreading/retraction motion of a droplet of a nematic LC, NPC-02, placed on a substrate plate modified with CRA-CM under the alternating photoirradiation.[29] UV-light irradiation leads to a lateral spreading of the droplet, since a sessile contact angle is changed from 24° to 11° after UV irradiation. The droplet was subsequently retracted by illumination with blue light, and the reversible changes were maintained even after 100 cycles of the alternating photoirradiation.

The surface-mediated motion of droplets was also observable for the other nematic LCs including 5CB and even for isotropic liquids such as olive oil, 1-methylnaphthalene, and 1,1,2,2-tetrachloroethane. However, droplets of some liquids, such as water, formamide, and ethylene glycol, display no shape change even under the same illumination conditions. In order to shed light on this kind of marked dependence of the nature of liquids on the droplet motion, hysteresis of contact angles[30] is an informative parameter to grasp dynamic interfacial phenomena.[31,32] In this context, contact-angle hystereses, defined as the difference between advancing (θ_{adv}) and receding (θ_{rec}) contact angles, were measured for the liquids before and after UV-light irradiation; they are summarized in Table 15.1. As expected, the liquids that show no surface-mediated motion exhibit larger contact-angle hystereses. On the contrary, the liquids displaying reversible changes in droplet shape under photoirradiation exhibit hystereses that are smaller than differences in contact angles before and after photoirradiation. Based on these results, an essential conclusion can be drawn. The reversible spreading/retraction motion is induced by light when θ_{rec} for E (*trans*)-isomer, θ_{rec}^{tr}, is larger than θ_{adv} for Z (*cis*)-isomer, θ_{adv}^{cis}.[29] Note that NPC-02, when compared to common organic liquids, exhibits both smaller hysteresis and larger photoinduced changes in contact angles upon photoisomerization (Table 15.1); therefore, NPC-02 is quite suitable for experiments of light-driven motion.

15.2.3 Displacement of Liquids

Because a gradient in surface tension induces a net mass transport of liquids, as mentioned in Section 2.2,[31,32] it was anticipated that spatially controlled photoirradiation of a CRA-CM–modified substrate results in differences in surface energies so that the motion of a liquid droplet can be guided by light.

TABLE 15.1   Contact angles of the CRA-CM monolayer for various liquids (deg)^a

	trans-rich		cis-rich		
	θ_{adv}	θ_{rec}	θ_{adv}	θ_{rec}	$(\theta_{rec}^{tr} - \theta_{adv}^{cis})^b$
no motion					
water	94	40	86	51	−46
formamide	68	17	62	19	−45
ethylene glycol	61	36	56	39	−20
poly(ethylene glycol)	42	37	38	31	−1
motion					
1-methylnaphthalene	26	24	10	18	4
1,1,2,2-tetrachloroethane	18	16	12	11	4
5CB	43	37	22	19	15
NPC-02	28	24	11	10	13
olive oil	29	25	17	13	8

^{a}The values were within experimental error of ± 2.
^{b}See the text.

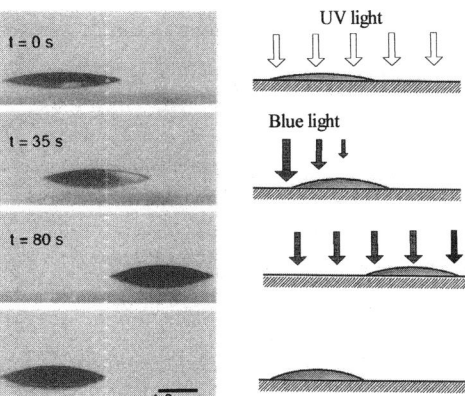

FIG. 15.3 Photographs of the light-driven motion of an olive-oil droplet on a silica substrate modified with CRA-CM. (Top three parts of this figure): the olive oil droplet on a *cis*-rich surface moved toward a surface with a higher surface energy upon asymmetrical irradiation with 436 nm light, whereas a sessile contact angle was changed from 18° to 25° by the irradiation. In the bottom part of the figure, the droplet moved in the opposite direction by switching the position of the photoirradiation.

This is the case, as shown in Figure 15.3. Olive oil is used here because it fulfills the necessary condition required for the liquid motion (Table 15.1). When a droplet of olive oil is placed on a Z-rich surface (Figure 15.3a), followed by asymmetrical illumination with blue light to recover *E*-isomer (Figure 15.3b), the droplet moves in order to escape the light irradiation (Figure 15.3c), whereas flood exposure with the blue light stops the displacement (Figure 15.3d). The displacement velocity of a droplet is influenced by steepness and intensity of the light gradient. A typical speed of 35 μm/sec was observed for the motion of a droplet of olive oil of approximately 2 μl.

The displacement of a liquid is realized according to the following mechanism. A difference in surface energy is generated between both edges of a liquid droplet placed on an *E*-rich surface and a Z-rich surface induced by asymmetrical photoirradiation. This situation leads to an asymmetrical drop shape causing a Laplace pressure gradient in the droplet.[33] The pressure gradient is then rapidly equalized to create a directional motion of the droplet.

The uniqueness of the surface-mediated movement of a liquid can be shown by the following examples. Figure 15.4 shows the light-driven motion of a glass bead trapped by a droplet of a nematic LC, NPC-02 by the alternating actions of spreading and dewetting the droplet. Because the asymmetrical spreading and dewetting of NPC-02 causes the deformation of the equilibrium meniscus between the liquid and the glass bead, a spatial inhomogeneity in capillary force occurs in the coalesced liquid. This imbalanced capillary force is responsible for the glass-bead displacement. The second example, shown in Figure 15.5, is featured by the movement of a liquid contained in a capillary glass tube. Because the self-assembling of a CRA-CM monolayer on a silica surface is performed simply by wetting the surface with an aminoalkylating reagent and subsequently with a CRA-CM solution, the inside wall of a capillary tube of silica glass can be readily subjected to the surface modification

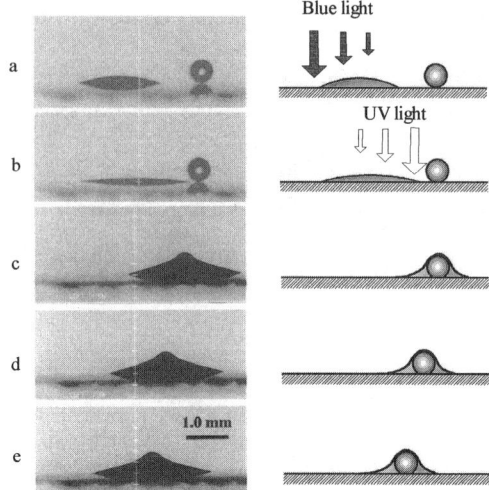

FIG. 15.4   Light-driven motion of a glass bead trapped in an NPC-02 droplet. The glass bead was captured under alternating UV and blue-light photoirradiation (a, b, and c) and conveyed by the alternating photoirradiation (d and e).

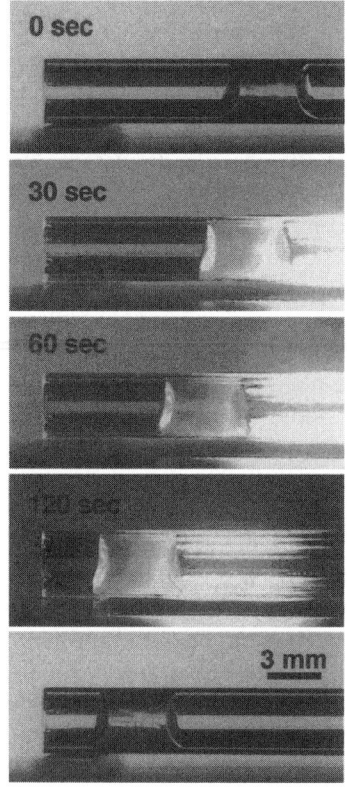

FIG. 15.5   Light-driven motion of an NPC-02 droplet filled in a CRA-CM–modified capillary tube of a 2.3-mm diameter. In photo a, droplet of NPC-02 was placed in the tube exposed to UV light. Subsequent irradiation with blue light at the right edge pushed the droplet to the left. The droplet was stopped by flood exposure to blue light (bottom).

with the photoresponsive monolayer. When a CRA-CM–modified capillary tube is irradiated with UV light to generate a Z-rich surface, the liquid in the tube starts to escape from blue-light irradiation so that the fluid motion in a glass tube can be controlled at will.

15.2.4 Outlook

This section dealt with the novel functionality of a photoresponsive monolayer to carry out the displacement of liquids simply by photoirradiation as a result of the photocontrol of surface energy. Self-assembled monolayers of CRA-CM are designed and fabricated to ensure sufficient E/Z photoisomerizability even in densely packed monomolecular layers, which are quite stable toward solvent treatments. The following critical condition should be met in order to realize the surface-mediated photomanipulation of liquid motion: $\theta_{rec}^{tr} < \theta_{adv}^{cis}$. In this context, contact-angle hysteresis plays an essential role in this kind of dynamic behavior of liquid displacement.

One of the potential applications of the surface-mediated photo-manipulation of liquid displacement is to achieve on-chip reactions of tiny amounts of chemical reagents. A light-guided transport of a liquid droplet containing fluorescamine was attained to make remote mixing with another droplet dissolving dodecylamine. The successful mixing was confirmed by the appearance of fluorescence resulting from the coupling reaction.[34] The improvement of performances of light-guided microfluidic systems using the present technique requires a photochromic molecular system that shows larger changes in surface-free energy upon illumination and a deeper understanding of the origin of contact-angle hysteresis in molecular levels to reduce the hysteresis. We believe that these approaches allow us to achieve rapid manipulation of the motion of liquids, even aqueous solutions, to find applications in a broad range of areas from on-chip synthesis to microfluidic devices.

15.3 PHOTOCONTROL OF POLYMER CHAIN ORGANIZATIONS

Thin polymer-chain organizations (mostly orientations) are arranged primarily by mechanical procedures such as stretching, rubbing, and frictional deposition. An alternative orientation is to use epitaxial processes via a transfer from a substrate surface. This section describes our attempts to arrange polymer films via transfer from an irradiated Az surface layer. Epitaxial control of polymer chains, especially conjugated ones, are fascinating in connection with photofunctions such as light absorption, emission, photoconduction, and so forth. We introduce two examples of the photocontrol of conjugated polymer chains in this section. Schematic illustrations are given in Figure 15.6.

15.3.1 Polydiacetylene[35,36]

The topochemical polymerization of diacetylenes proceeds with retention of molecular packing of the crystal structure. The reactivity strongly depends on

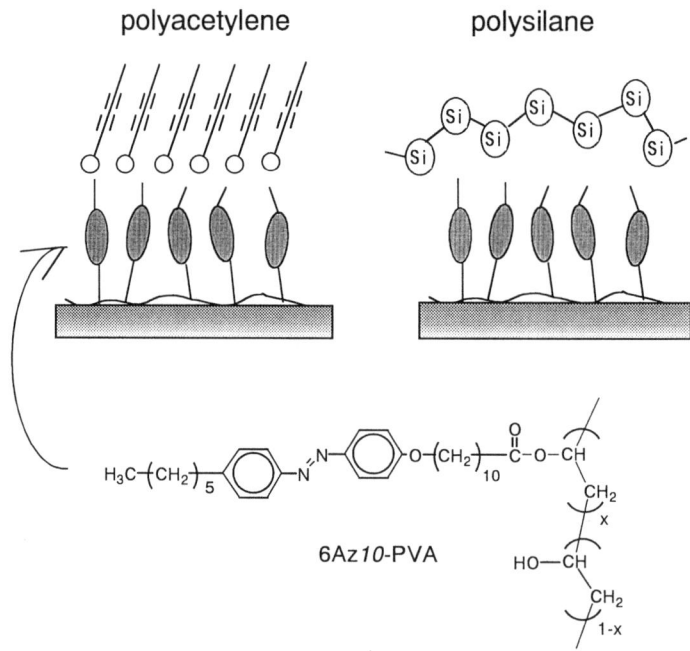

FIG. 15.6 Photocontrol of conjugated polymers (π-conjugated polydiacetylene and σ-conjugated polysilane) by surface Az monolayer.

the stacking distance of the monomers and the angle between the diacetylene rod and the stacking axis.[37] It is anticipated that the Az command layer can change such packing parameters, thus leading to alterations of the polymerization behavior. The motivation of this investigation using a heterolayered LBK system is attributed in part to the work of Peterson and coworkers,[38] which demonstrated that the packing structure of LBK layers of 22-tricosenoic acid developed epitaxially following the crystallite structure of an initially deposited monolayer to a large thickness. The initial monolayer can be exchanged with a photoisomerizable Az monolayer for the control of the packing state of an overlayered amphiphile.

On the top of a 6Az10-PVA monolayer on a substrate plate, an LBK film of 10,12-pentacosadiyonic acid (PDA) was prepared. Figure 15.7 indicates the changes in UV-visible absorption spectra upon 254 nm irradiation onto this bilayered LB film. Here, this procedure was made on the E- (a) and Z-Az surfaces (b). As seen in the figure, there was a marked difference in the conjugation state of the resulting polydiacetylene for the E- and Z-Az surfaces. The E- and Z-Az surfaces led to a blue- and red-colored polydiacetylene film, respectively. Thus, the Az command layer is able to control the π-conjugated state of polydiacetylene. An important issue involved in this process is that light absorbed by the Az in the UV region is transformed to the information in longer visible regions (blue or red in color).

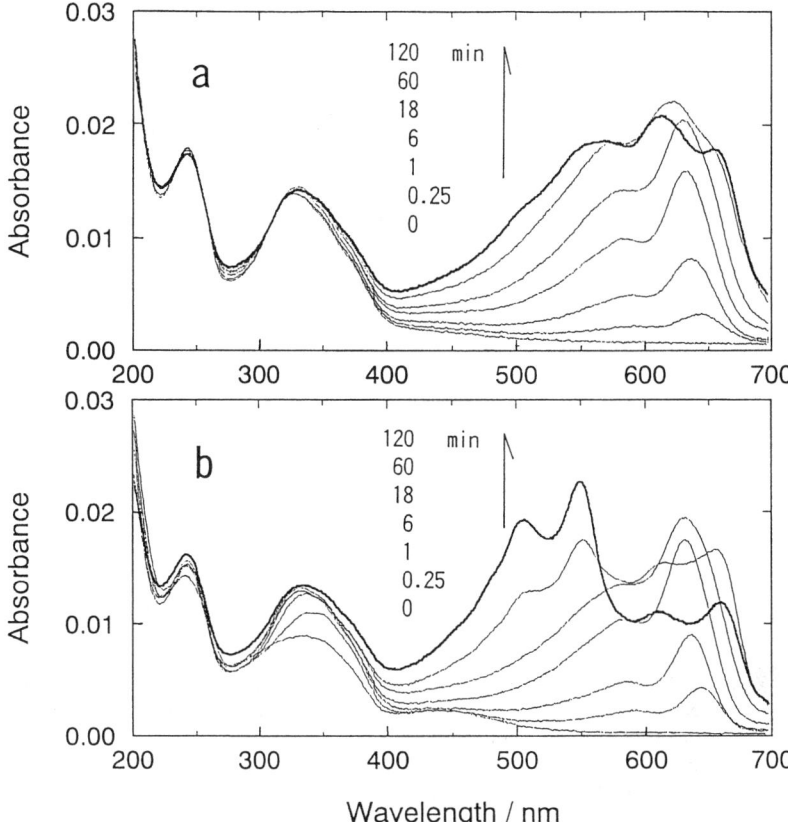

FIG. 15.7 Changes in UV-visible absorption spectra of polydiacetylene LB film on Az monolayer. *E*-and *Z*-Az surfaces give rise to blue (a) and red (b) films at a final stage. (From reference 36; copyright permission from the Society of Polymer Science, Japan.)

15.3.2 Polysilane[39]

Polysilanes (polysilylenes) consist of catenated silicon (Si)[39] backbone, which indicates unique optoelectric properties stemming from the delocalized σ-electron along the Si chain. This class of the polymer absorbs light strongly in the UV region, and this property is quite informative for detecting the polymer-chain organization. A spincast film of poly(di-*n*-hexylsilane) (PDHS), having a thickness below 100 nm, is prepared on the 6Az10-PVA monolayer. In this system, the photoisomerization of Az monolayer gives rise to controls of two aspects, the conformation and orientation.

The crystallization process of PDHS film was monitored. The spincast films of PDHS on a 6Az10-PVA monolayer were first heated to 100°C and then cooled. The enhancement of 370 nm band with concomitant decrease of 320 nm band reflects formation of the all-*trans*-zigzag Si chain conformation upon crystallization. The crystallization rate strongly depended on the initial isomerization state of the Az of the surface layer (see Figure 15.8). The crystallization process was accelerated on the *Z*-Az monolayer in comparison

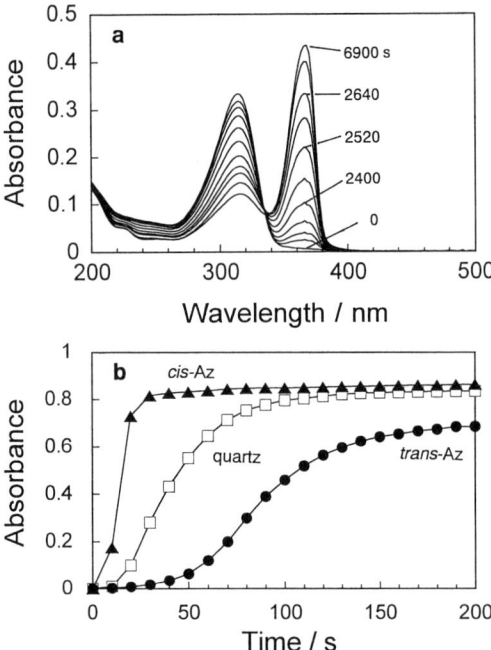

FIG. 15.8 Changes in UV-visible absorption spectrum of a poly(di-*n*-hexylsilane) spincast film in the course of crystallization to adopt an all-*trans*-zigzag conformation of the Si backbone at 15°C (a). Photocontrol of the crystallization rate as observed by the absorbance increases at 366 nm, corresponding to the increase of the all-*trans*-zigzag conformation of the Si chain (b). (From reference 39; copyright permission from the American Chemical Society.)

with that on a control quartz surface. In contrast, crystallization was retarded on the *E*-Az monolayer. Since the surface of the Az layer in both *E* and *Z* form is commonly more hydrophobic than the quartz surface, molecular topological requirements at the contacting interface should play an important role in the rate control. It seems that the nucleation process at the interface of PDHS film and the substrate is controlled by the photoisomerization, which then results in the crystallization control of the whole film.

In-plane anisotropic induction of the Az monolayer by linearly polarized light can orient the backbone of PDHS. In this experiment, the linearly polarized light is first irradiated with sufficient nonpolarized UV light and then with linearly polarized 436 nm light. This procedure generates a large dichroism orthogonal to the polarization vector in the π-π* absorption (long axis) of the Az monolayer. On the top of this oriented molecular layer, a PDHS spincast film is prepared. The PDHS film initially is in the disordered state, giving only the 320 nm absorption band. In this stage, the film shows no preferential orientation. As the crystallization proceeds for a few days at room temperature, the orientation starts to emerge. The orientation can be more pronounced by an additional annealing and cooling process (second crystallization). Figure 15.9 indicates a typical example of polarized UV-vis absorption spectra obtained after these procedures had been done. Light

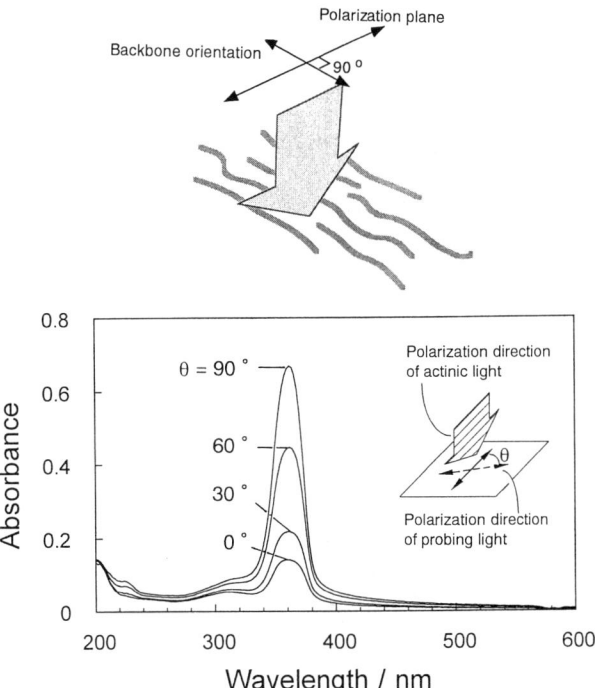

FIG. 15.9 Photo-orientation of the Si backbone of poly(di-*n*-hexylsilane) via transfer from an Az monolayer photoaligned via exposure to linearly polarized light. (From reference 39; copyright permission from the American Chemical Society.)

absorption takes place most strongly for perpendicularly polarized light and most weakly for the parallel one. Since the electric transition dipole moment of this band lies in the direction of backbone extension, the results obviously indicate that the polymer backbone is oriented in a manner orthogonal to the polarization plane of the actinic light, that is, parallel to the Az long-axis orientation.

Two-step orientational transfer from the Az monolayer to the PDHS film and then to nematic LCs can be further achieved.[40] The LC molecules align in the parallel direction of the photo-oriented PDHS chain. In the direct Az-LC transfer, the thermal back reaction of Az recovers the LC alignment to the original state. The two-step LC alignment, on the other hand, has a feature that the thermal process of the azobenzene chromophore no longer affects the orientation of LC. The two-step transfer is further used to align mesostructured sol-gel silicate formed with a template of organic surfactant aggregates.[41] In this attempt, the Az monolayer is too fragile, and the molecular orientation does not survive the conditions of the sol-gel process. The fragile orientation of Az is firmly fixed in the oriented polymer film, and then the photo-orientation of the mesostructured inorganic-organic hybrid is successfully performed. In this sense, the two-step orientational transfer is essential.

15.4 PHOTOINDUCED MOTIONS AND MASS MIGRATIONS

The energy conversion from light to mechanical response is one of the most fascinating targets in organic photochromic systems. Section 15.2 already indicated the light-driven droplet displacement. This section introduces the motion of Az polymers themselves. Rapid motions observed here might allow for applications of actuators of polymer-based micromachine systems. The photoinduced surface-relief generation described in the latter part of this section (Section 15.4.2) deals with much thicker films than monolayers. However, the mass migration in thicker films is closely related to the structure and properties of the monolayer systems.

15.4.1 Photomechanical Effects in Azobenzene Polymer Monolayers

15.4.1.1 Macrosize Effect

Photochromic molecules change their geometry and polarity by illumination and are regarded as light-switchable molecular machines.[42,43] When they are attached to a polymer chain, the photoisomerization triggers conformational changes of the main and side chains, resulting in the induction of secondary macrosized effects such as viscosity changes and induction of sol-gel transition in the solvated systems, changes in surface wetability and glass-transition temperature (Tg), and shape changes in a bulk state. In biological systems, macromolecular components (proteins) are hierarchically organized, and highly efficient and anisotropic motions are performed. One encounters great difficulties in assembling artificial polymer systems that fulfill two opposing concepts, namely, highly ordered structure and large mobility. As a prototype for such molecular systems, a monolayer system on water is quite appealing.[44,45,46,47] The water surface strongly aligns the amphiphilic molecules; nevertheless, the relatively low viscosity might provide high flexibility for large film deformations. Because of their molecularly arrayed arrangement, the macroscopic-film deformation almost exactly corresponds to the sum of the individual motion of a large number of existing molecules or polymer units (typically in the order of 10^{16} for film-balance experiments). This can be regarded as a typical example of the collective and active performance of supramolecules.[48]

We reported in 1993 that monolayers of poly(vinyl alcohol) derivatives bearing an Az side chain 6Azn-PVA and its homologues exhibit large photoinduced area changes.[49,50] Here, n denotes the length of the methylene spacer connecting PVA and Az. The area changes can be interpreted as the consequence of reversible on/off contact motion with the water surface, which is switched by the polarity change between the Az photoisomers as schematically illustrated in Figure 15.10. The expansion factors defined as the final area on UV-light irradiation divided by the initial area for 6Az1-PVA, 6Az5-PVA, and 6Az10-PVA were ca. 1.5, 2, and 3, respectively, when the area was monitored at a pressure of 2 mN m^{-1}.[49] To our knowledge, the 6Az10-PVA monolayer exhibits the largest film deformation in the monolayer system. Film expansion and contraction processes can be repeated many times with full reproducibility.

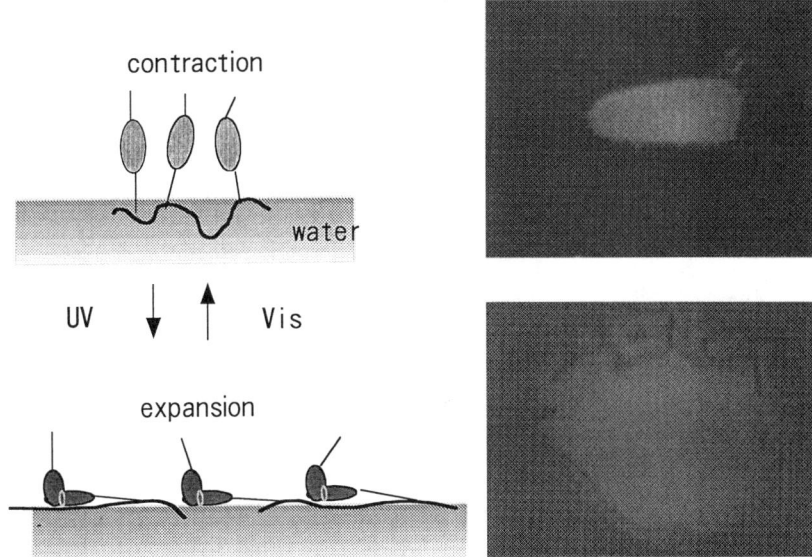

FIG. 15.10 Photoinduced expansion and contraction of 6Az10-PVA monolayer at the air-water interface. The area changes of an isolated domain can be visualized *in situ* by Brewster angle microscopy. (From reference 52; copyright permission from the American Chemical Society.)

15.4.1.2 Visualization by Microscopy

To date, almost all attention has been paid to the macroscopic behavior at a centimeter level (film area and surface pressure) obtained with a Langmuir film balance. The photoinduced response at a molecular level, on the other hand, is obtainable by UV-visible absorption spectroscopy and surface-potential measurements. In these contexts, our attention recently has been paid to microscopic observations that cover the ranges between the previously cited scale levels using 6Azn-PVA monolayers.

Static morphologies of 6Az10-PVA monolayers on water were first observed by Brewster angle microscopy (BAM) in correlation with the surface pressure-area curve. It was shown that the isomerization state of an Az unit greatly changes the morphological features and rheological properties.[51] The 6Az10-PVA in the E form showed rigid iceberg-like domain structures, and the Z form, in contrast, was a liquid-like monolayer.

We next attempted to observe the photoinduced film expansion and contraction for an isolated small domain under zero-pressure conditions.[52] BAM images at the expanded and contracted states are shown in Figure 15.10. New insights were obtained from this approach, as follows: (1) 6Az10-PVA monolayer at zero pressure exhibits four- to fivefold expansion that is even larger than that estimated with a macroscopic film balance experiment (ca. three fold); (2) the UV-light–induced expansion is a nonlinear process with respect to the progress of the $E \rightarrow Z$ photoisomerization, exhibiting an induction time at an early stage (Figure 5.11); and (3) upon visible-light irradiation, the film shows self-contraction without pressure. These experiments

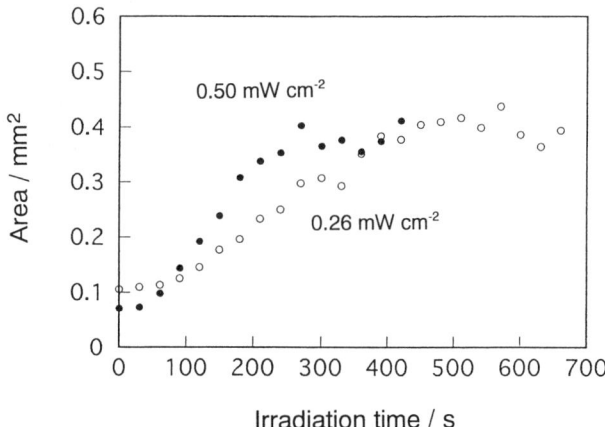

FIG. 15.11   Time course of area expansion of 6Az10-PVA monolayer as directly estimated by BAM. Note that induction time is observed before expansion. (From reference 52; copyright permission from the American Chemical Society.)

gave precise knowledge about the intrinsic film properties free from any distortions producible with an applied pressure.

The change in film thickness is detected by the intensity change of the reflected light in BAM. The decisive evidence was in fact obtained by *in situ* X-ray reflectometry.[53] Kiessig fringes were observed for specular measurement for 6Az10-PVA monolayers on a water surface. Analyzing the X-ray data indicated that the thickness of the monolayer became larger for the *E* form than for the *Z* form. The change in thickness of 0.2–0.3 nm was due to the *E-Z* configurational change in the hydrophobic side chain.

The morphologies on solid surfaces can be monitored by atomic force microscopy (AFM). AFM and BAM take images at a resolution of a nanometer (in height profiles) and micrometer level, respectively. Therefore, visualization by these two microscopic methods provides complementary information.

The photoinduced morphological changes were found to occur also on the solid surface of mica in a humidified atmosphere.[54] Clear-humidity dependence of the morphological change was observed for a film transferred at a lateral density of 1.2 nm^2 per Az. In a dry state (relative humidity [RH] below 25%), the flat morphology was maintained. At higher humidities, the monolayer underwent larger lateral movement to form a network structure, resulting in network morphology depending on the humidity. The film shrinks insufficiently at RH of 40%, but migrates at larger distances to form networks at 60 and 80% RH. The hydration of mica should play an important role in providing the high mobility of the monolayer.[55]

Depending on the lateral-packing density, the monolayers showed photodriven large and characteristic 2D and 3D morphological changes at a submicrometer scale. This might be the first visualization of the photomechanical response of a monolayered photochromic polymer taking place on a solid surface.[56]

First, the monolayers in the Z form were deposited onto mica with average areas per Az unit of ca. 0.4 (dense film) and 1.2 nm^2 (sparse film). After the sparse film was stored in the dark for four days, the film exhibited network-shaped morphologies due to a contraction of the film during the thermal induction of the Z to E isomerization. After UV-light irradiation, the network-shaped films exhibited marked 2D swelling. The swollen morphology virtually reverted to the original one after storing it in a dark, highly humid vessel for another four days.

In the case of the dense film, the first dark adaptation gave a sponge-shaped film instead of a network-shaped one. After UV-light irradiation, the film morphology exhibited a completely different change. Upon UV-light illumination, most of the defects disappeared, and instead huge shallow panlike protrusions of 200–300 nm in diameter and ca. 10 nm in height appeared. This 3D change was also a nearly reversible process. Matsumoto et al.[57] also observed related light-induced morphological inductions in a polyion-complex-type Az-containing LB film on mica.

15.4.1.3 Visualization at the Single-Chain Level

The morphological changes discussed in Section 15.4.1.2 were obtained for monolayers at assembled states in 2D. In such an assembled state, secondary effects such as generation of 3D collapse are accompanied on the solid surface. This should be ascribed to the limited allowance of lateral diffusion compared to the rate of the photoisomerization process. For observation of intrinsic photoresponse of the monolayer, separation of the polymer chain is highly desired, ideally on a single-chain level.

Segregation to the level of the single-polymer involvement can be achieved by high dilutions in solutions[58] or in monolayers.[59] For 6Az10-PVA monolayer, however, the lipophilic long side chain strongly assists aggregation and prohibits isolation to a single chain.[52] In such cases, the process of so-called skeletonization of the LBK monolayer is effective. Skeletonization involves selective removal of one component out of the two from a hybrid organization. For LBK films, this procedure was first achieved by Katherine Blodgett.[60]

For 2D dilution of 6Az10-PVA monolayer, the monolayer in the Z form was spread with a highly expanding stuffing amphiphile such as oleic acid (OA) and methyl oleate (MO) onto pure water.[61] The Z-6Az10-PVA monolayer, mixed with the stuffing amphiphile, was transferred onto a freshly cleaved mica by vertical lifting at a very low surface pressure (1 mN m^{-1}). The oily stuffing amphiphile component (OA or MO) was then removed by immersion into methanol for 1 min or by evaporation at 120°C for 2 h. The resulting film was then stored in a humidified atmosphere (relative humidity exceeding 95%) for a week, which allowed the complete thermal reversion to the E form of Az accompanied by the film contraction.

Noncontact mode AFM images were taken for perfectly isolated dot-shaped monolayered films after removal of MO from a mixture of R ([MO]/[Az]) = 9. In Figure 15.12, three completely separated dot films are observed in the 500×500 nm^2 field. The dot films in the E form of Az were 50–60 nm (uncorrected for the cantilever shape) in diameter and ca. 1.5 nm

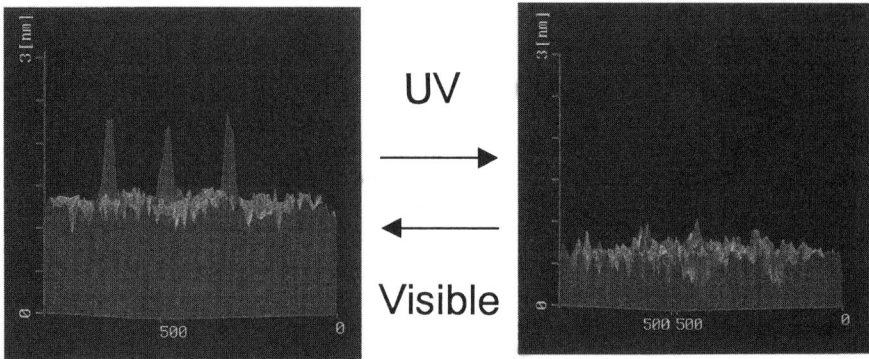

FIG. 15.12 Photoinduced morphological changes of 6Az10-PVA monolayer at an ultimately isolated dot state (most probably, single-chain separation) on mica upon light illumination. Dot films are obtained by the skeletonization method (see text). (From reference 61; copyright permission from the American Chemical Society.)

in height (left). UV-light illumination induced an expansion of the nano-film accompanying a considerable decrease in film thickness (right). The area showed an approximately fourfold expansion, with a reduction of the thickness to the level of 0.5 nm. Such behavior is reasonably understood by the assumed deformation model shown in Figure 15.10 and agrees with the BAM data obtained on the water surface.[52] Visible-light illumination reverted the initial dot shape.

The skeletonization from a 6Az10-PVA/OA mixture (R ([OA]/[Az]) = 3) with immersion into methanol provided a more crowded ensemble (Figure 15.13). In the initial state, the dot films were larger and were of irregular size. After one photocycle via expansion with UV light and contraction with visible light in a humidified atmosphere, the dots were separated into equal-sized smaller ones with the monolayer thickness upon subsequent visible-light illumination (left in Figure 15.13). For this more crowded ensemble, the expanded films contact each other to form continuous films (right); nevertheless, the dot films were precisely reproduced without fusion after shrinkage (left). Here again, the features and size in photoinduced changes were highly reproducible in many photocycles.

From the morphological features, it is inferred that the final dot films consist of one polymer chain. First, the diameter of the dots are of even size irrespective of the isolation procedures, namely skeletonization or photo-assisted segregation. Second, the smallest dot films retained their features in size and shape after many repeated photocycles. Such size stability is most likely explained by assuming the involvement of one polymer chain. Third, the average volume size was roughly estimated from the mean radius of the dot (from transmission electron microscopic data), and the height (AFM data) showed a reasonable coincidence with the supposed volume calculated from the molecular weight (Mw = 1×10^5) with an assumed density of 1 g dm^{-3}. These facts strongly support the interpretation of single-chain separation. Therefore, the nano-sized dot monolayer observed here can be regarded as the smallest light-driven nano-machine as the polymer material.

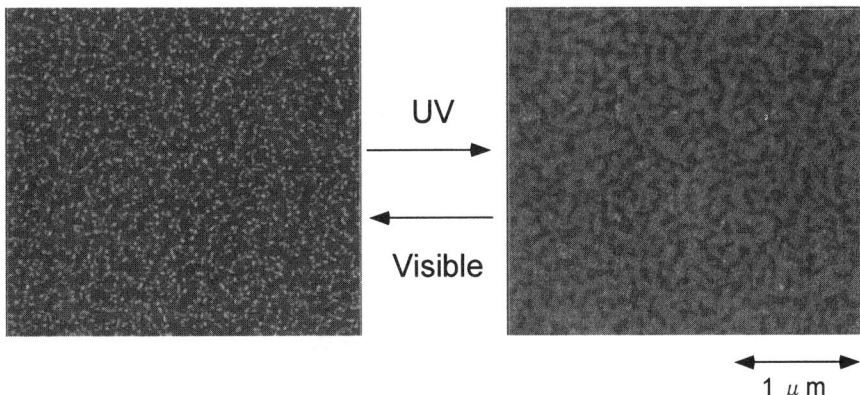

UV

Visible

1 μm

FIG. 15.13 Photoinduced morphological changes of 6Az10-PVA monolayer at dot state (most probably, single-chain separation) on mica upon light illumination. This is a more crowded ensemble compared to the case shown in Figure 15.12. (From reference 61; copyright permission from the American Chemical Society.)

15.4.2 Large Mass Migration

Surface-relief grating (SRG, regular topological surface modification) formed via irradiation with an interference pattern of coherent light has been demonstrated only recently,[62,63,64] and is perhaps the most attractive target in the current research of azo polymers[65] (Chapters 13 and 14 of this book focus on this subject). It is still unclear whether there are any common mechanisms involved between SRG formation and the photomechanical response observed in monolayers described in Section 15.4.1. Nevertheless, with regard to photoinduced motions in Az polymers, it is quite important to proceed with SRG experiments.

A great deal of data has been accumulated quite rapidly due to the basic phenomenological interest and attractive technological applications of SRG. This process has particular technological advantages because it offers a facile, all-optical, and single-step fabrication process that does not require a wet development procedure, and because the surface topology is erasable by application of circularly polarized light or heating above Tg, which realizes the repeatable utilization. There is no doubt that SRG is formed via large-scale polymer chain migration.

All attempts hitherto made for SRG experiments dealt with single-component Az polymers, including the side-chain and main-chain types. The efficiency of SRG formation can be tuned by changing the chemical structure of the Az polymer, but this requires tedious synthetic work. In this context, binary component host-guest systems in the SRG process seem appealing for tuning the film properties.[66] Incorporation of low-molecular-mass LC molecules is an intriguing candidate for this because LC molecules are expected to gain favorable fluidity for mass-migration with strong molecular cooperativity. The same Az polymer mentioned earlier (6Az10-PVA, host polymer) is mixed with 4'-pentyl-4-cyanobiphenyl (5CB, guest molecule). The chemical structures of the materials are shown in Figure 15.14. These two

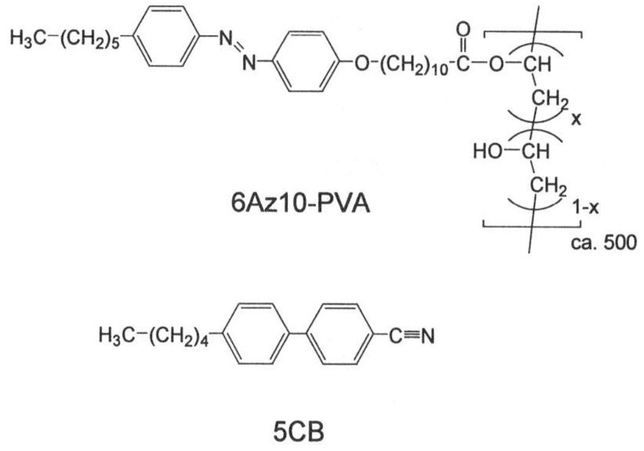

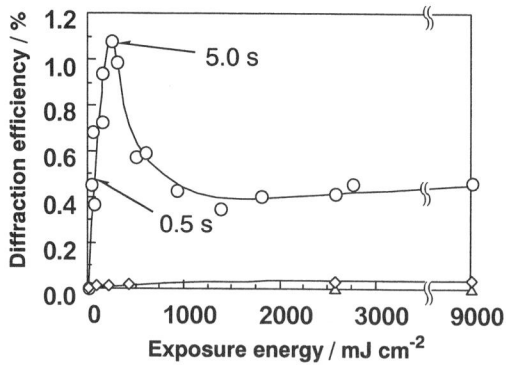

6Az10-PVA

5CB

FIG. 15.14   Molecular structure of the materials used in photogenerated surface-relief formation.

materials were originally combined to construct monolayer and LBK layers for the command surface model; one uses this model to obtain detailed information on the molecular motion and cooperativity by means of UV-visible and IR (Infrared) spectroscopy.[67,68]

The spincast films were prepared from chloroform solutions dissolving 6Az10-PVA and 5CB at $f = 0.5$, f being the molar fraction of 5CB ([5CB]/ ([Az unit] + [5CB])) on a glass or quartz substrate. The hybrid films were irradiated with nonpolarized UV (365 nm) light in advance to attain a Z-rich photoequilibrated state (UV-light treatment). Without this procedure, no SRG formation was performed. Therefore, exposure to UV light is a trigger step of SRG formation.

Figure 15.15 displays the growth profiles of the first-order diffraction efficiency evaluated with an He-Ne laser beam as a function of the total exposure energy. This figure contains the data obtained with the pure 6Az10-PVA film (diamonds) and the hybrid film of 6Az10-PVA/5CB at $f = 0.5$ with

FIG. 15.15   The diffraction efficiency originated from the formation of SRG by exposure to an interfered Ar ion layer as a function of exposure energy. Pure 6Az10-PVA (diamonds), 6Az10-PVA/5CB (1:1) hybrid after UV-light irradiation (circles) and without UV-light irradiation (triangles). (From reference 66; copyright permission from Wiley-VCH.)

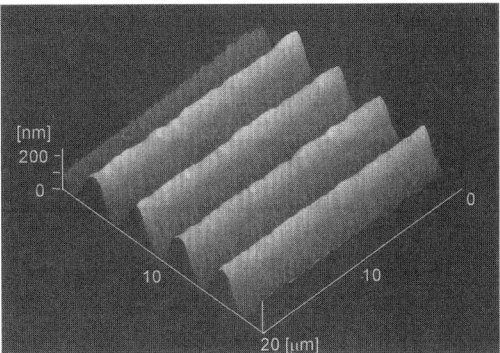

FIG. 15.16   Typical topological AFM image of surface relief gratings.

UV light treatment (circles) and without treatment (triangles). For the hybrid film with UV-light treatment, a sharp growth of diffraction efficiency was observed at an early exposure stage (circles). The diffraction efficiency reached a maximum of 1.1% in 250 mJ cm^{-2}, corresponding to ca. 5.0 s exposure at 50 mW cm^{-2}. Further exposure reduced the efficiency to give a constant value around 0.5%.

The increase in the diffraction efficiency synchronized with the growth of the surface geometrical modulation. A typical relief pattern observed by AFM is shown in Figure 15.16. It is stressed that the relief pattern was already observed after exposure to the interferometric Ar$^+$ laser (50 mJ cm^{-2}) at a noticeably short period of 0.5 s. Such rapid topological induction proceeded at room temperature. The depth from the peak to the bottom (Δh) was comparable to the initial film thickness. The photomodulated structure was stable and was unchanged for at least several months at room temperature. For the pure 6Az10-PVA film in the same procedure with UV-light treatment, no significant surface undulation was admitted.

The surface undulation can be erased by nonpolarized UV-light irradiation at a very low light doses (200 mJ cm^{-2}) at room temperature. Heating at 100°C (isotropic phase of the hybrid film) for 30 min also erased the surface structure.

Figure 15.17 indicates the most essential feature of the present system. This figure displays the diffraction efficiency (closed circles) and the surface modulation depth (open circles, Δh) of the hybrid films at various molar fractions performed after UV-light treatment. The diffraction efficiency showed a sudden increase (to 2.3%) at $f = 0.67$, which corresponds to the stoichiometry of two 5CB molecules per one Az unit. Below and above this ratio, the diffraction efficiency rapidly decreased. The profile of Δh almost followed that of the diffraction efficiency to give the maximum depth (100 nm) at $f = 0.67$. This 6Az10-PVA(1)/5CB(2) mixing ratio exactly coincides with the molecular compatibility (hybridization criterion) as revealed by DSC (Differential Scanning Colorimetry) and Langmuir monolayer experiments.[67] Due to the cooperative nature of this binary system, this can be dubbed a host-guest supramolecular SRG material.

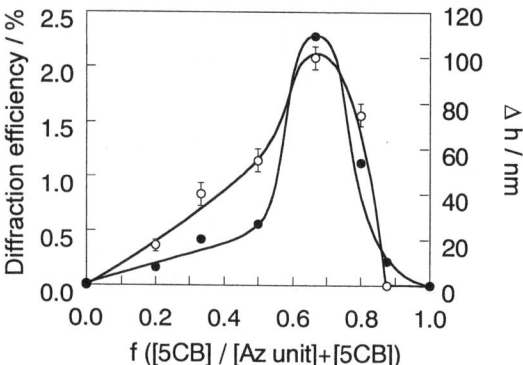

FIG. 15.17 Diffraction efficiency (filled circles) and top-to-valley height (open circles) estimated by AFM as a function of the molar fraction (f) of 5CB. (From reference 66; copyright permission from Wiley-VCH.)

Increased fluidity drastically enhances the efficiency of SRG formation; however, this essentially sacrifices the photogenerated shape stability. On the other hand, the shape stability in terms of long-term storage and durability at high temperatures is generally attained by use of rigid and high Tg polymers. Our recent approach revealed that introduction of oligo(ethylene oxide) (EO) side chain instead of LC incorporation in the Az polymer can overcome the above antimony.[69] The polymer is synthesized by radical copolymerization of an Az containing (meth)acrylate monomer and an EO containing meth(acrylate) monomer. Introduction of oligo(ethylene oxide) in the polymer provides an adequately fluid LC polymer. After a surface-relief structure is formed, the polymer is then subjected to chemical crosslinking via formalization (acetal formation with formaldehyde) between the hydroxyl group at the terminus of EO. Before the chemical crosslinking, the SRG was diminished below 90°C, but after formalization the shape was exactly maintained even at 250°C. Thus, soft and crosslinkable azo polymers provide a new class of materials for rapid surface-relief formation and persistent fixation.

15.5 CONCLUDING REMARKS

Our recent efforts and results on active and collective controls of soft materials by the E/Z photoisomerization of Az arranged in 2D and 3D are reviewed in this chapter. Large and efficient effects are actualized through precise alignment of molecular orientation in which the reaction vectors are not scattered but are favorably aligned. Such aligned molecular systems are most typically built via monolayer formation at interfaces by the LBK method or by self-assembling adsorption. The important breakthrough for realizing such dynamic systems is the use of polymer amphiphilic materials and spatially designed molecules as the Az derivatives that allow for the efficient photo-isomerization in the two dimensions. The data presented here clearly substantiate the validity of such an approach. The applicability of photoalignment of materials is not limited to fluid LCs, and recent work has shown wider opportunities for varieties of materials. Az polymers and molecular assemblies are generally regarded as useful media for optical memory, light switching, and modulations, and still more appealing applications are potentially involved. This chapter showed some new examples focusing on the motional functions that can provide many implications for creations of microactuator devices, directional microcatalytic media, downsized fluidic devices, and so forth. We believe that the photoresponsive Az polymers are still rapidly extending the horizons of the smart soft-materials science and technology.

REFERENCES

1. Ichimura, K., Suzuki, Y., Seki, T., Hosoki, A., and Aoki, K. Reversible change in alignment mode of nematic liquid crystals regulated photochemically by command surfaces modified with an azobenzene monolayer. *Langmuir*, **4**, 1422 (1988).
2. Ichimura, K. Photoalignment of liquid-crystal systems. *Chem. Rev.*, **100**, 1847 (2000).
3. Kidowaki, K., Fujiwara, T., and Ichimura, K. Surface-assisted photomanipulation of orientation of a polymer liquid crystal. *Chem. Lett.*, **643** (1999).
4. Ichimura, K., Momose, M., Kudo, K., Akiyama, H., and Ishizuki, N. Surface-assisted photolithography to form anisotropic dye layers as a new horizon of command surfaces. *Langmuir*, **11**, 2314 (1995).
5. Ichimura, K., Momose, M., and Fujiwara, T. Photoimages formed by lyotropic liquid crystals. *Chem. Lett*, 1022 (2000).
6. Ruslim, C., and Ichimura, K. Photocontrolled alignment of chiral nematic liquid crystals. *Adv. Mater.*, **13**, 37 (2001).
7. Ichimura, K., Furimi, S., Morino, S., Kidowaki, M., Nakagawa, M., Ogawa, M., and Nishiura, Y. Photocontrolled orientation of discotic liquid crystals. *Adv. Mater.*, **12**, 950 (2000).
8. Bain, C. D., and Whitesides, G. M. Depth sensitivity of wetting: monolayers of ω-mercapto ethers on gold. *J. Am. Chem. Soc.*, **110**, 5897 (1988).
9. Whitesides, G. M., and Laibinis, P. E. Wet chemical approaches to the characterization of organic surfaces: self-assembled monolayers, wetting, and the physical-organic chemistry of the solid-liquid interface. *Langmuir*, **6**, 87 (1990).
10. Aoki, K., Kawanishi, Y., Seki, T., Sakuragi, M., Tamaki, T., and Ichimura, K. Reversible alignment change of liquid crystals induced by photochromic molecular films: properties of azobenzene chromophores covalently attached to silica surfaces. *Liq. Cryst.*, **19**, 119 (1995).
11. Siewierski, L. M., Brittain, W. J., Petrash, S., and Foster, M. D. Photoresponsive monolayers containing in-chain azobenzene. *Langmuir*, **12**, 5838 (1996).
12. Möller, G., Harke, M., and Motschmann, H. Controlling microdroplet formation by light. *Langmuir*, **14**, 4955 (1998).
13. Wolf, M. O., and Fox, M. A. Photochemistry and surface properties of self-assembled monolayers of cis- and trans-4-cyano-4'-(10-thiodecoxy)stilbene on polycrystalline gold. *J. Am. Chem. Soc.*, **117**, 1845 (1995).
14. Wolf, M. O., and Fox, M. A. Photoisomerization and photodimerization in self-assembled monolayers of cis- and trans-4-cyano-4'-(10-mercaptodecoxy)stilbene on gold. *Langmuir*, **12**, 955 (1996).
15. Fox, M. A., and Wooten, M. D. Characterization, adsorption, and photochemistry of self-assembled monolayers of 10-thiodecyl 2-anthryl ether on gold. *Langmuir*, **13**, 7099 (1997).
16. Abbott, S., Ralston, J., Reynolds, G., and Hayes, R. Reversible wettability of photoresponsive pyrimidine-coated surfaces. *Langmuir*, **15**, 8923 (1999).
17. Ueda, M., Kim, H.-B., and Ichimura, K. Photocontrolled aggregation of colloidal silica. *J. Mater. Chem.*, **4**, 883 (1994).
18. Willner, I., and Rubin, S. Control of the structure and functions of biomaterials by light. *Angew. Chem. Int. Ed. Engl.*, **35**, 367 (1996).

19. Willner, I., and Blonder, R. Patterning of surfaces by photoisomerizable antibody-antigen monolayers. *Thin Solid Films*, **266**, 254 (1995).

20. Chaudhury, M. K., and Whitesides, G. M. How to make water run uphill. *Science*, **256**, 1539 (1992).

21. Dürr, H., and Bouas-Laurent, H. (Ed.). *Photochromism: Molecules and Systems*: Elsevier: Amsterdam, 1990.

22. Kurita, E., Fukushima, N., Fujimaki, M., Matsuzawa, Y., Kudo, K., and Ichimura, K. Macrocyclic amphiphiles. Part 2. Multi-point adsorptivitiy of the crown conformer of calix[4]resorcinarenes and their derivatives on surfaces of amorphous polar substrates. *J. Mater. Chem.*, **8**, 397 (1998).

23. Fujimaki, M., Kawahara, S., Matsuzawa, Y., Kurita, E., Hayashi, Y., and Ichimura, K. Macrocyclic amphiphiles — Monolayers of o-octacarboxymethoxylated calix[4]resorcinarenes with azobenzene residues exhibiting efficient photoisomerizability. *Langmuir*, **14**, 4495 (1998).

24. Ichimura, K., Fukushima, N., Fujimaki, M., Kawahara, S., Matsuzawa, Y., Hayashi, Y., and Kudo, K. Macrocyclic amphiphiles. 1. Properties of calix[4]resorcinarene derivatives substituted with azobenzenes in solutions and monolayers. *Langmuir*, **13**, 6780 (1997).

25. Oh, S.-K, Nakagawa, M., and Ichimura, K. Self-assembled monolayers derived from calix[4]resorcinarenes exhibiting excellent desorption-resistance and their applicability to surface energy photocontrol. *Chem. Lett.*, 349 (1999).

26. Ichimura, K., Oh, S.-K., Fujimaki, M., Matsuzawa, Y., and Nakagawa, M. Convenient preparation of self-assembled monolayers derived from calix[4]resorcinarene derivatives exhibiting resistance to desorption. *J. Inclusion Phenomena Macrocycl. Chem.*, **35**, 173 (1999).

27. Adamson, A. W., and Gast, A. P. *Physical Chemistry of Surfaces*, 6th ed.: John Wiley & Sons: New York, 1997.

28. Tillman, N., Ulman, A., Schildkraut, J. S., and Penner, T. L. Incorporation of phenoxy groups in self-assembled monolayers of trichlorosilane derivatives. Effects on films thickness, wettability, and molecular orientation. *J. Am. Chem. Soc.* **110**, 6136 (1998).

29. Ichimura, K., Oh, S.-K., and Nakagawa, M. Light-driven motion of liquids on a photoresponsive surface. *Science*, **288**, 1624 (2000).

30. Chaudhury, M. K. Interfacial interaction between low energy surfaces. *Mater. Sci. Eng.* **R16**, 97 (1996).

31. Chen, Y. L., Helm, C. A., and Israelachvili, J. N. Molecular mechanisms associated with adhesion and contact angle hysteresis of monolayer surfaces. *J. Phys. Chem.*, **95**, 10736 (1991).

32. Yoshizawa, H., Chen, Y.-L., and Israelachvili, J. N. Fundamental mechanisms of interfacial friction. 1. Relation between adhesion and friction. *J. Phys. Chem.*, **97**, 4128 (1993).

33. Brochard, F. Motions of droplets on solid surfaces induced by chemical or thermal gradients. *Langmuir*, **5**, 432 (1989).

34. Oh, S.-K., Nakagawa, M., and Ichimura, K. *J. Am. Chem. Soc.*, submitted for publication.

35. Seki, T., Tanaka, K., and Ichimura, K. Phototropic discrimination of the polymerization behavior of diacetylene Langmuir-Blodgett films by an azobenzene-containing monolayers. *Adv. Mater.*, **9**, 563 (1997).

36. Seki, T., Tanaka, K., and Ichimura, K. Photopolymerization of diacetylene Langmuir-Blodgett films on an azobenzene-containing monolayer. *Polym. J.*, **30**, 646 (1998).

37. Enkelmann, V. Structural aspects of the topochemical polymerization of diacetylenes. *Adv. Polym. Sci.*, **63**, 91 (1984).

38. Veale, G., and Peterson, I. R. Novel effects of counterions on Langmuir films of 22-tricosenoic acid. *J. Colloid Interface Sci.*, **103**, 178 (1985).

39. Seki, T., Fukuda, K., and Ichimura, K. Photocontrol of polymer chain organization using a photochromic monolayer. *Langmuir*, **15**, 5098 (1999).

40. Fukuda, K., Seki, T., Ichimura, K., and Komitov, L. Uniform planar alignment of nematics on photooriented linear polysilane. *Mol. Cryst. Liq. Crys.*, **368**, 535 (2001).

41. Kawashima, Y., Nakagawa, M., Seki, T., and Ichimura, K. Submitted for publication.

42. Irie, M. Photoresponsive polymers. *Adv. Polym. Sci.* **94**, 27 (1990).

43. Kumar, G. S., and Neckers, D. C. Photochemistry of azobenzene-containing polymers. *Chem. Rev.*, **89**, 1915 (1989).

44. Blair, H. S., and Pogue, H. I. Investigation of photoresponsive effects in polymer monolayers. *Polymer*, **20**, 99 (1979).

45. Gruder, H., Vilanove, R., and Rondelez, F. Reversible photochemical strain in Langmuir monolayers. *Phys. Rev. Lett.*, **44**, 590 (1980).

46. Malcolm, B. R., and Pieroni, O. The photoresponse of an azobenzene-containing poly(L-lysine) in the monolayer state. *Biopolymers*, **29**, 1121 (1990).

47. Menzel, H. Langmuir-Blodgett films of photochromic polyglutamates. 7. The photomechanical effect in monolayers of polyglutamates with azobenzene moieties in the side chains. *Macromol. Chem. Phys.*, **195**, 3747 (1994).

48. Seki, T., Sekizawa, H., Tanaka, K., Matsuzawa, Y., and Ichimura, K. Photoresponsive monolayers on water and solid surfaces. *Supramol. Sci.*, **5**, 373 (1998).

49. Seki, T., and Tamaki, T. Photomechanical effect in monolayers of azobenzene side chain polymers. *Chem. Lett.*, 1739 (1993).

50. Seki, T., Fukuda, R., Yokoi, M., Tamaki, T., and Ichimura, K. Photomechanical response of azobenzene containing monolayers on water surface. *Bull. Chem. Soc. Jpn.*, **69**, 2375 (1996).

51. Seki, T., Sekizawa, H., and Ichimura, K. Morphological changes in monolayer of a photosensitive polymer observed by Brewster angle microscopy. *Polym. Commun.*, **38**, 725 (1997).

52. Seki, T., Sekizawa, H., Morino, S., and Ichimura, K. Inherent and coopoerative photomechanical motions in monolayers of an azobenzene containing polymer at the air-water interface. *J. Phys. Chem. B.*, **102**, 5313 (1998).

53. Kago, K., Fürst, M., Matsuoka, H., Yamaoka, H., and Seki, T. Nanostructure of photochromic polymer/liquid crystal hybrid monolayer on water surface observed by in-situ x-ray reflectometry. *Langmuir*, **15**, 2237 (1999).

54. Seki, T., Tanaka, K., and Ichimura, K. Photomechanical response in monolayered polymer films on mica at high humidity. *Macromolecules*, **30**, 6401 (1997).

55. Chen, Y.-L., and Israelachvili, J. N. Effects of ambient conditions on adsorbed surfactant and polymer monolayers. *J. Phys. Chem.*, **96**, 7752 (1992).

56. Seki, T., Kajima, J., and Ichimura, K. Multifarious photoinduced morphologies in monomolecular films of azobenzene side chain polymer on mica. *Macromolecules*, **33**, 2709 (2000).

57. Matsumoto, M., Miyazaki, D., Tanaka, M., Azumi, R., Manda, E., Kondo, Y., Yoshino, N., and Tachibana, H. Reversible light-induced morphological change in Langmuir-Blodgett films. *J. Am. Chem. Soc.*, **120**, 1479 (1998).

58. Ebihara, K., Koshihara, S., Yoshimoto, M., Maeda, T., Ohnishi, T., Koinuma, H., and Fujiki, M. Direct observation of helical polysilane nanostructures by atomic force microscopy. *Jpn. J. Appl. Phys.*, **36**, L1211 (1997).

59. Kumaki, J., Nishikawa, Y., and Hashimoto, T. Visualization of single-chain conformations of a synthetic polymer with atomic force microscopy. *J. Am. Chem. Soc.*, **118**, 3321 (1996).

60. Blodgett, K., and Langmuir, I. Built-up films of barium stearate and their optical properties. *Phys. Rev.*, **51**, 964 (1937).

61. Seki, T., Kojima, J., and Ichimura, K. Light-driven dot films consisting of single polymer chain. *J. Phys. Chem. B*, **103**, 10338 (1999).

62. Rochon, P., Batalla, E., and Natansohn, A. Optically induced surface gratings on azo aromatic polymer films. *Appl. Phys. Lett.*, **66**, 136 (1995).

63. Kim, D. Y., Tripathy, S. K., Li, L., and Kumar, J. Laser-induced holographic surface relief gratings on nonlinear optical polymer films. *Appl. Phys. Lett.*, **66**, 1166 (1995).

64. Ramanujam, P. S., Holme, N. C. R., and Hvilsted, S. Atomic force and optical near-field microscopic investigations of polarization holographic gratings in a liquid crystalline azobenzene side-chain polyester. *Appl. Phys. Lett.*, **66**, 1166 (1995).

65. Viswanathan, N. K., Kim, D. Y., Shaping, Williams, J., Liu, W., Li, L., Samuelson, J., Kumar, J., and Tripathy, S. K. Surface relief structures on azo polymer films. *J. Mater. Chem.*, **9**, 1941 (1999).

66. Ubukata, T., Seki, T., and Ichimura, K. Surface relief gratings in host-guest supramolecular materials. *Adv. Mater.*, **12**, 1675 (2000).

67. Ubukata, T., Seki, T., and Ichimura, K. Modeling the interface region of command surface. Part 1. Structural evaluations of azobenzene/liquid crystal hybrid Langmuir monolayers. *J. Phys. Chem. B*, **104**, 4141 (2000).

68. Ubukata, T., Seki, T., Morino, S., and Ichimura, K. Modeling the interface region of command surface 2. Spectroscopic evaluations of azobenzene/liquid crystal hybrid Langmuir-Blodgett films under illumination. *J. Phys. Chem. B*, **104**, 4148 (2000).

69. Zettsu, N., Ubukata, T., Seki, T., and Ichimura, K. Soft crosslinkable azo polymer for rapid surface relief formation and persistent fixation. *Adv. Mater.*, **13**, 1693 (2001).

16

3D DATA STORAGE AND NEAR-FIELD RECORDING

YOSHIMASA KAWATA*
SATOSHI KAWATA[†]
Shizuoka University, Department of Mechanical Engineering, Johoku, Hamamatsu 432-8561, Japan
[†]*Osaka University, Department of Applied Physics, Suita, Osaka 565-0871, Japan*

ABSTRACT

In this chapter three-dimensional (3D) optical data storage and near-field recording techniques using photochromic materials for high-density memory are described. In 3D data storage, multilayered bit data are recorded in a 3D volume of photochromic materials. A reversible transformation of the photochromic molecules between two isomers with different absorption spectra can be stimulated by irradiation with light of appropriate wavelengths, allowing information to be stored. We present various optics for 3D memories using photochromic materials, and we show the near-field recording technique for high-density storage and its application to near-field microscopy for biology.

16.1 INTRODUCTION

Optical memory devices such as compact disks (CDs) and digital versatile disks (DVDs) are becoming essential items of audio and visual media as well as of external computer memory media. In these devices, a laser beam is used to record and read information. Because the laser spot can be focused to within a 1 μm scale, optical memory can access higher density and capacity than conventional magnetic memories can.

Optical memory density is ultimately limited by the diffraction of electromagnetic waves. Present techniques have almost reached this limitation in those optical memories that are commercially available as CDs or DVDs. Even with an infinitely large objective lens with high numerical-aperture (NA) value, the bit data resolution distance for recording and reading cannot be reduced to less than half of the beam wavelength.

Photochromic materials have been developed in order to dramatically increase the memory density.[1-5] These materials can be used for photon-mode recording, which is based on the photochemical reaction of the medium. In photon-mode recording, light characteristics such as wavelength, polarization, and phase can be multiplexed in data storage and thus can, in a potentially dramatic ways, increase the memory density.

Consequently, current efforts in optical memory devices are geared toward the development of durable short wavelength compact lasers that emit blue or green light.[6-8] Doubling the frequency (or halving the wavelength) of the laser reduces the beam-spot radius by two, thereby increasing the density by four when recorded in two dimensions. If one aims to increase the memory density by 100 times the current benchmark, laser diodes with output wavelengths 10 times shorter than those currently available have to be employed. This requirement is obviously impossible because neither the laser materials for such an ultra short wavelength nor optical components, particularly lenses, can be manufactured in the 70–80 nm wavelength range.

There are two ways to achieve higher density: the extension of the data-recording space in the axial direction, and the reduction of bit size. In this chapter, we describe a method for overcoming the density limit, namely, introduce an additional axial dimension in the recording process.[9-14] The z or longitudinal axis is used in addition to the surface dimension (x–y space) of conventional optical memory. The data are thus written not on the material surface but within the three-dimensional (3D) thick volume. The media that can be used are photochromic materials.[15,16]

In this chapter, we also describe the near-field recording technique. We can greatly reduce the bit size by using near-field optics.[17-21]

16.2 BIT-ORIENTED 3D MEMORY

Figure 16.1 shows a principle of bit-oriented three-dimensional (3D) optical memory. A laser beam is focused on a point in a recording medium. Chemical reactions of the medium should be induced at that spot because extremely high intensity is produced at the focus point. By 3D scanning of the focus

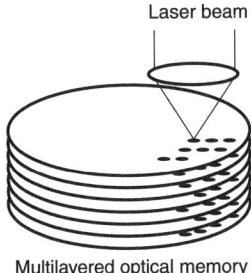

FIG. 16.1 Principle of bit-oriented 3D memory.

spot in the medium, we can record bit data in the medium in three dimensions.

Various materials such as photopolymers, photorefractive crystals, photochromic materials, and polarization sensitive materials can store data. Among these materials, photorefractive recording media in which data are stored as refractive-index change are the most promising in 3D memory because they have little absorption; hence, the light penetrates into the deep layers in the media.

Figure 16.2 shows a typical readout system of 3D memory. This is essentially a reflection confocal microscope configuration. Because the pinhole in front of a detector eliminated scattered light from the out-of-focus region, the system gives very good axial resolution and high contrast. The optical setup is very simple. We can also use other configurations, as readout systems; these are sensitive refractive-index variations such as phase contrast microscope, differential phase contrast microscope, and differential interference microscope.[22–24]

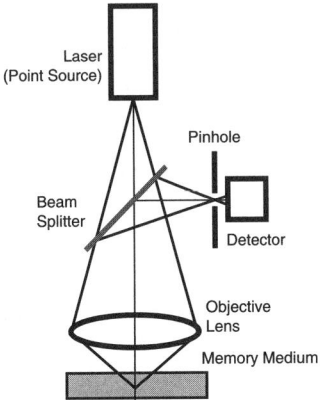

FIG. 16.2 A typical optical setup of reflection confocal microscope.

16.3 PHOTOCHROMIC MATERIALS FOR 3D OPTICAL MEMORY

A photochromic compound is characterized by its ability to alternate between two different chemical forms having different absorption spectra in response to light of appropriate wavelengths. Photochromic materials are promising as recording media for optical memory, because the media store erasable/rewritable data in photon mode. Because the data-recording mechanism is based on the photochemical reaction of each molecule, extremely high spatial resolution is expected.

The following properties are required for photochromic materials to be used as optical memory media:

1. thermal stability of both isomers;
2. resistance to fatigue during cyclic write and erase processes;
3. fast response;
4. high sensitivity;
5. nondestructive readout capability.

For 3D recording, the large two-photon absorption coefficient is preferred because 3D memory uses basically a two-photon process to access a point in a volumetric medium. In the following sections, typical photochromic compounds used for 3D memories are described.

16.3.1 Spirobenzopyran Derivatives

P. M. Rentzepis *et al.*[9] first demonstrated a bit-oriented 3D optical memory using a photochromic spirobenzopyran shown in Figure 16.3. Isomer A has an absorption band shorter than 450 nm; upon irradiation with ultraviolet (UV) light, it converts to isomer B, which has an absorption band around 600 nm, as shown in Figure 16.4. Isomer B gives fluorescence around 700 nm upon photoexcitation with 500–700 nm light.

Nondestructive readout by using a small difference in the refractive index of the spirobenzopyran isomers (Figure 16.3) in the near-infrared (IR) range was demonstrated.[15] Small refractive-index change was detected with a near-IR laser-scan differential phase-contrast microscope (Figure 16.5). The two isomers of the spirobenzopyran have very different absorption spectra, which suggests that the isomers also have different refractive indices. In the near-IR region around 800 nm, both isomers have negligible absorption. Therefore, near-IR light does not stimulate the photochromic photoreaction.

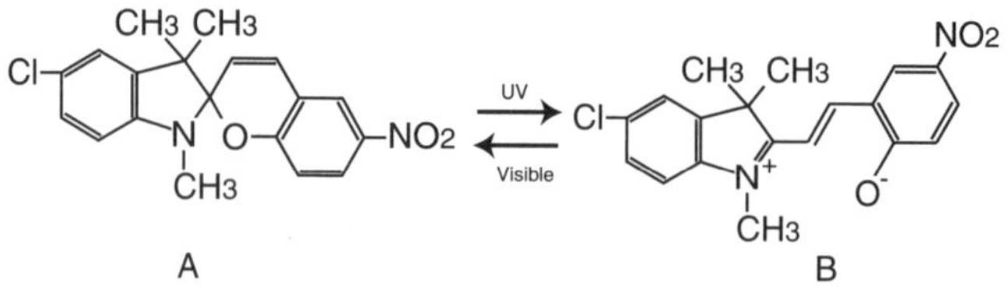

FIG. 16.3   Chemical structure of spirobenzopyran. Parts A and B are spirobenzopyran isomers.

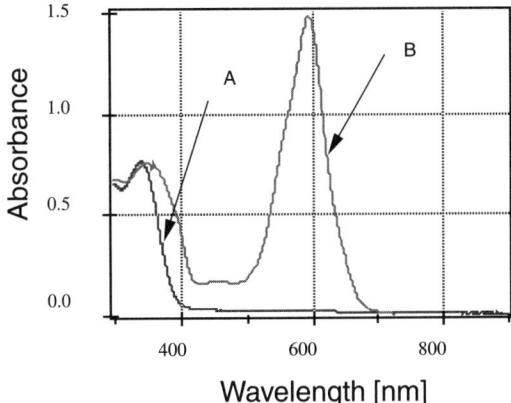

FIG. 16.4 Absorption spectra of polystyrene film containing spirobenzopyran before (A) and after (B) 442-nm light exposure.

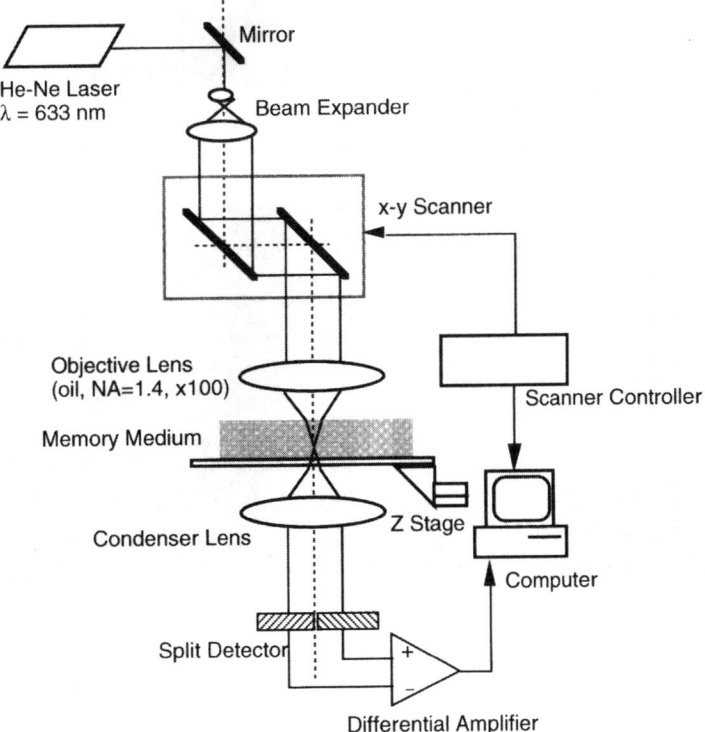

FIG. 16.5 Optical system for readout of 3D memory. For nondestructive readout, a near-IR differential phase-contrast microscope is used.

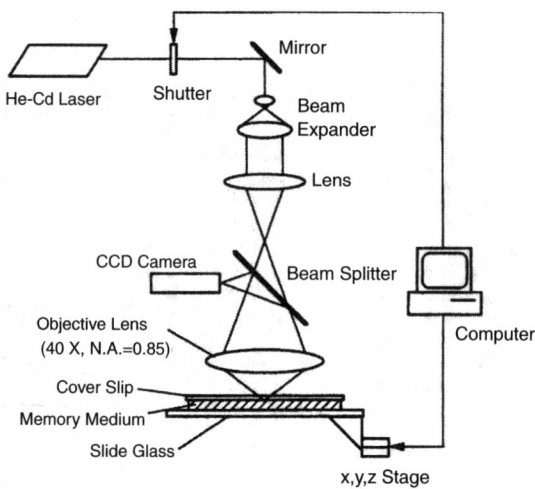

FIG. 16.6   Optical setup for recording 3D bit information in photochromic memory.

Data were recorded by focusing the laser light (441.6 nm) on two layers separated by as much as 70 μm using a laser-scanning microscope shown in Figure 16.6. The bit sequences with 24 × 24 bits per layer form the letters "A" and "B." The bit interval was 5 μm × 5 μm in each layer, and the separation distance between layers was 70 μm.

Figure 16.7 shows the readout of the two layers. The crosstalk between the two layers was small enough to read recorded data clearly by the differential phase-contrast microscope. Since the signal intensity was proportional to the derivative of the refractive index, the edges of the bits were slightly enhanced. The scanning rate was 1 layer per second, and the readout could be repeated more than 7000 times without destroying the recorded information. The differences in refractive index were measured to be 0.02 between the two isomers at 830 nm.[25] Spirobenzopyrans used in these systems have poor durability, and the photogenerated isomers are thermally unstable.

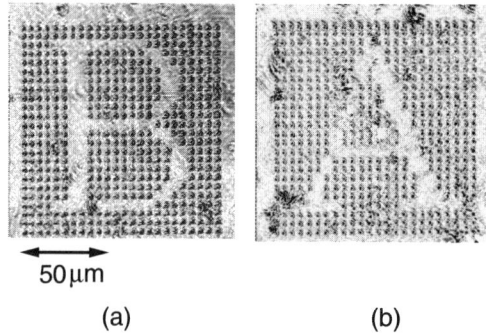

50 μm

(a)   (b)

FIG. 16.7   Bit patterns read from photochromic memory by near-IR laser-scanning differential phase-contrast microscopy; (a) first layer, (b) second layer. The bit interval is 5 μm, and the layer distance is 70 μm.[15]

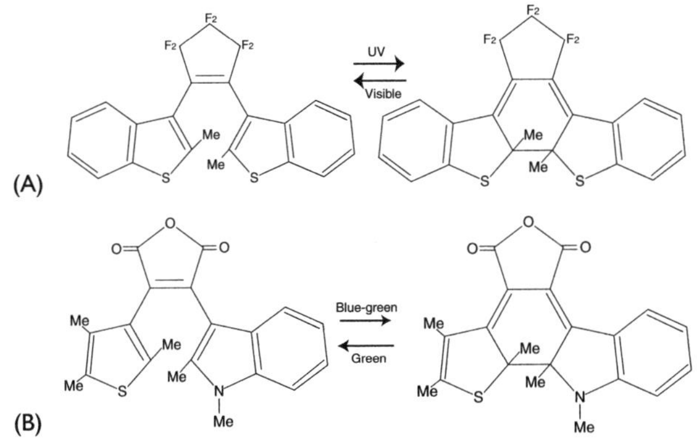

FIG. 16.8 Photochromic reactions of (a) 1,2-bis-(2-methyl-1-benzothiophen-3-yl)perfluorocyclopentene, and (b) 2-(1,2-dimethyl-3-indolyl)-3-(2,4,5-trimethyl-3-thienyl)maleic anhydride.

Spirobenzopyran derivatives were used as media of wavelength-multiplexed memory.[3,26] A narrow absorption spectrum is required for the wavelength multiplex. The narrow spectrum was obtained by using J-aggregates of spirobenzopyrans.[26] The J-aggregates have an absorption peak width of a few dozen nanometers. A five-wavelengths-multiplexed memory was demonstrated, but considerably large crosstalk between multiplexed channels was observed.[3]

16.3.2 Diarylethene Derivatives

Figure 16.8 shows one of the diarylethene derivatives with heterocyclic rings, or (a) 1,2-bis-(2-methyl-1-benzothiophen-3-yl) perfluorocyclopentene, and (b) 2-(1,2-dimethyl-3-indolyl)-3-(2,4,5-trimethyl-3-thienyl) maleic anhydride.[5,27] The compounds have no thermochromicity even at 200°C, and their colored closed-ring forms are stable for more than three months at 80°C. Furthermore, the cycle of cyclization/ring-opening reactions can be repeated more than 10^4 times while maintaining the photochromic performance. The diarylethene derivatives are the most promising photochromic compounds for high-density optical memory media.

Poly(methyl metacrylate) containing cis-1,2-diciano-1,2-bis(2,4,5-trimethyl-3-thienyl)ethene (B1536) was used as a photochromic optical memory medium.[16] Figure 16.9 shows the chemical structures of B1536 and the absorption spectra of two isomers.

B1536 is a yellow isomer that is converted into a red isomer upon irradiation with 380 nm light. The written data can be erased by irradiation with 543 nm light. This photochromic material (B1536) did not exhibit any apparent fatigue even after 100 write/erase cycles; the written data (i.e., the red isomer) were stable at 80°C for more than three months, and thermal back reaction (from the red isomer to the yellow isomer) did not occur even at 200°C.[28]

For recording, a two-photon process at the focused spot was used in order to access a point in a thick medium. A Ti:sapphire laser at 760 nm

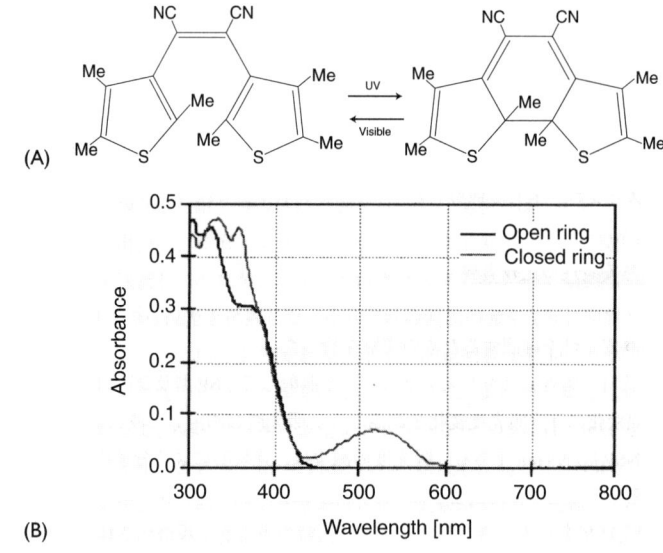

FIG. 16.9 (A) Chemical structure of cis-1,2-diciano-1,2-bis(2,4,5-trimethyl-3-thienyl)ethene (B1536) and (B) its absorption spectra.

in mode-locked pulsed laser operation was employed as the recording light source. Because the probability of the occurrence of two-photon absorption is proportional to the squared intensity of the incident light, photoisomerization can be induced only at the focal spot, where the intensity is very high.[9,11,12,29,30] This technique is attractive for recording data in an erasable medium, because it is possible to write information onto a particular layer without erasing the data already written on neighboring layers.

Figure 16.10 shows readout images of written bits in 26 consecutive layers. The bit and layer intervals were 2 μm and 5 μm, respectively. The data were read out with a reflection confocal microscope (RCM). The details of the readout system are discussed in the Section 4. With the RCM, the written data are clearly read without crosstalk.

16.3.3 Azobenzene Derivatives

Azobenzene derivatives constitute a family of dye molecules well known for their photochromic properties, which are due to the reversible cis-trans photoisomerization.[31–34] Figure 16.11 shows a photoisomerization process of azobenzene. Azobenzene derivatives have two geometric isomers: the trans and the cis forms. The isomerization reaction is a light- or heat-induced interconversion of the two isomers. Because the trans form is generally more stable, the thermal isomerization is generally in the direction of from cis to trans. Light induces transformations in both directions.

Three-dimensional optical memory using a new material, urethane-urea copolymer, which contains azo-dyes as a side chain, was demonstrated in 1998.[35] Figure16.12a shows the chemical structure of urethane-urea

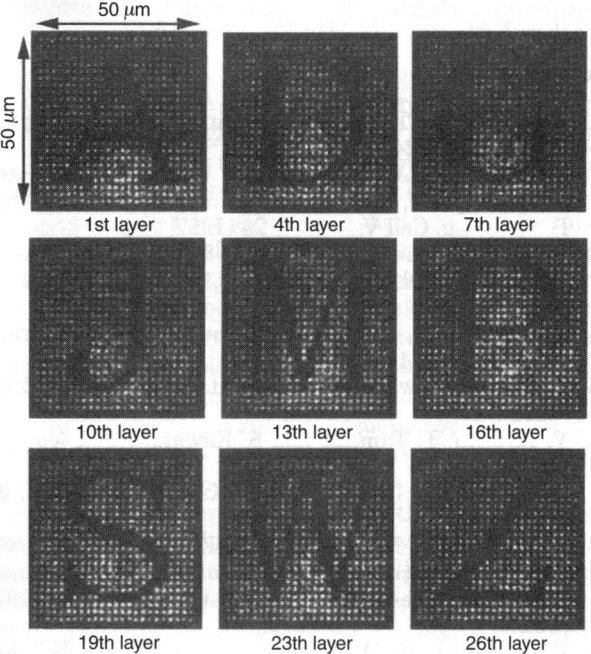

FIG. 16.10 Readout of bit patterns written into photochromic memory using a diarylethene derivative. The data were read out with a reflection confocal microscope.[16]

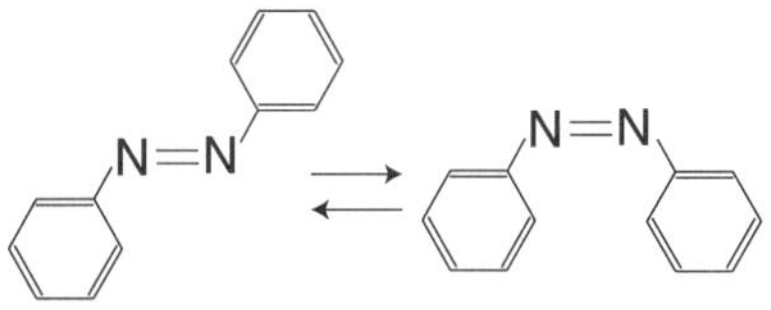

FIG. 16.11 Photoinduced isomerization of azobenzene.

copolymer. The urethane-urea copolymer was originally developed for a nonlinear optical waveguide.[36,37] The copolymer has comparatively high optical nonlinearity and is stable at room temperature. The copolymer has a side-chain structure, or azo-dye, to semifix chromophores as a photosensitizer for optical memory use.

The absorption spectrum of the copolymer shows a maximum at 476.3 nm and small absorption in the region longer than 600 nm (Figure 16.12b). When illuminated with blue light, the azo-dye undergoes cis-trans isomerization, producing a refractive-index change. The expected change in the refractive index is on the order of 10^{-2}.

A recording medium was developed in which photosensitive films (urethane-urea copolymer), nonphotosensitive (polyvinyl alcohol, PVA and poly(methyl methacrylate), PMMA) films, were coated on a glass substrate. Figure 16.13 shows the structure of the multilayered recording medium.

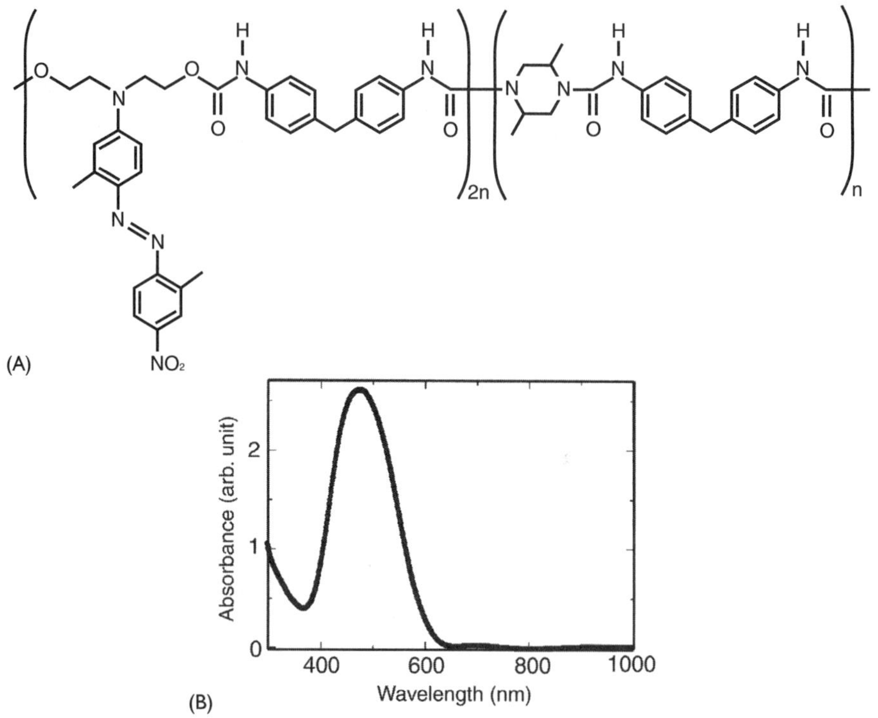

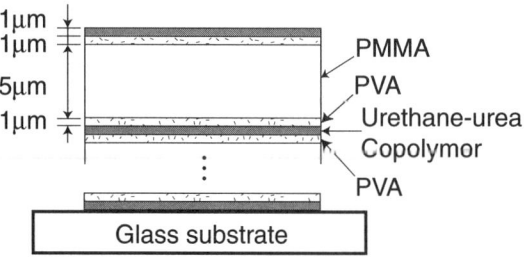

FIG. 16.12 (A) Chemical structure of urethane-urea copolymer and (B) its absorption spectrum.

FIG. 16.13 Multilayered recording medium in which photosensitive films and nonphotosensitive transparent PVA and PMMA films are stacked alternately.

Figure 16.14 shows the axial distribution of the recorded data. The figure was reconstructed from a set of many images captured when the focus plane was changed. The cross sections along the optical axis are also shown in Figure 16.14. The four recording layers were clearly detected, and the bit data were also clearly recognized. The side lobes in the cross section along the optical axis were due to the aberrations of the objective.

Figure 16.15 shows the readout result of the four-layered optical memory. The readout system was a reflection-type confocal microscope. For reading, a He-Ne laser was used as a light source, because the urethane-urea copolymer has little absorption for red light. The scattered light at the recorded bit data

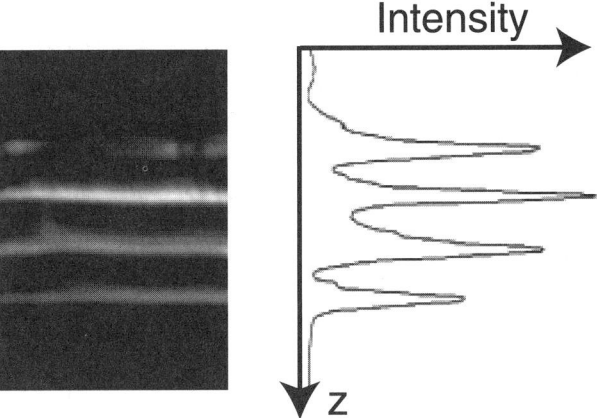

FIG. 16.14 Axial distribution of the four-layers recorded data. The data were read out with a reflection confocal microscope.[35]

FIG. 16.15 Recording and reading results of the four-layers data. The distance between neighboring bits is 1.5 μm, and the distance between layers is about 7 μm.

due to the refractive-index change was detected with a photo-multiplier tube (PMT). The pattern of "1," "3," and "4" was formed with bit sequences in the first, third, and fourth layers, respectively. The distance between bits in the layer is 1.5 μm, and the distance between layers was 7 μm.

The data shown in Figure 16.15 were recorded with a two-photon process. A Ti:sapphire laser was used as a light source of two-photon recording. The wavelength of laser light was 800 nm. Since urethane-urea copolymer has little absorption at the wavelength region longer than 600 nm, bit data should be recorded with a two-photon process.

16.4 RECORDING AND READOUT OPTICS

16.4.1 Single-Photon Recording

Although two-photon excitation is desirable for bit-data recording in 3D optical memories, single-photon recording also gives good separation between recorded planes. Figures 16.7 was data recorded with a single-photon process. A typical system for a single-photon recording is shown in Figure 16.6. A He-Cd laser of 441.6 nm wavelength was used for spirobenzopyran (Figure 16.3), and an Ar⁺ laser of 488.0 nm was used for urethane-urea copolymer (Figure 16.12a). The crosstalk was negligible.

Since single-photon recording does not require ultrashort-pulse lasers, conventional semiconductor lasers can be used in the 3D memories. The optics in CD and DVD devices can be easily applied for a recording system of 3D optical memory.

16.4.2 Single-Beam, Two-Photon Recording

Two-photon excitation is preferable in 3D optical memory because the crosstalk between two adjacent layers is much reduced. Another advantage of two-photon excitation is reduction in multiple scattering. This reduction occurs because of the use of an illumination beam at infrared wavelength.

A typical recording optics for a single-beam, two-photon recording system is similar to the one-photon recording system shown in Figure 16.6, except for the light source. A Ti:sapphire laser was used as the light source, because the cooperative nature of two-photon excitation requires a high-peak-power laser light to produce efficient excitation.[11,29] Systems have been developed that use a mode-locked Ti:sapphire laser,[16,30,38] a mode-locked dye-laser,[11] and a mode-locked yttrium aluminum garnet (YAG).[9,39] A two-photon process with continuous-wave light has also been investigated.[40,41] For the separation of layers, a higher numerical aperture lens is required, which makes a well-confirmed spot at a point in a thick medium.

The ultrashort-pulse lasers increase the cost of a recording device and make it difficult to produce a compact system, but a compact ultrashort pulse laser in which Er-doped fiber is used as a resonator has been developed.[42,43] Semiconductor lasers with mode-locked operations are also under investigation,[44,45] so the cost and size of ultrashort-pulse lasers will decrease in the future.

16.4.3 Right-Angle, Two Beams, Two-Photon Recording

P. M. Rentzepis *et al.* proposed and demonstrated 3D memory using right-angle two-beam in order to access a point in a volumetric medium, as shown in Figure 16.16.[9,39,46,47] These were excellent demonstrations to present the possibility of the bit-oriented 3D memory. For writing data in a spirobenzopyran, which requires excitation in the ultraviolet range, two-photon absorption of either a 1064 nm and a 532 nm photon (corresponding to 355 nm excitation), or two 532 nm photons (corresponding to 266 nm excitation), was used. At the intersection of the two beams, isomer A absorbs two photons simultaneously and photoisomerizes to isomer B.

For reading data, a similar two-photon process was employed. When isomer B molecules were excited by absorbing two 1064 nm photons, the excitation will cause only the written molecules to emit fluorescence. Although the fluorescence readout method is sensitive, the memory is erased during the reading process because the reverse reaction from the photoexcited isomer B to isomer A takes place, to some extent. Such a destructive readout

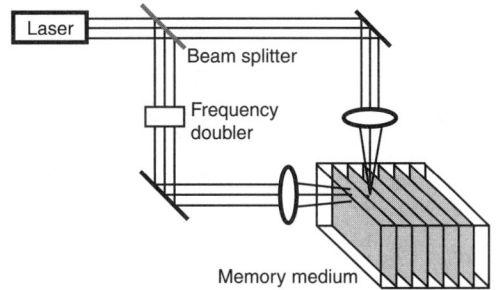

FIG. 16.16 Principle of 3D optical memories proposed by P. M. Rentzepis *et al.*[9]

method cannot be applied to practical use. It is essential that nondestructive readout methods be developed for practical use.[4,5,15,16] Among these methods, the most promising one is to read the memories by detecting the refractive-index changes along with the photoisomerization with long wavelength light, which does not induce photochromic reactions.

The disadvantages of the system, however, were the difficulty in aligning two beams so as to intersect at the same point and the limitation of the working distances of objective lenses. To achieve higher density storage, high NA objective lenses had to be used. Because the working distances of the lenses were not large, it was difficult to make right-angle configuration. In order to solve these problems, Rentzepis *et al.* proposed to use one beam as a plane addressing beam.

16.4.4 Reflection Confocal Reading

The reflection confocal microscope (RCM) is an attractive configuration as a readout system of multilayered optical memories because it has extremely high axial resolution and its configuration is substantially easier than that of transmission confocal microscopes. A typical RCM system is shown in Figure 16.2.

In transmission confocal microscopes that are equipped with phase-reading systems, the focus deviates from the pinhole because of the inhomogeneity of the refractive index and/or the thickness of the memory medium and substrate. As a result, the detected signal has a background owing to the local inhomogeneity of the medium and substrate.

The use of RCM alleviates the problems encountered when the transmission microscope is used, because the beam that is reflected at the data bit arrives at the pinhole even in the presence of inhomogeneity. RCMs, however, have not been used for reading 3D-stored data, primarily because the 3D spatial-frequency band of the reading optics of this configuration does not necessarily pick up the 3D band of the writing optics involved.

When one chooses the appropriate optical parameters, including the NAs of the objective lenses and the wavelengths of the writing and reading lights, there is a common spatial band for both writing and reading in three dimensions. Figures 16.17a and 16.17b show a data readout obtained with

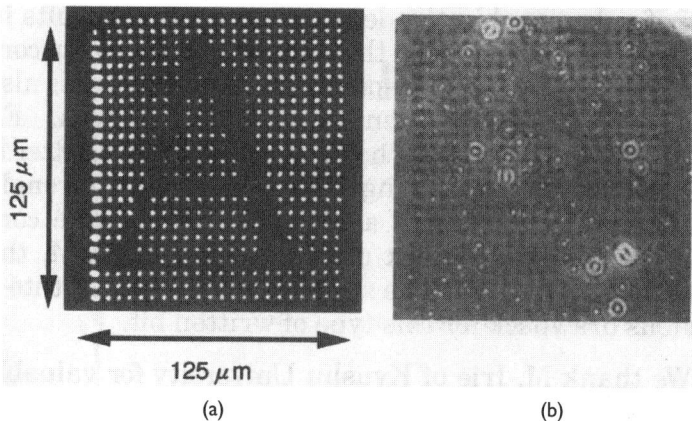

FIG. 16.17 Experimental comparison of data reading by (a) the RCM and (b) the transmission microscope.

an RCM and with a transmission phase-contrast microscope, respectively. The bit-data image obtained with the reflection readout system is much clearer than that obtained with the transmission readout system.

Figure 16.18 shows the spatial-frequency band of a written bit and the transfer-function band of the reading system.[48–51] This figure shows a vertical cut of the 3D band of the transfer function (k-vector) space that includes the axial frequency ($1/z$). $1/r$ is the transaxial spatial frequency of the polar coordinate $r = \sqrt{x^2 + y^2}$, where x and y are the two-dimensional coordinates of the transaxial plane. Shaded regions represent the spatial-frequency components of the two-photon writing optics and the 3D Fourier transform of a bit profile formed with the focused optics by the two-photon process. The ellipse in Figure 16.18 is a vertical cut of a bun-shaped band in three dimensions, which is obtained by 3D autocorrelation of a doughnut-shaped band that represents a 3D spatial-frequency band of a single-photon focused optics. The NAs of the objective lenses for writing and reading are given by $n \sin \alpha_\omega$ and

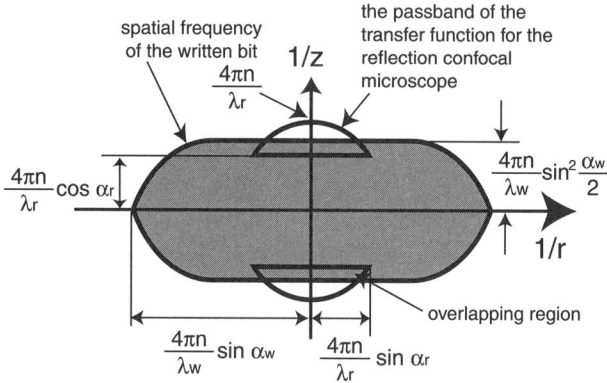

FIG. 16.18 Band of the writing optics for two-photon processes, and the band of the reading optics for RCM.

$n \sin \alpha_r$, respectively. n is the refractive index of the recording material, and α_ω and α_r are the maximum semi-angles of the objectives. λ_ω and λ_r are the wavelengths for writing and reading, respectively. The 3D spatial-frequency band expands when λ_ω decreases or when $n \sin \alpha_\omega$ increases. The parabolas (one concave up and one concave down) shown in Figure 16.18 are vertical cuts of the band of the reading optics of the RCM. The parabolas are semicylindrical in 3D space. The band of the reading optics is also given by the NA of the objective lens and the wavelength λ_r. The written data can be read when there is a common area (actually a common volume) in the 3D frequency band for both writing and reading. Appropriate $n \sin \alpha_\omega$, $n \sin \alpha_r$, λ_ω, and λ_r, which make the bun and the crescents overlap, can be selected. It is necessary to choose objective lenses with very high NAs to create an overlap between the bands of the writing and reading optical systems. For example, when NA=1.4 is used for writing, an NA higher than 1.2 is required for reading (λ_ω = 760 nm and λ_r = 633 nm).

16.4.5 Reflection Confocal Reading with Multilayered Medium

Another way to achieve the reading data with RCM is to use the multilayered recording medium in which photosensitive thin films and nonphotosensitive transparent films are alternately stacked. Since the photosensitive films are thinner than the depth of focus of the recording beam, the spatial-frequency distribution of the recorded-bit data is extended in the axial direction. The extended distribution overlaps the coherent optical transfer function of the RCM.

Figure 16.19 shows the spatial-frequency distributions of bit data recorded with focused laser beam and coherent optical transfer function (CTF) of reflection type confocal microscope.[48,49] Figure 16.19a shows a spatial-frequency distribution of bit datum recorded in very thick medium. This distribution coincides with the spatial-frequency distribution of the focused light to record the bit datum, because the bit is recorded with the focused beam. It is assumed that the NA of the objective lens is given by $n \sin \alpha$ and $k = 2\pi/\lambda$, where λ denotes the wavelength.

Figure 16.19b shows the spatial frequency distribution of a bit datum recorded in the thin layer, the thickness of which is same as the wavelength of recording light. Because the extension of the recorded bit in the axial direction is limited by the thickness of the photosensitive film, the spatial-frequency distribution of the bit datum is much extended in the axial direction. The distribution is calculated by the convolution between the spatial-frequency distribution of the focused spot and the distribution of thin photosensitive film.

Figure 16.19c shows the CTF of the reflection confocal microscope. As Wilson *et al.* pointed out,[51] the spatial distribution shown in Figure 16.19a has no overlap with the CTF of a reflection confocal microscope unless we use an extremely high NA lens.

The spatial distribution shown in Figure 16.19b easily has overlapping areas with the CTF, if we carefully select the thickness of the recording layer. As a conclusion, it is possible to read the data with the reflection confocal

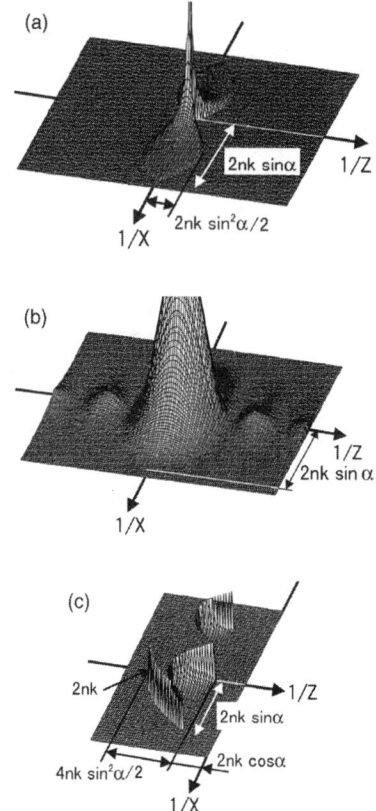

FIG. 16.19 Spatial-frequency distribution of the recorded bit datum (a) recorded in a thick medium and (b) recorded in a thin layer. The thickness of the recording layer is the same as the wavelength of the recording light. Images a and b are truncated at an arbitrary value to show the structure in the high spatial frequency region. Image c shows CTF of the reflection confocal microscope.

configuration by using the medium piled up in the thin recording layers and nonphotosensitive transparent layers alternately.

The structure of the multilayered recording medium is shown in Figure 16.13 in Section 3.3. In general, it is difficult to pile up organic films alternately without influencing the other films. In the medium, PVA is selected as the nonphotosensitive transparent film because water and pyridine as solvents for PVA and urethane-urea do not dissolve the urethane-urea film and the PVA film, respectively.

The process of making the medium was as follows. The urethane-urea copolymer was dissolved in pyridine at 6.5 wt%. The mixed solution was filtered through a 0.2 μm filter, and then it was spin-coated onto a glass plate at 1000 rpm for 10 s. The residual solvent was removed after heat treatment at 130°C for 12 h. Further drawing in a vacuum at 80°C for 20 h was done to remove the solvent completely. The film was gradually cooled to room temperature. The thickness was about 1 μm. After evaporating the pyridine solvent in an oven, the PVA and PMMA films were also spin-coated on the urethane-urea copolymer film. PMMA was used to control the distance between the photosensitive layers, but it interfered with the urethane-urea copolymer layer when it was spin-coated on the layer directly. To avoid such interference, PVA thin film was coated on a urethane-urea copolymer layer in order to protect it from PMMA. The thicknesses of the PVA and PMMA films were about 1 μm and 5 μm, respectively. The urethane-urea copolymer was spin-coated again on the PVA film.

The readout result with RCM optics is shown in Figure 16.15. The four recording layers are clearly detected. As a result, it has been determined that the reflection type confocal microscope configuration can be used as a read-out system of multilayered optical memory by using the recording medium in which the photosensitive thin films and transparent films are piled up alternately.

16.4.6 Confocal Phase-Contrast Microscope

As described in Section 16.4.5, the detection of refractive-index change between two isomers in a photochromic material is the most promising technique for nondestructive readout.[15] It is necessary to develop a readout system that is sensitive to refractive-index distribution. For this purpose, several methods, such as phase-contrast microscope, differential phase-contrast microscope, and reflection confocal microscope configurations have been proposed.

Figure 16.20 shows a laser-scanning confocal microscope for phase-contrast imaging. Since a point light source illuminates a point in a medium, it is possible to reduce unnecessary scattered light. The point detector of a confocal microscope detects only the light intensity from a specific point of interest in the thick sample. Only the light intensity in the conjugate pair of the point of interest in the thick volume (or the focused point of the laser

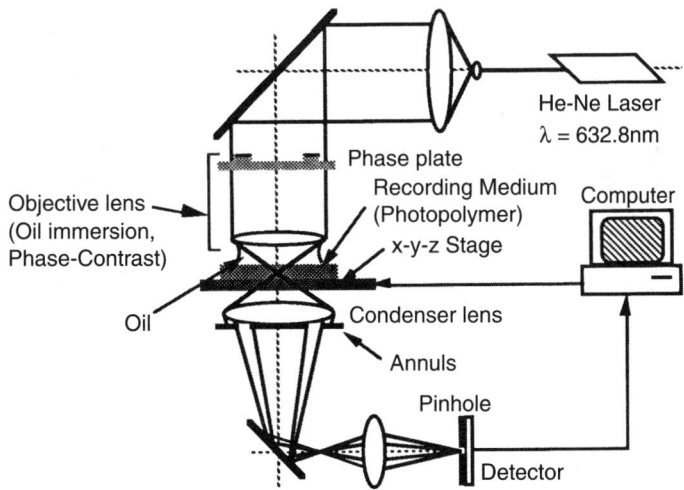

FIG. 16.20 Confocal phase-contrast microscope for 3D memory.

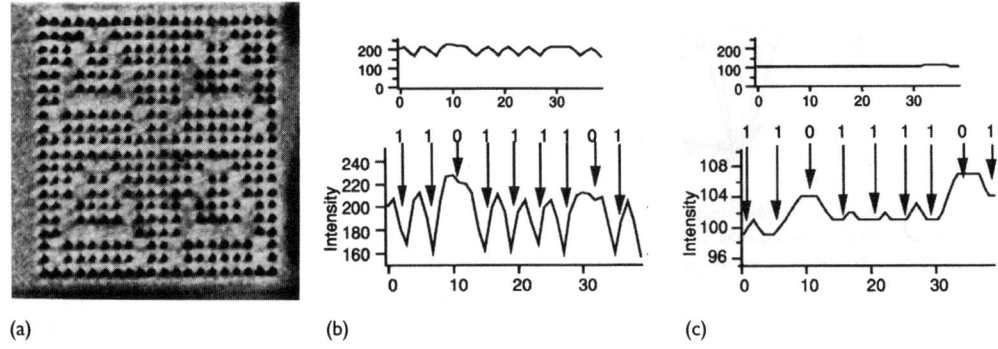

(a)   (b)   (c)

FIG. 16.21 (a) Readout data by confocal phase-contrast microscope and (b) its cross section. (c) Cross section of the same area read by a conventional phase-contrast microscope.

beam in the volume) is detected in a confocal microscope. The scattered light produced by other nonfocused points does not contribute to the detected signal. Hence, the signal contrast of the images is excellent, and the crosstalk between planes is negligible compared to images obtained using a nonconfocal microscope. Spatial resolution is also better because of the nonlinear spatial response (the product of illumination point-spread functions and the detection amplitude point-spread function).[52-54]

Figure 16.21a presents an example of bit data read using a confocal microscope. A He-Ne laser (632.8 nm) was used together with a phase-contrast objective and an annular pupil for phase-contrast (dark-field) imaging. Figure 16.21b shows a part of a cross section of the readout data. For comparison, the same segment of the data was read using a conventional microscope with the same objective lens (Figure 16.21c). The results demonstrated the advantages of confocal microscopy for high-contrast and high-resolution imaging of 3D structures.

16.4.7 Polarization Reading

It is well known that under irradiation with linearly polarized blue or green light, the azo-dye chromophores can undergo trans-cis photoisomerization to induce an anisotropic, or uniaxial, orientation of polymer side groups.[55] Figure 16.22 shows a schematic diagram of photoisomerization of the azo-dye of the urethane-urea copolymer (Figure 16.12a) by illumination with a polarized light. The azo-dye of the trans state shown in Figure 16.22a absorbs the light, whose polarization direction is parallel to the dipole moment of the azo-dye, and then flip-flops to the cis isomer, as shown in Figure 16.22b. The cis isomer also absorbs the blue and green light and photoisomerizes, again generating a trans state. There are two possibilities of cis-trans photoisomerization: One is a flip-flop process generating the same state before illumination, and the other is an inversion process generating a trans state whose axis is rotated about 90° from the initial state, as shown in Figure 16.22c.[56,57] The trans states show less absorption of the illuminated

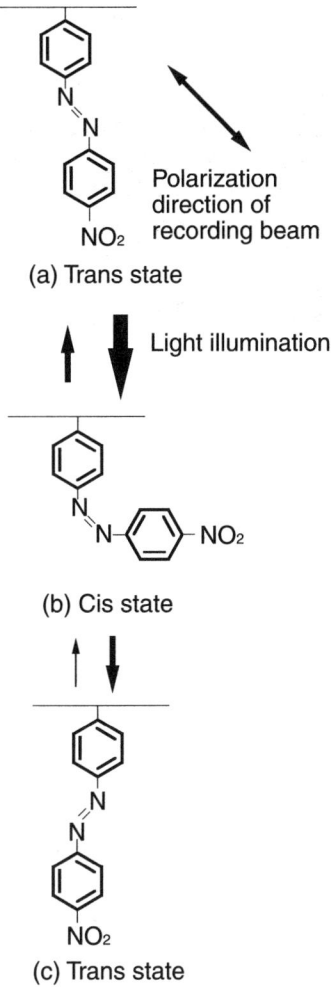

(a) Trans state

Polarization direction of recording beam

Light illumination

(b) Cis state

(c) Trans state

FIG. 16.22 Mechanism of the generation of anisotropy by photoisomerization of urethane-urea copolymer.

light, because the direction of the azo-group axis is nearly perpendicular to the polarization of the illuminated light. The trans chromophores, which return to the initial states through the flip-flop process, reabsorb the light and are reexcited to the cis isomers. As a result, the number of trans states, as shown in Figure 16.22a, decreases, and then the number of trans states, as shown in Figure 16.22c, increases by the exposure of the linearly polarized light. The increment of the number of azo-dyes whose axes are perpendicular to the direction of the illuminated polarization generates the optical anisotropic refractive-index distribution.[58,59]

Figure 16.23a shows the readout results of bit data that were recorded with the light of polarization angle from 0° to 180° with a pitch of 15°. The two vertical dots were recorded with the same polarization state. The data were read out with four polarization states. For recording, Ar$^+$ laser

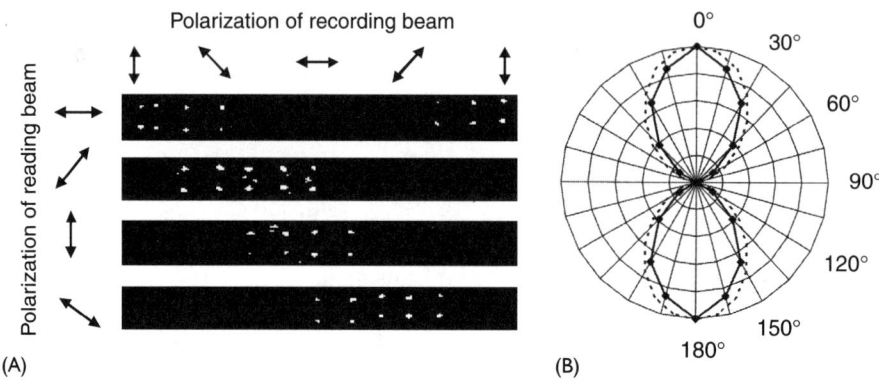

FIG. 16.23 Writing and reading with various polarization directions: (a) Readout results of data that were recorded with the polarization angles from 0° to 180° with a pitch of 15°. The data were read out with four polarization states. (b) Polar plot of readout intensity of data with the function of the polarization direction of the readout beam. Data were recorded with the horizontally polarized light in this figure.

(514.5 nm) was used, and readout was carried out using white light. It was noted that the reading intensity was highest when the polarization angle of the recording beam was perpendicular to that for reading, which confirmed that the azo-dye molecules were oriented perpendicular to the polarization of the Ar^+ laser light.

Figure 16.23b shows a polar plot of the readout intensity of data with the function of the polarization direction of the readout beam. The data were recorded with the horizontally polarized light in this figure. The data were read out with the highest contrast when the polarization of the reading light was vertical; namely, the polarization of the readout light was perpendicular to that of the recording beam.

The most interesting feature of the urethane-urea copolymer is the selective anisotropy for the particular direction. If the anisotropy is excited linearly to the direction perpendicular to the polarization direction of the writing beam, the readout intensity is expected to have the dependency of $\cos^2(\theta)$, as shown in Figure 16.23b with a dashed line. θ is an angle between the polarization directions of the reading beam and the axis of excited anisotropy. The angular dependence of the readout intensity in the material is narrower than the expected dependence. This is a promising property in polarization-multiplexed optical storage. Figure 16.23b shows that there will be little crosstalk between the recorded data when the polarization angle is approximately ±60°.

Figure 16.24 shows the polarization-multiplexed recorded data "X," "Y," and "Z" at the respective recording polarization angles 0°, 60°, and 120°. Each pattern was recorded with the same exposure time. The lateral distance between bits in the plane was 3 μm. The expected refractive-index change was estimated as about 0.01.

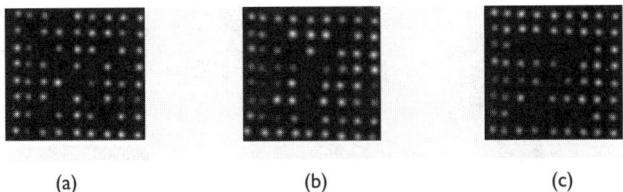

(a)   (b)   (c)

FIG. 16.24 Readout results of the polarization-multiplexed recorded data at the respective recording polarization angles 0°, 60°, and 120°.

16.4.8 Differential Phase-Contrast Microscope with a Split Detector

An alternative readout system is a scanning differential phase-contrast microscope with a split detector, as shown in Figure 16.5. The optical configuration is compact and easy to align. The memory medium, in which the data bits have been recorded, is located at the focus of an objective lens. The band limit of the optical transfer function (OTF) is the same as that of a conventional microscope with incoherent illumination.[60] The resolution, especially the axial resolution of the phase-contrast microscope, is similar to that obtained by Zernike's phase-contrast microscope. The contrast of the image is much improved compared to that of Zernike's phase-contrast microscope, however, because the nondiffracted components are completely eliminated by the subtraction of signals between two detectors. The readout system is therefore sensitive to small phase changes.

16.5 NEAR-FIELD RECORDING

16.5.1 Azobenzene Derivatives for Near-Field Recording

Near-field optics is another promising way to achieve high-density optical memory. Since near-field techniques overcome the diffraction limit of light by the contribution of evanescent wave,[61–63] it is possible to store a bit datum in a nanometric region. The nanometric resolution is achieved by the contribution of evanescent fields.

The near-field recording techniques were first demonstrated in photolithography applications.[64] A photomask with fine structures was placed directly on a photoresist and was illuminated with light. If the distance between the photomask and the photoresist was small enough, or if the distance was smaller than the decay length of evanescent field, it was possible to copy the fine structures of the photomask onto the photoresist film. Although this technique was proposed for nanofabrication with light, in other points of view it was recorded as the near-field recording of the optical-field distribution near the photomask in photosensitive materials.

Various materials have been examined for the media of near-field recording. Martin *el al.*[20] and Betzig *et al.*[19] used optomagnetic material, and Irie used diarylethene derivatives.[5]

Urethane-urea copolymer materials shown in Figure 16.12 are also promising for the application of near-field recording.[65–67] Figures 16.25a and

FIG. 16.25 The experimental results of near-field recording with (a) 500 nm particles and (b) 50 nm particles. The topographical distribution of the film surface was detected with an AFM. Some residual particles that were not removed by the washing were also observed in image a. It was found that urethane-urea copolymer has a high-enough resolution for near-field recording.

16.25b show the near-field recording results on a urethane-urea copolymer thin film. These images show the results of recording bit sizes of 500 nm and 50 nm, respectively. They showed the topographical change of the film surface detected with an atomic force microscope (AFM).

The 500 nm nits were recorded as follows. The urethane-urea copolymer was spin-coated on a glass substrate. Water containing the 500 nm polystyrene particles was dropped on the film and dried. In the drying process, the particles were arranged in a hexagonal structure by using the self-organization process discovered by Nagayama.[68]

The film was illuminated with a laser light of 488 nm wavelength and 30 mW power. Note that the laser light is near a resonant absorption peak of the urethane-urea film. After illumination of the laser light, the film was immersed in water and the particles were removed from the film. In Figure 16.25a the particles that were not removed by washing were also observed.

Because the light intensity under the particles was much stronger than that in the area between the particles caused by the multiscattering between the particles and the substrate,[69] the holes were produced by the light intensity at the areas under the particles. This was the same mechanism used to produce surface-relief grating with light illumination.[70]

For the recording of 50 nm bits, the 50 nm particles were prepared in the same manner. The particles were melted away with benzene.

The bit size of 50 nm was 10 times smaller than the wavelength of the illumination light. This was accomplished by the contribution of evanescent wave generated by the small particles. The results clearly show that the urethane-urea copolymer has a high enough resolution for the near-field recoding.

16.5.2 Application of Near-Field Recording to High-Resolution Imaging

In the near-field recording technique presented in Section 16.5.1, the evanescent field distributions near the particles were recorded as surface topography of a photosensitive film, and the topography was read out with an AFM. This is good for the application of high-resolution imaging, because the evanescent distribution near a specimen can be converted to the topographical change of the film surface and detected with an AFM.

Figure 16.26 shows the observed result of high-resolution imaging of a paramecium. This was the observed result of the film surface with an AFM. The scan area was 10 μm $\times$ 10 μm.

The near-field recording of a paramecium was done with contact recording technique. Water containing living paramecia was dropped on a urethane-urea copolymer film, and the specimen on the film was exposed to laser light

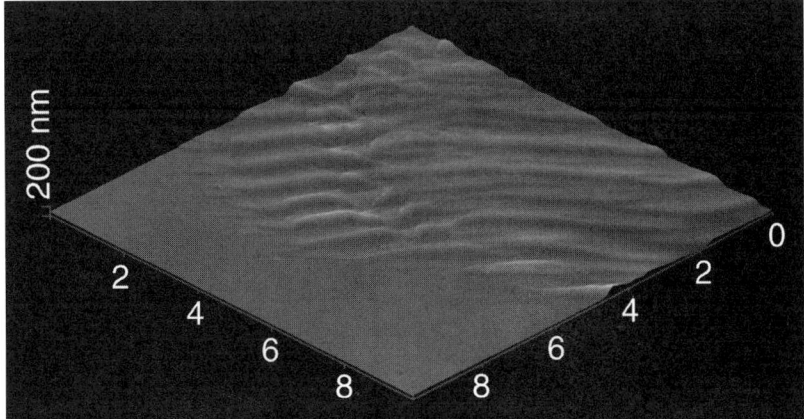

FIG. 16.26 The observed result of a paramecium. The surface of the photosensitive film was modulated by the intensity distribution of light induced by the fine structures of the specimen. The cilia of a paramecium are clearly observed, and it can be seen that the end of each cilium branches into two cilia.

of 488 nm wavelength. The power of the laser light was about 100 mW, and the specimen was exposed to direct light that had not passed through any lenses. The exposure time was 200 ms, and it was adjusted by a mechanical shutter. After the exposure, the film was immersed in water and the specimen was removed from the film.

It was found that the surface of the photosensitive film was modulated by the intensity distribution of light induced by the fine structures of the specimen. The cilia of a paramecium were clearly observed. One could recognize that the end of each cilium branched into two cilia. Because the spatial resolution of the observed result is smaller than 100 nm, the evanescent wave distribution generated by the fine structures of the specimens were imaged with the near-field recording technique.

The high-resolution imaging based on near-field recording can observe living and moving biological or very fast phenomena, because the whole optical field distribution near the surface of a specimen is recorded in a short time as the topographical change of a photosensitive film. This is a unique advantage of the near-field recording technique for high-resolution imaging because other near-field microscopes require the scanning time of a probe tip to acquire an image of a specimen.

The high-resolution imaging system has some additional advantages: (1) A high signal-to-noise ratio is expected because no small aperture is required, (2) the optical field is not disturbed due to the multiscattering between a probe tip and the specimen, and (3) by conversion of optical field to the topographical distributions of the film, it is possible to introduce extremely high resolution microscope systems—such as AFMs, scanning tunneling microscopes, and scanning electron microscopes—into the fields of the near-field microscope as a detection system.

Figure 16.27 shows the result of observing the movement of a euglena cell's tail. The cell was exposed three times in order to observe its sequential movements. Each exposure time was 50 ms, and the exposure interval was 500 ms.

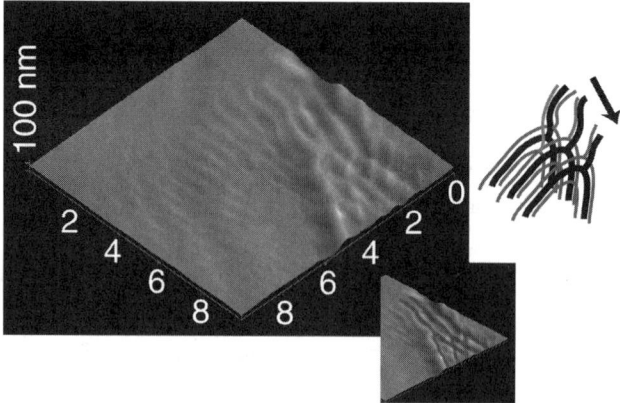

FIG. 16.27 The movement of a euglena cell. The cell was exposed three times in order to observe its sequential movement. The exposure time was 50 ms, and the exposure interval was 500 ms.

The inset portion of Figure 16.27 illustrates the movement of the euglena, which moved from the center to the right and down. The sequential movement of the euglena and the gracilis was observed, and the fine structures of the euglena and the gracilis were clearly imaged with subwavelength resolution.

16.6 CONCLUDING REMARKS

Various optics for 3D memories using photochromic materials and near-field recording for high-density memory were presented in this chapter. The major advantages of photochromic materials are erasable/rewritable capability, high resolution, and high sensitivity. As described here, some photochromic materials have adequate properties as recording media of 3D memory and near-field recording. Three-dimensional recording techniques do not conflict with other techniques for achieving high density, such as wavelength multiplexing, polarization multiplexing, and wavelength shortening. Three-dimensional recording techniques can be combined with these other techniques to increase density. Two-photon excitation is a key technique in 3D memory. It is required to develop photochromic materials with large two-photon absorption coefficient.[71] The multilayered memory is just an extension of z-direction of conventional optical memories, so scanning, tracking, and autofocusing techniques in the conventional memory systems can be useful with some modifications.

In the application of high-resolution imaging based on the near-field recording technique, it is interesting to observe various light properties such as polarization, absorption spectrum, fluorescence, and so forth, in near-field regions with the combination of functionalized films for the measurement of these properties. The system has the potential of being an application of the material science of thin films.

REFERENCES

1. Hiershberg, Y. *J. Am. Chem. Soc.* **1956**, *78*, 2304.
2. Irie, M., and Uchida, K. *Bull. Chem. Soc. Jpn.* **1998**, *71*, 985.
3. Hibino, J., Moriyama, K., Suzuki, M., and Kishimoto, K. *Thin Solid Films* **1992**, *210/211*, 562.
4. Uchida, M., and Irie, M. *J. Am. Chem. Soc.* **1993**, *115*, 6442.
5. *Photo-reactive Materials for Ultrahigh Density Optical Memory*, M. Irie, Ed. Amsterdam: Elsevier, 1994.
6. Nakamura, S. *Science* **1998**, *281*, 956.
7. Akiyana, T., Uno, M., Kitaura, H., Narumi, K., Kojima, R., Nishiuchi, K., and Yamada, N. *Jpn. J. Appl. Phys.* **2001**, *40*, 1598.
8. Aoki, I. *Jpn. J. Appl. Phys.* **2001**, *40*, 1817.
9. Parthenopoulos, D. A., and Rentzepis, P. M. *Science* **1989**, *24*, 843.
10. Parthenopoulos, D. A., and Rentzepis, P. M. *J. Appl. Phys.* **1990**, *68*, 5814.
11. Strickler, J. H., and Webb, W. W. *Opt. Lett.* **1991**, *16*, 1780.
12. Kawata, S., and Toriumi, A. *Proc. SPIE* **1997**, *3109*, 174.
13. Akimov, D. A., Fedotov, A. B., Koroteev, N. I., Levich, E. V., Magnitskii, S. A., Naumov, A. N., Sidorov-Biryukov, D. A., Sokolyuk, N. T., and Zheltikov, A. M. *Opt. Mem. Neural. Netw.* **1997**, *6*, 31.

14. Kawata, S., and Kawata, Y. *Chem. Rev.* **2000**, *100*, 1777.
15. Toriumi, A., Herrman, J. M., and Kawata, S. *Opt. Lett.* **1997**, *22*, 555.
16. Toriumi, A., Kawata, S., and Gu, M. *Opt. Lett.* **1998**, *23*, 1924.
17. Pohl, D. W., Denk, W., and Lanz, M. *Appl. Phys. Lett.* **1984**, *44*, 651.
18. Betzig, E., Lewis, A., Harrotunian, A., Isaacson, M., and Kratschmer, E. *Biophys. J.* **1986**, *49*, 269.
19. Betzig, E., Trautman, J. K., Wolfe, R., Gyorgy, E. M., and Finn, L. *Appl. Phys. Lett.* **1992**, *61*, 142.
20. Martin, Y., Rishton, S., and Wickramashinghe, H. K. *Appl. Phys. Lett.* **1997**, *71*, 1.
21. Tsujioka, T., and Irie, M. *Appl. Opt.* **1998**, *37*, 4419.
22. Kawata, Y., Ueki, H., Hashimoto, Y., and Kawata, S. *Appl. Opt.* **1995**, *34*, 4105.
23. Ueki, H., Kawata, Y., and Kawata, S. *Appl. Opt.* **1996**, *35*, 2457.
24. Kawata, Y., Tanaka, T., and Kawata, S. *Optical Memory and Neural Networks* **1999**, *8*, 1.
25. Kano, H., Wada, K., and Kawata, S. *Extended Abstracts of the 43rd Spring Meeting of the Japan Society of Applied Physics and Related Societies*, **1996**, p. 886.
26. Ando, E., Miyazaki, J., and Morimoto, K. *Thin Solid Films* **1985**, *133*, 21.
27. Hamano, M., and Irie, M. *Jpn. J. Appl. Phys.* **1996**, *35*, 1764.
28. Irie, M., and Mohri, M. *J. Org. Chem.* **1998**, *53*, 803.
29. Denk, W., Strickler, J. H., and Webb, W. W. *Science* **1990**, *248*, 73.
30. Kawata, Y., Ishitobi, H., and Kawata, S. *Opt. Lett.* **1998**, *23*, 756.
31. Sekkat, Z., and Knoll, W. *J. Opt. Soc. Am. B* **1995**, *12*, 1855.
32. Watanabe, O., Tsuchimori, M., and Okada, A. *J. Mater. Chem.* **1996**, *6*, 1487.
33. Egami, C., Suzuki, Y., Sugihara, O., Okamoto, N., Fujimura, H., Nakagawa, K., and Fujiwara, H. *Appl. Phys. B* **1997**, *64*, 471.
34. Wang, C., Fei, H., Yang, Y., Wei, Z., Qui, Y., and Chen, Y. *Opt. Commun.* **1999**, *159*, 58.
35. Ishikawa, M., Kawata, Y., Egami, C., Sugihara, O., Okamoto, N., Tsuchimori, M., and Watanabe, O. *Opt. Lett.* **1998**, *23*, 1781.
36. Watanabe, O., Tsuchimori, M., Okada, A., and Ito, H. *Appl. Phys. Lett.* **1997**, *71*, 750.
37. Tsuchimori, M., Watanabe, O., Ogata, S., and Okuda, A. *Jpn. J. Appl. Phys.* **1997**, *36*, 5518.
38. Day, D., and Gu, M. *Appl. Opt.* **1998**, *37*, 6299.
39. Dvornikov, A. S., and Rentzepis, P. M. *Opt. Commun.* **1997**, *136*, 1.
40. Hell, S., Booth, M., Wilms, S., Schnetter, C., Kirsh, A., Arndt-Jovin, D., and Jovin, T. *Opt. Lett.* **1998**, *23*, 1238.
41. Gu, M., and Day, D. *Opt. Lett.* **1999**, *24*, 288.
42. Fermann, M. E., Hofer, M., Haberl, F., and Schmidt, A. J. *Opt. Lett.* **1991**, *16*, 244.
43. Yamada, E., Yoshida, E., Kitoh, T., and Nakazawa, M. *Electron. Lett.* **1995**, *31*, 1324.
44. Hansen, P. B., Raybon, G., Koren, U., Miller, B. I., Young, M. G., Newkirk, M. A., Chien, M. D., Tell, B., and Burrus, C. A. *Electron. Lett.* **1993**, *29*, 739.
45. Arahira, S., Matsui, Y., Kunii, T., Oshiba, S., and Ogawa, Y. *IEEE Photon. Technol. Lett.* **1996**, *32*, 1211.
46. Wang, M. M., Esener, S. C., McCormik, F. B., Cokgor, I., Dvornikov, A. S., and Rentzepis, P. M. *Opt. Lett.* **1997**, *22*, 558.
47. Dvornikov, A. S., and Rentzepis, P. M. *Opt. Commun.* **1995**, *119*, 341.
48. Streibl, N. *J. Opt. Soc. Am. A* **1985**, *2*, 121.
49. Sheppard, C. J. R. *Optik* **1986**, *74*, 128.
50. Sheppard, C. J. R., Gu, M., and Mao, X. Q. *Opt. Commun.* **1991**, *81*, 281.
51. Wilson, T., Kawata, Y., and Kawata, S. *Opt. Lett.* **1996**, *21*, 1003.
52. Wilson, T., and Sheppard, C. J. R. *Theory and Practice of Scanning Optical Microscopy*. London: Academic Press, **1984**.
53. Nakamura, O., and Kawata, S. *J. Opt. Soc. Am. A* **1990**, *7*, 522.
54. Wilson, T. *Confocal Microscopy*. London: Academic Press, **1990**.
55. Sekkat, Z., Prêtre, P., Knoesen, A., Volksen, W., Lee, V. Y., Miller, R. D., Wood, J., and Knoll, W. *J. Opt. Soc. Am. B.* **1998**, *15*, 401.
56. Gibbson, W. M., Shannon, P. J., Sun, S. T., and Swetlin, B. J. *Nature* **1991**, *351*, 49.
57. Andderle, K. *Liquid Crystals* **1991**, *9*, 691.
58. Todorov, T., Nikolova, L., and Tomova, N. *Appl. Opt.* **1984**, *23*, 4309.

59. Ebralidze, T. D., and Mumladze, A. N. *Appl. Opt.* **1990**, *29*, 446.
60. Kawata, Y., Juškaitis, R., Tanaka, T., Wilson, T., and Kawata, S. *Appl. Opt.* **1996**, *35*, 2466.
61. Denk, W., and Pohl, D. W. *J. Vac. Sci. Technol. B* **1991**, *9*, 510.
62. Inouye, Y., and Kawata, S. *Opt. Lett.* **1994**, *19*, 159.
63. Kawata, Y., Xu, C., and Denk, W. *J. Appl. Phys.* **1999**, *85*, 1294.
64. Fischer, U. C., and Zingsheim, H. P. *J. Vac. Sci. Technol.* **1981**, *19*, 881.
65. Kawata, Y., Egami, C., Sugihara, O., Okamoto, N., Tsuchimori, M., Watanabe, O., and Nakamura, O. *Opt. Commun.* **1999**, *161*, 6.
66. Kawata, Y., Murakami, M., Egami, C., Sugihara, O., Okamoto, N., Tsuchimori, M., and Watanabe, O. *Appl. Phys. Lett.* **2001**, *78*, 2247.
67. Kitano, H., Murakami, M., Kawata, Y., Egami, C., Sugihara, O., Okamoto, N., Tsuchimori, M., and Watanabe, O. *J. Microscopy* **2001**, *202*, 162.
68. Nagayama, K. *Mater. Sci. Eng. C* **1994**, *1*, 87.
69. Inami, W., and Kawata, Y. *J. Appl. Phys.* **2001**, *89*, 5876.
70. Kim, D. Y., Li, L., Jiang, X. L., Shivshankar, V., Kumer, J., and Tripathy, S. K. *Macromolecules* **1995**, *28*, 8835.
71. Albota, M., Beljonne, D., Brédas, J. L., Ehrlich, J. E., Fu, J. Y., Heikal, A. A., Hess, S. E., Kogej, T., Levin, M. D., Marder, S. R., McCord-Maughon, D., Perry, J. W., Röckel, H., Rumi, M., Subramaniam, G., Webb, W. W., Wu, X. L., and Xu, C. *Science*, **1998**, *281*, 1653.

17

SYNTHESIS AND APPLICATIONS OF AMORPHOUS DIARYLETHENES

TSUYOSHI KAWAI*
MASAHIRO IRIE[†]
Department of Applied Chemistry; Graduate School of Engineering, Kyushu University, Hakozaki 6-10-1, Higashi-ku, Fukuoka 812-8581, Japan
[†]*CREST, Japan Science and Technology Corporation, JST, Hakozaki 6-10-1, Higashi-ku, Fukuoka 812-8581, Japan*

Photochromic molecular materials have attracted considerable attention from both fundamental as well as practical points of view.[1] Among a number of photochromic compounds, diarylethenes are the most promising for applications to optoelectronic devices, such as optical-memory media and switching devices, because of their thermal, irreversible, and fatigue-resistant photochromic performance.[2–12] Diarylethenes undergo cyclization and cycloreversion reactions, as shown in Figure 17.1.[2,3] For the applications, diarylethenes have to be dispersed in solid-film matrices, such as in polymers,[4,6,8–11,13] or in sol-gel glasses.[14] The concentration of diarylethenes in polymer matrices is usually limited below 30% because of bleeding from the matrices. There is a strong desire for an increase in the concentration for the applications, for which a large degree of photoinduced optical property changes is required. Some diarylethenes undergo photochromic reactions not only in polymer matrices but also in the single-crystalline phase.[15,16] Although the concentration in the single crystals is ideally high, it is not easy to fabricate the crystals to thin films. Amorphous photochromic materials satisfy both requirements of a high concentration and an easy fabrication to thin films.

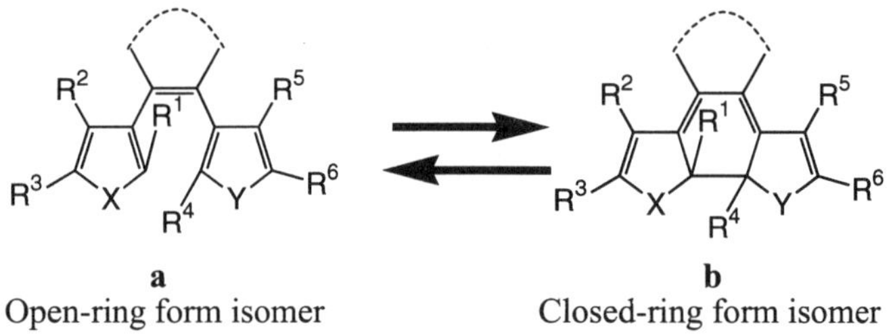

FIG. 17.1 Photochromic reactions of diarylethenes.

17.1 QUASI-STABLE AMORPHOUS DIARYLETHENES

The stability of the amorphous state of diarylethene **1**, of which the molecular structure is shown in Figure 17.2, has been studied for the first time.[17,18] Diarylethene **1** undergoes reversible photocoloration and photobleaching in the super-cooled isotropic bulk solid state, which is transparent and exhibits no marked light scattering in the visible wavelength region. Crystal growth, however, takes place at room temperature within a couple of days, and the sample becomes milky yellowish white. The crystal growth was suppressed by introducing diarylethene **1** into a thin layer cell in its molten state above Tm (melting temperature).

The thin layer cell was composed of two quartz plates as the windows and a polymer film as the spacer of 4 μm to 20 μm in thickness. Once molten **1** is introduced into the thin layer cell and cooled to room temperature, the amorphous phase becomes stable enough for practical use. The transparency was maintained for more than one year in the thin layer cell. X-ray diffraction study showed no diffraction peak, and polarized microscopy observation also indicated the amorphous state.

Upon irradiation with UV light, the open-ring form isomer **1a** converted to the closed-ring form isomer **1b**, which is red in color and shows an absorption spectral peak at 520 nm. Upon irradiation with visible light with a wavelength longer than 480 nm, the color was bleached, and isomer **1b** returned to the original open-ring form isomer **1a**. Because the photochromic reaction took place in the solid state and no marked molecular diffusion occurred, clear pattern formation could be achieved. The refractive-index change, Δn, was relatively small, $\Delta n_{max}=0.005$ at 633 nm. The isotropic bulk solid state was stable only in the thin layer cells. Slow crystal growth was difficult to avoid in thicker cells.[19]

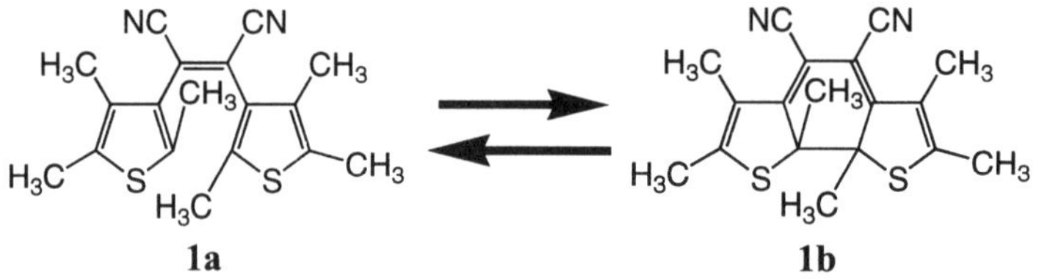

FIG. 17.2 Photochromic reaction of diarylethene 1.

17.2 THERMALLY STABLE AMORPHOUS DIARYLETHENES

In order to obtain stable amorphous states of organic materials, the glass transition temperature (Tg) of the materials should be higher than room temperature. Molecular motion is frozen at temperatures below Tg, and crystal growth is suppressed. The Tg of **1a** was found to be lower than room temperature. Although several photochromic amorphous materials containing azobenzene unit have been synthesized,[20] the practical use of amorphous azobenzenes is limited because of thermal instability of the cis-isomer and low reactivity in the solid state. Figure 17.3 shows some molecular structures

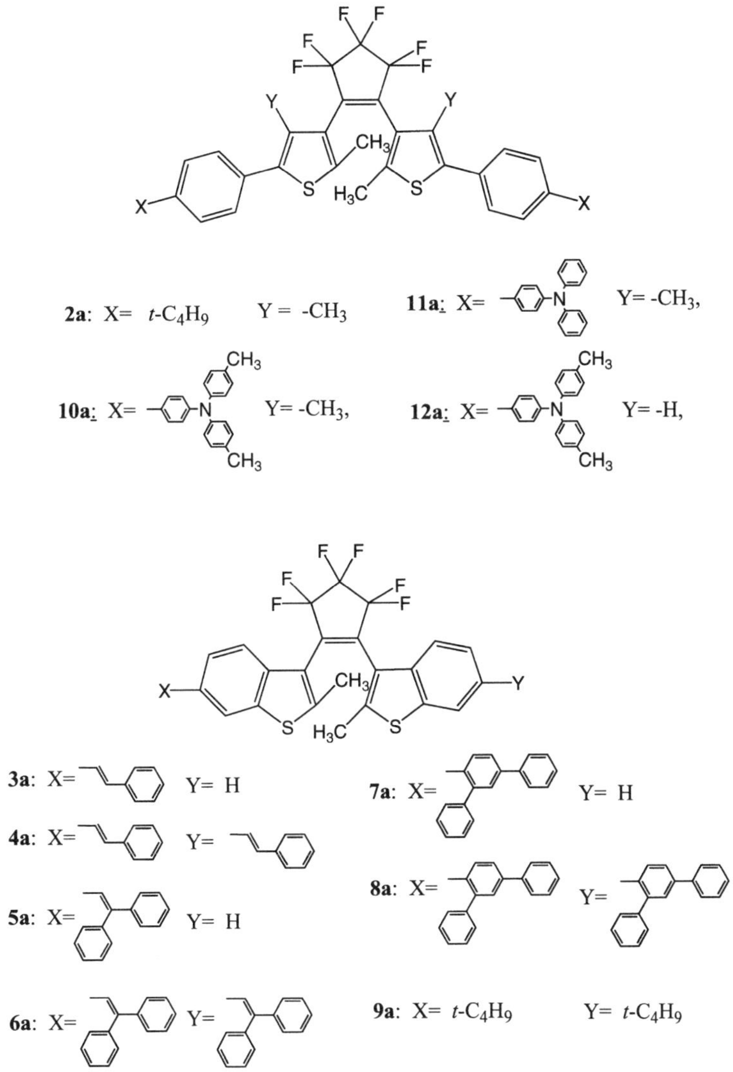

FIG. 17.3 Molecular structures of amorphous diarylethenes.

TABLE 17.1   Glass Transition Temperatures of Amorphous Diarylethenes

Compound	Tg /°C
2a	68
4a	56
5a	67
6a	92
7a	88
8a	127
9a	47
10a	103
10b	124
11a	94

of stable amorphous diarylethenes. Tg values are summarized in Table 17.1.[21–27] **2a** has bulky tert-butyl groups at both ends of the 1,2-bis(5-phenyl-3-thiophenyl)hexafluorocyclopentene.[21] Figure 17.4 shows a typical Differential Scanning Calorimetry (DSC) profile of **2a**. At about 70°C, a clear shift of the baseline was observed, indicating the glass-to-liquid transition. From the threshold temperature, Tg was evaluated to be 68°C. A small endothermal peak indicates weak intermolecular interactions in the glassy state. **2a** is the first example of a diarylethene with a Tg higher than room temperature.

One of the effective molecular design principles for increasing Tg is to introduce rigid phenyl or condensed phenyl side groups. Migration and rotation of large and rigid phenyl or condensed phenyl moieties are prohibited in the amorphous solid. Therefore, the rigid aromatic groups are easily frozen at temperatures higher than room temperature. From this point of view, bis(benzothienyl)ethene structure is favorable. The highest Tg, 127°C, was observed for **8a**, which has 2,4-diphenylbenzene substituents. In the case of substituted 1,2-bis(2-methyl-1-benzothiophen-3-yl) perfluorocyclopentenes,

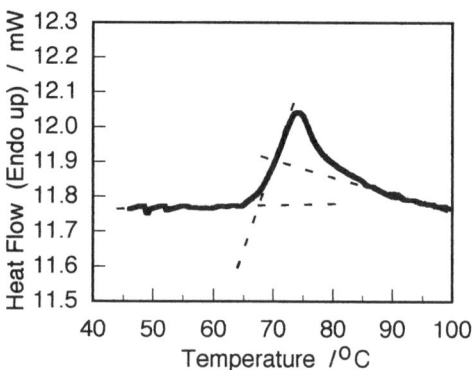

FIG. 17.4 DSC profile of diarylethene **2** at second and subsequent heating scans. The temperature scan rate is 10°C/min. The sample weight is 9.08 mg.

3–9, molecular symmetry does not play an important role for Tg. For example, mono-substituted compound **5a** showed Tg at 67°C, which is higher than that of symmetric di-substituted compound **4a**, in which the numbers of phenyl groups are the same.

The Tgs of diarylethenes bearing triphenylamino groups, **10a** and **12a**, are relatively high.[24,26] The absence of methyl groups at the 4 position of the thiophene rings in **12a** decreased Tg in comparison with that of **10a**.

The closed-ring form isomer **10b** showed higher Tg than that of the open-ring form isomer, **10a**. This difference is consistent with the melting-point difference between the open-ring and closed-ring isomers. The rigid and coplanar structure of the closed-ring isomers is responsible for increasing the melting point as well as the Tg.

17.3 OPTICAL PROPERTIES OF AMORPHOUS DIARYLETHENES

Thin films of the amorphous diarylethenes can be formed easily by a conventional spin-coating method on appropriate substrates with various solvents such as toluene, hexane, dichloromethane, and so forth. Both the open-ring and the closed-ring form isomers and their mixtures form thin films. Figure 17.5 shows the reversible absorption spectral changes of a bulk amorphous film of **2**.[21] The spectral changes are almost the same as those in the solution.

In Figure 17.6, **2b** film was irradiated with visible light for photobleaching, and then UV light (313 nm) was irradiated until the photostationary

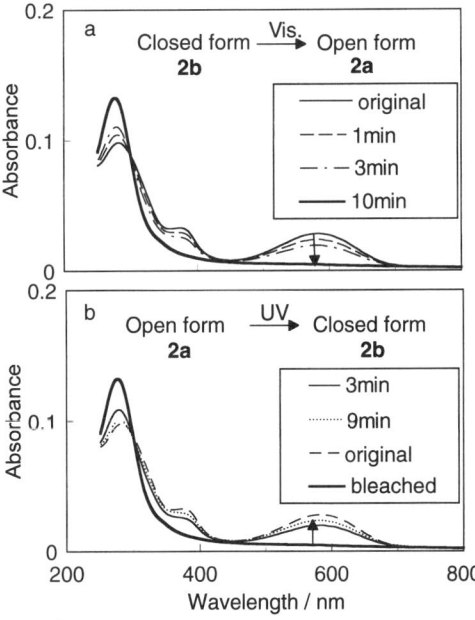

FIG. 17.5 Absorption spectral change of amorphous films of (a) diarylethene **2b** upon irradiation with visible light ($\lambda > 480$ nm and (b) diarylethene **2a** upon UV light ($\lambda = 313$ nm) irradiation.

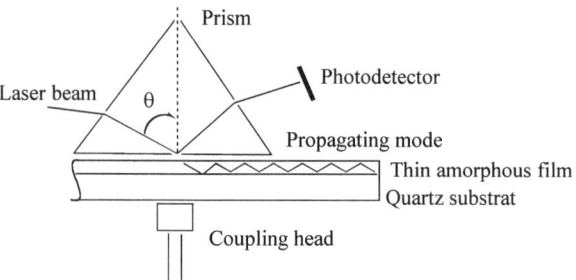

FIG. 17.6   Schematic illustration of prism-coupler setup used for refractive index measurement.

state was attained. After UV-light irradiation, about 80% of the absorbance at 580 nm was recovered. The conversion of 80% from **2a** to **2b** was almost the same as that in the solution phase. The conversion of the film prepared from **2a** solution was 40%, which is half of the conversion of the film prepared from a solution of the closed-ring form isomer. This difference in the maximum conversion to **2b** is caused by the conformation of the open-ring form isomers.[28,29] The isomer has two conformations, anti-parallel and parallel conformations. The former is photoactive whereas the latter is inactive, and half of the open-ring form isomers are in the inactive parallel conformations in solution. Half of **2a** molecules in the film prepared from **2a** solution are in the inactive parallel conformation. The conformational change is difficult in the amorphous film below Tg. In contrast, **2a** in the bleached film prepared from the solution of **2b** is in an anti-parallel conformation, and the maximum conversion to **2b** at the photostationary state is about two times larger than that in the film prepared from **2a** solution. A similar increase in the conversions in the film prepared from the closed-form isomer has been observed in amorphous diarylethenes, **3–10**. It should be noted that heat treatment of the bleached **7a** film at a temperature above Tg resulted in a decrease in the maximum conversion, which indicates that conformational change takes place at temperatures above Tg.

Optical refractive-index changes of amorphous films are useful for photonic applications. The prism-coupler setup, illustrated in Figure 17.6, has been used to evaluate the refractive indices of the bulk amorphous films. The refractive index of **2a** film was 1.550 at 817 nm, whereas **2b** film was 1.589 at the same wavelength. Upon visible-light irradiation, the refractive index decreased to a value close to that of the open-ring form isomer, as shown in Figure 17.7. The refractive-index changes could be repeated many times. Similar refractive-index changes were also observed in other amorphous diarylethenes, as summarized in Table 17.2.

The refractive index of **10a** film was increased from 1.619 to 1.635 upon UV-light irradiation, while that of the **10b** film decreased from 1.659 to 1.624 upon irradiation with visible light.[24] The refractive index of the bleached **10b** film was slightly higher than that of the **10a** film. This difference suggests that the conformation of **10a** molecules in the original **10a** film is different from that in the bleached **10b** film and, therefore, the density is

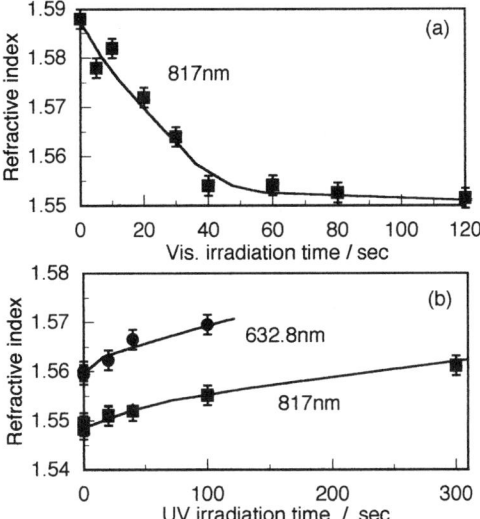

FIG. 17.7 Changes in refractive index of amorphous film of diarylethene **2** at 817 nm (squares) and 632.8 nm (circles) upon irradiation with (a) visible light and (b) UV light. The sample films (film thickness = 4 μm) were prepared from (a) closed-ring and (b) open-ring form isomers.

TABLE 17.2 Refractive Indices of Amorphous Diarylethenes

Compounds	Wavelength	Refractive Index
2a	633nm	1.561
	817nm	1.550
2b	817nm	1.588
9a	633nm	1.562
	817nm	1.551
9b	633nm	1.672
	817nm	1.615
10a	817nm	1.619
10b	817nm	1.659

slightly different. The refractive index of **10b** film in the photostationary state by irradiation with 313 nm light was about 0.015 larger than that of the bleached **10b** film, which is 43% of the refractive-index change of **10b** film. This ratio is roughly in accordance with the conversion ratio at the photostationary state, 48%, suggesting a linear relationship between composition and refractive index.

The dependence of refractive index on composition and wavelength has been studied for **9** in detail. The refractive index of **9** changed linearly with the concentration of the closed-ring form isomers in the film, as shown in Figure 17.8. [25] The experimental points could be extrapolated to evaluate the refractive index of the film of pure closed-ring form isomer **9b**, and the result is shown in Table 17.2. The difference of refractive index at 633 nm between the open-ring isomer **9a** and the closed-ring isomer **9b** is as large as 0.11,

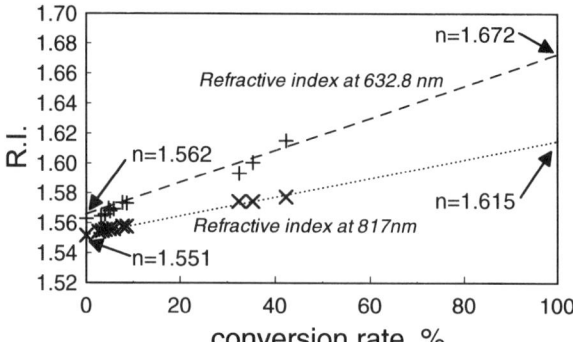

FIG. 17.8 Refractive-index changes of diarylethene **9** as a function of conversion rate measured at 632.8 nm and 817 nm.

which is due partly to a dispersion effect, as discussed next. At 817 nm, far from the absorption spectrum of **9b**, the variation of the refractive index was still very large, $\Delta n=0.064$, which is more than ten times larger than usual photorefractive effects.

Dispersion of the refractive index of the amorphous film can be analyzed based on the Sellmeier equation,[13,30] as follows:

$$n_\lambda{}^2 = n_\infty{}^2 + D\lambda_0{}^2/(\lambda^2-\lambda_0{}^2) \tag{17.1}$$

where D is related to the oscillator strength, λ is the monitoring wavelength, λ_0 is the wavelength of the absorption maximum, and n_∞ is the refractive index at the longest wavelength limit. The wavelength dependence of the refractive index simulated for amorphous diarylethene **9** is shown in Figure 17.9.[25] The parameters D and n_∞ are 0.38 and 1.5374 respectively for the open-ring form isomer film, and 0.18 and 1.5735 for the closed-ring form isomer film. This result shows that n_∞ for the colored closed-ring form isomer film is increased but D is decreased. Similar results have been reported by Kim *et al.*[13] At the longest wavelength distant from any dispersion effect due

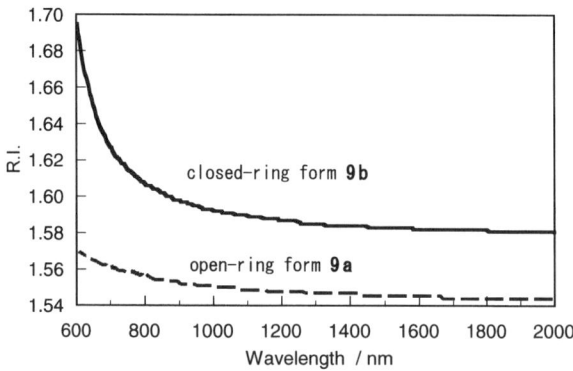

FIG. 17.9 Simulated Sellmeier plots for amorphous films of diarylethene **9a** and **9b**.

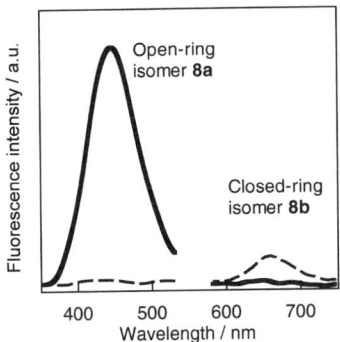

FIG. 17.10   Fluorescence spectral change of bulk amorphous film **8** excited with 280 nm by photo-irradiation: (------) closed-ring form isomer **8b**, and (——) open-ring form isomer generated by irradiation with visible light ($\lambda > 450$ nm).

to the absorption, the variation of n_∞ of the photochromic film was 0.052. This high value makes the sample a good candidate for holographic grating and Mach-Zehnder optical devices.

Fluorescence properties of amorphous diarylethenes have also been studied. Amorphous diarylethene **8a** exhibited relatively strong fluorescence in the solid state.[27] As shown in Figure 17.10, the amorphous film showed fluorescence from both the open-ring and closed-ring form isomers. Since the excitation spectra of the emissions agreed with the absorption bands of the open-ring and closed-ring form isomers, these emissions are assigned to the emissions from both isomers. The emission wavelength can be switched between 450 nm and 650 nm by alternative irradiation with UV and visible light.

17.4 CHARGE TRANSPORT IN AMORPHOUS DIARYLETHENE FILMS

Recently, amorphous diarylethene **11** was successfully applied to a junction type device shown in Figure 17.11.[12] An amorphous **11** film and a phothalo-cyanine film were used as a photochromic layer and a photoabsorbing layer, respectively. The ionization potential I_p of the vacuum-deposited amorphous film of **11** was found to decrease from 6.2 eV of the open-ring form isomer state to 5.7 eV of the closed-ring form isomer state. Since I_p of the phothalo-cyanine film ($I_p = 5.4$ eV) is close to that of the closed-ring form isomer state, the photogenerated holes in the phothalocyanine layer can be injected into the diarylethene film in the closed-ring isomer state and can move to the electrode under the biased field. When the film is in the open-ring isomer state, the film acts as the blocking layer against the photogenerated holes, because I_p of the open-ring isomer state is considerably larger than that of the phothalocyanine layer. Actually, the photocurrent response was not observed in the open-ring isomer state, whereas the finite photocurrent was observed in the closed-ring isomer state corresponding to the excitation light pulse, as shown in Figure 17.12. Because neither **11a** nor **11b** absorb the excitation

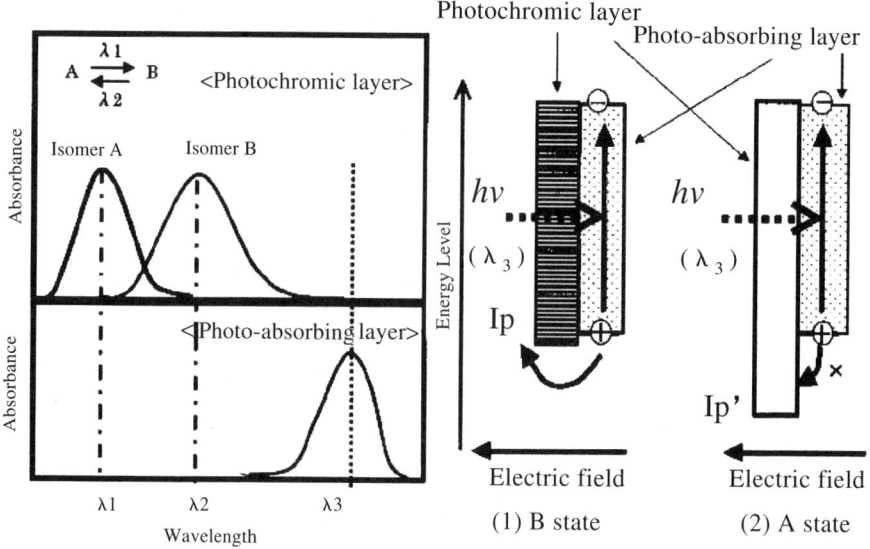

FIG. 17.11   Working principles of nondestructive readout method using photocurrent detection.

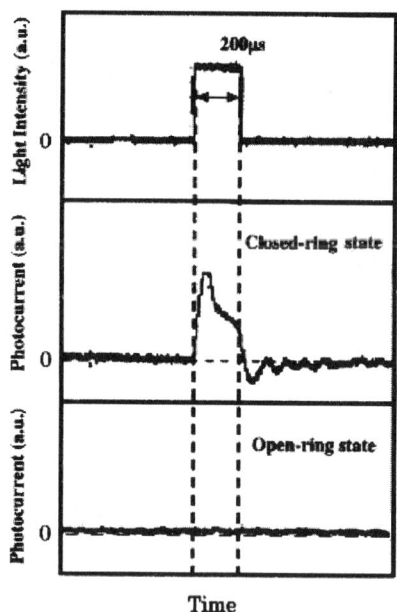

FIG. 17.12   Photocurrent waveform corresponding to irradiated light.

light, $\lambda = 780\,nm$, the photocurrent response could be observed many times without any photoisomerization. Therefore, this device acts as a photocurrent-mode photomemory with nondestructive readout capability. The proposed readout method using photocurrent detection has advantages compared to a readout method using absorption or fluorescence change detection. The method can be applied to this layer with near-field optical recording media because of high sensitivity and also has a potential ability of high readout resolution because of its use of an electrode probe with a sharp apex.

17.5 SUMMARY

Photochromic diarylethenes with appropriate substituent groups form stable amorphous thin films with Tgs higher than room temperature. These amorphous diarylethene films undergo reversible photochromic reactions and show relatively large changes in the refractive index and other optical and electrical properties. Various potential applications are possible, such as high-density optical recording and optical switching devices.

REFERENCES

1. Dürr, H., and Bouas-Laurent, H. *Photochromism: Molecules and Systems*. Amsterdam: Elsevier, 1990.
2. Irie, M., and Mohri, M. Thermally irreversible photochromic systems – reversible photocyclization of diarylethene derivatives. *J. Org. Chem.* 53, 803, 1988.
3. Irie, M. Diarylethenes for memories and switches. *Chem. Rev.* 100, 1685, 2000.
4. Tsujioka, T., Tatezono, F., Harada, T., Kuroki, K., and Irie, M. Recording sensitivity and superlow-power readout of photon-mode photochromic memory. *Jpn. J. Appl. Phys.* 33, 5788, 1994.
5. Hamano, M., and Irie, M. Rewritable near-field optical recording on photochromic thin films. *Jpn. J. Appl. Phys.* 35, 1764, 1996.
6. Toriumi, A., and Kawata, S. Reflection confocal microscope readout system for three-dimensional photochromic optical data storage. *Opt. Lett.* 23, 1924, 1998.
7. Fukaminato, T., Kobatake, S., Kawai, T., and Irie, M. Three-dimensional erasable optical memory using a photochromic diarylethene single crystal as the recording medium. *Proc. Jpn. Acad.* 77, B, 30, 2001.
8. Ebisawa, F., Hoshin, M., and, Sukegawa, K. Self-holding photochromic polymer machzehnder optical switch. *Appl. Phys. Lett.* 65, 2919, 1994.
9. Hoshino, M., Ebisawa, F., Yoshida, T., and Sukegawa, K. Refractive index change in photochromic diarylethene derivatives and its application to optical switching devices. *J. Photochem. Photobiol. A: Chem.* 1997, 105, 75.
10. Tanio, N., and Irie, M. Photooptical switching of polymer film wave-guide containing photochromic diarylethenes. *Jpn. J. Appl. Phys.* 33, 1550, 1994.
11. Tanio, N., and Irie, M. Refractive-index of organic photochromic dye amorphous polymer composites. *Jpn. J. Appl. Phys.* 33, 3942, 1994.
12. Tsujioka, T., Hamada, Y., Shibata, K., Taniguchi, A., and Fuyuki, T. Nondestructive readout of photochromic optical memory using photocurrent detection. *Appl. Phys. Lett.* 78, 2282 (2001).
13. Kim, E., Choi, Y.-K., and Lee, M.-H. Photoinduced refractive index change of a photochromic diarylethene polymer. *Macromolecules* 32, 4855, 1999.
14. Biteau, J., Chaput, F., Lahlil, K., Boilot, J.-P., Tsivgoulis, G., Lehn, J.-M., Darracq, B.,

Moris, C., and Levy, Y. Large and stable refractive index change in photochromic hybrid materials. *Chem. Mater.* 10, 1945, 1998.

15. Irie, M., Uchida, K., Eriguchi, T., and Tsuzuki, H. Photochromism of single-crystalline diarylethenes. *Chem. Lett.* 899, 1995.

16. Kobatake, S., Yamada, T., Uchida, K., Kato, N., and Irie., M. Photochromism of 1,2-bis(2,5-dimethyl-3-thienyl)perfluorocyclopentene in a single crystalline phase. *J. Am. Chem. Soc.* 121, 2380, 1999.

17. Kawai, T., Koshido, T., and Yoshino, K. Optical and dielectric-properties of photochromic dye in amorphous state and its application. *Appl. Phys. Lett.* 67, 795, 1995.

18. Koshido, T., Kawai, T., and Yoshino, K. Novel photomemory effects in an amorphous photochromic dye. *Jpn. J. Appl. Phys.* 34, L389, 1995.

19. Kaneuchi, Y., Kawai, T., Hamaguchi, M., Yoshino, K., and Irie, M. Optical properties of photochromic dyes in the amorphous state. *Jpn. J. Appl. Phys.* 36, 3736, 1997.

20. Shirota, Y., Moriwaki, K., Yoshikawa, S., Ujike, T., and Nakano, H. 4-[di(biphenyl-4-yl)amino]azobenzene and 4,4'-bis[bis(4'-tert-butylbiphenyl-4-yl)amino]azobenzene as a novel family of photochromic amorphous molecular materials. *J. Mater. Chem.* 8, 2579, 1998.

21. Kawai, T., Fukuda, N., Groschl, D., Kobatake, S., and Irie, M. Refractive index change of dithienylethene in bulk amorphous solid phase. *Jpn. J. Appl. Phys.* 38, 1194, 1999.

22. Kim, M.-S., Kawai, T., and Irie, M. Synthesis and photochromism of amorphous diarylethene having styryl substituents. *Mol. Cryst. Liq. Cryst.* 345, 251, 2000.

23. Kim, M.-S., Kawai, T., and Irie, M. Synthesis of fluorescent amorphous diarylethenes. *Chem. Lett.* 1188, 2000.

24. Fukudome, M., Kamiyama, K., Kawai, T., and Irie, M. Photochromism of a dithienylethene having diphenylamino side groups in the bulk amorphous phase25 refractive index change of an amorphous bisbenzothienylethene. *Chem. Lett.* 70, 2001.

25. Chauvin, J., Kawai, T., and Irie, M. *Jpn. J. Appl. Phys.*, 40, 2518, 2001.

26. Utsumi, H., Nagahama, D., Nakano, H., and Shirota, Y. A novel family of photochromic amorphous molecular materials based on dithienylethene. *J. Mater. Chem.* 10, 2436, 2000.

27. Kim, M.-S., Kawai, T., and Irie, M. Synthesis of fluorescent amorphous diarylethenes. *Chem. Lett.* 702, 2001.

28. Irie, M., Miyatake, O., Uchida, K., and Eriguchi, T. Photochromic diarylethenes with intralocking arms. *J. Am. Chem. Soc.* 116, 9984, 1994.

29. Irie, M., Sakemura, K., Okinaka, M., and Uchida, K. Photochromism of dithienylethenes with electron-donating substituents. *J. Org. Chem.* 60, 8305, 1995.

30. Smith, D., Riccius, H., and Edwin, R. Refractive index of lithium-niobate. *Opt. Commun.* 17, 332, 1976.

■ INDEX